MATRIX STRUCTURAL ANALYSIS

MATRIX STRUCTURAL ANALYSIS

J. L. MEEK

Reader of Civil Engineering
University of Queensland

McGRAW-HILL BOOK COMPANY

New York St. Louis San Francisco Düsseldorf Johannesburg
Kuala Lumpur London Mexico Montreal New Delhi
Panama Rio de Janeiro Singapore Sydney Toronto

MATRIX STRUCTURAL ANALYSIS

Library of Congress Catalog Card Number 77-139559

07-041316-9

1 2 3 4 5 6 7 8 9 0 MAMM 7 9 8 7 6 5 4 3 2 1

This book was set in Laurel and News Gothic by York Graphic Services, Inc., and printed on permanent paper and bound by The Maple Press Company. The designer was Elliot Epstein; the drawings were done by John Cordes, J. & R. Technical Services, Inc. The editors were Roland S. Woolson, B. J. Clark, and J. W. Maisel. Stuart Levine supervised production.

CONTENTS

PREFACE

Now the engine, from its capability of performing by itself all these purely material operations, spares intellectual labour, which may be more profitably employed. Thus the engine may be considered as a real manufactory of figures, which will lend its aid to those useful sciences and arts that depend on numbers. Again who can foresee the consequences of such an invention? In truth, how many precious observations remain practically barren for the progress of the sciences, because there are not powers sufficient for computing the results! And what discouragement does the perspective of a long and arid computation cast into the mind of a man of genius, who demands time exclusively for meditation, and who beholds it snatched from him by the material routine of operations. . . . Thus the idea of constructing an apparatus capable of aiding human weakness in such researches is a conception which, being realized, would mark a glorious epoch in the history of the sciences.

Charles Babbage and his Calculating Engineers—A Commentary
L. F. MENABREA OF TURIN, OFFICER OF THE MILITARY ENGINEERS
October, 1842.

This book has been written for the student as well as the practicing engineer. It attempts to outline the fundamental principles of modern structural analysis and also give program sketches that should allow the theory to be converted into useful computer programs. Some emphasis has been placed on algorithms for solution of linear equations because these form the essential core of any system for large-scale problems. That is, there is more to a successful program than being able to generate equilibrium equations or member-stiffness matrices. Because the number of courses in matrix algebra has proliferated in the last decade, this text is in contrast with many published works on the subject in that it assumes that the reader has some knowledge of matrix algebra. Toward structural analysis, however, it adopts the viewpoint that the reader has been only superficially exposed to the modern matrix approach. Essential building blocks are established for transformation of force and displacement quantities and then applied to problems of equilibrium and deflection calculation. An important aspect of this is that continuous as well as discrete systems are considered.

A matrix interpretative program, SMIS, for all standard matrix operations (except eigenvalue calculation) has been included so that students may have rapid access to the computer for working class assignments.

Over several years of undergraduate and postgraduate teaching, the writer has found the use of this program to be a major factor in the interest of students in the subject. In fact, the attraction of the student to the subject then has, in many cases, a quality that can only be described as magical. In this the author is indebted to Professors R. W. Clough and E. Wilson of the University of California, Berkeley, for allowing the use of the SMIS program in its original form, and to Mr. R. Nilsson of the University of Queensland for the writing of the free-field input subroutine for the version reproduced herein (see Appendix A).

It appears, at least to the writer, that it will be inevitable that the matrix-tensor concepts will creep forward into the undergraduate curriculum until most, if not all, current teaching is replaced. It is hoped that this book is a forerunner of many that will pioneer the way for such changes. The breadth and scope of such a training should produce university graduates with a structural awareness not even dreamed possible today.

The author would like to express his sincere appreciation for the inspiration given him during various stays at the University of California, Berkeley, by Professor R. W. Clough and for many ideas in the finite-element and large-displacement analyses given by Dr. C. Felippa. Thanks, too, go to Misses E. Hoare and V. Howard for their careful typing of the manuscript.

J. L. Meek

MATRIX STRUCTURAL ANALYSIS

1
CONCEPTS OF STRUCTURAL ANALYSIS

1.1 INTRODUCTION

The purpose of this text is to present some of the ideas of modern structural analysis. Since the introduction of the automatic digital computer in the early 1950s, there has been a virtual revolution in the approach, not only to the computational methods, but also to the fundamental theorems of structural analysis. From the numerous published research reports of the last decade, it is intended to attempt a collation of common underlying principles that will allow the reader to approach the subject of structural analysis as a program developer as well as an informed user of the service bureau library routines. To this end, numerous program sketches are dispersed throughout the theory. Although economy must always dictate the use of general-purpose programs, the writer feels strongly that the important man in the structural team will be the innovator capable of rapid program development based on a thorough knowledge of structural theory and computer techniques. From teaching experience over several years, the conclusion has been reached that a high proportion of students with honors potential can become proficient in these fields in approximately 2 years of postgraduate training. This time may be further reduced when the concepts of vectors, tensors, and matrices percolate to the lower years of the engineering undergraduate curriculum.

The writer would like to emphasize that this is not a text on matrix theory, but one on practical structural analysis. Matrix theory is a branch of pure mathematics, and it is indeed fortunate that such a highly organized body of algebra should form one of the basic tools of the new approach to structural mechanics.

It seems that once the initial step has been taken to teach statics and deflection theory in matrix form, the student's perception of the physical situations involved will be considerably heightened. This is particularly true if he has access to an automatic digital computer because then the range and number of problems that can be solved in a given time is vastly increased.

The course of study proposed in this text follows a path rather similar in many respects to that suggested by our present-day courses in classical structural mechanics. Now, however, matrix theory will form the basic framework for our discussions. With this in mind, the order of presentation will be:

1 Force and displacement quantities expressed as vectors with components given in matrix form; transformation of vector components for change in coordinate axes and conditions of equilibrium for a rigid body.

2 The solution of equilibrium problems for determinate framed structures using the matrix transformations introduced in 1.

3 The theorems related to calculation of deflections of elastic structures and their application to statically determinate structures.

4 The conditions by which a structure may be classified as stable or unstable and determinate or indeterminate. Thence means are devised for the solution of statically indeterminate structures. It will be shown that the process involves both the static and deflection analyses developed in 2 and 3.

5 The alternative expression for equilibrium equations based on displacement transformations and the formalization of this procedure as the stiffness method of analysis.

6 The extension of the theory, applied so far to framed structures, to include elements of more than one dimension, referred to as *finite elements.*

7 The limitations of the elastic analysis and its adaptation to the calculation of load-deflection relationships for structures in which nonlinear behavior is exhibited from both material and geometric causes. In the latter case it is found that buckling phenomena can be predicted.

It is our intention then, in this introduction, to discuss these topics briefly so that the reader may gain insight into the problems that are to be encountered. It should be noted that the matrix notation formalizes the algebra in a way most suitable for computer coding. An important aspect not mentioned above is that at the present time computer memory capacities are limited so that economical use of storage is needed in order to produce programs that may handle problems of a reasonable size.

1.2 FORCE AND DISPLACEMENT QUANTITIES CONSIDERED AS VECTORS; MATRIX NOTATION

A suitable starting point for our discussion will be a study of force and displacement quantities as vectors and the resulting transformations of their components for various changes (that is, rotation and translation) in the coordinate axes. In addition, stress and strain components are studied. It is

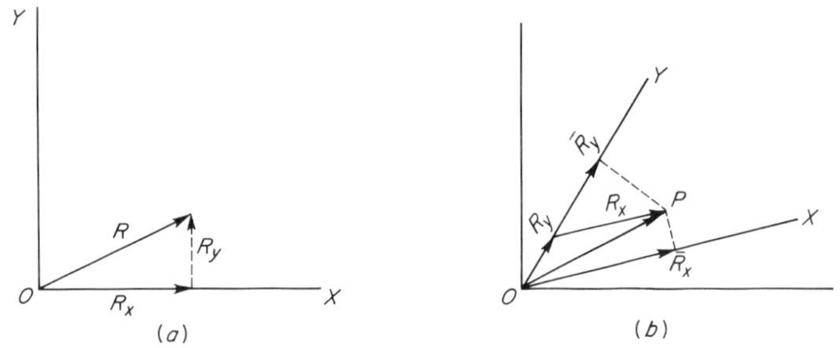

Figure 1.1 Coordinate systems. (*a*) Orthogonal coordinate axes, angle $XOY = 90°$; (*b*) skew coordinate axes, angle $XOY \neq 90°$.

found that these are different from force and displacement quantities, and it will be shown that they belong to a class of physical phenomena known mathematically as *rank* 2 *tensors* (see pp. 00ff).

The coordinate axes chosen will be a right-handed cartesian set. Then transformations associated with change in axes will always have consistent signs. Such a set of axes is shown in the plane in Fig. 1.1*a*. In the XY coordinate system shown, the vector R has components R_X and R_Y expressed as projections on the X and Y axes such that:

1 The vector law of addition (parallelogram rule) is

$$R = R_X + R_Y$$

2 The magnitude of R, written mod R or simply $|R|$, is expressed by the pythagorean rule

$$|R|^2 = R_X{}^2 + R_Y{}^2$$

The matrix representation of the vector R will be

$$R = \begin{bmatrix} R_X \\ R_Y \end{bmatrix}$$

In passing it will be noticed that the expression for mod R in matrix notation is

$$|R|^2 = R^T R$$

It must be realized that the above representation of a vector in a rectangular cartesian coordinate system is neither unique nor always the most useful one. Thus, for example, a set of skew coordinate axes is shown in Fig. 1.1*b*. Notice in Fig. 1.1*b* that two representations of the vector R are possible in the skew coordinates. The first of these is given by the components R_X, R_Y that obey the parallelogram rule (rule 1 above) for vector addition. These are parallel to the coordinate axes OX and OY respectively. Alternatively, R

may be expressed by the components R_X, R_Y measured along the axes to the perpendiculars drawn from P to the axes OX and OY. It is evident that these components do not obey rules 1 and 2 above. However, both these sets of components play important parts in the general theory of nonorthogonal coordinate systems and the transformations therein of vector components. Since this text is considered a relatively elementary one, only orthogonal coordinate systems (and their transformations) are discussed and the reader is referred to the various works on skew structures for the more difficult but nevertheless extremely fascinating skew transformations (see Refs. 1 and 27).

The essence of the vector concept expressed in matrix form is an understanding of the effect on the components of the vector if the axes XY suffer a rotation through the angle θ to $\overline{X}\overline{Y}$, measured positive as shown in Fig. 1.2. The components R_X, R_Y are obtained by projecting R_X and R_Y onto the $O\overline{X}$ and $O\overline{Y}$ axes respectively, so that

$$\overline{R}_X = R_X \cos\theta + R_Y \sin\theta$$

$$\overline{R}_Y = -R_X \sin\theta + R_Y \cos\theta$$

Alternatively, in matrix form,

$$\begin{bmatrix} \overline{R}_X \\ \overline{R}_Y \end{bmatrix} = \begin{bmatrix} \cos\theta & \sin\theta \\ -\sin\theta & \cos\theta \end{bmatrix} \begin{bmatrix} R_X \\ R_Y \end{bmatrix}$$

The use of this and related transformations is explored in considerable detail in Chaps. 2 and 3. Therein, it will be shown that the basic transformations for force and displacement quantities are:

1 A rotation of the coordinate axes (as above)

2 A parallel shift of axes from one origin to another

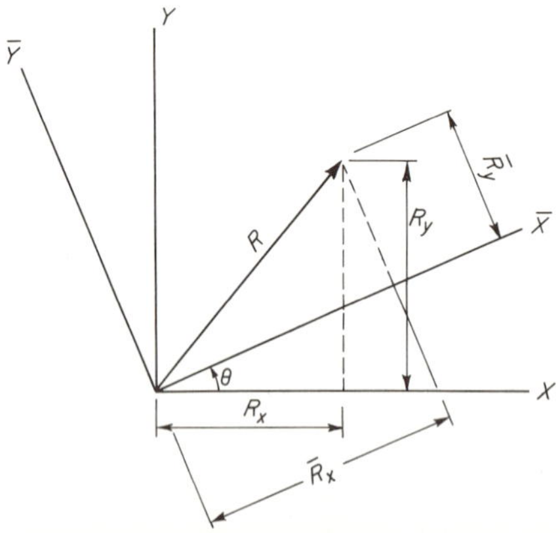

Figure 1.2 Rotation of coordinate axes.

Both three-dimensional and planar situations are discussed because in the latter it is usually inconvenient to carry the null components of the three-dimensional case through the calculations.

1.3 FORCE TYPES; JOINT EQUATIONS OF EQUILIBRIUM

After establishing the necessary theory for the transformation of force components in Chap. 2, it is then our task to apply this to the calculation of forces in statically determinate structures. These are considered first since it appears that they will be the simplest to analyze. Many clues are discovered therein on how analyses are best set up in the computer for more complex situations.

For the statically determinate structure, the elastic properties of the members are not required in order to calculate member forces. The details of how computer input can be minimized are outlined in Sec. 1.4.

In the process of analyzing the structure, three types of forces are distinguishable.

1 *Node Forces* These are externally applied forces. No restriction is placed on where the loads may be applied. If they act along a member, then means are devised for transferring them to be statically equivalent node forces (see Sec. 3.9). Node forces are reduced so that their components are expressed in a common, or global, coordinate system.

2 *Member Forces* These are forces applied through the nodes to the members of the structure. They may be expressed with components in the global coordinate system or in the local coordinates for the member, or alternatively they may be given simply as axial force, twist, and bending moments. The latter interpretation will be necessary for the designer proportioning the size of the member.

In determinate structures, member forces are caused by node forces only, whereas in indeterminate structures, it will be found that other effects such as temperature variation, foundation settlement, etc., can produce member forces. It is important to realize that the reactive forces from the members acting on the nodes are in equilibrium with the external forces (including reactions).

3 *Reactive Forces* These are supplied by the supporting system and give the whole structure, considered as a rigid body, its stability. That is, these forces form an equilibrium set when combined with group 1 forces. In order to clarify these definitions, let us examine the forces acting on the nodes and on the members of the portal frame shown in Fig. 1.3*a*. The portal frame has been dissected into a number of free bodies in Fig. 1.3*b*.

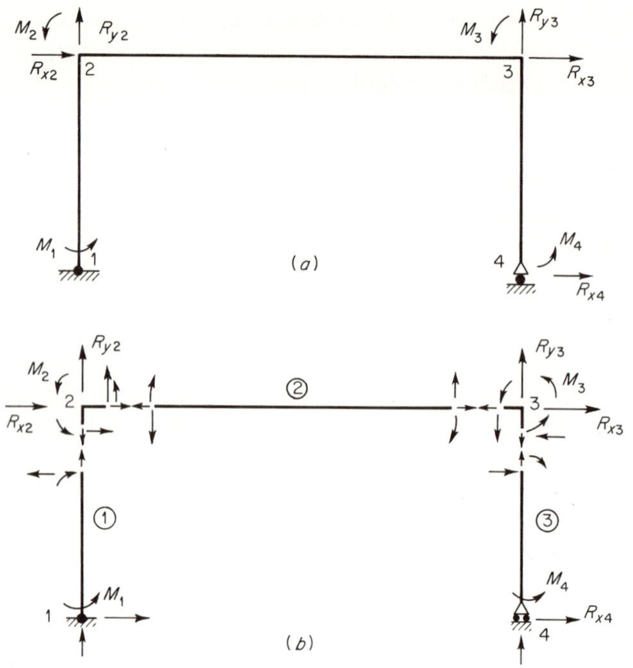

Figure 1.3 (*a*) Portal frame; node forces. (*b*) Node, member, and reactive forces.

Notice in Fig. 1.3 that three types of equilibrium are observed:

a Each joint is in equilibrium.

b Each member is in equilibrium.

c The structure as a whole is in equilibrium.

Since the joint equilibrium equations include both member forces and reactive forces, their solution will yield all unknown forces associated with the statically determinate structure. It will be shown in Chap. 3, that the equations of equilibrium are most readily set out in the computer in the global coordinate system and to do this the transformations outlined in Sec. 1.2 (and developed in Chap. 2) are employed to transform member forces. Notice the difference here between this and the *hand calculation* approach in which the calculator may employ several methods alternately in the one structure to facilitate the solution.

Statically Equivalent Forces In the discussion of node forces the important concept of statically equivalent forces was introduced. By definition, two force systems are said to be *statically equivalent* if when they are transferred to a common coordinate system (and origin), they have identical components.

Notice that if two force systems are statically equivalent, the negative of one will necessarily equilibrate the other. Now, since member forces with signs reversed equilibrate the node forces in a structure, it follows that member forces are statically equivalent to the applied node forces.

Our intent in Chap. 3 is to study the determinate structure so that programs may be evolved for its analysis. It is emphasized that thorough knowledge of the transformations used therein and of the solution techniques employed will allow the student to program more complex situations with relative ease.

In order that the joint equilibrium equations may be set up automatically with a minimum of input details, it will be useful to study the structure topology, that is, how the members of the structure are connected to form the whole system of joints, members, and loops.

1.4 STRUCTURE TOPOLOGY

When a line diagram of the structure is drawn, as in Fig. 1.4, and members and joints numbered, the resulting graph has properties that are independent of the geometry or material properties. This forms part of a relatively new mathematical development called *linear graph theory* (see Ref. 4).

From the node and member numbering it will be possible to relate the forces in a member to the joint equilibrium equations in which they participate. A member will have its positive sense directed from its smaller to its larger node number (this is used in defining a positive member coordinate sense). The connectivity of the structure in Fig. 1.4 is expressed in the table:

Member No.	*End 1*	*End 2*
1	1	2
2	2	3
3	3	4

It is found that there are many other useful properties of linear graphs associated with the nodes and closed loops it contains. Some of these may be used in statically indeterminate analysis. Also, the topological properties of

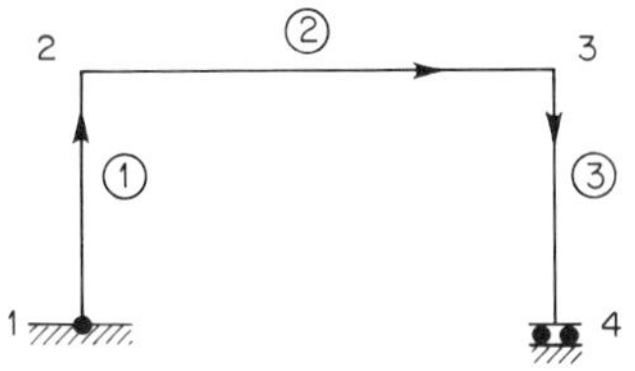

Figure 1.4 Structure topology.

structural systems are similar to those of other network problems encountered in such diverse fields as transportation engineering, pipe flow, and electrical circuit theory. For example, one problem in statically indeterminate analysis of a structure is in what manner the structure shall be cut so that the remaining determinate system will be as stiff as possible. Consider the rigid frame shown in Fig. 1.5*a*, and enquire in which way the structure shall be cut so that the remaining determinate system will be as stiff as possible. One such possibility is shown in Fig. 1.5*b* and has been obtained by cutting all the beam members.

Just how stiff the determinate structure will be must depend on the relative flexibilities of the beams and columns and on the assumption that the foundation is rigid. Also we have excluded the possibility of making only a hinge release at a section. That is, all releases must be complete cuts.

The equivalent transportation problem is given in Fig. 1.6*a*, in which it is required to find the shortest paths between an origin *A* and the points 1, 2, and 3. Again, depending on the relative lengths assigned to the individual branches, a possible solution is that shown in Fig. 1.6*b*. The similarity should be noted between the graphs in 1.5*b* and 1.6*b*.

The minimum input data required for the solution of the statically determinate structure will be:

1 The structure topology, comprising a table of numbers giving the details of how the members are connected together. Included with this should be a member type list. That is, a member may be pin-jointed, contain an internal hinge, and so on. Such internal releases in the member will affect the number of independent member forces contributing to the joint equilibrium equations.

2 The joint coordinates expressed in the global system.

3 The joint supports.

4 The external load system.

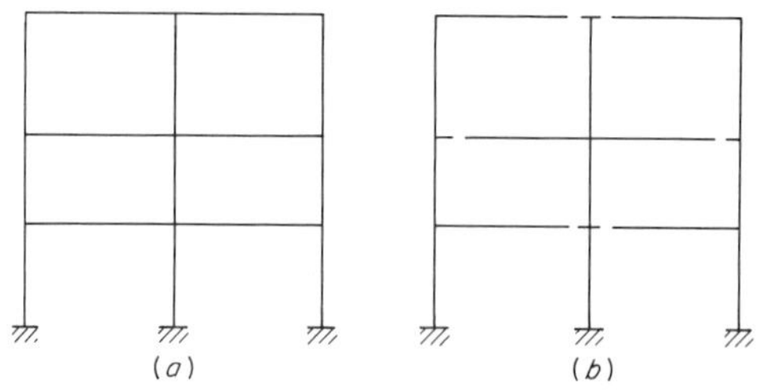

Figure 1.5 (*a*) Continuous structure; (*b*) determinate or *cut* structure.

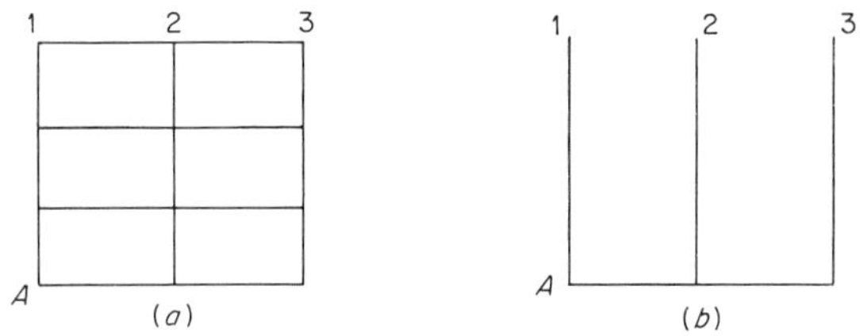

Figure 1.6 (*a*) Transportation network; (*b*) shortest paths.

In Chap. 3, these topics are discussed in relation to the development of computer programs. For class tuition, however, it is useful to introduce a matrix interpretative system that will handle standard matrix operations, such as addition, multiplication, inversion, etc. In this way the matrices associated with class problems may be generated by hand and then manipulated by the computer.

1.5 MATRIX INTERPRETATIVE SYSTEM

When the joint equilibrium equations are set up as outlined in the previous sections, it is soon found that the number of numerical operations involved for their direct solution becomes quite large even for small structures. For example, consider the analysis of the pin-jointed truss shown in Fig. 1.7 by the method of joint analysis. In the matrix presentation, this involves solving all equations simultaneously (two per joint), and if reactions are included, there will be a total of 14 equations.

What will be required for class work is a language for handling matrices that will be accepted by the computer. This could be a type of compiler language of a higher level than languages such as Fortran, that could accept messages dealing with matrix operations. Alternatively, a program could be developed that simply recognizes standard matrix algebra commands such as add, subtract, and multiply and acts accordingly. It would be anticipated that the latter course of action would be simpler than developing a compiler language. This approach was followed by several research pioneers in the field. One such program, called *SMIS* (*symbolic matrix interpretative system*), was developed by Professors Wilson and Clough at the University of California, Berkeley, and a modification of this system with the use of a free-field input

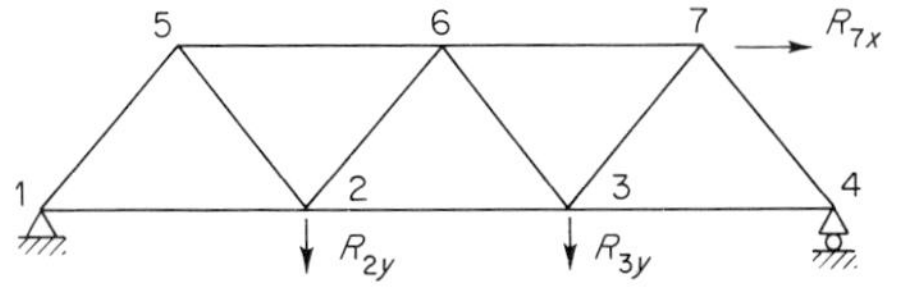

Figure 1.7 Truss analysis.

and without some of the routines used in dynamic analysis is available for use in conjunction with this text. This system consists of a main program (or monitor) with a set of subroutines that not only carry out matrix operations but also allocate storage and keep track of names assigned to the various matrices by the user. See Appendix A, for the complete list and operating instructions.

Example 1.1 The use of SMIS

It is required to multiply two 6×6 matrices and print out the result. For convenience the three matrices will be labeled A, B, and C so that the operation is written

$C = A \times B$

From Appendix A it is found that the following information must be punched on cards:

```
LOAD     A    6    6
```
cards with elements of A in free field

```
LOAD     B    6    6
```
cards with elements of B in free field

```
MULT     A    B    C
PRINT    C    *    RESULT C = A × B
```

The reader will be able to follow the several examples in the text in which structural problems have been solved by using SMIS.

1.6 DEFLECTIONS; FLEXIBILITY AND STIFFNESS

When forces are applied to a structure, deflections occur, or alternatively, when displacements are prescribed, node forces are necessary to produce them. The magnitudes of these effects (displacements or forces) will depend on the load-deflection relationships for the component parts of the structure as well as on how they are connected together. Structure load-deflection characteristics are usually described under the definitions of flexibility and stiffness.

Flexibility and Stiffness

The flexibility of a structure is a property associated with the application of forces to the structure. That is, a flexibility coefficient is the value of the deflection at a point of the structure, in a given direction, due to a unit load applied at a second point and in a second direction. If the material properties of the structure are elastic and displacements small, the flexibility coefficients will have unique values. When the above conditions are not satisfied, the coefficients will vary with load and deflection, and the calculation of deflections presents a formidable task. Aspects of nonlinearity are discussed in Sec. 1.10 and again in more detail in Chap. 9.

Member distortions will arise from a variety of stress patterns. In framed structures, axial and bending stresses are usually considered in their effect on member distortions (see Chap. 4). In addition the shear distortions may be approximated. In more complex elements (see Sec. 1.9), force fields may vary in a prescribed way over the element.

As an example of flexibility and stiffness relationships, consider the most elementary case of a structure of two nodes connected by springs 1 and 2 as shown in Fig. 1.8. The springs have flexibilities f_1 and f_2 respectively, and when loads are applied to the nodes, distortions in the springs cause deflections of the node points.

The coefficient f_i gives the deflection of spring i for a unit force in the spring. When loads P_1 and P_2 are applied, the node deflections are

$$y_1 = f_1(P_1 + P_2)$$
$$y_2 = f_1(P_1 + P_2) + f_2P_2$$

In matrix form,

$$\begin{bmatrix} y_1 \\ y_2 \end{bmatrix} = \begin{bmatrix} f_1 & f_1 \\ f_1 & f_1 + f_2 \end{bmatrix} \begin{bmatrix} P_1 \\ P_2 \end{bmatrix}$$

Thus, the flexibility coefficients for joints 1 and 2 are f_1 and $f_1 + f_2$, respectively.

The stiffness of the structure, in contrast to its flexibility, is associated with the application of displacements to the structure. A stiffness coefficient is the value of the force produced in a given direction at a point owing to a unit displacement applied at a second point and in a second direction.

The stiffness coefficients will thus vary depending on how many displacements are prescribed, since implicit in the above definition is the condition that while one displacement is given a unit value, all other displacements are held equal to zero. Again, the stiffness coefficient will be unique (for the same displacement set) if the structure is elastic and displacements are small, but will vary in a nonlinear manner when these conditions are not satisfied.

For the springs shown in Fig. 1.8, the individual spring stiffnesses are given as the reciprocal of the corresponding flexibility. That is,

$$k_i = \frac{1}{f_i}$$

The stiffness coefficients for joint 1 of the structure are now calculated with joint 2 both fixed and free. Firstly, if node 2 is unrestrained, the stiffness at node 1 is simply k_1 since spring 2 then suffers a rigid-body displacement. If,

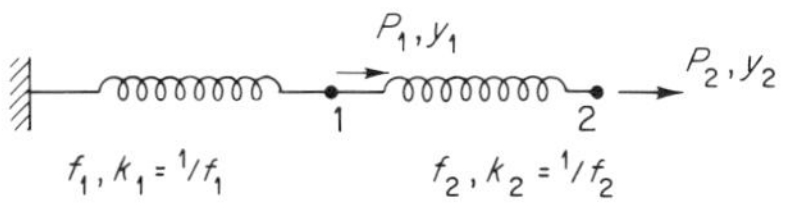

Figure 1.8 Structure deflections.

however, node 2 is considered and consequently restrained when node 1 is given its unit displacement, the stiffness for node 1 obviously involves both spring stiffnesses and is equal to $k_1 + k_2$.

The calculation of node deflections from member distortions is developed in detail in Chap. 4.

1.7 STATICALLY INDETERMINATE STRUCTURES

When the equations of joint equilibrium are set up for the general structure, as outlined in Sec. 1.3, it will be found that, except for the particular case of static determinacy, there will be more equations than unknowns. This means that providing the structure is stable, a number of the member forces or reactions can be given arbitrary values and then the remaining forces obtained in terms of those assumed and the external loads. When such is the case, the structure is said to be statically indeterminate. For example, in the portal frame shown in Fig. 1.9 there are four joints, and hence in its planar condition, there are $4 \times 3 = 12$ equations of joint equilibrium. If each member has end moments M_i, M_j and axial force F, there is a total of 9 member forces, and together with the 6 reactions, these make a total of 15 unknowns. Then the structure is said to be $15 - 12 = 3$ times statically indeterminate. It will be shown now that a solution of a statically indeterminate problem requires a knowledge of the flexibilities of the members.

Thus, suppose the structure is cut as shown in Fig. 1.9*b*. The 10-kip load applied at the top of the left-hand column will cause the part of the structure on the left side of the cut to deflect as shown, so that now there is a lack of compatibility at the cut. That is, the structure is no longer continuous across the cut section. For the particular case shown in Fig. 1.9*b*, there will be three different types of discontinuity present, namely, (1) axial, (2) transverse, and (3) rotation. To restore continuity, the unknown force quantities X_1, X_2, and X_3 are applied. In this way three simultaneous equations are evolved that take into consideration the flexibility properties of the structure as well as the geometry and external load system. We see that the indeterminate analysis requires:

1 The calculation of force transformations in the "cut" structure, and a classification of the force types into known node forces and indeterminate forces

2 The deflection analysis for both known and unit values of the unknown forces

3 The return from the cut-discontinuous structure to the actual structure by applying the required magnitudes of the forces at the cuts

In Chap. 5 the statically indeterminate structure is studied in detail and algorithms given for the computerization of its analysis. It is important to realize that considerable effort is saved if the generation of force transforma-

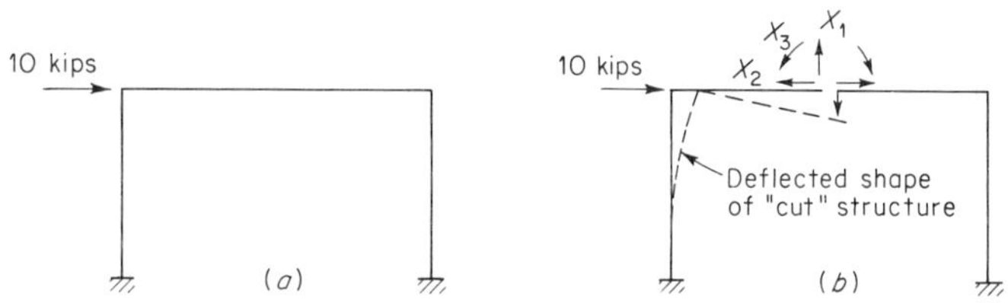

Figure 1.9 Indeterminate analysis. (*a*) Statically indeterminate structure; (*b*) cut-determinate structure.

tion matrices in 1 above and the selection of the unknown forces are carried out automatically by the computer program.

When more forces are available in the structure than the minimum number required for static equilibrium, interesting member force phenomena occur that are in equilibrium with zero external forces. Thus, for example, temperature strains, lack of fit, plastic yield, and foundation settlement will all cause member forces in the indeterminate structure.

1.8 KINEMATIC DETERMINACY; DISPLACEMENT TRANSFORMATIONS

Thus far, we have explored only the ideas of force transformations, static determinacy, flexibility, etc., concepts, all bound to the study of forces applied to structures. This approach follows logically from the assumption that firstly all structures must be reduced to a statically determinate system. In this way, in the determination of member forces it will be required to solve many equations, but by clever manipulations such as sectioning, using free-body diagrams of parts of the structure, and so on, the process may be reduced, for manual purposes, to the solution of one or two of the equations at each step.

In the alternative approach, referred to as the displacement or stiffness method of analysis, the joint displacements are considered degrees of freedom, and if all these were known, the structure would be kinematically determinate. Notice that some assumption must be made regarding the displacements of a member between its node values as a function of these values, since if the member displacement cannot be expressed in this way, the structure would have an infinite number of degrees of kinematic indeterminacy. For example, a member of uniform cross section with a constant axial force will have linear displacements, whereas the same member under a linear moment field will have transverse displacements given by a cubic function.

The first problem, then, in kinematic analysis is to calculate member and node forces, given node displacements.

Consider, for example, the structure (Fig. 1.10) composed of three springs, whose force-deflection relationships are given by the stiffnesses k_1, k_2, and k_3. The structure has one node common to the three springs, and the coordinate

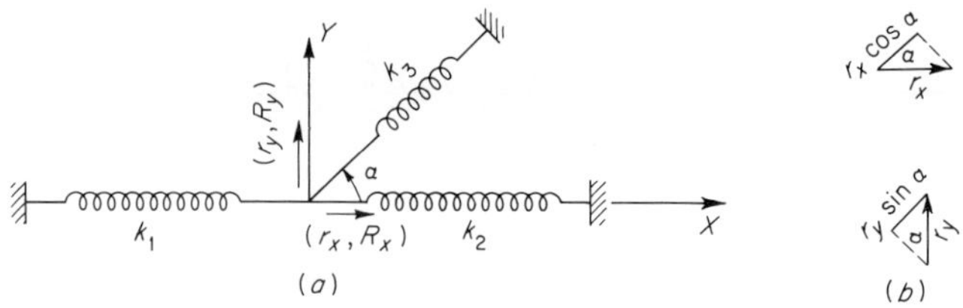

Figure 1.10 (*a*) Stiffness of a structure node; (*b*) displacement patterns $r_x = 1$ and $r_y = 1$.

axes are chosen so that springs 1 and 2 coincide with the X axis and spring 3 is inclined to these by the angle α.

From Fig. 1.10*b* the elongations of the springs, Δ_1, Δ_2, and Δ_3, are given in terms of the node displacements r_x, r_y by the equations

1 r_x, with $r_y = 0$:

$$\Delta_1 = r_x \qquad \Delta_2 = -r_x \qquad \Delta_3 = -r_x \cos\alpha$$

2 $r_x = 0$, with r_y:

$$\Delta_1 = 0 \qquad \Delta_2 = 0 \qquad \Delta_3 = -r_y \sin\alpha$$

These six equations may be expressed in the single matrix equation,

$$\begin{bmatrix} \Delta_1 \\ \Delta_2 \\ \Delta_3 \end{bmatrix} = \begin{bmatrix} 1 & 0 \\ -1 & 0 \\ -\cos\alpha & -\sin\alpha \end{bmatrix} \begin{bmatrix} r_x \\ r_y \end{bmatrix} \tag{i}$$

In terms of the member displacements Δ_1, Δ_2, Δ_3, the member forces F_1, F_2, F_3 are given by the stiffness coefficients, such that

$$F_1 = k_1 \Delta_1$$
$$F_2 = k_2 \Delta_2$$
$$F_3 = k_3 \Delta_3$$

The node forces R_X, R_Y produced by the displacements r_x, r_y may be calculated by projecting the member forces on the X and Y axes, so that

$$R_X = F_1 - F_2 - F_3 \cos\alpha$$
$$R_Y = -F_3 \sin\alpha$$

Substituting for F_1, F_2, and F_3 in terms of the spring distortions and putting the equations in matrix form, we have

$$\begin{bmatrix} R_X \\ R_Y \end{bmatrix} = \begin{bmatrix} k_1 & -k_2 & -k_3 \cos \alpha \\ 0 & 0 & -k_3 \sin \alpha \end{bmatrix} \begin{bmatrix} \Delta_1 \\ \Delta_2 \\ \Delta_3 \end{bmatrix}$$

The required stiffness expression relating the node forces to the node displacements is obtained by substituting for Δ_1, Δ_2, Δ_3 from Eq. (*i*), so that

$$\begin{bmatrix} R_X \\ R_Y \end{bmatrix} = \begin{bmatrix} k_1 & -k_2 & -k_3 \cos \alpha \\ 0 & 0 & -k_3 \sin \alpha \end{bmatrix} \begin{bmatrix} 1 & 0 \\ -1 & 0 \\ -\cos \alpha & -\sin \alpha \end{bmatrix} \begin{bmatrix} r_x \\ r_y \end{bmatrix}$$

That is,

$$\begin{bmatrix} R_X \\ R_Y \end{bmatrix} = \begin{bmatrix} k_1 + k_2 + k_3 \cos^2 \alpha & k_3 \cos \alpha \sin \alpha \\ k_3 \cos \alpha \sin \alpha & k_3 \sin^2 \alpha \end{bmatrix} \begin{bmatrix} r_x \\ r_y \end{bmatrix} \qquad (ii)$$

When r_x and r_y are known, Eq. (*ii*) may be used to obtain the node forces. However, if the node forces R_X and R_Y are known, Eq. (*ii*) must be solved in order to determine the node deflections. Then Eq. (*i*) may be used to determine the member distortions and hence their forces. Thus we see that even when the known quantities are the node forces, a kinematic analysis may be used to calculate member forces. Now, however, member elasticity properties must be assigned before starting the analysis, since deflections must be calculated.

Evidently the structure with all nodes restrained against displacement and rotation is kinematically determinate. This proves to be a convenient starting point for any analysis because then the node deflections may be calculated by applying node forces equal and opposite to the restraining forces, to the structure. For example, it is required to calculate the end deflection of the cantilever beam shown in Fig. 1.11*a*. From elementary strength of materials,

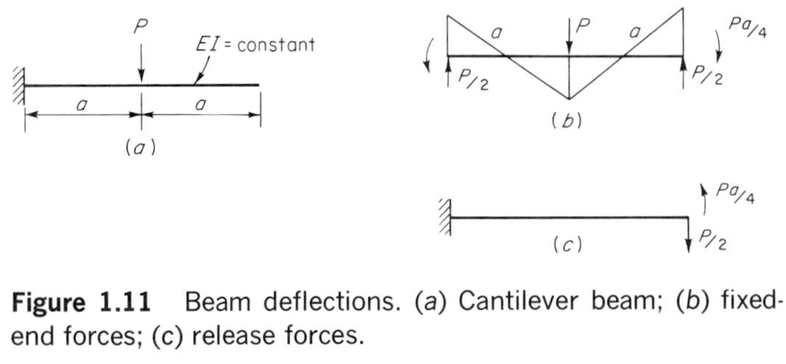

Figure 1.11 Beam deflections. (*a*) Cantilever beam; (*b*) fixed-end forces; (*c*) release forces.

$$\delta_{\text{end}} = \frac{Pa^3}{3EI} + \frac{Pa^3}{2EI} = \frac{5Pa^3}{6EI}$$

The end of the cantilever will be the only node, so that the structure with this node restrained is shown in Fig. 1.11*b*. The forces to release the end are taken as the reverse of those in Fig. 1.11*b* and are shown as the moment and concentrated force in Fig. 1.11*c*. Then,

$$\delta_{\text{end}} = \frac{(P/2)(2a)^3}{3EI} - \frac{(Pa/4)(2a)^2}{2EI} = \frac{5Pa^3}{6EI}$$

The two results agree. It should be noted that the method of analyzing a structure from a kinematically determinate base for which all node displacements are zero may be used either in the displacement method or in the force method based on the selection of redundant forces.

It will be clear that the calculation of member-stiffness matrices and fixed-end forces is an important part of the displacement method. The member stiffness may be calculated by inverting the corresponding flexibility matrix if force patterns are assumed in the member. For the finite elements discussed in Sec. 1.9, displacement patterns are used to describe the member distortions, and in this case the member stiffness may be built up directly. The stiffness method is found to be possibly the most useful method of analyzing many structural types, particularly those composed of finite elements. The displacement patterns mentioned above, due to one node's being given a unit displacement with all others held equal to zero, are easily generated since they are highly localized and affect only a few members.

For a comparison of the force and displacement methods used in framed structures one should look at the direct stiffness method given in Chap. 7 and the structure-cutter approach in Chap. 5. These will give programs whose internal structures are vastly different but which require the same input data, namely, geometry, connectivity, elasticity, support conditions, and loading.

1.9 STRUCTURAL ELEMENTS OF MORE THAN ONE DIMENSION; THE FINITE ELEMENT

Perhaps one of the most interesting and useful developments that has arisen from the impetus given to research in structural mechanics by the digital computer has been in the application of the methods of analysis originally reserved for framed structures to problems of continuum mechanics. The progress in this field has been breathtaking in the new possibilities that have been revealed.

The problem posed is relatively simple, that is: Is it practical to devise an element that has two- or even three-dimensional characteristics instead of the one-dimensional nature of the line element used in framed structures? If such elements can be devised, then the continuum may be built up of a number of elements connected together in some prescribed manner. Finally both deflections and stress values will be produced. Many satisfactory elements have now been tested against known analytic solutions, and remarkable agreement observed. The elements in common use to date have generally been polygonal figures such as the triangle, tetrahedron, rectangle, and parallelogram. Recently, elements with curved sides have been devised and these seem to offer many advantages over the simple elements in specific applications. This use of elements of more than one dimension is now commonly referred to as the *finite-element method of analysis*.

Although the finite-element idea can be developed intuitively by analogy with the line element, it is important to realize that it can also be based on the minimization principles (see Ref. 38). Needless to say, the elements in use have been constructed with varying degrees of refinement and today plane stress (strain), plate bending, shell, and even general three-dimensional solids may be analyzed, once suitable programs are devised for handling the considerable amount of arithmetic that is involved.

To make a detailed comparison of the finite-element method with other numerical techniques such as the finite-difference method would take considerable effort. Our basis for a reasonably thorough study of the finite-element method in Chap. 8 will be that at the present time it seems to be the best engineering approach to the analysis of a great variety of complex stress problems. Some of its very useful features are listed below:

1 Displacement boundary conditions are readily accommodated.

2 Body forces and temperature strains may be included.

3 Material properties may vary throughout the continuum. Stratified material (such as found in rock mechanics problems) may be analyzed.

4 The actual stress-strain relationship for the material can be used within the element, and flow rules such as those in plasticity theory incorporated.

5 Buckling loads may be predicted and large displacements included.

6 Many different element types may be included in one program. For example, in plate problems, beam elements can be programmed, and in plane-stress analysis, truss elements can be included so that a material such as reinforced concrete is approximated.

Because classical structural analysis has used the principles of statics in the calculation of the primary forces in structures assumed to be statically determinate, the line element has most often had its elastic properties deduced in terms of known force fields. Two such force fields are shown in Fig. 1.12*a*. For a truss element, the force field is simply member force equal to S, a constant. For the beam element with terminal moments M_1, M_2 only, the moment field will be linear. The corresponding displacement fields (for prismatic members) are given in Fig. 1.12*b*. It is seen that for the truss ele-

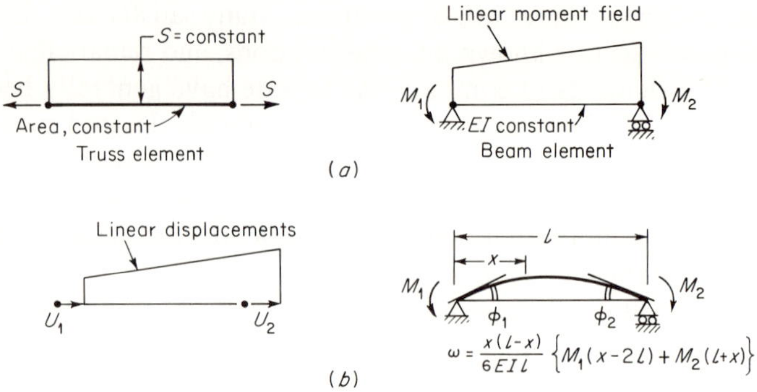

Figure 1.12 (*a*) Force fields; (*b*) displacement fields.

ment the displacement field is linear, whereas the beam deflection field, obtained from the linear moment field by double integration, is a cubic function. Evidently an alternative approach to assuming the force fields will be to assume the displacement fields and obtain the forces by suitable differentiation. Notice that if the member is nonuniform, the assumption of the displacement fields will not necessarily lead to the same force fields as for the prismatic member.

In finite-element analysis both force and displacement fields have been used, although the latter type has considerable vogue for use in compatible elements. Compatible elements are those for which displacements along a boundary are completely specified by the nodal point displacements contained in this boundary. When this is the case, no interference occurs between adjacent elements when they are strained. It should be noted, however, that noncompatible elements can still be useful. For example, in some plate bending elements the deflections in adjacent elements match, but normal slopes do not. Some of the criteria for constructing satisfactory elements are discussed in Chap. 8.

In order to focus our attention on the use of elements of more than one dimension, consider the thin cantilever beam of rectangular cross section loaded as shown in Fig. 1.13. The thickness h is considered small so that the beam is in a state of plane stress. The stress distributions near the load point P and at the built-in support will be complicated by the boundary conditions there. However, in the main portion of the beam, provided that $d < \frac{1}{10}l$, the stresses will be substantially those given by the linear bending-stress distribution for σ_x and the parabolic shear-stress distribution for τ_{xy}.

In Fig. 1.13*b* the beam is shown subdivided into a number of triangular elements in each of which the strain fields will be specified. It is most important to realize that Fig. 1.13*b* is not a physical model of the real structure in which the small triangles of material are pinned together in some way. Such a model would be altogether too flexible because of high stress concentrations at the pins. To return from this physical model to the finite-element model

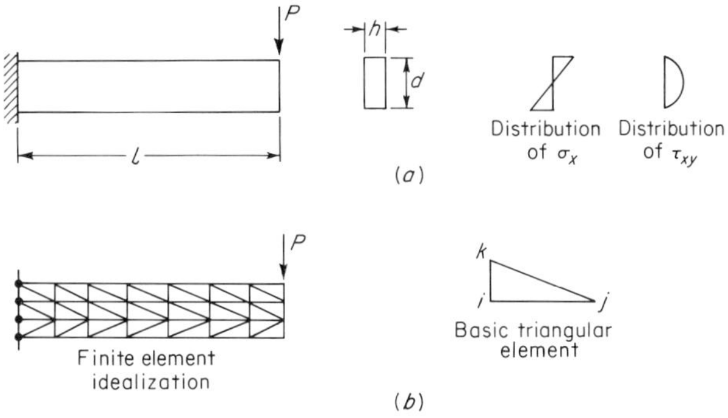

Figure 1.13 Finite-element idealization. (*a*) Cantilever beam; (*b*) idealized triangular subdivision.

in which strain fields are specified would require surface forces of unknown magnitude to be applied along the surfaces of each element.

Given that the triangles have the specified strain patterns, the triangles in the whole assemblage are simply constrained to have the same displacements at the nodes. Two triangle types and their node systems are shown in Fig. 1.14. The most elementary model that may be constructed for the triangle will have constant strains and hence linear displacements. That is, ε_x, ε_y, and γ_{xy} will each be constant, and hence if the thickness is constant, so also will σ_x, σ_y, and τ_{xy} be constant within the triangle. The displacements u and v in the x and y coordinate directions respectively will be linear functions, such that

$$u = \alpha_1 + \alpha_2 x + \alpha_3 y$$

$$v = \alpha_4 + \alpha_5 x + \alpha_6 y$$

It is now observed that if u and v are given at the nodes i, j, k, the α coefficients may be determined and hence u and v within the triangle. It will be noticed from Fig. 1.13*b* that the constant stresses within each triangle will not approximate the normal- and shear-stress distributions very accurately unless there are a reasonable number (for example, 10) of triangles across any section.

The second triangle shown in Fig. 1.14 is assumed to have a linear distribution of strains and hence for constant thickness, a linear distribution of

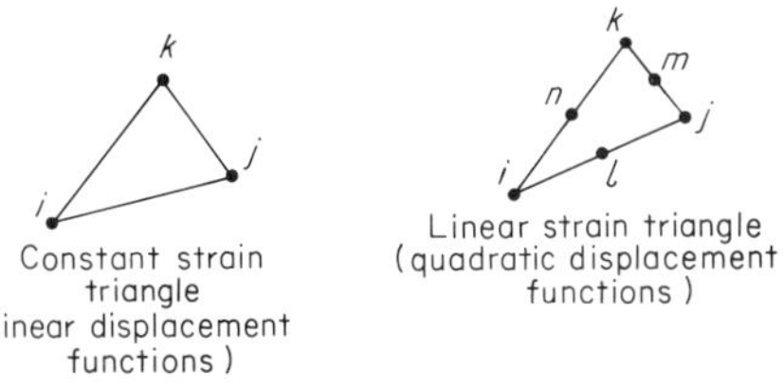

Figure 1.14

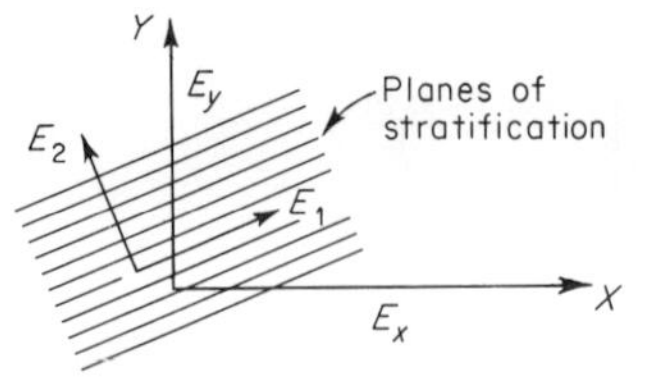

Figure 1.15 Stratified material.

stresses. On integration of the strain functions the displacements u, v are given as quadratic functions such that

$$u = \alpha_1 + \alpha_2 x + \alpha_3 y + \alpha_4 x^2 + \alpha_5 xy + \alpha_6 y^2$$
$$v = \alpha_7 + \alpha_8 x + \alpha_9 y + \alpha_{10} x^2 + \alpha_{11} xy + \alpha_{12} y^2$$

Now there are 12 parameters to be determined, and they may be calculated if the node displacements u, v are known at 6 nodes taken to be at the apices and midpoints of sides.

It is anticipated (and such is the case) that this element gives a better means of approximating the stresses in the beam.

The process of using the finite-element models is quite complicated. Briefly, it involves finding node forces statically equivalent to the stress fields in the element and also finding the elastic relationship between these and the corresponding node displacements. The theory is discussed in detail in Chap. 8. The finite-element idea has been introduced here in order to delineate further areas of study that are necessary preliminaries to the study of structural mechanics.

Firstly the multidimensional model has introduced stresses σ_x, σ_y, τ_{xy}, etc., and strains ε_x, ε_y, and ε_{xy}. These are different from the components of force and displacement vectors, and a prime concern will be to define their transformation theories for a change in coordinate axes, taken here to mean simply a rotation from one cartesian system to another. It is found that stresses and strains are tensor quantities and their transformation laws are different from those for vectors. At the same time it will be of interest to formulate Hooke's law in matrix form and examine the effect thereon of a rotation of axes.

For example, a material may be stratified as shown in Fig. 1.15 in a direction that differs from the cartesian reference axes. It will be of considerable importance to be able to write down Hooke's law in the reference axes if it is known in the stratified axes. The displacement functions are in effect polynomials that express the interpolation for the function within the element. The use of xy coordinates is clumsy for the triangle, and in the theory of interpolation presented in Chap. 8, it is found that a new set of coordinates, called *natural* (or *area*) *coordinates,* leads to much simpler expression of the interpolation.

1.10 STRUCTURAL BEHAVIOR; NONLINEAR EFFECTS; AXIAL-FORCE EFFECTS AND BUCKLING

Traditionally, structural analysis has meant the investigation of the behavior of line structures (trusses, frames, and grids) under loads whose magnitudes are such that:

1 The effects of geometry changes are small and may be neglected in their influence on linear theory.

2 The material properties are everywhere elastic.

It might be argued that the linear elastic range, being the operational zone of the structure, is of greatest interest. Such may be the case, but nevertheless, the inelastic range will be of importance too when factors of safety against collapse and deflections for overload conditions are being estimated. Of course, for materials such as some plastics and soil, the nonlinear characteristics may be too important to ignore at any stage. It is found also, that in the response of buildings to seismic loading (a topic beyond the scope of this text), some members will yield, which will greatly increase the absorption of energy by the structure.

The load-deflection curve for the portal frame shown in Fig. 1.16*a*, (assumed to be braced against out-of-plane buckling) may be as shown in Fig. 1.16*b*. It is seen that the curve is highly nonlinear with the small deflection-elastic range's being only a small proportion of the total curve. However, at any load P_1 in the load-deflection history, a tangent may be drawn to the curve so that over a limited range of load, the behavior is approximately linear. Although the slope of this tangent differs from the initial elastic slope, the basic ideas of the linear theory still apply. In this way, the whole curve may be approximated by loading with a sequence of incremental loads and adjusting the geometry and member properties at each step. Some aspects of this type of nonlinear analysis are discussed in detail in Chap. 9.

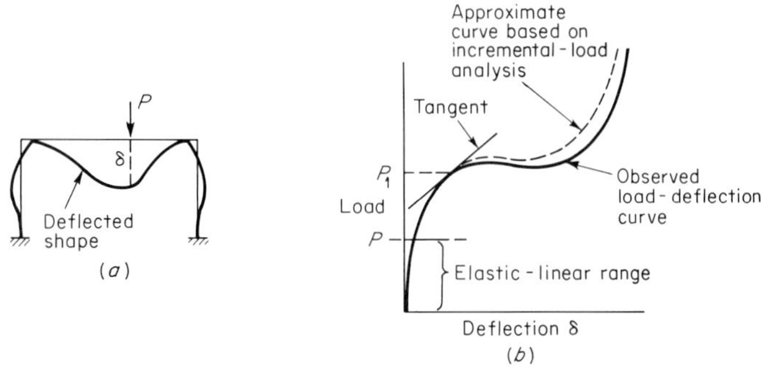

Figure 1.16 Structure deflections. (*a*) Structure geometry; (*b*) load-deflection curve.

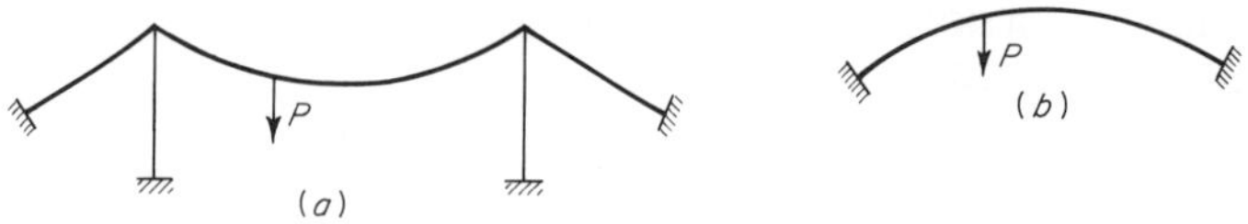

Figure 1.17 Structures with large axial forces. (*a*) Suspension cable; (*b*) arch rib.

Another interesting feature of classical linear structural analysis is that it overlooks the effect that axial forces in the structure members have on stiffness, even though their effect may be linear (i.e., the tangent stiffness discussed above). The two obvious structures of this type are the suspension cable and the arch rib (Fig. 1.17). In both cases the dead load of the structure may greatly influence the initial elastic slope of the load-deflection curve for P. Thus in Fig. 1.18*a* the effect of axial tension in the cable is to produce a stiffening of the structure, whereas in the arch, the axial compression renders the structure more flexible as shown in Fig. 1.18*b*. The effect of axial force on structure equilibrium (stiffness) can be visualized readily by considering a tight wire, shown in Fig. 1.19, of length l, carrying a tension T. In the initially straight position, the elastic support available for the transverse load P is virtually zero, since the bending inertia of the wire is very small. However, the wire may carry the load P provided that the deflection δ is small, according to the resolution of forces at the point A,

$$P = \frac{4T}{l}\delta$$

It is clear that this is a linear relationship between load and deflection. Such a process is of course well known to the prestress-concrete designer who uses deflected, high-tensile strands in tension to carry transverse loads. In this case the deflected cables are in effect frozen in position by the adjacent concrete.

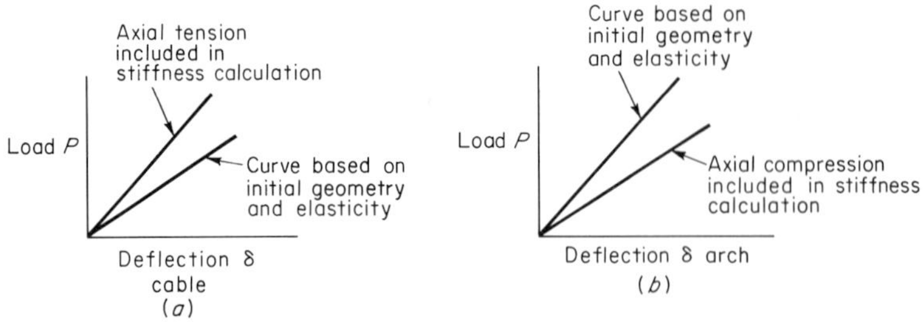

Figure 1.18 Initial load-deflection curves.

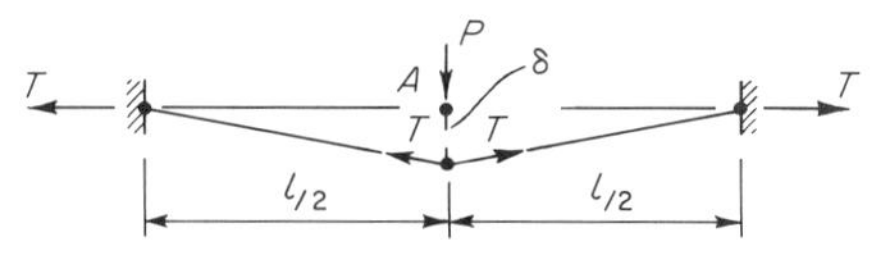

Figure 1.19 Deflected cable.

Stability Analysis; Euler Loads

It will be shown in Chap. 9 that this idea of geometric stiffness (i.e., the change in joint stiffness due to the presence of axial forces in members) may be used to good effect in the calculation of buckling loads of highly complicated structures. It suffices here to illustrate the principle by considering the rigid column, shown in Fig. 1.20, hinged at one end and supported at its free end in the transverse direction by a spring whose force per unit deflection is k. The column of length l carries an axial force S that may be tensile or compressive and a transverse force P at the free end. With the member in the deflected position in Fig. 1.20a, equilibrium of the joint A requires that the sum of the forces in the transverse direction be equal to zero. That is,

$$P - S\frac{\delta}{l} - k\delta = 0$$

It is now asked: For what member force S will the column remain in the deflected position when P is equal to zero? In this case,

$$S\frac{\delta}{l} = -k\delta$$

That is, the deflection δ is indeterminate and

$$\frac{S}{l} = -k$$

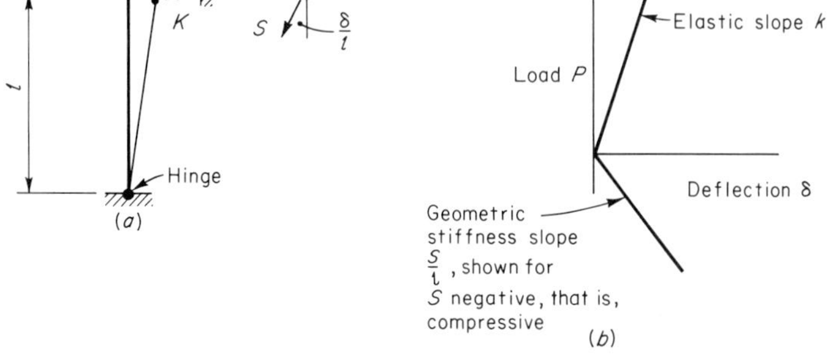

Figure 1.20 Column stability.

This condition is graphed in Fig. 1.20*b* in which the two stiffness slopes shown are equal and opposite. The condition is also recognized as the bifurcation of the equilibrium when $S = -kl$, as shown in Fig. 1.21. The horizontal line drawn in Fig. 1.21 is, of course, only a mathematical fiction. In reality the structure performance will probably be as shown by the dotted line, and the meaning of the bifurcation point is simply that from this point on, the theory must consider large displacements. It is interesting to note that the effect of the displacement δ can be treated in two ways. In the first instance, the column can be considered to be in its original position as far as equilibrium is concerned and a fictitious force $-S(\delta/l)$ applied to allow for the displacement δ. That is, the new joint load will be $P - S(\delta/l)$ in the transverse direction.

The alternative approach is to consider the stiffness k modified by the factor S/l so that the equilibrium equation at the joint is written

$$P - \left(\frac{S}{l} + k\right)\delta = 0$$

That is, the effective stiffness of the joint in the transverse direction is

$$k_{\text{tangent}} = \frac{S}{l} + k$$

When the above theory is applied to structures with many joints and hence many degrees of freedom, the resulting equations take the form

$$AX = \lambda X$$

In matrix algebra this equation is recognized as a characteristic or eigenvalue equation, and solutions exist for the determinantal equation,

$$|A - \lambda I| = 0$$

Recourse may be had to many methods in order to solve this equation. In the buckling problem only the lowest critical load may be of interest, and in this case some iterative process may be used to calculate λ_1 and the corresponding buckling mode X_1. Perhaps the most convenient way to solve the

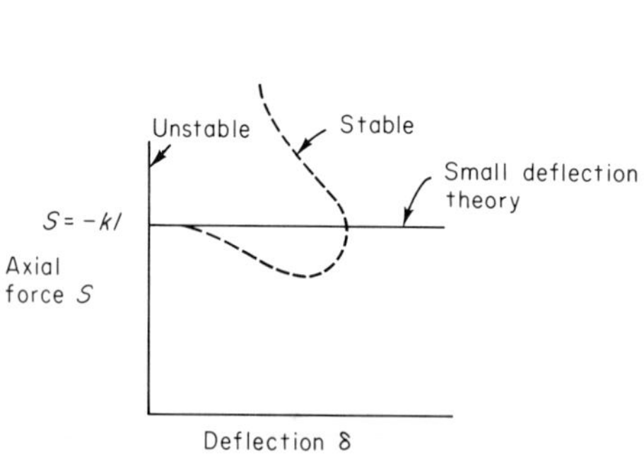

Figure 1.21 Large-deflection effects.

characteristic equation is to use the appropriate subroutine obtained from the computer center library. This will be almost certain to have the facility for extracting the required maximum or minimum root. Notice that just as δ was indeterminate in the simple column problem, so also will be the magnitude of the mode X_1. That is, the shape of the buckling mode only has been determined. It is usual to normalize the vector X_1 so that either its largest amplitude is equal to unity or, alternatively, *mod* X_1 is equal to unity.

It is shown in Chap. 2, that the strains associated with a given displacement have a linear part and a quadratic part. The linear terms produce the familiar deflections of the linear elasticity theory.

The quadratic terms are found to be involved in the geometric stiffness concept. These terms are derived in Chap. 2 for later use in Chap. 9 on nonlinear analysis.

The large-displacement analysis will have many important applications in buckling because it may be used to predict the postbuckling behavior and strength of a structure. In the case of plates this reserve in strength may be considerable.

2 TRANSFORMATIONS. VECTORS, TENSORS

It may be said without exaggeration that the metamorphosis in our mental processes arising from the digital computer, possibly a more profound one than that due to atomic physics and engineering, is still not appreciated sufficiently in many places. As yet we have merely started drawing upon the inherent capabilities of the electronic computing devices, and the vision formed in the mind of a man like John Von Neumann is only to a very small degree fulfilled.

JOHN H. ARGYRIS AND S. KELSEY
Modern Fuselage Analysis and the Elastic Aircraft

2.1 FORCES AND DISPLACEMENTS AS VECTOR QUANTITIES. COMPONENTS GIVEN IN MATRIX FORM

In Chap. 1, we have outlined some of the thoughts behind the development of modern structural analysis. It was mentioned therein, that although the theory may be employed in cases where nonlinear behavior is encountered, the basic principles involve linear analysis. Fundamental to the treatment of this subject will be a study of force and displacement quantities with their vector components expressed in matrix form.

The parallelogram law for the addition of force quantities, proved by simple experiment, provides the relation between the physical phenomena and the mathematical abstraction of the concept of force. The law, too, is a useful computational tool allowing for easy addition of force quantities by graphic means. In addition, it leads to fundamental conditions of equilibrium expressed in a graphic and perhaps one could say even a pictorial form. Thus for the equilibrium of forces at a point, their force polygon must close. In addition, for a general force system in a plane, the funicular polygon must close. This condition ensures that the moment of the forces about any point in the plane is zero for equilibrium.

Historically, it seems that in the first stages of the development of the study of statics, this graphic presentation had not only a useful purpose in aiding the student's perception of the new science, but also immense calculational advantages. In fact it removed the necessity for numerical calculation and placed reliance on accurate draftsmanship. Even in today's world of high-speed calculators and computers, it is apparent that such a graphic presentation should be given in a first course on theory of structures. The lack of mechanical equipment of sufficient speed and adaptability to handle the arithmetic process led to a whole era of the development of the art of graphic statics. Much of this, for example, in truss analysis, is still in vogue today. The extent of development may be judged by the treatises on strength of materials in which continuous beams are solved by funicular polygon[3] or in which moments of inertia of irregular sections are calculated by graphic means.

Early investigators were not slow to realize that displacements could also be added by the parallelogram law and that in some way, forces and displacements exhibited similar mathematical properties. Thus both forces and displacements are prescribed by a magnitude and direction of action or movement. Moments and rotations as well exhibit these same *vector* properties.

Naturally, the analytic treatment of structural analysis has suffered with this insistence on graphic presentation. The present age of computer utilization has freed the subject of structural analysis from this bond with graphics, and there has been a corresponding increase in the depth of understanding of fundamentals probably not thought conceivable a few decades ago. Anticipating later chapters in this book, one may cite the contragradient nature of force and associated displacement quantities in linear transformation as an important example of the new theory. It will be found that such a concept necessarily requires that all components of force and displacement quantities be considered simultaneously and thus is completely disguised if only single components are studied. This might be considered analogous to an attempt to explain the inverse of a matrix by using the reciprocal of a single integer as an example.

Developed probably in the initial stages as an exercise in pure mathematics, matrix algebra has become an increasingly important tool of structural analysis, until one might speculate that now a unified matrix theory of structures is available. All previous methods of analysis have simply been incomplete variations of this single central theory.

In the matrix presentation of forces and displacements, the components of the vector quantity become all important and the laws of vector algebra have their equivalent statements in matrix algebra. To start a discussion on force and displacement systems, it will be necessary to define the coordinate systems in which they may be defined. There may be only one, overall set of coordinates for the whole system, sometimes designated *global* or *frame coordinates*. In addition, there may be separate coordinate systems for each component or region of the structure. These will be denoted as *member* or *local coordinates*. Usually, the global or frame coordinates will be a right-handed, rectangular cartesian set in which the complete geometry of the structure is expressed. The member or local coordinates may also be rectangular cartesian sets, but sometimes it will be found useful to define a new type of natural coordinates that have advantages in expressing the particular member properties.[10,38]

Vector or matrix properties will be related to the coordinate system under consideration. Thus for two force vectors to be additive, they must be of the same form (distinguishing between force and moment vectors) and have components expressed in the same coordinate system. If, however, these forces act at different points on a body, their total effect will not be given simply by the addition of components, but the action of one force must be transferred to a point on the line of action of the other (or alternatively they may both be transferred to one point) and this will involve generation of moment components as well.

The theory will be largely concerned with how these force (or displacement) quantities transform, that is, how their components expressed as row (or column) matrices transform from one set of coordinate axes to another and from one point to another in the system.

Now, force and displacement systems for a deformable body are not usually independent. Forces are applied and generate a set of displacements (or vice versa). If the material is assumed to be linear elastic (or hookean), there will be a linear relationship between the forces and the displacements, provided that the displacements are not of such magnitude as to cause a gross distortion of the body so that the statics of the structure is significantly changed. If, however, the loads are of sufficient magnitude, it may be that the material exhibiting nonlinear properties above a certain stress level will deform plastically and the load-deflection relationship will no longer be linear. When a structure is assumed to behave in a linear elastic manner for both material and geometric properties, the principle of superposition is valid. This means that the total effect of several force systems is the sum of their separate effects.

The various possibilities of linear and nonlinear characteristics for the simple slender column with an eccentric load are shown in Fig. 2.1. Notice in Fig. 2.1 that even if the material is considered nonyielding, the behavior along line *B* is nonlinear. This will be accentuated when plastic distortions occur, line *C*, since the deflections are then larger and the geometry change more pronounced. The actual, observed deflection will be curve *D*. The first portion of all the curves *B* to *D* has been drawn to coincide with *A* to show that for a load below a specified level the structure behaves in a linear fashion.

Given then the elastic structure, a linear relationship exists between forces and displacements. This relationship will be specified in a certain coordinate system, and one may enquire just how this is effected in a transformation of components to another coordinate system. It is found in this case that the transformation is slightly more complicated than that for the vector quantities, it being in the nature of a tensor transformation.

It will be most convenient to begin with components of forces, moments, displacements, and rotations, expressed in a right-handed, rectangular sys-

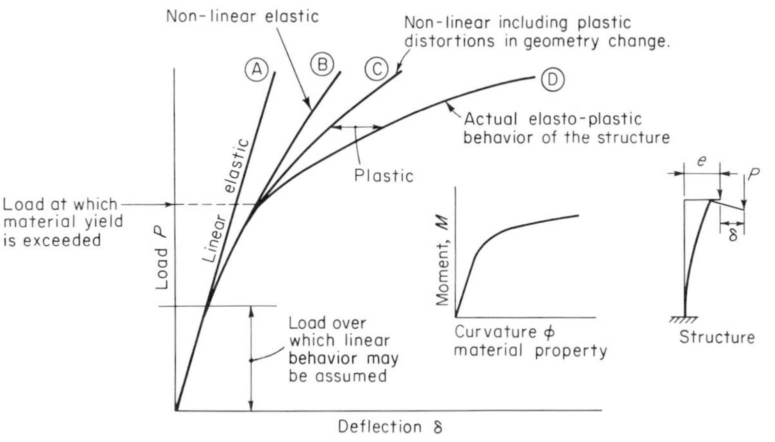

Figure 2.1

tem of cartesian coordinates. This right-handed system will be used almost exclusively since it leads to a consistency with the sign conventions of coordinate geometry theory. In Fig. 2.2, the sign convention for positive force and displacement components is given. Notice the double-headed-arrow notation used for the graphic representation of moment and rotation vectors. Special note should be made of the positive sense of moment and rotation vectors. The positive sense will be in a clockwise direction when looking along the arrow toward the head. This is shown pictorially in Fig. 2.2*b* and *d*.

Consider two simple cases of force and moment components given in Fig. 2.3*a* and *b*. In Fig. 2.3*a* I, the member is in pure tension; in Fig. 2.3*a* II, a pure shear is applied to the end. In Fig. 2.3*b* I, the member has a torque applied, whereas in Fig. 2.3*b* II, it is subjected to pure bending moment causing tension on its top fiber.

The components of a vector given as in Fig. 2.2 completely define its magnitude and direction. This may be demonstrated by defining unit vectors $\hat{i}, \hat{j}$, and $\hat{k}$ along the coordinate axes OX, OY, OZ respectively and then using the parallelogram law to add first the X and Y components and then this resultant with the Z component. Thus, if the force vector F has components (F_x, F_y, F_z),

$$F = (F_x\hat{i} + F_y\hat{j}) + F_z\hat{k} \tag{2.1}$$

In matrix notation the forces F and M and the displacements v and θ are defined by Eqs. (2.2) and (2.3).

Force and moment quantities:

$$F = \begin{bmatrix} F_x \\ F_y \\ F_z \end{bmatrix} \qquad M = \begin{bmatrix} M_x \\ M_y \\ M_z \end{bmatrix} \tag{2.2}$$

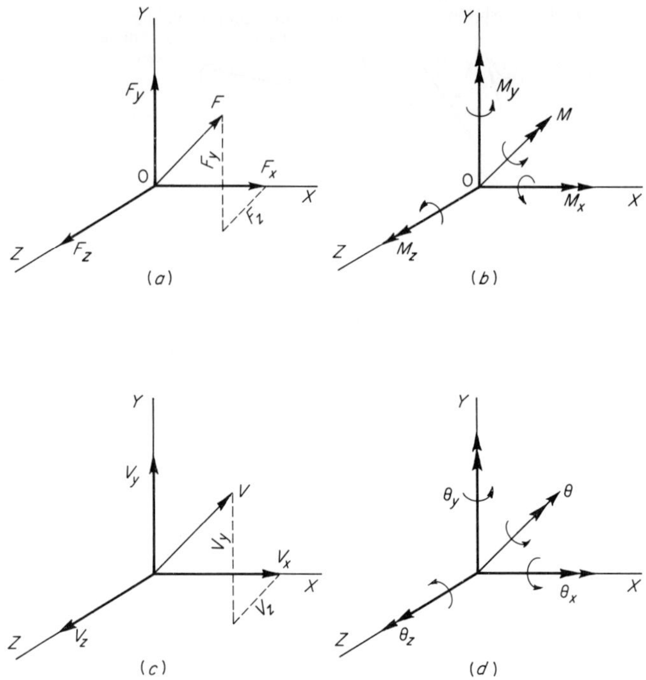

Figure 2.2 Positive forces and displacements. (*a*) Forces, (*b*) moments, (*c*) displacements, (*d*) rotations.

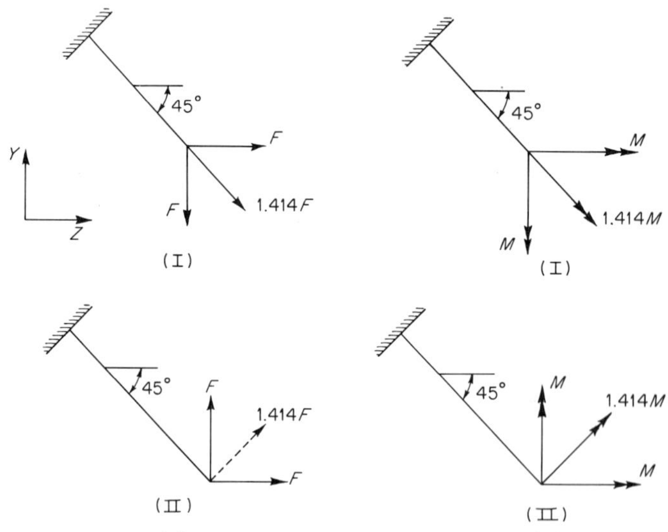

Figure 2.3

Displacement and rotation quantities:

$$v = \begin{bmatrix} v_x \\ v_y \\ v_z \end{bmatrix} \qquad \theta = \begin{bmatrix} \theta_x \\ \theta_y \\ \theta_z \end{bmatrix} \tag{2.3}$$

For convenience in writing it is often useful to make the definitions in transposed form. That is, for example,

$$F^T = \langle F_x \quad F_y \quad F_z \rangle \tag{2.2a}$$

In the example given in Fig. 2.3, the various force vectors will be:

Fig. 2.3*a* I: $F^T = \langle F \quad -F \quad 0 \rangle = F\langle 1 \quad -1 \quad 0 \rangle$

Fig. 2.3*a* II: $F^T = \langle F \quad F \quad 0 \rangle = F\langle 1 \quad 1 \quad 0 \rangle$

Fig. 2.3*b* I: $M^T = \langle M \quad -M \quad 0 \rangle = M\langle 1 \quad -1 \quad 0 \rangle$

Fig. 2.3*b* II: $M^T = \langle M \quad M \quad 0 \rangle = M\langle 1 \quad 1 \quad 0 \rangle$

It is not always necessary to subdivide force and moment (or displacement and rotation) quantities in this way, and often the force or displacement quantity referred to, for example, at one end of a member in a three-dimensional structure, will actually be a vector of six components defined by the combination of Eqs. (2.2) and (2.3). Thus define R and r as force and displacement matrices by Eqs. (2.4*a*) and (2.4*b*).

$$R^T = \langle F^T \quad M^T \rangle = \langle F_x \quad F_y \quad F_z \; M_x \quad M_y \quad M_z \rangle \tag{2.4a}$$

$$r^T = \langle v^T \quad \theta^T \rangle = \langle v_x \quad v_y \quad v_z \quad \theta_x \quad \theta_y \quad \theta_z \rangle \tag{2.4b}$$

It should be emphasized that the combined vectors of Eqs. (2.4) will be used for algebraic convenience and that the matrices (vectors) R and r should be considered composed of two vector quantities each with three components.

An immediate advantage of the rectangular cartesian coordinate system is seen when the magnitude of the vector is calculated, since all the vector additions in Eq. (2.1) are composed of right-angle triangles. Magnitudes are thus given in each addition by use of Pythagoras' theorem. That is, if magnitudes are being considered, Eq. (2.1) becomes

$$F^2 = (F_x{}^2 + F_y{}^2) + F_z{}^2 \tag{2.5}$$

In matrix or vector notation, the magnitude of the vector as defined in Eqs. (2.2) and (2.3) is given by the square root of the inner product of the vector with itself. The inner product in matrix notation is found to be the premultiplication of the vector by its transpose.

If the magnitude of the vector is denoted mod $F = |F|$,

$$|F| = \sqrt{F^T F} = \left\{ [F_x \quad F_y \quad F_z] \begin{bmatrix} F_x \\ F_y \\ F_z \end{bmatrix} \right\}^{1/2} = (F_x{}^2 + F_y{}^2 + F_z{}^2)^{1/2} \tag{2.6}$$

The components (F_x, F_y, F_x) of the vector are its direction numbers. That is, they specify its directional properties in addition to defining its magnitude as in Eq. (2.6). These may be made nondimensional by dividing by the vector magnitude $|F|$, and these nondimensional parameters are the direction cosines of the vector with reference to the X, Y, Z coordinate axes. These direction cosines are

$$l_x = \frac{F_x}{|F|} \qquad l_y = \frac{F_y}{|F|} \qquad l_z = \frac{F_z}{|F|} \tag{2.7}$$

That is,

$$l^T = \langle l_x \quad l_y \quad l_z \rangle = \frac{1}{|F|} F^T \tag{2.8}$$

Equation (2.8) may be used in a variety of ways, for example,

$$F^T = |F| \langle l_x \quad l_y \quad l_z \rangle \tag{2.8a}$$

It should be noticed even at this stage that the matrix notation in Eqs. (2.6) and (2.8*a*) leads directly to the arithmetic process. Since it is the intention to rely entirely on computer methods, this ready translation to arithmetic makes the matrix notation ideal for program coding.

The parallelogram law for the addition of two vectors may be expressed simply as the addition of their corresponding components. That is,

$$F = F_1 + F_2 = \begin{bmatrix} F_{1x} \\ F_{1y} \\ F_{1z} \end{bmatrix} + \begin{bmatrix} F_{2x} \\ F_{2y} \\ F_{2z} \end{bmatrix} = \begin{bmatrix} F_{1x} + F_{2x} \\ F_{1y} + F_{2y} \\ F_{1z} + F_{2z} \end{bmatrix} \tag{2.9}$$

Let the vector $F_1{}^T = \langle 10 \quad 12 \quad 0 \rangle$; $F_2{}^T = \langle 5 \quad 7 \quad 0 \rangle$. By Eq. (2.9) $F^T = \langle 15 \quad 19 \quad 0 \rangle$. This addition is verified in Fig. 2.4.

2.2 THE DIRECTED LINE ELEMENT

In line with our policy to work firstly with simple structures and to generalize them for complicated systems, the first structures to be considered will be composed of straight members connecting certain specified nodes. These will, of course, be truss or frame structures considered either in the plane or more generally in three dimensions. A node is defined as the junction point of two or more members or as the connection point of the structure with the foundation media. To designate these, it will be convenient, and in fact nec-

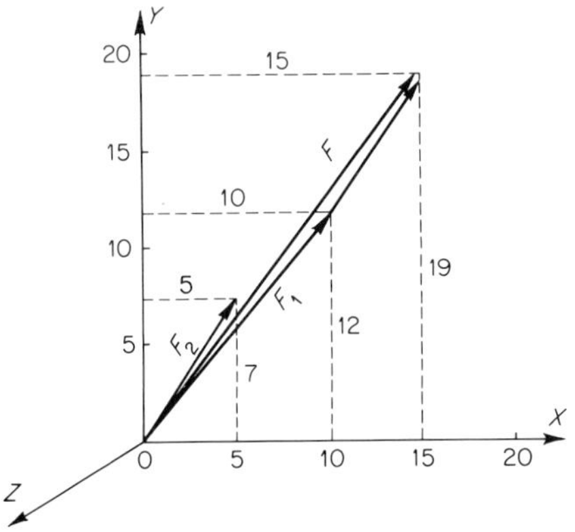

Figure 2.4

essary for efficient computer solution, to number the nodes and to specify member position in the structure by the node numbers at its ends. This configuration of members and nodes gives the topology of the structure and may be shown to exhibit fundamental properties independent of the dimensions or material properties of the structure[4,9]. For each node, a set of coordinates (x,y,z) is specified, and these (for straight prismatic members) completely define the geometry of the structure.

Consider the plane truss shown in Fig. 2.5. The nodes have been numbered 1, 2, . . . , 5 and the members (1), (2), . . . , (7).

In the truss the general member will be designated i, j and the positive sense of the member given in the direction i to j. This means that the member has been given vector status, and this vector is used later as one of the local coordinates for the member.

In some computer programs it is necessary to make the member node numbering such that the node number i is always less than node number j. This convention is followed in Table 2.1, in which node and member properties for the truss in Fig. 2.5 are given.

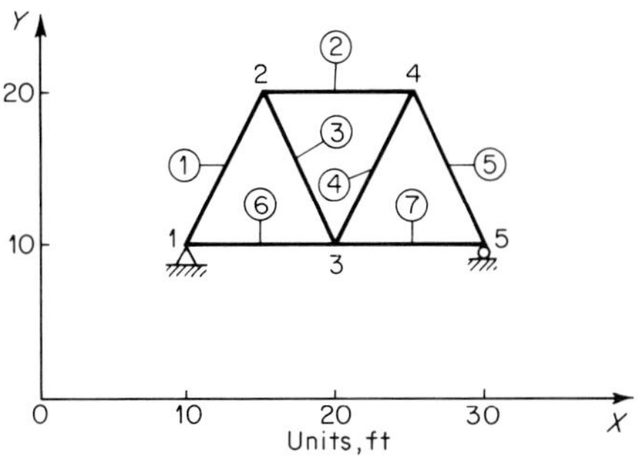

Figure 2.5 Line elements.

TABLE 2.1 (*a*) **Node coordinates**

Node No.	*X* coordinate	*Y* coordinate
1	10.0	10.0
2	15.0	20.0
3	20.0	10.0
4	25.0	20.0
5	30.0	10.0

(*b*) **Member connectivity**

Member No.	Node *i*	Node *j*
1	1	2
2	2	4
3	2	3
4	3	4
5	4	5
6	1	3
7	3	5

In this text it has been found convenient to group structures according to whether their behavior is of a two- or three-dimensional nature. In the two-dimensional group there is the additional subdivision into plane frames and grid structures. Both are considerably less complex than the general three-dimension group. There will be other structures such as high-rise rigid frames and pile groups with rigid headstocks that deserve special consideration.

The member node numbers given in Table 2.1*b* can be extracted from the table and used as coordinate indices in Table 2.1*a*. That is, let I, J be node numbers for member N of Table 2.1*b*. Then the coordinates of the ends of member N are extracted from Table 2.1*a*, as $X(I)$, $Y(I)$, $X(J)$, and $Y(J)$

Let the vector

$$M^T = \langle [X(J) - X(I)] \quad [Y(J) - Y(I)] \rangle \tag{2.10}$$

Then the length of member N is given as

$$p = |M| = (M^T M)^{1/2}$$

For the structure in three dimensions the same definition applies except that

$$M^T = \langle [X(J) - X(I)] \quad [Y(J) - Y(I)] \quad [Z(J) - Z(I)] \rangle \tag{2.11}$$

The direction cosines of the member are given, according to Eq. (2.8), as

$$l^T = \langle l_x \quad l_y \quad l_z \rangle = \frac{1}{|M|} M^T \tag{2.12}$$

Expansion of Eq. (2.12) shows that $l_x = [X(J) - X(I)]/|M|$, etc.

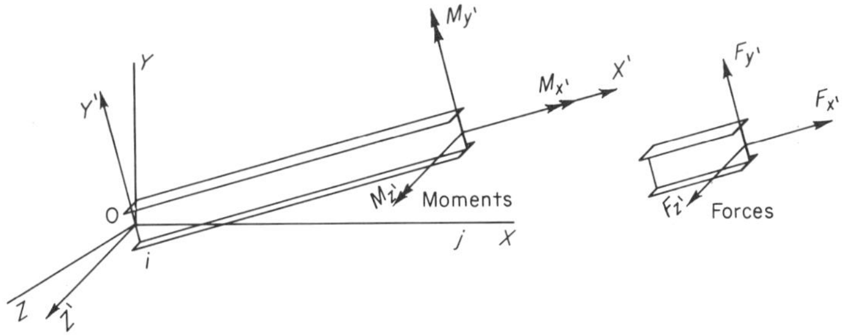

Figure 2.6 Local and global coordinate axes.

The above definition of member axes is adequate for the analysis of pin-jointed trusses. However, for rigid-jointed structures, the moments and shears occur at right angles to the member length and it is necessary to define a complete set of local coordinate axes in which to reference member force and distortion components. Because the distortions of a member due to applied bending moments are most readily calculated if principal axes are used to express their components, the local coordinate axes will be taken as the member direction for the OX' axis, and the principal axes of the cross section in a positive, right-handed sense to OX' will be taken as the OY' and OZ' axes. The axes for an I section are shown in Fig. 2.6. Force and moment vectors $(F,M)_j$ are shown acting on end j and are in the $X'Y'Z'$ system.

If all the member Z' axes of a given planar structure are perpendicular to the plane and all loads applied in the plane, then all member force components out of the plane are zero except M_z and these moments produce in-plane distortions. The behavior of the structure is then described completely in two dimensions.

Alternatively, if all members lie in the XZ plane and all Y' axes of members are perpendicular to this plane and all loads applied in the Y direction, the structure again behaves in a two-dimensional manner even though deflections occur at right angles to the plane of the structure.

The planar and grid types are shown in Fig. 2.7*a* and *b*. Note that for the planar structure, the nonzero force components are (F_x,F_y,M_z) with corresponding displacements (v_x,v_y,θ_z). In the grid structure these are altered to (M_x,M_z,F_y) and (θ_x,θ_z,v_y) respectively. Thus in either case, for any member or joint only three force and three deflection quantities need be considered.

2.3 TRANSFORMATIONS OF FORCE AND DISPLACEMENT COMPONENTS FOR ROTATION OF COORDINATE AXES; LOCAL AND GLOBAL COORDINATE SYSTEMS; TWO-DIMENSIONAL STRUCTURES

In Sec. 2.2 it has been shown that, in general, more than one set of coordinates may be used in describing the geometry of a structure, its member properties, force systems, and node deflections. It is convenient to express all

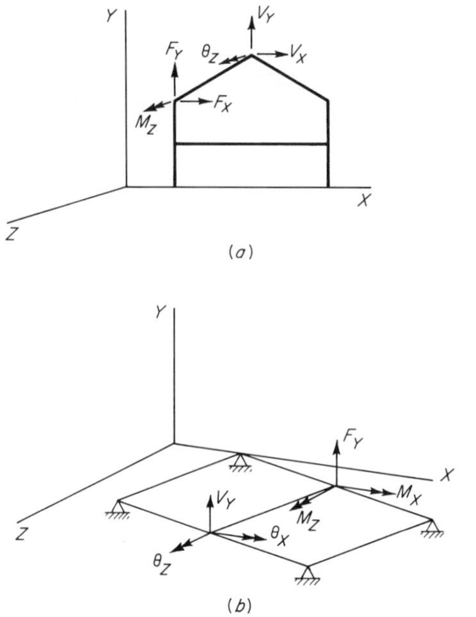

Figure 2.7 Planar structures. (*a*) Plane frame; (*b*) grid structure.

node force and displacement quantities in the global system. Member properties and forces are more readily (and usefully) expressed each in its own particular local set of coordinates. It is fundamental to the theory of structures to obtain the transformations relating the set of variables in one system to the equivalent set in a second system. From Fig. 2.6 it is seen that the member coordinate system $X'Y'Z'$ is obtained from XYZ by a rotation about the point O. In this section the fundamental transformation laws for relating vector components under rotation of coordinate axes will be obtained for plane frame and grid structures, that is, for two-dimensional coordinate systems. The first will be for XY components for rotations of the axes about OZ, and the second for XZ components for rotations about the OY axis. It should be apparent that in the first instance, components in the Z direction are unaltered and in the second, components in the Y direction are unaltered.

Consider, in the first instance, the plane frame with forces and displacements whose components, except of course for M_z and θ_z, are all in the XY plane. It is required to calculate the new components of the force $F^T = \langle F_x \quad F_y \rangle$ when the axes are rotated, as shown in Fig. 2.8*a*, about OZ through the angle α in the positive sense. Since the Z component remains unaltered, it may be neglected and the rotation of axes is then shown diagrammatically in Fig. 2.8*b*. In Fig. 2.8*b* the two sets of components $\langle F_x \quad F_y \rangle$ and $\langle F'_x \quad F'_y \rangle$ are shown for the force F.

The components $\langle F'_x \quad F'_y \rangle$ are obtained from $\langle F_x \quad F_y \rangle$ by projecting each

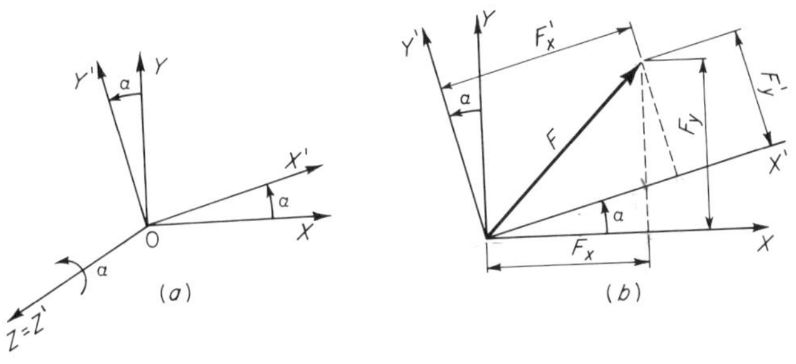

Figure 2.8 (*a*) Rotation about *OZ*; (*b*) components of *F* in *X'Y'* systems.

in turn onto the axes OX' and OY'. Thus,

$$\begin{aligned} F'_x &= F_x \cos \alpha + F_y \sin \alpha \\ F'_y &= -F_x \sin \alpha + F_y \cos \alpha \end{aligned} \tag{2.13}$$

These two equations may be expressed succinctly in matrix form as Eq. (2.14),

$$\begin{bmatrix} F'_x \\ F'_y \end{bmatrix} = \begin{bmatrix} \cos \alpha & \sin \alpha \\ -\sin \alpha & \cos \alpha \end{bmatrix} \begin{bmatrix} F_x \\ F_y \end{bmatrix} \tag{2.14}$$

On further examination it is seen that the trigonometric terms in the coefficient matrix on the right-hand side of Eq. (2.13) may be interpreted as the direction cosines of OX' and OY' with respect to the X, Y axes. Thus for OX',

$$\langle l_{11} \quad l_{12} \rangle = \langle \cos \alpha \quad \sin \alpha \rangle$$

and for OY',

$$\langle l_{21} \quad l_{22} \rangle = \langle -\sin \alpha \quad \cos \alpha \rangle$$

Hence,

$$\begin{bmatrix} \cos \alpha & \sin \alpha \\ -\sin \alpha & \cos \alpha \end{bmatrix} = \begin{bmatrix} l_{11} & l_{12} \\ l_{21} & l_{22} \end{bmatrix} = L^T \tag{2.15}$$

Substitution of Eq. (2.15) in (2.14) yields

$$\begin{bmatrix} F'_x \\ F'_y \end{bmatrix} = \begin{bmatrix} l_{11} & l_{12} \\ l_{21} & l_{22} \end{bmatrix} \begin{bmatrix} F_x \\ F_y \end{bmatrix} \tag{2.16}$$

Finally, Eq. (2.16) may be written by using the symbolic form for the matrix and force quantities, that is,

$$F' = L^T F \tag{2.17}$$

It should be noted that the direction cosines of the X', Y' axes in fact define unit vectors along these axes and hence

$$l_{11}^2 + l_{12}^2 = 1$$
$$l_{21}^2 + l_{22}^2 = 1 \tag{2.18}$$

Further, since the vectors so defined are orthogonal,

$$l_{11}l_{21} + l_{12}l_{22} = 0 \tag{2.19}$$

The results of Eqs. (2.18) and (2.19) may be expressed in matrix form by saying that the matrix L is orthogonal. That is, that the product of L with its transpose is equal to the unit matrix I. That is,

$$LL^T = I \tag{2.20}$$

It is easily shown by expansion that the terms of the product LL^T are equal to the left-hand sides in Eqs. (2.18) and (2.19), and hence (2.20) is true.

Expressed differently, the orthogonality relationship states that the transpose of L is its inverse. That is,

$$L^T = L^{-1} \tag{2.21}$$

It is established in Eq. (2.21) that the inverse of the matrix L exists and hence the transformation in Eq. (2.17) is reversible. That is, if the components of the force vector F' are known in the $X'Y'$ coordinate system, those in the XY system may be obtained by multiplying both sides of the equation by $(L^T)^{-1} = L$, namely,

$$F = LF' \tag{2.22}$$

Example 2.1 **Calculation of cantilever-beam deflections in global coordinates**

The cantilever beam shown in Fig. 2.9 has material properties $EI = 10^4$ in.-kip units, and $EA = 5 \times 10^2$ in.-kip units. The end j is subjected to the force vector $F^T = \langle 10 \quad 8 \rangle$. It is required to calculate the deflections $v^T = \langle v_x \quad v_y \rangle$ of end j in the global coordinates.

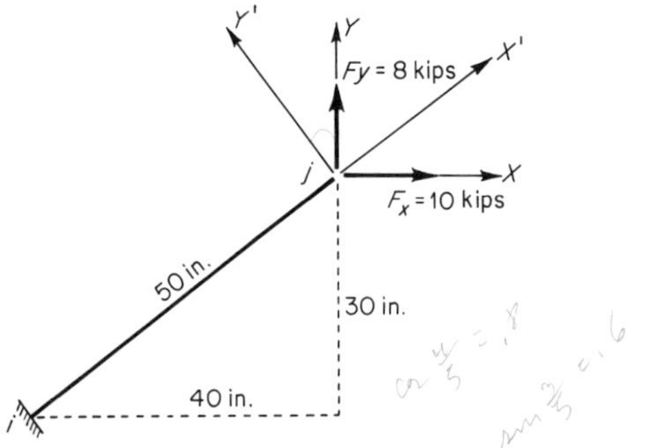

Figure 2.9

The L^T matrix for the member coordinates is calculated to be

$$L^T = \begin{bmatrix} 0.8 \\ -0.6 \end{bmatrix} \begin{bmatrix} 0.6 \\ 0.8 \end{bmatrix}$$

Therefore,

$$F' = L^T F = \begin{bmatrix} 0.8 & 0.6 \\ -0.6 & 0.8 \end{bmatrix} \begin{bmatrix} 10 \\ 8 \end{bmatrix} = \begin{bmatrix} 12.80 \\ 0.40 \end{bmatrix}$$

Deflections of the cantilever end j may now be calculated by using the F' force components and the simple formulas from strength of materials. Thus,

$$v' = \begin{bmatrix} v'_x \\ v'_y \end{bmatrix} = \begin{bmatrix} \dfrac{F'_x l}{EA} \\ \dfrac{F'_y l^3}{3EI} \end{bmatrix} = \begin{bmatrix} 0.04 \\ 1.06667 \end{bmatrix} \text{in.}$$

Finally, Eq. (2.22) may be used to transform the displacement components to the XY coordinates. That is,

$$v = Lv' = \begin{bmatrix} 0.8 & -0.6 \\ 0.6 & 0.8 \end{bmatrix} \begin{bmatrix} 0.04 \\ 1.0667 \end{bmatrix} = \begin{bmatrix} -0.608 \\ 0.8773 \end{bmatrix}$$

Transformations for Grid Structures

Consider this second class of two-dimensional systems in which member moments and torques are produced by loads applied at right angles to the plane of the grid. The basic transformation of coordinates from frame to member coordinates is a rotation about the OY axis, shown in its positive sense in Fig. 2.10a.

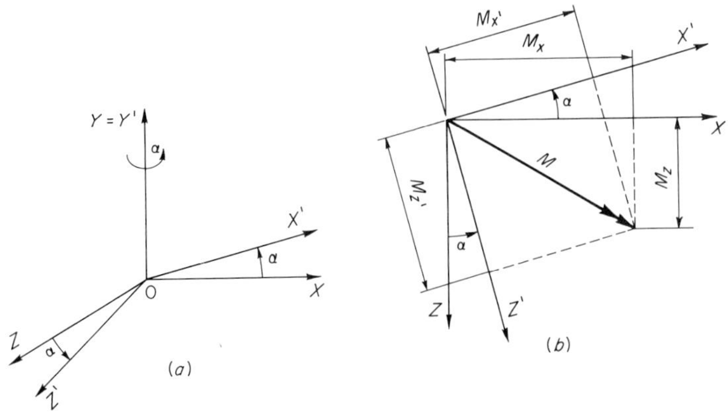

Figure 2.10 (a) Basic rotation of coordinates; (b) components of M in XZ and $X'Z'$ systems.

Proceeding in the same way as in Eq. (2.13), it is seen that the components $\langle M'_x \quad M'_z \rangle$ of the moment vector are obtained from $\langle M_x \quad M_z \rangle$ by projecting each in turn onto the axes OX', OZ' and summing terms. That is,

$$\begin{aligned} M'_x &= M_x \cos\alpha - M_z \sin\alpha \\ M'_z &= M_x \sin\alpha + M_z \cos\alpha \end{aligned} \tag{2.23}$$

In matrix form,

$$\begin{bmatrix} M'_x \\ M'_z \end{bmatrix} = \begin{bmatrix} \cos\alpha & -\sin\alpha \\ \sin\alpha & \cos\alpha \end{bmatrix} \begin{bmatrix} M_x \\ M_z \end{bmatrix} \tag{2.24}$$

The transformation matrix of Eq. (2.24) is not the same as that of Eq. (2.14), and it is convenient to change the coordinate axes to make them agree. To do this, consider the structure to lie in the XY plane and to be loaded by forces parallel to the Z axis. This new set of coordinates is shown in Fig. 2.11*a* and *b*.

The projection of (M_x, M_y) components onto the X', Y' axes in Fig. 2.11 leads to Eqs. (2.25).

$$\begin{aligned} M'_x &= M_x \cos\alpha + M_y \sin\alpha \\ M'_y &= -M_x \sin\alpha + M_y \cos\alpha \end{aligned} \tag{2.25}$$

That is,

$$\begin{bmatrix} M'_x \\ M'_y \end{bmatrix} = \begin{bmatrix} \cos\alpha & \sin\alpha \\ -\sin\alpha & \cos\alpha \end{bmatrix} \begin{bmatrix} M_x \\ M_y \end{bmatrix} \tag{2.26}$$

The matrix L^T may now be defined in an identical manner with that in Eq. (2.15), and substitution in Eq. (2.26) yields

$$M' = L^T M \tag{2.27}$$

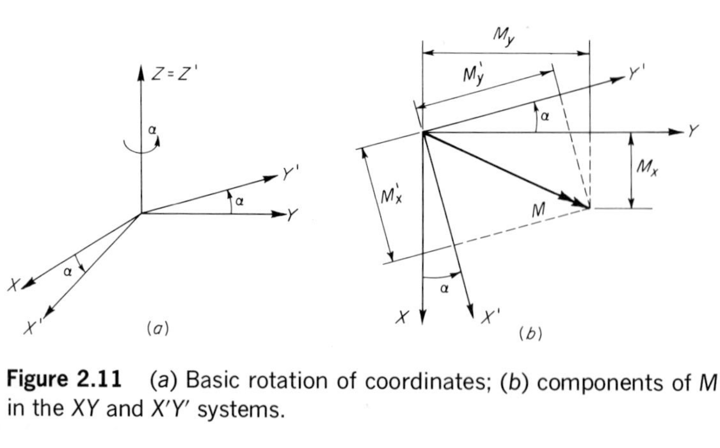

Figure 2.11 (*a*) Basic rotation of coordinates; (*b*) components of *M* in the *XY* and *X'Y'* systems.

It follows that the inverse transformation is again valid. Hence,

$$M = LM' \tag{2.28}$$

Example 2.2 **Calculation of cantilever-beam end rotations and deflections**

The cantilever beam shown in Fig. 2.12 has material properties $EI = 10^5$ in.-kip units and $EI_P = 2 \times 10^5$ in.-kip units. The force (moment) vector at end j is given as

$$M^T = \langle 200 \quad 600 \rangle \text{ in.-kip units}$$

It is required to calculate $\langle \theta_x \quad \theta_y \rangle$ and also v_z, the deflection of j.

The L^T matrix for the member coordinates is calculated, noting that the rotation α is negative for this case. That is,

$$L^T = \begin{bmatrix} 0.8 & -0.6 \\ 0.6 & 0.8 \end{bmatrix}$$

Then,

$$\begin{bmatrix} M'_x \\ M'_y \end{bmatrix} = \begin{bmatrix} 0.8 & -0.6 \\ 0.6 & 0.8 \end{bmatrix} \begin{bmatrix} 200 \\ 600 \end{bmatrix} = \begin{bmatrix} -200 \\ 600 \end{bmatrix} \text{ in.-kips}$$

From strength of materials,

$$v_z = \frac{M'_x l^2}{2EI} = \frac{-200 \times 50^2}{2 \times 10^5} = -2.5 \text{ in.}$$

The rotations about the X', Y' axes are given by

$$\theta' = \begin{bmatrix} \theta'_x \\ \theta'_y \end{bmatrix} = \begin{bmatrix} \dfrac{M'_x l}{EI} \\ \dfrac{M'_y l}{EI_P} \end{bmatrix} = \begin{bmatrix} -0.1 \\ 0.15 \end{bmatrix} \text{rad}$$

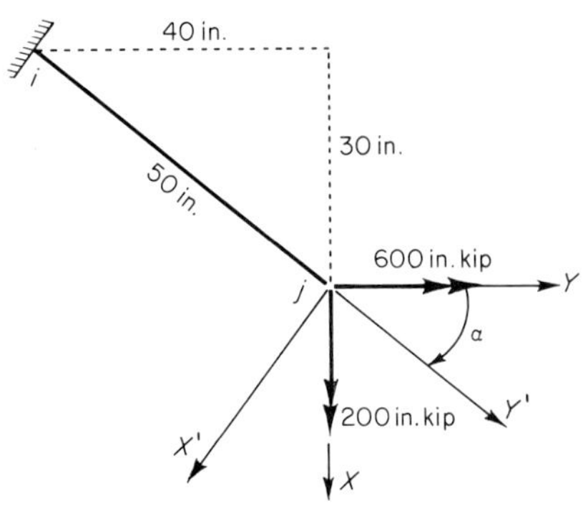

Figure 2.12

Finally $\theta^T = \langle \theta_x \quad \theta_y \rangle$ is obtained by using the transformation equation (2.28) to transform from the $X'Y'$ components to XY components. That is,

$$\theta = \begin{bmatrix} \theta_x \\ \theta_y \end{bmatrix} = L\theta' = \begin{bmatrix} 0.8 & 0.6 \\ -0.6 & 0.8 \end{bmatrix} \begin{bmatrix} -0.1 \\ 0.15 \end{bmatrix} = \begin{bmatrix} 0.01 \\ 0.18 \end{bmatrix} \text{rad}$$

All the transformations for rotation of axes in grid and plane frame structures associated with the four types of vector quantities M, F, θ, and v are summarized in Eqs. (2.29).

$$F' = L^T F \qquad F = LF' \tag{2.29a}$$

$$v' = L^T v \qquad v = Lv' \tag{2.29b}$$

$$M' = L^T M \qquad M = LM' \tag{2.29c}$$

$$\theta' = L^T \theta \qquad \theta = L\theta' \tag{2.29d}$$

2.4 TRANSFORMATIONS OF FORCE AND DISPLACEMENT VECTORS FOR COORDINATE AXES IN THREE-DIMENSIONAL SPACE

In Sec. 2.3, the transformation equations for vector components have been developed for plane frame and grid structures, for which the structural behavior can be described in a two-dimensional sense. These transformations are really only special cases of the general transformation equation for rotation of coordinates about the three axes in XYZ space. They are useful because in such cases they make a significant reduction in the size of the problem to be solved.

In the general case developed in this section, the rotated coordinate axes X', Y', Z' will be described relative to the original set X, Y, Z by their direction cosines l_{ij}. In the notation l_{ij}, the first index indicates the axes X', Y', Z' and has the range of 1 to 3, and the second index, also ranging from 1 to 3, indicates the axes X, Y, Z. Thus,

$$\begin{aligned} l_{1j} &= (l_{11}, l_{12}, l_{13}) && \text{direction cosines of } X' \\ l_{2j} &= (l_{21}, l_{22}, l_{23}) && \text{direction cosines of } Y' \\ l_{3j} &= (l_{31}, l_{32}, l_{33}) && \text{direction cosines of } Z' \end{aligned} \tag{2.30}$$

The axes are shown in Fig. 2.13, and by definition of direction cosines,

$$l_{11} = \cos\alpha \qquad l_{12} = \cos\beta \qquad l_{13} = \cos\gamma \tag{2.30a}$$

Similar expressions define l_{2j} and l_{3j}.

Consider the force $F^T = \langle F_x \quad F_y \quad F_z \rangle$ shown in Fig. 2.13, and project its components on the OX' axis. Then,

$$F'_x = F_x \cos\alpha + F_y \cos\beta + F_z \cos\gamma \tag{2.31}$$

That is,

$$F'_x = F_x l_{11} + F_y l_{12} + F_z l_{13} \tag{2.32}$$

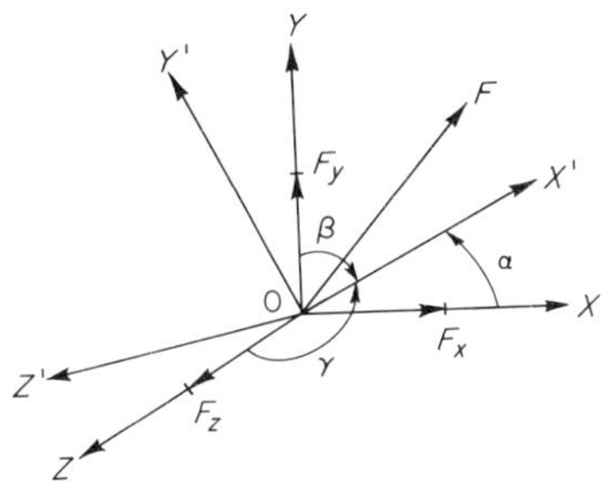

Figure 2.13 Rotation of axes, three dimensions.

By similar reasoning it may be shown that

$$\begin{aligned} F'_y &= F_x l_{21} + F_y l_{22} + F_z l_{23} \\ F'_z &= F_x l_{31} + F_y l_{32} + F_z l_{33} \end{aligned} \tag{2.33}$$

These three equations may be written in matrix form, as Eq. (2.34),

$$\begin{bmatrix} F'_x \\ F'_y \\ F'_z \end{bmatrix} = \begin{bmatrix} l_{11} & l_{12} & l_{13} \\ l_{21} & l_{22} & l_{23} \\ l_{31} & l_{32} & l_{33} \end{bmatrix} \begin{bmatrix} F_x \\ F_y \\ F_z \end{bmatrix} \tag{2.34}$$

Thus, if L is the matrix whose columns are the direction cosines of $X'Y'Z'$ with respect to XYZ axes,

$$L = \begin{bmatrix} l_{11} & l_{21} & l_{31} \\ l_{12} & l_{22} & l_{32} \\ l_{13} & l_{23} & l_{33} \end{bmatrix} \tag{2.35}$$

The transformation law in Eq. (2.34) may thus be written

$$F' = L^T F \tag{2.36}$$

It should be noted that this is identical in matrix form with Eq. (2.17).

The direction cosines of the $X'Y'Z'$ axes are unit orthogonal vectors, and thus

$$\begin{aligned} &\sum_{j=1,3} l_{ij} l_{ij} = 1 \qquad i = 1, 2, 3 \\ &\sum_{j=1,3} l_{ij} l_{kj} = 0 \qquad i = 1, 2, 3 \\ &\qquad\qquad\qquad\quad\; k = 1, 2, 3 \\ &\qquad\qquad\qquad\quad\; i \neq k \end{aligned} \tag{2.37}$$

Equations (2.37) may be written in more compact form if the tensor index notation is used. That is, repeated indices indicate that the expression should be summed over the range of the indices. The Kronecker delta δ_{ij} is defined so that $\delta_{ij} = 1$, $i = j$, and $\delta_{ij} = 0$, $i \neq j$.

With this notation, Eq. (2.37) becomes

$$l_{ij}l_{kj} = \delta_{ik} \tag{2.38}$$

Thus, for example, if $i = k = 1$,

$$l_{1j}l_{1j} = l_{11}^2 + l_{12}^2 + l_{13}^2 = \delta_{11} = 1$$

and if $i = 1$ and $k = 2$,

$$l_{1j}l_{2j} = l_{11}l_{21} + l_{12}l_{22} + l_{13}l_{23} = \delta_{12} = 0$$

By direct multiplication of terms it can be shown that the expression $l_{ij}l_{kj}$ actually generates the nine terms of L^TL as i, k range from 1 to 3, and in that case δ_{ik} generates the unit matrix I.

That is, Eq. (2.38) is equivalent to

$$L^TL = I \tag{2.39}$$

It follows that, as in the two-dimensional problem,

$$L^{-1} = L^T$$

The reverse transformation to Eq. (2.36) is thence

$$F = LF' \tag{2.40}$$

It should be noted that the two-dimensional transformations may be obtained from Eq. (2.35) if the Z' axis remains parallel to OZ and hence $l_{33} = 1$ and $l_{13} = l_{31} = l_{23} = l_{32} = 0$. That is, in this case,

$$L = \begin{bmatrix} l_{11} & l_{21} & 0 \\ l_{12} & l_{22} & 0 \\ 0 & 0 & 1 \end{bmatrix} \tag{2.40a}$$

Furthermore, all the transformations for three-dimensional forces and displacements are given by Eqs. (2.29) if the appropriate matrices and vectors are substituted therein.

Finally, for the vectors R and r as defined in Eqs. (2.4a) and (2.4b), the transformation laws are written

$$R' = \begin{bmatrix} F' \\ M' \end{bmatrix} = \begin{bmatrix} L^T & 0 \\ 0 & L^T \end{bmatrix} \begin{bmatrix} F \\ M \end{bmatrix} = L_D{}^T r \tag{2.41}$$

$$r' = \begin{bmatrix} v' \\ \theta' \end{bmatrix} = \begin{bmatrix} L^T & 0 \\ 0 & L^T \end{bmatrix} \begin{bmatrix} v \\ \theta \end{bmatrix} = L_D{}^T r \tag{2.42}$$

The matrix $L_D = \begin{bmatrix} L & 0 \\ 0 & L \end{bmatrix}$ is a 6×6 matrix with L as diagonal blocks. It is called a *superdiagonal matrix*.

2.5 TRANSFORMATION OF FORCE AND DISPLACEMENT COMPONENTS FROM ONE SET OF COORDINATE AXES TO A PARALLEL SET

The transformation of force and displacement components for parallel sets of coordinate axes manifests itself in many ways in elementary mechanics. Such transformations occur in the calculation of rigid-body reactions, in the calculation of shear-force, bending-moment, and axial-thrust diagrams, and in the calculation of beam deflections. The transformations type is particularly significant because it exhibits the contragradient nature of force and displacement transformations that are not revealed in Eqs. (2.29) simply because the L matrix of the transformations is orthogonal.

The discussion may be commenced with the statement of two apparently unrelated problems:

1 Given the force vector $R^T = \langle F^T \quad M^T \rangle$ acting at a point $P(x,y,z)$, calculate the equivalent force vector $R_O{}^T = \langle F_O{}^T \quad M_O{}^T \rangle$ acting at the origin of coordinates, O.

2 Given the displacement vector $r_O{}^T = \langle v_O{}^T \quad \theta_O{}^T \rangle$ of a rigid body at the origin O, calculate the equivalent displacement vector $r^T = \langle v^T \quad \theta^T \rangle$ at the point $P(x,y,z)$, with the assumption that displacements are small so that displacements due to rotations can be linearized.

Transformation of Force Components

A force $R^T = \langle F_x \quad F_y \quad F_z \quad M_x \quad M_y \quad M_z \rangle$ is shown in Fig. 2.14 acting at the point $P(x,y,z)$. It is required to replace R with an equivalent force R_O acting through the origin O.

The process of transferring (F_x,F_y,F_z) may be accomplished by placing

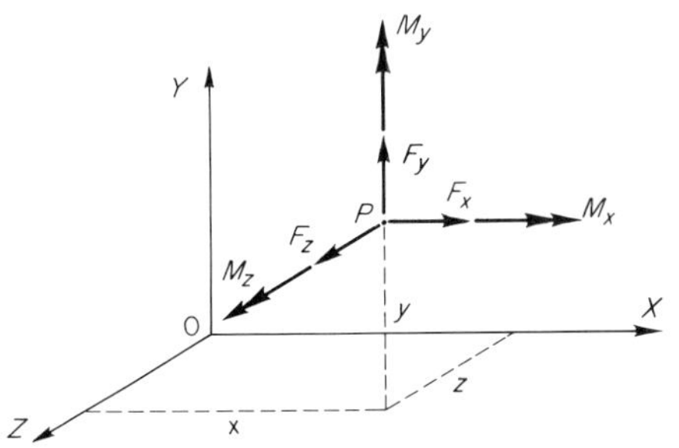

Figure 2.14 Transfer of force components.

$-F_x$ at P and $\pm F_x$ at O, etc. It will be seen that this produces F_x at O together with couples about the axes OY and OZ. The moments M_x, M_y, M_z produce effects that are independent of their point of application. The forces at O that are equivalent to those applied at P are thus given by Eq. (2.43).

$$\begin{aligned}
F_{xO} &= F_{xP} \\
F_{yO} &= F_{yP} \\
F_{zO} &= F_{zP} \\
M_{xO} &= M_{xP} - F_y z + F_z y \\
M_{yO} &= M_{yP} + F_x z - F_z x \\
M_{zO} &= M_{zP} - F_x y + F_y x
\end{aligned} \tag{2.43}$$

The structure of the transformation in Eq. (2.43) becomes apparent when the six equations are combined in matrix form as Eq. (2.44).

$$\begin{bmatrix} F_x \\ F_y \\ F_z \\ M_x \\ M_y \\ M_z \end{bmatrix}_O = \left[\begin{array}{ccc|ccc} 1 & 0 & 0 & 0 & 0 & 0 \\ 0 & 1 & 0 & 0 & 0 & 0 \\ 0 & 0 & 1 & 0 & 0 & 0 \\ \hline 0 & -z & y & 1 & 0 & 0 \\ z & 0 & -x & 0 & 1 & 0 \\ -y & x & 0 & 0 & 0 & 1 \end{array}\right] \begin{bmatrix} F_x \\ F_y \\ F_z \\ M_x \\ M_y \\ M_z \end{bmatrix} \tag{2.44}$$

For algebraic manipulation of Eq. (2.44) it will be more convenient to write the equation in the form

$$R_O = TR_P \tag{2.45}$$

The matrix T may be partitioned, as shown by the dashed lines in Eq. (2.44), into four 3×3 submatrices. That is,

$$T = \begin{bmatrix} I & 0 \\ W & I \end{bmatrix} \tag{2.46}$$

The diagonal blocks I are 3×3 unit matrices, and W is an antisymmetric matrix with off-diagonal terms generated from the coordinates of P. Such antisymmetric matrices are always encountered not only for transformation of moment terms, but also in the transformation of linearized displacement terms due to rotations.

The inverse of T, namely, T^{-1}, may be written

$$T^{-1} = \begin{bmatrix} A & B \\ C & D \end{bmatrix}$$

Then,

$$TT^{-1} = \begin{bmatrix} I & 0 \\ W & I \end{bmatrix} \begin{bmatrix} A & B \\ C & D \end{bmatrix} = \begin{bmatrix} I & 0 \\ 0 & I \end{bmatrix} \tag{2.47}$$

On multiplication of the matrices in Eq. (2.47), it is seen that

$$A = I$$
$$B = 0$$

Hence,

$$D = I$$
$$W + C = 0$$

That is,

$$C = -W = W^T$$

Finally, the inverse of T is given by

$$T^{-1} = \begin{bmatrix} I & 0 \\ W^T & I \end{bmatrix} \tag{2.48}$$

The transformation of forces from the origin O to P is then given by

$$R_P = T^{-1}R_O \tag{2.49}$$

The matrix T^{-1} could also be obtained directly by writing down the equations equivalent to (2.43) for transformation of forces from O to P.

The transformation of forces from P to O may be generalized one further step by supposing that the force at P is given with components of $R_P{}^T = \langle F'^T \quad M'^T \rangle_P$ expressed in the coordinate system X', Y', Z' obtained from the axes $\bar{X}$, $\bar{Y}$, $\bar{Z}$, which are parallel to X, Y, Z but with origin at P, by rotation of axes. In this case the transformation to O is performed in two stages. Firstly, the inverse relationship to Eq. (2.41) is used to transform R'_P to the parallel set of axes through P, and then these components are transformed to O by using Eq. (2.45). See Fig. 2.15. That is,

$$R_P = (L_D{}^T)^{-1}R'_P = L_D R'_P \tag{2.50}$$

$$R_O = TL_D R'_P \tag{2.51}$$

Let

$$b = TL_D \tag{2.52}$$

The matrix b will be defined as the *force transformation matrix* for transformation of the force R'_P, expressed in local coordinates at P, to the global components at O.

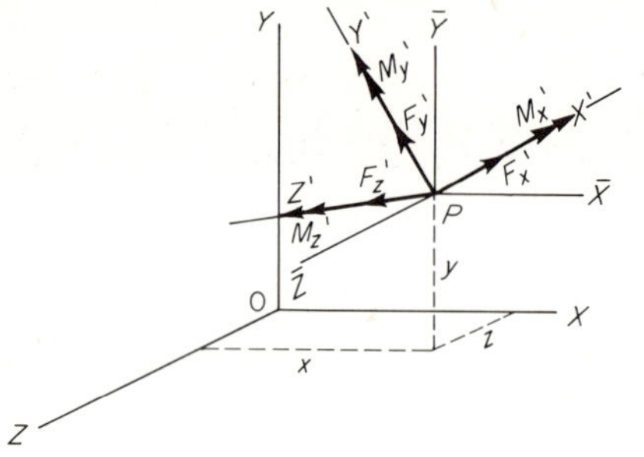

Figure 2.15

The Matrix Equations of Static Equilibrium for a Rigid Body

Consider the rigid body shown in Fig. 2.16, which is subject to the force systems $R_1, R_2, \ldots, R_n$, where $R_i^T = \langle F_i^T \quad M_i^T \rangle$, that act at points $P_1, P_2, \ldots, P_n$ of the body. The coordinates of every $P_i(x_i,y_i,z_i)$ are assumed to be known. If the forces are given in local coordinates at P_i, then in addition to the coordinates, the direction cosine matrices L_i for each system must be given. From the coordinates of P_i and L_i, the matrix b_i may be generated, since from Eq. (2.52).

$$b_i = T_i L_{Di}^T \tag{2.53}$$

The ith force R_i is then transformed to O. When all such transformations to O have been performed, the sum of the corresponding $\tilde{R}$ components gives the resultant force at O equivalent to all the force systems $R_1, R_2, \ldots, R_n$ acting on the body. That is,

$$\tilde{R} = \sum_{i=1,n} b_i R_i \tag{2.54}$$

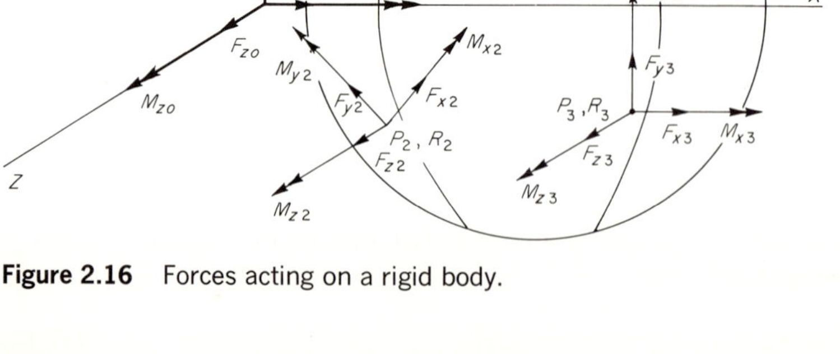

Figure 2.16 Forces acting on a rigid body.

For equilibrium, $\tilde{R} = 0$. That is,

$$\sum_{i=1,n} b_i R_i = 0 \tag{2.55}$$

These equations may be used not only to test the equilibrium of a system of forces acting on a body, but also for calculating statically determinate reactions. Suppose, for example, that point a is fixed and the reactions there are R_a, expressed in any coordinate system. Then by using Eq. (2.55) for equilibrium,

$$b_a R_a + \sum_{i=1,n} b_i R_i = 0 \tag{2.56}$$

Hence,

$$R_a = -b_a^{-1} \sum_{i=1,n} b_i R_i \tag{2.57}$$

If the six components necessary to fix the body in space do not occur at the one point, Eq. (2.57) is complicated by the fact that the columns of b_a must be extracted from the various b matrices and combined in b_a. There must be at least six equations for static determinacy and stability, but it is possible that the six support forces chosen do not satisfy this last condition. If the system of supports is such that the body is unstable, then b_a will be singular. This condition will be detected when it is attempted to calculate the inverse.

Calculation of Bending Moment and Shear Force; Simple Beams

The treatment given in elementary texts on the construction of bending moment, shear force, and axial-thrust diagrams for simple beams is only a special application of Eqs. (2.57) and (2.44). That is, Eq. (2.57) is applied first to calculate the reactions, and then Eq. (2.44) is used to transform the appropriate forces to the point under consideration.

The beam ij is shown in Fig. 2.17 to coincide with the X axis, the end i

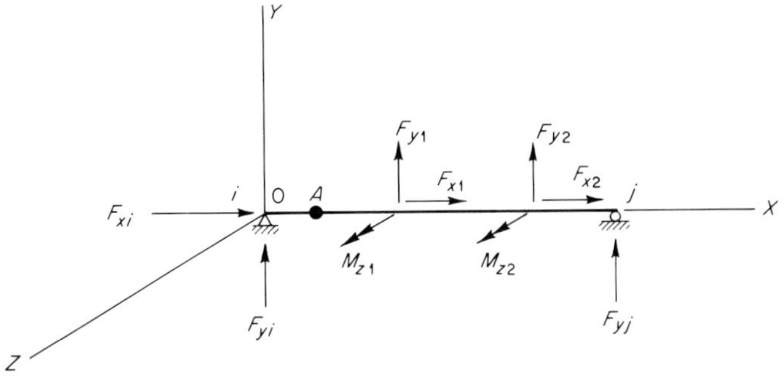

Figure 2.17 Simple beam forces and reactions.

being also the origin O of coordinates. Only force components in the XY plane are considered so that the general force R is given by Eq. (2.58).

$$R^T = \langle F_x \quad F_y \quad M_z \rangle \tag{2.58}$$

The beam has hinged support at i and roller at j, so that the composite reaction vector is given by

$$R_a{}^T = \langle F_{xi} \quad F_{yi} \quad F_{yj} \rangle \tag{2.59}$$

All forces will be transferred to O. The transformation for R_a is given by

$$R_{aO} = \begin{bmatrix} 1 & 0 & 0 \\ 0 & 1 & 1 \\ 0 & 0 & l \end{bmatrix} \begin{bmatrix} F_{xi} \\ F_{yi} \\ F_{yj} \end{bmatrix} = b_a R_a$$

The transformation matrix for force R_k at coordinate x_k on the beam is

$$T_k = \begin{bmatrix} 1 & 0 & 0 \\ 0 & 1 & 0 \\ 0 & x_k & 1 \end{bmatrix}$$

Substitution of these expressions in Eq. (2.57) yields

$$R_a = -b_a{}^{-1} \sum_{k=1,N} T_k R_k$$

$$= \frac{-1}{l} \begin{bmatrix} l & 0 & 0 \\ 0 & l & -1 \\ 0 & 0 & 1 \end{bmatrix} \sum_{k=1,N} T_k R_k$$

If the beam is loaded with a central concentrated load $R^T = \langle 0 \quad 10 \quad 0 \rangle$ and $l = 10.0$, $x_k = 5.0$,

$$R_a = \begin{bmatrix} F_{xi} \\ F_{yi} \\ F_{yj} \end{bmatrix} = \frac{1}{10} \begin{bmatrix} -10 & 0 & 0 \\ 0 & -10 & 1 \\ 0 & 0 & -1 \end{bmatrix} \begin{bmatrix} 1 & 0 & 0 \\ 0 & 1 & 0 \\ 0 & 5.0 & 1 \end{bmatrix} \begin{bmatrix} 0 \\ 10 \\ 0 \end{bmatrix} = \begin{bmatrix} 0 \\ -5 \\ -5 \end{bmatrix}$$

Stress Resultant

The member shown in Fig. 2.18 has been cut by a plane normal to the local OX' axis for that member. For equilibrium of both parts of the structure, equal and opposite sets of forces must be placed on both sides of the cut. These forces will be designated $S^T = \langle S_x \quad S_y \quad S_z \quad M_x \quad M_y \quad M_z \rangle$ and act parallel

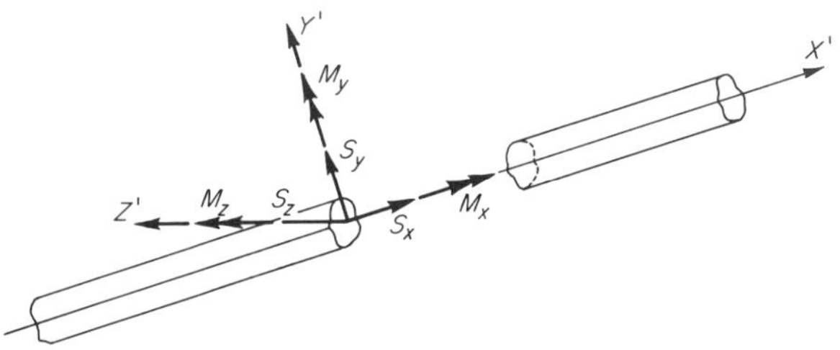

Figure 2.18 Stress resultants.

with the axes X', Y', Z' as shown in Fig. 2.18. From this figure, it will be seen that the stress resultants acting on the left-hand side of the cut will be equal to the algebraic sum of the forces on the right-hand portion of the member, expressed in local coordinates, and similarly the stress resultant acting on the right-hand side of the cut will be equal to the algebraic sum of the forces on the left-hand portion of the member.

With the above definition of stress resultant it is now possible to calculate the stress resultant at point A of the beam shown in Fig. 2.17. To do this, define the transformation matrix T_{Ak} as that matrix necessary to transform the force R_k to an equivalent set at A. By definition,

$$T_{Ak} = \begin{bmatrix} 1 & 0 & 0 \\ 0 & 1 & 0 \\ 0 & x_k - x_A & 1 \end{bmatrix}$$

For the simple beam

$$S_A{}^T = \langle S_x \quad S_y \quad M_z \rangle_A$$

Then

$$S_A = \sum_{\substack{\text{forces to} \\ \text{right of } A}} T_{Ak} R_k$$

If $x_A = 2.5$ and if $R_k{}^T = \langle 0 \quad 10 \quad 0 \rangle$ is the single concentrated load at the center of the beam length 10,

$$S_A = \begin{bmatrix} S_x \\ S_y \\ M_z \end{bmatrix} = \begin{bmatrix} 1 & 0 & 0 \\ 0 & 1 & 0 \\ 0 & 2.5 & 1 \end{bmatrix} \begin{bmatrix} 0 \\ 10 \\ 0 \end{bmatrix} + \begin{bmatrix} 1 & 0 & 0 \\ 0 & 1 & 0 \\ 0 & 10 & 1 \end{bmatrix} \begin{bmatrix} 0 \\ -5 \\ 0 \end{bmatrix} = \begin{bmatrix} 0 \\ -5 \\ -25 \end{bmatrix}$$

Transformation of Displacements

So far, only the transformation of force components from P to O (Fig. 2.15) has been considered, and it is noted that whereas there is a one-to-one correspondence between the direct-force components, the moment components at P are augmented at O by the couples of the force components at P about the coordinate axes.

It would be possible to proceed in a similar manner by considering displacements at P and their effect at O. However, it is more convenient to start with the direct and rotation components at O and then calculate the corresponding displacements at $P(x,y,z)$. It will be possible then to transfer these components into any given local coordinate system at P. Consider the rigid body shown in Fig. 2.19, and suppose that the origin of coordinates, O, is given the displacements $r_O{}^T = \langle v_x \quad v_y \quad v_z \quad \theta_x \quad \theta_y \quad \theta_z \rangle$. It is seen that r_O consists of the direct-displacement vector $\langle v_x \quad v_y \quad v_z \rangle$ and the rotation-displacement vector $\langle \theta_x \quad \theta_y \quad \theta_z \rangle$. For any point $P(x,y,z)$, the direct displacements will be produced by a combination of the two effects:

1 The displacements $\langle v_x \quad v_y \quad v_z \rangle$ at O

2 The displacement components produced by rotations, $\langle \theta_x \quad \theta_y \quad \theta_z \rangle$

The rotations are assumed to be small so that the resulting displacements are linearized. Thus, for example, if P is the point (x,y,O) and the rotation vector is θ_z, that is, (O,O,θ_z), the displacements of P are (see Fig. 2.20),

$$v_x = -l\theta_z \sin \phi = -ly$$
$$v_y = l\theta_z \cos \phi = lx$$

For the point $P(x,y,z)$, all such linear displacement terms for the rotations $\langle \theta_x \quad \theta_y \quad \theta_z \rangle$ may be calculated, thus, by considering each in turn.

The displacements at P due to those at O are then given by Eqs. (2.60).

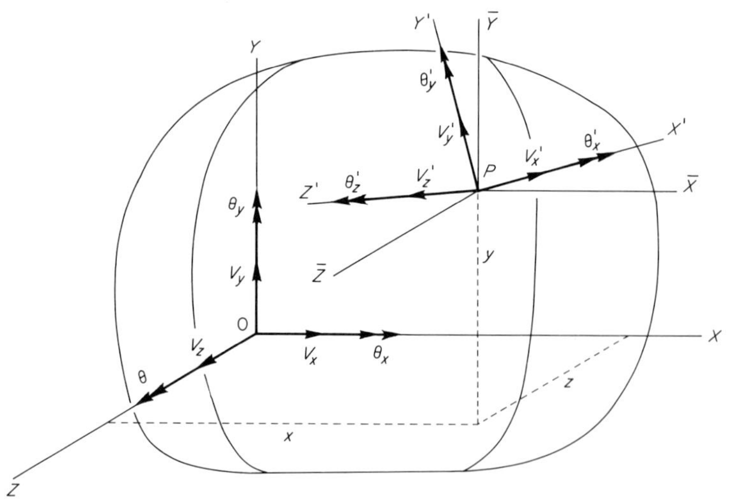

Figure 2.19 Displacement components at O and P.

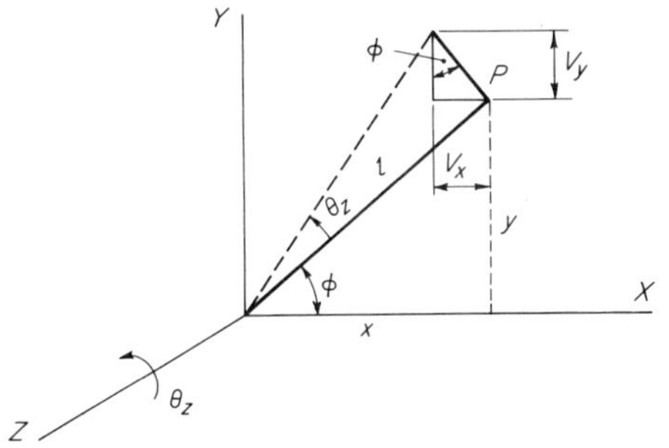

Figure 2.20 Displacement components $\theta_z \neq 0$.

$$
\begin{aligned}
v_{xP} &= v_{xO} + \theta_y z - \theta_z y \\
v_{yP} &= v_{yO} - \theta_x z + \theta_z x \\
v_{zP} &= v_{zO} + \theta_x y - \theta_y x \\
\theta_{xP} &= \theta_{xO} \\
\theta_{yP} &= \theta_{yO} \\
\theta_{zP} &= \theta_{zO}
\end{aligned}
\tag{2.60}
$$

These equations are written in matrix form as Eq. (2.61). The matrix of coefficients on the right-hand side is called a *displacement transformation matrix*.

$$
\begin{bmatrix} v_x \\ v_y \\ v_z \\ \theta_x \\ \theta_y \\ \theta_z \end{bmatrix}_P = \left[\begin{array}{ccc|ccc} 1 & 0 & 0 & 0 & z & -y \\ 0 & 1 & 0 & -z & 0 & x \\ 0 & 0 & 1 & y & -x & 0 \\ \hline 0 & 0 & 0 & 1 & 0 & 0 \\ 0 & 0 & 0 & 0 & 1 & 0 \\ 0 & 0 & 0 & 0 & 0 & 1 \end{array}\right] \begin{bmatrix} v_x \\ v_y \\ v_z \\ \theta_x \\ \theta_y \\ \theta_z \end{bmatrix}_O
\tag{2.61}
$$

Comparison of Eq. (2.61) with Eq. (2.44) shows that this transformation matrix is the transpose of the force transformation matrix. Thence, Eq. (2.61) may be written

$$r_P = T^T r_O \tag{2.62}$$

This property of the T matrices will be shown in Chap. 3 to be a particular statement of the more general contragradient law for displacement and force transformations.

Once the displacement components at P have been obtained through Eq. (2.62), in global coordinates, Eq. (2.42) may be used to calculate the components r'_P in any local coordinates at P. That is,

$$r'_P = L_D{}^T r_P = L_D{}^T T^T r_O \tag{2.63}$$

The coefficient matrix $L_D{}^T T^T$ in Eq. (2.63) will be used to define the displacement transformation matrix a. That is,

$$a = L_D{}^T T^T \tag{2.64}$$

From Eq. (2.52), it is seen that

$$a = b^T \tag{2.65}$$

Once again this is a statement of the contragradient nature of force and displacement transformations.

Because the matrices $L_D{}^T$ and T^T are nonsingular, it is possible to find the inverse transformation to the transformation of Eq. (2.63). That is, if the displacements of a rigid body are known in local coordinates at P, those at O may be calculated. Thus,

$$\begin{aligned} r_O &= (L_D{}^T T^T)^{-1} r'_P \\ &= (T^T)^{-1} L_D r'_P \end{aligned} \tag{2.66}$$

The inverse of T^T is obtained from Eq. (2.48), since $(T^T)^{-1} = (T^{-1})^T$. That is,

$$(T^T)^{-1} = \begin{bmatrix} I & W \\ 0 & I \end{bmatrix} \tag{2.67}$$

2.6 TRANSFORMATION OF THE INERTIA MATRIX; CARTESIAN TENSOR TRANSFORMATIONS FOR ROTATION OF AXES; THE PARALLEL-AXES THEOREM

The transformation matrices developed for forces and displacements in the previous sections may be used generally for any vector quantity. There are, however, physical phenomena such as inertia, stress, and strain, whose components cannot be expressed in vector form and whose transformation laws are different from that for vectors. For example, the inertia (or second moment of area) quantities associated with a plane area in the XY coordinate system are I_x, I_y, and I_{xy}. These do not form the components of a vector.

In this section tensor transformations are introduced to obtain expressions for transformations of inertia components under rotation of axes. The term *inertia* is used here in the loose sense to be synonymous with second moment of area. If the term inertia is used in the dynamic sense, then the quantity defined herein if multiplied by density and thickness would be the mass inertia for a thin disk.

Definition of Moments of Inertia

Consider the plane area shown in Fig. 2.21, lying in the YZ plane, whose normal coincides with the X axis. For a prismatic member whose centroidal

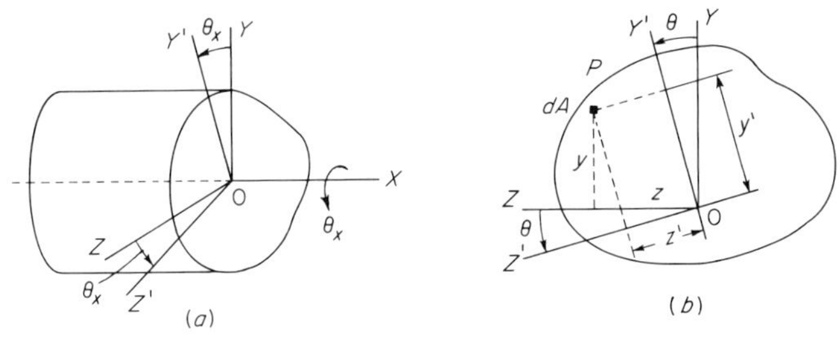

Figure 2.21 (*a*) Member cross-section coordinates; (*b*) coordinates of area *dA* in *XY* and *X'Y'* systems.

axis coincides with OX, the area will be the member cross-sectional area. The second moment of the area, shown in Fig. 2.21*b*, about the axis OZ is defined by the integral

$$I_{zz} = \int_{\text{area}} y^2 \, dA \tag{2.68}$$

Similarly, the second moment of the area about OY is given by

$$I_{yy} = \int_{\text{area}} z^2 \, dA \tag{2.69}$$

These terms alone are not sufficient to define the quadratic functions of the area associated with the axes OY, OZ, and it is necessary to introduce the product of inertia defined in Eq. (2.70),

$$I_{yz} = \int_{\text{area}} yz \, dA \tag{2.70}$$

Note that the term I_{zy} is also equal to the integral of Eq. (2.70). The four quantities so defined may be grouped in a 2×2 matrix that is symmetric and will be defined as the moment of inertia matrix J. That is,

$$J = \begin{bmatrix} I_{zz} & I_{zy} \\ I_{yz} & I_{yy} \end{bmatrix} \tag{2.71}$$

The construction of the J matrix may be undertaken in a different way by using the point P to define the vector **OP**, given by its components in Eq. (2.72),

$$Y = \begin{bmatrix} y \\ z \end{bmatrix} \tag{2.72}$$

The terms z^2, y^2, and yz contributing to the integrals I_{zz}, I_{yy}, and I_{yz} may be formed as a square matrix by the postmultiplication of Y by its transpose. That is,

$$YY^T = \begin{bmatrix} y^2 & yz \\ zy & z^2 \end{bmatrix} \tag{2.73}$$

The integral of a matrix over the area is formed by integrating term by term, so that J is obtained from YY^T by Eq. (2.74),

$$J = \int_{\text{area}} YY^T \, dA \tag{2.74}$$

Transformation of Inertia Matrix

The transformation law for the vectors Y and Y' for rotation of axes is known, and from this it is possible to determine the transformation of J. In the rotated coordinate system shown in Fig. 2.21b, the coordinates of the point P are given by $Y'^T = \langle y' \quad z' \rangle$. The inertia matrix relative to the $X'Y'$ axes is then calculated by substitution of Y' in Eq. (2.74). That is,

$$J' = \int_{\text{area}} Y'Y'^T \, dA \tag{2.75}$$

From Eq. (2.29) it is seen that the vector Y' is related to Y through the transformation

$$Y' = L^T Y \tag{2.76}$$

Substitution for Y' and Y'^T from Eq. (2.76) in Eq. (2.75) gives the expression for J' in terms of the YZ coordinates and the rotation matrix L. Since L is constant for all points in the cross section, it may be taken outside the integral. That is,

$$J' = \int_{\text{area}} L^T YY^T L \, dA = L^T \int_{\text{area}} YY^T \, dA \, L = L^T J L \tag{2.77}$$

The transformation law for rotation of axes is thus

$$J' = L^T J L \tag{2.78}$$

The matrix L is nonsingular, $L^{-1} = L^T$ [Eq. (2.21)], so that the transformation of the inertia matrix J' to J is given by Eq. (2.79),

$$J = LJ'L^T \tag{2.79}$$

Notice that the transformation in Eq. (2.78) is different from that for vector quantities and involves a postmultiplication by L and a premultiplication by L^T. The pre- and postmultiplication is a feature of rank-2 tensor transformations. The transformation in Eq. (2.78) is now expanded in order to examine the individual terms of J'. That is,

$$J' = \begin{bmatrix} \cos\theta & \sin\theta \\ -\sin\theta & \cos\theta \end{bmatrix} \begin{bmatrix} I_{zz} & I_{zy} \\ I_{yz} & I_{yy} \end{bmatrix} \begin{bmatrix} \cos\theta & -\sin\theta \\ \sin\theta & \cos\theta \end{bmatrix}$$

$$= \begin{bmatrix} \cos\theta & \sin\theta \\ -\sin\theta & \cos\theta \end{bmatrix} \left[\begin{array}{c|c} I_{zz}\cos\theta + I_{zy}\sin\theta & -I_{zz}\sin\theta + I_{zy}\cos\theta \\ \hline I_{yz}\cos\theta + I_{yy}\sin\theta & -I_{yz}\sin\theta + I_{yy}\cos\theta \end{array}\right]$$

Finally,

$$J' = \left[\begin{array}{c|c} I_{zz}\cos^2\theta + 2I_{zy}\sin\theta\cos\theta + I_{yy}\sin^2\theta & (I_{yy} - I_{zz})\sin\theta\cos\theta + I_{zy}(\cos^2\theta - \sin^2\theta) \\ \hline (I_{yy} - I_{zz})\sin\theta\cos\theta + I_{zy}(\cos^2\theta - \sin^2\theta) & I_{zz}\sin^2\theta + 2I_{zy}\sin\theta\cos\theta + I_{yy}\cos^2\theta \end{array}\right] \tag{2.80}$$

Thus, for example,

$$I'_{zz} = I_{zz}\cos^2\theta + 2I_{zy}\sin\theta\cos\theta + I_{yy}\sin^2\theta \tag{2.81}$$

This and similar expressions are found in elementary texts on strength of materials.

Principal Axes; Principal Moments of Inertia

The principal axes of the cross section are defined as those axes for which the product of inertia vanishes. They are also the axes for which the moments of inertia have maximum and minimum values. When the inertia matrix is expressed in principal axes, it will be a diagonal matrix. That is, if $\bar{J}$ is the inertia matrix in the principal axes,

$$\bar{J} = \begin{bmatrix} \bar{I}_{zz} & 0 \\ 0 & \bar{I}_{yy} \end{bmatrix} \tag{2.82}$$

The problem then is to find the transformation that reduces J given in Eq. (2.71) to diagonal form.

Let θ be the angle giving the required rotation of axes. From Eqs. (2.80) and (2.82),

$$\bar{I}_{zy} = 0 = (I_{yy} - I_{zz})\sin\theta\cos\theta + I_{zy}(\cos^2\theta - \sin^2\theta) \tag{2.83}$$

Hence

$$\tan 2\bar{\theta} = \frac{2I_{yz}}{I_{zz} - I_{yy}} \tag{2.84}$$

The solution of Eq. (2.84) gives two values of $\bar{\theta}$, 90° apart. It is then possible to calculate $\bar{I}_{zz}$ and $\bar{I}_{yy}$ directly from Eq. (2.80) or to form the L matrix and produce $\bar{I}$ by using Eq. (2.79). An interesting alternative approach is to use the properties of the characteristic equation

$$JX = \lambda IX \tag{2.85}$$

Equation (2.85) may be put into homogeneous form by transposing the right-hand side, so that

$$(J - \lambda I)X = 0 \tag{2.86}$$

These are a set of homogeneous linear equations and may be solved only if the determinant of coefficients is equal to zero. That is,

$$|J - \lambda I| = 0 \tag{2.87}$$

In full, Eq. (2.87) is written

$$\begin{vmatrix} I_{zz} - \lambda & I_{zy} \\ I_{yz} & I_{yy} - \lambda \end{vmatrix} = 0 \tag{2.88}$$

On expansion, the determinant of Eq. (2.88) leads to a quadratic equation for λ, Eq. (2.89),

$$\lambda^2 - \lambda(I_{zz} + I_{yy}) - I_{zy}^2 + I_{zz}I_{yy} = 0 \tag{2.89}$$

Hence,

$$\lambda_{1,2} = \frac{(I_{zz} + I_{yy}) \pm \sqrt{(I_{zz} + I_{yy})^2 - 4(I_{zz}I_{yy} - I_{zy}^2)}}{2} \tag{2.90}$$

$$\lambda_{1,2} = \frac{(I_{zz} + I_{yy}) \pm \sqrt{(I_{zz} - I_{yy})^2 + 4I_{zy}^2}}{2} \tag{2.91}$$

Corresponding to the roots λ_1 and λ_2, Eq. (2.85) will yield solutions X_1 and X_2 that are now shown to be orthogonal. That is,

$$X_1{}^T X_2 = 0 \tag{2.92}$$

From Eq. (2.86) it will be seen that only one of the terms of X is independent, so that solutions may be normalized. Then,

$$X_1{}^T X_1 = X_2{}^T X_2 = 1 \tag{2.93}$$

Substitute with λ_1, λ_2 in Eq. (2.85)

$$JX_1 = \lambda_1 X_1 \tag{2.85a}$$

$$JX_2 = \lambda_2 X_2 \tag{2.85b}$$

premultiply Eq. (2.85*a*) by $X_2{}^T$ and (2.85*b*) by $X_1{}^T$. Thence from (2.85*a*),

$$X_2{}^T J X_1 = \lambda_1 X_2{}^T X_1 \tag{2.85c}$$

and from (2.85*b*),

$$X_1{}^T J X_2 = \lambda_2 X_1{}^T X_2 \tag{2.85d}$$

Transpose both sides of Eq. (2.85*d*);

$$X_2{}^T J X_1 = \lambda_2 X_2{}^T X_1 \tag{2.85e}$$

The substitution of (2.85*c*) from (2.85*e*) gives

$$0 = (\lambda_2 - \lambda_1)X_2{}^T X_1$$

Since it is assumed that $\lambda_2 \neq \lambda_1$, it follows that

$$X_2{}^T X_1 = 0 \tag{2.94}$$

Now from matrix Q composed of X_1 and X_2 as columns,

$$Q = [X_1 \quad X_2]$$

Then,

$$Q^T J Q = \begin{bmatrix} X_1{}^T \\ X_2{}^T \end{bmatrix} J[X_1 \quad X_2] = \begin{bmatrix} X_1{}^T J X_1 & X_1{}^T J X_2 \\ X_2{}^T J X_1 & X_2{}^T J X_2 \end{bmatrix}$$

$$= \begin{bmatrix} \lambda_1 X_1{}^T X_1 & \lambda_2 X_1{}^T X_2 \\ \lambda_2 X_2{}^T X_1 & \lambda_2 X_2{}^T X_2 \end{bmatrix}$$

Hence

$$Q^T J Q = \begin{bmatrix} \lambda_1 & 0 \\ 0 & \lambda_2 \end{bmatrix} \tag{2.95}$$

It is seen that λ_1 and λ_2 are the principal moments of inertia and Q is the required rotation matrix to produce the principal axes.

Geometrical Interpretation of Tensor Transformation and Principal Axes

The above transformations of the inertia tensor have a very useful graphic interpretation generally called *Mohr's circle of inertia.* It is found that Mohr's circle conveniently expresses the transformation of any tensor of rank 2 in two dimensions.

If the points (I_{zz}, I_{zy}) and $(I_{yy}, -I_{yz})$ are plotted with the coordinate axes as shown in Fig. 2.22, it is found that the values of λ coincide with the ordinates where the circle cuts the X axis. From Fig. 2.22, the radius R of the circle is given

$$R = \sqrt{\frac{(I_{zz} - I_{yy})^2}{2} + I_{zy}{}^2}$$

The values I_1 and I_2 where the circle cuts the X axis are

$$I_1 = \frac{I_{zz} + I_{yy}}{2} + R$$

$$I_2 = \frac{I_{zz} + I_{yy}}{2} - R$$

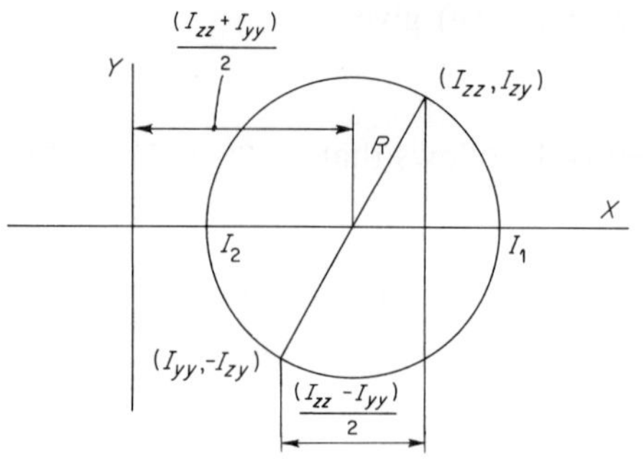

Figure 2.22 Mohr's circle of inertia.

These equations are identical with Eqs. (2.90) and (2.91).

The Parallel-axes Theorem

The parallel-axes theorem for the transformation of the inertia matrix is derived by application of the transformation of coordinates to Eq. (2.74). Consider the coordinate axes Y', Z' passing through the centroid O of the plane area shown in Fig. 2.23. The origin of the parallel set of axes is displaced from O by the coordinates (v,w). The relationship between the two sets of coordinates is given by Eq. (2.96),

$$\begin{bmatrix} y \\ z \end{bmatrix}' = \begin{bmatrix} y \\ z \end{bmatrix} - \begin{bmatrix} v \\ w \end{bmatrix} \tag{2.96}$$

Let $\mathbf{00'}$ be the vector with components $C^T = \langle v \quad w \rangle$. Then by Eqs. (2.72) and (2.96),

$$Y' = Y - C \tag{2.97}$$

For the axes Y', Z', the J' matrix is given as

$$J' = \int_{\text{area}} Y'Y'^T \, dA \tag{2.98}$$

Substitute in Eq. (2.98) for Y' from Eq. (2.97). That is,

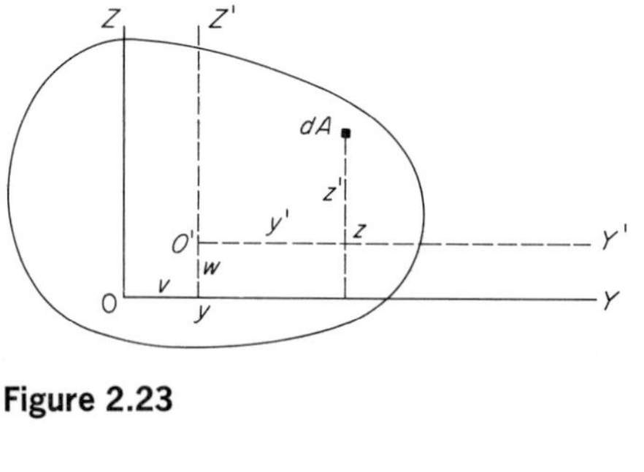

Figure 2.23

$$J' = \int_{\text{area}} (Y - C)(Y^T - C^T)\, dA$$

On multiplication,

$$J' = \int YY^T\, dA + \int Y\, dA\, C^T + C \int Y^T\, dA + C \int dA\, C^T \tag{2.99}$$

The terms $\int Y\, dA$ and $\int Y^T\, dA$ in Eq. (2.99) are each equal to zero, since Y is measured from axes through the centroid and $\int y\, dA = \int z\, dA = 0$ for such a case. Also, $\int dA = A$, the area. Finally,

$$J' = J + CC^TA \tag{2.100}$$

In expanded form, Eq. (2.100) is written

$$\begin{bmatrix} I'_{zz} & I'_{zy} \\ I'_{yz} & I'_{yy} \end{bmatrix} = \begin{bmatrix} I_{zz} & I_{zy} \\ I_{yz} & I_{yy} \end{bmatrix} + A \begin{bmatrix} v^2 & uw \\ uw & w^2 \end{bmatrix} \tag{2.101}$$

Equating terms of the matrices on both sides of Eq. (2.101) gives the usual expressions for the inertias about the parallel axes, for example,

$$I'_{zz} = I_{zz} + Av^2 \tag{2.102}$$

Example 2.3 Calculation of moments of inertia; principal axes and principal moments of inertia

It is required to calculate J for the angle section shown in Fig. 2.24. The axes XOY are parallel to the sides of the angle. The centroid is located at O. Equation (2.100) will be used first to calculate J' and then to shift to J at XOY.

Centroidal Axes

Area No.	A	y′	Ay′	x′	Ax′
1	6	3	18	0.5	3.0
2	5	5.5	27.5	3.5	17.5

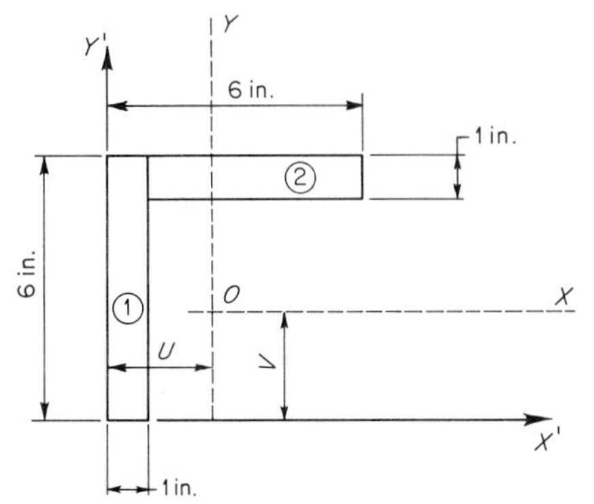

Figure 2.24 Plane area moments of inertia.

$\Sigma A = 11 \qquad \Sigma Ay' = 45.5 \qquad \Sigma Ax' = 20.5$

$$v = \frac{-45.5}{11} \qquad u = \frac{-20.5}{11}$$

$$= -4.1363 \qquad = -1.8636$$

Calculation of J' The moments of inertia J_1' and J_2' of areas 1 and 2 shown in Fig. 2.24 will be calculated by a shift from their own local centroidal axes. Thus,

$$J_1' = \begin{bmatrix} \frac{6 \times 1^3}{12} & 0 \\ 0 & \frac{1 \times 6^3}{12} \end{bmatrix} + 6 \begin{bmatrix} \left(\frac{-1}{2}\right)^2 & \left(\frac{-1}{2}\right)(-3) \\ \left(\frac{-1}{2}\right)(-3) & (-3)^2 \end{bmatrix} = \begin{bmatrix} 0.5 & 0 \\ 0 & 18 \end{bmatrix} + 6 \begin{bmatrix} 0.25 & 1.5 \\ 1.5 & 9.0 \end{bmatrix}$$

$$= \begin{bmatrix} 2.0 & 9.0 \\ 9.0 & 72.0 \end{bmatrix}$$

Similarly,

$$J_2' = \begin{bmatrix} \frac{1 \times 5^3}{12} & 0 \\ 0 & \frac{5 \times 1^3}{12} \end{bmatrix} + 5 \begin{bmatrix} (-3.5)^2 & (-3.5)(-5.5) \\ (-3.5)(-5.5) & (-5.5)^2 \end{bmatrix}$$

$$= \begin{bmatrix} 10.4167 & 0 \\ 0 & 0.4167 \end{bmatrix} + \begin{bmatrix} 61.25 & 96.25 \\ 96.25 & 151.25 \end{bmatrix}$$

$$= \begin{bmatrix} 71.667 & 96.25 \\ 96.25 & 151.667 \end{bmatrix}$$

$$J' = J_1' + J_2' = \begin{bmatrix} 73.667 & 105.25 \\ 105.25 & 223.667 \end{bmatrix}$$

Shift from $X'Y'$ to XY

$$J = J' - ACC^T = \begin{bmatrix} 73.667 & 105.25 \\ 105.25 & 223.667 \end{bmatrix} - 11 \begin{bmatrix} 1.8636^2 & 4.1363 \times 1.8636 \\ 4.1363 \times 1.8636 & 4.1363^2 \end{bmatrix}$$

$$= \begin{bmatrix} 35.46 & 20.46 \\ 20.46 & 35.46 \end{bmatrix}$$

Principal Axes The directions of the principal axes are obtained from Eq. (2.84).

$$\tan 2\theta = \frac{2 \times 20.46}{0} = \infty$$

That is,

$$2\theta = 90^\circ \quad \text{or} \quad 270^\circ$$

$$\theta = 45^\circ \quad \text{or} \quad 135^\circ$$

From Eq. (2.15),

$$L = \frac{1}{\sqrt{2}}\begin{bmatrix} 1.0 & 1.0 \\ -1.0 & 1.0 \end{bmatrix}$$

From Eqs. (2.82) and (2.78),

$$\bar{J} = \tfrac{1}{2}\begin{bmatrix} 1.0 & -1.0 \\ 1.0 & 1.0 \end{bmatrix}\begin{bmatrix} 35.46 & 20.46 \\ 20.46 & 35.46 \end{bmatrix}\begin{bmatrix} 1.0 & 1.0 \\ -1.0 & 1.0 \end{bmatrix}$$

That is,

$$\bar{J} = \begin{bmatrix} 15.0 & 0 \\ 0 & 55.92 \end{bmatrix}$$

2.7 TRANSFORMATION OF THE STRESS TENSOR; TWO-DIMENSIONAL STRESS STATE AND ROTATION OF AXES

A two-dimensional stress state with stress components $(\sigma_x,\sigma_y,\tau_{xy})$ acts on the infinitesimal cube *dx*, *dy*, *dz* as in Fig. 2.25*a*. These stress components will be represented in the *XOY* coordinate axes. An equivalent set of stress components $(\sigma'_x,\sigma'_y,\tau'_{xy})$ given in the $X'OY'$ coordinate system (Fig. 2.25*b*) may also be used to describe the state of stress at point *O*. The axes X', O, Y' have been obtained from X, O, Y by the rotation $\theta_z = \alpha$.

The sign convention used for stress components is that positive stresses act on positive faces of the cube in positive directions and on negative faces in negative directions. For σ_x and σ_y this is equivalent to saying tension is positive, compression negative. The positive sense of all stress components is shown in Fig. 2.25*a*.

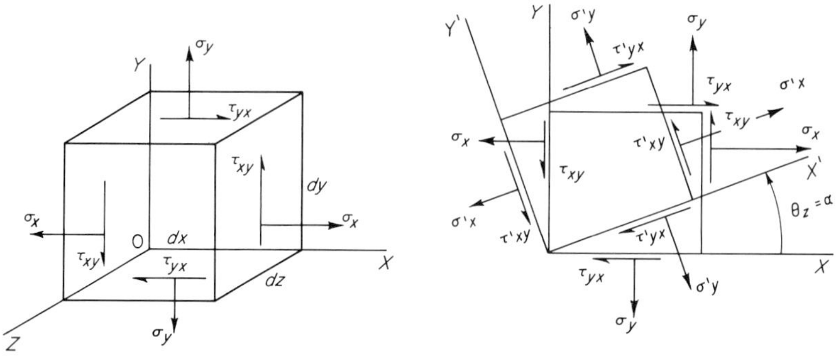

Figure 2.25 (*a*) Stress on infinitesimal cube in two-dimensional stress state; (*b*) stress systems σ and σ' on original and rotated coordinate systems.

The two stress systems σ and σ' representing the stress at point O are shown in Fig. 2.25*b*. It should be emphasized that these two representations are identical and one may be obtained from the other by suitable transformation.

The stress matrices (or stress tensors) σ and σ' are now defined in the two coordinate systems as the 2×2 matrices with components σ_x, σ_y, and $\tau_{xy} = \tau_{yx}$. Thus,

$$\sigma = \begin{bmatrix} \sigma_x & \tau_{xy} \\ \tau_{yx} & \sigma_y \end{bmatrix} \qquad \sigma' = \begin{bmatrix} \sigma'_x & \tau'_{xy} \\ \tau'_{yx} & \sigma'_y \end{bmatrix} \tag{2.103}$$

The stress matrix is symmetric, since by taking moments about the center of the cube it is shown that for equilibrium,

$$\tau_{xy} = \tau_{yx} \tag{2.104}$$

Since the matrix σ' is obtained from σ only by a rotation of coordinate axes, it is required to find the transformation relationship as a function of this rotation α. It is found that the stress matrix or tensor behaves just like the inertia tensor under coordinate transformations.

The relationship is provided by considering the small wedge of material at O, shown in Fig. 2.26, and writing its equations of equilibrium in terms of $(\sigma_x,\sigma_y,\tau_{xy})$ and the stresses (p_x,p_y) on the wedge surface. The normal to the inclined surface, $\hat{n}$, subtends the angle α with the positive OX direction. The stresses (p_x,p_y) are defined as follows. Let F_x be equal to the sum of all forces resolved across the wedge in the X direction. Then,

$$p_x = \frac{F_x}{A} \tag{2.105}$$

In Eq. (2.105), A is the area of the inclined face. Similarly,

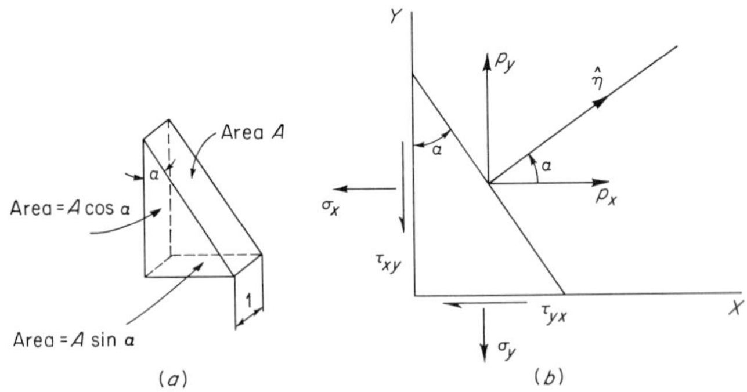

Figure 2.26 (*a*) Wedge of unit thickness; (*b*) stress components acting on wedge.

$$p_y = \frac{F_y}{A} \tag{2.106}$$

The stress components (p_x, p_y) may be expressed in terms of the stress tensor σ, and the direction cosines of $\hat{n}$ by writing the equations of equilibrium $\Sigma F_x = 0$ and $\Sigma F_y = 0$. That is,

$$\Sigma F_x = 0 \qquad \text{therefore} \qquad p_x A - \sigma_x A \cos\alpha - \tau_{yx} A \sin\alpha = 0 \tag{i}$$

$$\Sigma F_y = 0 \qquad \text{therefore} \qquad p_y A - \tau_{xy} A \cos\alpha - \sigma_y A \sin\alpha = 0 \tag{ii}$$

In matrix form Eqs. (*i*) and (*ii*) are combined to give

$$\begin{bmatrix} p_x \\ p_y \end{bmatrix} A = \begin{bmatrix} \sigma_x & \tau_{yx} \\ \tau_{xy} & \sigma_y \end{bmatrix} \begin{bmatrix} \cos\alpha \\ \sin\alpha \end{bmatrix} A \tag{2.107}$$

Remember that it is required to express σ' in the rotated coordinates in terms of σ, and so the vector terms of Eq. (2.107) are transformed to components in the $X'Y'$ axes. Notice that there are two vectors in Eq. (2.107). Firstly, the force vector $\langle p_x \quad p_y \rangle A$ and secondly the outward normal to the area A, $\langle \cos\alpha \quad \sin\alpha \rangle A$. Both of these vectors obey the transformation laws established in Eq. (2.29). That is,

$$\begin{bmatrix} p_x \\ p_y \end{bmatrix} A = L \begin{bmatrix} p'_x \\ p'_y \end{bmatrix} A \tag{2.108}$$

and

$$\begin{bmatrix} \cos\alpha \\ \sin\alpha \end{bmatrix} A = L \begin{bmatrix} \cos\alpha' \\ \sin\alpha' \end{bmatrix} A \tag{2.109}$$

Notice that the angle α' is measured from OX' to the outward normal $\hat{n}$ and Eq. (2.109) is also an expression of the fact that $\alpha = \alpha' + \theta$ and hence,

$$\begin{aligned} \cos\alpha &= \cos(\alpha' + \theta) = \cos\theta\cos\alpha' - \sin\theta\sin\alpha' \\ \sin\alpha &= \sin(\alpha' + \theta) = \sin\theta\cos\alpha' + \cos\theta\sin\alpha' \end{aligned} \tag{2.110}$$

Substitute Eqs. (2.108) and (2.109) in Eq. (2.107). That is,

$$L \begin{bmatrix} p'_x \\ p'_y \end{bmatrix} A = \begin{bmatrix} \sigma_x & \tau_{yx} \\ \tau_{xy} & \sigma_y \end{bmatrix} L \begin{bmatrix} \cos\alpha' \\ \sin\alpha' \end{bmatrix} A \tag{2.111}$$

Premultiply both sides of Eq. (2.111) by L^T. Thence,

$$\begin{bmatrix} p'_x \\ p'_y \end{bmatrix} A = L^T \begin{bmatrix} \sigma_x & \tau_{yx} \\ \tau_{xy} & \sigma_y \end{bmatrix} L \begin{bmatrix} \cos\alpha' \\ \sin\alpha' \end{bmatrix} A \tag{2.112}$$

If Eq. (2.107) is examined, it is seen that it is not an expression unique to the XOY coordinate system, but in fact relates stresses on the wedge face to the stress tensor in any coordinate system. That is, it is also valid to use σ' and α' with stress components $\langle p'_x \quad p'_y \rangle$.

From Eq. (2.107), it follows

$$\begin{bmatrix} p'_x \\ p'_y \end{bmatrix} A = \begin{bmatrix} \sigma'_x & \tau'_{yx} \\ \tau'_{xy} & \sigma'_y \end{bmatrix} \begin{bmatrix} \cos \alpha' \\ \sin \alpha' \end{bmatrix} A \tag{2.113}$$

The Eqs. (2.112) and (2.113) have identical left-hand sides, and hence their right-hand sides must also be equal. That is,

$$\begin{bmatrix} \sigma'_x & \tau'_{yx} \\ \tau'_{xy} & \sigma'_y \end{bmatrix} = L^T \begin{bmatrix} \sigma_x & \tau_{yx} \\ \tau_{xy} & \sigma_y \end{bmatrix} L \tag{2.114}$$

or

$$\sigma' = L^T \sigma L \tag{2.115}$$

It is seen immediately that this is the transformation law connecting σ' and σ and also that its form is identical with Eq. (2.78). It follows that the theory on principal inertias and principal axes must also have its counterpart in the stress tensor. There will be principal stresses and principal directions. Since by definition, $\tau_{xy} = \tau_{yx} = 0$ on the principal planes, these are planes of zero shear stress. Again the same graphic presentation of the transformation is possible and is known as *Mohr's circle of stress.*

The angle of rotation to principal axes is given [see Eq. (2.84)] by

$$\tan 2\bar{\theta} = \frac{2\tau_{yz}}{\sigma_x - \sigma_y}$$

Three-dimensional Stress State and Rotation of Axes

The stress state in three-dimensional space requires the six independent stress quantities σ_x, τ_{yx}, τ_{zx}, σ_y, τ_{zy}, σ_z, shown in Fig. 2.27, for its complete specifi-

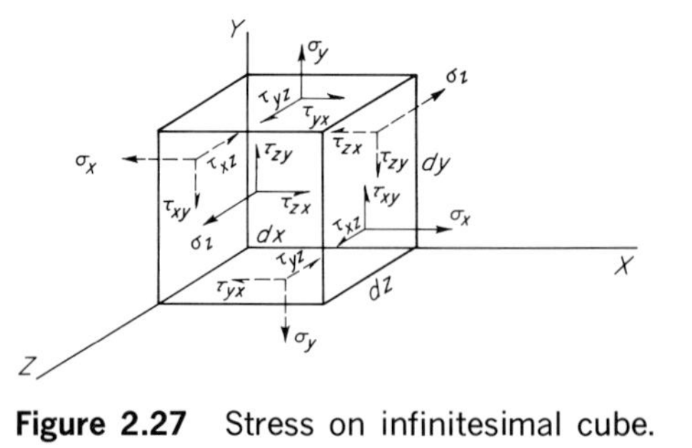

Figure 2.27 Stress on infinitesimal cube.

cation. By taking moments about the center of the infinitesimal cube it is shown that

$$\tau_{xy} = \tau_{yx}$$
$$\tau_{yz} = \tau_{zy}$$
$$\tau_{zx} = \tau_{xz}$$

That is, the stress matrix (or tensor) σ as defined in Eq. (2.116) is symmetric,

$$\sigma = \begin{bmatrix} \sigma_x & \tau_{yx} & \tau_{zx} \\ \tau_{xy} & \sigma_y & \tau_{zy} \\ \tau_{xz} & \tau_{yz} & \sigma_z \end{bmatrix} \tag{2.116}$$

As for the two-dimension stress state, it is required to obtain the transformation of the components of the stress tensor for rotation of axes.

Consider the infinitesimal tetrahedron shown in Fig. 2.28 with stresses (p_x, p_y, p_z) acting on the inclined face and the normal PN having direction cosines $n^T = \langle l_{n1} \quad l_{n2} \quad l_{n3} \rangle$. For the tetrahedron the three equations of equilibrium, $\Sigma F_x = 0$, $\Sigma F_y = 0$, and $\Sigma F_z = 0$, must be satisfied by forces acting on its sides.

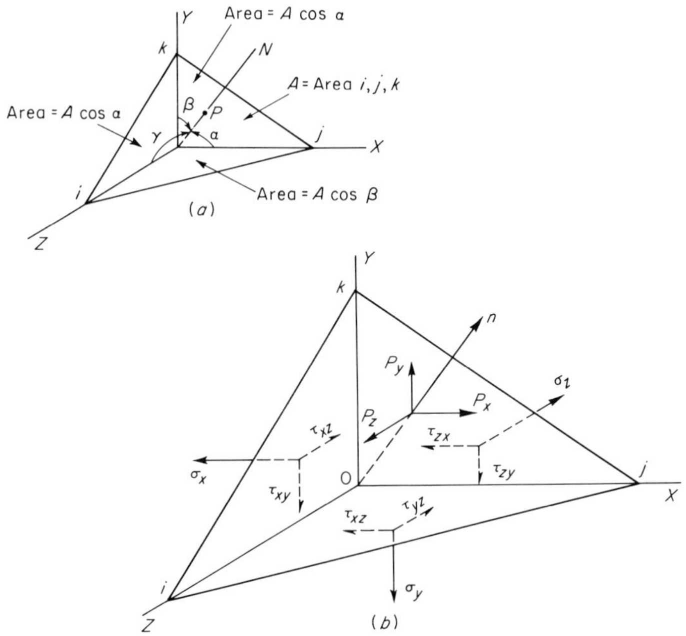

Figure 2.28 (*a*) Geometry of tetrahedron; (*b*) stress components on faces of tetrahedron.

From Fig. 2.28*a* it is seen that

$$n^T = \langle l_{n1} \quad l_{n2} \quad l_{n3} \rangle = \langle \cos\alpha \quad \cos\beta \quad \cos\gamma \rangle$$

From the equations of equilibrium and Eq. (2.107), it is seen that the relationship between $\langle p_x \quad p_y \quad p_z \rangle$ and σ is given by Eq. (2.117),

$$\begin{bmatrix} p_x \\ p_y \\ p_z \end{bmatrix} A = \begin{bmatrix} \sigma_x & \tau_{yx} & \tau_{zx} \\ \tau_{xy} & \sigma_y & \tau_{zy} \\ \tau_{xz} & \tau_{yz} & \sigma_z \end{bmatrix} \begin{bmatrix} l_{n1} \\ l_{n2} \\ l_{n3} \end{bmatrix} A \tag{2.117}$$

Apply the transformation law established for three-dimension vector components in Eq. (2.40) to express components $\langle p_x \quad p_y \quad p_z \rangle$ and $\langle l_{n1} \quad l_{n2} \quad l_{n3} \rangle$ in terms of $\langle p'_x \quad p'_y \quad p'_z \rangle$ and $\langle l'_{n1} \quad l'_{n2} \quad l'_{n3} \rangle$ in the rotated coordinate system. Then,

$$L \begin{bmatrix} p'_x \\ p'_y \\ p'_z \end{bmatrix} A = \begin{bmatrix} \sigma_x & \tau_{yx} & \tau_{zx} \\ \tau_{xy} & \sigma_y & \tau_{zy} \\ \tau_{xz} & \tau_{yz} & \sigma_z \end{bmatrix} L \begin{bmatrix} l'_{n1} \\ l'_{n2} \\ l'_{n3} \end{bmatrix} A \tag{2.118}$$

Premultiply both sides of Eq. (2.118) by L^T. Then,

$$\begin{bmatrix} p'_x \\ p'_y \\ p'_z \end{bmatrix} A = L^T \begin{bmatrix} \sigma_x & \tau_{yx} & \tau_{zx} \\ \tau_{xy} & \sigma_y & \tau_{zy} \\ \tau_{xz} & \tau_{yz} & \sigma_z \end{bmatrix} L \begin{bmatrix} l'_{n1} \\ l'_{n2} \\ l'_{n3} \end{bmatrix} A \tag{2.119}$$

However, it is recalled that the stress state at point O may equally well be expressed by the stress matrix σ' with components in the rotated coordinate axes. The equilibrium of the tetrahedron is then given by Eq. (2.120).

$$\begin{bmatrix} p'_x \\ p'_y \\ p'_z \end{bmatrix} A = \begin{bmatrix} \sigma'_x & \tau'_{zx} & \tau'_{zx} \\ \tau'_{xy} & \sigma'_y & \tau'_{zy} \\ \tau'_{xz} & \tau'_{yz} & \sigma'_z \end{bmatrix} \begin{bmatrix} l'_{n1} \\ l'_{n2} \\ l'_{n3} \end{bmatrix} A \tag{2.120}$$

By comparing Eqs. (2.119) and (2.120) it is seen that the transformation law for transformation of the stress tensor under rotation of axes is given by

$$\sigma' = L^T \sigma L \tag{2.121}$$

Pre- and postmultiplication of Eq. (2.121) by L and L^T respectively gives the inverse transformation

$$\sigma = L \sigma L^T \tag{2.122}$$

The Stress Quadric

The normal stress σ_n on the plane *ijk*, in Fig. 2.28*b*, may be found by resolving the force components $\langle p_x \quad p_y \quad p_z \rangle A$ on the plane in the direction of the normal $\hat{n}$ and dividing by the area. That is,

$$\sigma_n A = [l_{n1} \quad l_{n2} \quad l_{n3}] \begin{bmatrix} p_x \\ p_y \\ p_z \end{bmatrix} A \tag{2.123}$$

From Eq. (2.117) substitute for $\langle p_x \quad p_y \quad p_z \rangle$. Then,

$$\sigma_n = [l_{n1} \quad l_{n2} \quad l_{n3}] \sigma \begin{bmatrix} l_{n1} \\ l_{n2} \\ l_{n3} \end{bmatrix} \tag{2.124}$$

However, $\hat{n}^T = \langle l_{n1} \quad l_{n2} \quad l_{n3} \rangle$ is the unit vector normal to the plane *ijk*; thence Eq. (2.124) is written

$$\sigma_n = \hat{n}^T \sigma_n \hat{n} \tag{2.125}$$

Consider the vector R, whose components define the point P, coordinates (x,y,z), such that

$$R^T = \sigma_n{}^{-1/2} \hat{n}^T = \langle x \quad y \quad z \rangle \tag{2.126}$$

Also, from Eq. (2.126),

$$\hat{n} = \sigma_n{}^{1/2} R \tag{2.127}$$

Substitute now for $\hat{n}$ in Eq. (2.125).

$$\sigma_n = R^T \sigma R \sigma_n$$

Hence,

$$R^T \sigma R = 1 \tag{2.128}$$

When Eq. (2.128) is expanded in terms of σ_x, τ_{xy}, etc., and $\langle x \quad y \quad z \rangle$, the expression obtained is the equation for an ellipsoid. (Note that for two dimensions this reduces to an ellipse.) That is,

$$R^T \sigma R = x^2 \sigma_x + y^2 \sigma_y + z^2 \sigma_z + 2xy\tau_{xy} + 2yz\tau_{yz} + 2zx\tau_{zx} = 1 \tag{2.129}$$

If the vector $\overline{OP} = \overline{R}$ is plotted from the origin O, then P lies on the quadratic surface given by Eq. (2.129). The ellipse of stress is shown in Fig. 2.29 The principal axes of the quadratic surface are such that for these axes, $\overline{R}$ is also the normal to the surface. By a rotation of axes, σ may be transferred to the principal axes, for which the shear stresses are zero.

The equation to the quadratic surface in the principal axes is then given by

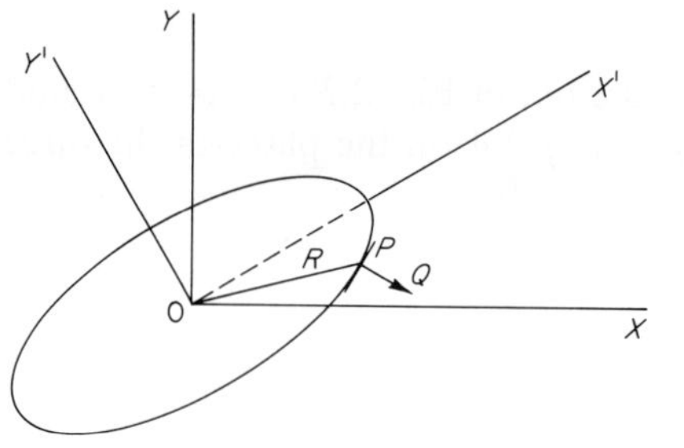

Figure 2.29 Ellipse of stress.

$$x^2\sigma_1 + y^2\sigma_2 + z^2\sigma_3 = 1 \tag{2.130}$$

Let L be the rotation matrix for the transformation to the principal axes. Then from Eq. (2.130), the vector $\bar{R}$ in the principal axes is given

$$\bar{R} = L^T R$$

and hence

$$R = L\bar{R}$$

Substitution for R in Eq. (2.128) gives the quadric expressed in principal axes.

$$\bar{R}^T L^T \sigma L \bar{R} = 1 \tag{2.131}$$

From Eqs. (2.130) and (2.131), it is seen that $L^T\sigma L$ must be equal to $\bar{\sigma}$ for L to be the required transformation matrix.

From Eqs. (2.85)ff it is seen that L has columns composed of solutions to the homogeneous equations,

$$\sigma X = \lambda X \tag{2.132}$$

The roots $\lambda_1, \lambda_2, \lambda_3$ are the characteristic values, or eigenvalues, of Eq. (2.132) and are in fact the principal stresses $\sigma_1, \sigma_2, \sigma_3$. The solution to Eq. (2.132) requires that $|\sigma - \lambda I| = 0$ for nontrivial solutions. On expansion this determinant gives the cubic equation in λ. In full,

$$|\sigma - \lambda I| = \begin{vmatrix} \sigma_x - \lambda & \tau_{yx} & \tau_{zx} \\ \tau_{xy} & \sigma_y - \lambda & \tau_{zy} \\ \tau_{xz} & \tau_{yz} & \sigma_z - \lambda \end{vmatrix} \tag{2.133}$$

Thence,

$$\lambda^3 - I_1\lambda^2 + I_2\lambda - I_3 = 0 \tag{2.134}$$

The coefficients I_1, I_2, I_3 in Eq. (2.134) are constants, or invariants, and are obtained from the expansion of Eq. (2.133). That is,

$$
\begin{aligned}
I_1 &= \sigma_x + \sigma_y + \sigma_z = \sigma_1 + \sigma_2 + \sigma_3 \\
I_2 &= \sigma_x\sigma_y + \sigma_y\sigma_z + \sigma_z\sigma_x - \tau_{xy}^2 - \tau_{yz}^2 - \tau_{zx}^2 = \sigma_1\sigma_2 + \sigma_2\sigma_3 + \sigma_3\sigma_1 \qquad (2.135) \\
I_3 &= |\sigma| = \sigma_1\sigma_2\sigma_3
\end{aligned}
$$

The first invariant I_1 is called the *trace* of matrix σ and is written

$$I_1 = \text{trace } \sigma = \text{tr } \sigma \qquad (2.136)$$

2.8 DEFINITION OF STRAIN COMPONENTS IN CARTESIAN COORDINATES; TRANSFORMATION OF THE STRAIN TENSOR

When forces are applied to a body or the temperature changes, deformations occur. These deformations are not necessarily elastic in the hookean sense. That is, the stress-strain curve may be not a straight line, and the deformations not recoverable. It is usual to assume in the classical theory, however, that the deformations are small. This requirement of small deformations ensures that the equations of equilibrium written in terms of the undistorted geometry will not be in serious error. In the theory developed in this section on strain it is found that certain nonlinear, quadratic terms occur that together form a quadratic strain tensor. It is also discovered that this quadratic strain tensor is useful in the analysis of bodies whose displacements are large (see Chap. 9).

The displacements of a body, general point P in Fig. 2.30, are examined in the neighborhood of P, and the definitions of strain are developed. Let P be the point $\langle x \quad y \quad z\rangle$, and Q a neighboring point $\langle x + \Delta x \quad y + \Delta y \quad z + \Delta z\rangle$. The vector $\langle \Delta x \quad \Delta y \quad \Delta z\rangle$ is assumed small by the mathematical concept of neighboring point. When the body is deformed, let P move to P_D defined by displacements $\langle v_x \quad v_y \quad v_z\rangle$ from P. The coordinates of P_D are then

$$\langle x + v_x \quad y + v_y \quad z + v_z\rangle$$

Then Q will be displaced to Q_D by this same amount plus the vector that gives

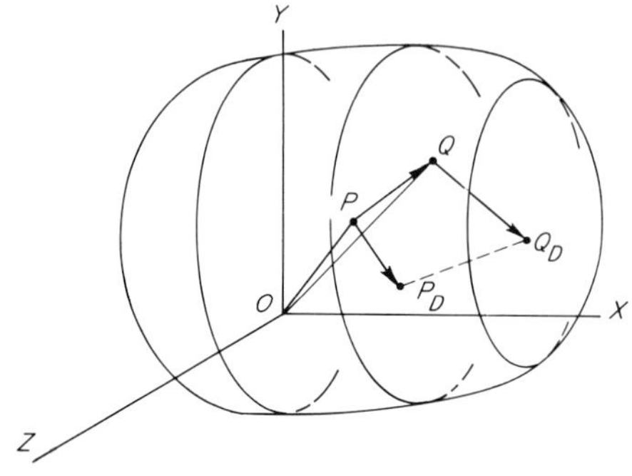

Figure 2.30 Initial and final states of deformed body.

the variation in $\langle v_x \quad v_y \quad v_z \rangle$ because of the distance $\langle \Delta x \quad \Delta y \quad \Delta z \rangle$ that Q is from P. Notice that this implies that the strain field is continuous and, as is shown also, the first derivatives with respect to x, y, z are also continuous. Let $\langle \delta v_x \quad \delta v_y \quad \delta v_z \rangle$ be the variation in the displacement field from P to Q.

That is, the coordinates of Q_D are given by

$$\langle x + \Delta x + v_x + \delta v_x \quad y + \Delta y + v_y + \delta v_y \quad z + \Delta z + v_z + \delta v_z \rangle$$

The vector **PQ** is given as the difference of the vectors **OP** and **OQ** and is designated Ω. That is,

$$\mathbf{PQ} = \Omega = \mathbf{Q} - \mathbf{P} = \begin{bmatrix} x + \Delta x \\ y + \Delta y \\ z + \Delta z \end{bmatrix} - \begin{bmatrix} x \\ y \\ z \end{bmatrix} = \begin{bmatrix} \Delta x \\ \Delta y \\ \Delta z \end{bmatrix} \tag{2.137}$$

After straining, the vector $\mathbf{P}_D\mathbf{Q}_D$ is designated Ω_D and again is obtained by subtraction of $\mathbf{OP}_D$ from $\mathbf{OQ}_D$. Thence,

$$\mathbf{P}_D\mathbf{Q}_D = \Omega_D = \begin{bmatrix} \Delta x + \delta v_x \\ \Delta y + \delta v_y \\ \Delta z + \delta v_z \end{bmatrix} \tag{2.138}$$

The difference between **PQ** and $\mathbf{P}_D\mathbf{Q}_D$ is a measure of the strain at P and is given by

$$\delta\Omega = \Omega_D - \Omega = \begin{bmatrix} \delta v_x \\ \delta v_y \\ \delta v_z \end{bmatrix} \tag{2.139}$$

Since P and Q are neighboring points and it has been stated that the displacement functions and their derivatives are continuous, the variations $\langle \delta v_x \quad \delta v_y \quad \delta v_z \rangle$ may be written as linear functions of $\langle \Delta x \quad \Delta y \quad \Delta z \rangle$ and derivatives of the displacements at P. That is, for Q close to P,

$$\delta v_x = \frac{\partial v_x}{\partial x} \Delta x + \frac{\partial v_x}{\partial y} \Delta y + \frac{\partial v_x}{\partial z} \Delta z \tag{2.140}$$

Similar expressions may be written for δv_y and δv_z and all three combined in the matrix equation (2.141). Note that $v_{x,x}$ denotes partial differentiation of v_x with respect to x, etc.

$$\begin{bmatrix} \delta v_x \\ \delta v_y \\ \delta v_z \end{bmatrix} = \begin{bmatrix} v_{x,x} & v_{x,y} & v_{x,z} \\ v_{y,x} & v_{y,y} & v_{y,z} \\ v_{z,x} & v_{z,y} & v_{z,z} \end{bmatrix} \begin{bmatrix} \Delta_x \\ \Delta_y \\ \Delta_z \end{bmatrix} \tag{2.141}$$

If M is used to denote the matrix of partial differential coefficients in Eq. (2.141), it may be written

$$\delta\Omega = M\Omega \tag{2.142}$$

It will be now shown that M may be decomposed into a symmetrical matrix that gives the classical definition of strain components and an antisymmetrical matrix that gives a rotation effect that is neglected in the nonrotational theory of elasticity. That is, this matrix causes rotation without strain. Thus, define the *direct* strains by

$$\varepsilon_x = v_{x,x} \qquad \varepsilon_y = v_{y,y} \qquad \varepsilon_{z,z} = v_{z,z} \tag{2.143}$$

and the shear strains by

$$\begin{aligned} \varepsilon_{xy} &= \varepsilon_{yx} = \tfrac{1}{2}(v_{x,y} + v_{y,x}) \\ \varepsilon_{yz} &= \varepsilon_{zy} = \tfrac{1}{2}(v_{y,z} + v_{z,y}) \\ \varepsilon_{zx} &= \varepsilon_{xz} = \tfrac{1}{2}(v_{z,x} + v_{x,z}) \end{aligned} \tag{2.144}$$

The rotation matrix terms are defined by

$$\begin{aligned} 2\theta_x &= v_{z,y} - v_{y,z} \\ 2\theta_y &= v_{x,z} - v_{z,x} \\ 2\theta_z &= v_{y,x} - v_{x,y} \end{aligned} \tag{2.145}$$

The matrix M is then decomposed into the strain matrix ε and rotation matrix θ, so that

$$M = \varepsilon + \theta = \begin{bmatrix} \varepsilon_x & \varepsilon_{xy} & \varepsilon_{xz} \\ \varepsilon_{yx} & \varepsilon_y & \varepsilon_{yz} \\ \varepsilon_{zx} & \varepsilon_{zy} & \varepsilon_z \end{bmatrix} + \begin{bmatrix} 0 & -\theta_z & \theta_y \\ \theta_z & 0 & -\theta_x \\ -\theta_y & \theta_x & 0 \end{bmatrix} \tag{2.146}$$

Substitution in Eq. (2.142) gives

$$\delta\Omega = (\varepsilon + \theta)\Omega \tag{2.147}$$

From Eqs. (2.138) and (2.142)

$$\Omega_D = \Omega + M\Omega \tag{2.148}$$

The relationship between Ω and Ω_D is shown graphically in Fig. 2.31. It is seen that $\delta\Omega$ is a measure of total strain at P. The direct strain along the line

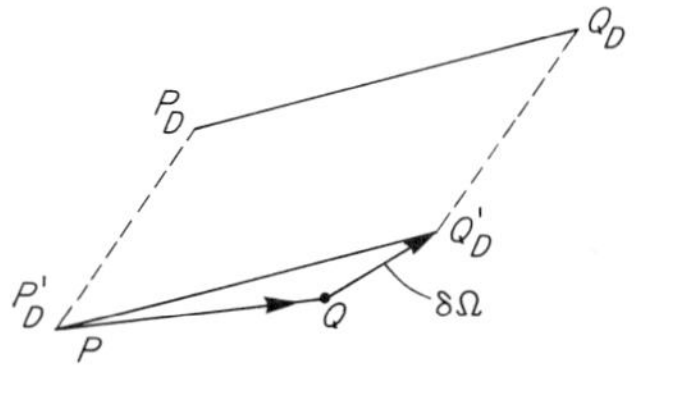

Figure 2.31 Deformation at P.

PQ will be measured in the difference in the vector magnitudes $|\Omega_D|$ and $|\Omega|$.

From Eq. (2.6), the magnitude of Ω, namely, $|\Omega|$, is given by

$$|\Omega|^2 = \Omega^T\Omega = \Delta x^2 + \Delta y^2 + \Delta z^2 \tag{2.149}$$

The similar expression for the strained length $\mathbf{P}_D\mathbf{Q}_D$ is

$$|\Omega_D|^2 = \Omega_D{}^T\Omega_D = (\Omega + M\Omega)^T(\Omega + M\Omega) \tag{2.150}$$

Multiplying the matrices on the right-hand side of Eq. (2.150),

$$|\Omega_D|^2 = \Omega^T\Omega + \Omega^T M\Omega + \Omega^T M^T\Omega + \Omega^T M^T M\Omega \tag{2.151}$$

For small displacements the term $\Omega^T M^T M\Omega = |\delta\Omega|^2$ may be neglected in comparison with the linear terms. It will be used later, however, in the discussion on the quadratic (nonlinear) strain tensor. By multiplication of terms it may be shown that $\Omega^T M^T\Omega = \Omega^T M\Omega$; hence,

$$|\Omega_D|^2 = \Omega^T\Omega + 2\Omega^T M\Omega \tag{2.152}$$

Hence for small strains,

$$|\Omega_D|^2 = |\Omega|^2 + 2\Omega^T M\Omega \tag{2.153}$$

That is, the change in length of $\mathbf{PQ}$ in straining to $\mathbf{P}_D\mathbf{Q}_D$ is given by

$$|\Omega_D|^2 = |\Omega|^2\left(1 + \frac{2\Omega^T M\Omega}{|\Omega|^2}\right)$$

and hence,

$$|\Omega_D| - |\Omega| = \frac{\Omega^T M\Omega}{|\Omega|} \tag{2.154}$$

From Eq. (2.146), M may be expressed in terms of ε and θ, so that

$$|\Omega_D| - |\Omega| = \frac{1}{|\Omega|}(\Omega^T\varepsilon\Omega + \Omega^T\theta\Omega) \tag{2.155}$$

On multiplication of the second term it is found that $\Omega^T\theta\Omega = 0$, a result that is expected for rotation without extension. Finally then,

$$|\Omega| - |\Omega| = \frac{1}{|\Omega|}\Omega^T\varepsilon\,\Omega \tag{2.156}$$

The Eq. (2.156) may be further simplified if it is written in terms of its magnitude $|\Omega|$ and the unit vector $t = (l_{s1}, l_{s2}, l_{s3})$ defining its direction [see Eq. (2.8*a*)]. Then

$$\Omega^T = |\Omega|\langle l_{s1} \quad l_{s2} \quad l_{s3}\rangle = |\Omega|t^T \tag{2.157}$$

Substitute in Eq. (2.157) and use $\Delta|\Omega|$ to signify the change in $|\Omega|$. Then,

$$\Delta|\Omega| = |\Omega_D| - |\Omega| = |\Omega|t^T\varepsilon t \tag{2.158}$$

The strain in the direction t is defined as the $\mathbf{P} \xrightarrow{lt} \mathbf{Q}$ $(\Delta|\Omega|)/|\Omega|$ and is given by Eq. (2.159)

$$\varepsilon_t = \mathbf{P} \xrightarrow{lt} \mathbf{Q} \frac{\Delta|\Omega|}{|\Omega|} = t^T \varepsilon t \tag{2.159}$$

Shear Strain at the point P

In Fig. 2.32, the displacements for P and Q have been plotted for the two-dimensional space XY, so that the only strains occurring are ε_x, ε_y, and $\varepsilon_{xy} = \varepsilon_{yx}$. It will be seen in the figure that the change in the right angle from XOY to $X'O'Y'$ in the deformed state is given by

Change in right angle $= v_{x,y} + v_{y,x}$

However, the term $v_{x,y} + v_{y,x}$ is simply twice the shear strain defined in Eq. (2.144). It is seen then that the shear strain at P associated with the direction defined by t, will be twice the change in the angle between t and its normal t_1. The two vectors t and t_1 at the point P in two-dimensional space are shown in Fig. 2.33.

Let Ω_1 be a vector $\mathbf{PQ}'$ measured along t_1 at right angles to $\mathbf{PQ}$. Then,

$$t_1{}^T t = t^T t_1 = 0 = \Omega_1{}^T \Omega$$

since t_1 and t are orthogonal. Also, from Eqs. (2.139) and (2.142),

$$\Omega_{1D} - \Omega_1 = \delta\Omega_1 = M\Omega_1 = \varepsilon\Omega_1$$

for small strains. As for Ω_D,

$$\Omega_{D1} = \Omega_1 + \varepsilon\Omega_1$$

Let β be the angle between the vectors Ω_D and Ω_{D1} in the deformed state. Then,

$$\Omega_{D1}{}^T \Omega_D = |\Omega_{D1}||\Omega_D| \cos\beta \tag{2.160}$$

That is,

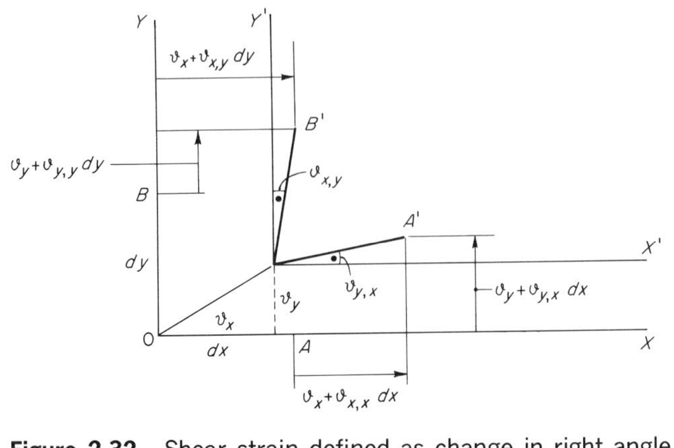

Figure 2.32 Shear strain defined as change in right angle.

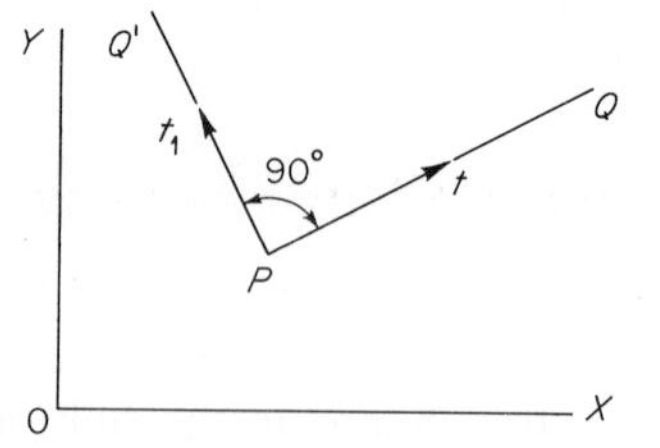

Figure 2.33 Axes for determining shear strain at *P*.

$$|\Omega_{D1}||\Omega_D| \cos \beta = (\Omega_1 + \varepsilon\Omega_1)^T(\Omega + \varepsilon\Omega) \tag{2.161}$$

On multiplication of the matrices on the right-hand side,

$$|\Omega_{1D}||\Omega_D| \cos \beta = \Omega_1{}^T\Omega + \Omega_1{}^T\varepsilon\Omega + \Omega_1{}^T\varepsilon^T\Omega + \Omega_1{}^T\varepsilon^T\varepsilon\Omega \tag{2.162}$$

In Eq. (2.162), $\Omega_1{}^T\Omega = 0$ since these vectors are parallel with the orthogonal vectors t_1 and t. The term $\Omega_1{}^T\varepsilon^T\varepsilon\Omega$ is a second-order term and is neglected here for linear elasticity.

For small changes in angle,

$$\cos \beta = \sin\left(\frac{\pi}{2} - \beta\right) \approx \frac{\pi}{2} - \beta = 2\varepsilon_{1\text{-}2} \tag{2.163}$$

By definition $\varepsilon_{1\text{-}2} = \varepsilon_{2\text{-}1}$ is the shear strain associated with the lines t_1 and t_2.

Finally, from Eq. (2.162)

$$|\Omega_{1D}||\Omega|2\varepsilon_{1\text{-}2} = |\Omega_{1D}||\Omega_D|(2t_1{}^T\varepsilon t)$$

That is,

$$\varepsilon_{1\text{-}2} = t_1{}^T\varepsilon t \tag{2.164}$$

Transformation of the Strain Tensor ε; Rotation of Axes

The transformation of the strain tensor ε to components ε' in the coordinate system $X'Y'Z'$ follows immediately from Eq. (2.159), in which it is recognized that the vector t obeys the usual transformation rules of Eq. (2.29). That is, if t' is the unit vector referred to the $X'Y'Z'$ axes and L the usual rotation matrix of direction cosines,

$$t = Lt' \tag{2.165}$$

Substitute for t in Eq. (2.159)

$$\varepsilon_t = t'^TL^T\varepsilon Lt' \tag{2.166}$$

However ε_t may also be written directly in terms of ε' and t' by following the reasoning leading to Eq. (2.159), so that

$$\varepsilon_t = t'^T\varepsilon't \tag{2.167}$$

Since the left-hand sides of Eqs. (2.165) and (2.167) are identical in value, it must follow that

$$\varepsilon' = L^T \varepsilon L \tag{2.168}$$

Thus the strain quantity ε obeys the same law of transformation as the inertia matrix J and the stress matrix σ.

The principal strain directions are defined as those directions for which the shear strains are zero. Since when the shear stresses are zero, so also are the shear strains, it follows that the principal stress and strain directions coincide.

Alternatively, solutions of the equations

$$\varepsilon X = \lambda X$$

yield the principal strains ε_1, ε_2, ε_3 as the eigenvalues. The solutions (X_1, X_2, X_3) for the values ε_1, ε_2, ε_3 will be the principal-axes' direction cosines.

The Strain Quadric

Associated with the strain ε_t (direction t), define the position vector R whose components define the point $P\langle x \quad y \quad z\rangle$ such that

$$R^T = \varepsilon_t^{-1/2} t^T = \langle x \quad y \quad z \rangle \tag{2.169}$$

Then,

$$t^T = \varepsilon_t^{1/2} R^T \tag{2.170}$$

Substitution of Eq. (2.170) in Eq. (2.159) leads to the quadratic function associated with the strain matrix ε. That is,

$$R^T \varepsilon R = 1 \tag{2.171}$$

The similarity of Eqs. (2.171) and (2.128) should be noticed.

Expand the terms $R^T \varepsilon R$:

$$R^T \varepsilon R = x^2 \varepsilon_x + y^2 \varepsilon_y + z^2 \varepsilon_z + 2xy\varepsilon_{xy} + 2yz\varepsilon_{yz} + 2zx\varepsilon_{zx} = 1 \tag{2.172}$$

As for the principal stress quadric, a suitable rotation of axes reduces Eq. (2.172) to the form

$$x^2 \varepsilon_1 + y^2 \varepsilon_2 + z^2 \varepsilon_z = 1 \tag{2.173}$$

Since the axes of principal stress and strain coincide, so also do the principal axes of the stress and strain quadrics.

Quadratic Strain Tensor

The discussion on strain is completed by examining the strain ε_t when the quadratic term in Eq. (2.151) is not considered of second-order magnitude. It will be found that these quadratic terms are useful in forming the geometric stiffness matrices in Chap. 9.

When the term $\Omega^T M^T M \Omega$ in Eq. (2.149) is retained, the expression for the magnitude of the strained length $|\Omega_D|$ becomes

$$|\Omega_D|^2 = |\Omega|^2 + 2\Omega^T M\Omega + \Omega^T M^T M\Omega \tag{2.174}$$

When substitution for Ω is made, $\Omega = |\Omega|t$,

$$|\Omega_D|^2 = |\Omega|^2(1 + 2t^T Mt + t^T M^T Mt)$$

That is,

$$|\Omega_D| - |\Omega| = |\Omega|(t^T Mt + \tfrac{1}{2}t^T M^T Mt) \tag{2.175}$$

In the term $t^T M^T$, M may be replaced by $\varepsilon = \theta$, and thence $t^T Mt = t^T \varepsilon t$ as before. Then,

$$\varepsilon_t = \frac{|\Omega_D| - |\Omega|}{|\Omega|} = t^T(\varepsilon + \tfrac{1}{2}M^T M)t \tag{2.176}$$

The tensor $\varepsilon + \tfrac{1}{2}M^T M$ is the strain tensor with quadratic terms. Define η, the quadratic strain tensor,

$$\eta = \tfrac{1}{2}M^T M \tag{2.177}$$

so that

$$\varepsilon_t = t^T(\varepsilon + \eta)t \tag{2.178}$$

Shear Strain, Large Displacements

In an analogous manner to that for the calculation of small shear strains, the change in the right angle between t_1 and t is given from the expression

$$\Omega_{1D}{}^T \Omega_D = |\Omega_{1D}||\Omega_D| \cos\beta = (\Omega_1 + M\Omega_1)^T(\Omega + M\Omega)$$

That is,

$$|\Omega_{1D}||\Omega_D| \cos\beta = 2\Omega_1{}^T M\Omega + \Omega_1{}^T M^T M\Omega \tag{2.179}$$

When $M = \varepsilon + \theta$ is substituted in the first term on the right-hand side of Eq. (2.179) and it is noted that $\Omega_1{}^T\theta\Omega = 0$, it is found that

$$\cos\beta = \sin\left(\frac{\pi}{2} - \beta\right) = 2\varepsilon_{1\text{-}2} = 2t_1{}^T(\varepsilon + \tfrac{1}{2}M^T M)t \tag{2.180}$$

that is,

$$\varepsilon_{1\text{-}2} = t_1{}^T(\varepsilon + \eta)t \tag{2.181}$$

The same tensor $\varepsilon + \eta$ is thus able to give both direct and shear strain terms when the quadratic terms are included.

On expansion of $\tfrac{1}{2}M^T M = \eta$, it is seen that

$$\begin{aligned}
\eta_{11} &= \tfrac{1}{2}(v_{x,x}^2 + v_{y,x}^2 + v_{z,x}^2) \\
\eta_{22} &= \tfrac{1}{2}(v_{x,y}^2 + v_{y,y}^2 + v_{z,y}^2) \\
\eta_{33} &= \tfrac{1}{2}(v_{x,z}^2 + v_{y,z}^2 + v_{z,z}^2) \\
\eta_{12} &= \tfrac{1}{2}(v_{x,x}v_{x,y} + v_{y,x}v_{y,y} + v_{z,x}v_{x,z}) \\
\eta_{13} &= \tfrac{1}{2}(v_{x,x}v_{x,z} + v_{y,x}v_{y,z} + v_{z,x}v_{z,z}) \qquad \text{etc.}
\end{aligned} \tag{2.182}$$

PROBLEMS

2.1 Calculate the reaction at O for the beams shown in Fig. P2.1*a* and *b*, using the matrix transformation theory. Draw the member-force diagrams in each case.

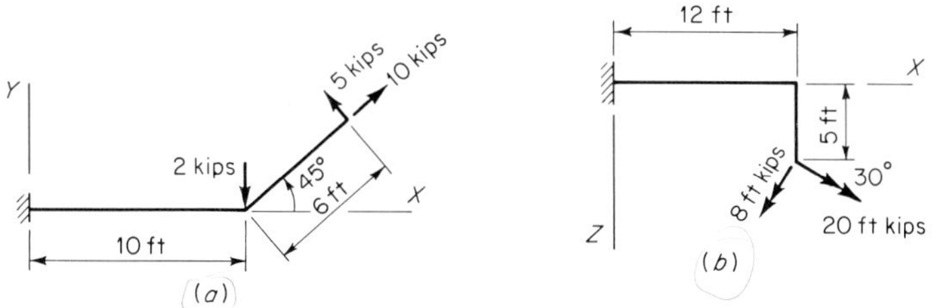

Figure P2.1.

2.2 Draw stress-resultant diagrams, moment, shear, and axial force for the beam shown in Fig. P2.2.

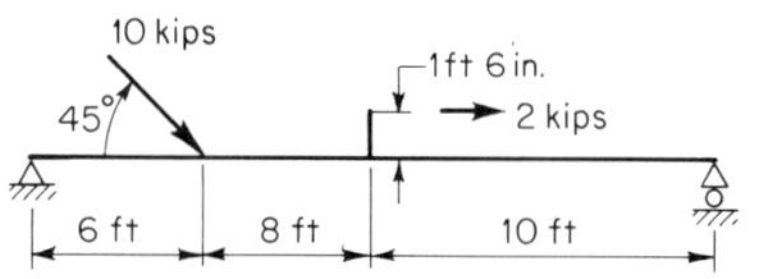

Figure P2.2.

2.3 The force systems 1, 2, and 3 act in the XY plane on the rigid plate shown in Fig. P2.3. Calculate the static equivalent of these forces at O.

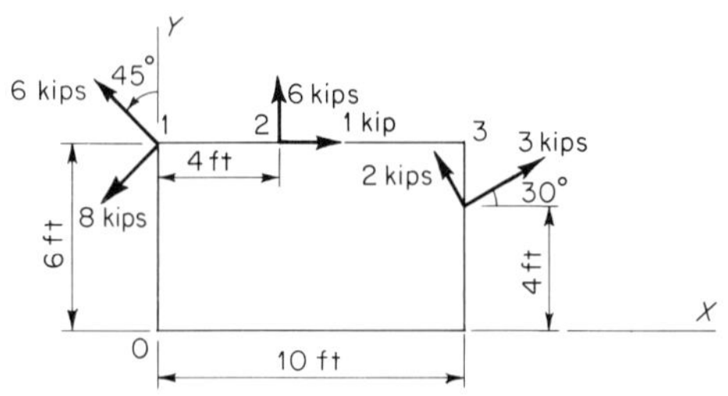

Figure P2.3.

2.4 The plate shown in Fig. P2.3 is given displacements at O

$$\begin{bmatrix} u \\ v \\ \theta_z \end{bmatrix} = \begin{bmatrix} 1.5 \text{ in.} \\ 0.3 \text{ in.} \\ 0.01 \text{ rad} \end{bmatrix}$$

Calculate the displacements at points 1, 2, and 3 in the coordinate directions shown.

2.5 Calculate the moment of inertia tensors in XY components for the plane areas shown in Fig. P2.4. Then calculate the principal axes and the principal moments of inertia.

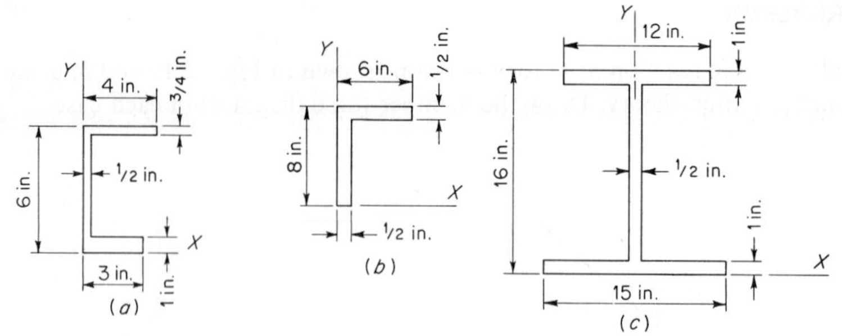

Figure P2.4.

2.6 Calculate the components of the stress tensor at the point A of the beam shown in Fig. P2.5.

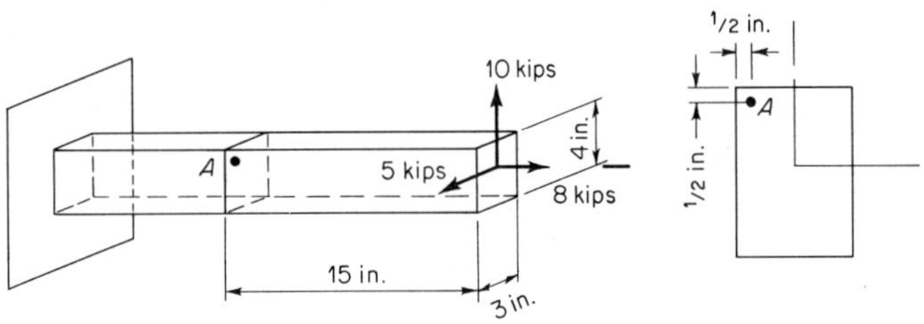

Figure P2.5.

2.7 For the point A of the beam shown in Fig. P2.5, calculate the strain tensor if $E = 30 \times 10^6$ in.-lb units and $\mu = 0.25$.

2.8 Calculate the principal-stress values and their directions for the stress tensor

$$\sigma = \begin{bmatrix} 20.0 & 5.0 \\ 5.0 & 20.0 \end{bmatrix}$$

3
STRUCTURE IDEALIZATION. VIRTUAL DISPLACEMENTS; DETERMINATE STRUCTURES

3.1 STRUCTURAL IDEALIZATION

In Chap. 2, it has been the intention to set up the fundamental concepts of force and displacement vectors and their transformations that may be applied generally to the analysis of structures irrespective of which method of analysis we choose to use. It was found that these transformations could be used to transform forces to a common origin and hence could be applied not only to problems of equilibrium, but also to the calculation of bending-moment, shear, and axial-force diagrams for members.

In this chapter we shall use the transformations to set up the equations of equilibrium for joints of a structure. The idea of a structure joint is most commonly applied in truss structures, in which the term *pin-jointed* conjures up the picture of members attached at their extremities by pins that drop nicely into place. In most cases this is certainly not true since the members are either monolithic (as in concrete structures) or riveted or welded to large gusset plates. In some large-span trusses in which large pin connections are used ostensibly to reduce secondary bending trusses, one wonders how much effect friction forces must have.

Nevertheless, the mathematical model of the pin located at the intersection of the centroidal axes of members common to the node is a useful one since it allows the equilibrium equations for a point to be used there. For the truss these will be

$$\Sigma F_x = 0 \qquad \Sigma F_y = 0 \qquad \Sigma F_z = 0$$

In the same way, the concept of a joint can be applied in the analysis of framed structures, except that now bending moments occur in members so that in addition to the force equilibrium conditions above, the moment conditions for joint equilibrium will be

$$\Sigma M_x = 0 \qquad \Sigma M_y = 0 \qquad \Sigma M_z = 0$$

In this case the joint may be considered a slice of the structure, infinitesimally small in its longitudinal dimensions and completely rigid so that it will not influence either the static or deflection calculations for the structure. The two types of joint are shown in Fig. 3.1. The idea of a joint and joint equilibrium has not been widely used in the calculation of member forces in determinate framed structures, mainly because it is generally more convenient to calculate member forces in framed structures, manually, by the method of sections. That is, free bodies of larger portions of the structure than a single joint are used.

When computers are used, however, for the static analysis of a determinate structure, it is most convenient to work in terms of the joint equilibrium equations because the geometry and topology of the structure are given at these points. Furthermore, these same ideas may be used in the displacement method of analysis. In this way, then, it becomes essential that we generalize the definitions of a joint (or node) and a structural member so that they may be used over the whole range of structural analysis.

Definition of Structure

A structure is considered composed of rigid nodes connected by flexible members supported from a rigid foundation so that the whole system is statically and geometrically stable. This definition implies that the model to be analyzed is only a mathematical one of the real structure and the approximation will be improved by suitable refining of the model. The structural elements may be finite or infinitesimal in dimensions. They may be straight prismatic members or curved in elevation, plane, or three-dimensional space. They may be two dimensional in form as in planar stress or plate bending finite-element analysis or three dimensional as in general stress analysis. The important feature of an element is that its force-displacement relationships must be known. That is, the stress field may be expressed in terms of several generalized (or member) forces. Then the force-displacement relationships at the points of application of the member forces are calculable. Alternatively,

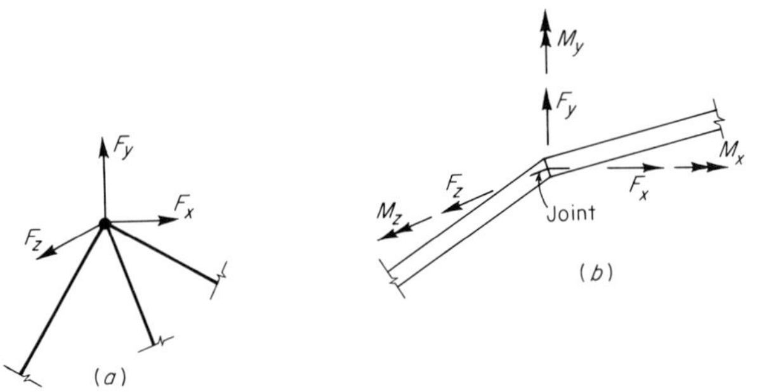

Figure 3.1 Joints for truss and framed structures. (*a*) Pin-joint, idealized truss structure; (*b*) joint in frame structure.

the displacement fields in the element may be decided. Then the member forces are calculable if the member displacements are known.

Example 3.1 Force and displacement fields

The prismatic cantilever beam *ij* in Fig. 3.2 may have its force field, here considered bending moment only, completely specified by the load P and moment M_j acting, as shown, on the free end j. These will be the member forces, and together they produce the linear force field M_x.

$$M_x = P(l - x) + M_j \tag{3.1}$$

From elementary strength of materials it is known that such a force field produces transverse displacement v_x that is a cubic function of the distance x along the beam. Thus, for a comparison of the two types of specification for a member, force and displacement, it is assumed that the displacements are cubic in nature and given by Eq. (3.2).

$$y = Ax^3 + Bx^2 + Cx + D \tag{3.2}$$

Then the slope is

$$y' = \theta = 3Ax^2 + 2Bx + C \tag{3.3}$$

and the curvature is

$$y'' = 6Ax + 2B \tag{3.4}$$

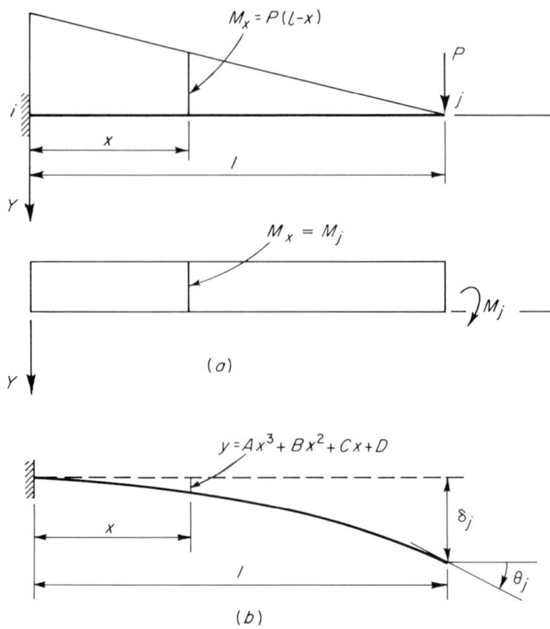

Figure 3.2 Force and displacement fields. (*a*) Linear force field; (*b*) cubic displacement field.

From the strength of materials it is known that for P and M the deflection and rotation at end j are given by Eq. (3.5).

$$\delta_j = \frac{Pl^3}{3EI} + \frac{M_j l^2}{2EI} \qquad (3.5)$$

$$\theta_j = \frac{Pl^2}{2EI} + \frac{M_j l}{EI}$$

In matrix form,

$$\begin{bmatrix} \delta_j \\ \theta_j \end{bmatrix} = \frac{l}{EI}\begin{bmatrix} \frac{l^2}{3} & \frac{l}{2} \\ \frac{l}{2} & 1 \end{bmatrix}\begin{bmatrix} P \\ M_j \end{bmatrix} \qquad (3.6)$$

The inverse deflection-force relationship is obtained by solving Eq. (3.6). Then

$$\begin{bmatrix} P \\ M_j \end{bmatrix} = \frac{2EI}{l}\begin{bmatrix} \frac{6}{l^2} & -\frac{3}{l} \\ -\frac{3}{l} & 2 \end{bmatrix}\begin{bmatrix} \delta_j \\ \theta_j \end{bmatrix} \qquad (3.7)$$

If the force and displacement fields are to be identical, then Eq. (3.7) should also be obtained by consideration of the displacement field.

For the displacement field the constants A, B, C, D are obtained by substitution in the kinematic boundary conditions. These are that when $x = 0$, then $y = 0$ and therefore $D = 0$, and when $x = 0$, then $y' = 0$ and therefore $C = 0$.

When $x = l$, then $y = \delta_j$ and $y' = \theta_j$; thence

$$\delta_j = Al^3 + Bl^2$$

$$\theta_j = 3Al^2 + 2Bl$$

In matrix form,

$$\begin{bmatrix} \delta_j \\ \theta_j \end{bmatrix} = l\begin{bmatrix} l^2 & l \\ 3l & 2 \end{bmatrix}\begin{bmatrix} A \\ B \end{bmatrix}$$

Solving for A and B,

$$\begin{bmatrix} A \\ B \end{bmatrix} = \frac{-1}{l^3}\begin{bmatrix} 2 & -l \\ -3l & l^2 \end{bmatrix}\begin{bmatrix} \delta_j \\ \theta_j \end{bmatrix} \qquad (3.8)$$

Substitution of Eq. (3.8) in Eq. (3.4) gives the expression

$$y'' = 6\left(-\frac{2}{l^3}\delta_j + \frac{1}{l^2}\theta_j\right)x + 2\left(\frac{3}{l^2}\delta_j - \frac{1}{l}\theta_j\right) \qquad (3.9)$$

The bending moment is

$$M_x = EIy'' = 2EI\left[\left(\frac{-6}{l^3}\delta_j + \frac{3}{l^2}\theta_j\right)x + \left(\frac{3}{l^2}\delta_j - \frac{1}{l}\theta_j\right)\right] \tag{3.10}$$

The forces on end j required to produce the cubic displacement field may now be obtained from the force boundary conditions,

1 When $x = 0$:

$$M = Pl + M_j = 2EI\left(\frac{3}{l^2}\delta_j - \frac{1}{l}\theta_j\right)$$

2 When $x = l$:

$$M = M_j = 2EI\left(-\frac{3}{l^2}\delta_j + \frac{2}{l}\theta_j\right)$$

In matrix form,

$$\begin{bmatrix} l & 1 \\ 0 & 1 \end{bmatrix}\begin{bmatrix} P \\ M_j \end{bmatrix} = 2EI\begin{bmatrix} \frac{3}{l^2} & -\frac{1}{l} \\ -\frac{3}{l^2} & \frac{2}{l} \end{bmatrix}\begin{bmatrix} \delta_j \\ \theta_j \end{bmatrix}$$

Premultiply both sides by the matrix $(1/l)\begin{bmatrix} 1 & -1 \\ 0 & l \end{bmatrix}$; then,

$$\begin{bmatrix} P \\ M_j \end{bmatrix} = \frac{2EI}{l}\begin{bmatrix} \frac{6}{l^2} & -\frac{3}{l} \\ -\frac{3}{l} & 2 \end{bmatrix}\begin{bmatrix} \delta_j \\ \theta_j \end{bmatrix} \tag{3.11}$$

Equations (3.7) and (3.11) are identical and show that the choice of displacement and force fields for the member give the same member-force-displacement relationships.

The nodes of the idealized structure may be of infinitely small dimensions, in which case they are used in the mathematical sense of a point to write the equations of equilibrium. However, they may be assumed to be of finite dimensions. For example, in the analysis of a very stiff pile headstock on slender piles, the headstock may be considered node. Various types of structures are shown in Fig. 3.3.

The structure shown in Fig. 3.3*a* has members (1), · · · , (8) and nodes 1, · · · , 8. The nodes in this structure are shown by the dark lines |, across the member and are considered infinitely thin. In this case the node has no effect on the structure properties. The structure typifies the idealizations made in analyzing frames and grids.

For the pile-headstock structure of Fig. 3.3*b*, the assumption may be made that the headstock is rigid. This rigid body has six equations of equilibrium, that may be written at the common origin O. It will require transformations

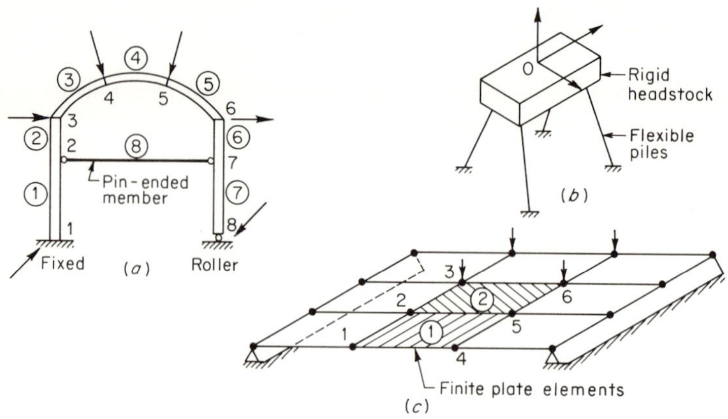

Figure 3.3 Structural idealization.

of the type given in Eq. (2.63) to relate displacements at O to those at the top of each pile member.

For both the structures given in Fig. 3.3*a* and *b*, it is usual to assume that the structural member may be considered a line element, that is, of infinitely small transverse dimensions, while still exhibiting the elastic properties of the actual member. This assumption simplifies displacement compatibility requirements where members intersect. The structure in Fig. 3.3*c* represents a different type of idealization. In this case the elements (1) and (2), for example, are two dimensional and so exhibit different properties from the line element. The plate bending elements shown in Fig. 3.3*c* have force fields characterized by bending and twisting moments. Displacements are matched at the nodes 1, 2, etc. This will generally imply that certain displacement compatibilities are also satisfied between adjacent elements along lines such as 1-2-3 and 2-5.

3.2 MEMBER FORCES AND DISPLACEMENTS

From the definition of a structural member, or element, it is axiomatic that only a finite number of forces or displacements may be used to completely specify the element force and displacement fields. The actual number of force or displacement components will vary depending on the coordinate system chosen, but the actual minimum number will be called a *fundamental set*. The beam shown in Fig. 3.2, for example, requires the two values (P,M_j) to specify its moment field completely. This is, however, not a unique set, because (M_i,M_j) could equally well be used to describe the same force field. Similarly, (δ_j,θ_j) completely define the member displacement field [Eqs. (3.2) and (3.8)]. Again this is not a unique set since (ϕ_i,ϕ_j) also define the dis-

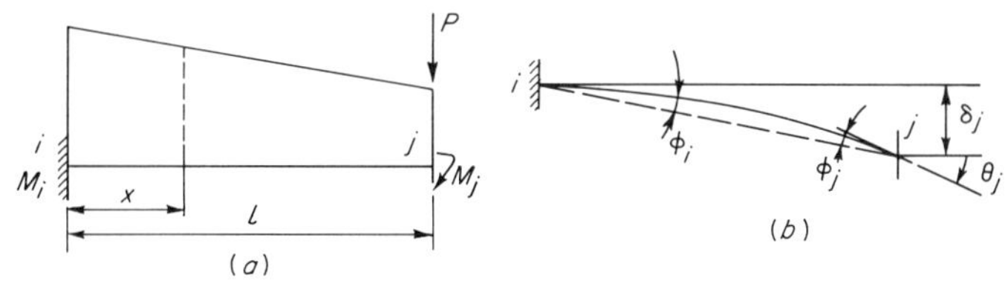

$M_x = P(L - x) + M_j$,

or $M_x = M_i(1 - x/L) + M_j\, x/L$

Figure 3.4 Member forces and distortions.

torted shape, these values being the rotations of the tangents at the ends of the member relative to the chord in the distorted position (see Fig. 3.4*b*).

In this chapter methods are obtained for the calculation of member forces for framed (or beam) and truss structures that are statically determinate, that is, structures whose member forces may be determined by statics alone. The transformations given in Chap. 2 are found to be useful in setting up the necessary joint equilibrium equations. It must be emphasized that in these statically determinate systems, the material properties of the members do not influence the member force values, unless deflections are so large as to change the statics of the structure. These large-deflection problems will be considered in Chap. 9. However, member material properties and dimensions are important when the deflections of the structure are considered. For example, a beam member may be haunched and thus have variable moments of inertia. This will affect the bending distortions of the member for given applied moments. It is usually found convenient to express the member deflection-load properties in the fundamental set of forces and deflections. On the other hand, the equilibrium equations are most naturally expressed in frame or global components. It is important to distinguish between frame and local coordinate components and the basic force system. In Fig. 3.5 the force systems have been set out for the pin-jointed truss member and the frame or beam member with linearly varying moment (constant shear) and constant axial force.

It will be found that the transformations developed in Chap. 2 are used in transforming force system I to system II and to system III.

The fundamental member distortions and joint displacements, corresponding to the force systems of Fig. 3.5, are shown in Fig. 3.6. Notice that to obtain fundamental distortions from joint displacements, rigid-body displacements of the member must be eliminated.

Finally member-node forces will be defined as those forces acting on the nodes from a member. Since action and reaction are equal and opposite, the member-node forces are opposite in sign to the member forces. That is, they are opposite in sign to those shown in Fig. 3.5*a* and *b*, II and III. It will be usual in this case to express member-node forces in global or frame coordinates.

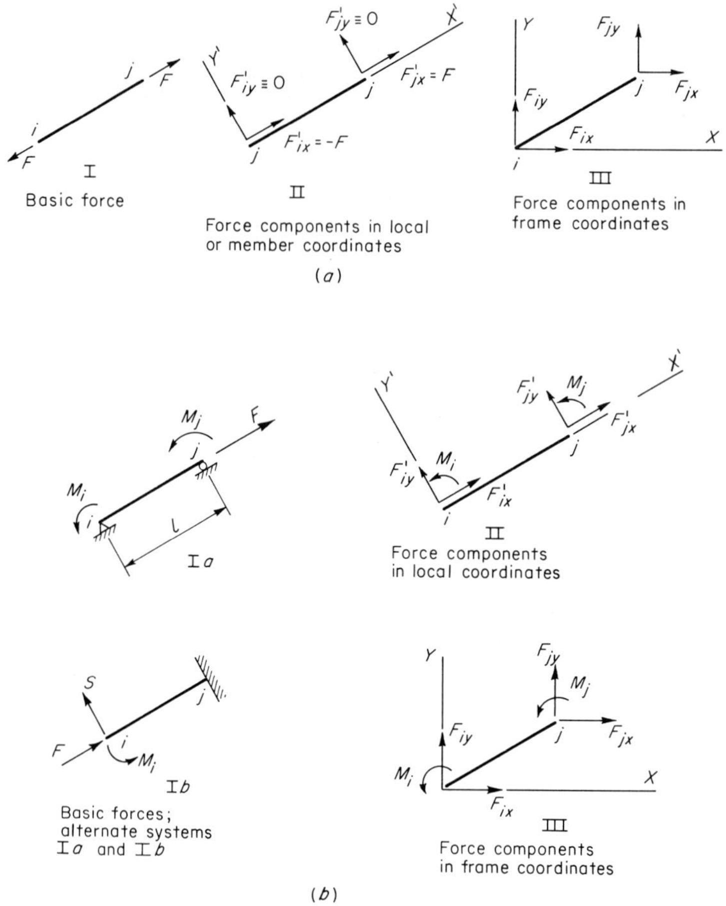

Figure 3.5 (*a*) Pin-jointed truss member; (*b*) frame or beam member with linear moment variation.

3.3 PRINCIPLE OF VIRTUAL DISPLACEMENTS AND THE CONTRAGRADIENT LAW

The principle of virtual displacements (or virtual work) is one of several work and energy theorems useful not only in the derivation of equilibrium conditions, but also in calculation of structure deflections. It leads to a very useful expression for transformations between compatible sets of displacements, given the transformation between the corresponding statically equivalent force systems.

The equilibrium form of the principle of virtual displacements is expressed as follows:

If a body in equilibrium under any system of forces whatsoever is given an arbitrary virtual (that is, not necessarily the actual) rigid-body displacement,

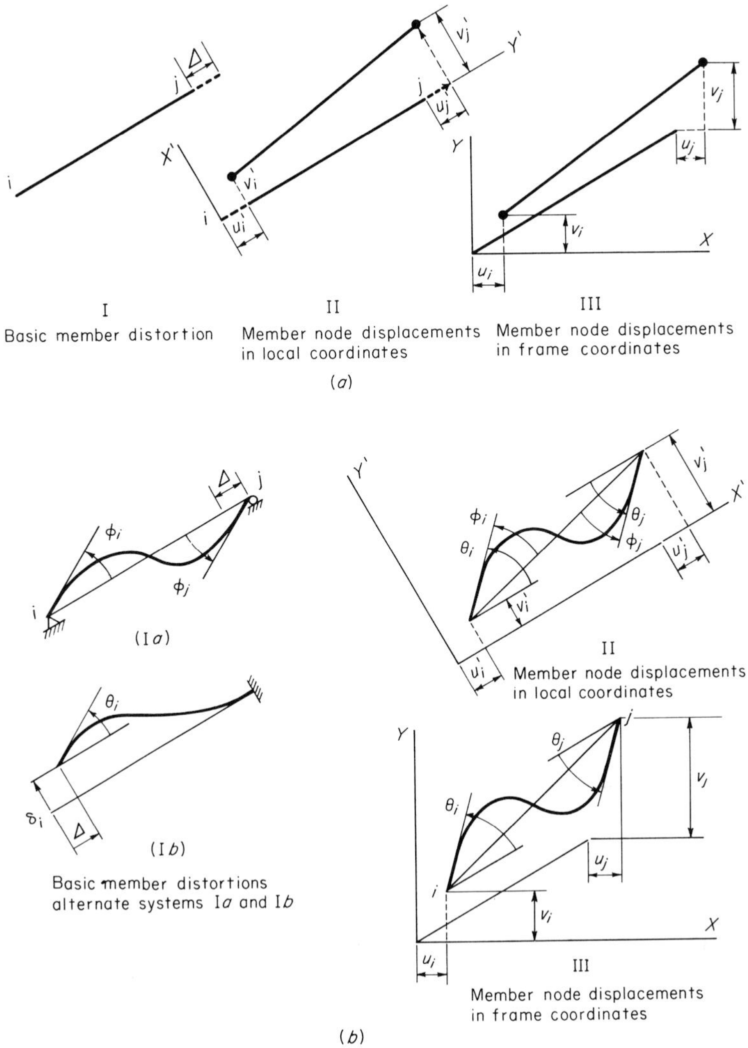

Figure 3.6 (*a*) Pin-jointed truss member; (*b*) frame or beam member.

the work done by the forces will be zero. Conversely, if the work done vanishes, then the force system is in equilibrium. The theorem is proved by application of Eqs. (2.51) and (2.52). Consider the system of forces R_1, $R_2, \ldots, R_n$ acting on the rigid body (for example, that shown in Fig. 2.16). These forces may be transformed to a statically equivalent system R_O acting at the origin. That is,

$$R_O = b_1R_1 + b_2R_2 + \cdots + b_nR_n \tag{3.12}$$

Give the rigid body a virtual displacement such that the displacement at O is

r_O. The displacements $r_1, r_2, \ldots, r_n$ are related to r_O by Eqs. (2.64) and (2.65), so that

$$r_i = a_i r_O = b_i^T r_O \tag{3.13}$$

The work done by the forces $R_1, R_2, \ldots, R_n$ is

$$W = R_1^T r_1 + R_2^T r_2 + \cdots + R_n^T r_n \tag{3.14}$$

Substitution of Eq. (3.13) gives

$$W = R_1^T b_1^T r_O + R_2^T b_2^T r_O + \cdots + R_n^T b_n^T r_O$$

or

$$W = (b_1 R_1 + b_2 R_2 + \cdots + b_n R_n)^T r_O$$

The expression in parentheses is, however, from Eq. (3.12), only the resultant force at O, namely, R_O; hence

$$W = R_O^T r_O \tag{3.15}$$

The system of forces $R_1, R_2, \ldots, R_n$ is, by hypothesis, in equilibrium, and thence the resultant force at O must be identically zero. That is,

$$W = R_O^T r_O = 0 r_O = 0 \tag{3.16}$$

The theorem is proved. Conversely, if the work done as given by Eq. (3.15) is equal to zero, the resultant force R_O must be equal to zero and hence the body is in equilibrium.

The second form of the principle of virtual displacements considers the structure composed of members and nodes acted on by an external force system. It is argued that since each of the nodes is in equilibrium under the action of its external loads and internal member-node forces, the work done must vanish for any virtual displacement. In this case the virtual displacement includes distortions of the body (but maintains displacement boundary-condition compatibility). The work done by member-node forces must be equal and opposite to that done by member forces. The work done by external forces on all nodes plus work done by member-node forces on all nodes is equal to zero for equilibrium of the nodes.

Let

W_e = external work done by applied forces

W_i = internal work done by member forces

Then,

$-W_i$ = work done by the member-node forces

Then for joint equilibrium,

$$W_e - W_i = 0$$

or

$$W_e = W_i \tag{3.17}$$

Notice that whereas the member-node forces are in equilibrium with the applied forces, the member forces are statically equivalent to the applied forces.

Although Eq. (3.17) may be used to calculate deflections and also to express node equilibrium directly, it is more useful in the present context of matrix analysis to generalize Eq. (3.17) and express it in matrix form. In doing this the contragradient nature of force and displacement transformations is revealed.

The Contragradient Law

Consider the two statically equivalent force systems Q and P connected by the linear transformation,

$$Q = BP \tag{3.18}$$

In certain circumstances this may be a reversible transformation, that is, $P = B^{-1}Q$. This is, however, an unnecessary restriction, and it will be assumed that in general there will be more terms in Q than in P. This implies that there are solutions to Q for which P is zero. For example, in the truss shown in Fig. 3.7, there is a relationship connecting the member forces $\langle F_1 \quad F_2 \quad F_3 \rangle = Q^T$ and $R^T = \langle R_x \quad R_y \rangle$. However, there will be solutions for which $\langle F_1 \quad F_2 \quad F_3 \rangle$ have zero-external-force components since the three member forces alone may satisfy the conditions of joint equilibrium, $\Sigma\, F_x = 0$ and $\Sigma\, F_y = 0$. This corresponds, of course, to a temperature change of one or more of the members.

Associated with the force systems Q and P will be the sets of displacements q and p. These are compatible displacements so that if the displacements in the directions of Q are q, those in the directions of P are p. From Eq. (3.17) it is seen that if two statically equivalent force systems are given compatible displacements, the work done in each system is the same. That is,

$$P^T p = Q^T q \tag{3.19}$$

However from Eq. (3.18),

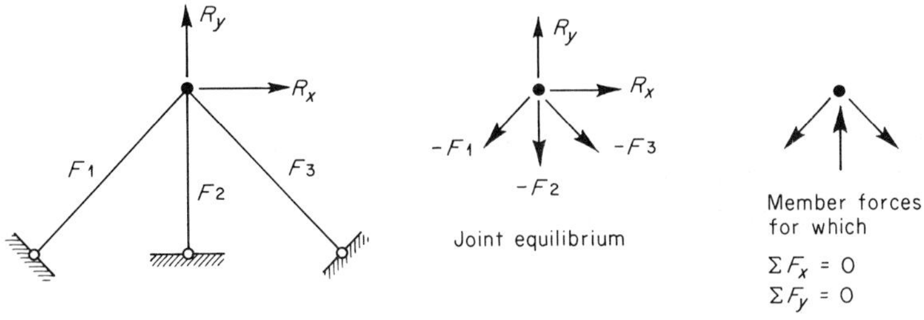

Figure 3.7 Force transformation matrices for a structure.

$$Q^T = P^T B^T$$

Hence,

$$P^T p = P^T B^T q$$

That is, since P and Q are arbitrary,

$$p = B^T q \tag{3.20}$$

It will be found that Eqs. (3.18) and (3.20) form a most important set of fundamental relationships that will be used continuously throughout this text.

The contragradient law appears in a second form in which Q and P are statically equivalent forces and the associated sets of displacements q and p are related by the linear transformation

$$q = Cp \tag{3.21}$$

Then from Eq. (3.19),

$$P^T p = Q^T C p$$

so that

$$P = C^T Q \tag{3.22}$$

Under certain circumstances the transformations (3.18) and (3.22) are identical. This occurs when the transformation (3.18) is reversible. The implication is then that Q and P are vectors of the same dimension and B is a square nonsingular matrix. Then,

$$P = B^{-1} Q \tag{3.23}$$

Comparison of Eqs. (3.22) and (3.23) shows that in this case

$$C^T = B^{-1} \tag{3.24}$$

Alternatively,

$$B = (C^T)^{-1} = (C^{-1})^T \tag{3.25}$$

It will be shown that the above reversible transformation occurs in the relationship between the node and member forces and between the node and member displacements in statically determinate structures. Then P is the node forces, and Q the member forces. The Eq. (3.25) provides a convenient way of obtaining the B matrix.

3.4 FORCE TRANSFORMATION MATRICES FOR STRUCTURES; FORCE–INFLUENCE MATRICES; SIGN CONVENTIONS FOR BEAM ELEMENTS IN BENDING

Consider the structure composed of rigid nodes and flexible members, external node forces R and member forces S. The member forces S are related to R by a linear *force transformation* matrix. As mentioned in Chap. 2, this matrix may be determined from static-equilibrium equations based on initial structure geometry alone for the statically determinate structure. The linear relationship between member and node forces is expressed in matrix form by Eq. (3.26).

$$S = bR \tag{3.26}$$

The symbol b has been used to be consistent with the definition of force transformation matrices in Eq. (2.52). The remainder of Chap. 3 will discuss ways of obtaining the b matrix for statically determinate structures by computer methods. The member types considered will be those of Fig. 3.5a I and b I.

Example 3.2 Force transformation matrix for pin-jointed truss

The truss shown in Fig. 3.8 has members (1), (2), . . . , (7) and nodes 1, 2, 3, . . . , 5 and is under the action of the forces R_1, R_2, R_3 parallel to the Y axis. The member forces may be calculated by the method of joints or method of sections. If in Eq. (3.26) the force R_1 is set equal to unity and $R_2 = R_3 = 0$, it is found that the member forces S are identical with the first column of the b matrix, and similarly for $R_2 = 1$, etc. From this it is seen that the b matrix may be obtained by successive applications of a unit load at the required node points.

It must be emphasized that this will not in general be the most effective way to calculate the b matrix, but that it does bring out the force influence property of the rows of b. The numerical values of the b matrix for the truss in Fig. 3.8 are given in Eq. (3.27).

$$\begin{bmatrix} S_1 \\ S_2 \\ S_3 \\ S_4 \\ S_5 \\ S_6 \\ S_7 \end{bmatrix} = \begin{bmatrix} -0.433 & -0.144 & -0.288 \\ -0.144 & -0.433 & -0.288 \\ 0.866 & 0.288 & 0.577 \\ 0.288 & -0.288 & -0.577 \\ -0.288 & 0.288 & -0.577 \\ 0.288 & 0.866 & 0.577 \\ 0.288 & 0.288 & 0.577 \end{bmatrix} \begin{bmatrix} R_1 \\ R_2 \\ R_3 \end{bmatrix} \tag{3.27}$$

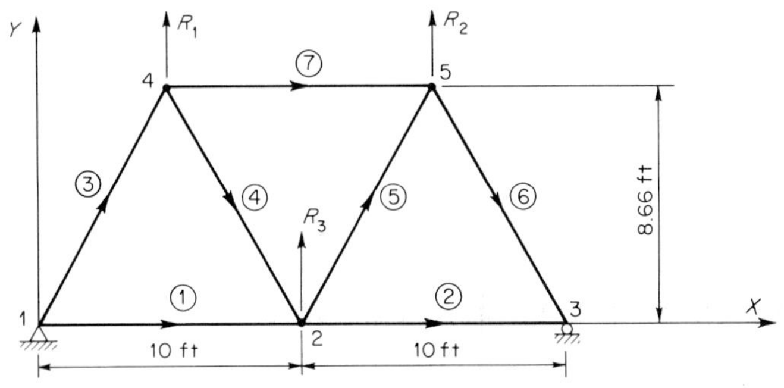

Figure 3.8 Determinate pin-jointed truss.

Since the columns of b are produced by application of unit loads, the rows of b give the value of a particular force as the unit load assumes various positions on the structure. Hence the rows of b are influence functions for member forces, and the b force transformation matrix may also be termed a *force-influence* matrix.

It is observed, also, that in Eq. (3.27) the b matrix is not a square matrix but is actually 7×3. This appears to be in conflict with the statement in Sec. 3.3 that the b matrix is nonsingular for statically determinate structures.

In the present example of the truss in Fig. 3.8, not all possible node loads have been included. When this is done and horizontal forces are placed at nodes 2, 3, 4, and 5, the number of node forces is then seven and b is a square matrix that is found to be nonsingular. This property is demonstrated in the next example on beam structures.

Example 3.3 Force transformation matrix for beam

The beam structure shown in Fig. 3.9 is included in the definition of structure given in Sec. 3.1 if the nodes at points 1, 2, . . . , 5 are considered infinitely thin, rigid portions of the beam. In this way they will influence neither the geometry nor the elastic properties of the beam. The beam is supported by a pin at node 1 and a roller at node 4. The loads R_1, R_2, and R_3 are applied in the positive Y direction at nodes 2, 3, and 5. The structure has beam elements (1), (2), (3), and (4) whose fundamental member forces will be taken to be (M_i, M_j) as shown in Fig. 3.5*b* I*a*. Then for the beam, Eq. (3.26) is given by

$$\begin{bmatrix} M_{1i} \\ M_{1j} \\ M_{2i} \\ M_{2j} \\ M_{3i} \\ M_{3j} \\ M_{4i} \\ M_{4j} \end{bmatrix} = \begin{bmatrix} 0 & 0 & 0 \\ -4.2 & -1.8 & +3.0 \\ +4.2 & +1.8 & -3.0 \\ -1.8 & -4.2 & +7.0 \\ +1.8 & +4.2 & -7.0 \\ 0 & 0 & +10.0 \\ 0 & 0 & -10.0 \\ 0 & 0 & 0 \end{bmatrix} \begin{bmatrix} R_1 \\ R_2 \\ R_3 \end{bmatrix} \tag{3.28}$$

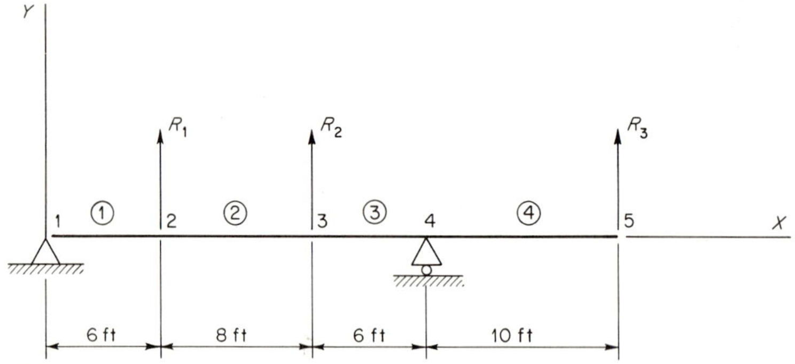

Figure 3.9 Determinate beam structure.

The b matrix in Eq. (3.28) is not square and is evidently not invertible. If, however, at nodes 1 to 5, moment forces are included, the number of forces becomes equal to 8 and the b matrix is then square and nonsingular. The complete b matrix is written in Eq. (3.29).

$$\begin{bmatrix} M_{1i} \\ M_{1j} \\ M_{2i} \\ M_{2j} \\ M_{3i} \\ M_{3j} \\ M_{4i} \\ M_{4j} \end{bmatrix} = \begin{bmatrix} 0 & 0 & 0 & 1.0 & 0 & 0 & 0 & 0 \\ -4.2 & -1.8 & +3.0 & -0.7 & 0.3 & 0.3 & 0.3 & 0.3 \\ +4.2 & +1.8 & -3.0 & 0.7 & 0.7 & -0.3 & -0.3 & -0.3 \\ -1.8 & -4.2 & +7.0 & -0.3 & -0.3 & 0.7 & 0.7 & 0.7 \\ +1.8 & +4.2 & -7.0 & 0.3 & 0.3 & 0.3 & -0.7 & -0.7 \\ 0 & 0 & +10.0 & 0 & 0 & 0 & 1.0 & 1.0 \\ 0 & 0 & -10.0 & 0 & 0 & 0 & 0 & -1.0 \\ 0 & 0 & 0 & 0 & 0 & 0 & 0 & 1.0 \end{bmatrix} \begin{bmatrix} R_1 \\ R_2 \\ R_3 \\ M_1 \\ M_2 \\ M_3 \\ M_4 \\ M_5 \end{bmatrix} \tag{3.29}$$

Now the b matrix in Eq. (3.29) may be inverted, and the inverse relationship is given in Eq. (3.30).

$$\begin{bmatrix} R_1 \\ R_2 \\ R_3 \\ M_1 \\ M_2 \\ M_3 \\ M_4 \\ M_5 \end{bmatrix} = \begin{bmatrix} -0.16667 & -0.16667 & +0.125 & +0.125 & 0 & 0 & 0 & 0 \\ 0 & 0 & -0.125 & -0.125 & +0.16667 & +0.16667 & 0 & 0 \\ 0 & 0 & 0 & 0 & 0 & 0 & -0.10 & -0.10 \\ 1.0 & 0 & 0 & 0 & 0 & 0 & 0 & 0 \\ 0 & 1.0 & 1.0 & 0 & 0 & 0 & 0 & 0 \\ 0 & 0 & 0 & 1.0 & 1.0 & 0 & 0 & 0 \\ 0 & 0 & 0 & 0 & 0 & 1.0 & 1.0 & 0 \\ 0 & 0 & 0 & 0 & 0 & 0 & 0 & 1.0 \end{bmatrix} \begin{bmatrix} M_{1i} \\ M_{1j} \\ M_{2i} \\ M_{2j} \\ M_{3i} \\ M_{3j} \\ M_{4i} \\ M_{4j} \end{bmatrix} \tag{3.30}$$

It was noted that whereas the columns of the b matrix are member-force diagrams for unit loads, the rows of b are influence functions. Thus to generate the b matrix, it is possible to proceed by either of these methods. Typical unit-load and influence diagrams are shown in Fig. 3.10. Notice that influence lines are required for unit moment moving across the beam, as well as for unit load. It is interesting to examine the coefficient matrix in Eq. (3.30), since this transforms the member forces in the beam into the joint equilibrium equations. For example, the moment equilibrium equation at joint 3 will be

$$-M_{2j} - M_{3i} + M_3 = 0 \tag{3.31}$$

That is,

$$\begin{bmatrix} 1 & 1 \end{bmatrix} \begin{bmatrix} M_{2j} \\ M_{3i} \end{bmatrix} = M_3 \tag{3.32}$$

It will be observed that this relationship is given by line six in Eq. (3.30).

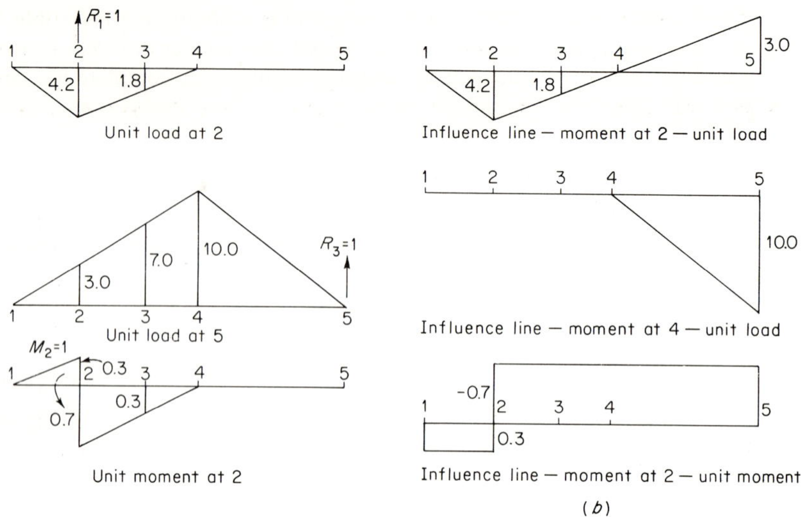

Figure 3.10 Typical unit-force and influence diagrams. (a) Unit-force diagrams; (b) influence diagrams.

Force Transformation Matrix for Frame Structures; Alternative Sign Convention for Moments

The member-moment sign convention used in the previous section is quite useful in computer solutions because it leads to a consistent sign convention between member and joint moments in the joint equilibrium equations. However, in beam and frame problems it has been the practice in noncomputer methods of solution to use the engineer's sign convention for member moments. In this convention, for the member ij shown in Fig. 3.11 the moments M_i, M_j at the ends will be positive when they produce tension on the side of the beam nearest the viewer if the end i is on the left hand. When this sign convention is used, the force transformation in Eq. (3.29) becomes

$$\begin{bmatrix} M_{1i} \\ M_{1j} \\ M_{2i} \\ M_{2j} \\ M_{3i} \\ M_{3j} \\ M_{4i} \\ M_{4j} \end{bmatrix} = \begin{bmatrix} 0 & 0 & 0 & -1.0 & 0 & 0 & 0 & 0 \\ 4.2 & 1.8 & -3.0 & -0.7 & 0.3 & 0.3 & 0.3 & 0.3 \\ 4.2 & 1.8 & -3.0 & -0.7 & -0.7 & 0.3 & 0.3 & 0.3 \\ 1.8 & 4.2 & -7.0 & -0.3 & -0.3 & 0.7 & 0.7 & 0.7 \\ 1.8 & 4.2 & -7.0 & -0.3 & -0.3 & -0.3 & 0.7 & 0.7 \\ 0 & 0 & -10.0 & 0 & 0 & 0 & 1.0 & 1.0 \\ 0 & 0 & -10.0 & 0 & 0 & 0 & 0 & 1.0 \\ 0 & 0 & 0 & 0 & 0 & 0 & 0 & 1.0 \end{bmatrix} \begin{bmatrix} R_1 \\ R_2 \\ R_3 \\ M_1 \\ M_2 \\ M_3 \\ M_4 \\ M_5 \end{bmatrix} \tag{3.33}$$

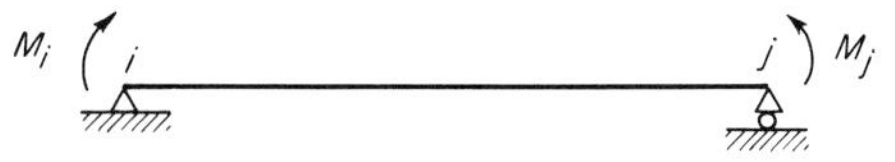

Figure 3.11 Beam-element-moment sign convention.

Inversion of the coefficient matrix in Eq. (3.33) gives the joint equilibrium equations. That is,

$$\begin{bmatrix} R_1 \\ R_2 \\ R_3 \\ M_1 \\ M_2 \\ M_3 \\ M_4 \\ M_5 \end{bmatrix} = \begin{bmatrix} -0.16667 & 0.16667 & 0.125 & -0.125 & 0 & 0 & 0 & 0 \\ 0 & 0 & -0.125 & 0.125 & -0.16667 & -0.16667 & 0 & 0 \\ 0 & 0 & 0 & 0 & 0 & 0 & -0.1 & 0.1 \\ -1.0 & 0 & 0 & 0 & 0 & 0 & 0 & 0 \\ 0 & 1.0 & -1.0 & 0 & 0 & 0 & 0 & 0 \\ 0 & 0 & 0 & 1.0 & -1.0 & 0 & 0 & 0 \\ 0 & 0 & 0 & 0 & 0 & 1.0 & -1.0 & 0 \\ 0 & 0 & 0 & 0 & 0 & 0 & 0 & 1.0 \end{bmatrix} \begin{bmatrix} M_{1i} \\ M_{1j} \\ M_{2i} \\ M_{2j} \\ M_{3i} \\ M_{3j} \\ M_{4i} \\ M_{4j} \end{bmatrix} \tag{3.34}$$

3.5 DISPLACEMENT TRANSFORMATION MATRICES

In Sec. 3.3, it has been shown [Eqs. (3.21) and (3.22)] that if a displacement transformation of the type of Eq. (3.35) connects two displacement systems, the transformation of corresponding forces is Eq. (3.36).

$$q = Cp \tag{3.35}$$

$$P = C^T Q \tag{3.36}$$

In the present usage, q will be replaced by v, the vector of fundamental member displacements, and p by r, the node displacements in frame or global coordinates. By analogy with Eq. (2.64), the displacement transformation matrix is designated a, so that in terms of node and member displacements, Eq. (3.35) is written

$$v = ar \tag{3.37}$$

In terms of node and member forces, Eq. (3.36) becomes

$$R = a^T S \tag{3.38}$$

However, Eq. (3.38) expresses the fact that R is statically equivalent to S. That is,

$$R - a^T S = 0 \tag{3.39}$$

expresses the equations of equilibrium of the nodes. If R is expressed in frame coordinates, then Eqs. (3.39) are the node equilibrium equations in frame coordinates.

If a is nonsingular, then

$$S = (a^T)^{-1}R \tag{3.39a}$$

By comparison of Eq. (3.39a) with Eq. (3.26), it is deduced that for such a condition,

$$b = (a^T)^{-1} \tag{3.40}$$

It is evident then that Eq. (3.40) provides a method for calculation of the b matrix for the statically determinate structure. General expressions for a are derived in Sec. 3.6.

Example 3.4 Displacement transformation for pin-jointed truss

The three-bar truss shown in Fig. 3.12, with members (1), (2), and (3) and nodes 1, 2, and 3, has three possible node displacements $\langle r_{2x} \quad r_{2y} \quad r_{3x}\rangle$. The a matrix expressing member elongations $\langle \Delta_1 \quad \Delta_2 \quad \Delta_3\rangle$ as functions of the node displacements may be calculated by considering the transformation from global to local axes at each joint. Transform $\langle r_{2x} \quad r_{2y}\rangle$ into local member coordinates at node 2 for both members (1) and (2), the positive sense of the X' axes for each member being given by the arrows in the figure.

From Eq. (2.15) for member i:

$$L_i = \begin{bmatrix} \cos\alpha_i & -\sin\alpha_i \\ \sin\alpha_i & \cos\alpha_i \end{bmatrix}$$

For $i = 1$, $\cos\alpha_1 = 0.6$, $\sin\alpha_1 = 0.8$:

$$L_1 = \begin{bmatrix} 0.6 & -0.8 \\ 0.8 & 0.6 \end{bmatrix}$$

For $i = 2$, $\cos\alpha_2 = 0.6$, $\sin\alpha_2 = -0.8$:

$$L_2 = \begin{bmatrix} 0.6 & 0.8 \\ -0.8 & 0.6 \end{bmatrix}$$

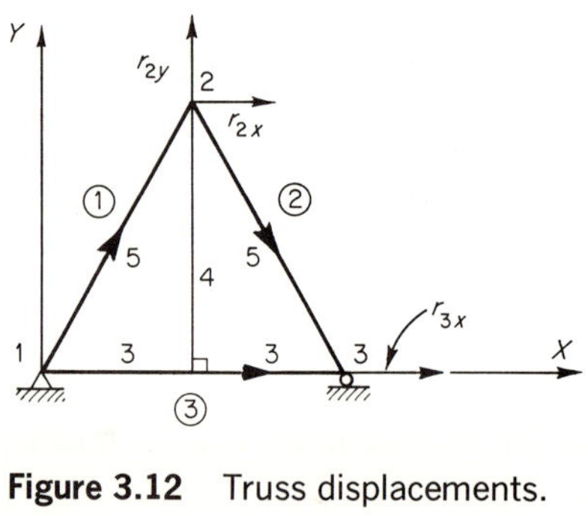

Figure 3.12 Truss displacements.

Only the X' components of member distortions in local coordinates need be considered, since the Y' components do not contribute to elongation of the members. Then,

$$r'_{1x} = \Delta_1 = [0.6 \quad -0.8]\begin{bmatrix} r_{2x} \\ r_{2y} \end{bmatrix} \tag{i}$$

$$r'_{2x} = -\Delta_2 = [0.6 \quad 0.8]\begin{bmatrix} r_{2x} \\ r_{2y} \end{bmatrix} \tag{ii}$$

For joint 3, only the displacement r_{3x} need be considered.

For member (2):

$$r'_{2x} = \Delta_2 = [0.6 \quad 0.8]\begin{bmatrix} r_{3x} \\ r_{3y} \end{bmatrix} \tag{iii}$$

For member (3):

$$L_3 = \begin{bmatrix} 1 & 0 \\ 0 & 1 \end{bmatrix}$$

Therefore,

$$r'_{3x} = \Delta_3 = [1 \quad 0]\begin{bmatrix} r_{3x} \\ r_{3y} \end{bmatrix} \tag{iv}$$

Equations (*i*) through (*iv*) may be combined, omitting $r_{3y} = 0$. That is,

$$\begin{bmatrix} \Delta_1 \\ \Delta_2 \\ \Delta_3 \end{bmatrix} = \begin{bmatrix} 0.6 & -0.8 & 0 \\ -0.6 & -0.8 & 0.6 \\ 0 & 0 & 1 \end{bmatrix}\begin{bmatrix} r_{2x} \\ r_{2y} \\ r_{3x} \end{bmatrix}$$

This is the equation

$v = ar$

Then

$$a^T = \begin{bmatrix} 0.6 & -0.6 & 0 \\ -0.8 & -0.8 & 0 \\ 0 & 0.6 & 1 \end{bmatrix}$$

$$(a^T)^{-1} = \begin{bmatrix} 0.833 & -0.833 & 0.5 \\ 0.625 & 0.625 & 0.375 \\ 0 & 0 & -1.0 \end{bmatrix}$$

Just as for the force transformation matrix, the columns of the a matrix are produced by successively applying unit joint displacements. The rows of a are displacement-influence functions. That is, they give the values of a particular member displacement as a unit displacement is produced at successive nodes.

3.6 TRANSFORMATIONS FOR BASIC MEMBER FORCE AND DISPLACEMENT COMPONENTS

In the process of developing the automatic procedures for generation of the b and a matrices, it is necessary to reconsider the basic (fundamental) member force and displacement components and to derive the relationships between these and the components expressed firstly in the local member coordinates and secondly in the frame or global coordinates.

A Force Transformations

The force transformation matrices are derived from the basic force quantities shown in Fig. 3.5*a* and *b*. If S_{xy} designates the member forces with components in frame coordinates, then the member node forces are $-S_{xy}$. That is, the equations of joint equilibrium are in the form

$$-S_{xy} + R = 0 \tag{3.41}$$

Hence,

$$S_{xy} = R \tag{3.42}$$

The actual step of writing the equations of joint equilibrium in the form $\Sigma F_x = 0$, etc., is bypassed, and instead, the equality in Eq. (3.42) is used. This saves negating member forces as in Eq. (3.41). Force transformations will be developed for truss and frame members.

1 Pin-jointed Truss Member

The pin-jointed truss member affords the simplest example of force transformation. The pin-jointed member has a force field completely described by the forces $\pm F$ acting on the member ends, and evidently, for equilibrium of the member these are equal and opposite. The relationship between F and the force components in the local $X'Y'$ coordinates is given in Eq. (3.43)

$$\begin{bmatrix} F'_{ix} \\ F'_{iy} \\ F'_{jx} \\ F'_{jy} \end{bmatrix} = \begin{bmatrix} -1 \\ 0 \\ 1 \\ 0 \end{bmatrix} F \tag{3.43}$$

Define the rotation transformation matrix L, as in Eq. (2.15). That is,

$$L = \begin{bmatrix} l_{11} & l_{21} \\ l_{12} & l_{22} \end{bmatrix}$$

Then also $L_D = \begin{bmatrix} L & 0 \\ 0 & L \end{bmatrix}$ is the 4×4 matrix with L as diagonal blocks. The matrix L_D transforms the force components $\langle F'_{ix} \quad F'_{iy} \quad F'_{jx} \quad F'_{jy} \rangle$ to global components. That is,

$$\begin{bmatrix} F_{ix} \\ F_{iy} \\ F_{jx} \\ F_{jy} \end{bmatrix} = L_D \begin{bmatrix} F'_{ix} \\ F'_{iy} \\ F'_{jx} \\ F'_{jy} \end{bmatrix} = \left[\begin{array}{cc|cc} \cos\alpha & -\sin\alpha & 0 & 0 \\ \sin\alpha & \cos\alpha & 0 & 0 \\ \hline 0 & 0 & \cos\alpha & -\sin\alpha \\ 0 & 0 & \sin\alpha & \cos\alpha \end{array}\right] \begin{bmatrix} -1 \\ 0 \\ 1 \\ 0 \end{bmatrix} F \tag{3.44}$$

Since the two coefficient matrices on the right-hand side of Eq. (3.44) are so simple, it is convenient to perform the matrix multiplication and Eq. (3.44) reduces to Eq. (3.45).

$$\begin{bmatrix} F_{ix} \\ F_{iy} \\ F_{jx} \\ F_{jy} \end{bmatrix} = \begin{bmatrix} -\cos\alpha \\ -\sin\alpha \\ \cos\alpha \\ \sin\alpha \end{bmatrix} F \tag{3.45}$$

Equation (3.45) is used in Sec. 3.7 to develop an automatic analysis program for the solution of member forces in statically determinate trusses.

2 Frame Members, Type Ia

In Fig. 3.5*b* I, two alternative basic systems are given that describe the same force field in the frame member, that is, constant axial force and linearly varying moment. Each has its particular transformation, and both of these are given in detail. Consider first the basic set of forces $\langle M_i \quad M_j \quad F \rangle$ shown in Fig. 3.5*b* I*a* and again in Fig. 3.13. The sign convention used is that given in Fig. 2.2 and for the present two-dimensional structure, reduces to the more familiar rule that anticlockwise moments on the member (as viewed from the positive Z' axis on the $X'Y'$ plane) are positive. With the moments M_i and M_j acting as shown, the beam reactions are

$$R_A = \frac{M_i + M_j}{l}$$

$$R_B = \frac{-(M_i + M_j)}{l}$$

These reactions together with the moments (M_i, M_j) and the axial force F are

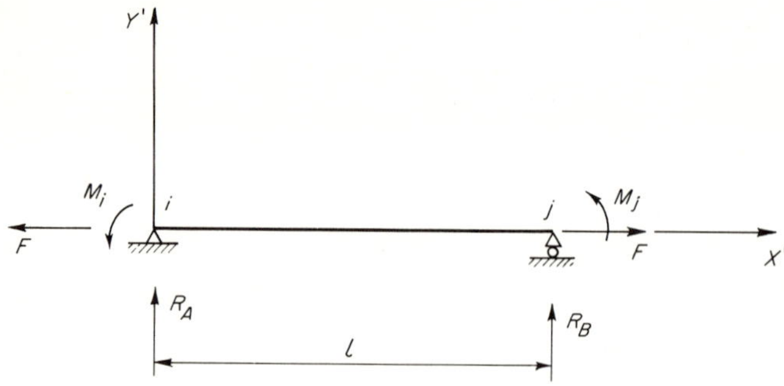

Figure 3.13 Basic force system Ia.

sufficient to define all the $X'Y'$ components on the ends i and j of the member.

$$\begin{bmatrix} F'_{ix} \\ F'_{iy} \\ M'_i \\ F'_{jx} \\ F'_{jy} \\ M'_j \end{bmatrix} = \begin{bmatrix} 0 & 0 & -1 \\ \frac{1}{l} & \frac{1}{l} & 0 \\ 1 & 0 & 0 \\ \hline 0 & 0 & 1 \\ \frac{-1}{l} & \frac{-1}{l} & 0 \\ 0 & 1 & 0 \end{bmatrix} \begin{bmatrix} M_i \\ M_j \\ F \end{bmatrix} \tag{3.46}$$

This transformation is written

$$F' = TS \tag{3.47}$$

The transformation matrix from local to global components at ends i and j is given by Eq. (2.40a), and thus,

$$L_D = \begin{bmatrix} L & 0 \\ 0 & L \end{bmatrix} = \left[\begin{array}{ccc|ccc} l_{11} & l_{21} & 0 & 0 & 0 & 0 \\ l_{12} & l_{22} & 0 & 0 & 0 & 0 \\ 0 & 0 & 1 & 0 & 0 & 0 \\ \hline 0 & 0 & 0 & l_{11} & l_{21} & 0 \\ 0 & 0 & 0 & l_{12} & l_{22} & 0 \\ 0 & 0 & 0 & 0 & 0 & 1 \end{array}\right] \tag{3.48}$$

By using Eq. (3.48), the transformation is given from F' to F and hence [with Eq. (3.47)] from S to F. That is,

$$F = L_D F' = L_D T S \tag{3.49}$$

This transformation is used in Example 3.3, which shows the generation of the b matrix, applicable also for the gable frame given in Fig. 3.15.

3 Frame Members, Type I*b*

The second type of basic member-force designation given here is shown in Fig. 3.14. The force field in the member is described in terms of the three force quantities at end i of the member $\langle M_i \quad S_i \quad F \rangle$. The forces at end j of the member are obtained by simple transformation.

$$S_j = -S_i$$
$$M_j = -M_i + S_i l$$

The transformation to $X'Y'$ components is now given by Eq. (3.50).

$$\begin{bmatrix} F'_{ix} \\ F'_{iy} \\ M'_i \\ F'_{jx} \\ F'_{jy} \\ M'_j \end{bmatrix} = \begin{bmatrix} 0 & 0 & 1 \\ 0 & 1 & 0 \\ 1 & 0 & 0 \\ \hdashline 0 & 0 & -1 \\ 0 & -1 & 0 \\ -1 & l & 0 \end{bmatrix} \begin{bmatrix} M_i \\ S_i \\ F \end{bmatrix} \tag{3.50}$$

Again, this equation is written

$$F' = T_1 S \tag{3.51}$$

The transformation in global coordinates is identical with Eq. (3.48), so that

$$F = L_D F' = L_D T_1 S \tag{3.52}$$

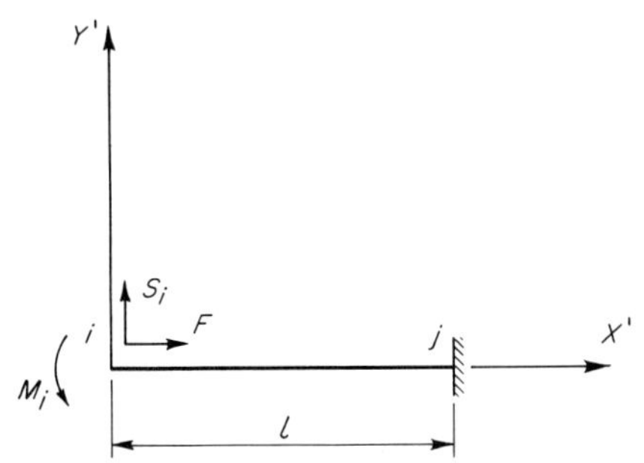

Figure 3.14 Basic member force system I*b*.

Example 3.5 The determination of the force transformation matrix b for the gable frame shown in Fig. 3.15

The gable frame shown in Fig. 3.15 has a pin support at node 1 and a horizontal roller (vertical support) at node 5. It is both stable and determinate. There are 12 node force quantities, apart from the 3 reactions, and these will require 12 independent equilibrium equations. However, there are 4 members, each with 3 force components, so that there are sufficient member force quantities to complete the joint equilibrium equations. The arrow on each member shows its positive sense. For this structure, member forces are taken as $\langle M_i \quad M_j \quad F\rangle$, so that the transformation equation (3.49) applies. The matrices L and T and the product $L_D T$ are formed for each member.

Member 1

$$L = \begin{bmatrix} 0 & -1 & 0 \\ 1 & 0 & 0 \\ 0 & 0 & 1 \end{bmatrix} \qquad T = \begin{bmatrix} 0 & 0 & -1 \\ 0.05 & 0.05 & 0 \\ 1 & 0 & 0 \\ \hline 0 & 0 & 1 \\ -0.05 & -0.05 & 0 \\ 0 & 1 & 0 \end{bmatrix} \qquad L_D T = \begin{bmatrix} -0.05 & -0.05 & 0 \\ 0 & 0 & -1 \\ 1 & 0 & 0 \\ \hline 0.05 & 0.05 & 0 \\ 0 & 0 & 1 \\ 0 & 1 & 0 \end{bmatrix}$$

Member 2

$$L = \begin{bmatrix} 0.94868 & -0.31622 & 0 \\ 0.31622 & 0.94868 & 0 \\ 0 & 0 & 1 \end{bmatrix} \qquad T = \begin{bmatrix} 0 & 0 & -1 \\ 0.03162 & 0.03162 & 0 \\ 1 & 0 & 0 \\ \hline 0 & 0 & 1 \\ -0.03162 & -0.03162 & 0 \\ 0 & 1 & 0 \end{bmatrix}$$

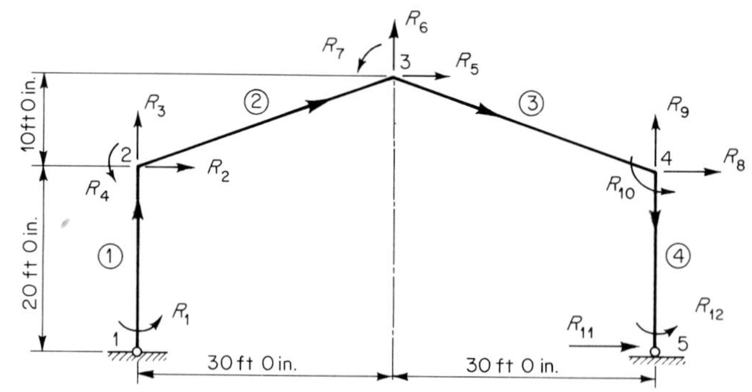

Figure 3.15 Gable-frame dimensions—joint loads.

$$L_DT = \begin{bmatrix} -0.009999 & -0.009999 & -0.94868 \\ 0.029995 & 0.029995 & -0.31622 \\ 1.0 & 0 & 0 \\ \hline 0.009999 & 0.009999 & 0.94868 \\ -0.029997 & -0.029997 & 0.31622 \\ 0 & 1.0 & 0 \end{bmatrix}$$

Member 3

$$L = \begin{bmatrix} 0.94868 & 0.31622 & 0 \\ -0.31622 & 0.94868 & 0 \\ 0 & 0 & 1 \end{bmatrix} \qquad T = \begin{bmatrix} 0 & 0 & -1 \\ 0.03162 & 0.03162 & 0 \\ 1 & 0 & 0 \\ \hline 0 & 0 & 1 \\ -0.03162 & -0.03162 & 0 \\ 0 & 1 & 0 \end{bmatrix}$$

$$L_DT = \begin{bmatrix} 0.009999 & 0.009999 & -0.94868 \\ 0.029997 & 0.029997 & 0.31622 \\ 1 & 0 & 0 \\ \hline -0.009999 & -0.009999 & 0.94868 \\ -0.029997 & -0.029997 & -0.31622 \\ 0 & 1 & 0 \end{bmatrix}$$

Member 4

$$L = \begin{bmatrix} 0 & 1 & 0 \\ -1 & 0 & 0 \\ 0 & 0 & 1 \end{bmatrix} \qquad T = \begin{bmatrix} 0 & 0 & -1 \\ 0.05 & 0.05 & 0 \\ 1 & 0 & 0 \\ \hline 0 & 0 & 1 \\ -0.05 & -0.05 & 0 \\ 0 & 1 & 0 \end{bmatrix} \qquad L_DT = \begin{bmatrix} 0.05 & 0.05 & 0 \\ 0 & 0 & 1 \\ 1 & 0 & 0 \\ \hline -0.05 & -0.05 & 0 \\ 0 & 0 & -1 \\ 0 & 1 & 0 \end{bmatrix}$$

From each of the L_DT matrices the appropriate parts are now extracted for each joint equilibrium equation.

For joint 3, for example,

$$\begin{bmatrix} 0.009999 & -0.009999 & 0.94868 \\ -0.029997 & -0.029997 & 0.31622 \\ 0 & 1 & 0 \end{bmatrix} \begin{bmatrix} M_i \\ M_j \\ F \end{bmatrix}_2 + \begin{bmatrix} 0.009999 & 0.009999 & -0.94868 \\ 0.029997 & 0.029997 & 0.31622 \\ 1 & 0 & 0 \end{bmatrix} \begin{bmatrix} M_i \\ M_j \\ F \end{bmatrix}_3 = \begin{bmatrix} R_5 \\ R_6 \\ R_7 \end{bmatrix}$$

This equation could be written

$$(L_DT)_{2j}S_2 + (L_DT)_{3i}S_3 = R_3$$

All the joint equations are combined in Eq. (3.53). The joint numbers and the type of force equilibrium equation, that is, F_x, F_y, or M, are shown down the left-hand side, and the member numbers 1, 2, 3, 4 are shown across the top.

The inverse to Eq. (3.53) is shown in Eq. (3.54) (see pages 107 and 108, respectively); the coefficient matrix is the required b matrix. Its columns give the values of the various member forces for unit values of the node loads.

Thus, from column 11, the bending-moment diagram has been plotted in Fig. 3.16a, for a unit load R_{11}. A similar diagram for R_7 is plotted in part b of the figure. In Fig. 3.16c the axial-force diagram is plotted from column 6, for $R_6 = 1$.

Example 3.6 Truss analysis; derivation of the force transformation matrix

In Example 3.1, Sec. 3.4, the force transformation matrix for the truss shown in Fig. 3.8 is given without derivation [Eq. (3.27)]. Here it is shown how Eq. (3.45) may be used to set up the joint equilibrium equations. In Example 3.2 the joint equilibrium equations in which the reactions occur are omitted. When the coefficient matrix of the member forces is inverted, the complete b matrix results. Notice that in Eq. (3.27), only part of the b matrix is given and in fact is reproduced as columns 2, 5, and 7 in Eq. (3.56).

The positive sense of members is shown by the arrows on the members in Fig. 3.8 so that the direction cosines are given as follows:

Members 1, 2, and 7:

$\cos\alpha = 1.0 \qquad \sin\alpha = 0.0$

Members 3 and 5:

$\cos\alpha = 0.5 \qquad \sin\alpha = 0.86603$

Members 4 and 6:

$\cos\alpha = 0.5 \qquad \sin\alpha = -0.86603$

Only seven joint equilibrium equations are written. Those at the pin-support, node 1, and that for the Y direction at the roller support, node 3, are omitted. The rows and columns in Eq. (3.55) are labeled so that the member entries, from use of Eq. (3.45), are clearly defined.

$$\begin{array}{ccc} \textit{Node} \downarrow & \textit{Member} \rightarrow & \\ 2 & X & \\ & Y & \\ 3 & X & \\ 4 & X & \\ & Y & \\ 5 & X & \\ & Y & \end{array}
\begin{array}{c} \begin{array}{ccccccc} 1 & 2 & 3 & 4 & 5 & 6 & 7 \end{array} \\
\begin{bmatrix} 1.0 & -1.0 & 0 & 0.5 & -0.5 & 0 & 0 \\ 0 & 0 & 0 & -0.86603 & -0.86603 & 0 & 0 \\ 0 & 1.0 & 0 & 0 & 0 & 0.5 & 0 \\ 0 & 0 & 0.5 & -0.5 & 0 & 0 & -1.0 \\ 0 & 0 & 0.86603 & 0.86603 & 0 & 0 & 0 \\ 0 & 0 & 0 & 0 & 0.5 & -0.5 & 1.0 \\ 0 & 0 & 0 & 0 & 0.86603 & 0.86603 & 0 \end{bmatrix} \end{array}
\begin{bmatrix} F_1 \\ F_2 \\ F_3 \\ F_4 \\ F_5 \\ F_6 \\ F_7 \end{bmatrix} = \begin{bmatrix} R_{2x} \\ R_{2y} \\ R_{3x} \\ R_{4x} \\ R_{4y} \\ R_{5x} \\ R_{5y} \end{bmatrix} \qquad (3.55)$$

$$
\begin{array}{cc}
\text{Joint} & \\
\downarrow & \\
1\{ & M \\
 & X \\
2 & Y \\
 & M \\
 & X \\
3 & Y \\
 & M \\
 & X \\
4 & Y \\
 & M \\
5 & X \\
 & M
\end{array}
\overset{\textit{Member}\longrightarrow}{}
\left[\begin{array}{ccc|ccc|ccc|ccc}
\multicolumn{3}{c|}{1} & \multicolumn{3}{c|}{2} & \multicolumn{3}{c|}{3} & \multicolumn{3}{c}{4} \\
1.0 & 0 & 0 & 0 & 0 & 0 & 0 & 0 & 0 & 0 & 0 & 0 \\
0.05 & 0.05 & 0 & -0.009999 & -0.009999 & -0.94868 & 0 & 0 & 0 & 0 & 0 & 0 \\
0 & 0 & 1.0 & 0.029997 & 0.029997 & -0.31622 & 0 & 0 & 0 & 0 & 0 & 0 \\
0 & 1.0 & 0 & 1.0 & 0 & 0 & 0 & 0 & 0 & 0 & 0 & 0 \\
0 & 0 & 0 & 0.009999 & 0.009999 & 0.94868 & 0.009999 & 0.009999 & -0.94868 & 0 & 0 & 0 \\
0 & 0 & 0 & -0.029997 & -0.029997 & 0.31622 & 0.029997 & 0.029997 & 0.31622 & 0 & 0 & 0 \\
0 & 0 & 0 & 0 & 1.0 & 0 & 1.0 & 0 & 0 & 0 & 0 & 0 \\
0 & 0 & 0 & 0 & 0 & 0 & -0.009999 & -0.009999 & 0.94868 & 0.05 & 0.05 & 0 \\
0 & 0 & 0 & 0 & 0 & 0 & -0.029997 & -0.029997 & -0.31622 & 0 & 0 & 1.0 \\
0 & 0 & 0 & 0 & 0 & 0 & 0 & 1.0 & 0 & 1.0 & 0 & 0 \\
0 & 0 & 0 & 0 & 0 & 0 & 0 & 0 & 0 & -0.5 & -0.5 & 0.0 \\
0 & 0 & 0 & 0 & 0 & 0 & 0 & 0 & 0 & 0 & 1.0 & 0.0
\end{array}\right]
\begin{bmatrix} M_{1i} \\ M_{1j} \\ F_1 \\ M_{2i} \\ M_{2j} \\ F_2 \\ M_{3i} \\ M_{3j} \\ F_3 \\ M_{4i} \\ M_{4j} \\ F_4 \end{bmatrix}
=
\begin{bmatrix} R_1 \\ R_2 \\ R_3 \\ R_4 \\ R_5 \\ R_6 \\ R_7 \\ R_8 \\ R_9 \\ R_{10} \\ R_{11} \\ R_{12} \end{bmatrix}
\tag{3.53}
$$

$$
\begin{array}{cccccccccccc}
1 & 2 & 3 & 4 & 5 & 6 & 7 & 8 & 9 & 10 & 11 & 12
\end{array}
$$

$$
\begin{bmatrix}
1.0 & 0 & 0 & 0 & 0 & 0 & 0 & 0 & 0 & 0 & 0 & 0 \\
-1.0 & 20.0 & 0 & 0 & 20.0 & 0 & 0 & 20.0 & 0 & 0 & 20.0 & 0 \\
-0.01666 & 0.3333 & 1.0 & -0.01666 & 0.49996 & 0.5 & -0.01666 & 0.3333 & 0 & -0.01666 & 0 & -0.01666 \\
1.0 & -20.0 & 0 & 1.0 & -20.0 & 0 & 0 & -20.0 & 0 & 0 & -20.0 & 0 \\
-0.5 & 10.0 & 0 & -0.5 & 15.0 & -15.0 & 0.5 & 20.0 & 0 & 0.5 & 30.0 & 0.50 \\
-0.00527 & 0.1054 & 0 & -0.00527 & 1.10679 & 0.15811 & -0.00527 & 1.0541 & 0 & -0.00527 & 0.9487 & -0.00527 \\
0.5 & -10.0 & 0 & 0.5 & -15.00 & 15.0 & 0.5 & -20.0 & 0 & -0.5 & -30.0 & -0.50 \\
0 & 0 & 0 & 0 & 0 & 0 & 0 & 0 & 0 & 1.0 & 20.0 & 1.00 \\
0.00527 & -0.1054 & 0 & 0.00527 & -0.1581 & 0.15811 & 0.00527 & 0.8433 & 0 & 0.00527 & 0.9487 & 0.00527 \\
0 & 0 & 0 & 0 & 0 & 0 & 0 & 0 & 0 & 0 & -20.0 & -1.0 \\
0 & 0 & 0 & 0 & 0 & 0 & 0 & 0 & 0 & 0 & 0 & 1.0 \\
0.01666 & -0.3333 & 0 & 0.01666 & -0.49996 & 0.5 & 0.01666 & -0.3333 & 1.0 & 0.01666 & 0 & 0.01666
\end{bmatrix}
\begin{bmatrix} R_1 \\ R_2 \\ R_3 \\ R_4 \\ R_5 \\ R_6 \\ R_7 \\ R_8 \\ R_9 \\ R_{10} \\ R_{11} \\ R_{12} \end{bmatrix}
=
\begin{bmatrix} M_{1i} \\ M_{1j} \\ F_1 \\ M_{2i} \\ M_{2j} \\ F_2 \\ M_{3i} \\ M_{3j} \\ F_3 \\ M_{4i} \\ M_{4j} \\ F_4 \end{bmatrix}
\tag{3.54}
$$

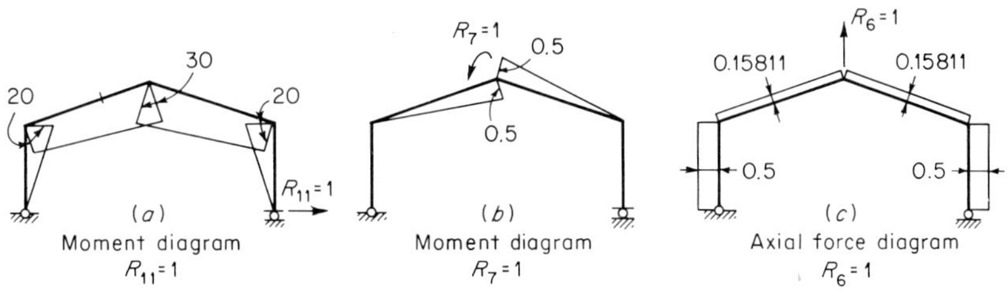

Figure 3.16 Unit-load diagrams.

The solution of Eq. (3.55) gives the member forces in terms of the node forces. That is, $S = bR$.

$$\begin{bmatrix} F_1 \\ F_2 \\ F_3 \\ F_4 \\ F_5 \\ F_6 \\ F_7 \end{bmatrix} = \begin{bmatrix} 1.0 & -0.2887 & 1.0 & 0.75 & -0.4330 & 0.75 & -0.1443 \\ 0 & -0.2887 & 1.0 & 0.25 & -0.1443 & 0.25 & -0.4330 \\ 0 & 0.5773 & 0 & 0.5 & 0.8660 & 0.5 & 0.2887 \\ 0 & -0.5773 & 0 & -0.5 & 0.2887 & -0.5 & -0.2887 \\ 0 & -0.5773 & 0 & 0.5 & -0.2887 & 0.5 & 0.2887 \\ 0 & 0.5773 & 0 & -0.5 & 0.2887 & -0.5 & 0.8660 \\ 0 & 0.5773 & 0 & -0.5 & 0.2887 & 0.5 & 0.2887 \end{bmatrix} \begin{bmatrix} R_{2x} \\ R_{2y} \\ R_{3x} \\ R_{4x} \\ R_{4y} \\ R_{5x} \\ R_{5y} \end{bmatrix} \tag{3.56}$$

4 Frame Members Ia; Intermediate Hinge

When a frame member, such as shown in Fig. 3.17, has a hinge P at a point between i and j, then the condition $M = 0$ at the hinge necessitates that the bending moment at P be zero for any applied (M_i, M_j). When there are no loads on the member, this means that a linear relationship exists between M_i and M_j. Consider the case when $A \neq 0$ (the distance from i to the hinge), and define M_i as the basic member-force quantity. From Fig. 3.17, it is seen that

$$M_j = \frac{B}{A} M_i \tag{3.57}$$

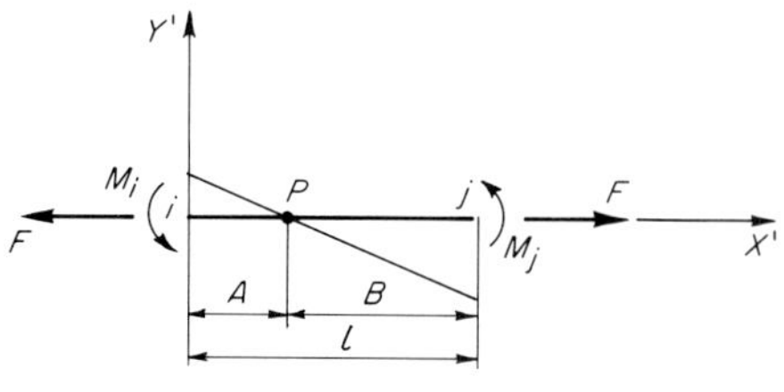

Figure 3.17 Basic member forces—member with hinge.

If this value for M_j is substituted in Eq. (3.46), the resulting transformation relating $\langle M_i \quad F \rangle$ to node forces is given in Eq. (3.58).

$$\begin{bmatrix} F'_{ix} \\ F'_{iy} \\ M'_i \\ \hline F'_{jx} \\ F'_{jy} \\ M'_y \end{bmatrix} = \begin{bmatrix} 0 & -1 \\ \dfrac{1}{A} & 0 \\ 1 & 0 \\ \hline 0 & 1 \\ \dfrac{-1}{A} & 0 \\ \dfrac{B}{A} & 0 \end{bmatrix} \begin{bmatrix} M_i \\ F \end{bmatrix} \tag{3.58}$$

The transformation from local to global components is again given by Eq. (3.49). It must be noted that an arbitrary addition of hinges along members may result in an unstable structure, a condition that will be recognized by a singular matrix in the equilibrium equations.

Example 3.7 Analysis of continuous beam; member with internal hinge

The continuous beam shown in Fig. 3.18 has two members and three supports. The hinge in member 1 makes the structure statically determinate. It will be analyzed for joint moments only. That is, axial force components may be neglected. There are only three basic member force components $\langle M_{1i} \quad M_{2i} \quad M_{2j} \rangle$ that enter into the three equations of joint equilibrium. These equations are written

$$\begin{bmatrix} 1 & 0 & 0 \\ 2 & 1 & 0 \\ 0 & 0 & 1 \end{bmatrix} \begin{bmatrix} M_{1i} \\ M_{2i} \\ M_{2j} \end{bmatrix} = \begin{bmatrix} M_1 \\ M_2 \\ M_3 \end{bmatrix} \tag{3.59}$$

When the coefficient matrix in Eq. (3.59) is inverted, the b matrix is obtained. That is,

$$\begin{bmatrix} M_{1i} \\ M_{2i} \\ M_{2j} \end{bmatrix} = \begin{bmatrix} 1 & 0 & 0 \\ -2 & 1 & 0 \\ 0 & 0 & 1 \end{bmatrix} \begin{bmatrix} M_1 \\ M_2 \\ M_3 \end{bmatrix} \tag{3.60}$$

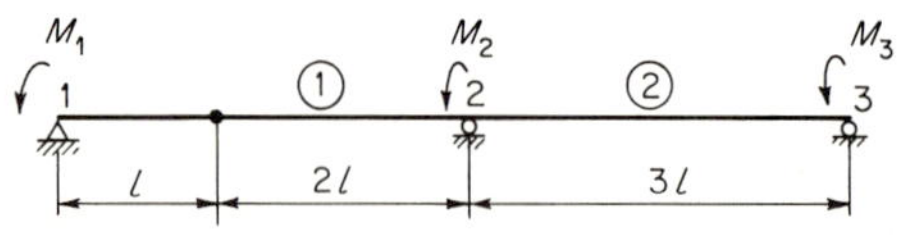

Figure 3.18 Member with hinge.

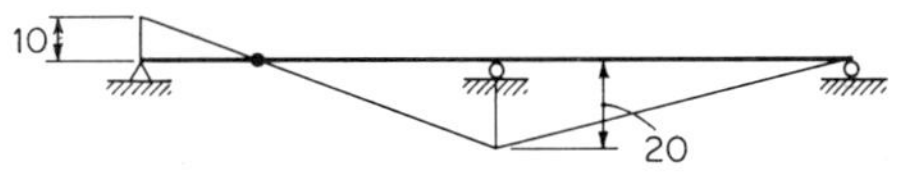

Figure 3.19 Bending-moment diagram—continuous beam.

If then the joint moments are given by $\langle M_1 \quad M_2 \quad M_3 \rangle = \langle 10, \quad 0 \quad 0 \rangle$, the bending-moment diagram for the beam is given in Fig. 3.19.

Example 3.8 Analysis of portal frame; member with hinge

The statically determinate portal frame $ABCD$ shown in Fig. 3.20 has pin supports at A and D and a hinge at P in the center of member BC. For the portal frame there are eight possible node forces (excluding the four reactions), and also eight basic member forces. Then the equations of statics should be sufficient to determine member forces. The transformation matrices $L_D T$ used in the joint equilibrium equations are given below for each of the three members.

Member 1

$$L = \begin{bmatrix} 0 & -1 & 0 \\ 1 & 0 & 0 \\ 0 & 0 & 1 \end{bmatrix} \qquad T = \left[\begin{array}{ccc} 0 & 0 & -1 \\ 0.05 & 0.05 & 0 \\ 1 & 0 & 0 \\ \hline 0 & 0 & 1 \\ -0.05 & -0.05 & 0 \\ 0 & 1 & 0 \end{array}\right] \qquad L_D T = \left[\begin{array}{ccc} -0.05 & -0.5 & 0 \\ 0 & 0 & -1 \\ 1 & 0 & 0 \\ \hline 0.05 & 0.05 & 0 \\ 0 & 0 & 1 \\ 0 & 1 & 0 \end{array}\right]$$

Member 2

$$L = \begin{bmatrix} 1 & 0 & 0 \\ 0 & 1 & 0 \\ 0 & 0 & 1 \end{bmatrix} \qquad T = \left[\begin{array}{cc} 0 & -1 \\ 0.04 & 0 \\ 1 & 0 \\ \hline 0 & 1 \\ -0.04 & 0 \\ 1 & 0 \end{array}\right] \qquad L_D T = \left[\begin{array}{cc} 0 & -1 \\ 0.04 & 0 \\ 1 & 0 \\ \hline 0 & 1 \\ -0.04 & 0 \\ 1 & 0 \end{array}\right]$$

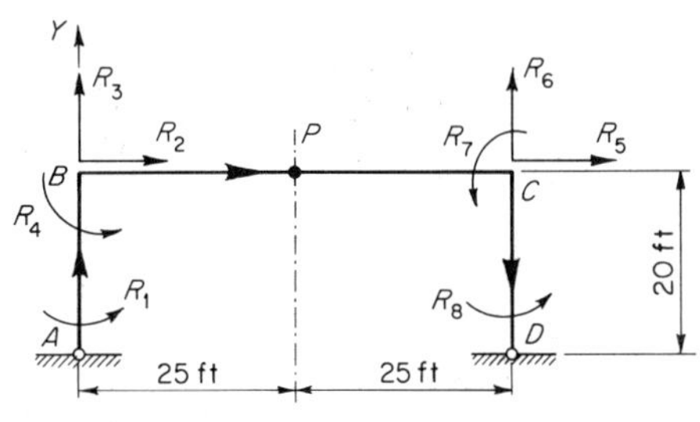

Figure 3.20 Portal-frame analysis.

Member 3

$$L = \begin{bmatrix} 0 & 1 & 0 \\ -1 & 0 & 0 \\ 0 & 0 & 1 \end{bmatrix} \qquad T = \left[\begin{array}{ccc} 0 & 0 & -1 \\ 0.05 & 0.05 & 0 \\ 1 & 0 & 0 \\ \hdashline 0 & 0 & 1 \\ -0.05 & -0.05 & 0 \\ 0 & 1 & 0 \end{array}\right] \qquad L_D T = \left[\begin{array}{ccc} 0.05 & 0.05 & 0 \\ 0 & 0 & 1 \\ 1 & 0 & 0 \\ \hdashline -0.05 & -0.05 & 0 \\ 0 & 0 & -1 \\ 0 & 1 & 0 \end{array}\right]$$

The joint equilibrium equations are written in Eq. (3.61)

$$\begin{array}{cc} \text{Node} & \\ 1 & M \\ 2 & F_x \\ & F_y \\ & M \\ 3 & F_x \\ & F_y \\ & M \\ 4 & M \end{array} \overset{\text{Member} \rightarrow \quad 1 \qquad\qquad 2 \qquad\qquad 3}{\begin{bmatrix} 1.0 & 0 & 0 & 0 & 0 & 0 & 0 & 0 \\ 0.05 & 0.05 & 0 & 0 & -1 & 0 & 0 & 0 \\ 0 & 0 & 1 & 0.04 & 0 & 0 & 0 & 0 \\ 0 & 1 & 0 & 1 & 0 & 0 & 0 & 0 \\ 0 & 0 & 0 & 0 & 1 & 0.05 & 0.05 & 0 \\ 0 & 0 & 0 & -0.04 & 0 & 0 & 0 & 1 \\ 0 & 0 & 0 & 1 & 0 & 1 & 0 & 0 \\ 0 & 0 & 0 & 0 & 0 & 0 & 1 & 0 \end{bmatrix}} \begin{bmatrix} M_{1i} \\ M_{1j} \\ F_1 \\ M_{2i} \\ F_2 \\ M_{3i} \\ M_{3j} \\ F_3 \end{bmatrix} = \begin{bmatrix} R_1 \\ R_2 \\ R_3 \\ R_4 \\ R_5 \\ R_6 \\ R_7 \\ R_8 \end{bmatrix} \qquad (3.61)$$

On inverting Eq. (3.61), the force transformation matrix is obtained [Eq. (3.62)].

$$\begin{bmatrix} M_{1i} \\ M_{1j} \\ F_1 \\ M_{2i} \\ F_2 \\ M_{3i} \\ M_{3j} \\ F_3 \end{bmatrix} = \begin{bmatrix} 1.0 & 0 & 0 & 0 & 0 & 0 & 0 & 0 \\ -0.5 & 10.0 & 0 & 0.5 & 10.0 & 0 & -0.5 & -0.5 \\ -0.02 & 0.4 & 1.0 & -0.02 & 0.4 & 0 & -0.02 & -0.02 \\ 0.5 & -10.0 & 0 & 0.5 & -10.0 & 0 & 0.5 & 0.5 \\ 0.025 & -0.5 & 0 & 0.025 & 0.5 & 0 & -0.025 & -0.025 \\ -0.5 & 10.0 & 0 & -0.5 & 10.0 & 0 & 0.5 & -0.5 \\ 0 & 0 & 0 & 0 & 0.0 & 0 & 0 & 1.0 \\ 0.02 & -0.4 & 0 & 0.02 & -0.4 & 1.0 & 0.02 & 0.02 \end{bmatrix} \begin{bmatrix} R_1 \\ R_2 \\ R_3 \\ R_4 \\ R_5 \\ R_6 \\ R_7 \\ R_8 \end{bmatrix} \qquad (3.62)$$

From the matrix of coefficients in Eq. (3.62), the force diagrams for various unit loads may be constructed. Thus for $R_1 = 1$ and $R_2 = 1$, the moment diagrams are shown in Fig. 3.21*a* and *b* respectively. For member 2, the moment M_j is obtained from M_i, since $M_j = (B/A)M_i = \frac{1}{2}M_i$ in this case.

B Displacement Transformations

In Eqs. (3.39) and (3.40), it has been shown that $b = (a^T)^{-1}$ for the statically determinate structure. It is seen that, instead of using force transformations to express basic forces in terms of their global components, the

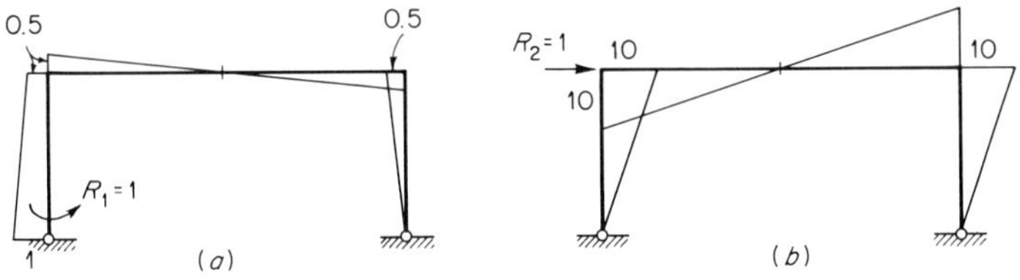

Figure 3.21 Diagrams for unit loads.

a matrix may be set up and inverted. The displacement transformation from node displacements to member distortions may be obtained from equations such as (3.44) and (3.49) by applying the contragradient law. That is, these force transformations may be considered of the form of Eq. (3.22),

$$P = C^T Q \tag{3.63}$$

The relationship between basic member distortions and node displacements in frame coordinates is then given,

$$q = Cp \tag{3.64}$$

For the pin-jointed truss member in Fig. 3.6*a*, Eq. (3.64) gives the member distortion

$$\Delta = [-\cos\alpha \quad -\sin\alpha \quad \cos\alpha \quad \sin\alpha]\begin{bmatrix} v_{ix} \\ v_{iy} \\ v_{jx} \\ v_{jy} \end{bmatrix} \tag{3.65}$$

For the frame member of Fig. 3.6*b*, by using Eq. (3.49), it is seen that

$$v = T^T L_D{}^T r \tag{3.66}$$

In Eq. (3.66)

$$r^T = \langle v_{ix} \quad v_{iy} \quad \theta_i \quad v_{jx} \quad v_{jy} \quad \theta_j \rangle$$

and (3.67)

$$v^T = \langle \phi_i \quad \phi_j \quad \Delta \rangle$$

Although the above application of the contragradient law is considered a sufficient demonstration of the displacement-transformation relationships, it is useful to work through the detailed geometrical reasoning given below, in which each node is given a unit displacement in turn, while the remainder are kept equal to zero. That is, displacement-influence functions are constructed.

Pin-jointed Truss Member

The unit displacements $v_{ix} = 1, v_{iy} = 1$, etc., are shown in Fig. 3.22 (I to IV). From the four diagrams in Fig. 3.22 in turn it is seen that

$$\Delta = [-\cos\alpha \quad -\sin\alpha \quad \cos\alpha \quad \sin\alpha]\begin{bmatrix} v_{ix} \\ v_{iy} \\ v_{jx} \\ v_{jy} \end{bmatrix}$$

Beam Member; Basic Distortions $\langle \phi_i \quad \phi_j \quad \Delta \rangle$

The six unit displacements are shown in Fig. 3.23, this time in member coordinates. It has already been shown that for rotation of axes [see Eq. (2.63)],

$$r' = L_D{}^T r$$

From Fig. 3.23, it is seen that

$$\begin{bmatrix} \phi_i \\ \phi_j \\ \Delta \end{bmatrix} = \left[\begin{array}{ccc|ccc} 0 & \frac{1}{l} & 1 & 0 & \frac{-1}{l} & 0 \\ 0 & \frac{1}{l} & 0 & 0 & \frac{-1}{l} & 1 \\ -1 & 0 & 0 & 1 & 0 & 0 \end{array}\right] \begin{bmatrix} v'_{ix} \\ v'_{iy} \\ \theta'_i \\ v'_{jx} \\ v'_{jy} \\ \theta'_j \end{bmatrix} \tag{3.68}$$

Comparison of Eq. (3.68) with (3.46) shows that this is in fact

$$v = T^T r' \tag{3.69}$$

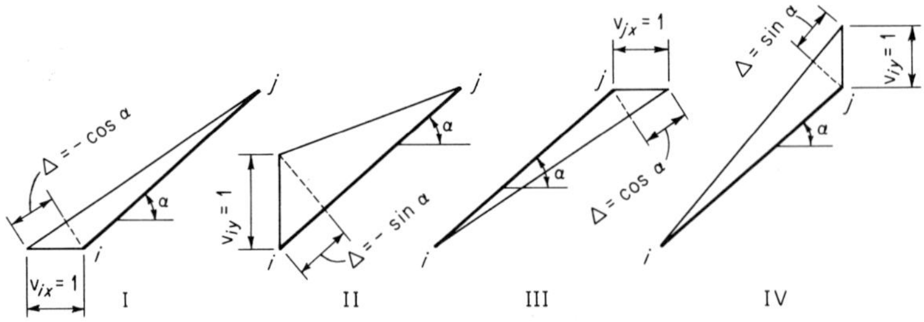

Figure 3.22 Truss member—node displacements.

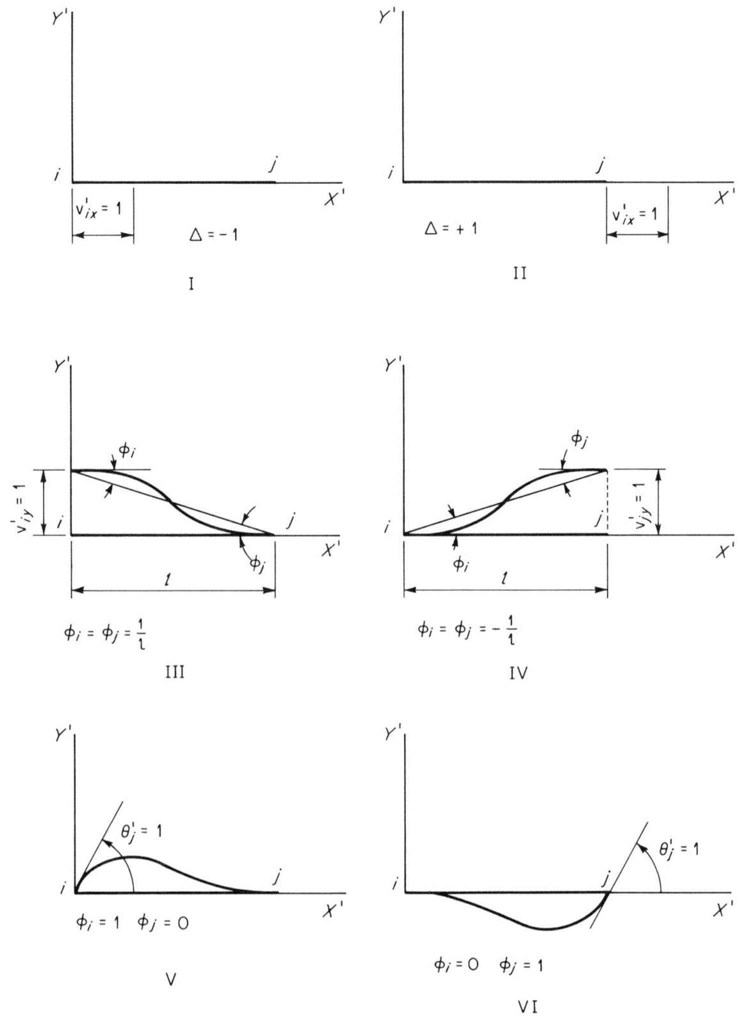

Figure 3.23 Node displacements.

Thus, finally,

$$v = T^T L_D{}^T r \tag{3.70}$$

3.7 COMPUTER PROGRAMS FOR FORCE TRANSFORMATION MATRICES IN DETERMINATE STRUCTURES

The matrices developed in Sec. 3.6 [Eqs. (3.45), (3.49), etc.] for setting up the joint equilibrium equations may be generated automatically in the computer if members and nodes are numbered sequentially and each member treated as a directed-line element as in Sec. 2.2. Each member is described by the node numbers *IP*, *IQ* at its ends. These numbers indicate its position in the structure and its positive sense. In addition the coordinates of *IP* and *IQ* give the member length and direction cosines. Herein, frame and truss structures are considered separately, although it should become apparent that they may both be included in a single general program.

A Truss Structures

If all equations for nodes are written down for the truss structure [member equation (3.45)], it will be seen that each node generates two rows and each member one column of the resulting matrix of coefficients [see Eq. (3.55)]. Since the number of members plus the number of reactions (equals three for simple trusses), is equal to twice the number of nodes, the matrix so formed will not be square, having more rows than columns. It has been shown that one method of handling this is to delete the equations corresponding to the external reactions. Alternatively, if the reactions are included, their values are obtained from the analysis. In this case one more column must be added for each reaction, so that finally a square matrix is obtained. Note that a roller support produces one reaction component perpendicular to the plane of the roller, whereas a pin reaction in a planar structure has two independent components assumed to be parallel to the global coordinate axes. Since a roller may be on any plane, the reaction direction is specified by the direction cosines of the normal to the plane. All reactions will be specified in this same way, that is, by the node of application, together with two direction cosines that specify the direction of application.

For example, the reactions on the truss in Fig. 3.8 are a pin at node 1 and a horizontal roller at node 3. These reactions are specified:

Node number	cos α	sin α
1	1	0
1	0	1
3	0	1

Equation (3.45) (for member force contribution to node equations) is now written for the general member k and nodes *IP*, *IQ* in the form of Eq. (3.71) to indicate how this member force contributes to the coefficient matrix.

$$\begin{aligned} \begin{bmatrix} F_{ix} \\ F_{iy} \end{bmatrix}_{IP} &= \begin{bmatrix} -\cos\alpha_k \\ -\sin\alpha_k \end{bmatrix}_{IP} F_k \\ \begin{bmatrix} F_{jx} \\ F_{jy} \end{bmatrix}_{IQ} &= \begin{bmatrix} \cos\alpha_k \\ \sin\alpha_k \end{bmatrix}_{IQ} F_k \end{aligned} \tag{3.71}$$

Let the statics matrix be designated B. Then the matrix $\langle -\cos\alpha_k \quad -\sin\alpha_k \rangle_{IP}$

is loaded into column k of B in the rows $2 \times (IP - 1)$ and $2 \times IP$ respectively, and the matrix $\langle \cos \alpha_k \quad \sin \alpha_k \rangle_{IQ}$ is loaded into column k, rows $2 \times (IQ - 1)$ and $2 \times IQ$. The positions of these terms in B are shown schematically in Fig. 3.24. In this figure, three reaction columns R_L, R_M, R_N have been shown. For any reaction R_T, the direction cosines will be $\langle \cos \alpha_T \; \sin \alpha_T \rangle$, and these values are loaded into rows $2 \times (T - 1)$ and $2 \times T$ respectively.

Fortran Coding to Generate the B Matrix

The complete coding showing input and output details is given in Appendix C. The coding given below illustrates the method of loading the coefficients into the B matrix.

Notation

X,Y	*node coordinate arrays*
IP,IQ	*member node arrays*
RCOS, RSIN	*direction cosines of reaction vector*
IR	*reaction node number array*
NEMBS	*number of members*
NJNTS	*number of joints*
NRCT	*number of reactions*
DIMENSION	B(60,60), X(25), Y(25), IP(50), IQ(50), RCOS(10) RSIN(10), IR(10)

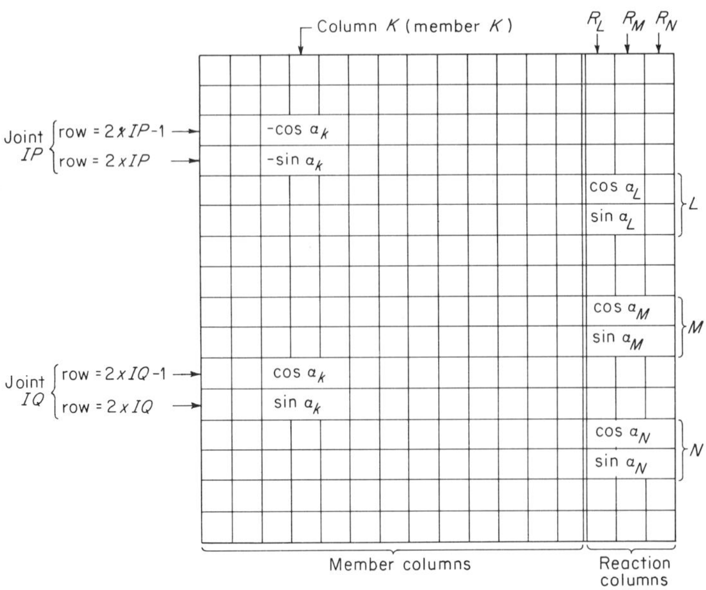

Figure 3.24 Construction of statics matrix.

```
C        CLEAR B TO ZERO
         DO 50 I = 1,60
         DO 50 J = 1,60
    50   B(I,J) = 0.0

C        LOAD MEMBER FORCE COEFFICIENTS
         DO 100 II = 1, NEMBS
         I = IP(II)
         J = IQ(II)
         XDIF = X(J) -X(I)
         YDIF = Y(J) -Y(I)
         AL = SQRT(XDIF**2 + YDIF**2)
         COS = XDIF/AL
         SIN = YDIF/AL
         M = 2*I
         B(M-1,II) = -COS
         B(M,II) = -SIN
         M = 2*J
         B(M-1,II) = COS
   100   B(M,II) = SIN

C        LOAD IN REACTION COEFFICIENTS
         DO 110 II = 1,NRCT
         I = 2*IR(II)
         M = NEMBS + II
         B(I-1,M) = RCOS (II)
   110   B(I,M) = RSIN(II)
```

B Frame Structures

The same procedure is followed for frame structures as for truss structures. For the planar structure, each joint has three equations of equilibrium, $\Sigma F_x = 0$, $\Sigma F_y = 0$, and $\Sigma M_z = 0$. A planar member generates two or three columns, depending on whether or not the member has a hinge along its length, that is, whether the number of basic force quantities is two or three. Since two types of members are possible, a member must have a distinguishing parameter, or flag, to indicate its type. The program for generating B is given below. The student should use the program in Appendix C, together with the coding below to produce his own determinate-frame program.

In Fig. 3.25 the statics matrix is shown for the portal frame with four members and five joints. Member 2 has an intermediate hinge, so that only two columns are required for its force quantities. In Fig. 3.25 the locations into which the coefficients of member 3 forces are loaded are shown with the appropriate indexing. Notice that the presence of the hinge in member 2 requires that four external reactions be provided for equilibrium.

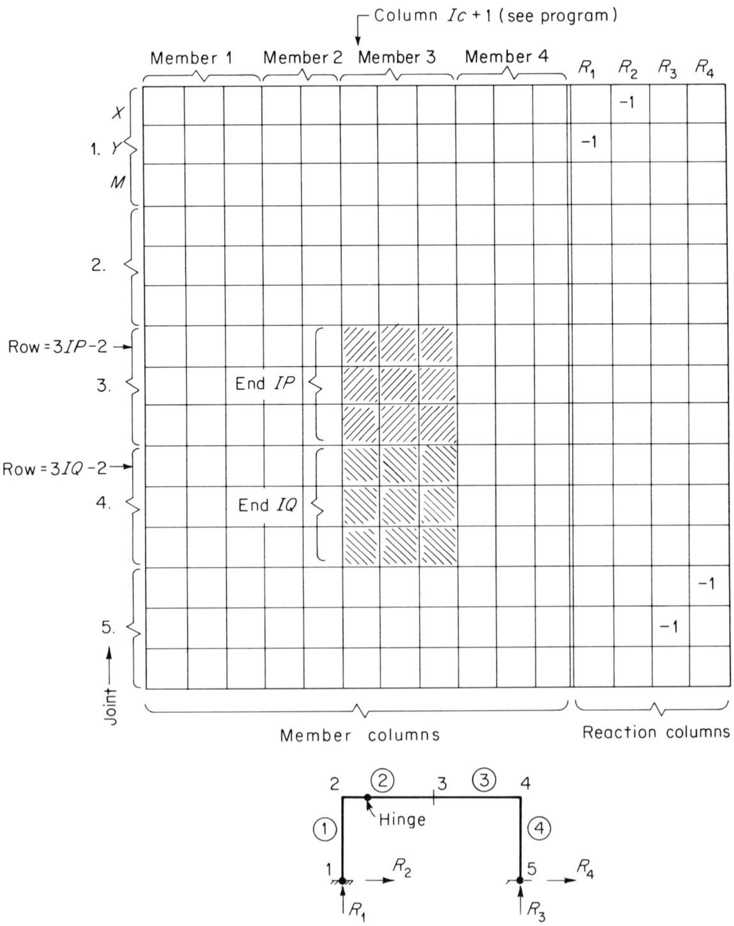

Figure 3.25 Construction of statics matrix; loading of coefficients for member 3.

Program to generate joint equilibrium equations for frame structures

```
*         SUBROUTINE TO CLEAR A BLOCK OF MEMORY TO ZERO
          SUBROUTINE CLEAR (A,N)
          DIMENSION A(1)
          DO 100 = I,N
    100   A(I) = 0.0
          RETURN
          END

*         MAIN PROGRAM NODE COORDINATES
*         ARE LOCATED IN ARRAYS X,Y. MEMBER NODE NUMBERS
*         ARE IN IP AND IQ
```

```
*       MEMBER MAY HAVE AN INTERMEDIATE HINGE
*       TYPE (N) = 0 STANDARD MEMBER
*       TYPE (N) = 1 MEMBER WITH HINGE
        DIMENSION B(60,60), X(25), Y(25), IP(25), IQ(25),
        TYPE (25), L(3,3), T(6,3), LT(6,3), A(10), NRCT(10), REACT(10,3)
        INTEGER TYPE
        REAL L, LT
        CALL CLEAR (B, 3600)
        CALL CLEAR (L,9)
        L(3,3) = 1.0
        ID = 0
        IC = 0
        IH = 1
*       IH RECORDS HINGE MEMBERS LIMITED IN
*       NUMBER BY DIMENSION STATEMENT A(10)

        DO 100 II = 1, NEMBS
        CALL CLEAR (T,18)
        IT = IP(II)
        JT = IQ(II)
        XDIF = X(JT) - X(IT)
        YDIF = Y(JT) - Y(IT)
        AL = SQRT (XDIF**2+YDIF**2)
        COS = XDIF/AL
        SIN = YDIF/AL
        BL = 1.0/AL
*       FORM L MATRIX
        L(1,1) = COS
        L(2,2) = COS
        L(1,2) = -SIN
        L(2,1) = SIN
*       FORM T MATRIX TWO CASES
        IF (TYPE (II).EQ.1) GO TO 120
*       TYPE = 1 MEMBER WITH HINGE AT 120
*       NO HINGE FOLLOWS
        T(1,3) = -1.0
        T(2,1) = BL
        T(2,2) = BL
        T(5,1) = -BL
        T(5,2) = -BL
        T(3,1) = 1.0
        T(4,3) = 1.0
        T(6,2) = 1.0
*       ID IS COUNTER ON NUMBER OF COLUMNS INT
        ID = 3
```

```
          GO TO 125

     120  C = 1.0/A(IH)
          T(1,2) = -1.0
          T(2,1) = C
          T(3,1) = 1.0
          T(4,2) = 1.0
          T(5,1) = -C
          T(6,1) = (AL - A(IH))/C
          ID = 2
          IH = IH + 1

     125  CONTINUE
          CALL CLEAR (LT,18)
          DO 128 I = 1,3
          DO 128 J = 1,ID
          DO 128 K = 1,3

     128  LT(I,J) = LT (I,J) + L(I,K)*T(K,J)
          DO 129 I = 1,3
          DO 129 J = I,ID
          DO 129 K = 1,3

     129  LT(I+3,J) = LT(I+3,J) + L(I,K)*T(K+3,J)
*         LOAD LT INTO THE APPROPRIATE B LOCATIONS
          M = 3*IT-3
          N = 3*JT-3
          DO 150 J = 1,3
          DO 150 K = 1,ID
          JJ = IC+K
          MM = M+J
          NN = N+J
          B(MM,JJ) = B(MM,JJ) + LT(J,K)
     150  B(NN,JJ) = B(NN,JJ) + LT(J+3,K)
          IC = IC + ID
     100  CONTINUE
*         WHEN ALL MEMBERS ARE PROCESSED LOAD
*         IN COEFFICIENTS FOR REACTIONS
*         NR = NUMBER OF REACTIONS
          DO 180 = I = 1,NR
          IC = IC + 1
          II = NRCT (NR)*3-3
          DO 180 J = 1,3
     180  B(II+J,IC) = REACT(I,J)
```

3.8 TREE STRUCTURES

Tree structures are a special class of statically determinate structure arrived at by considering the structure with all its supports fixed and with no internal releases (see Fig. 3.26*a*). The structure then consists of a number of closed loops. When all these loops have been broken by removing one member from each loop in such a way that the remaining structure is always stable, the final residual structure is a tree. (More details of tree structures are contained in Chaps. 5 and 6.) From Fig. 3.26*a* and *b* it should be evident that for a particular configuration of members in a structure, many different trees may be formed. The tree structure formed may be actually several separate trees, and these may be single or many branched.

The question that arises is that given a tree structure, what special techniques are available to generate the force transformation matrix for member forces from node forces. It will be found that the process of solving joint equilibrium equations [Eq. (3.39), for example] may be reduced to inverting a lower triangular matrix. This may be done directly without reverting to a standard invert routine.

Consider the single-branch tree with members (1), . . . , (4) and nodes 1, . . . , 4 shown in Fig. 3.27. The tree may be in three-dimensional space. The forces acting on the nodes are the general force-moment type [see Eq. (2-4)] and are represented then for the Nth node by

$$R_N^T = \langle F^T \quad M^T \rangle = \langle F_{NX} \quad F_{NY} \quad F_{NZ} \quad M_{NX} \quad M_{NY} \quad M_{NZ} \rangle$$

The force R_1 at node 1 may be transformed to an equivalent force R_{2-1} at node 2 by the use of Eq. (2.45) with the transformation matrix now designated U_1. That is,

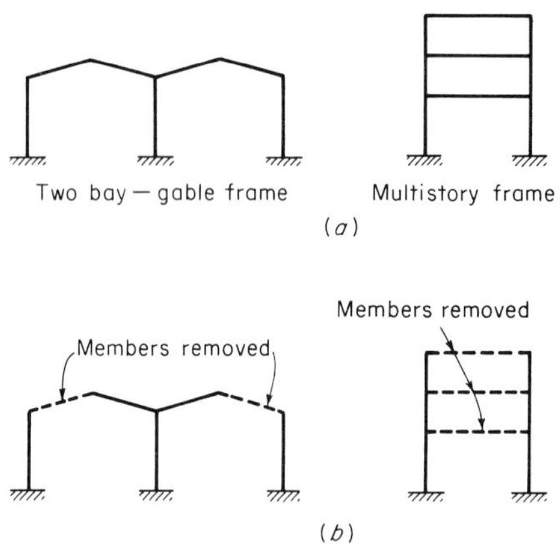

Figure 3.26 Tree structures. (*a*) Structures without releases; (*b*) structures when members are removed.

$$R_{2-1} = U_1 R_1 \tag{3.72}$$

From Eq. (2.44),

$$U_1 = \left[\begin{array}{ccc|ccc} 1 & 0 & 0 & 0 & 0 & 0 \\ 0 & 1 & 0 & 0 & 0 & 0 \\ 0 & 0 & 1 & 0 & 0 & 0 \\ \hline 0 & -z_{12} & y_{12} & 1 & 0 & 0 \\ z_{12} & 0 & -x_{12} & 0 & 1 & 0 \\ -y_{12} & x_{12} & 0 & 0 & 0 & 1 \end{array}\right] \tag{3.73}$$

In Eq. (3.73),

$$z_{12} = z_1 - z_2 \qquad y_{12} = y_1 - y_2 \qquad x_{12} = x_1 - x_2 \tag{3.74}$$

The force R_{2-1} may be transformed to the local coordinates for member (1) at node 2 by using the L_D direction-cosine matrix. Denoting member components by primed quantities,

$$R'_{2-1} = L_D{}^T U_1 R_1 \tag{3.75}$$

The member force at node 2, member (1) must be equal and opposite to R'_{2-1}. Member forces are described as in Fig. 3.5 I*b*. Therefore,

$$S_1 = -R'_{2-1} = -L_{D1}{}^T U_1 R_1 \tag{3.76}$$

This process may be repeated by transforming R_{2-1} and R_2 to node 3. Then member force for member (2) is given by

$$S_2 = -R'_{3-2} = -L_{D2}{}^T(U_2 R_2 + U_2 U_1 R_1) \tag{3.77}$$

Similarly,

$$S_3 = -L_{D3}{}^T(U_3 R_3 + U_3 U_2 R_2 + U_3 U_2 U_1 R_1) \tag{3.78}$$

$$S_4 = -L_{D4}{}^T(U_4 R_4 + U_4 U_3 R_3 + U_4 U_3 U_2 R_2 + U_4 U_3 U_2 U_1 R_1) \tag{3.79}$$

The force transformation matrix b for the tree structure may be developed from Eqs. (3.76) to (3.79) by combining them in matrix form. That is,

$$\begin{bmatrix} S_1 \\ S_2 \\ S_3 \\ S_4 \end{bmatrix} = -\begin{bmatrix} L_{D1}{}^T U_1 & 0 & 0 & 0 \\ L_{D2}{}^T U_2 U_1 & L_{D2}{}^T U_2 & 0 & 0 \\ L_{D3}{}^T U_3 U_2 U_1 & L_{D3}{}^T U_3 U_2 & L_{D3}{}^T U_3 & 0 \\ L_{D4}{}^T U_4 U_3 U_2 U_1 & L_{D4}{}^T U_4 U_3 U_2 & L_{D4}{}^T U_4 U_3 & L_{D4}{}^T U_4 \end{bmatrix} \begin{bmatrix} R_1 \\ R_2 \\ R_3 \\ R_4 \end{bmatrix} \tag{3.80}$$

Equation (3.80) is simplified if it is written as the product of L_D and U matrices, as in Eq. (3.81).

$$\begin{bmatrix} S_1 \\ S_2 \\ S_3 \\ S_4 \end{bmatrix} = - \begin{bmatrix} L_{D1} & 0 & 0 & 0 \\ 0 & L_{D2} & 0 & 0 \\ 0 & 0 & L_{D3} & 0 \\ 0 & 0 & 0 & L_{D4} \end{bmatrix}^T \begin{bmatrix} U_1 & 0 & 0 & 0 \\ U_2U_1 & U_2 & 0 & 0 \\ U_3U_2U_1 & U_3U_2 & U_3 & 0 \\ U_4U_3U_2U_1 & U_4U_3U_2 & U_4U_3 & U_4 \end{bmatrix} \begin{bmatrix} R_1 \\ R_2 \\ R_3 \\ R_4 \end{bmatrix} \tag{3.81}$$

The b matrix in Eqs. (3.80) and (3.81) has been obtained directly, instead of by the inversion processes in Sec. 3.6ff. If the joint equilibrium equations are written down, it is possible to see how this inversion has taken place.

As the basic member-force system, use that on end i as given in Fig. 3.14, generalized if necessary to three dimensions. This member force (member N) will be designated S_{Ni}. Firstly, using Eq. (2.42), transform S_{Ni} to global coordinates. Then,

$$F_{Ni} = L_{DN}S_{Ni} \tag{3.82}$$

Furthermore, F_{Ni} may be transformed to the equivalent force at end j, and the negative of this transformed force must be F_{Nj}, the member force at end j, since the member is in equilibrium. That is,

$$F_{Nj} = -U_N^{-1}L_{DN}S_{Ni} \tag{3.83}$$

For the joints 1, 2, 3, 4 of the structure in Fig. 3.27, the equilibrium equations are written by using Eqs. (3.82) and (3.83)

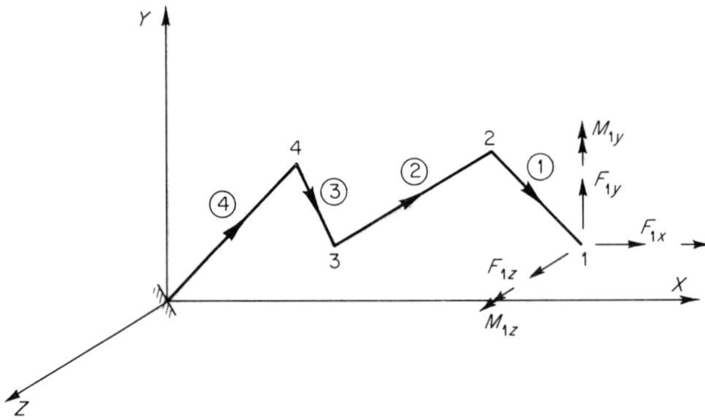

Figure 3.27 Single-branch tree.

$$\begin{bmatrix} R_1 \\ R_2 \\ R_3 \\ R_4 \end{bmatrix} = \begin{bmatrix} -U_1^{-1}L_{D1} & 0 & 0 & 0 \\ L_{D1} & -U_2^{-1}L_{D2} & 0 & 0 \\ 0 & L_{D2} & -U_3^{-1}L_{D3} & 0 \\ 0 & 0 & L_{D3} & U_4^{-1}L_{D4} \end{bmatrix} \begin{bmatrix} S_{1i} \\ S_{2i} \\ S_{3i} \\ S_{4i} \end{bmatrix} \tag{3.84}$$

Decomposing the coefficient matrix on the right-hand side of Eq. (3.84),

$$\begin{bmatrix} R_1 \\ R_2 \\ R_3 \\ R_4 \end{bmatrix} = \begin{bmatrix} -U_1^{-1} & 0 & 0 & 0 \\ I & -U_2^{-1} & 0 & 0 \\ 0 & I & -U_3^{-1} & 0 \\ 0 & 0 & I & -U_4^{-1} \end{bmatrix} \begin{bmatrix} L_{D1} & 0 & 0 & 0 \\ 0 & L_{D2} & 0 & 0 \\ 0 & 0 & L_{D3} & 0 \\ 0 & 0 & 0 & L_{D4} \end{bmatrix} \begin{bmatrix} S_{1i} \\ S_{2i} \\ S_{3i} \\ S_{4i} \end{bmatrix} \tag{3.85}$$

The unit matrix I will be of the same order as L_D and U. Thus, for the three-dimensional structure, I will be I_6, and for the two-dimensional structure, it will be I_3.

The member forces are obtained by inverting Eq. (3.85), remembering that $L_D^{-1} = L_D^T$.

$$\begin{bmatrix} S_{1i} \\ S_{2i} \\ S_{3i} \\ S_{4i} \end{bmatrix} = \begin{bmatrix} L_{D1} & 0 & 0 & 0 \\ 0 & L_{D2} & 0 & 0 \\ 0 & 0 & L_{D3} & 0 \\ 0 & 0 & 0 & L_{D4} \end{bmatrix}^T \begin{bmatrix} -U_1^{-1} & 0 & 0 & 0 \\ I & -U_2^{-1} & 0 & 0 \\ 0 & I & -U_3^{-1} & 0 \\ 0 & 0 & I & -U_4^{-1} \end{bmatrix}^{-1} \begin{bmatrix} R_1 \\ R_2 \\ R_3 \\ R_4 \end{bmatrix} \tag{3.86}$$

A comparison of Eqs. (3.86) and (3.81) shows that the first matrices composed of the L terms are identical and hence the U matrices must each be the inverse of the other. The inverse of a lower triangular matrix is also a lower triangular matrix, so that the U matrix in Eq. (3.86) multiplied by its inverse may be written as shown in Eq. (3.87).

$$\begin{bmatrix} -U_1^{-1} & 0 & 0 & 0 \\ I & -U_2^{-1} & 0 & 0 \\ 0 & I & -U_3^{-1} & 0 \\ 0 & 0 & I & -U_4^{-1} \end{bmatrix} \begin{bmatrix} A & 0 & 0 & 0 \\ B & C & 0 & 0 \\ D & E & F & 0 \\ G & H & K & M \end{bmatrix} = \begin{bmatrix} I & 0 & 0 & 0 \\ 0 & I & 0 & 0 \\ 0 & 0 & I & 0 \\ 0 & 0 & 0 & I \end{bmatrix} \tag{3.87}$$

It is easily shown that $A = -U_1$, $C = -U_2$, etc., for the diagonal terms. To obtain B, multiply the second row of the first matrix by the first row of the second. That is,

$$\begin{aligned} IA + B(-U_2^{-1}) &= 0 \\ B = U_2A &= -U_2U_1 \end{aligned} \tag{3.88}$$

It is seen that the general off-diagonal block (i,j), $i > j$, is obtained by premultiplying block $(i-1,j)$ by block (i,i). This is a most useful result; see Chap. 6 where it is used in greater detail for tree structures of more than one branch and primary tree. The only requirement is that the nodes be numbered so that the joint equilibrium matrix is a lower triangle.

3.9 STRUCTURES WITH DISTRIBUTED LOADS

It only remains in this chapter to consider structures in which loads occur along members, that is, at points other than node points.

These loads may be easily accommodated by adding the analyses of two structures. The first has all joints fixed against translation and rotation. Fixed-end forces may be calculated for this structure. The second structure has the negative of these fixed forces applied at its nodes.

Thus, consider the simple beam shown in Fig. 3.28, loaded with the distributed force w per unit length over the half-span. For members (1) and (2) in Fig. 3.28c, the analysis for the joint forces proceeds normally. Thus,

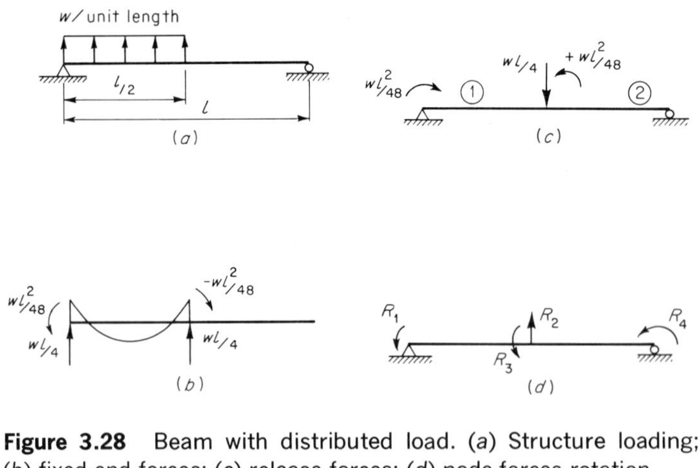

Figure 3.28 Beam with distributed load. (a) Structure loading; (b) fixed-end forces; (c) release forces; (d) node-forces rotation.

$$\begin{bmatrix} M_{1i} \\ M_{1j} \\ M_{2i} \\ M_{2j} \end{bmatrix} = \begin{bmatrix} 1 & 0 & 0 & 0 \\ -\frac{1}{2} & \frac{l}{4} & \frac{1}{2} & \frac{1}{2} \\ \frac{1}{2} & -\frac{l}{4} & \frac{1}{2} & -\frac{1}{2} \\ 0 & 0 & 0 & 1 \end{bmatrix} \begin{bmatrix} R_1 \\ R_2 \\ R_3 \\ R_4 \end{bmatrix}$$

Now from Fig. 3.28,

$$\langle R_1 \quad R_2 \quad R_3 \quad R_4 \rangle = wl \left\langle -\frac{l}{48} \quad \frac{1}{4} \quad \frac{l}{48} \quad 0 \right\rangle$$

Then,

$$\begin{bmatrix} M_{1i} \\ M_{1j} \\ M_{2i} \\ M_{2j} \end{bmatrix}_{\text{release}} = wl \begin{bmatrix} 1 & 0 & 0 & 0 \\ -\frac{1}{2} & \frac{l}{4} & \frac{1}{2} & \frac{1}{2} \\ \frac{1}{2} & -\frac{l}{4} & \frac{1}{2} & -\frac{1}{2} \\ 0 & 0 & 0 & 1 \end{bmatrix} \begin{bmatrix} -\frac{l}{48} \\ \frac{1}{4} \\ \frac{l}{48} \\ 0 \end{bmatrix} = wl \begin{bmatrix} -\frac{l}{48} \\ \frac{l}{12} \\ -\frac{l}{16} \\ 0 \end{bmatrix}$$

The fixed-end forces on the members, from Fig. 3.28*b*, are given by

$$\begin{bmatrix} M_{1i} \\ M_{1j} \\ M_{2i} \\ M_{2j} \end{bmatrix}_{\text{fixed}} = wl \begin{bmatrix} \frac{l}{48} \\ -\frac{l}{48} \\ 0 \\ 0 \end{bmatrix}$$

The final solution is given by

$$\begin{bmatrix} M_{1i} \\ M_{1j} \\ M_{2i} \\ M_{2j} \end{bmatrix} = wl \begin{bmatrix} -\frac{l}{48} \\ \frac{l}{12} \\ -\frac{l}{16} \\ 0 \end{bmatrix} + wl \begin{bmatrix} \frac{l}{48} \\ -\frac{l}{48} \\ 0 \\ 0 \end{bmatrix} = wl \begin{bmatrix} 0 \\ -\frac{l}{16} \\ \frac{l}{16} \\ 0 \end{bmatrix}$$

PROBLEMS

3.1 Set up the equilibrium equations at the joints of the trusses shown in Fig. P3.1. Then use SMIS to calculate member forces and reactions.

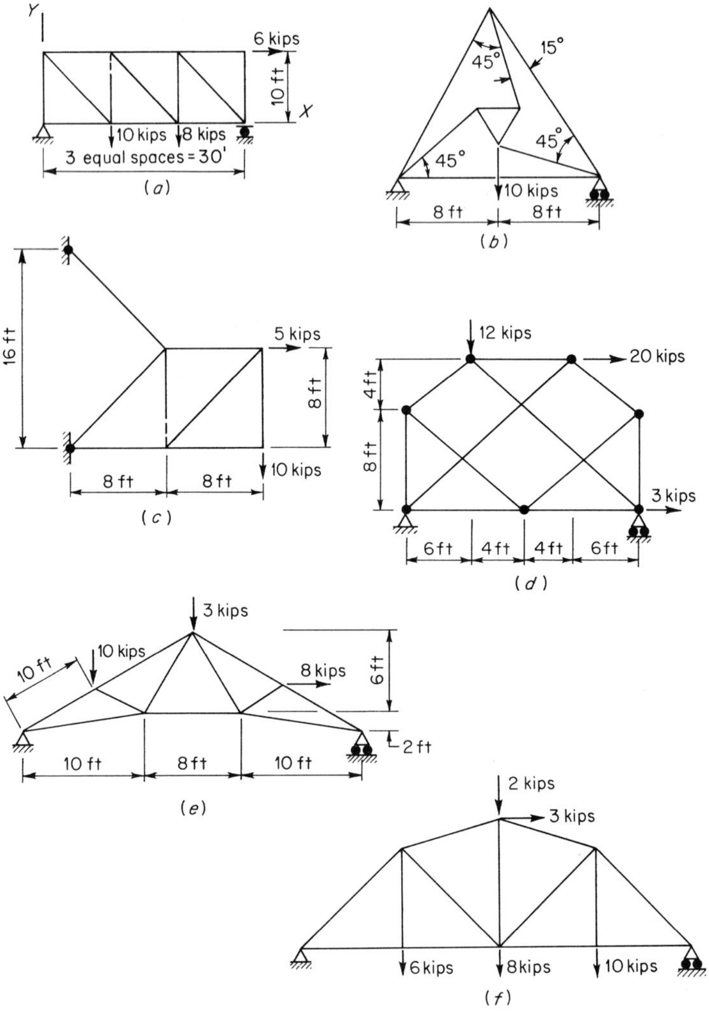

Figure P3.1.

3.2 Use the beam member-force definition given in Fig. 3.12 to set up the equilibrium equations of the joints of the structures shown in Fig. P3.2. Use SMIS to calculate these member forces for the loads shown. Then draw bending-moment and shear-force diagrams.

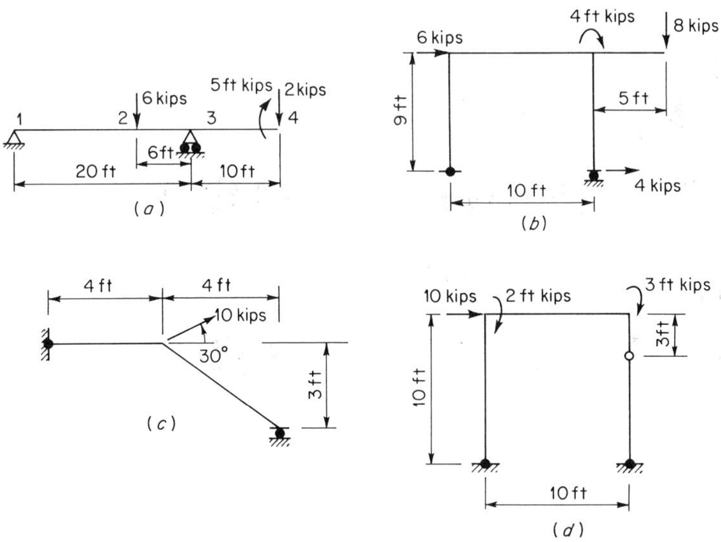

Figure P3.2.

3.3 Repeat the calculations for the structures shown in Fig. P3.2, using the member-force description given in Fig. 3.13.

3.4 Use the displacement-transformation theory in Eq. (3.70) to set up the equilibrium equations for the joints of the structures given in Fig. P3.2. Show these are identical with those calculated in Prob. 3.2.

3.5 Using the theory developed in Sec. 3.9 to reduce the distributed forces on the structure members shown in Fig. P3.3, calculate fixed-end and joint release forces. Thence, set up the equations of joint equilibrium, and solve these, using SMIS.

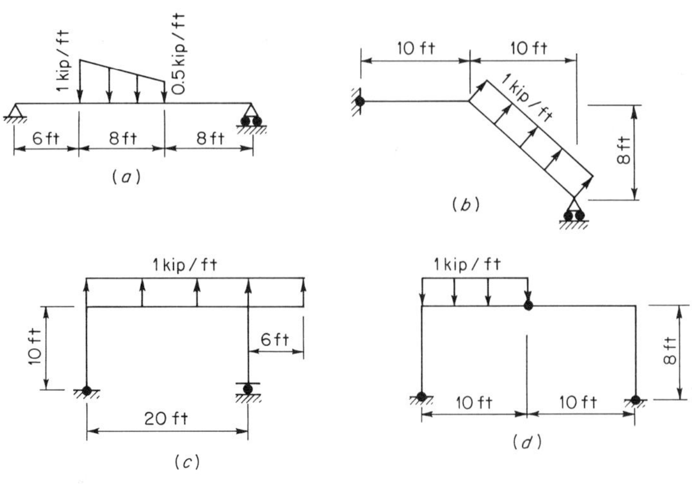

Figure P3.3.

4 DEFLECTIONS. DETERMINATE STRUCTURES. FINITE AND INFINITESIMAL SYSTEMS

4.1 NODE DEFLECTIONS FROM MEMBER DISTORTIONS

In Chaps. 2 and 3, methods for the analysis of determinate structures have been considered. The actual mathematical requirements for stability, determinacy, and indeterminacy were not considered, however, since they require a discussion of the concepts of static indeterminacy and release systems that are not introduced until Chap. 5.

The statically determinate structure could really be considered a collection of rigid bodies (members) connected in such a way that their mutual forces may be calculated from the equations of statics alone. That is, deflection calculations are not necessary for force calculation in statically determinate structures. Of course, deflection calculations are important in such structures since the service performance may be dictated by limiting deflections. It must be restated that for the structural idealization given in Sec. 3.1 (Fig. 3.3), the member distortions can be determined when the member forces are known (and vice versa). The most usual practice is to assume that the member materials are linearly elastic. In such a case there will be a linear relationship between member distortions and applied member forces.

However, to obtain the relationship between the node deflections and member distortions it is unnecessary to assume any member properties. Member distortions could arise from plastic flow, shrinkage, or temperature changes. All that is important is that the member distortions and the subsequent node displacements are compatible and do not violate the kinematic boundary conditions at supports.

The member forces may be obtained from Eq. (3.26), that is,

$$S = bR \tag{4.1}$$

Suppose now that the member distortions v (in the directions of S) produce node deflections r in the directions of R. To obtain the relationship between r and v, it is sufficient only to evoke the contragradient law by using Eqs. (3.18) and (3.20). Thus,

$$r = b^T v \tag{4.2}$$

Alternatively Eq. (4.2) may be obtained by using the equivalence of internal and external work under compatible displacements [Eq. (3.17)]. Consider the force $R_i = 1$, which produces a set of member forces S_i, with all other forces equal to zero. Then from Eq. (4.1),

$$\langle S_{1i} \quad S_{2i} \quad \cdots \quad S_{ki}\rangle = S_i{}^T = \langle b_{1i} \quad b_{2i} \quad \cdots \quad b_{ki}\rangle \tag{4.3}$$

That is, S_i is the ith column of the b matrix. For the displacements (v,r), the external work W_e and internal work W_i may now be calculated. Thus,

$$W_e = 1 \times r_i \tag{4.4}$$

and

$$W_i = [b_{1i} \quad b_{2i} \quad \cdots \quad b_{ki}]\begin{bmatrix} v_1 \\ v_2 \\ \vdots \\ v_k \end{bmatrix} = S_i{}^T v \tag{4.5}$$

Since [from Eq. (3.17)]

$$\begin{aligned} W_e &= W_i \\ r_i &= S_i{}^T v = \{b_i\}^T v \end{aligned} \tag{4.6}$$

The notation $\{b_i\}$ indicates the ith column of b. For all node points $i = 1, 2, \ldots, N$, thus r_i may be calculated, and all such equations are written in Eq. (4.7).

$$\begin{bmatrix} r_1 \\ r_2 \\ \vdots \\ r_N \end{bmatrix} = \begin{bmatrix} b_{11} & b_{21} & \cdots & b_{k1} \\ b_{12} & b_{22} & \cdots & b_{k2} \\ \cdots & \cdots & \cdots & \cdots \\ b_{1N} & b_{2N} & \cdots & b_{kN} \end{bmatrix} \begin{bmatrix} v_1 \\ v_2 \\ \vdots \\ v_k \end{bmatrix} \tag{4.7}$$

It is evident that the coefficient matrix on the right-hand side of Eq. (4.7) is simply b^T. That is, Eq. (4.7) is the statement of Eq. (4.2) in expanded form.

Example 4.1 The truss shown in Fig. 3.8 has member length (7) shortened by 1 in. Calculate the vertical deflections at nodes 2, 4, and 5. Since the matrix in Eq. (3.27) is the b matrix for loads in the vertical directions at 2, 4, and 6, Eq. (4.2) gives deflections in these directions as

$$\begin{bmatrix} r_1 \\ r_2 \\ r_3 \end{bmatrix} = \begin{bmatrix} -0.433 & -0.144 & 0.866 & 0.288 & -0.288 & 0.288 & 0.288 \\ -0.144 & -0.433 & 0.288 & -0.288 & 0.288 & 0.866 & 0.288 \\ -0.288 & -0.288 & 0.577 & -0.577 & -0.577 & 0.577 & 0.577 \end{bmatrix} \begin{bmatrix} v_1 \\ v_2 \\ v_3 \\ v_4 \\ v_5 \\ v_6 \\ v_7 \end{bmatrix} \tag{4.8}$$

Thus, for $v_7 = 1$ in. and all other $v_i = 0$ with $i = 1, 6$, the deflections are given as

$$\begin{bmatrix} r_1 \\ r_2 \\ r_3 \end{bmatrix} = \begin{bmatrix} 0.288 \\ 0.288 \\ 0.577 \end{bmatrix} \text{in.}$$

For linear elastic members, there is a linear relationship between member forces and member distortions. Thus for member i, let the member force components be $\langle S_{i1} \quad S_{i2} \quad \cdots \quad S_{iM} \rangle$. Then the corresponding member distortions will be $\langle v_{i1} \quad v_{i2} \quad \cdots \quad v_{iM} \rangle$. When each of the S_{ij} values is set equal to unity in turn, the remaining $M - 1$ values being set to zero, the deflections so obtained form the columns of the member-flexibility matrix. This matrix will be designated f, and the term f_{pq} will give the distortion v_{ip} for unit load S_{iq}. The member flexibility relationship is written in Eq. (4.9).

$$\begin{bmatrix} v_1 \\ v_2 \\ \vdots \\ v_M \end{bmatrix}_i = \begin{bmatrix} f_{11} & f_{12} & \cdots & f_{1M} \\ f_{21} & f_{22} & \cdots & f_{2M} \\ \cdots & \cdots & \cdots & \cdots \\ f_{M1} & f_{M2} & \cdots & f_{MM} \end{bmatrix}_i \begin{bmatrix} S_1 \\ S_2 \\ \vdots \\ S_M \end{bmatrix}_i \tag{4.9}$$

Note that the subscript i on the matrices in Eq. (4.9) indicates that member i is being considered. When the general expression is considered, Eq. (4.9) is written

$$v_i = f_i S_i \tag{4.10}$$

Now, $S^T = \langle S_1 \quad S_2 \quad \cdots \quad S_k \rangle$ and $v^T = \langle v_1 \quad v_2 \quad \cdots \quad v_k \rangle$, so that if f is defined as the matrix with the f_i as diagonal blocks, all equations such as (4.10) may be combined to give

$$\begin{bmatrix} v_1 \\ v_2 \\ \vdots \\ v_k \end{bmatrix} = \begin{bmatrix} f_1 & & & \mathbf{0} \\ & f_2 & & \\ & & \ddots & \\ \mathbf{0} & & & f_k \end{bmatrix} \begin{bmatrix} S_1 \\ S_2 \\ \vdots \\ S_k \end{bmatrix} \tag{4.11}$$

That is,

$$v = fS \tag{4.12}$$

It will be shown that f_i is a symmetrical matrix, so that $f_i{}^T = f_i$ and thence f is also symmetrical. The coefficients in f_i may be considered influence coefficients for deflections of member i. That is, whereas columns of f_i give deflections for a unit value of a particular member force, rows of f_i give values of particular deflections for unit values of each member force. This indicates that f_i is a deflection-influence coefficient matrix for the member as well as a member-flexibility matrix.

In Eq. (4.12), a relationship is established between member distortions v and member forces S. Since the latter are related to node forces by Eq. (4.1), so also are v. That is,

$$v = fS = fbR \tag{4.13}$$

Finally, the node deflections produced by the loads R are calculated from Eq. (4.2) to be

$$r = b^T v = b^T f b R \tag{4.14}$$

Equation (4.14) is an important and powerful equation for the calculation of node deflections. If there are N node force values, then $b^T f b$ is an $N \times N$ matrix and is defined as the structure-flexibility matrix F. That is,

$$F = b^T f b \tag{4.15}$$

$$r = FR \tag{4.16}$$

From Eq. (4.16) it is seen that F is the deflection-influence coefficient matrix for the whole structure. Its rows are deflection-influence coefficients. It will be shown also that $F = F^T$. That is, F is a symmetrical matrix. It should be noted that if f is symmetrical then the multiplication in Eq. (4.15) automatically produces a symmetrical matrix.

4.2 STRESS–STRAIN RELATIONSHIPS FOR LINEARLY ELASTIC MATERIALS

The deformation laws for structural members are based on the laws of elasticity, relating stresses and their corresponding strains. These stress-strain relationships may be written in many useful ways, of which the tensor forms developed in this section are the most fundamental.

For the material that is homogeneous and isotropic, two elastic constants are sufficient to relate stress and strain components.

For a uniaxial stress σ_x, the strains produced are

$$\varepsilon_x = \frac{\sigma_x}{E}$$

$$\varepsilon_y = \frac{-\nu\sigma_x}{E}$$

$$\varepsilon_z = \frac{-\nu\sigma_x}{E}$$

The elastic constants are E, Young's modulus, and ν, Poisson's ratio. When all three stress components are present, the sum of the contributions to the strains gives

$$\begin{aligned} \varepsilon_x &= \frac{1}{E}(\sigma_x - \nu\sigma_y - \nu\sigma_z) \\ \varepsilon_y &= \frac{1}{E}(-\nu\sigma_x + \sigma_y - \nu\sigma_z) \\ \varepsilon_z &= \frac{1}{E}(-\nu\sigma_x - \nu\sigma_y + \sigma_z) \end{aligned} \tag{4.17}$$

The shear-strain components depend only on the shear stresses and are related to these by the shear modulus of elasticity G. This is not an independent constant, but is related to E and ν by

$$G = \frac{E}{2(1 + \nu)} \tag{4.18}$$

Then,

$$\begin{aligned} \varepsilon_{xy} = \varepsilon_{yx} &= \frac{\tau_{xy}}{2G} = (1 + \nu)\frac{\tau_{xy}}{E} \\ \varepsilon_{yz} = \varepsilon_{zy} &= \frac{\tau_{yz}}{2G} = (1 + \nu)\frac{\tau_{yz}}{E} \\ \varepsilon_{zx} = \varepsilon_{xz} &= \frac{\tau_{zx}}{2G} = (1 + \nu)\frac{\tau_{zx}}{E} \end{aligned} \tag{4.19}$$

In Eqs. (4.17) and (4.19) all the relationships have been written down relating the terms of the stress tensor ε to those of the strain tensor σ.

In tensor form these are written

$$\varepsilon = \frac{1 + \nu}{E}\sigma - \frac{\nu}{E} I_3 \operatorname{tr} \sigma \tag{4.20}$$

In Eq. (4.20),

$$\operatorname{tr} \sigma = \operatorname{trace} \sigma = \sigma_x + \sigma_y + \sigma_z$$

Equations (4.17) may be solved for direct stresses in terms of direct strains. Then,

$$\begin{aligned}\sigma_x &= \frac{\lambda}{\nu}[(1-\nu)\varepsilon_x + \nu\varepsilon_y + \nu\varepsilon_z]\\ \sigma_y &= \frac{\lambda}{\nu}[\nu\varepsilon_x + (1-\nu)\varepsilon_y + \nu\varepsilon_z]\\ \sigma_z &= \frac{\lambda}{\nu}[\nu\varepsilon_x + \nu\varepsilon_y + (1-\nu)\varepsilon_z]\end{aligned} \tag{4.21}$$

In Eqs. (4.21),

$$\lambda = \frac{\nu}{(1+\nu)(1-2\nu)} \tag{4.22}$$

Also,

$$\begin{aligned}\tau_{xy} &= \tau_{yx} = 2G\varepsilon_{xy}\\ \tau_{yz} &= \tau_{zy} = 2G\varepsilon_{yz}\\ \tau_{zx} &= \tau_{xz} = 2G\varepsilon_{zx}\end{aligned} \tag{4.23}$$

The constants λ and G are known as the *Lamé constants of elasticity*. Equations (4.21) and (4.23) may be written in terms of the tensors ε and σ so that

$$\sigma = 2G\varepsilon + \lambda I_3 \operatorname{tr} \varepsilon \tag{4.24}$$

For two-dimensional elasticity, Eqs. (4.17) and (4.21) are written

$$\begin{bmatrix}\varepsilon_x\\ \varepsilon_y\end{bmatrix} = \frac{1}{E}\begin{bmatrix}1 & -\nu\\ -\nu & 1\end{bmatrix}\begin{bmatrix}\sigma_x\\ \sigma_y\end{bmatrix} \tag{4.25}$$

$$\begin{bmatrix}\sigma_x\\ \sigma_y\end{bmatrix} = \frac{E}{1-\nu^2}\begin{bmatrix}1 & \nu\\ \nu & 1\end{bmatrix}\begin{bmatrix}\varepsilon_x\\ \varepsilon_y\end{bmatrix} \tag{4.26}$$

Equations (4.20) and (4.24) are useful in transformations of the tensor components in various coordinate systems, but for matrix structures analysis, more useful forms are developed in Sec. 4.3.

4.3 INFINITESIMAL SYSTEMS

Although the fundamental stress-strain relationships and their laws of transformation are provided by the stress and strain tensors, not all the components of the stress tensor are necessary to define the fundamental deformations of the infinitesimal cube $dx\ dy\ dz$ since $\varepsilon_{xy} = \varepsilon_{yx}$, etc. The fundamental deformations for the two-dimensional element are shown in Fig. 4.1.

Thus for the element $dx\ dy\ dz$, the fundamental strain quantities are

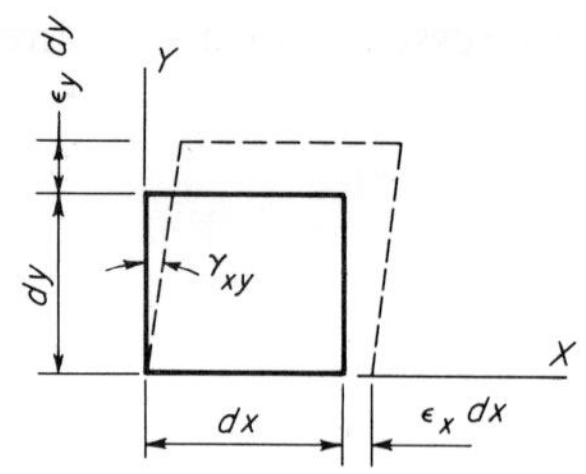

Figure 4.1 Basic element distortions.

$$\varepsilon^T = \langle \varepsilon_x \quad \varepsilon_y \quad \varepsilon_z \quad \gamma_{xy} \quad \gamma_{yz} \quad \gamma_{zx} \rangle \tag{4.27}$$

The corresponding stress quantities are

$$\sigma^T = \langle \sigma_x \quad \sigma_y \quad \sigma_z \quad \tau_{xy} \quad \tau_{yz} \quad \tau_{zx} \rangle \tag{4.28}$$

Then the relationship between ε and σ is given by

$$\varepsilon = f\sigma \tag{4.29}$$

where f is given by Hooke's law to be

$$f = \frac{1}{E}\begin{bmatrix} 1 & -\nu & -\nu & 0 & 0 & 0 \\ -\nu & 1 & -\nu & 0 & 0 & 0 \\ -\nu & -\nu & 1 & 0 & 0 & 0 \\ 0 & 0 & 0 & \Phi & 0 & 0 \\ 0 & 0 & 0 & 0 & \Phi & 0 \\ 0 & 0 & 0 & 0 & 0 & \Phi \end{bmatrix} \tag{4.30}$$

In Eq. (4.30),

$$\Phi = 2(1 + \nu) \tag{4.31}$$

From Eq. (4.29) it is possible to derive an equation similar to Eq. (4.14), in infinitesimal form. Consider the element $dx\ dy\ dz$, and write the element forces and distortions in terms of σ and ε.

Define the matrix

$$T_1 = \begin{bmatrix} dy\,dz & \cdot & \cdot & \cdot & \cdot & \cdot \\ \cdot & dz\,dx & \cdot & \cdot & \cdot & \cdot \\ \cdot & \cdot & dx\,dy & \cdot & \cdot & \cdot \\ \cdot & \cdot & \cdot & dy\,dz & \cdot & \cdot \\ \cdot & \cdot & \cdot & \cdot & dz\,dx & \cdot \\ \cdot & \cdot & \cdot & \cdot & \cdot & dx\,dy \end{bmatrix} \tag{4.32}$$

Then the stress resultants on the element will be

$$S^T = \langle F_x \quad F_y \quad F_z \quad F_{xy} \quad F_{yz} \quad F_{zx} \rangle \tag{4.33}$$

$$S = T_1\sigma \tag{4.34}$$

Then

$$\sigma = T_1^{-1}S \tag{4.35}$$

Let

$$T_2 = \begin{bmatrix} dx & \cdot & \cdot & \cdot & \cdot & \mathbf{0} \\ \cdot & dy & \cdot & \cdot & \cdot & \cdot \\ \cdot & \cdot & dz & \cdot & \cdot & \cdot \\ \cdot & \cdot & \cdot & dx & \cdot & \cdot \\ \cdot & \cdot & \cdot & \cdot & dy & \cdot \\ \mathbf{0} & \cdot & \cdot & \cdot & \cdot & dz \end{bmatrix} \tag{4.36}$$

Then the deformations of the elements are given by

$$v = T_2\varepsilon \tag{4.37}$$

Substitution for ε and σ from Eqs. (4.29) and (4.35) respectively gives

$$v = T_2 f T_1^{-1} S \tag{4.38}$$

That is, for the element $dx\,dy\,dz$, the flexibility matrix is given by

$$\bar{\bar{f}} = T_2 f T_1^{-1} \tag{4.39}$$

Suppose that the stress distribution throughout the member is known in terms of the coordinates x, y, z. That is, let

$$\sigma = b_{xyz}R \tag{4.40}$$

Then,

$$S = T_1 b_{xyz} R \tag{4.41}$$

The matrix $\bar{\bar{b}} = T_1 b_{xyz}$ is the force transformation matrix relating the forces S to the loads R. That is,

$$S = \bar{\bar{b}} R \tag{4.42}$$

The deflections δr at the load points R due to the distortions v may now be calculated by using the contragradient law. That is,

$$\delta r = \bar{\bar{b}}^T v = \bar{\bar{b}}^T \bar{\bar{f}} \bar{\bar{b}} R \tag{4.43}$$

Substitute for $\bar{\bar{b}}$ and $\bar{\bar{f}}$ from Eqs. (4.39) and (4.43), to give

$$\delta r = b^T T_1{}^T T_2 f T_1{}^{-1} T_1 b R \tag{4.44}$$

On multiplication, it is seen that

$$T_1{}^T T_2 = I \, dV \tag{4.45}$$

where $dV = dx\, dy\, dz$ is the volume of the infinitesimal element. Then,

$$\delta r = b^T f b \, dV \, R \tag{4.46}$$

When all such contributions are summed over the volume of the body, the total deflections r are obtained.

$$r = \int_{\text{vol}} b^T f b \, dV \, R \tag{4.47}$$

Equation (4.47) provides a useful method of deflection calculation when the stress distribution is known (or assumed) in the structure. This is particularly the case for members where the node forces produce specially restricted stress patterns. Several examples of these stress patterns are given in this section.

An alternative form for Eq. (4.47) exists when deflections are required at points other than the loaded points R. In such a case, let $\bar{b} = \bar{b}(x,y,z)$, namely, transform forces at points other than the load points. Then the strains are still given by Eq. (4.38), and use of the contragradient law now gives

$$\bar{r} = \int_{\text{vol}} \bar{b}^T f b \, dV \, R \tag{4.48}$$

Examples of calculation of member deflections by using Eq. (4.47) and (4.48)

Example 4.2 The prismatic beam that has area A and length l and is fixed at one end is loaded axially with a force distributed linearly over the length l (Fig. 4.2). The intensity of loading varies from O at the free end to R at the fixed end. It is required to calculate the deflection at the free end. From Fig. 4.2, it is seen that the stress σ_x at section x is the only stress on an element of the cross section and

$$\sigma_x = \frac{Rx}{lA} \tag{i}$$

Therefore,

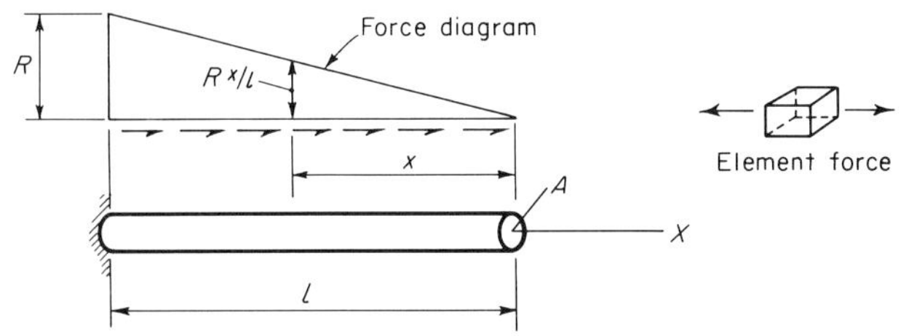

Figure 4.2 Prismatic-member axial force.

$$b = \frac{x}{Al} \tag{ii}$$

The stress-strain flexibility relationship reduces to

$$\varepsilon_x = \frac{1}{E}\sigma_x \tag{iii}$$

Thence

$$f = \frac{1}{E} \tag{iv}$$

Place a unit load on the free end so that

$$\bar{b} = \frac{1}{A} \tag{v}$$

Then from Eq. (4.48),

$$\bar{r} = \int_{\text{vol}} \bar{b}^T f b \, dV \, R = \int_o^l \int_{\text{area}} \frac{1}{A}\frac{1}{E}\frac{x}{Al} \, dy \, dz \, dx \, R$$

$$= \frac{R}{AEl}\int_o^l x \, dx = \frac{Rl}{2AE}$$

That is,

$$\bar{r} = \frac{l}{AE} x \text{ average force}$$

Example 4.3 The beam element shown in Fig. 4.3 is of length dx and is subjected to the uniform moment M, as shown, that produces stresses given by the engineer's theory of bending. It is required to calculate the relative rotations of the two faces of the infinitesimal slice.

The stress distribution is

$$\sigma_x = -\frac{My}{I_z} \tag{i}$$

Therefore,

$$b = -\frac{y}{I_z} \tag{ii}$$

Only axial stresses σ_x are produced, therefore, as in Example 4.2,

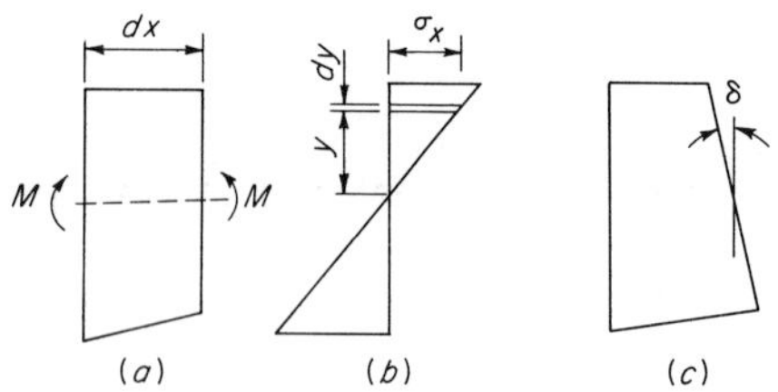

Figure 4.3 (*a*) Beam element in pure bending; (*b*) assumed stress distribution; (*c*) relative displacement of cross sections.

$$f = \frac{1}{E} \tag{iii}$$

Hence, using Eq. (4.47),

$$\delta\theta = \left[\int_{\text{area}} \left(-\frac{y}{I_z}\right)\left(\frac{1}{E}\right)\left(-\frac{y}{I_z}\right) dA\right] dx\ M \tag{iv}$$

However, $I_z = \int_{\text{area}} y^2\, dA$, so that

$$\delta\theta = \frac{dx}{EI_z} M \tag{v}$$

Equation (*v*) is the flexibility relationship for a thin-slice dx subject to a pure bending moment. Since a member of finite length is composed of an infinite number of such slices, Eq. (*v*) may be used to calculate bending deflections by integrating over the member length.

Example 4.4 A cantilever beam Fig. 4.4 of variable moment of inertia $I_z(x)$ is subjected to bending moments $M = M_z(x)$ produced by a load system defined by a single parameter R. The inertia and moment functions are assumed continuous over the length of the beam.

Equation (*v*) in Example 4.2 is used together with Eq. (4.48) and numerical integration to calculate the deflection at the free end of the cantilever. Because $M(x)$ may be expressed in terms of the single parameter R and the coordinate x, the equation $S = bR$, for the element dx, becomes

$$M(x) = mR \tag{i}$$

That is,

$$b \equiv m$$

For $\bar{R} = 1$ on the free end,

$$\bar{b} = \bar{m} = -(l - x) \tag{ii}$$

From Eq. (*v*), Example 4.2,

$$f = \frac{dx}{EI} \tag{iii}$$

The deflection at the free end is given by

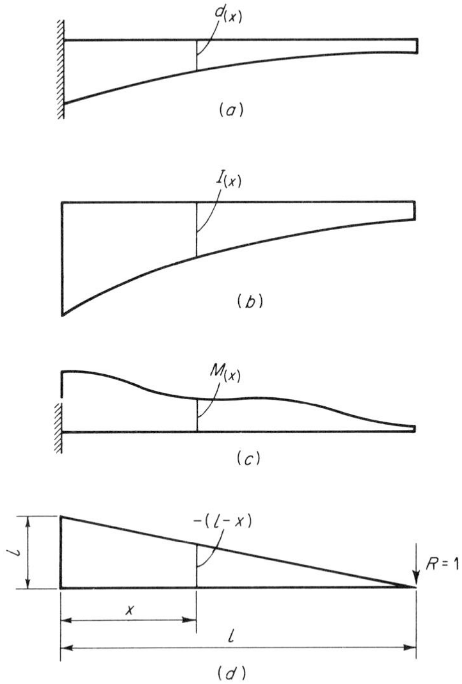

Figure 4.4 (*a*) Beam profile; (*b*) moment of inertia; (c) bending moment; (*d*) bending moment for unit load on free end.

$$\bar{r} = \int_0^l \bar{b}^T f b \, dx \, R \qquad (iv)$$

That is,

$$\bar{r} = \int_0^l \frac{\bar{m}m}{EI} \, dx \, R \qquad (v)$$

Suppose, now, that $\bar{m}$, m, and I have been calculated at five equal intervals along the beam, as shown in Fig. 4.5, and further, let

$$\beta_i = \frac{I_1}{I_i} \qquad (vi)$$

Applying Simpson's rule to Eq. (v) gives

$$\bar{r} = \frac{h}{3EI_1}(\bar{m}_1 m_1 + 4\bar{m}_2 m_2 \beta_2 + 2\bar{m}_3 m_3 \beta_3 + 4\bar{m}_4 m_4 \beta_4 + \bar{m}_5 m_5 \beta_5)R \qquad (vii)$$

On using the values of $\bar{m}$ given in Fig. 4.5, with $h = l/4$,

$$\bar{r} = \frac{l^2}{12EI_1}(m_2\beta_2 + m_3\beta_3 + 3m_4\beta_4 + m_5\beta_5)R \qquad (viii)$$

Suppose that the moment m is produced by the uniform load w per unit length, so that

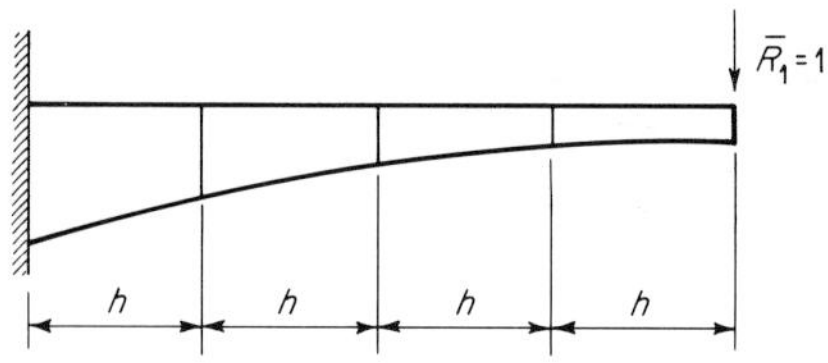

Inertia values	I_5	I_4	I_3	I_2	I_1
Relative $1/I$ values	β_5	β_4	β_3	β_2	1
Moment values	M_5	M_4	M_3	M_2	M_1
$\bar{m}$ values	$-l$	$-3l/4$	$-l/2$	$-l/4$	0
Simpson's coefficient	1	4	2	4	1

Figure 4.5 Parameters for numerical integration.

$$m_x = \frac{-w(l-x)^2}{2} \qquad \text{or} \qquad m = -\frac{(l-x)^2}{2} \qquad R = w$$

Then,

$$\langle m_2 \quad m_3 \quad m_4 \quad m_5 \rangle = \langle -\frac{l^2}{32} - \frac{l^2}{8} - \frac{9l^2}{32} - \frac{l^2}{2} \rangle$$

Further, let the beam be of constant section so that $I_1 = I =$ constant. Then,

$$\bar{r} = \frac{-l_4}{12EI}\left(\frac{1}{32} + \frac{1}{8} + \frac{27}{32} + \frac{1}{2}\right)w = \frac{-wl_4}{8EI}$$

Example 4.5 Beam deflections including shear distortions

In Example 4.4, it was assumed that the applied loads R produced distortions that were a function of σ_x only. However, shear stresses act in the cross section of the beam, and these produce additional deflection.

The cantilever beam shown in Fig. 4.6 is loaded in its plane of symmetry so that the distribution of shear and normal stresses given by the engineer's theory of bending are as shown in 4.6*c*. In this case the infinitesimal element used in deflection calculations will be the slice shown by cross section $t\,dy$ of length dx in Fig. 4.6*b*.

The stresses on the element will be

$$\sigma = -\frac{my}{I} \qquad \tau = \frac{VQ}{It} \tag{i}$$

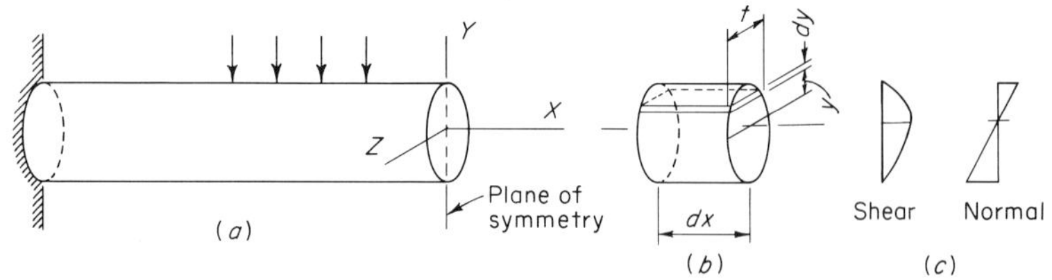

Figure 4.6 (*a*) Cantilever beam including bending and shear distortions; (*b*) element of beam; (*c*) stress distribution.

Then,

$$\begin{bmatrix} \sigma_x \\ \tau \end{bmatrix} = \begin{bmatrix} \dfrac{-y}{I} & 0 \\ 0 & \dfrac{Q}{It} \end{bmatrix} \begin{bmatrix} m \\ V \end{bmatrix} \tag{ii}$$

For the cantilever-beam load R on the free end,

$m = xR \qquad V = R$

Therefore,

$$\begin{bmatrix} \sigma_x \\ \tau \end{bmatrix} = \begin{bmatrix} \dfrac{-xy}{I} \\ \dfrac{Q}{IT} \end{bmatrix} R \tag{iii}$$

The flexibility relationship for the infinitesimal element is given by

$$\begin{bmatrix} \varepsilon_x \\ \gamma_{xy} \end{bmatrix} = \begin{bmatrix} \dfrac{1}{E} & 0 \\ 0 & \dfrac{1}{G} \end{bmatrix} \begin{bmatrix} \sigma_x \\ \tau \end{bmatrix} \tag{iv}$$

Hence, using Eq. (4.47)

$$r = \int_{\text{vol}} \begin{bmatrix} \dfrac{-xy}{I} & \dfrac{Q}{It} \end{bmatrix} \begin{bmatrix} \dfrac{1}{E} & 0 \\ 0 & \dfrac{1}{G} \end{bmatrix} \begin{bmatrix} \dfrac{-xy}{I} \\ \dfrac{Q}{It} \end{bmatrix} dV\, R \tag{v}$$

The volume dV is given from Fig. 4.6b,

$$dV = t\, dy\, dx \tag{vi}$$

Thus,

$$r = \int_o^l \int_{-c_1}^{c_2} \begin{bmatrix} \dfrac{-xy}{I} & \dfrac{Q}{It} \end{bmatrix} \begin{bmatrix} \dfrac{1}{E} & 0 \\ 0 & \dfrac{1}{G} \end{bmatrix} \begin{bmatrix} \dfrac{-xy}{I} \\ \dfrac{Q}{It} \end{bmatrix} t\, dy\, dx\, R \tag{vii}$$

On multiplication of the terms in (*vii*) and integration over the cross section, remembering that $I = \int_{-c_1}^{c_2} y^2 t\, dy$,

$$r = \left(\int_O^l \frac{x^2}{EI}\, dx + \frac{1}{GI^2} \int_o^l \int_{-c_2}^{c_1} \frac{Q^2}{t}\, dy\, dx \right) R \tag{viii}$$

It is seen immediately that the first term in (*viii*) is the normal bending deflection term, and on integration,

$$r_{\text{bending}} = \frac{Rl^3}{3EI} \tag{ix}$$

The shear term depends on the distribution of shear stress in the cross section, that is, on Q^2/t. For the rectangular beam, this is parabolic as shown in Fig. 4.7, and for the I beam, it includes the cross shear in the flanges. Thus for the rectangle,

$$Q = \frac{1}{2}t\left(\frac{d^2}{4} - y^2\right)$$

$$Q^2 = \frac{1}{4}t^2\left(\frac{d^4}{16} - \frac{d^2y^2}{2} + y^4\right)$$

$$I^2 = \frac{t^2d^6}{144}$$

The shear-deflection term becomes

$$r_{\text{shear}} = \int_o^l \frac{1}{GI^2}\frac{t}{4}\left[2\int_o^{d/2}\left(\frac{d^4}{16} - \frac{d^2y^2}{2} + y^4\right)dy\right]dx\,R$$

$$= \int_o^l \frac{6}{5Gt\,d}\,dx\,R$$

Thus for length l,

$$r = \frac{Rl^3}{3EI} + \frac{6Rl}{5Gt\,d}$$

For a material with $G \approx \frac{1}{2}E$, then

$$r = \frac{Rl}{Et\,d}\left(\frac{4l^2}{d^2} + \frac{12}{5}\right)$$

The ratio,

$$\frac{r_{\text{shear}}}{r_{\text{bending}}} = \frac{3}{5}\frac{d^2}{l^2}$$

Thus, if $d = \frac{1}{10}l$,

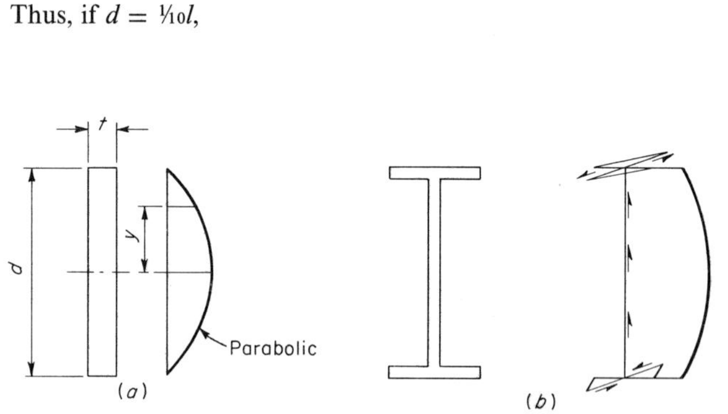

Figure 4.7 Distribution of shear stresses. (*a*) Rectangular cross section; (*b*) I section with shear stress.

$$r_{\text{shear}} = 0.6\% \, r_{\text{bending}}$$

if $d = \frac{1}{5}l$,

$$r_{\text{shear}} = 2.4\% \, r_{\text{bending}}$$

For I sections, shear distortions will be a larger percentage of bending distortions, and it is usual to include shear distortions in computer programs that calculate frame-structure deflections.

In such cases the effective shear area is defined by

$$A_{\text{eff}} = \frac{I^2}{\int_{\text{area}} (Q^2/t) \, dy}$$

Then by Eq. (*viii*),

$$\int_{\text{area}} b^T f b = \frac{x^2}{EI} + \frac{1}{GA_{\text{eff}}}$$

Example 4.6 Deflections of a beam in axes other than the principal axes

The problem is posed by the angle member shown in Fig. 4.8, which is loaded at the intersection of the angle legs. This point is also the shear center, so that the member deflects without twisting. The angle has its lower flange tapered, so that the principal axes of the cross sections change with distance along the member. From the theory of Sec. 2.6, it is determined that the inertia tensor transforms under rotation of coordinate axes according to the law [see Eq. (2.78)]

$$J' = L^T J L$$

where

$$L = \begin{bmatrix} \cos\theta & -\sin\theta \\ \sin\theta & \cos\theta \end{bmatrix} \quad \text{and} \quad J = \begin{bmatrix} I_{zz} & I_{zy} \\ I_{yz} & I_{yy} \end{bmatrix}$$

Let the inertia tensor in principal axes be $\bar{J}$, so that

$$\bar{J} = \begin{bmatrix} \bar{I}_{zz} & 0 \\ 0 & \bar{I}_{yy} \end{bmatrix} \tag{4.49}$$

The bending deformations of the angle are associated with $\bar{J}^{-1}$, and it is useful to

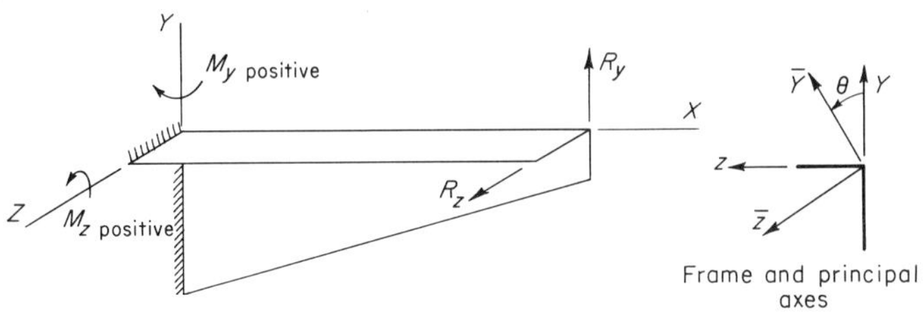

Figure 4.8 Clipped-angle cantilever beam.

determine the transformation law for this matrix. Thus, if J is the inertia matrix in the XY system,

$$\bar{J} = L^T J L$$

L being that transformation matrix to diagonalize J. Then,

$$\bar{J}^{-1} = (L^T J L)^{-1} = L^{-1} J^{-1} (L^T)^{-1} = L^T J^{-1} L \tag{4.50}$$

That is, J^{-1} transforms in the same way as J. On inverting J, it is found that

$$J^{-1} = \frac{1}{I_{yy} I_{zz} - I_{yz}^2} \begin{bmatrix} I_{yy} & -I_{zy} \\ -I_{yz} & I_{zz} \end{bmatrix} \tag{4.51}$$

For the angle in Fig. 4.8, the b matrix is now determined in order to transform the forces $\langle R_y \quad R_z \rangle$ into moments about the principal axes at section x.

Firstly, in frame coordinates,

$$\begin{bmatrix} M_z \\ M_y \end{bmatrix} = \begin{bmatrix} l - x & 0 \\ 0 & (l - x) \end{bmatrix} \begin{bmatrix} R_y \\ R_z \end{bmatrix} \tag{4.52}$$

Write this equation

$$M_{zy} = b_1 R \tag{4.53}$$

The moments M_{zy} transform into $\overline{M}_{zy}$ components in the principal axes according to Eq. (2.29), noting the sign change in Fig. 4.8 for M_y and the reversed order of M_z and M_y. That is,

$$\overline{M}_{zy} = L^T M_{zy} = L^T b_1 R \tag{4.54}$$

Then

$$b = L^T b_1 \tag{4.55}$$

$$\overline{M}_{zy} = bR \tag{4.56}$$

The rotations produced by the principal-axes moments $\overline{M}_{zy}$ are uncoupled. That is, $\overline{M}_z$ produces deflections in the $\bar{y}$ plane only, and vice versa. For the infinitesimal length dx, deflection $\delta\theta_{zy}$ is given by

$$\delta\bar{\theta}_{zy} = \frac{1}{E} \bar{J}^{-1} \overline{M}_{zy} \, dx \tag{4.57}$$

Apply the contragradient law to Eq. (4.56) connecting the displacement δr at the free end to $\delta\bar{\theta}_{zy}$. That is,

$$\delta r = \begin{bmatrix} \delta_{ry} \\ \delta_{rz} \end{bmatrix} = b^T \, \delta\bar{\theta}_{zy} \tag{4.58}$$

Combining Eqs. (4.56), (4.57), and (4.58),

$$\delta r = b^T \frac{1}{E} \bar{J}^{-1} b \, dx \, R \tag{4.59}$$

Substitute for b and $\bar{J}^{-1}$ from Eqs. (4.50) and (4.55). Then,

$$\delta r = \frac{1}{E} b_1{}^T L L^T J^{-1} L L^T b_1 R \tag{4.60}$$

However,

$$LL^T = I$$

Hence,

$$\delta r = b_1{}^T \frac{1}{E} J^{-1} b_1 \, dx \, R \tag{4.61}$$

It is seen in Eq. (4.61) that it is unnecessary to perform the transformations of moments and inertias to the principal axes and that $(1/E)J^{-1}\,dx$ is the flexibility matrix for the angle element of length dx under the moments $\langle M_z \quad M_y \rangle$.

For the cantilever length l, the sum of all such contributions is given

$$r = \frac{1}{E} \int_o^l b_1{}^T J^{-1} b_1 \, dx \, R \tag{4.62}$$

Notice that for the nonprincipal axes, the flexibilities are coupled. That is, a moment M_z produces rotations about both the z and y axes.

4.4 ELASTIC STRAIN ENERGY; MAXWELL'S AND BETTI'S THEOREMS; DERIVATIVES OF THE STRAIN ENERGY FUNCTION WITH RESPECT TO NODE FORCES

When a set of forces is applied to an elastic body, work is done by the forces that may be recovered when the forces are removed. That is, the body returns to its original shape. This strain energy may be calculated in terms of the applied forces and the elastic moduli. Thus for the simple tension member in Fig. 4.9, let the load P be incremented by the amount δP so that the extension of the bar increases by δe. Then from the shaded area of Fig. 4.9, work done is

$$\delta W = P \,\delta e + \tfrac{1}{2}\delta P \,\delta e \tag{4.63}$$

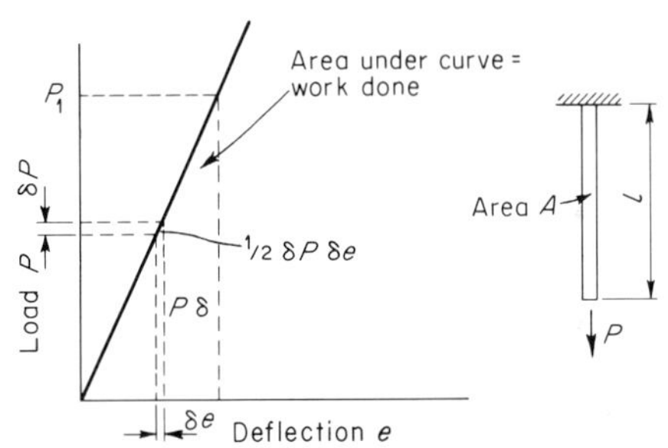

Figure 4.9 Elastic strain energy in bar in simple tension.

Integrating from O to P_1 and carrying the process in Eq. (4.63) to the limit (that is, second-order terms are ignored),

$$W = \int_o^{P_1} P\,\delta e \tag{4.64}$$

For the elastic stress state,

$$\delta e = \frac{l}{AE}\,\delta P$$

Hence,

$$W = \frac{l}{AE}\int_o^{P_1} P\,dP = \frac{P^2 l}{2AE} \tag{4.65}$$

Now $l/AE = f$, the member flexibility; hence,

$$W = \tfrac{1}{2}P^2 f$$

This process may be generalized for any elastic structure or member under the action of more than one force. Since member properties are considered fundamental, strain energy expressions are first developed for members and then summed for all members in order to give the strain energy in the whole structure.

Maxwell's Theorem

Consider the structural element whose elastic-member force distortion law is known to be

$$v_i = f_i S_i \tag{4.66}$$

If there are only two force and displacement components, Eq. (4.66) is written in full as

$$\begin{bmatrix} v_1 \\ v_2 \end{bmatrix} = \begin{bmatrix} f_{11} & f_{12} \\ f_{21} & f_{22} \end{bmatrix} \begin{bmatrix} S_1 \\ S_2 \end{bmatrix} \tag{4.67}$$

Calculate the elastic strain energy U in the element due to the forces S_1 and S_2 as follows:

1 Apply S_1 alone. Then

$$U_1 = \tfrac{1}{2}f_{11}S_1^2$$

Now apply S_2. Then the additional stored energy must be

$$U_{12} = \tfrac{1}{2}f_{22}S_2^2 + f_{12}S_2S_1$$

Thence,

$$U = U_1 + U_{12} = \tfrac{1}{2}f_{11}{}^2S_1^2 + \tfrac{1}{2}f_{22}S_2^2 + f_{12}S_2S_1 \tag{4.68}$$

2 If the forces are removed and applied with S_2 first, a new expression is obtained for the strain energy. Thus, apply S_2; then

$$U_2 = \tfrac{1}{2} f_{22} S_2^2$$

Now apply S_1. The additional stored energy must be

$$U_{21} = \tfrac{1}{2} f_{11} S_1^2 + f_{21} S_1 S_2$$

That is,

$$U = U_2 + U_{21} = \tfrac{1}{2} f_{22} S_2^2 + \tfrac{1}{2} f_{11} S_1^2 + f_{21} S_1 S_2 \tag{4.69}$$

However, the law of conservation of energy states that U must be independent of the order of application of S_1 and S_2. It follows that

$$f_{12} = f_{21} \tag{4.70}$$

The important result has been proved, that the member-flexibility matrix is symmetric. The structure-flexibility matrix is obtained from the equation

$$F = b^T f b \tag{4.71}$$

The matrix f is evidently symmetric since it has been shown to be composed of diagonal blocks of member flexibilities that are symmetric. Thence,

$$F^T = (b^T f b)^T = b^T f^T b = b^T f b \tag{4.72}$$

That is, if the member-flexibility matrices are symmetric, so also is the structure-flexibility matrix.

The relationship of symmetry expressed in Eq. (4.70) is known as *Maxwell's reciprocal theorem.*

The expression for strain energy U in Eq. (4.69) may be written in symmetrical form by writing $f_{21} = \tfrac{1}{2}(f_{12} + f_{21})$. That is,

$$U = \tfrac{1}{2} f_{11} S_1^2 + \tfrac{1}{2} f_{12} S_1 S_2 + \tfrac{1}{2} f_{21} S_1 S_2 + \tfrac{1}{2} f_{22} S_2^2 \tag{4.73}$$

In matrix form,

$$U = \tfrac{1}{2}[S_1 \quad S_2] \begin{bmatrix} f_{11} & f_{12} \\ f_{21} & f_{22} \end{bmatrix} \begin{bmatrix} S_1 \\ S_2 \end{bmatrix} \tag{4.74}$$

That is,

$$U = \tfrac{1}{2} S_i^T f_i S_i \tag{4.75}$$

The expression in Eq. (4.75) is for one member only, but is now generalized for the whole structure by use of S and f.

Let the members in the structure be designated a, b, c, etc. Then the strain energy for the whole structure is given as the sum of all member values.

$$U_{\text{total}} = \tfrac{1}{2} S_a^T f_a S_a + \tfrac{1}{2} S_b^T f_b S_b + \tfrac{1}{2} S_c^T f_c S_c + \cdots \tag{4.76}$$

Thence,

$$U_{\text{total}} = \tfrac{1}{2}[S_a \quad S_b \quad S_c \quad \cdots]\begin{bmatrix} f_a & & \mathbf{0} & \\ & f_b & & \\ \mathbf{0} & & f_c & \\ & & & \ddots \end{bmatrix}\begin{bmatrix} S_a \\ S_b \\ S_c \\ \cdot \end{bmatrix} \tag{4.77}$$

However, from Eq. (4.11), $[S_a \quad S_b \quad S_c \quad \cdots] = S$, etc. That is,

$$U_{\text{total}} = \tfrac{1}{2}S^T fS \tag{4.78}$$

The expression for strain energy may now be expressed in terms of the external load vector, since by Eq. (3.26),

$$S = bR$$

Substitute for S in Eq. (4.78). Then,

$$U_{\text{total}} = \tfrac{1}{2}R^T b^T f bR \tag{4.79}$$

However by similar reasoning to that employed to obtain S_i, the strain energy of a member, the strain energy for the whole structure may be written as a function of R and the structure-flexibility matrix F. That is,

$$U_{\text{total}} = \tfrac{1}{2}R^T FR \tag{4.80}$$

By comparing Eqs. (4.79) and (4.80) it is seen that a definition has been obtained for $F = b^T f b$ and furthermore that this agrees with the expression obtained previously, Eq. (4.15).

If the member under consideration in Eq. (4.75) is the infinitesimal element, this equation may be written by using the notation of Sec. 4.3. For the elemental member $dx\ dy\ dz$,

$$\begin{aligned} \delta U = \tfrac{1}{2}S^T \bar{\bar{f}} S &= \tfrac{1}{2}\sigma^T T_1{}^T T_2 f T_1{}^{-1} T_1 \sigma \\ &= \tfrac{1}{2}\sigma^T f\sigma\, dV \end{aligned}$$

For the whole body,

$$U = \tfrac{1}{2}\int_{\text{vol}} \sigma^T f\sigma\, dV \tag{4.81}$$

However,

$$\sigma = b(x,y,z)R$$

Thence,

$$U = \tfrac{1}{2}R^T \int_{\text{vol}} b^T f b\, dV\, R \tag{4.82}$$

From Eq. (4.82) it is seen that an alternative definition of F is given by

$$F = \int_{\text{vol}} b^T f b \, dV \tag{4.83}$$

Betti's Theorem

Betti's theorem is a relationship between two force systems plus their related displacements on a linearly elastic structure that is readily derived from energy considerations.

Consider the two force systems R_1 and R_2 acting on the elastic structure. Let r_{21} be the deflections in the R_2 directions produced by R_1 and r_{12} be those deflections in the R_1 directions produced by R_2. Then,

$$r_{21} = F_{21}R_1 \tag{4.84}$$

$$r_{12} = F_{12}R_2 \tag{4.85}$$

Similarly,

$$r_{11} = F_{11}R_1 \qquad \text{and} \qquad r_{22} = F_{22}R_2$$

Then the whole flexibility matrix for the structure is written

$$\begin{bmatrix} r_1 \\ r_2 \end{bmatrix} = \begin{bmatrix} F_{11} & F_{12} \\ F_{21} & F_{22} \end{bmatrix} \begin{bmatrix} R_1 \\ R_2 \end{bmatrix} \tag{4.86}$$

It is apparent from the symmetry of the F matrix that

$$F_{21}{}^T = F_{12} \tag{4.87}$$

Then multiply both sides of Eq. (4.84) by $R_2{}^T$ and both sides of Eq. (4.85) by $R_1{}^T$.

$$R_2{}^T r_{21} = R_2{}^T F_{21} R_1$$

$$R_1{}^T r_{12} = R_1{}^T F_{12} R_2$$

In the second of these equations, note that

$$R_1{}^T r_{12} \equiv r_{12}{}^T R_1 = R_2{}^T F_{12}{}^T R_1$$

Since Eq. (4.87) is valid,

$$R_1{}^T r_{12} = R_2{}^T F_{21} R_1$$

That is,

$$R_1{}^T r_{12} = R_2{}^T r_{21} \tag{4.88}$$

This expression is known as *Betti's theorem.*

If the symmetry of the F matrix has not been demonstrated, then Betti's theorem can be shown to be true from energy considerations. This may be regarded as a proof of the symmetry relationship $F_{12} = F_{21}{}^T$. Apply the R_1 load system. The strain energy is

$$U_1 = \tfrac{1}{2} R_1{}^T F_{11} R_1$$

Now apply R_2. The additional strain energy is

$$U_{12} = \tfrac{1}{2}R_2{}^T F_{22} R_2 + R_1{}^T F_{12} R_2$$

Thus

$$U_{\text{total}} = U_1 + U_{12} = \tfrac{1}{2}R_1{}^T F_{11} R_1 + \tfrac{1}{2}R_2{}^T F_{22} R_2 + R_1{}^T F_{12} R_2$$

If the order of loading is reversed, the expression for U_{total} is found to be

$$U_{\text{total}} = \tfrac{1}{2}R_1{}^T F_{11} R_1 + \tfrac{1}{2}R_2{}^T F_{22} R_2 + R_2{}^T F_{21} R_1$$

From the law of conservation of energy, it follows that

$$R_1{}^T F_{12} R_2 = R_2{}^T F_{21} R_1$$

or

$$R_1{}^T r_{12} = R_2{}^T r_{21}$$

Derivatives of the Strain Energy Function

When the strain energy expression U is differentiated partially with respect to one of the loads R_i, the resulting value is found to be the deflection at that load point and in that direction. This is known as *Castigliano's second theorem*.

Differentiate Eq. (4.80) partially with respect to the load R_i.

$$\frac{\partial U}{\partial R_i} = \frac{1}{2}\frac{\partial R^T}{\partial R_i} FR + \frac{1}{2} R^T F \frac{\partial R}{\partial R_i} \tag{4.89}$$

Now

$$R^T = \langle R_1 \quad R_2 \quad \cdots \quad R_i \quad \cdots \quad R_N \rangle \tag{4.90}$$

Hence,

$$\frac{\partial R^T}{\partial R_i} = \langle 0 \quad 0 \quad \cdots \quad 0 \quad \underset{\substack{\uparrow \\ i\text{th column}}}{1} \quad 0 \quad \cdots \quad 0 \rangle = \left(\frac{\partial R}{\partial R_i}\right)^T \tag{4.91}$$

Thus when FR is premultiplied by $\partial R^T/\partial R_i$, only the ith row of F remains, and similarly when $R^T F$ is postmultiplied by $\partial R/\partial R_i$, only the ith column of F remains. Thence,

$$\frac{\partial U}{\partial R_i} = \tfrac{1}{2}(F_{i1}R_1 + F_{i2}R_2 + \cdots + F_{ii}R_i + \cdots + F_{iN}R_N)$$
$$+ \tfrac{1}{2}(F_{1i}R_1 + F_{2i}R_2 + \cdots + F_{ii}R_i + \cdots + F_{Ni}R_N)$$

However, $F_{ij} = F_{ji}$ (Maxwell's symmetry relationship). Therefore,

$$\frac{\partial U}{\partial R_i} = F_{i1}R_1 + F_{i2}R_2 + \cdots + F_{ii}R_i + \cdots + F_{iN}R_N = r_i \tag{4.92}$$

The result is Castigliano's second theorem.

4.5 MEMBER-FLEXIBILITY MATRICES; SIMPLE CASES

The fundamental expression $r = b^T f b R$ for the calculation of node deflections requires a knowledge not only of the construction of the force transformation matrix b, but also of the means to construct the member-flexibility matrix f. The simplest flexibility is for the prismatic member that is under axial force only and is of length l and area A, for which

$$f = \frac{l}{AE} \tag{4.93}$$

For the straight prismatic beam under linearly varying end moments, the alternative basic forces are shown in Fig. 4.10. For the forces in Fig. 4.10a the distortions $\langle \phi_i \quad \phi_j \rangle$ of the member relative to the chord i, j are given as

$$\begin{bmatrix} \phi_i \\ \phi_j \end{bmatrix} = \frac{l}{6EI} \begin{bmatrix} 2 & 1 \\ 1 & 2 \end{bmatrix} \begin{bmatrix} M_i \\ M_j \end{bmatrix} \tag{4.94}$$

That is, the member-flexibility matrix is

$$f = \frac{l}{6EI} \begin{bmatrix} 2 & 1 \\ 1 & 2 \end{bmatrix} \tag{4.95}$$

More complete details of flexibility matrices are given in Sec. 4.10.

With the alternative sign convention in Fig. 4.10b,

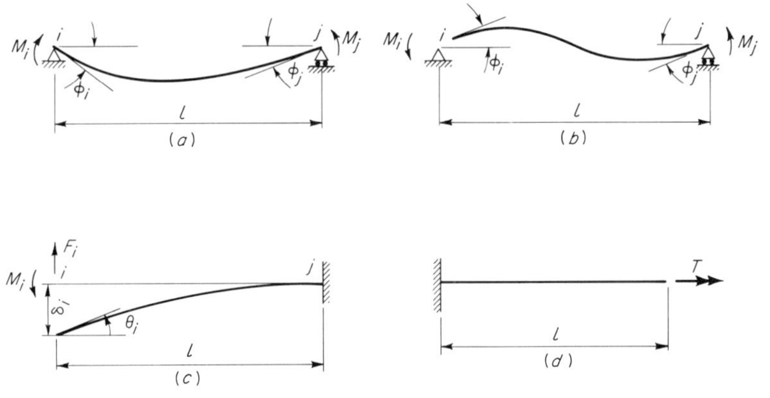

Figure 4.10 Prismatic-member basic-force quantities. (a) Sign convention +ve for moment tension on lower fiber; (b) sign convention +ve for anticlockwise moment; (c) forces defined on one end, i, of member; (d) prismatic-member torque.

$$f = \frac{l}{6EI}\begin{bmatrix} 2 & -1 \\ -1 & 2 \end{bmatrix} \tag{4.96}$$

When the basic forces considered are $\langle M_i \quad S_i \rangle$ (Fig. 4.10c), the deflections at end i are given as

$$\begin{bmatrix} \theta_i \\ \delta_i \end{bmatrix} = \frac{l}{EI}\begin{bmatrix} 1 & -\dfrac{l}{2} \\ -\dfrac{l}{2} & \dfrac{l^2}{3} \end{bmatrix}\begin{bmatrix} M_i \\ S_i \end{bmatrix}$$

In this case then,

$$f = \frac{l}{EI}\begin{bmatrix} 1 & -\dfrac{l}{2} \\ -\dfrac{l}{2} & \dfrac{l^2}{3} \end{bmatrix} \tag{4.97}$$

For the prismatic member in torsion (Fig. 4.10d), the flexibility calculation is simple only when the cross section is circular. In this case,

$$f = \frac{l}{GI_P} \tag{4.98}$$

The term I_P is the polar moment of inertia of the cross section.

For the rectangular section the following values have been taken from Ref. 33.

$$f = \frac{l}{Gk_1(2a)^3 2b}$$
$$k_1 = \frac{1}{3}\left(1 - \frac{192}{\pi^5}\frac{a}{b}\sum_{n=1,3,5}\frac{1}{n^5}\tanh\frac{n\pi b}{2a}\right) \tag{4.99}$$

$\dfrac{b}{a}$:	1	1.2	1.5	2.0	2.5	3	4	5	10	α
k_1:	0.1406	0.166	0.196	0.229	0.249	0.263	0.281	0.291	0.302	0.333

Calculation of Structure Deflections

Either of the Eqs. (4.100) or (4.101) may be used.

$$r = b^T f b R \tag{4.100}$$
$$\bar{r} = \bar{b}^T f b R \tag{4.101}$$

If the flexibility matrix F is required, then it is necessary to perform the multiplication $b^T f b$. When only the deflections r, for one or two load cases or points of the structure, are required, it is more economical in computation time to multiply R by b, which produces the vector bR, then to multiply this vector by f to produce the vector fbR, and finally to multiply by $\bar{b}^T$. In this way three multiplications of a vector by a matrix are required. If the process is executed by multiplying $\bar{b}^T fb$, then this involves two multiplications of matrices by matrices.

Example 4.7 Calculation of truss deflections

The truss in Fig. 4.11 has vertical forces of $2.7k$ and $3.2k$ at nodes 2 and 3, as shown. It is required to calculate the vertical deflections at the load points and, in addition, the horizontal deflection at node 5.

The truss analysis program in Appendix C gives the member forces $S = bR$ and the unit forces $\bar{b}$ at nodes 2, 3, and 5 respectively, as follows:

$$S = \begin{bmatrix} 4.095 \\ 5.056 \\ 0.0 \\ 0.0 \\ -4.999 \\ 3.276 \\ -1.120 \\ 0.672 \\ 5.652 \\ -3.033 \\ 0.0 \\ -4.116 \\ -5.081 \end{bmatrix} \qquad \bar{b} = \begin{array}{c} \begin{array}{ccc} 2 & 3 & 5 \end{array} \\ \begin{bmatrix} -0.952 & -0.476 & 0.762 \\ -0.556 & -1.111 & 0.444 \\ 0.0 & 0.0 & 0.0 \\ 0.0 & 0.0 & 0.100 \\ 1.163 & 0.581 & 0.291 \\ -0.762 & -0.381 & -0.190 \\ -0.463 & 0.740 & 0.370 \\ 0.278 & -0.444 & 0.222 \\ -0.621 & -1.242 & 0.497 \\ 0.333 & 0.667 & -0.267 \\ 0.0 & 0.0 & -1.005 \\ 0.957 & 0.479 & -0.766 \\ 0.558 & 1.117 & -0.447 \end{bmatrix} \end{array}$$

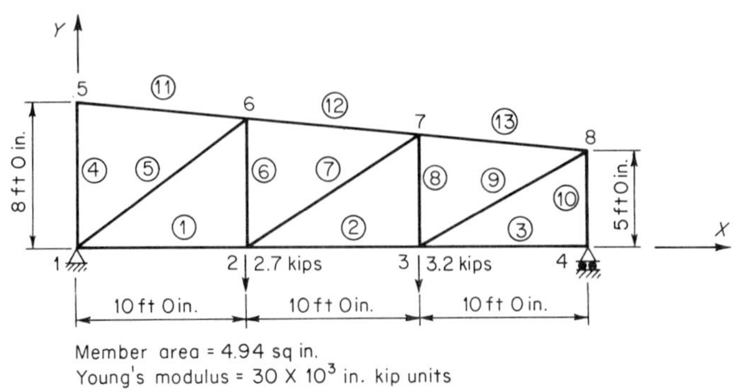

Figure 4.11 Statically determinate truss deflections.

The matrix of member flexibilities is given by

$$f = \frac{1}{30 \times 10^3 \times 4.94} \times \begin{bmatrix} 120.0 & & & & & & & & & & & & \mathbf{0} \\ & 120.0 & & & & & & & & & & & \\ & & 120.0 & & & & & & & & & & \\ & & & 96.0 & & & & & & & & & \\ & & & & 146.48 & & & & & & & & \\ & & & & & 84.0 & & & & & & & \\ & & & & & & 139.95 & & & & & & \\ & & & & & & & 72.0 & & & & & \\ & & & & & & & & 134.25 & & & & \\ & & & & & & & & & 60.0 & & & \\ & & & & & & & & & & 120.6 & & \\ & & & & & & & & & & & 120.6 & \\ \mathbf{0} & & & & & & & & & & & & 120.6 \end{bmatrix}$$

The deflection calculations, by using the matrix interpretative program, are given on pages 160 and 161.

Example 4.8 Calculation of beam deflections

The same procedure may be used for beams and frames as for truss members, except that special provision must be made for distributed loads. When calculations are made by desk calculator or by semiautomated methods, it is usual to neglect axial distortions and consider only bending effects. Example 4.8 shows how the beam type of member is used in the construction of a flexibility matrix for a cantilever beam with two concentrated loads.

The cantilever beam shown in Fig. 4.12 has forces R_1 and R_2 applied at nodes 1 and 2. The b matrix used to generate the member forces M_i, M_j is constructed from the unit-load diagrams $R_1 = 1$ and $R_2 = 1$ shown in Fig. 4.12*b* and *c*.

The calculation of F for the forces $\langle R_1 \quad R_2 \rangle$ will be carried out for both the alternative sign conventions in Fig. 4.10.

Case 1 In the first instance the sign convention given in Fig. 4.10*b* will be used for member moments. That is, anticlockwise moments on the member ends are positive.

Then member-flexibility matrices are given as in Eq. (4.96),

$$f = \frac{l}{6EI}\begin{bmatrix} 2 & -1 \\ -1 & 2 \end{bmatrix}$$

The force transformation matrix b is constructed from Fig. 4.12*b* and *c*. That is,

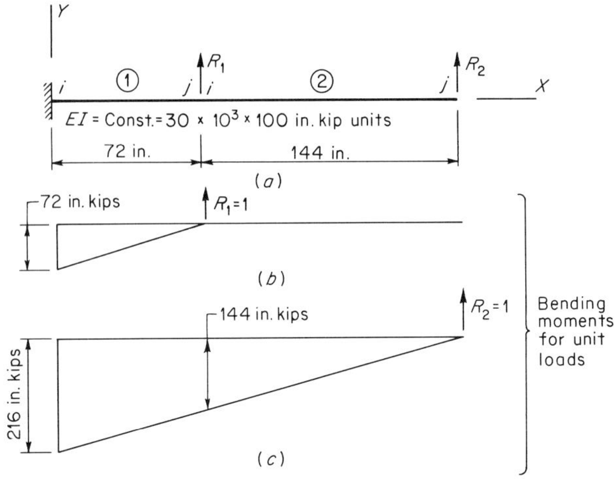

Figure 4.12 Cantilever-beam members and forces.

$$\begin{bmatrix} M_{1i} \\ M_{1j} \\ M_{2i} \\ M_{2j} \end{bmatrix} = \begin{bmatrix} -72 & -216 \\ 0 & 144 \\ 0 & -144 \\ 0 & 0 \end{bmatrix} \begin{bmatrix} R_1 \\ R_2 \end{bmatrix}$$

The matrix of member flexibilities is given by

$$f = 10^{-5} \begin{bmatrix} 0.8 & -0.4 & 0 & 0 \\ -0.4 & 0.8 & 0 & 0 \\ 0 & 0 & 1.6 & -0.8 \\ 0 & 0 & -0.8 & 1.6 \end{bmatrix}$$

The calculation of the flexibility matrix for the forces $\langle R_1 \quad R_2 \rangle$ is given below. The multiplier 10^{-5} has been omitted from the calculation.

```
START

LOAD B 4 2
     4 ROWS,        2 COLUMNS
           **********
 FORCE TRANSFORMATION MATRIX B
           **********

PRINT B

                    1                2
     1          72.00000        216.00000
     2           0.00000        144.00000
     3           0.00000        144.00000
     4           0.00000          0.00000
```

```
LOAD F 4 4
   4 ROWS,        4 COLUMNS
         **********
  FLEXIBILITY MATRIX
         **********

PRINT F

                  1              2              3              4
     1         0.80000        0.40000        0.00000        0.00000
     2         0.40000        0.80000        0.00000        0.00000
     3         0.00000        0.00000        1.60000        0.80000
     4         0.00000        0.00000        0.80000        1.60000

MULT F B C

TRANS B D

MULT D C E
         **********
  STRUCTURE FLEXIBILITY MATRIX*10(5)
         **********

PRINT E

                  1              2
     1       4147.19999    16588.80003
     2      16588.80003   111974.40038
```

Case 2 In this case, the sign convention used is that tension on the lower fibers is produced by positive moment (Fig. 4.10*a*). In the member *ij*, this may be interpreted as tension on the side nearest the viewer when the end *i* is on the left hand. Then,

$$f = \frac{l}{6EI}\begin{bmatrix} 2 & 1 \\ 1 & 2 \end{bmatrix}$$

The *b* matrix now has terms numerically equal to the previous values. However, the signs will be changed for every M_i row. That is,

$$\begin{bmatrix} M_{1i} \\ M_{1j} \\ M_{2i} \\ M_{2j} \end{bmatrix} = \begin{bmatrix} 72 & 216 \\ 0 & 144 \\ 0 & 144 \\ 0 & 0 \end{bmatrix}\begin{bmatrix} R_1 \\ R_2 \end{bmatrix}$$

The member-flexibility matrix f is now given by using Eq. (4.95),

$$f = 10^{-5}\begin{bmatrix} 0.8 & 0.4 & 0 & 0 \\ 0.4 & 0.8 & 0 & 0 \\ 0 & 0 & 1.6 & 0.8 \\ 0 & 0 & 0.8 & 1.6 \end{bmatrix}$$

The beam-flexibility calculation is given below, and it is seen that the results obtained are the same as in case 1. The important difference in the sign conventions is that in the second case, it is usual to consider deflections positive in the direction that a force acts, whereas in the b matrix obtained from equilibrium equations, it is implied that forces and displacements are positive in the directions of the positive coordinate axes. It is found that the type 2 sign convention has been the most usual in the approach to statically indeterminate analysis, although as will be shown, type 1 is quite acceptable.

```
START

LOAD B 4 2
    4 ROWS,       2 COLUMNS
         **********
 FORCE TRANSFORMATION MATRIX B
         **********

PRINT B

              1                2
    1    -72.00000       -216.00000
    2      0.00000        144.00000
    3      0.00000       -144.00000
    4      0.00000          0.00000

LOAD F 4 4
    4 ROWS,       4 COLUMNS
         **********
 FLEXIBILITY MATRIX
         **********

PRINT F

              1                2               3               4
    1      0.80000        -0.40000         0.00000         0.00000
    2     -0.40000         0.80000         0.00000         0.00000
    3      0.00000         0.00000         1.60000        -0.80000
    4      0.00000         0.00000        -0.80000         1.60000

MULT F B C

TRANS B D

MULT D C E
         **********
 STRUCTURE FLEXIBILITY MATRIX*10(5)
         **********

PRINT E

              1                2
    1     4147.20000     16588.80005
    2    16588.80003    111974.40038
```

```
START

        DEFLECTION CALCULATIONS FOR TRUSS

LOAD S 13 1
   13 ROWS,      1 COLUMNS
        MEMBER FORCES

PRINT S

              1
    1         4.09500
    2         5.05600
    3         0.00000
    4         0.00000
    5        -4.99900
    6         3.27600
    7        -1.12000
    8         0.67200
    9         5.65200
   10        -3.03300
   11         0.00000
   12        -4.11600
   13        -5.08100

LOAD F 13 13
   13 ROWS,     13 COLUMNS
        MEMBER LENGTHS=(FLEXIBILITY FOR AREA=1)

PRINT F

              1              2              3              4              5
    1     120.00000        0.00000        0.00000        0.00000        0.00000
    2       0.00000      120.00000        0.00000        0.00000        0.00000
    3       0.00000        0.00000      120.00000        0.00000        0.00000
    4       0.00000        0.00000        0.00000       96.00000        0.00000
    5       0.00000        0.00000        0.00000        0.00000      146.48000
    6       0.00000        0.00000        0.00000        0.00000        0.00000
    7       0.00000        0.00000        0.00000        0.00000        0.00000
    8       0.00000        0.00000        0.00000        0.00000        0.00000
    9       0.00000        0.00000        0.00000        0.00000        0.00000
   10       0.00000        0.00000        0.00000        0.00000        0.00000
   11       0.00000        0.00000        0.00000        0.00000        0.00000
   12       0.00000        0.00000        0.00000        0.00000        0.00000
   13       0.00000        0.00000        0.00000        0.00000        0.00000

              9             10             11             12             13
    1       0.00000        0.00000        0.00000        0.00000        0.0000
    2       0.00000        0.00000        0.00000        0.00000        0.0000
    3       0.00000        0.00000        0.00000        0.00000        0.0000
    4       0.00000        0.00000        0.00000        0.00000        0.0000
    5       0.00000        0.00000        0.00000        0.00000        0.0000
    6       0.00000        0.00000        0.00000        0.00000        0.0000
    7       0.00000        0.00000        0.00000        0.00000        0.0000
    8       0.00000        0.00000        0.00000        0.00000        0.0000
    9     134.25000        0.00000        0.00000        0.00000        0.0000
   10       0.00000       60.00000        0.00000        0.00000        0.0000
   11       0.00000        0.00000      120.60000        0.00000        0.0000
   12       0.00000        0.00000        0.00000      120.60000        0.0000
   13       0.00000        0.00000        0.00000        0.00000      120.6000

MULT F S V

LOAD BT 3 13
    3 ROWS,     13 COLUMNS
        TRANSPOSE OF FORCE TRANSFORMATION MATRIX
```

6	7	8
0.00000	0.00000	0.00000
0.00000	0.00000	0.00000
0.00000	0.00000	0.00000
0.00000	0.00000	0.00000
0.00000	0.00000	0.00000
34.00000	0.00000	0.00000
0.00000	139.95000	0.00000
0.00000	0.00000	72.00000
0.00000	0.00000	0.00000
0.00000	0.00000	0.00000
0.00000	0.00000	0.00000
0.00000	0.00000	0.00000
0.00000	0.00000	0.00000

```
PRINT BT

          1           2           3           4           5
 1    -0.95200    -0.55600     0.00000     0.00000     1.16300
 2    -0.47600    -1.11100     0.00000     0.00000     0.58100
 3     0.76200     0.44400     0.00000     0.10000     0.29100

          9          10          11          12          13
 1    -0.62100     0.33300     0.00000     0.95700     0.55800
 2    -1.24200     0.66700     0.00000     0.47900     1.11700
 3     0.49700    -0.26700    -1.00500    -0.76600    -0.44700

MULT BT V DEF

SCALE DEF .000006746
 SCALAR=   0.6746000E-05
        NODE DEFLECTIONS 2,3,5 IN INCHES

PRINT DEF

          1
 1    -0.02111
 2    -0.02403
 3     0.00937
```

Example 4.9 Deflections of a frame structure

The portal frame shown in Fig. 4.13 has a 20-kip load applied horizontally at the beam level.

Deflections are calculated for bending distortions only and also with axial distortions of members included, and the results are compared. The flexibility matrix for member (2) must be developed for the basic force M_i. The member is shown in Fig. 4.14, and the rotation corresponding to $M_i = 1$ is calculated by using the principle of virtual displacements. By integrating the moment diagram in Fig. 4.14 multiplied by itself, it is found that

$$\phi_i = \frac{A}{3EI}\left(1 + \frac{B^3}{A^3}\right) = f_i \tag{4.102}$$

From Eq. (3.62) for load R_5 alone,

$$\begin{bmatrix} M_{1i} \\ M_{1j} \\ F_1 \\ M_{2i} \\ F_2 \\ M_{3i} \\ M_{3j} \\ F_3 \end{bmatrix} = \begin{bmatrix} 0 \\ 10.0 \\ 0.4 \\ -10.0 \\ 0.5 \\ 10.0 \\ 0.0 \\ -0.4 \end{bmatrix} [R_5] \tag{i}$$

The member-flexibility matrices are given by (note $6I/A = 150.0$)

6	7	8
-0.76200	-0.46300	0.27800
-0.38100	0.74000	-0.44400
-0.19000	0.37000	-0.22200

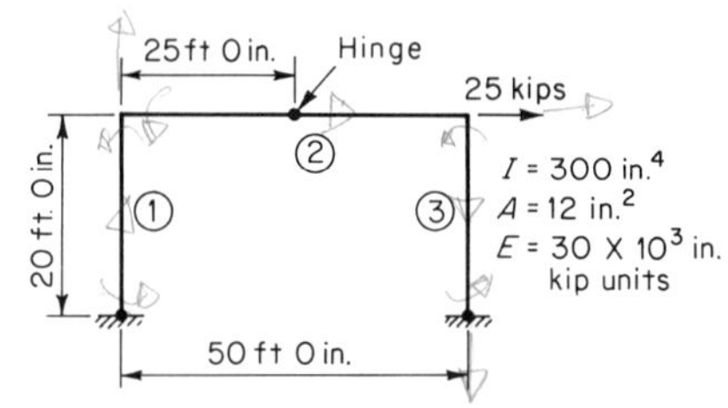

Figure 4.13

$$f_1 = f_3 = \frac{0.4}{9} \times 10^{-4} \begin{bmatrix} 2 & 1 & 0 \\ 1 & 2 & 0 \\ 0 & 0 & 150.0 \end{bmatrix}$$

$$f_2 = \frac{1}{9} \times 10^{-4} \begin{bmatrix} 2 & 0 \\ 0 & 150 \end{bmatrix}$$

Case 1 *Deflection Calculation When Axial Forces and Distortions Are Included* For $R_5 = 25k$, the member forces are given by

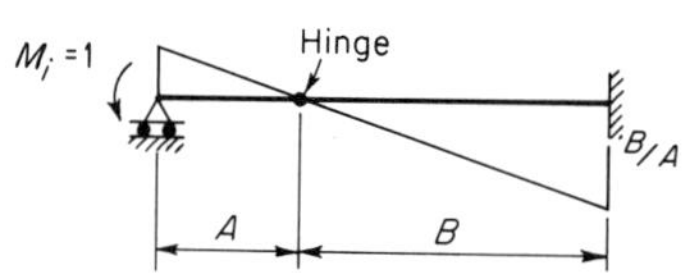

Figure 4.14 Flexibility coefficient for member with hinge.

$$S^T = \langle 0 \quad 250.0 \quad 10.0 \quad -250.0 \quad 12.5 \quad 250.0 \quad 0 \quad -10.0\rangle$$

The member-flexibility matrix f is

$$f = \frac{1}{9} \times 10^{-4} \begin{bmatrix} 0.8 & -0.4 & 0 & 0 & 0 & 0 & 0 & 0 \\ -0.4 & 0.8 & 0 & 0 & 0 & 0 & 0 & 0 \\ 0 & 0 & 60.0 & 0 & 0 & 0 & 0 & 0 \\ 0 & 0 & 0 & 2 & 0 & 0 & 0 & 0 \\ 0 & 0 & 0 & 0 & 150 & 0 & 0 & 0 \\ 0 & 0 & 0 & 0 & 0 & 0.8 & -0.4 & 0 \\ 0 & 0 & 0 & 0 & 0 & -0.4 & 0.8 & 0 \\ 0 & 0 & 0 & 0 & 0 & 0 & 0 & 60.0 \end{bmatrix}$$

The deflection calculation is coded on pages 166 and 167.

Case 2 *Deflection Calculation, Axial Distortions Neglected* Now

$$S^T = \langle 0 \quad 250.0 \quad -250.0 \quad 250.0 \quad 0\rangle$$

and

$$f = \frac{1}{9} \times 10^{-4} \begin{bmatrix} -0.8 & -0.4 & 0 & 0 & 0 \\ -0.4 & 0.8 & 0 & 0 & 0 \\ 0 & 0 & 2 & 0 & 0 \\ 0 & 0 & 0 & 0.8 & -0.4 \\ 0 & 0 & 0 & -0.4 & 0.8 \end{bmatrix}$$

Again, the deflection calculation is coded on page 168.

Example 4.10 Deflections of a structure including bending and torsion flexibilities

In this example, deflections are calculated for loads on a simple three-dimensional structure. The structure is a cantilever composed of a ¼-in. mild-steel rod with two 90° bends, as shown in Fig. 4.15*a*. The members (1), (2), and (3) are parallel with the X, Z, and Y axes respectively. Member properties must be described in terms of the Y, Z axes in the member cross sections. In this case, for the circular cross section, $I_z = I_y$ and so the member axes may be chosen to be always parallel with the global axes since member directions are parallel to one of these in every case.

Deflections will be calculated at the end of the cantilever for a force, on the end, whose components in the directions X, Y, Z are each equal to 1 lb. For each member there will be four bending moments $(M_i, M_j)_z$ about axis OZ' for the member and $(M_i, M_j)_y$ about axis OY' for the member. Positive values are shown in Fig. 4.15. The b matrix is then constructed from the unit-load diagrams shown in Fig. 4.15*c*.

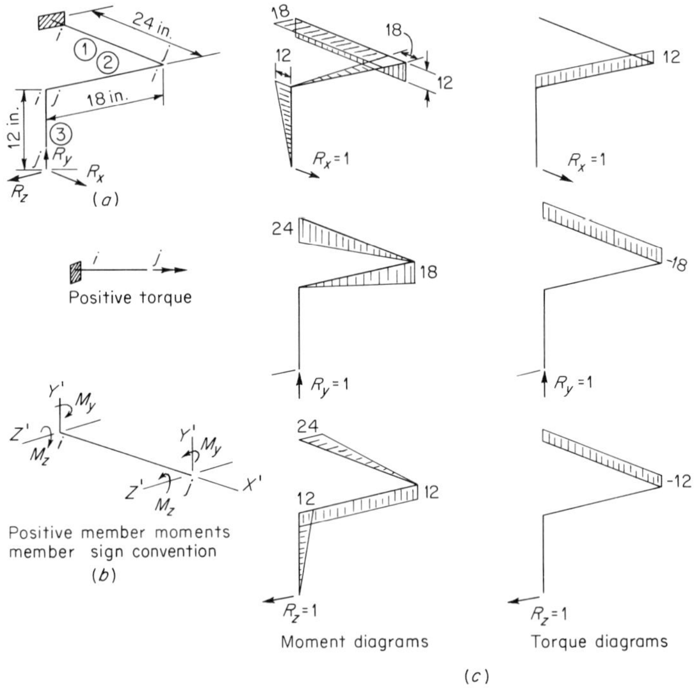

Figure 4.15

$$
\begin{bmatrix} M_{1iz} \\ M_{1jz} \\ M_{1iy} \\ M_{1jy} \\ T_1 \\ M_{2iz} \\ M_{2jz} \\ M_{2iy} \\ M_{2jy} \\ T_2 \\ M_{3iz} \\ M_{3jz} \\ M_{3iy} \\ M_{3jy} \\ T_3 \end{bmatrix} = \begin{bmatrix} 12 & 24 & 0 \\ 12 & 0 & 0 \\ 18 & 0 & -24 \\ 18 & 0 & 0 \\ 0 & -18 & -12 \\ 0 & 18 & 12 \\ 0 & 0 & 12 \\ 18 & 0 & 0 \\ 0 & 0 & 0 \\ 12 & 0 & 0 \\ 12 & 0 & 0 \\ 0 & 0 & 0 \\ 0 & 0 & -12 \\ 0 & 0 & 0 \\ 0 & 0 & 0 \end{bmatrix} \begin{bmatrix} R_X \\ R_Y \\ R_Z \end{bmatrix}
$$

```
START
    CALCULATION OF PORTAL FRAME DEFLECTIONS

LOAD S 8 1
    8 ROWS,      1 COLUMNS

PRINT S  * MEMBER FORCES

              1
    1          0.00000
    2        250.00000
    3         10.00000
    4       -250.00000
    5         12.50000
    6        250.00000
    7          0.00000
    8        -10.00000

LOAD F 8 8
    8 ROWS,      8 COLUMNS

PRINT F  *  MEMBER FLEXIBILITIES

              1              2              3              4              5
    1          0.80000       -0.40000        0.00000        0.00000        0.00000
    2         -0.40000        0.80000        0.00000        0.00000        0.00000
    3          0.00000        0.00000       60.00000        0.00000        0.00000
    4          0.00000        0.00000        0.00000        2.00000        0.00000
    5          0.00000        0.00000        0.00000        0.00000      150.00000
    6          0.00000        0.00000        0.00000        0.00000        0.00000
    7          0.00000        0.00000        0.00000        0.00000        0.00000
    8          0.00000        0.00000        0.00000        0.00000        0.00000

LOAD B 1 8
    1 ROWS,      8 COLUMNS

PRINT B  *  FORCE TRANSFORMATION MATRIX B[TRANSPOSE]

              1              2              3              4              5
    1          0.10000       10.00000        0.40000      -10.00000        0.50000

MULT F S D

MULT B D R

SCALE R 0.11111
 SCALAR=   0.1111100E+00

SCALE R .0001
 SCALAR=   0.1000000E-03

PRINT R  * DEFLECTIONS

              1
    1          0.11575
```

6	7	8
0.00000	0.00000	0.00000
0.00000	0.00000	0.00000
0.00000	0.00000	0.00000
0.00000	0.00000	0.00000
0.00000	0.00000	0.00000
0.80000	-0.40000	0.00000
-0.40000	0.80000	0.00000
0.00000	0.00000	60.00000

6	7	8
10.00000	0.00000	-0.40000

```
START

LOAD S 5 1
    5 ROWS,       1 COLUMNS

PRINT S MEMBER FORCES MOMENTS ONLY

               1
    1        0.00000
    2      250.00000
    3     -250.00000
    4      250.00000
    5        0.00000

LOAD F 5 5
    5 ROWS,       5 COLUMNS

PRINT F *  MEMBER FLEXIBILITIES MOMENTS ONLY

               1             2             3             4             5
    1        0.80000      -0.40000       0.00000       0.00000       0.00000
    2       -0.40000       0.80000       0.00000       0.00000       0.00000
    3        0.00000       0.00000       2.00000       0.00000       0.00000
    4        0.00000       0.00000       0.00000       0.80000      -0.40000
    5        0.00000       0.00000       0.00000      -0.40000       0.80000

LOAD B 1 5
    1 ROWS,       5 COLUMNS

PRINT B  * FORCE TRANSFORMATION MATRIX B(TRANSPOSE)

               1             2             3             4             5
    1        0.00000      10.00000     -10.00000      10.00000       0.00000

MULT F S D

MULT B D R

SCALE R 0.11111
 SCALAR=  0.1111100E+00

SCALE R .0001
 SCALAR=  0.1000000E-03

PRINT R  *DEFLECTION NO AXIAL STRAINS

               1
    1        0.10000
```

The member-flexibility matrices are now 5 × 5, having four moment terms and one torque term. Since the I and I_P values are the same for all members, the member flexibilities will vary according to their lengths. For the ¼-in.-diameter rod, $I = 0.00019$ and $I_P = 0.00038$. Use $E = 30 \times 10^6$ in.-lb units and $\nu = 0.25$.

$$f_1 = \frac{24}{6EI}\begin{bmatrix} 2 & 1 & 0 & 0 & 0 \\ 1 & 2 & 0 & 0 & 0 \\ 0 & 0 & 2 & 1 & 0 \\ 0 & 0 & 1 & 2 & 0 \\ 0 & 0 & 0 & 0 & \dfrac{6EI}{GI_P} \end{bmatrix} = \frac{1}{19 \times 10^2}\begin{bmatrix} 2.6667 & 1.3333 & 0 & 0 & 0 \\ 1.3333 & 2.6667 & 0 & 0 & 0 \\ 0 & 0 & 2.6667 & 1.3333 & 0 \\ 0 & 0 & 1.3333 & 2.6667 & 0 \\ 0 & 0 & 0 & 0 & 10.0 \end{bmatrix}$$

Then, $f_2 = {}^{18}\!/_{24} f_1$, so that

$$f_2 = \frac{1}{19 \times 10^2} \begin{bmatrix} 2 & 1 & 0 & 0 & 0 \\ 1 & 2 & 0 & 0 & 0 \\ 0 & 0 & 2 & 1 & 0 \\ 0 & 0 & 1 & 2 & 0 \\ 0 & 0 & 0 & 0 & 7.5 \end{bmatrix}$$

Also $f_3 = {}^{12}\!/_{24} f_1$, so that

$$f_3 = \frac{1}{19 \times 10^2} \begin{bmatrix} 1.3333 & 0.6667 & 0 & 0 & 0 \\ 0.6667 & 1.3333 & 0 & 0 & 0 \\ 0 & 0 & 1.3333 & 0.6667 & 0 \\ 0 & 0 & 0.6667 & 1.3333 & 0 \\ 0 & 0 & 0 & 0 & 5.0 \end{bmatrix}$$

The deflection calculations are listed on pages 172 to 174, the final answers being in inches.

Foundation Settlement

Foundation settlement in a statically determinate structure causes rigid-body movements of the structure. If reactions S_R have been calculated from the equation

$$S_R = b_R R \tag{4.103}$$

then node deflections are obtained from the foundation movements v_R by using the contragradient law. That is,

$$v = -b_R{}^T v_R \tag{4.104}$$

The negative sign occurs because the S_R are the equilibrium forces on the structure.

The simple beam in Fig. 4.16 has load nodes 1, 2 as shown, so that the reactions are given by

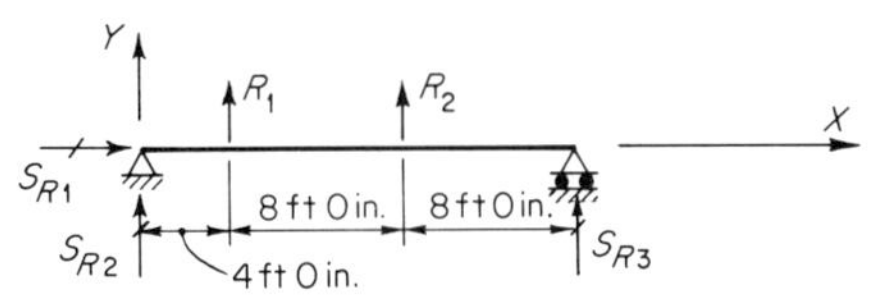

Figure 4.16 Foundation settlement of beam.

$$\begin{bmatrix} S_{R1} \\ S_{R2} \\ S_{R3} \end{bmatrix} = \begin{bmatrix} 0 & 0 \\ -0.8 & -0.2 \\ -0.2 & -0.6 \end{bmatrix} \begin{bmatrix} R_1 \\ R_2 \end{bmatrix}$$

If, then, v_{R3} settles by -2 in., the node deflections are given by

$$\begin{bmatrix} r_1 \\ r_2 \end{bmatrix} = -\begin{bmatrix} 0 & -0.8 & -0.2 \\ 0 & -0.2 & -0.6 \end{bmatrix} \begin{bmatrix} 0 \\ 0 \\ -2 \end{bmatrix} = \begin{bmatrix} -0.4 \\ -1.2 \end{bmatrix} \text{in.}$$

4.6 NATURAL FORCES AND DISPLACEMENTS; UNCOUPLED FLEXIBILITY ELASTIC CENTER

In Sec. 4.3, Example 4.6, it has been shown that when the deflections of a beam are calculated in terms of moment components expressed in the principal axes, these deflections are uncoupled. Thus a moment $\overline{M}_Z$ causes rotation $\bar{\theta}_z$ only, and thence $\bar{\theta}_y = 0$. It is useful to enquire if the process of diagonalization can be carried out for the general member- or structure-flexibility matrix. It was seen in Example 4.6, that the process of diagonalization can be achieved by calculating the matrix of eigenvectors.

The member force-deflection relationship is given by

$$v = fS \tag{4.105}$$

That is, in full for f, a 3×3 matrix,

$$\begin{bmatrix} v_1 \\ v_2 \\ v_3 \end{bmatrix} = \begin{bmatrix} f_{11} & f_{12} & f_{13} \\ f_{21} & f_{22} & f_{23} \\ f_{31} & f_{32} & f_{33} \end{bmatrix} \begin{bmatrix} S_1 \\ S_2 \\ S_3 \end{bmatrix} \tag{4.106}$$

The off-diagonal terms f_{ij}, $i \neq j$, are the cross-flexibility coefficients in the element. The enquiry then is into the conditions for which these off-diagonal terms may be made to vanish. That is, $f_{ij} = 0$ for $i \neq j$.

Let U be the eigenvectors of f. That is, solutions of Eq. (4.107)

$$fX = \lambda X \tag{4.107}$$

Then let $S = Us$ be a force transformation that is obviously reversible. By using the contragradient law, the displacements corresponding to s will be given by

$$\zeta = U^T v \tag{4.108}$$

Solving Eq. (4.108) for v,

$$v = (U^T)^{-1}\zeta \tag{4.109}$$

Substitute for v and S in Eq. (4.105),

$$(U^T)^{-1}\zeta = f\,Us$$

Hence,

$$\zeta = U^T f\,Us \tag{4.110}$$

The flexibility matrix

$$f = U^T f\,U \tag{4.111}$$

is a diagonal matrix. That is, from (4.107),

$$fU = U\lambda$$

Premultiply by U^T,

$$U^T f\,U = \lambda U^T U = I\lambda \tag{4.112}$$

From Eq. (4.112) it is seen that $f = \lambda$ with eigenvectors as diagonal terms. The forces $s = U^{-1}S$ will be the natural forces for the member or structure. The natural forces and displacements may now be calculated for the simple beam element (Fig. 4.17).

For the beam,

$$f = \frac{l}{6EI}\begin{bmatrix} 2 & 1 \\ 1 & 2 \end{bmatrix}$$

Equation (4.107) is now

$$\frac{l}{6EI}\begin{bmatrix} 2 & 1 \\ 1 & 2 \end{bmatrix} X = \lambda X$$

Hence the determinantal equation is

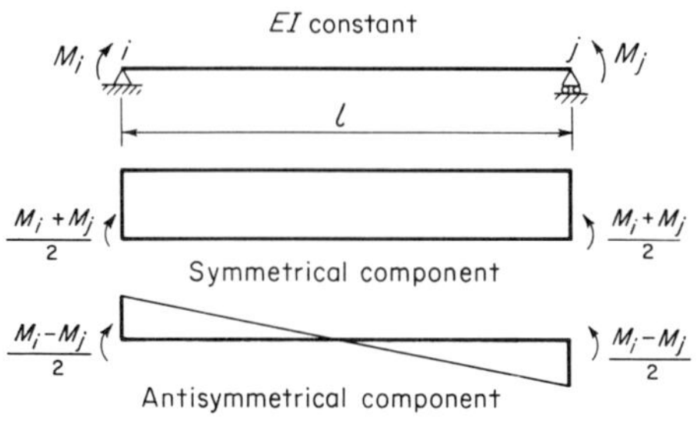

Figure 4.17 Beam element—natural-force components.

```
START
     DEFLECTION OF STRUCTURE INCLUDING TORSION AND BENDING DISTORTIONS

LOAD B 15 3
   15 ROWS,      3 COLUMNS

PRINT B   * FORCE TRANSFORMATION MATRIX

                 1              2              3
     1       12,00000       24,00000        0,00000
     2       12,00000        0,00000        0,00000
     3       18,00000        0,00000      -24,00000
     4       18,00000        0,00000        0,00000
     5        0,00000      -18,00000      -12,00000
     6        0,00000       18,00000       12,00000
     7        0.00000        0,00000       12,00000
     8       18,00000        0,00000        0,00000
     9        0,00000        0,00000        0,00000
    10       12,00000        0,00000        0,00000
    11       12,00000        0,00000        0,00000
    12        0,00000        0,00000        0,00000
    13        0,00000        0,00000      -12,00000
    14        0,00000        0.00000        0,00000
    15        0,00000        0,00000        0,00000

LOAD R 3 1
    3 ROWS,      1 COLUMNS

PRINT R    *  LOAD VECTOR

                 1
     1        1,00000
     2        1,00000
     3        1,00000

MULT B R S

LOAD F 15 15
   15 ROWS,     15 COLUMNS

PRINT F   * MEMBER FLEXIBILITY MATRIX

                 1              2              3              4              5
     1        2,00000        1,00000        0,00000        0,00000        0,00000
     2        1,00000        2,00000        0,00000        0,00000        0.00000
     3        0,00000        0,00000        2,00000        1,00000        0,00000
     4        0,00000        0,00000        1,00000        2,00000        0,00000
     5        0,00000        0,00000        0,00000        0,00000        7,50000
     6        0,00000        0,00000        0,00000        0,00000        0,00000
     7        0,00000        0,00000        0,00000        0,00000        0,00000
     8        0,00000        0,00000        0,00000        0.00000        0,00000
     9        0,00000        0,00000        0,00000        0,00000        0,0000
    10        0,00000        0,00000        0,00000        0,00000        0,0000
    11        0,00000        0,00000        0,00000        0,00000        0,0000
    12        0,00000        0,00000        0,00000        0,00000        0,0000
    13        0,00000        0,00000        0,00000        0,00000        0,0000
    14        0.00000        0,00000        0,00000        0,00000        0,0000
    15        0,00000        0,00000        0,00000        0.00000        0,0000
```

	7	8
.00000	0.00000	0.00000
.00000	0.00000	0.00000
.00000	0.00000	0.00000
.00000	0.00000	0.00000
.00000	0.00000	0.00000
.66667	1.33330	0.00000
.33330	2.66667	0.00000
.00000	0.00000	2.66667
.00000	0.00000	1.33330
.00000	0.00000	0.00000
.00000	0.00000	0.00000
.00000	0.00000	0.00000
.00000	0.00000	0.00000
.00000	0.00000	0.00000
.00000	0.00000	0.00000

```
              9           10          11          12          13
     1    0,00000     0,00000     0,00000     0,00000     0,00000
     2    0,00000     0,00000     0,00000     0,00000     0,00000
     3    0,00000     0,00000     0,00000     0,00000     0,00000
     4    0,00000     0,00000     0,00000     0,00000     0,00000
     5    0,00000     0,00000     0,00000     0,00000     0,00000
     6    0,00000     0,00000     0,00000     0,00000     0,00000
     7    0,00000     0,00000     0,00000     0,00000     0,00000
     8    1,33330     0,00000     0,00000     0,00000     0,00000
     9    2,66667     0,00000     0,00000     0,00000     0,00000
    10    0,00000    10,00000     0,00000     0,00000     0,00000
    11    0,00000     0,00000     1,33330     0,66670     0,00000
    12    0,00000     0,00000     0,66670     1,33330     0,00000
    13    0,00000     0,00000     0,00000     0,00000     1,33330
    14    0,00000     0,00000     0,00000     0,00000     0,66670
    15    0,00000     0,00000     0,00000     0,00000     0,00000

 MULT F S V

 TRANS B BT

 MULT BT V DEF

 SCALE DEF 0,000526
  SCALAR=  0,5260000E-03

 PRINT DEF   * DEFLECTIONS OF FREE END (INCHES) GLOBAL COORDINATES

              1
     1    2,56267
     2    4,09964
     3    2,50585

ALL CARDS PROCESSED  END OF JOB
```

$$\begin{vmatrix} 2-\lambda & 1 \\ 1 & 2-\lambda \end{vmatrix} = 0$$

On expansion, the resulting quadratic equation is given by

$$\lambda^2 - 4\lambda + 3 = 0$$

Hence,

$$\lambda_1 = 3 \qquad \lambda_2 = 1$$

The matrix is

$$U = \begin{bmatrix} 1 & 1 \\ 1 & -1 \end{bmatrix}$$

Substitute in Eq. (4.111)

$$\bar{f} = \frac{l}{6EI}\begin{bmatrix} 1 & 1 \\ 1 & -1 \end{bmatrix}\begin{bmatrix} 2 & 1 \\ 1 & 2 \end{bmatrix}\begin{bmatrix} 1 & 1 \\ 1 & -1 \end{bmatrix} = \frac{l}{6EI}\begin{bmatrix} 6 & 0 \\ 0 & 2 \end{bmatrix}$$

14	15
0,00000	0.00000
0,00000	0.00000
0,00000	0.00000
0.00000	0.00000
0,00000	0.00000
0,00000	0.00000
0,00000	0.00000
0,00000	0.00000
0,00000	0.00000
0,00000	0.00000
0,00000	0.00000
0,00000	0.00000
0,66670	0.00000
1,33330	0.00000
0,00000	5,00000

Finally,

$$\bar{f} = \frac{l}{3EI}\begin{bmatrix} 3 & 0 \\ 0 & 1 \end{bmatrix} \tag{4.113}$$

It remains now to give the physical interpretation of the forces s.

$$s = \begin{bmatrix} s_1 \\ s_2 \end{bmatrix} = U^{-1}S = \tfrac{1}{2}\begin{bmatrix} 1 & 1 \\ 1 & -1 \end{bmatrix}\begin{bmatrix} M_i \\ M_j \end{bmatrix}$$

Then,

$$\begin{aligned} s_1 &= \tfrac{1}{2}(M_1 + M_2) \\ s_1 &= \tfrac{1}{2}(M_1 - M_2) \end{aligned} \tag{4.114}$$

The distortion terms are

$$\begin{bmatrix} \zeta_1 \\ \zeta_2 \end{bmatrix} = U^T v = \begin{bmatrix} 1 & 1 \\ 1 & -1 \end{bmatrix}\begin{bmatrix} \phi_i \\ \phi_j \end{bmatrix} = \begin{bmatrix} \phi_i + \phi_j \\ \phi_i - \phi_j \end{bmatrix} \tag{4.115}$$

From (4.114) and (4.115), it is seen that the moments and distortions are the symmetric and antisymmetric components of the force and displacement fields, as shown in Fig. 4.17.

The Elastic Center

The idea of the elastic center arises in the consideration of the flexibility of single-bay plane frames. The cantilever shown in Fig. 4.18 lies in the XY plane and is loaded with force components $\langle R_x \quad R_y \quad M_z \rangle$ at the free end, which is taken to be the origin of coordinates. The 3×3 flexibility matrix for the cantilever can be calculated at O for the forces $\langle R_x \quad R_y \quad M_z \rangle$ for bending distortions only by using Eq. (v) in Example 4.3 and Eq. (4.47). That is, for the element ds in Fig. 4.18,

$$\delta\theta = \frac{ds}{EI} M \qquad f = \frac{ds}{EI} \tag{4.116}$$

$$F = \int_0^l b^T f b \, ds \tag{4.117}$$

If moment-causing tension on the lower fiber is positive, then for the axes shown, the bending moments for unit values of the forces $\langle R_x \quad R_y \quad M_z \rangle$ are, from Fig. 4.18,

$$m_x = y \qquad m_y = -x \qquad m_M = 1$$

That is,

$$m = [y \quad -x \quad 1] \begin{bmatrix} R_x \\ R_y \\ M_z \end{bmatrix} \tag{4.118}$$

This is the equation $S = bR$. Substitute then in Eq. (4.117) to obtain F.

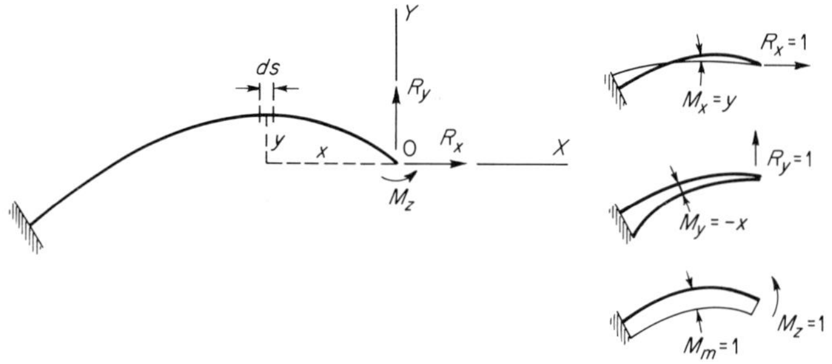

Figure 4.18 Elastic center of curved-member flexibilities.

$$F = \int_o^l \frac{1}{EI} \begin{bmatrix} y \\ -x \\ 1 \end{bmatrix} [y \quad -x \quad 1]\, ds$$

That is,

$$F = \int_o^l \frac{1}{EI} \begin{bmatrix} y^2 & -xy & y \\ -xy & x^2 & -x \\ y & -x & 1 \end{bmatrix} ds \tag{4.119}$$

Equation (4.119) is an interesting result because it shows similarity with Eq. (2.73)ff. In fact the first 2×2 submatrix is the moment of inertia of a cross section with the outline of the cantilever and with thickness everywhere equal to $1/EI$ of the cantilever. Furthermore, the terms $\int(1/EI)\, y\, ds$ and $\int(-x/EI)\, ds$ are the first moments of this cross section about the axes OX and OY. It is evident that these integrals are zero if O is the centroid of the cross section. This point is called the *elastic center*. Evidently, the area of the analogous section A is given by

$$A = \int_o^l \frac{1}{EI}\, ds \tag{4.120}$$

The coordinates $\overline{O}$, the elastic center, are

$$\overline{x} = \frac{\int_o^l (x/EI)\, ds}{A} \qquad \overline{y} = \frac{\int_o^l (y/EI)\, ds}{A} \tag{4.121}$$

Let $\overline{F}$ be the flexibility matrix referred to the elastic center. That is, extend a rigid arm from O to $\overline{O}$, and place forces at $\overline{O}$ instead of at O (see Fig. 4.19). Then,

$$\overline{F} = \int_o^l \frac{1}{EI} \begin{bmatrix} \overline{y}^2 & -\overline{x}\overline{y} & 0 \\ -\overline{x}\overline{y} & \overline{x}^2 & 0 \\ 0 & 0 & 1 \end{bmatrix} ds \tag{4.122}$$

The flexibility matrix is further simplified by a rotation of axes to $\overline{\overline{x}}$, $\overline{\overline{y}}$ (Fig. 4.19) through the angle θ.

By analogy with Eq. (2.84)

$$\tan 2\theta = \frac{2\overline{I}_{yx}}{\overline{I}_{yy} - \overline{I}_{xx}} \tag{4.123}$$

where the $\overline{I}$ values refer to the $\overline{OX}, \overline{OY}$ axes. Then from Eq. (2.78), the

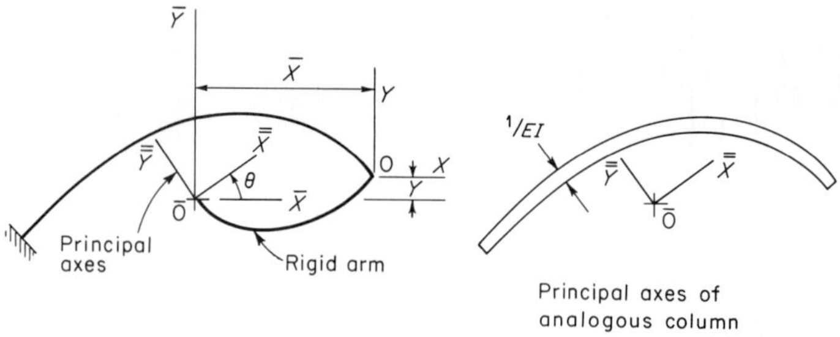

Figure 4.19 Elastic center.

flexibilities in the principal axes are given as

$$\bar{\bar{F}} = \int_o^l \frac{1}{EI} \begin{bmatrix} \bar{\bar{y}}^2 & 0 & 0 \\ 0 & \bar{\bar{x}}^2 & 0 \\ 0 & 0 & 1 \end{bmatrix} ds = L^T \bar{F} L \tag{4.124}$$

In Eq. (4.124),

$$L = \begin{bmatrix} \cos\theta & -\sin\theta & 0 \\ \sin\theta & \cos\theta & 0 \\ 0 & 0 & 1 \end{bmatrix} \tag{4.125}$$

It should be noted that the idea of elastic center is a useful one when frames and arches are being analyzed by hand methods but loses appeal in computer methods.

Consider now the simple beam in Fig. 4.17, shown as an analogous column in Fig. 4.20. By inspection it is seen that the elastic center must be at the beam center at $l/2$ from the end and that $\bar{x}$ is parallel with the beam axis. In addition, R_x causes zero moments and can be omitted. Then the flexibility for R_y is given as

$$f_{yy} = \int_{-l/2}^{+l/2} \frac{1}{EI} y^2\, dx = \frac{l^3}{12EI}$$

and for M_z,

$$f_M = \int_{-l/2}^{+l/2} \frac{1}{EI}\, dx = \frac{l}{EI}$$

From Fig. 4.20, it will be seen that R_y gives the antisymmetric-moment diagram and M_z the symmetric-moment component. The term f_{yy} is the flexibility for a moment of $l/2$ at the ends of the beam. That is, if $R_y = 2/l$,

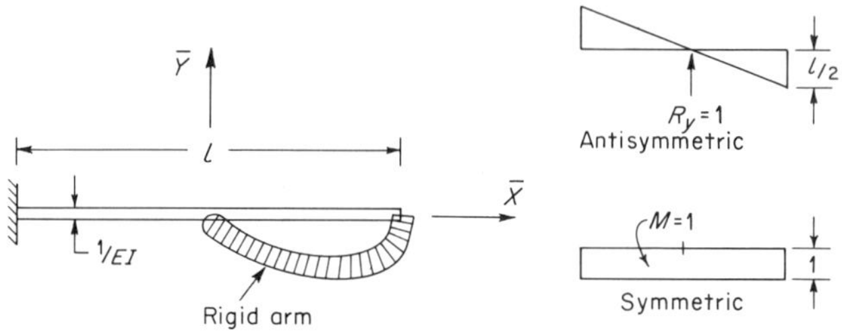

Figure 4.20 Analogous column for beam.

$$f'_{yy} = \frac{l^3}{12EI}\frac{4}{l^2} = \frac{l}{3EI}$$

Then f'_{yy} agrees with that for $s_1 = \tfrac{1}{2}(M_1 - M_2)$, Eq. (4.113).

4.7 GENERALIZED FORCES AND GENERALIZED DISPLACEMENTS

In Sec. 4.6, it has been found advantageous to treat forces in groups rather than as single quantities. That is, the linear variation of moments along the beam could be described in terms of two generalized or natural forces $\langle s_1 \quad s_2 \rangle$, where $s_1 = \tfrac{1}{2}(M_i + M_j)$ and $s_2 = \tfrac{1}{2}(M_i - M_j)$. This idea of describing a force system as composed of combinations of groups of forces rather than of single forces has many uses and will be discussed further in this section.

As the first example, the beam shown in Fig. 4.21*a* is loaded at points 1 and 2 with loads $\langle R_1 \quad R_2 \rangle$. The node displacements will then also be described by the discrete deflection values $\langle r_1 \quad r_2 \rangle$. In Fig. 4.21*b* and *c*, it is seen that the same loading is produced by the symmetric and antisymmetric load patterns $\langle Q_1 \quad Q_2 \rangle$, where Q_1 and Q_2 are related to R_1 and R_2 by Eq. (*i*).

$$\begin{bmatrix} Q_1 \\ Q_2 \end{bmatrix} = \begin{bmatrix} \tfrac{1}{2} & \tfrac{1}{2} \\ \tfrac{1}{2} & -\tfrac{1}{2} \end{bmatrix} \begin{bmatrix} R_1 \\ R_2 \end{bmatrix} \qquad (i)$$

The simple beam with end moments M_i and M_j and the uniformly distributed load w per unit length in Fig. 4.22 has three independent forces. These may be described in several ways. For example, $W = wl$ is the total load on the span, and let M_c be the bending moment at the beam center. The forces on the beam may be described by any one of the sets of generalized forces, $\langle M_i \quad w \quad M_j \rangle$, $\langle M_i \quad W \quad M_j \rangle$, or $\langle M_i \quad M_c \quad M_j \rangle$.

Given the generalized forces $\langle Q_1 \quad Q_2 \quad \cdots \quad Q_m \rangle$, it is possible to obtain a meaningful interpretation of the generalized displacements $\langle q_1 \quad q_2 \quad \cdots \quad q_m \rangle$ and to derive the flexibility relationship between Q and q.

Consider the force system R and displacements r with the known flexibility

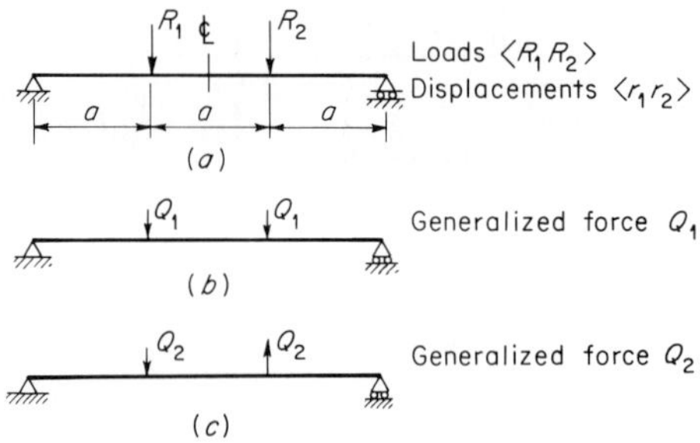

Figure 4.21 Symmetric and antisymmetric components.

relationship

$$r = fR \tag{4.126}$$

The generalized forces Q and generalized displacements q also describe the same force and displacement systems for the structure or member. Since Q describes the same force system as R, it can be related to R by the reversible transformation

$$R = BQ \tag{4.127}$$

and hence, also,

$$Q = B^{-1}R \tag{4.128}$$

Then the displacements q are related to those r, the relationship being given by the contragradient law so that

$$q = B^T r \tag{4.129}$$

The flexibility matrix for the Q forces may be obtained from Eq. (4.126) by substitution for $\langle r \quad R \rangle$ in terms of $\langle q \quad Q \rangle$. Thus,

$$q = B^T r = B^T f R = B^T f B Q$$

The flexibility matrix F_Q for the Q forces has been obtained,

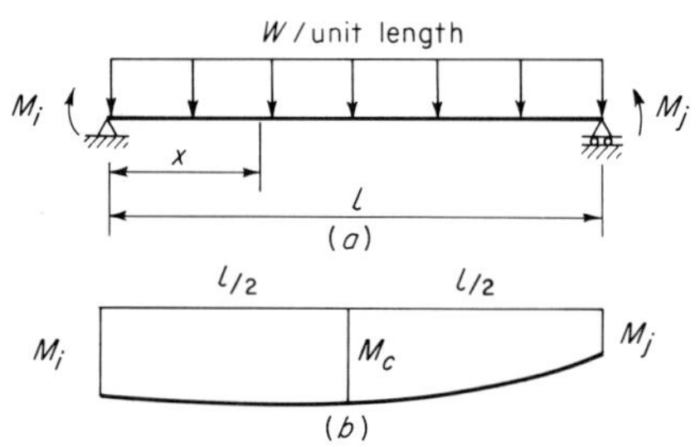

Figure 4.22 Beam-element uniform load.

$$F_Q = B^T f B \tag{4.130}$$

Example 4.11 The flexibility relationship for the beam in Fig. 4.21a is

$$\begin{bmatrix} r_1 \\ r_2 \end{bmatrix} = \frac{a^3}{54EI}\begin{bmatrix} 24 & 21 \\ 21 & 24 \end{bmatrix}\begin{bmatrix} R_1 \\ R_2 \end{bmatrix}$$

From Eq. (i), Sec. 4.7,

$$B = \tfrac{1}{2}\begin{bmatrix} 1 & 1 \\ 1 & -1 \end{bmatrix}$$

Therefore from Eq. (4.130),

$$F_Q = \frac{a^3}{72EI}\begin{bmatrix} 1 & 1 \\ 1 & -1 \end{bmatrix}\begin{bmatrix} 8 & 7 \\ 7 & 8 \end{bmatrix}\begin{bmatrix} 1 & 1 \\ 1 & -1 \end{bmatrix}$$

Therefore,

$$F_Q = \frac{a^3}{72EI}\begin{bmatrix} 30 & 0 \\ 0 & 2 \end{bmatrix}$$

Notice that

$$q_1 = \frac{5a^3}{12EI}Q_1$$

$$q_2 = \frac{a^3}{36EI}Q_2$$

Example 4.12 The simple beam shown in Fig. 4.22 will be analyzed with generalized forces $\langle M_i \quad w \quad M_j \rangle$. The generalized displacements will be calculated from Castigliano's second theorem

$$\frac{\partial U}{\partial Q_i} = q_i$$

For the member, the bending moment at section x is given in terms of M_i, M_j, and w,

$$M_x = \frac{wx}{2}(l - x) + M_i\left(1 - \frac{x}{l}\right) + M_j\frac{x}{l}$$

The strain energy due to bending is

$$U = \tfrac{1}{2}\int_o^l \frac{M_x^2}{EI}\,dx$$

Differentiate U with respect to Q_i so that

$$\frac{\partial U}{\partial Q_i} = \int_o^l \frac{M_x}{EI}\frac{\partial M_x}{\partial Q_i}\,dx$$

In this case, $\langle Q_1 \quad Q_2 \quad Q_3 \rangle \equiv \langle M_i \quad w \quad M_j \rangle$, hence

$$q_{M_i} = \frac{\partial U}{\partial M_i} = \frac{1}{EI}\int_o^l \left(1 - \frac{x}{l}\right)\left[\frac{wx}{2}(l-x) + M_i\left(1 - \frac{x}{l}\right) + M_j\frac{x}{l}\right]dx$$
$$= \frac{wl^3}{24EI} + \frac{M_il}{3EI} + \frac{M_jl}{6EI}$$

Similarly,

$$q_w = \frac{\partial U}{\partial w} = \frac{1}{EI}\int_o^l \frac{x(l-x)}{2}\left[\frac{wx}{2}(l-x) + M_i\left(1 - \frac{x}{l}\right) + M_j\frac{x}{l}\right]dx$$
$$= \frac{wl^5}{120EI} + \frac{M_il^3}{24EI} + \frac{M_jl^3}{24EI}$$

and

$$q_{M_j} = \frac{\partial U}{\partial M_j} = \frac{1}{EI}\int_o^l \frac{x(l-x)}{2}\left[\frac{wx}{2}(l-x) + M_i\left(1 - \frac{x}{l}\right) + M_j\frac{x}{l}\right]dx$$
$$= \frac{wl^3}{24EI} + \frac{M_il}{6EI} + \frac{M_jl}{3EI}$$

In matrix form,

$$\begin{bmatrix} q_{M_i} \\ q_w \\ q_{M_j} \end{bmatrix} = \frac{l}{6EI}\begin{bmatrix} 2 & \frac{l^2}{4} & 1 \\ \frac{l^2}{4} & \frac{l^4}{20} & \frac{l^2}{4} \\ 1 & \frac{l^2}{4} & 2 \end{bmatrix}\begin{bmatrix} M_i \\ w \\ M_j \end{bmatrix} \tag{4.131}$$

The interpretation of Eq. (4.131) is easily undertaken, since the moment terms $\begin{bmatrix} 2 & \cdots & 1 \\ \cdots & \cdots & \cdots \\ 1 & \cdots & 2 \end{bmatrix}$ are simply the linear terms in the 2×2 moment-flexibility matrix and the terms $l^3/(24EI)$ are the end rotations due to w. The term $l^5/(120EI)$ is the average deflection in the span multiplied by l.

Suppose now that (M_i, M_c, M_j) are expressed as in Eq. (4.128) by a linear transformation from (M_i, w, M_j). That is,

$$\begin{bmatrix} M_i \\ M_c \\ M_j \end{bmatrix} = \begin{bmatrix} 1 & 0 & 0 \\ \frac{1}{2} & \frac{l^2}{8} & \frac{1}{2} \\ 0 & 0 & 1 \end{bmatrix}\begin{bmatrix} M_i \\ w \\ M_j \end{bmatrix} \tag{4.132}$$

Then,

$$B^{-1} = \begin{bmatrix} 1 & 0 & 0 \\ \frac{1}{2} & \frac{l^2}{8} & \frac{1}{2} \\ 0 & 0 & 1 \end{bmatrix}$$

On inversion,

$$B = \begin{bmatrix} 1 & 0 & 0 \\ -\frac{4}{l^2} & \frac{8}{l^2} & -\frac{4}{l^2} \\ 0 & 0 & 1 \end{bmatrix}$$

The generalized flexibility matrix F_Q for (M_i, M_c, M_j) is then given as

$$F_Q = B^T F B = \frac{l}{30EI} \begin{bmatrix} 4 & 2 & -1 \\ 2 & 16 & 2 \\ -1 & 2 & 4 \end{bmatrix} \tag{4.133}$$

The generalized flexibility matrices given in Eqs. (4.131) to (4.133) may be used for the calculation of node deflections of structures with uniform loading between nodes.

Example 4.13 Cantilever-beam deflection

The beam in Fig. 4.23 has uniform EI and is loaded uniformly with w per length. The deflection of the free end is calculated by using the equation

$$r = \bar{b}^T f b R$$

The generalized member forces may be expressed in terms of the single force quantity w.

$$\begin{bmatrix} M_i \\ M_c \\ M_j \end{bmatrix} = -\begin{bmatrix} \frac{l^2}{2} \\ \frac{l^2}{8} \\ 0 \end{bmatrix} w$$

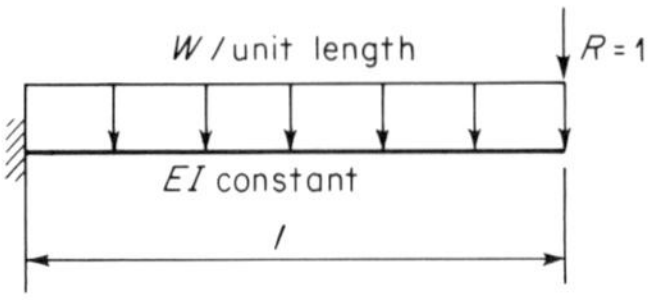

Figure 4.23 Cantilever-beam deflection.

Then,

$$fbR = \frac{-l}{30EI}\begin{bmatrix} 4 & 2 & -1 \\ 2 & 16 & 2 \\ -1 & 2 & 4 \end{bmatrix}\begin{bmatrix} \frac{l^2}{2} \\ \frac{l^2}{8} \\ 0 \end{bmatrix} w = \frac{-l}{30EI}\begin{bmatrix} \frac{9l^2}{4} \\ 3l \\ 0 \end{bmatrix}$$

For a unit load on the free end,

$$\bar{b} = -\begin{bmatrix} l \\ \frac{l}{2} \\ 0 \end{bmatrix}$$

Thence,

$$r = \bar{b}^T fbR = \frac{l}{30EI}\begin{bmatrix} l & \frac{l}{2} & 0 \end{bmatrix}\begin{bmatrix} \frac{9l^2}{4} \\ 3l^2 \\ 0 \end{bmatrix} = \frac{wl^4}{8EI}$$

Additional Generalized Flexibility Matrices

Generalized flexibility matrices may be formed for a variety of member load conditions:

1 Prismatic beam of length l, with EI constant, end moments M_i and M_j, concentrated load P, and distance xl from the left-hand end of the beam (Fig. 4.24a).

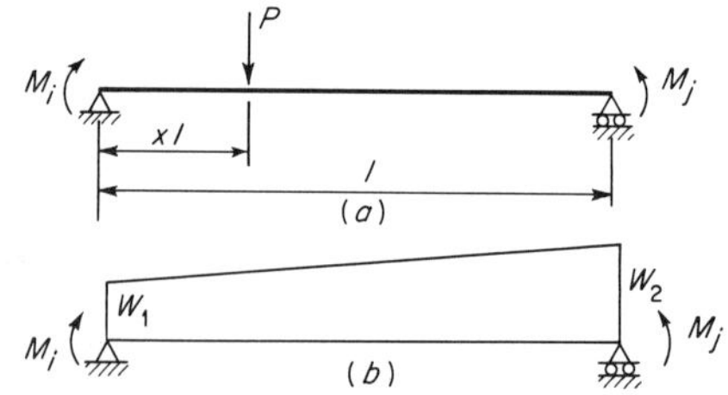

Figure 4.24 Generalized force systems.

$$\begin{bmatrix} q_i \\ q_j \\ q_P \end{bmatrix} = \frac{l}{6EI} \begin{bmatrix} 2 & 1 & (1-x)x(2-x)l \\ 1 & 2 & x(1-x)(1+x)l \\ (1-x)x(2-x)l & x(1-x)(1+x)l & 2x^2(1-x)^2l^2 \end{bmatrix} \begin{bmatrix} M_i \\ M_j \\ P \end{bmatrix} \tag{4.134}$$

2 Prismatic beam of length l, with EI constant, end moments M_i and M_j, linearly varying load, and end values w_1 and w_2 (Fig. 4.24*b*).

$$\begin{bmatrix} q_i \\ q_j \\ w_1 \\ w_2 \end{bmatrix} = \frac{l}{6EI} \begin{bmatrix} 2 & 1 & \dfrac{2l^2}{15} & \dfrac{7l^2}{60} \\ 1 & 2 & \dfrac{7l^2}{60} & \dfrac{2l^2}{15} \\ \dfrac{2l^2}{15} & \dfrac{7l^2}{60} & \dfrac{8l^4}{6 \times 105} & \dfrac{31l^4}{6 \times 420} \\ \dfrac{7l^2}{60} & \dfrac{2l^2}{15} & \dfrac{31l^4}{6 \times 420} & \dfrac{8l^4}{6 \times 105} \end{bmatrix} \begin{bmatrix} M_i \\ M_j \\ w_1 \\ w_2 \end{bmatrix} \tag{4.135}$$

Example 4.14 Node deflections of a uniform beam

This example makes use of the generalized flexibility matrices in Eqs. (4.134) and (4.135). The beam has members (1), (2), and (3), in Fig. 4.25, member (1) being loaded with linearly varying force, and member (3) with concentrated force. Deflections are calculated at nodes 1 and 2. For the beam, the equation $F = bR$ becomes

$$\begin{bmatrix} M_{1i} \\ M_{1j} \\ w_1 \\ w_2 \\ M_{2i} \\ M_{2j} \\ M_{3i} \\ M_{3j} \\ P \end{bmatrix} = \begin{bmatrix} 0 & 0 & 0 \\ 12 & 24 & -5 \\ 1 & 0 & 0 \\ 0 & 1 & 0 \\ 12 & 24 & -5 \\ 0 & 0 & -10 \\ 0 & 0 & -10 \\ 0 & 0 & 0 \\ 0 & 0 & 1 \end{bmatrix} \begin{bmatrix} w_1 \\ w_2 \\ P \end{bmatrix} \tag{i}$$

```
START
 CALCULATION OF BEAM DEFLECTIONS GENERALIZED FLEXIBILITY MATRICES

LOAD B 9 3
    9 ROWS,     3 COLUMNS

PRINT B    * FORCE TRANSFORMATION MATRIX

                 1             2              3
    1         0.00000       0.00000        0.00000
    2        12.00000      24.00000       -5.00000
    3         1.00000       0.00000        0.00000
    4         0.00000       1.00000        0.00000
    5        12.00000      24.00000       -5.00000
    6         0.00000       0.00000      -10.00000
    7         0.00000       0.00000      -10.00000
    8         0.00000       0.00000        0.00000
    9         0.00000       0.00000        1.00000

LOAD R 3 1
    3 ROWS,     1 COLUMNS

PRINT R   * LOAD VECTOR

                 1
    1         1.00000
    2         1.50000
    3        10.00000

MULT B R S

ZERO F 9 9
    9 ROWS,     9 COLUMNS

LOAD F1 4 4
    4 ROWS,     4 COLUMNS

PRINT F1   * MEMBER FLEXIBILITY MATRIX  1

                 1             2              3              4
    1         2.00000       1.00000       19.20000       16.80000
    2         1.00000       2.00000       16.80000       19.20000
    3        19.20000      16.80000        1.00000        1.00000
    4        16.80000      19.20000        1.00000        1.00000

SCALE F1  2.0
 SCALAR=  0.2000000E+01

ADDSM F1 F 1 1

LOAD F1 2 2
    2 ROWS,     2 COLUMNS

PRINT F1   *MEMBER FLEXIBILITY MATRIX 2

                 1             2
    1         2.00000       1.00000
    2         1.00000       2.00000

SCALE F1  2.0
 SCALAR=  0.2000000E+01

ADDSM F1 F 5 5

LOAD F1 3 3
    3 ROWS,     3 COLUMNS

PRINT F1   *MEMBER FLEXIBILITY MATRIX  3

                 1             2              3
    1         2.00000       1.00000        7.50000
    2         1.00000       2.00000        7.50000
    3         7.50000       7.50000        1.00000
```

```
SCALEI F1 3.3333
 SCALAR=   0.3333300E+01

ADDSMI F1 F 7 7

MULT F S V

LOAD BT 2 9
    2 ROWS,        9 COLUMNS

PRINT BT   *FORCE TRANSFORMATION MATRIX FOR DEFLECTIONS

                1              2              3              4              5
    1        0.00000        6.00000        0.00000        0.00000        6.00000
    2        0.00000      -10.00000        0.00000        0.00000      -10.00000

                9
    1        0.00000
    2        0.00000

MULT BT V DEF

PRINT DEF   *DEFLECTIONS MULTIPLIED BY EI

                1
    1      -748.80000
    2     17661.25001
```

The flexibility matrices for members 1, 2, and 3 are

$$f_1 = \frac{2}{EI}\begin{bmatrix} 2 & 1 & 19.2 & 16.8 \\ 1 & 2 & 16.8 & 19.2 \\ 19.2 & 16.8 & 1.0 & 1.0 \\ 16.8 & 19.2 & 1.0 & 1.0 \end{bmatrix}$$

$$f_2 = \frac{2}{EI}\begin{bmatrix} 2 & 1 \\ 1 & 2 \end{bmatrix} \qquad f_3 = \frac{3.33}{EI}\begin{bmatrix} 2 & 1 & 7.5 \\ 1 & 2 & 7.5 \\ 7.5 & 7.5 & 1 \end{bmatrix}$$

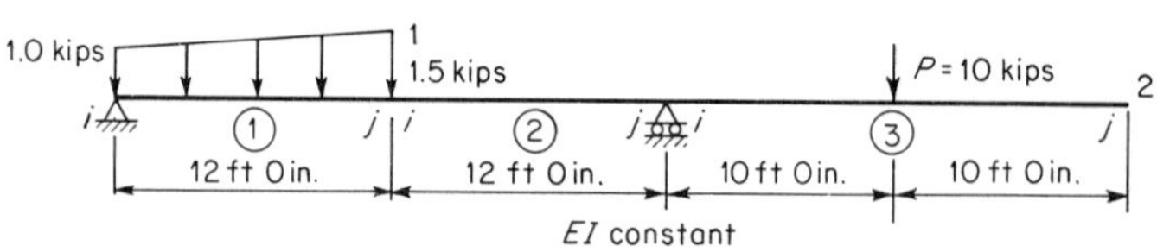

Figure 4.25 Beam loading.

```
    6               7               8
 0.00000         0.00000         0.00000
-20.00000       -20.00000        0.00000
```

Notice that terms such as $f_{w1,w1}$ and f_{PP} are not required to calculate deflections of nodes and may be set equal to unity.

$$\bar{b}^T = \begin{bmatrix} 0 & 6 & 0 & 0 & 6 & 0 & 0 & 0 & 0 & 0 \\ 0 & -10 & 0 & 0 & -10 & 0 & -20 & -20 & 0 & 0 \end{bmatrix}$$

4.8 LOADS ON MEMBERS BETWEEN NODES; CALCULATION OF NODE DEFLECTIONS

In Sec. 4.7, generalized forces and displacements are used to calculate structure deflections when loads other than node forces are applied to the structure. They could in fact form a basis for the treatment of all such conditions. However, because of the variety of load conditions that may be realized with prismatic or nonprismatic members, the universal application of generalized flexibility matrices would seem to be cumbersome for computer application. There are exceptions, however, in special structures (such as folded plates) where load patterns are restricted to a few types (see Ref. 7). In addition, they form a useful basis for discussion in the analysis of highly redundant structures by the force method (see Chap. 5). Two additional alternative methods that may be used to calculate the effects of loads along members are discussed in this section. The first method is similar to the generalized force method in that the effects of member loads are incorporated in the distortions of the member. That is, in addition to distortions due to the linear-moment diagram (described by end moments M_i, M_j), the distortions from member loads are included. The second method continues the idea formulated in Sec. 3.9. Therein, the bending moments and axial forces in statically determinate frame structures are obtained by first locking all joints and then calculating the node forces for this condition. The node deflections in this condition are thus all zero. The node deflections are calculated by applying the reverse of these joint forces (the release forces) to the nodes.

Definition of Member Forces

Forces acting along a member may be applied either in the global coordinate system or in the local member system of axes. The first type of loading is typical of gravity forces for which the gravity force is resolved into components $\rho \cos \beta$, $\rho \sin \beta$ (Fig. 4.26). The member loading in local member coordinates is typical of wind or general fluid pressures that act perpendicular to, or parallel with, the member. In Fig. 4.26*a* and *c*, the member forces may be concentrated or distributed.

It has been noted in this chapter, Sec. 4.3, that member distortions are most easily calculated in member coordinates (Fig. 4.26*c*). On the other hand, from Chaps. 2 and 3, it is seen that structures are best analyzed for node forces in the frame coordinate system. Thus, for the analysis of member forces, two transformations are generally necessary:

1 Transformation of force components along the member to local force components. This is achieved by use of Eq. (2.29), written here in Eqs. (*i*) and (*ii*). That is,

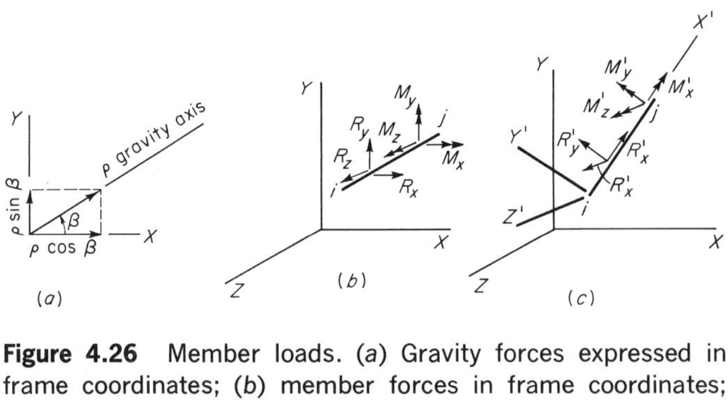

Figure 4.26 Member loads. (*a*) Gravity forces expressed in frame coordinates; (*b*) member forces in frame coordinates; (*c*) member forces in local coordinates.

$$R' = L^T R \qquad M' = L^T M \qquad \text{concentrated forces and moments} \qquad (i)$$

$$w' = L^T w \qquad m' = L^T m \qquad \text{distributed forces and moments} \qquad (ii)$$

2 Transformation of resulting distortions or fixed-end forces back to the relevant node quantities, either

$$r = L^T v \qquad \text{distortions} \qquad (iii)$$

$$R = LR' \qquad M = LM' \qquad \text{forces} \qquad (iv)$$

Although when the above transformations are used in (i) and (ii), six force components result, it is convenient to consider only one of these, one that is equivalent to M'_{iz}, M'_{jz} for the simple beam. The force components considered will thus be R'_y or w'_y.

Method 1 Deflections due to temperature change, member lack of fit, or loads along the member

It is found that member distortions due to member loads between nodes are simply one of the class that includes those due to temperature change and member lack of fit (e.g., due to errors in fabrication).

In Eq. (4.2), it is shown that for member distortions v, the node deflections produced are

$$r = b^T v$$

These member distortions v may be caused by effects other than the elastic strains due to the forces applied to the member at the nodes. For example, a member may undergo temperature changes that produce strains designated $\bar{v}$. These temperature strains are calculated here for truss and plane frame members.

Case 1 *Truss Member* A truss member of length l with coefficient of thermal expansion α undergoes a temperature change t. The change in length is

$$\Delta l = \bar{v} = l\alpha t \tag{4.136}$$

and thence,

$$r = b^T\bar{v} = b^T l\alpha t$$

Case 2 *Beam Member* The frame member shown in Fig. 4.27 suffers a temperature change t_1 on its upper side and another, t_2, on its lower side, and the varia-

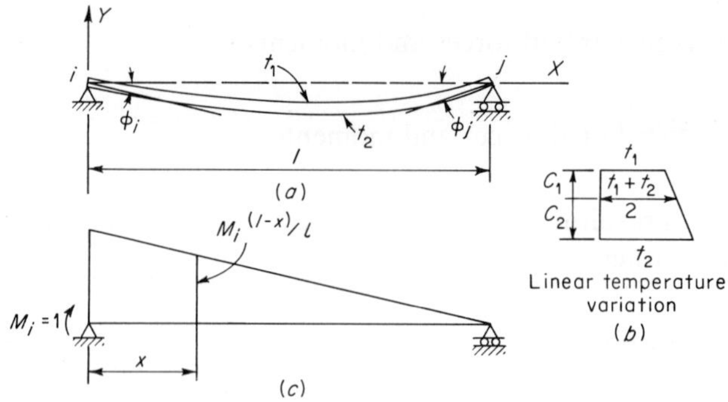

Figure 4.27 Beam-element temperature distortions.

tion through the cross section is assumed to be linear (Fig. 4.27*b*). From the figure it is seen that the change in the length of the member is

$$\Delta l = \frac{t_1 + t_2}{2} l\alpha$$

To calculate the end-rotation terms $\bar{\phi}_i$ and $\bar{\phi}_j$, apply unit moments $M_i = 1$ and $M_j = 1$ in turn, and use Eq. (4.48). That is,

$$\delta r = \int_{c_2}^{c_1} \bar{b}^T v$$

In this case,

$$v = \varepsilon_x\, dx = \frac{\alpha}{c_1 - c_2}[(t_1 - t_2)y + (c_1 t_2 - c_2 t_1)]\, dx$$

For $M_i = 1$,

$$\bar{b}^T = -\frac{(l - x)}{l}\frac{y}{I} d\, dy$$

Here d is the cross-section width at y. The rotation at end i is given by

$$\delta r = \frac{-dxy\, dy}{I} dx \frac{\alpha}{c_1 - c_2}[(t_1 - t_2)y + (c_1 t_2 - c_2 t_1)]\left(1 - \frac{x}{l}\right)$$

To obtain the total effect of all such increments, integrate first across the cross section and secondly over the length of the member. Note that

$$\int_{c_2}^{c_1} y\, dy = 0$$

$$\int_{c_2}^{c_1} dy^2\, dy = I$$

Hence,

$$\delta r = \frac{\alpha(t_2 - t_1)}{c_1 - c_2}\left(1 - \frac{x}{l}\right) dx$$

Integrating from *o to l*,

$$r = \bar{\phi}_i = \int_o^l \frac{\alpha(t_2 - t_1)}{c_1 - c_2}\left(1 - \frac{x}{l}\right) dx$$
$$= \frac{\alpha(t_2 - t_1)}{c_1 - c_2}\frac{l}{2}$$

Similarly,

$$\bar{\phi}_j = \frac{\alpha(t_2 - t_1)}{c_1 - c_2}\frac{l}{2}$$

These two equations may be combined to give

$$\begin{bmatrix} \phi_i \\ \phi_j \end{bmatrix} = \frac{\alpha l}{2(c_1 - c_2)}\begin{bmatrix} -1 & 1 \\ -1 & 1 \end{bmatrix}\begin{bmatrix} t_1 \\ t_2 \end{bmatrix} \tag{4.137}$$

It has now been shown that member distortions may be produced by causes other than the member forces.

For the member with flexibility f, member forces S, and temperature or lack-of-fit strains $\bar{v}$,

$$v = fS + \bar{v} = fbR + \bar{v} \tag{4.138}$$

The induced node deflections are calculated as

$$r = b^T v = b^T(fbR + \bar{v}) \tag{4.139}$$

If forces act along a member between its nodes, they will produce these $\bar{v}$ strains in addition to producing member forces S.

Consider, for example, a prismatic member of length l with a transverse load w per unit length. Then the contributions $\bar{v}$ to the end rotations are given by

$$\begin{bmatrix} \bar{\phi}_i \\ \bar{\phi}_j \end{bmatrix} = \frac{l^2}{24EI}\begin{bmatrix} 1 \\ 1 \end{bmatrix} wl$$

In the general case this will be written

$$\bar{v} = \delta R \tag{4.140}$$

Then,

$$v = fbR + \delta R = (fb + \delta)R \tag{4.141}$$

Then the deflections at the nodes are given as

$$r = b^T(fb + \delta)R \tag{4.142}$$

When the points at which the deflections are required do not coincide with the load points (such must always be the case with distributed loads), then Eq. (4.142) becomes

$$r = \bar{b}^T(fb + \delta)R \tag{4.143}$$

Example 4.15 Calculate the deflection at the center of the beam shown in Fig. 4.28. Equation (4.143) will be used. For the load w,

$$\begin{bmatrix} M_{1i} \\ M_{1j} \\ M_{2i} \\ M_{2j} \end{bmatrix} = \begin{bmatrix} 0 \\ \frac{l^2}{16} \\ \frac{l^2}{16} \\ 0 \end{bmatrix} w$$

The δ terms are nonzero for member (1) only and are given by

$$\begin{bmatrix} \delta_{1i} \\ \delta_{1j} \\ \delta_{2i} \\ \delta_{2j} \end{bmatrix} = \frac{l^2}{192EI} \begin{bmatrix} 1 \\ 1 \\ 0 \\ 0 \end{bmatrix}$$

The member-flexibility matrix is

$$f = \frac{l}{12EI} \begin{bmatrix} 2 & 1 & 0 & 0 \\ 1 & 2 & 0 & 0 \\ 0 & 0 & 2 & 1 \\ 0 & 0 & 1 & 2 \end{bmatrix}$$

For the vertical load at 2,

$$\bar{b}^T = \left\langle 0 \quad \frac{l}{4} \quad \frac{l}{4} \quad 0 \right\rangle$$

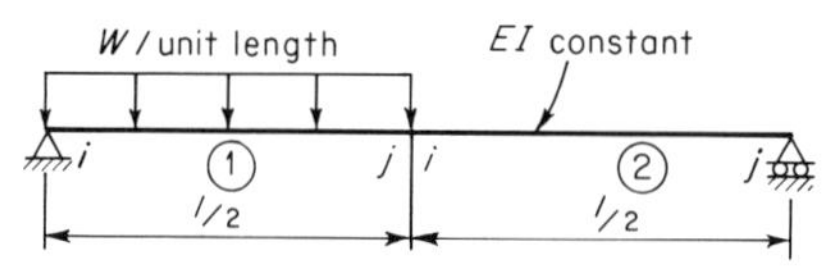

Figure 4.28 Beam deflections.

Then,

$$fb + \delta = \frac{l^3}{192EI}\begin{bmatrix}1\\2\\2\\1\end{bmatrix} + \frac{l^3}{192EI}\begin{bmatrix}1\\1\\0\\0\end{bmatrix} = \frac{l^3}{192EI}\begin{bmatrix}2\\3\\2\\1\end{bmatrix}$$

Finally,

$$r = \begin{bmatrix}0 & \frac{l}{4} & \frac{l}{4} & 0\end{bmatrix}\begin{bmatrix}2\\3\\2\\1\end{bmatrix}\frac{l^3}{192EI}w = \frac{5wl^4}{768EI}$$

It appears that method 1 gives a practical method for the inclusion of loads along members. For general loading and variable moment of inertia, numerical integration may be used to obtain δ. Of course, if the general loading as shown in Fig. 4.26 is applied to the member, rotation terms must be calculated for loads in each of the $X'Y'$ and $X'Z'$ planes.

Method 2 The second method now given has wide application and is most suitable for computer coding since it reduces all structures to the node loads only for analysis. Thus (returning at the present time to simple planar frames), the structure shown in Fig. 4.29 is analyzed in two separate stages. The complete solution is the sum of the two analyses. The first, node-fixed condition is analyzed for the forces acting on each member (Fig. 4.29*b*). The node displacements are thus, necessarily, zero. If all the forces required to fix every member are aggregated in global components at each joint, the equivalent effect to releasing all joints is to apply the negative of the joint fixity forces to the structure (Fig. 4.29*c*).

This method of analysis follows closely that used in the displacement method (Chap. 6), and details on the calculation of fixed-end forces are

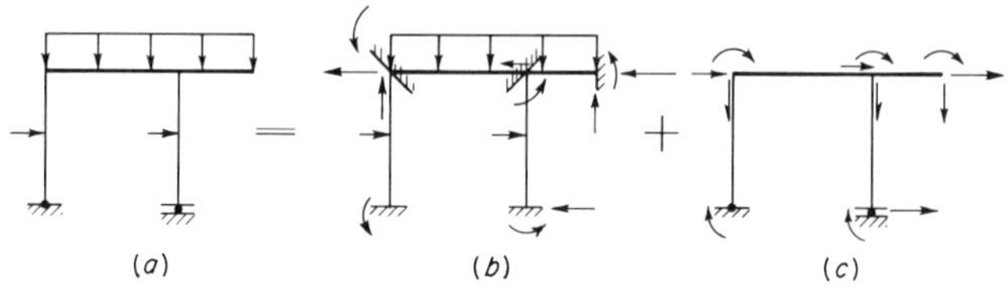

Figure 4.29 Analysis with release forces. (*a*) Structure with loads on members; (*b*) fixed-node structure; (*c*) structure with release-force node loads only.

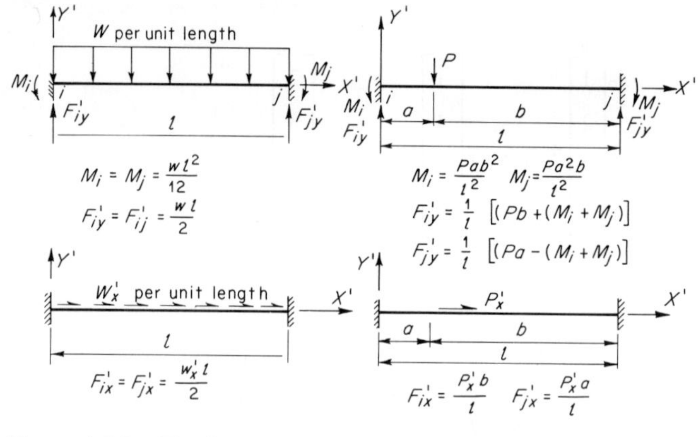

Figure 4.30 Fixed-end forces—simple cases.

given there. A table of fixed-end forces for prismatic members is also given in Appendix B. Some simple cases are given in Fig. 4.30. Note that here the sign convention for joints is used.

Example 4.16 Calculate the deflection at the free end of the cantilever beam, loaded w per unit length (see Fig. 4.31). The fixed-end condition is shown in Fig. 4.31*b*, and from this the release node forces are

$$\left\langle -\frac{wl}{2} \quad -\frac{wl^2}{12} \right\rangle$$

(See Fig. 4.31*c*.)

The deflection at end j in Fig. 4.31*c* may be written down by inspection. That is,

$$r = \frac{(wl/2)l^3}{3EI} - \frac{(wl^2/12)l^2}{2EI} = \frac{wl^4}{8EI}$$

Example 4.17 Calculate the deflection at the center of the beam shown in Fig. 4.28. For the beam in Fig. 4.28, the release forces to be applied to the all-fixed condition are shown separately with their resulting-moment diagrams in Fig. 4.32. For $\langle R_1 \quad R_2 \quad R_3 \rangle$ positive as shown in Fig. 4.32,

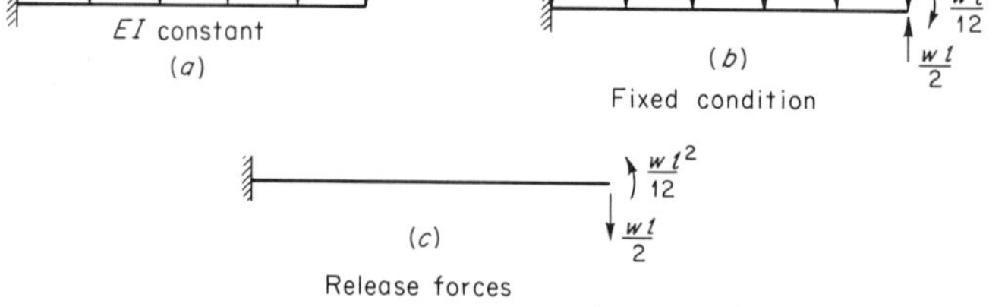

Figure 4.31

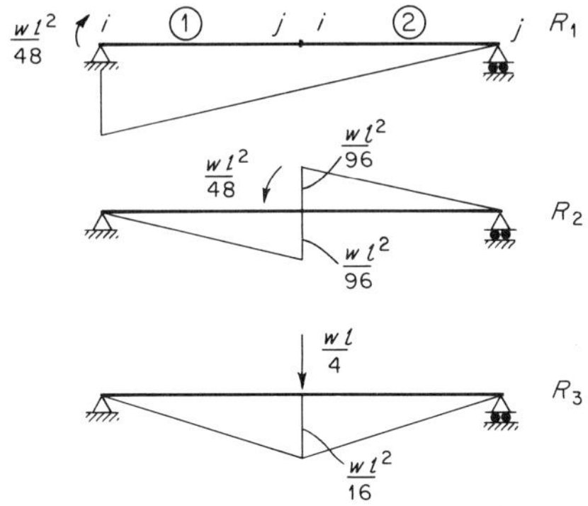

Figure 4.32 Release-force diagrams.

$$S = \begin{bmatrix} M_{1i} \\ M_{1j} \\ M_{2i} \\ M_{2j} \end{bmatrix} = \begin{bmatrix} 1 & 0 & 0 \\ \frac{1}{2} & \frac{1}{2} & \frac{l}{4} \\ \frac{1}{2} & -\frac{1}{2} & \frac{l}{4} \\ 0 & 0 & 0 \end{bmatrix} \begin{bmatrix} R_1 \\ R_2 \\ R_3 \end{bmatrix}$$

Hence,

$$r = \bar{b}^T f b R = \begin{bmatrix} 0 & \frac{l}{4} & \frac{l}{4} & 0 \end{bmatrix} \frac{l}{12EI} \begin{bmatrix} 2 & 1 & 0 & 0 \\ 1 & 2 & 0 & 0 \\ 0 & 0 & 2 & 1 \\ 0 & 0 & 1 & 2 \end{bmatrix} \begin{bmatrix} 1 & 0 & 0 \\ \frac{1}{2} & \frac{1}{2} & \frac{l}{4} \\ \frac{1}{2} & -\frac{1}{2} & \frac{l}{4} \\ 0 & 0 & 0 \end{bmatrix} \begin{bmatrix} \frac{1}{3} \\ \frac{1}{3} \\ \frac{4}{l} \end{bmatrix} \frac{wl^2}{16}$$

On multiplication it is found that

$$r = \begin{bmatrix} 0 & \frac{l}{4} & \frac{l}{4} & 0 \end{bmatrix} \begin{bmatrix} 2 \\ 3 \\ 2 \\ 1 \end{bmatrix} \frac{wl^3}{192EI} = \frac{5wl^4}{768EI}$$

4.9 DEFLECTION CALCULATIONS FOR CURVED MEMBERS

A curved member may always be approximated by a series of straight members, the junction of two members becoming a node in the model structure, as shown in Fig. 4.33. Evidently, the more straight segments are incorporated,

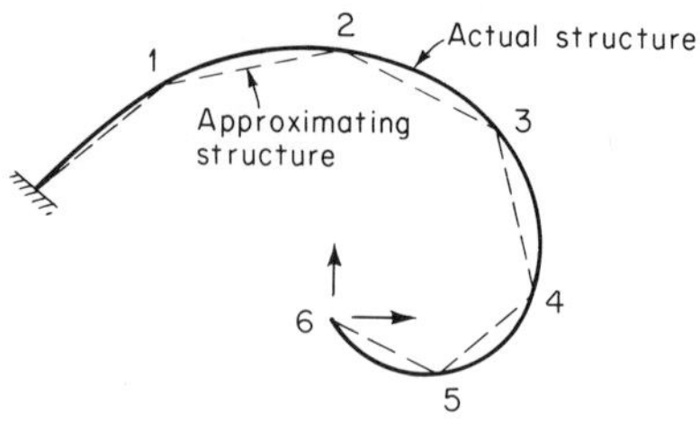

Figure 4.33

the smaller is their size and the closer do the properties approach those of the actual structure. The above procedure will be useful if the curve to the member is of difficult, or perhaps even nonmathematical, form.

In this section two classes of curved members are considered that fit into the categories of frame and grid structures. The two classes are thus:

1 Members curved in a plane and loaded in this plane. They possess a principal plane in the load plane and are usually referred to as *arch structures.*

2 Members curved in a plane and loaded perpendicular to this plane. Again a principal axis will be assumed to be parallel to the load plane. These members are often referred to as *beams curved in plan.*

There is an analogy between the forces and displacements of the two systems, and once one is analyzed, the other follows very simply. The detailed discussion here is restricted to arch members for which displacements are considered small.

1 Arch Members

The curved arch member is shown in Fig. 4.34. In the first instance the effects throughout the arch will be calculated for the forces $\langle R_{xO} \quad R_{yO} \quad M_{zO} \rangle$ acting at the free end. The calculation of the distribution of forces throughout the arch becomes an application of the transformation equations (2.49) and (2.29). That is, transfer $\langle R_{xO} \quad R_{yO} \quad M_{zO} \rangle$ first from O to P in axes parallel to XOY, and then rotate into the set X', P, Y' at right angles to the member cross section at P. Thence,

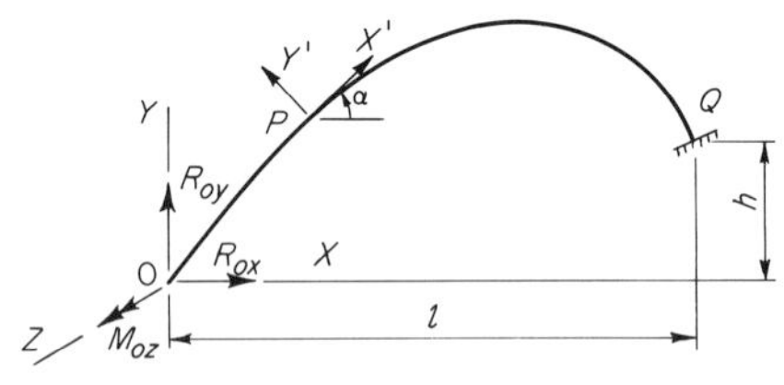

Figure 4.34 Arch member.

$$\begin{bmatrix} R_x \\ R_y \\ M_z \end{bmatrix}_P = \begin{bmatrix} 1 & 0 & 0 \\ 0 & 1 & 0 \\ y & -x & 1 \end{bmatrix} \begin{bmatrix} R_x \\ R_y \\ M_z \end{bmatrix}_O \tag{4.144}$$

or

$$R_P = b_1 R_O \tag{4.145}$$

Now transfer R_P to components in the local axes X', Y'.

From Eq. (2.29),

$$R'_P = L^T R_P = L^T b_1 R_O \tag{4.146}$$

Here

$$L^T = \begin{bmatrix} \cos\alpha & \sin\alpha & 0 \\ -\sin\alpha & \cos\alpha & 0 \\ 0 & 0 & 1 \end{bmatrix} \tag{4.147}$$

In Eq. (4.146), the relationship in Eq. (4.148) has been developed.

$$R'_P = bR_O \tag{4.148}$$

For the element length ds at P, the term R'_P is the member force. The flexibility of the element is given by

$$f = \begin{bmatrix} \dfrac{ds}{EA} & \cdot & \cdot \\ \cdot & \dfrac{ds}{GA_{\text{eff}}} & \cdot \\ \cdot & \cdot & \dfrac{ds}{EI} \end{bmatrix} \tag{4.149}$$

For numerical integration it is convenient to use equal increments of x rather than ds. Now, $dx = ds \cos\alpha$, so that $ds = dx \sec\alpha$. Thence,

$$f = \begin{bmatrix} \dfrac{\sec\alpha}{EA} & \cdot & \cdot \\ \cdot & \dfrac{\sec\alpha}{GA_{\text{eff}}} & \cdot \\ \cdot & \cdot & \dfrac{\sec\alpha}{EI} \end{bmatrix} dx \tag{4.150}$$

For the element ds, the distortion is given by

$$\delta v'_p = fR'_P \tag{4.151}$$

By the contragradient law, the resulting deflection at O is given by

$$\delta r_o = b^T \, \delta v'_P = b^T f b R_O \tag{4.152}$$

For all such elements, integrate over the arch length from $x = O$ to $x = l$. Therefore,

$$r_o = \int_o^l b^T f b \, dx \, R_O \tag{4.153}$$

The function $b^T f b$ may be evaluated in several ways. For example, if the curve is given as a set of discrete values with y and x measured from graphic presentation of the arch profile, then sec α must be determined by numerical methods.

Generally, however, the curve will be given in differentiable form,

$$y = g(x)$$

Matrix Integration

The matrix $b^T f b$ is integrated over the range O to l, term by term, so that the integral of all terms may be expressed in matrix form. That is, for the intervals h, shown in Fig. 4.35,

$$\int_o^l b^T f b \, dx = \frac{h}{3}[(b^T f b)_1 + 4(b^T f b)_2 + 2(b^T f b)_3 + \cdots + (b^T f b)_N] \tag{4.154}$$

When the deflection is required at O owing to a load at another point on the arch, the Eq. (4.153) is written

$$r_{o1} = \int_{x_1}^l \bar{b}^T f b \, dx \, R_1 \tag{4.155}$$

The integral takes place only over the region common to the diagrams b and $\bar{b}$. It is convenient in arch analysis to subdivide the arch into an even number of intervals, for example, 40, and then limit the load R_1 to the even points, so that no discontinuity occurs at R_1 in the application of Simpson's integration (see Fig. 4.35).

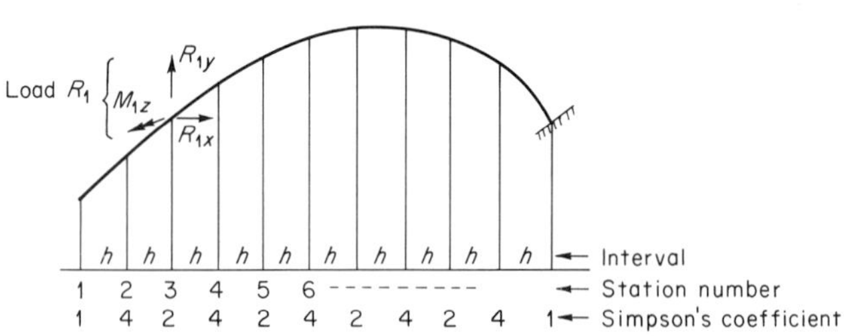

Figure 4.35 Numerical integration for arch.

For dead-load calculations on the arch, Eq. (4.155) is written

$$r_{ow} = \int_{o}^{l} \bar{b}^T f R'_w \, dx \tag{4.156}$$

The forces R'_w at any section are calculated from the arch material density w and area $A(x)$. The calculation of R'_w requires an integration process, and this can be accomplished by proceeding from point to point on the arch. For this purpose, suppose that w and $A(x)$ are known at equal intervals along the arch. Then suppose that R_w and R'_w are known at section n and further that $A(x)$ is given by A_n, A_{n+1}, A_{n+2}. Then R_w at $n + 2$ may be determined as a function of these quantities. A recurrence relationship is thus obtained that allows R'_w to be determined at every second interval. From Fig. 4.36*c*, the gravity axis is inclined at angle β with the X axis. Thence,

$$\begin{bmatrix} w_x \\ w_y \end{bmatrix} = \rho \begin{bmatrix} \cos\beta \\ \sin\beta \end{bmatrix} \tag{4.157}$$

From Fig. 4.36, given $[R_x \quad R_y \quad M_z]_n$, calculate $[R_x \quad R_y \quad M_z]_{n+2}$. Let

$\Omega_i = A_i \sec \alpha_i$

Then,

$$\begin{aligned} R_{x_{n+2}} &= R_{x_n} + \frac{w_x h}{3}(\Omega_n + 4\Omega_{n+1} + \Omega_{n+2}) \\ R_{y_{n+2}} &= R_{y_n} + \frac{w_y h}{3}(\Omega_n + 4\Omega_{n+1} + \Omega_{n+2}) \\ M_{z_{n+2}} &= M_{z_n} - R_{y_n} 2h - \frac{w_y h}{3}(2h\Omega_n + 4h\Omega_{n+1}) \\ &\quad + R_{x_n}(y_1 + y_2) + \frac{w_x h}{3}[(y_1 + y_2)\Omega_n + 4y_2\Omega_{n+1}] \end{aligned} \tag{4.158}$$

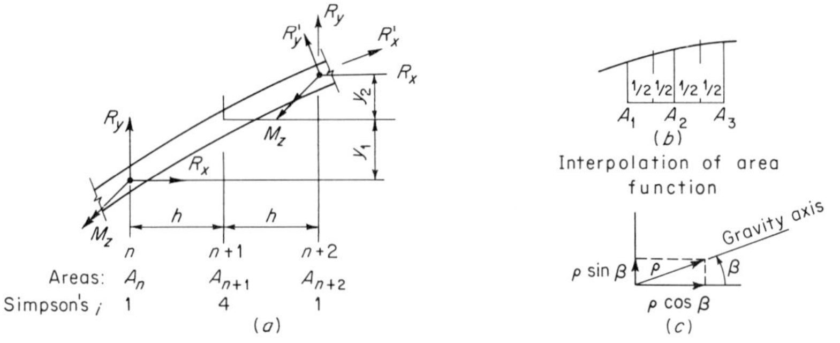

Figure 4.36 Force in rib—dead-load calculations.

Alternatively, in matrix form,

$$\begin{bmatrix} R_x \\ R_y \\ M_z \end{bmatrix}_{n+2} = \begin{bmatrix} 1 & 0 & 0 \\ 0 & 1 & 0 \\ y_1 + y_2 & -2h & 1 \end{bmatrix} \begin{bmatrix} R_x \\ R_y \\ M_z \end{bmatrix}_n + \frac{w_x h}{3} \begin{bmatrix} 1 & 4 & 1 \\ 0 & 0 & 0 \\ y_1 + y_2 & 4y_2 & 0 \end{bmatrix} \begin{bmatrix} \Omega_n \\ \Omega_{n+1} \\ \Omega_{n+2} \end{bmatrix} + \frac{w_y h}{3} \begin{bmatrix} 0 & 0 & 0 \\ 1 & 4 & 1 \\ -2h & -4h & 0 \end{bmatrix} \begin{bmatrix} \Omega_n \\ \Omega_{n+1} \\ \Omega_{n+2} \end{bmatrix} \tag{4.159}$$

In the particular case where the gravity axis coincides with the Y axis,

$w = w_y \qquad \text{and} \qquad w_x = 0$

Then Eq. (4.159) becomes

$$\begin{bmatrix} R_x \\ R_y \\ M_z \end{bmatrix}_{n+2} = \begin{bmatrix} 1 & 0 & 1 \\ 0 & 1 & 0 \\ y_1 + y_2 & -2h & 1 \end{bmatrix} \begin{bmatrix} R_x \\ R_y \\ M_z \end{bmatrix}_n + \frac{w_y h}{3} \begin{bmatrix} 0 & 0 & 0 \\ 1 & 4 & 1 \\ -2h & -4h & 0 \end{bmatrix} \begin{bmatrix} \Omega_n \\ \Omega_{n+1} \\ \Omega_{n+2} \end{bmatrix} \tag{4.160}$$

In local coordinates at section $n + 2$,

$$R'_{n+2} = L^T_{n+2} R_{n+2} \tag{4.161}$$

With the above recurrence relationships, the forces R' may be calculated by starting at the end O and proceeding along the member. If there are $2n + 1$ points of subdivision R', values are calculated at $n + 1$ points. If the area values A have been given at $n + 1$ coordinates, the intermediate values may be obtained at the half-intervals by quadratic interpolation.

Thus in Fig. 4.36*b*, let

$$A = Cx^2 + Dx + E$$

Solve for C, D, and E.

$$E = A_1$$
$$C = \tfrac{1}{2}(A_1 - 2A_2 + A_3)$$
$$D = \tfrac{1}{2}(4A_2 - 3A_1 - A_3)$$

Then,

$$A_{1/2} = \tfrac{1}{4}C + \tfrac{1}{2}D + E$$
$$A_{1\text{-}1/2} = \tfrac{9}{4}C + \tfrac{3}{2}D + E$$

Member-flexibility Matrices

In Eq. (4.153) the integral gives the flexibility matrix in global coordinates at end O. That is,

$$F_O = \int_0^l b^T f b \, dx \tag{4.162}$$

Transfer the force at O to an equivalent force at Q. Then,

$$\begin{bmatrix} R_x \\ R_y \\ M_z \end{bmatrix}_Q = \begin{bmatrix} 1 & 0 & 0 \\ 0 & 1 & 0 \\ h & -l & 1 \end{bmatrix} \begin{bmatrix} R_x \\ R_y \\ M_z \end{bmatrix}_O \tag{4.163}$$

This equation is written

$$R_Q = BR_O \tag{4.164}$$

The flexibility matrix for end Q is then given by

$$F_Q = B^T F_O B \tag{4.165}$$

It is seen that if the flexibility matrix at one end is known, that at the other end may be obtained from the arch geometry in Eq. (4.165).

2 Beams Curved in Plan

Only beams whose centroid and shear center coincide are considered. The beam is curved in one of its principal planes and the loads applied in the other principal plane. The axes chosen are given in Fig. 2.11 and are shown again in Fig. 4.37. It is seen that the XY plane is the plane in which the beam is curved, and the load is applied as a force F_z and moments M_X and M_Y. For the beam the origin of coordinates O has been taken at the fixed end.

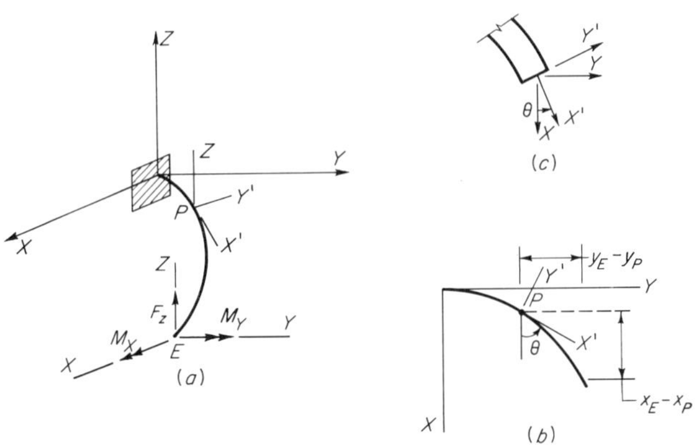

Figure 4.37 Beam curved in plan.

The coordinates of the free end will be (x_E, y_E), and those of the general point P, (x_P, y_P). Local and global coordinates at P are shown in Fig. 4.37*a* and *b*. The distances α_x and α_y are defined by

$$\alpha_x = x_E - x_P$$

$$\alpha_y = y_E - y_P$$

The distribution of forces throughout the beam for the forces $\langle F_Z \quad M_X \quad M_Y \rangle$ on the free end is obtained by application of Eqs. (2.49) and (2.29). The transformation from E to P in global components is given by Eq. (4.166).

$$\begin{bmatrix} F_z \\ M_x \\ M_y \end{bmatrix}_P = \begin{bmatrix} 1 & 0 & 0 \\ \alpha_y & 1 & 0 \\ -\alpha_x & 0 & 1 \end{bmatrix} \begin{bmatrix} F_z \\ M_x \\ M_y \end{bmatrix}_E \tag{4.166}$$

That is,

$$F_P = b_1 F_E \tag{4.167}$$

The transformation from F_P to local member components at P, that is, F_P', is given by Eq. (4.168).

$$\begin{bmatrix} F_z \\ M_x \\ M_y \end{bmatrix}' = \begin{bmatrix} 1 & 0 & 0 \\ 0 & \cos\theta & \sin\theta \\ 0 & -\sin\theta & \cos\theta \end{bmatrix} \begin{bmatrix} F_z \\ M_x \\ M_y \end{bmatrix}_P \tag{4.168}$$

That is,

$$F_P' = L^T F_P \tag{4.169}$$

Then, combined with Eq. (4.167),

$$F_P' = L^T b_1 F_E \tag{4.170}$$

or

$$F_P' = b F_E \tag{4.171}$$

For the principal axes shown in Fig. 4.37*c*, the flexibility matrix is given by

$$f = \begin{bmatrix} \dfrac{ds}{GA_{\text{eff}}} & \cdot & \cdot \\ \cdot & \dfrac{ds}{GI_P} & \cdot \\ \cdot & \cdot & \dfrac{ds}{EI_x} \end{bmatrix} \tag{4.172}$$

The torsion-stiffness term I_P may be calculated from St. Venant's torsion theory. From Fig. 4.37*b*,

$$ds = dx \sec \theta$$

Then

$$f = \begin{bmatrix} \frac{\sec \theta}{GA_{\text{eff}}} & \cdot & \cdot \\ \cdot & \frac{\sec \theta}{GI_P} & \cdot \\ \cdot & \cdot & \frac{\sec \theta}{EI_x} \end{bmatrix} dx \tag{4.173}$$

For the element ds at P, the element distortions due to F_P' are given by

$$\delta v_P' = fF_P' = fbF_E$$

By the contragradient law, the resulting displacement at E is given by

$$\delta r_E = b^T \, \delta v_P' = b^T f b F_E \tag{4.174}$$

For all such elements, the total deflection at E is obtained by integrating δr_E from $x = O$ to $x = x_E$. That is,

$$r_E = \int_0^{x_E} b^T f b \, dx \, F_E \tag{4.175}$$

This equation may be integrated in the same way as that of the arch problem [see Eq. (4.154)].

4.10 CALCULATION OF MEMBER-FLEXIBILITY MATRICES; NONUNIFORM MEMBERS, SHEAR DISTORTIONS, AXIAL LOAD EFFECTS

For general application of the equation $r = b^T f b R$ to the calculation of deflections of structures, means must be found to generate f for members other than those simple cases already discussed in Sec. 4.5. Thus in Sec. 4.9 the flexibility matrix has been developed for arches and beams curved in plan. In this section, the theory is extended by considering further calculations of f matrices. Perhaps the most common member form found in frame structures is that with nonuniform cross section and consequently with nonuniform moment of inertia. Three types of variable member inertia are shown in Fig. 4.38.

Special note should be made of the type in Fig. 4.38*b*, if it is to be included in three-dimensional structures, because in this case the curved centroidal axis will cause extra complications for forces in the Z' direction. However, in plane frame or bridge structures, this difficulty generally does not arise. In

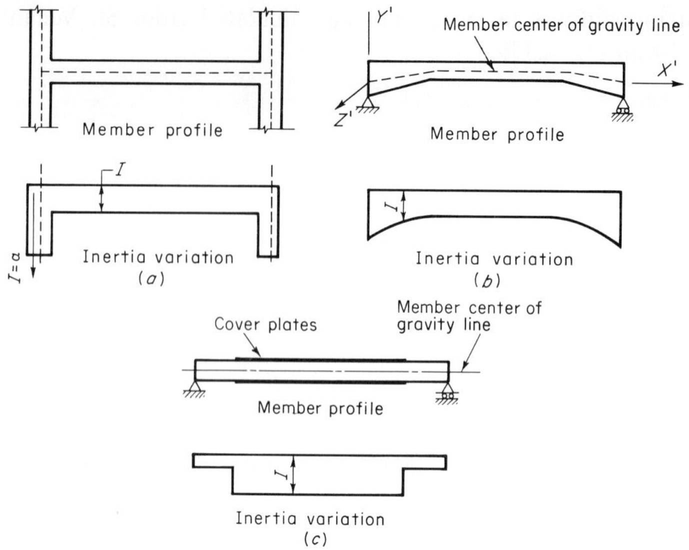

Figure 4.38 Member types with variable inertia.

this section then, any variation in member properties will be assumed to be symmetrical about the member centroid when used in three-dimensional analyses.

1 Member with Infinite End Inertia

This type of member is typical of rigid-frame structures in which columns and beam thicknesses are taken into consideration. The member is shown in Fig. 4.39 and is considered composed of three members (1), (2), and (3) of lengths A, l_1, and B respectively. Members (1) and (3) have infinite inertias; member (2) has inertia I.

The structure forces (M_i, M_j) applied at the ends i and j produce member forces in (1), (2), and (3), of which those of member (2) alone produce distortions. From Fig. 4.39, the b matrix for member (2) is given by

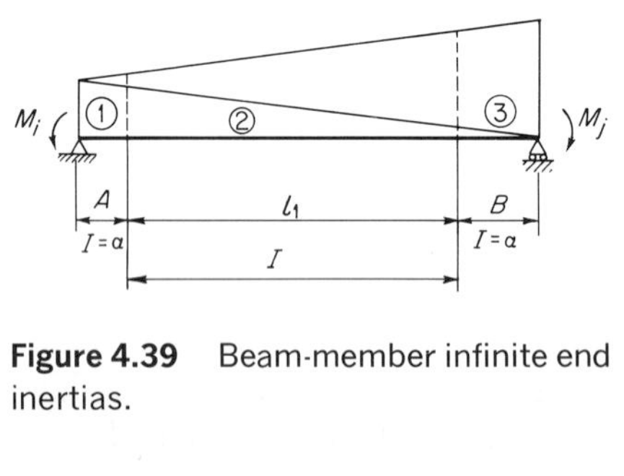

Figure 4.39 Beam-member infinite end inertias.

$$\begin{bmatrix} M_{2i} \\ M_{2j} \end{bmatrix} = \begin{bmatrix} 1 - \dfrac{A}{l} & \dfrac{A}{l} \\ \dfrac{B}{l} & 1 - \dfrac{B}{l} \end{bmatrix} \begin{bmatrix} M_i \\ M_j \end{bmatrix}$$

The flexibility matrix for member (2) is

$$f_2 = \frac{l_1}{6EI} \begin{bmatrix} 2 & 1 \\ 1 & 2 \end{bmatrix}$$

Hence,

$$f = \frac{l_1}{6EI} \begin{bmatrix} 1 - \dfrac{A}{l} & \dfrac{B}{l} \\ \dfrac{A}{l} & 1 - \dfrac{B}{l} \end{bmatrix} \begin{bmatrix} 2 & 1 \\ 1 & 2 \end{bmatrix} \begin{bmatrix} 1 - \dfrac{A}{l} & \dfrac{A}{l} \\ \dfrac{B}{l} & 1 - \dfrac{B}{l} \end{bmatrix} \tag{4.176}$$

The multiplication of matrices in Eq. (4.176) may be undertaken in algebraic form, but it is convenient in computer analysis to leave it in the above form.

2 Member with Variable Area and Inertia

The member is shown in Fig. 4.40*a*, and it will be assumed that the area and inertia values are continuous functions of x. Discontinuities may be accommodated either by considering each region separately or alternatively by subdividing the span into a sufficiently large number of intervals to isolate the discontinuity in only one small region. The member-flexibility matrix is calculated by

$$F = b^T f b$$

Now, however, f is expressed for the infinitesimal element dx, so that F is calculated by matrix integration similar to that in Eq. (4.47). Hence,

$$F = \int_o^l b^T f b \, dx$$

Consider the axial force and moment terms independently. Simpson's rule is used in the integration, and for illustration the member is divided into four equal intervals. In practice, the interval of subdivision may be finer for computer programs. Then,

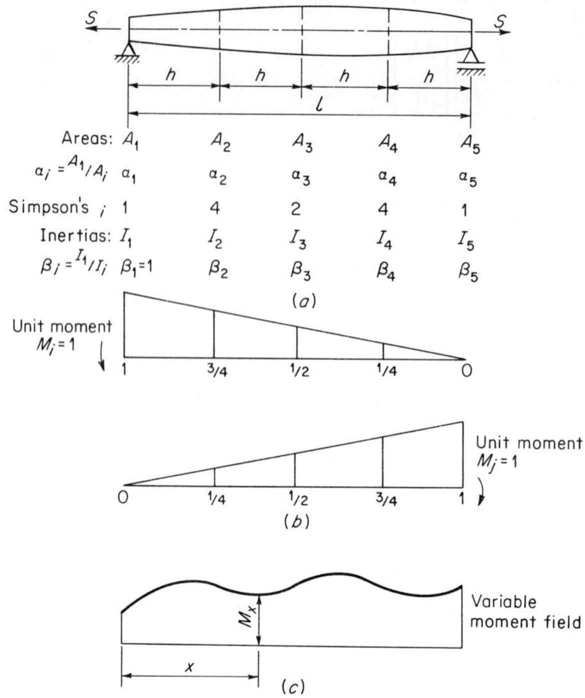

Figure 4.40 Integration scheme for nonuniform member.

$$
\begin{aligned}
f_{\text{axial}} &= \int_o^l 1 \frac{1}{EA} 1 dx \\
&= \frac{1}{EA_1} \int_o^l \alpha \, dx \qquad \alpha = \frac{A_1}{A} \\
&= \frac{h}{3EA_1} W^T \alpha
\end{aligned}
\tag{4.177}
$$

In Eq. (4.177),

$$
W^T = \langle 1 \quad 4 \quad 2 \quad 4 \quad 1 \rangle \qquad \alpha^T = \langle \alpha_1 \quad \alpha_2 \quad \alpha_3 \quad \alpha_4 \quad \alpha_5 \rangle \tag{4.178}
$$

For moment flexibility,

$$
\begin{aligned}
M_x &= M_i\left(1 - \frac{x}{l}\right) + M_j \frac{x}{l} \\
&= \left[1 - \frac{x}{l} \quad \frac{x}{l}\right] \begin{bmatrix} M_i \\ M_j \end{bmatrix} = bR
\end{aligned}
$$

Then,

$$f_{\text{moment}} = \begin{bmatrix} f_{11} & f_{12} \\ f_{21} & f_{22} \end{bmatrix} = \int_o^l \begin{bmatrix} 1 - \dfrac{x}{l} \\ \dfrac{x}{l} \end{bmatrix} \frac{1}{EI} \begin{bmatrix} 1 - \dfrac{x}{l} & \dfrac{x}{l} \end{bmatrix} dx$$

Let $\beta = I_1/I$; then

$$f_{\text{moment}} = \frac{1}{EI_1} \int_o^l \begin{bmatrix} 1 - \dfrac{x}{l} \\ \dfrac{x}{l} \end{bmatrix} \beta \begin{bmatrix} 1 - \dfrac{x}{l} & \dfrac{x}{l} \end{bmatrix} dx$$

$$= \frac{1}{EI_1} \int_o^l \beta \begin{bmatrix} \left(1 - \dfrac{x}{l}\right)^2 & \dfrac{x}{l}\left(1 - \dfrac{x}{l}\right) \\ \dfrac{x}{l}\left(1 - \dfrac{x}{l}\right) & \left(\dfrac{x}{l}\right)^2 \end{bmatrix} dx$$

For the sections 1 to 5,

$$\frac{x}{l} = (0, \tfrac{1}{4}, \tfrac{2}{4}, \tfrac{3}{4}, 1) = \zeta_1{}^T$$

$$1 - \frac{x}{l} = (1, \tfrac{3}{4}, \tfrac{2}{4}, \tfrac{1}{4}, 0) = \zeta_2{}^T$$

Define the matrix multiplication

$$A * B = \begin{bmatrix} a_{11}b_{11} & a_{12}b_{12} \\ a_{21}b_{21} & a_{22}b_{22} \end{bmatrix} = C$$

that is, a multiplication such that the terms of C are simply the products of corresponding terms of A and B. Then,

$$\left(\frac{x}{l}\right)^2 = \zeta_1 * \zeta_1$$

$$\left(1 - \frac{x}{l}\right)^2 = \zeta_2 * \zeta_2$$

$$\frac{x}{l}\left(1 - \frac{x}{l}\right) = \zeta_1 * \zeta_2$$

Then, define $\beta = \text{diag}\,\langle \beta_1 \quad \beta_2 \quad \beta_3 \quad \beta_4 \quad \beta_5 \rangle$. Thence,

$$f_{11} = W^T \beta \zeta_2 * \zeta_2$$
$$f_{12} = f_{21} = W^T \beta \zeta_1 * \zeta_2$$
$$f_{22} = W^T \beta \zeta_1 * \zeta_1$$

$$\begin{bmatrix} f_{11} & f_{12} \\ f_{21} & f_{22} \end{bmatrix} = \frac{1}{EI_1} \begin{bmatrix} W^T & \cdot \\ \cdot & W^T \end{bmatrix} \begin{bmatrix} \beta & \cdot \\ \cdot & \beta \end{bmatrix} \begin{bmatrix} \delta_2 * \delta_2 & \delta_1 * \delta_2 \\ \delta_2 * \delta_1 & \delta_1 * \delta_1 \end{bmatrix} \tag{4.179}$$

On expansion,

$$f_{11} = \frac{h}{3EI_1}[1 + 4\beta_2(3/4)^2 + 2\beta_3(2/4)^2 + 4\beta_4(1/4)^2]$$

$$f_{12} = f_{21} = \frac{h}{3EI_1}[4\beta_2 1/4 3/4 + 2\beta_3(2/4)^2 + 4\beta_4 3/4 1/4] \tag{4.180}$$

$$f_{22} = \frac{h}{3EI_1}[4\beta_2(1/4)^2 + 2\beta_3(2/4)^2 + 4\beta_4(3/4)^2 + \beta_5]$$

End Rotations

If a variable moment field m_x, shown in Fig. 4.40, is applied to the member, the end rotations $\langle \phi_i \quad \phi_j \rangle$ may be calculated by using Eq. (4.43), in which case

$m_x = bR$

Then,

$$\begin{bmatrix} \phi_i \\ \phi_j \end{bmatrix} = \frac{1}{EI_1} \int_o^l \begin{bmatrix} 1 - \dfrac{x}{l} \\ \dfrac{x}{L} \end{bmatrix} \beta m_x \, dx$$

Let

$m^T = \langle m_1 \quad m_2 \quad m_3 \quad m_4 \quad m_5 \rangle$

Finally,

$$\begin{bmatrix} \phi_i \\ \phi_j \end{bmatrix} = \frac{h}{3EI_1} \begin{bmatrix} W^T & \cdot \\ \cdot & W^T \end{bmatrix} \begin{bmatrix} \beta & \cdot \\ \cdot & \beta \end{bmatrix} \begin{bmatrix} \zeta_2 * m \\ \zeta_1 * m \end{bmatrix} \tag{4.181}$$

Expanding Eq. (4.181),

$$\phi_i = \frac{h}{3EI_1}(m_1 + 4\beta_2 3/4 m_2 + 2\beta_3 2/4 m_3 + 4\beta_4 1/4 m_4)$$

and,

$$\phi_j = \frac{h}{3EI_1}(4\beta_2 1/4 m_2 + 2\beta_3 2/4 m_3 + 4\beta_4 3/4 m_4 + m_5\beta_5) \tag{4.182}$$

Example 4.18 End rotations with uniform beam

Let m_x be produced by constant load w per unit length, so that

$$m_1 = m_5 = 0 \qquad m_2 = m_4 = \frac{3wl^2}{32} \qquad m_3 = \frac{wl^2}{8}$$

In this case, for four intervals, $h = l/4$, so that

$$\begin{aligned}\phi_i &= \frac{l}{12EI}\left[4\left(\frac{3}{4}\right)\frac{3wl^2}{32} + 2\left(\frac{2}{4}\right)\frac{wl^2}{8} + 4\left(\frac{1}{4}\right)\frac{3wl^2}{32}\right] \\ &= \frac{wl^2}{24EI} = \phi_j\end{aligned}$$

Equations (4.181) and (4.182) are readily incorporated in computer programs either for direct deflection calculation as in Sec. 4.8, method 1, or later in Chap. 6 for calculation of fixed-end moments.

3 Members for Which Shear Distortions Are Included

Once again, for calculation of the effects of shear distortions due to the end moments $\langle M_i \quad M_j \rangle$, Eq. (4.47) is used. The beam is shown in Fig. 4.41, and the shear-stress distribution due to transverse-stress resultant V is

$$\tau = \frac{VQ}{It}$$

However,

$$V = \begin{bmatrix} -\frac{1}{l} & \frac{1}{l} \end{bmatrix} \begin{bmatrix} M_i \\ M_j \end{bmatrix}$$

Therefore

$$\tau = \frac{Q}{It}\begin{bmatrix} -\frac{1}{l} & \frac{1}{l} \end{bmatrix} \begin{bmatrix} M_i \\ M_j \end{bmatrix}$$

For the infinitesimal element, the shear distortion is given as

$$\gamma = \frac{\tau}{G}$$

Thence,

$$r = \int_0^l b^T f b \, dV \, R$$

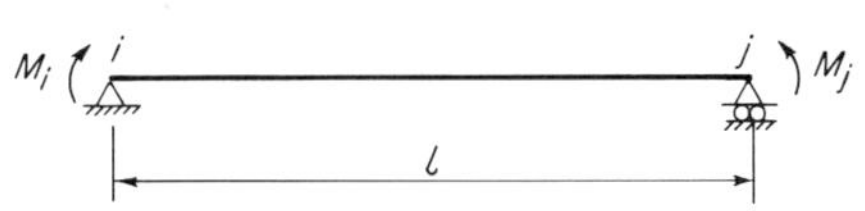

Figure 4.41

$$= \frac{1}{G}\int_o^l \begin{bmatrix} -\frac{1}{l} \\ \frac{1}{l} \end{bmatrix} \left(\frac{Q}{It}\right)^2 \begin{bmatrix} -\frac{1}{l} & \frac{1}{l} \end{bmatrix} dV\, R$$

$$= \frac{1}{G}\frac{1}{I^2}\int_o^l\int_{c_1}^{c_2} \frac{Q^2}{t^2}\frac{1}{l^2}\begin{bmatrix} 1 & -1 \\ -1 & 1 \end{bmatrix} t\,dy\,dx\,R$$

$$= \frac{1}{G}\frac{1}{I^2}\frac{1}{l^2}\begin{bmatrix} 1 & -1 \\ -1 & 1 \end{bmatrix}\int_o^l\int_{c_1}^{c_2}\frac{Q^2}{t}dy\,dx\,R$$

Let A_{eff} be the effective shear area considered for shear deflections, such that

$$A_{\text{eff}} = \frac{I^2}{\int_{c_1}^{c_2}\frac{Q^2}{t}dy}$$

Thence

$$\begin{bmatrix} \phi_i \\ \phi_j \end{bmatrix} = \frac{1}{GA_{\text{eff}}}\frac{1}{l}\begin{bmatrix} 1 & -1 \\ -1 & 1 \end{bmatrix}\begin{bmatrix} M_i \\ M_j \end{bmatrix}$$

That is,

$$f_{\text{shear}} = \frac{1}{GA_{\text{eff}}}\frac{1}{l}\begin{bmatrix} 1 & -1 \\ -1 & 1 \end{bmatrix}$$

Notice that f_{shear} is a singular matrix. This implies that a beam with shear resistance to deformation alone cannot carry the moments $\langle M_i \quad M_j \rangle$.

The complete member-flexibility matrix including shear and bending distortions is written

$$f = \frac{l}{6EI}\begin{bmatrix} 2+\alpha & 1-\alpha \\ 1-\alpha & 2+\alpha \end{bmatrix} \qquad \alpha = \frac{6EI}{GA_{\text{eff}}l^2} \tag{4.183}$$

4 Bending–flexibility Matrices for Prismatic Members, Axial–force Effects Included

Case 1 *Axial Compression* The effect of axial compression is clearly to increase bending flexibility, since the deflections produced by (M_i,M_j) cause P to induce moments that further increase the deflections and thus (ϕ_i,ϕ_j). The prismatic member of length l and with moments (M_i,M_j) and axial compression P is shown in Fig. 4.42. The moments M_x along the member are given by

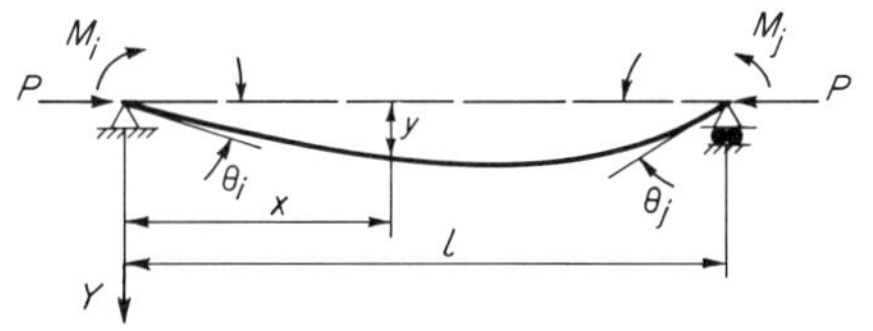

Figure 4.42 Prismatic-member axial compression and end moments.

$$M_x = \left(1 - \frac{x}{l}\right)M_i + \frac{x}{l}M_j + Py$$

The differential equation to the deflected shape is thence

$$EIy'' = -M_x = -\left(1 - \frac{x}{l}\right)M_i - \frac{x}{l}M_j - Py$$

Let

$$k^2 = \frac{P}{EI}$$

Then,

$$y'' + k^2y = -\frac{1}{EI}\left[\left(1 - \frac{x}{l}\right)M_i + \frac{x}{l}M_j\right] \tag{4.184}$$

The complementary function for Eq. (4.184) is

$$y_c = A \cos kx + B \sin kx$$

and the particular solution

$$y_P = \frac{-1}{k^2EI}\left[\left(1 - \frac{x}{l}\right)M_i + \frac{x}{l}M_j\right]$$

The complete solution is the sum of the complementary function and particular integral.

$$y = y_c + y_P = A \cos kx + B \sin kx - \frac{1}{k^2EI}\left[\left(1 - \frac{x}{l}\right)M_i + \frac{x}{l}M_j\right] \tag{4.185}$$

The constants A and B are determined from the boundary conditions, (1) when $x = 0, y = 0$, and (2) when $x = l, y = 0$.

1 $0 = A - \dfrac{M_i}{k^2EI} \qquad A = \dfrac{M_i}{k^2EI}$

2 $0 = -\dfrac{M_i}{k^2EI} \cos kl + B \sin kl - \dfrac{M_j}{k^2EI}$

$$B = \frac{1}{\sin kl k^2EI}(M_j - M_i \cos kl)$$

Substitute for A and B in Eq. (4.185).

$$y = \frac{1}{k^2EI}\left[M_i \cos kx + \frac{1}{\sin kl}(M_j - M_i \cos kl) \sin kx - \left(1 - \frac{x}{l}\right)M_i - \frac{x}{l}M_j\right] \tag{4.186}$$

On differentiation of Eq. (4.186) with respect to x, the equation to the slope of the deflected shape is given by

$$y' = \frac{1}{k^2EI}\left[-M_ik \sin kx + \frac{k}{\sin kl}(M_j - M_i \cos kl) \cos kx + \frac{1}{l}M_i - \frac{1}{l}M_j\right] \tag{4.187}$$

The end slopes ϕ_i and ϕ_j are determined by substituting $x = 0$ and $x = l$ respectively in Eq. (4.187). Thence,

$$\begin{aligned} y'_{(0)} = \phi_i &= \frac{1}{k^2EI}\left[M_i\left(\frac{1}{l} - \frac{k \cos kl}{\sin kl}\right) - M_j\left(\frac{1}{l} - \frac{k}{\sin kl}\right)\right] \\ -y'_{(l)} = \phi_j &= \frac{1}{k^2EI}\left[-M_i\left(\frac{1}{l} - \frac{k}{\sin kl}\right) + M_j\left(\frac{1}{l} - \frac{k \cos kl}{\sin kl}\right)\right] \end{aligned} \tag{4.188}$$

The notation is now changed slightly, so that

$$\beta = \frac{l}{2}\sqrt{\frac{P}{EI}} \qquad \beta = \frac{kL}{2}$$

and thus

$$\cos kl = \cos 2\beta = c$$
$$\sin kl = \sin 2\beta = s$$

With this change in notation, Eqs. (4.188) are rewritten in matrix form,

$$\begin{bmatrix} \phi_i \\ \phi_j \end{bmatrix} = \frac{1}{Pls}\begin{bmatrix} s - 2\beta c & -(s - 2\beta) \\ -(s - 2\beta) & s - 2\beta c \end{bmatrix}\begin{bmatrix} M_i \\ M_j \end{bmatrix} \tag{4.189}$$

If now the ϕ functions used in Ref. 21 are introduced,

$$\phi_1 = \beta \cot \beta$$

$$\phi_2 = \frac{1}{3}\frac{\beta^2}{1 - \beta \cot \beta}$$

$$\phi_3 = \frac{3\phi_2}{4} + \frac{\phi_1}{4}$$

$$\phi_4 = \frac{3\phi_2}{2} - \frac{\phi_1}{2}$$

$$\phi_5 = \phi_1\phi_2$$

after some algebra,

$$\begin{bmatrix}\phi_i\\ \phi_j\end{bmatrix}=\frac{l}{6EI}\begin{bmatrix}2\left(\dfrac{3}{4\phi_1}+\dfrac{1}{4\phi_2}\right) & \dfrac{3}{2\phi_1}-\dfrac{1}{2\phi_2}\\ \dfrac{3}{2\phi_1}-\dfrac{1}{2\phi_2} & 2\left(\dfrac{3}{4\phi_1}+\dfrac{1}{4\phi_2}\right)\end{bmatrix}\begin{bmatrix}M_i\\ M_j\end{bmatrix} \tag{4.190}$$

Case 2 Axial Tension When the member is subjected to axial tension, the effect is the reverse of that for axial compression. That is, the flexibility of the member is decreased. In this case, the equation for the deflected shape is

$$y''-k^2y=\frac{-1}{EI}\left[\left(1-\frac{x}{l}\right)M_i+\frac{x}{l}M_j\right] \tag{4.191}$$

Now the solution is

$$y=-\frac{M_i}{P}\cosh kx+\frac{\sinh kx}{P\sinh kl}(M_i\cosh kl-M_j)+\frac{1}{Pl}(-M_i+M_j) \tag{4.192}$$

and hence the equation for member slope is

$$y'=-\frac{kM_i}{P}\sinh kx+\frac{k\cosh kx}{P\sinh kl}(M_i\cosh kl-M_j)+\frac{1}{Pl}(-M_i+M_j) \tag{4.193}$$

Finally, let $s=\sinh kl$, $c=\cosh kl$ and $\beta=kl/2$. The end rotations are obtained from y'_o and y'_l. That is,

$$\begin{bmatrix}\phi_i\\ \phi_j\end{bmatrix}=\frac{1}{P}\begin{bmatrix}\dfrac{kc}{s}-\dfrac{1}{l} & \dfrac{1}{l}-\dfrac{k}{s}\\ \dfrac{1}{l}-\dfrac{k}{s} & \dfrac{kc}{s}-\dfrac{1}{l}\end{bmatrix}\begin{bmatrix}M_i\\ M_j\end{bmatrix} \tag{4.194}$$

Now, with

$$\phi_1=\beta\coth\beta \qquad \phi_2=\frac{1}{3}\,\frac{\beta^2}{\beta\coth\beta-1}$$

$$\phi_3=\tfrac{1}{4}(\phi_1+3\phi_2) \qquad \phi_4=\tfrac{1}{2}(\phi_1-3\phi_2)$$

after some algebra,

$$\begin{bmatrix}\phi_i\\ \phi_j\end{bmatrix}=\frac{l}{6EI}\begin{bmatrix}2\left(\dfrac{3}{4\phi_1}+\dfrac{1}{4\phi_2}\right) & \dfrac{3}{2\phi_1}-\dfrac{1}{2\phi_2}\\ \dfrac{3}{2\phi_1}-\dfrac{1}{2\phi_2} & 2\left(\dfrac{3}{4\phi_1}+\dfrac{1}{4\phi_2}\right)\end{bmatrix}\begin{bmatrix}M_i\\ M_j\end{bmatrix} \tag{4.195}$$

It should be noted that when Eqs. (4.190) and (4.195) are used, the effect of P on bending flexibility has been taken into consideration, whereas the effects of $\langle M_i \quad M_j \rangle$ on axial stiffness have not been. When the structure is statically determinate, the ϕ functions in Eqs. (4.190) and (4.195) may be calculated directly from the external loads. Later it will be seen that this is not possible in statically indeterminate structures in which the member flexibilities influence the values of the indeterminate forces, which in turn affect member forces and hence member flexibilities.

4.11 CALCULATION OF DEFLECTIONS ALONG A MEMBER

Deflections at any point of a statically determinate structure may be calculated by using Eq. (4.47) or (4.48) and simply including that point as a node of the structure. Thus for the portal frame shown in Fig. 4.43*a*, the deflections at point B are calculated by using nodes 1 to 5 and members (1) to (4) as shown.

However, it will be realized that every point added as a node will necessarily increase the size of the statics matrix to be solved (see Chap. 3) and also the sizes of the f and b matrices in Eq. (4.47), etc. Since the size of the statics matrix increases by 3 in plane frames and 6 in three-dimensional frames for each node added and the storage required for the whole matrix is a function of the square of its size, it will be appreciated that this process of adding nodes soon becomes prohibitive in computer storage. For example, in Fig. 4.43*b*, the three members (1), (2), and (3) have 15 nodes for which deflections are required, and this necessitates a 45×45 matrix for these nodes alone. In addition, if member (2)-(3)-(4) in Fig. 4.43*a* is nonprismatic, it will be inconvenient to calculate inertias as equal intervals over (2)-(3) and over a different interval in member (3)-(4) for flexibility-matrix generation. It is more practical to give inertias at equal intervals (for example, 10) over the whole length (2) to (4). It is convenient then to define members by their natural physical configuration. Thus they may be terminated at changes in structure geometry as, for example, in Fig. 4.43*b*, where the convenient number of members for the structure is five. When such a structure is analyzed for node forces, then node deflections and member bending moments are obtained. From the terminal member moments and loads on the member, bending moments may be calculated at those sections for which the inertias are given. Deflections may be calculated at points along the member in the local member coordinates, given the node deflections,

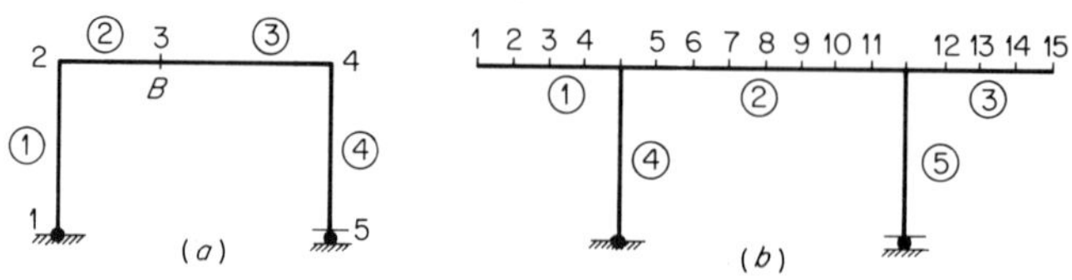

Figure 4.43 Member deflections.

using numerical quadrature on the M/EI diagram to obtain deflection relative to the end values. If Simpson's rule is used as the quadrature formula, deflections will be calculated for every second point of beam subdivision. Thus, for example, if M/EI is given at the tenth points along the member and deflections are required at all these tenth points, it is necessary to interpolate M/EI first to obtain values at the twentieth points. In the program given here, quadratic interpolation is used to obtain these half-interval values. Furthermore, special provision must be made for members that have a hinge along their length.

It is assumed, then, that displacements (u_i, v_i, θ_i) and (u_j, v_j, θ_j) have been obtained for member i, j in local coordinates. This involves a transformation such as those in Eq. (2.29) from the node displacements calculated in global coordinates.

Since the displacements (u_i, u_j) in the OX member coordinates (Fig. 4.44) will generally differ by only a small amount (the member distortions being small in its axial direction), the displacement u_x at distance x from i may be taken to be

$$u_x = u_i + (u_j - u_i)\frac{x}{l}$$

The v displacements will be of larger magnitude, so that v_x may differ significantly from (v_i, v_j), and are shown in Fig. 4.44b for a member with a hinge at A. Notice that a slope discontinuity occurs at the hinge.

Interpolation

Suppose that $\alpha = M/EI$ is given at the intervals n, $n + 2$, $n + 4$, etc. To calculate deflections at these points, the values of α are interpolated at $n + 1$, $n + 3$, etc. Thus, let

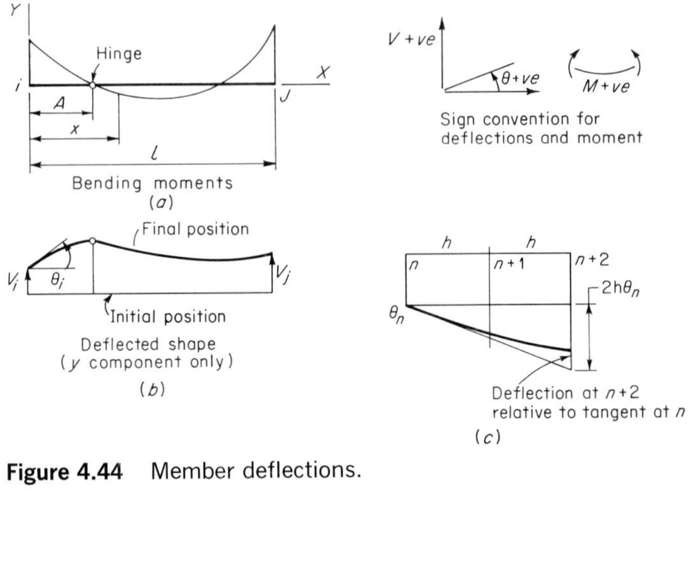

Figure 4.44 Member deflections.

$$\alpha = Ax^2 + Bx + C \tag{i}$$

The values are given by

$\alpha(n-1) \quad x = -1$

$\alpha(n) \quad x = 0$

$\alpha(n+1) \quad x = 1$

From these values of α are calculated A, B, and C as

$$A = ½[\alpha(n-1)] \quad -2\alpha(n) \quad +\alpha(n+1)$$
$$B = ½[\alpha(n+1)] \quad -\alpha(n-1)$$
$$C = \alpha(n)$$

To obtain $\alpha(n - ½)$, $\alpha(n + ½)$ substitute $x = \pm ½$ in Eq. (i).

A small segment of the beam is shown in Fig. 4.44c. It is seen that the slope and deflection at station $n + 2$ may be written in terms of the various quantities at n, $n + 1$, and $n + 2$.

Assuming that α is parabolic in its variation over the interval n to $n + 2$, the slope at $n + 2$ is given by

$$\theta_{n+2} = \theta_n + \frac{h}{3}(\alpha_n + 4\alpha_{n+1} + \alpha_{n+2}) \tag{4.196}$$

The deflection is given by

$$v_{n+2} = v_n + \theta_n 2h + \frac{h}{3}(\alpha_n 2h + 4\alpha_{n+1}h) \tag{4.197}$$

By using these two equations it is possible to progress from the left-hand end, which has slope θ_i and deflection v_i, to the slope $\bar{\theta}_j$ and deflection $\bar{v}_j$ at the right-hand end of the beam. Of course, because numerical quadrature has been used, $\bar{v}_j$ will be in general slightly different from v_j.

However, when a hinge is present, $\bar{v}_j$ will differ significantly from v_j because of the slope discontinuity there. When a hinge is present, an adjustment is made to deflections for $x > A$, the term A being the distance from i to the hinge. The correction will be

$$x > A \qquad v_x = \bar{v}_x + (v_j - \bar{v}_j)\frac{(x - A)}{(l - A)} \tag{4.198}$$

Finally, when all (u_x, v_x) have been calculated, they must be transformed back to global components [Eq. (2.29)].

Example 4.19 Calculate deflections at one-tenth points of member AC shown in Fig. 4.45

The member shown in Fig. 4.45 is prismatic with $I = 100$ in. and $E = 30{,}000$ in.-kip units. The deflections at A and C are calculated to be

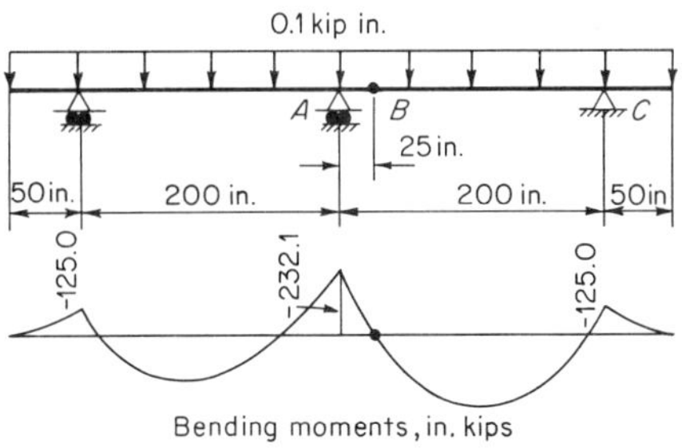

Figure 4.45 Deflection of structure—member with hinge.

$$\theta_A = -0.004586 \text{ rad}$$

$$v_A = 0$$

$$v_C = -1.75 \text{ in.}$$

From the program given in Appendix E, the deflections at the one-tenth points in member AC are:

Distance x/l from A	*Deflection*
0	0.0
0.1	0.08024
0.2	−0.15194
0.3	−0.47216
0.4	−0.76348
0.5	−1.01649
0.6	−1.22709
0.7	−1.39649
0.8	−1.53128
0.9	−1.64337
1.0	−1.75

4.12 TRANSFORMATION OF FLEXIBILITY MATRICES OF MEMBERS TO FRAME COORDINATES

In this section the flexibility matrices of the type used for members in tree structures are transformed to global coordinates. In Sec. 3.8, Eq. (3.80), it has been shown that member forces may be calculated by the multiplication of a rotation transformation matrix and a displacement-of-axes type of transformation matrix. If member forces are expressed in global coordinates, then the rotation transformation is unnecessary. However, in deflection calculations the flexibility matrix must then be expressed in global coordinates. The basic member forces have been given in Fig. 3.5*b* and are given again here in Fig. 4.45 for a plane frame member. For the forces $S'^T = \langle F'_{ix} \quad F'_{iy} \quad M'_i \rangle$, the flexibility matrix is given by

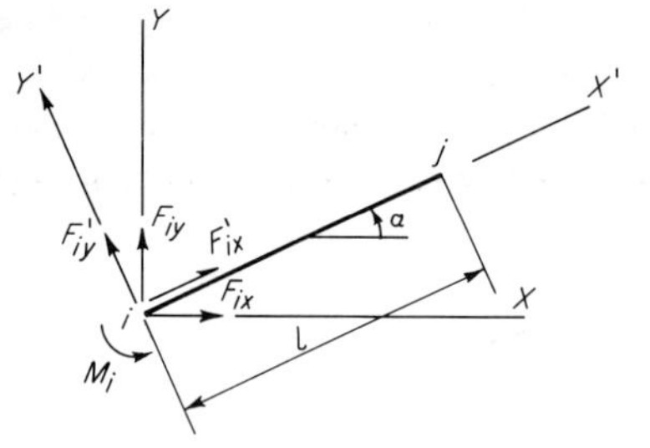

Figure 4.46 Member forces—global and local coordinates.

$$f' = \begin{bmatrix} f'_{11} & 0 & 0 \\ 0 & f'_{22} & f'_{23} \\ 0 & f'_{32} & f'_{33} \end{bmatrix} \tag{4.199}$$

The transformations for member quantities are given from Eqs. (2.29)

$$S' = L^T S \tag{4.200}$$

$$v' = L^T v \tag{4.201}$$

In this case,

$$L^T = \begin{bmatrix} \cos\alpha & \sin\alpha & 0 \\ -\sin\alpha & \cos\alpha & 0 \\ 0 & 0 & 1 \end{bmatrix} \tag{4.202}$$

Using Eq. (4.199),

$$v' = f'S' \tag{4.203}$$

Substituting Eqs. (4.200) and (4.201),

$$L^T v = f'L^T S \tag{4.204}$$

Hence,

$$v = Lf'L^T S \tag{4.205}$$

The flexibility matrix in the global coordinates is thus

$$f = Lf'L^T \tag{4.206}$$

Then,

$$v = fS \tag{4.207}$$

The transformation in Eq. (4.206) is used for f in global coordinates and when used, is similar in approach to the direct-stiffness method (see Chap. 7). It will be reencountered in tree-structure analysis in Chap. 5.

PROBLEMS

4.1 The trusses shown in Fig. P4.1 below have been analyzed in Prob. 3.1. Calculate the node deflections by using the equation $r = b^T f b R$. Let $EA = 1$ for all members.

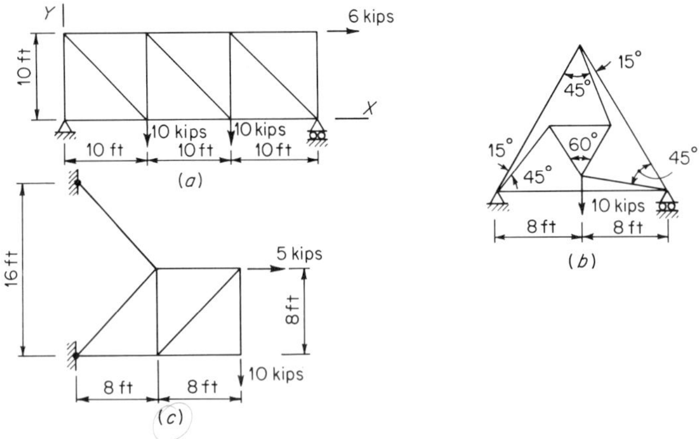

Figure P4.1

4.2 Calculate the deflections (displacements only) at the nodes of the beam and frame structures shown in Fig. P4.2 by using Eq. (4.14).

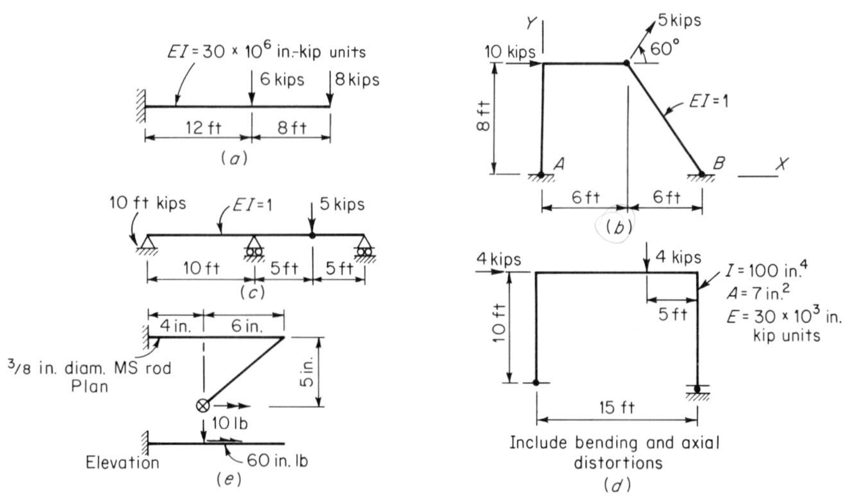

Figure P4.2

4.3 Calculate the Y and Z deflection components at the end of the clipped angle shown in Fig. P4.3 by using Eq. (4.62) and numerical integration.

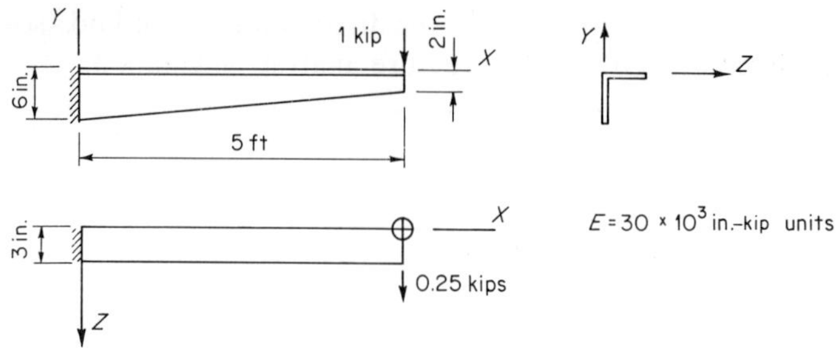

Figure P4.3

4.4 Recalculate the node deflections of the frame (*b*), shown in Fig. P4.2, if support *B*:

a Has a vertical movement of -1.0 in.

b Can be considered supported elastically by two springs in the vertical and horizontal directions, such that:

(*i*) $f_v = 0.2$ in.-kip

(*ii*) $f_h = 0.1$ in.-kip

4.5 Calculate the position of the elastic center for the frame shown in Fig. P4.4.

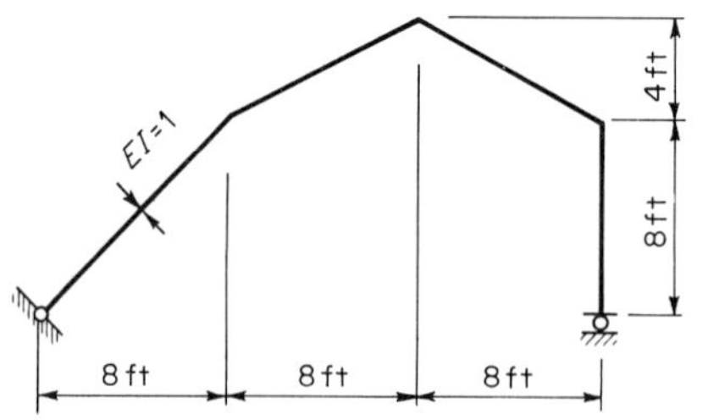

Figure P4.4

4.6 Calculate the deflection at point *A* of the beams shown in Fig. P4.5, using:

a Generalized flexibility matrices for the member with distributed load

b The lack-of-fit method given in Eq. (4.142)

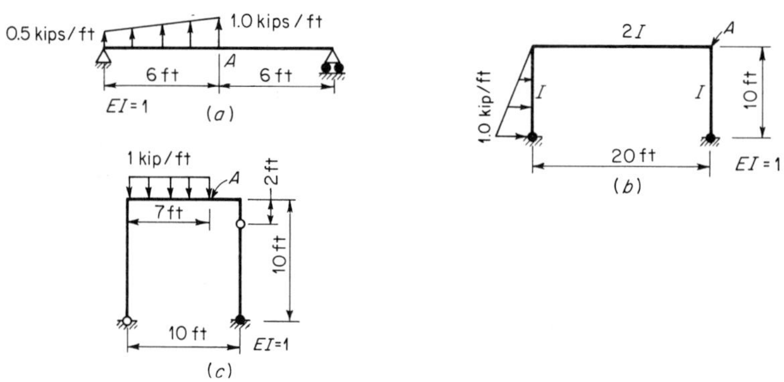

Figure P4.5

4.7 Calculate the generalized flexibility matrices for the prismatic beam subjected to end moments M_i, M_j and the sinusoidal loading shown in Fig. P4.6.

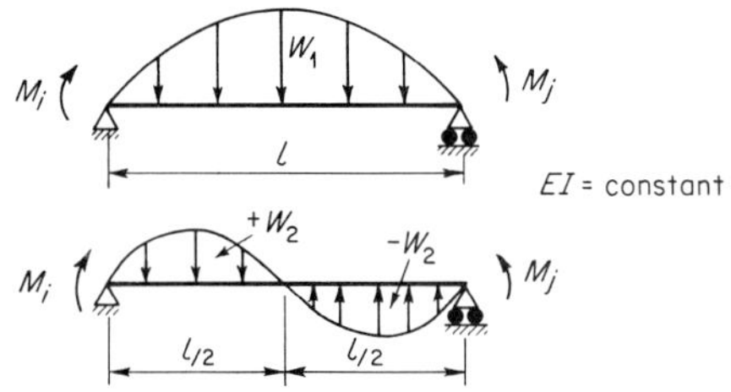

Figure P4.6

4.8 Recalculate the deflection at point A of the structures shown in Fig. P4.5, using the idea of fixed-end and release forces given in Sec. 4.8, method 2.

4.9 Calculate the deflection at the one-fourth points of the members AB of the structures shown in Fig. 4.8a and b, using the numerical quadrature given in Sec. 4.11.

5 STATICALLY INDETERMINATE STRUCTURES

5.1 CONDITIONS FOR STABILITY AND STATIC DETERMINACY OR INDETERMINACY OF STRUCTURES; SOME ELEMENTS OF LINEAR GRAPH THEORY; CONSISTENCY OF EQUATIONS IN TERMS OF RANK OF THE AUGMENTED COEFFICIENT MATRIX

This far, only stable, statically determinate structures have been discussed. For simple structures stability may be assessed on an intuitive basis. Of course, if analysis is attempted for an unstable structure, it will be found that in the generation of the force transformation matrix, singularity is encountered. That is, a matrix with zero diagonal term will be generated when the joint equations of equilibrium are being solved for member forces. This in fact forms a convenient method for testing for stability. It is informative, however, to examine some elements of linear graph theory and to use these to express the general conditions of stability and indeterminacy of structures.

Elements of Linear Graph Theory

Nodes, Branches, and Linear Graphs

Consider any configuration of curves in space, such as those shown in Fig. 5.1. Choose a set of points, called *nodes,* that define a decomposition of the graph into curves, called *branches,* such that:

1 Each branch is terminated by two different nodes.

2 No two branches have a point in common that is not a node.

A terminal point such as A in Fig. 5.1 will always be a node.

It must be realized that such a choice of nodes and branches is not unique; two such decompositions of members and nodes for the same graph are shown in Fig. 5.1*a* and *b*. Such a collection of nodes and branches is called a *linear graph.* It should be evident that any line representation of a frame structure

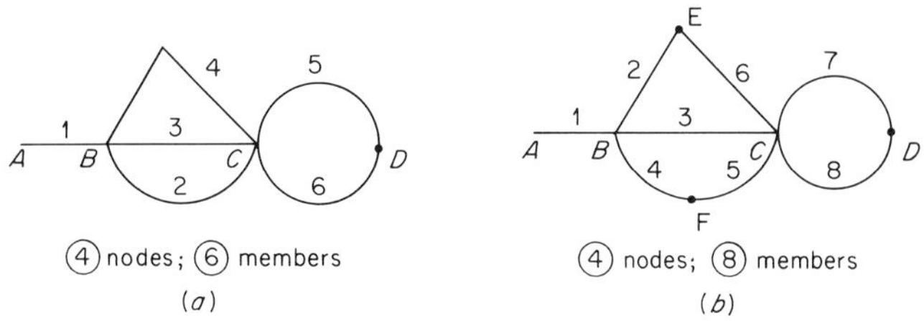

Figure 5.1 Linear graphs.

with its nodes and members is a linear graph. This linear graph has properties that are independent of the geometry and the deflection properties of the structure members.

Paths, Meshes, and Connected Graphs

A path is defined by an ordered sequence of letters and numbers denoting nodes and branches traversed in that order. Only simple paths in which each branch is traversed in one sense are considered. A path is called *open* if defined by a sequence that does not contain the same node twice and called *closed* if it does. A closed path that contains only one repeated node and begins and ends at that node is called a *mesh.*

Junctions, Loops, and Minimal Graphs

There are many branch-node configurations for any graph, among which there will be one with a minimum number of nodes and branches. This particular graph will be called a *minimal graph.*

Any point, such as C in Fig. 5.1, that is crossed along more than one path and must therefore be the terminal point of at least three branches and necessarily a node is called a *juncture.* Any mesh having at the most one juncture is called a *loop*. In a loop, at least one point other than a juncture must be chosen as a node.

Let L be the number of loops in a graph; J, the number of junctures; E, the number of ends; N, the number of nodes; B, the number of branches. Let B_m and N_m be the number of branches and the number of nodes of a minimal graph representing the same configuration. Then, except for an isolated loop,

$$N_m = L + J + E \tag{5.1}$$

For the configuration in Fig. 5.1*a*,

$$L = 1 \qquad J = 2 \qquad E = 1$$

Therefore, $N_m = 4$ in this case. It will be seen then that Fig. 5.1*a* is a minimal graph. For each node added to a minimal graph, an additional branch is added. Therefore,

$$B = B_m + (N - N_m) \tag{5.2}$$

Then,

$$B - N = B_m - N_m$$

However, B_m and N_m are unique for a given configuration. Therefore the quantity I, defined as their difference, is unique. That is,

$$B - N = I \tag{5.3}$$

A tree is a graph containing no meshes. A tree is shown in Fig. 5.2 and exhibits the general tree properties:

1 In any tree there is a unique open path from each node to every other node.

2 In any tree the number of branches is one less than the number of nodes. Denote the tree quantities B_T and N_T. Then

$$B_T - N_T = -1 = I_T \tag{5.4}$$

3 In any graph it is always possible to find one tree connecting every node.

Chords and Basic Meshes

In any arbitrary graph, choose one tree connecting all its nodes. Those branches that are not part of the tree will be called *chords,* corresponding to that choice of tree. Two such choices of trees and chords are shown in Fig. 5.2*b* and *c*. Call the mesh defined by any one chord and the unique open path in the tree connecting the terminal nodes of this chord a *basic mesh.*

The number of branches B in a linear graph is equal to the number of branches B_T in the tree connecting all its N nodes plus the number of chords removed in constructing the tree. That is,

$$B = B_T + M \tag{5.5}$$

The number of basic meshes M is thus given as

$$M = B - B_T = B - N_T + 1 \tag{5.6}$$

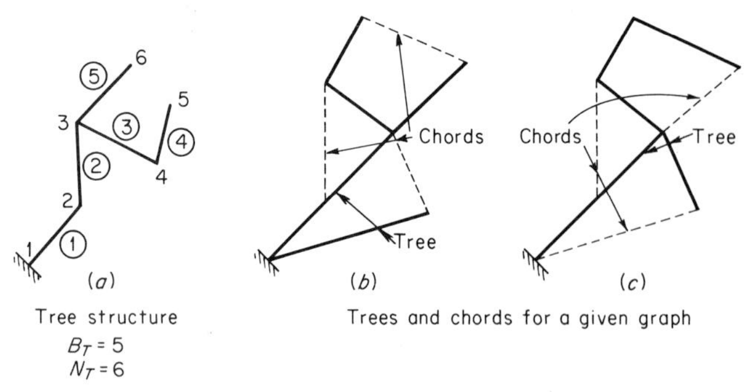

Figure 5.2 Graphs and trees.

However, $N_T = N$, so that

$$M = B - N + 1 = I + 1 \tag{5.7}$$

Since I is unique, so also is M. For a tree,

$$M_T = 0 \tag{5.8}$$

Topological Properties of a Graph

It will be see… the graph properties outlined so far depend only on the branch-n… ration and are completely independent of the graph deforma… tely equivalent linear graphs are shown in Fig. 5.3.

P… Continuous Structure, … Structure

… for a member t is defined as the six force … $M_x \quad M_y \quad M_z\rangle$ in the member, section t. … near combination of the stress resultants … e, the three releases

… incapable of transmitting bending moment. … equations, independent of joint equilibrium equations, … *of releases*. If no releases exist in a structure, it is called … *ucture*. The continuous structure with the configuration iden… hat of a structure with releases is called its *corresponding con… structure.*

…tically Equivalent Structure

Replace the ground support of a structure by any linear graph system that is a tree, connecting all the support points, including free ends but not physical cuts, so that the release system is not affected. Such tree structures representing supports are shown by the wavy lines in Fig. 5.4*a* and *b*.

Definitions of Determinacy, Indeterminacy, and Stability

A structure is statically determinate if under any applied loads it is possible to uniquely determine a stress-resultant matrix S_t that satisfies all release

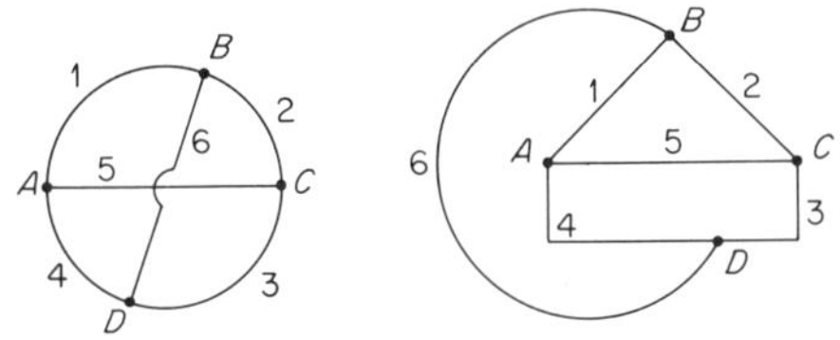

Figure 5.3 Pictorial representations of one linear graph.

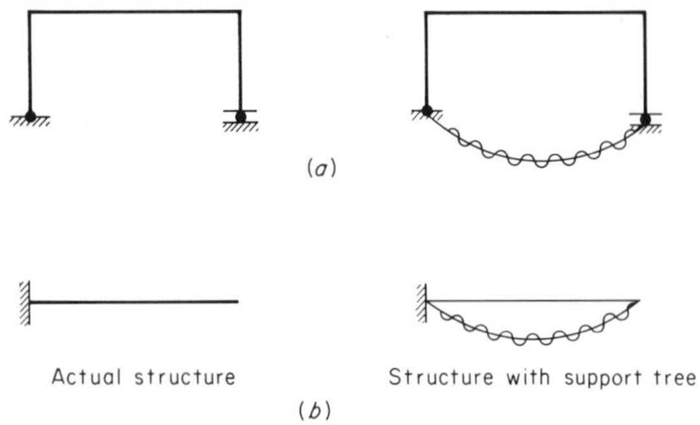

Figure 5.4 Support structures.

conditions by using only the equations of equilibrium. If, however, under any applied loads, it is possible only to determine the stress resultants everywhere as a linear function of k stress resultants with the release conditions' being satisfied, then the structure is statically indeterminate to the kth degree.

If, however, a release condition is necessarily violated, the structure is statically unstable.

These conditions are now expressed mathematically in terms of matrix rank. It should be noted that when any continuous structure, of which the linear graph is a tree, is acted on by a system of forces in equilibrium, the stress resultants are uniquely determined everywhere by equilibrium equations only (see Sec. 3.8).

Formulation of Necessary and Sufficient Mathematical Conditions for Determinacy and (or) Stability

Consider any structure $\mathbb{S}'$ with r releases, expressible by $X_i = 0$, $i = 1, 2, 3, \ldots, r$. The equation $X_i = 0$ represents some linear combination of stress resultants at a particular section. The structure $\mathbb{S}'$ is acted on by an arbitrary set of loads denoted by R'.

Let $\mathbb{S}$ denote a statically equivalent structure acted on by an equilibrium system of external loads R that are composed of R' acting on the part $\mathbb{S}'$ of $\mathbb{S}$ and $R - R'$ (a system self-equilibrating to R') acting anywhere on the added tree $\mathbb{S} - \mathbb{S}'$. Then as far as determinacy is concerned, $\mathbb{S}$ acted on by R may be considered instead of $\mathbb{S}'$ acted on by R'. Thus, for example, the beam in Fig. 5.5a is the structure $\mathbb{S}'$ and has loads R'. The added tree $\mathbb{S} - \mathbb{S}'$ connects the two supports and also has a branch to the end of the cantilever C. On this tree the reactions $\langle R_1 \quad R_2 \quad R_3 \rangle = R - R'$, satisfy the necessary equilibrium equations.

Let $\mathbb{S}_c$ be the corresponding continuous structure to $\mathbb{S}$, and consider it acted on by R. Choose any linear graph to represent $\mathbb{S}_c$ having N nodes and B branches and a tree connecting all nodes. For any such choice there are M basic meshes and chords, and from Eq. (5.7),

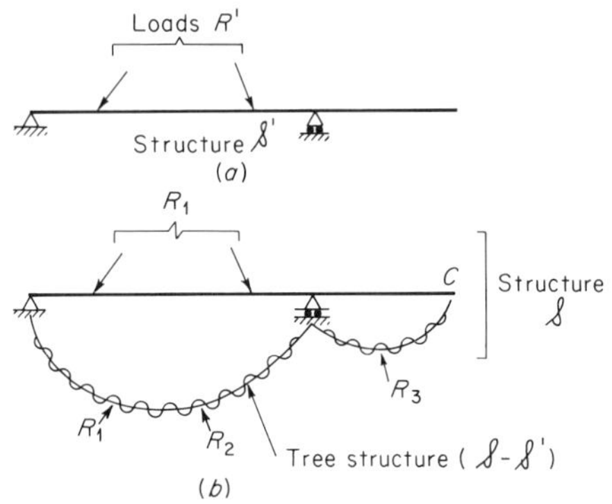

Figure 5.5 Statically equivalent structure.

$$M = B - N + 1 = I + 1$$

Let $S^T = \langle S_1{}^T \quad S_2{}^T \quad \cdots \quad S_M{}^T \rangle$ denote the stress-resultant vector composed of M subvectors each of dimension 6 in the three-dimension structure. Then S is a $6M \times 1$ matrix.

Imagine physical cuts at the M chord sections, and consider the stress resultants on the cut sections to be externally applied. Because the structure remaining after the cuts are made is a tree, it is now possible to express uniquely the stress resultants in $\mathbb{S}_c$ as linear combinations of the S chord forces and the external loads R' by using only the equilibrium equations of the joints. Since no release conditions are included, the stress resultants at these points will be nonzero (in general) and will be related to S and R' by the force transformation matrices denoted b_{xc} and $b_{xR'}$ respectively. Then, if $X^T = \langle X_1 \quad X_2 \quad \cdots \quad X_r \rangle$, there being r releases,

$$X = b_{xc}S + b_{xR'}R' \tag{5.9}$$

The matrix b_{xc} is of order $r \times 6M$ and depends only on

1 The geometry of the structure

2 The location and the type of releases

3 The choice of the chord sections chosen to induce determinacy in $\mathbb{S}_c$

The vector $b_{xR'}R' = P$ is a vector of r rows and must be completely arbitrary in its terms, since for stability this condition is necessary.

To return from $\mathbb{S}_c$ to the actual structure $\mathbb{S}$ (or $\mathbb{S}'$ since these are equivalent), it is necessary and sufficient to set X equal to zero. That is,

$$b_{xc}S + P = 0 \tag{5.10}$$

Equation (5.10) is a set of r linear simultaneous equations in $6M$ unknowns S. If for an arbitrary vector P, there is a unique solution S, $\mathbb{S}'$ is statically determinate, because all release conditions have been satisfied and the stress resultants (that is, member forces) have been determined from the equations of joint equilibrium alone. Several cases arise, depending on whether $r \lesseqgtr 6M$, and all these must be considered in the complete expression of stability, determinacy, and (or) indeterminacy. These conditions are expressed in terms of the ranks of the matrices $[b_{xc}]$ and $[b_{xc} \mid P]$. The second matrix is the coefficient matrix augmented with the extra column composed of the arbitrary force vector. The mathematical expression for the solution of Eq. (5.10) is given at the end of this section [see Eqs. (5.11)ff].

The Expression of Determinacy and Stability in Terms of Matrix Rank

Let ρ be the rank of b_{xc}, the $r \times 6M$ matrix of the coefficients of S, and ρ_A the rank of the augmented $r \times 6M + 1$ matrix that consists of $[b_{xc} \mid P]$. Then five cases arise as follows:

1 If $r > 6M$, shown diagrammatically. In this case, $\rho_A > \rho$ because by adding an

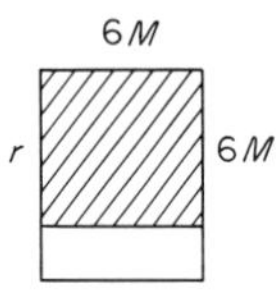

arbitrary column P, it is possible to increase the rank by 1. The structure $\mathbb{S}'$ is unstable, and no solution exists for R'.

2 If $r = 6M$ and $\rho = r$, then $\rho_A = \rho$, since $r \times r$ is the largest square matrix

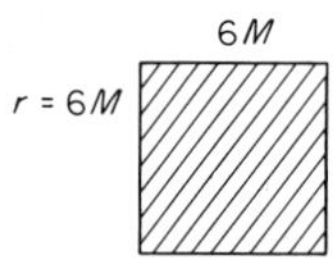

possible. In this case, b_{xc} is nonsingular, and a unique solution exists. The structure $\mathbb{S}'$ is statically determinate.

3 If $r = 6M$ and $\rho < r$, then $\rho_A > \rho$. This case is similar to case 1 in that the arbitrary column added to b_{xc} to form the augmented matrix may be chosen to increase the rank so that $\rho_A = \rho + 1$. As in 1, the structure is unstable.

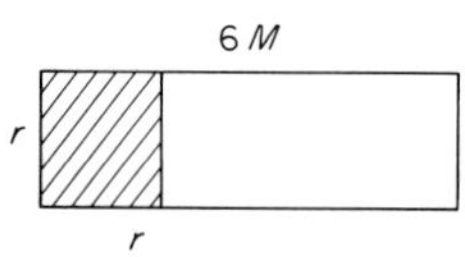

4 If $r < 6M$ and $\rho < r$, then, as in 1 and 3 above, $\rho_A = \rho + 1$ and the structure $\mathbb{S}'$ is unstable.

5 If $r < 6M$ and $\rho = r$, then, as in 2 above, $\rho_A = \rho$. Then, $6M - r$ stress resultants may be chosen arbitrarily. When these have been chosen, the re-

maining r stress resultants are uniquely determined. The structure $\mathcal{S}'$ is statically indeterminate to the kth $= 6M - r$ degree.

In the solution of the joint equilibrium equations encountered in Chap. 3, certain observations are apparent. Firstly, the solution by these equations alone is possible only if the coefficient matrix is square. If the number of equations, r, is greater than the number of unknowns, no solution exists. If the number of unknowns, $6M$, is greater than the number of equations, r, the situation warrants further investigation. It will be this class of structure, $r < 6M$, that is studied in this chapter.

Example 5.1 Unstable structure

The structure with three pins in a straight line shown in Fig. 5.6 is a well-known case of instability, since the joint equation of equilibrium $\Sigma F_y = 0$ is indeterminate at node 2. The release conditions are $M_1 = M_2 = M_3 = 0$, so that Eq. (5.10) is written

$$\begin{bmatrix} M_1 \\ M_2 \\ M_3 \end{bmatrix} = \begin{bmatrix} 0 \\ 0 \\ 0 \end{bmatrix} = \begin{bmatrix} 0 & 1 & l_1 \\ 0 & 1 & -l_2 \\ 0 & 1 & -(l_2 + l_3) \end{bmatrix} \begin{bmatrix} R_1 \\ R_2 \\ R_3 \end{bmatrix} + \begin{bmatrix} P_1 \\ P_2 \\ P_3 \end{bmatrix}$$

Since b_{xc} has a zero column, its rank is equal to 2. That is, $\rho = 2$. It is evident that the rank of $\rho_A = 3$, since P is arbitrary.

The structure is unstable.

Example 5.2 Statically determinate structure

The simple beam with one end cantilevered and its corresponding continuous structure are shown in Fig. 5.7. In this case, $M = 2$, so that there are six chord forces S as shown. The release conditions are also six in number and are

$$M_A = M_B = H_B = M_C = H_C = V_C = 0$$

The expression for the forces X at the releases is determined in terms of

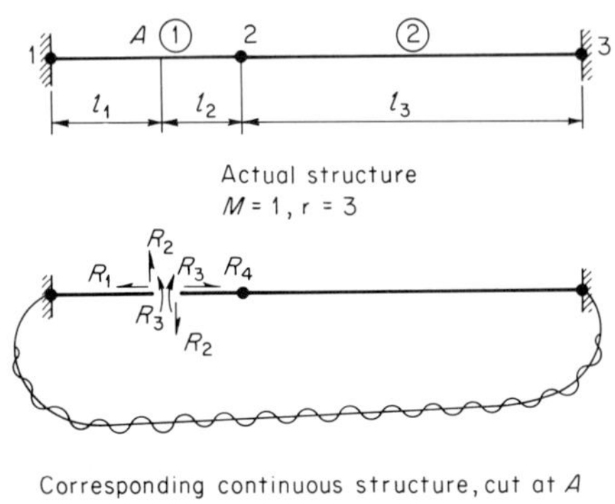

Figure 5.6 Unstable structure.

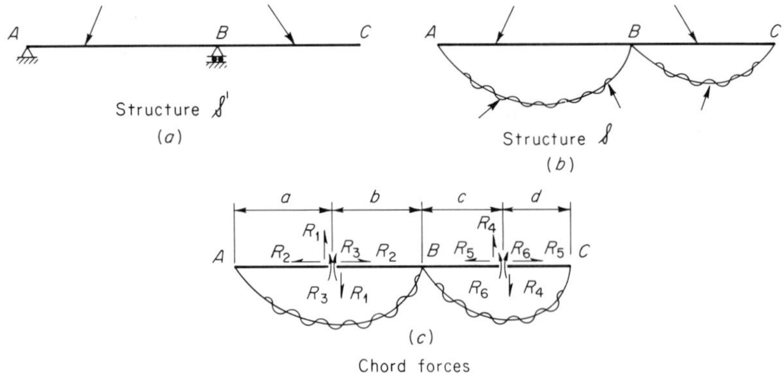

Figure 5.7 Determinate structure.

$\langle R_1 \quad R_2 \quad \cdots \quad R_6 \rangle$.

$$\begin{bmatrix} M_A \\ M_B \\ H_B \\ M_C \\ H_C \\ V_C \end{bmatrix} = \begin{bmatrix} 0 \\ 0 \\ 0 \\ 0 \\ 0 \\ 0 \end{bmatrix} = \begin{bmatrix} -a & 0 & -1 & 0 & 0 & 0 \\ -b & 0 & 1 & -c & 0 & -1 \\ 0 & 1 & 0 & 0 & -1 & 0 \\ 0 & 0 & 0 & -d & 0 & 1 \\ 0 & 0 & 0 & 0 & 1 & 0 \\ 0 & 0 & 0 & -1 & 0 & 0 \end{bmatrix} \begin{bmatrix} R_1 \\ R_2 \\ R_3 \\ R_4 \\ R_5 \\ R_6 \end{bmatrix} + \begin{bmatrix} P_1 \\ P_2 \\ P_3 \\ P_4 \\ P_5 \\ P_6 \end{bmatrix} = 0$$

It is evident that to determine the rank of this coefficient matrix by inspection is not an easy task. If, however, it is inverted, it will be found to be nonsingular. That is, its rank is 6. Then $\rho = \rho_A = 6$, and the structure in Fig. 5.7*a* is stable and determinate. Notice that in the inversion, each row must be searched to choose a nonzero value since in some columns the pivotal (diagonal) value is zero. In the equations for Example 5.1, the first column contained all zeros.

Example 5.3 Statically indeterminate structure

The portal frame shown in Fig. 5.8*a* has two release conditions, $M_A = M_B = 0$. The number of meshes, $M = 1$, and the number of chord forces = 3. The rank of the coefficient matrix is 2, and so the structure is statically indeterminate to the $k = 1$ degree. Then,

$$\begin{bmatrix} M_A \\ M_B \end{bmatrix} = \begin{bmatrix} 0 \\ 0 \end{bmatrix} = \begin{bmatrix} -a & -h & -1 \\ -b & h & 1 \end{bmatrix} \begin{bmatrix} R_1 \\ R_2 \\ R_3 \end{bmatrix} + \begin{bmatrix} P_1 \\ P_2 \end{bmatrix}$$

If, for example, $R_2 = 10$ kips and $P_1 = P_2 = 0$, then R_1 and R_3 are uniquely determined, since

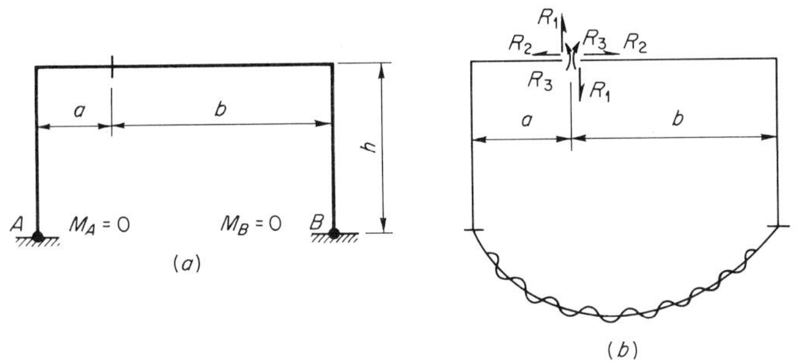

Figure 5.8 Statically indeterminate structure (stable). (*a*) Actual structure; (*b*) continuous structure.

$$\begin{bmatrix} -a & -1 \\ -b & 1 \end{bmatrix} \begin{bmatrix} R_1 \\ R_3 \end{bmatrix} = -h \begin{bmatrix} -10 \\ 10 \end{bmatrix}$$

$$\begin{bmatrix} R_1 \\ R_2 \end{bmatrix} = \frac{h}{l} \begin{bmatrix} -1 & b \\ 1 & -a \end{bmatrix} \begin{bmatrix} -10 \\ 10 \end{bmatrix}$$

Solution of Sets of Linear Simultaneous Equations

In this section the general conditions for the solution of m equations in n unknowns are derived in terms of the rank of the coefficient and augmented coefficient matrix.

The rank r of the matrix A is defined as the order of the submatrix of A of the highest order that has a nonvanishing determinant.

Thus, consider the m equations with n unknowns, with coefficient matrix A written in full as Eqs. (5.11) and in matrix form as Eq. (5.12).

$$\begin{bmatrix} a_{11} & a_{12} & \cdots & a_{1n} \\ a_{21} & a_{22} & \cdots & \\ \cdots & \cdots & \cdots & \cdots \\ a_{m1} & a_{m2} & \cdots & a_{mn} \end{bmatrix} \begin{bmatrix} d_1 \\ d_2 \\ \vdots \\ d_n \end{bmatrix} = \begin{bmatrix} b_1 \\ b_2 \\ \vdots \\ b_n \end{bmatrix} \tag{5.11}$$

or

$$A\delta = \beta \tag{5.12}$$

Assume that the equations are arranged so that the leading square submatrix A_{11} is of order r and is nonsingular. The rank of A is taken to be r. Write Eq. (5.12) in partitioned form:

$$\begin{bmatrix} A_{11} & A_{12} \\ A_{21} & A_{22} \end{bmatrix} \begin{bmatrix} \delta_1 \\ \delta_2 \end{bmatrix} = \begin{bmatrix} \beta_1 \\ \beta_2 \end{bmatrix} \tag{5.13}$$

Premultiply both sides of Eq. (5.13) by the matrix

$$P = \begin{bmatrix} I & 0 \\ -A_{21}A_{11}^{-1} & I \end{bmatrix} \tag{5.14}$$

Notice that this is possible since $|A_{11}| \neq 0$, a necessary condition for A_{11}^{-1} to exist. Then,

$$\begin{bmatrix} A_{11} & A_{12} \\ 0 & A_{22} - A_{21}A_{11}^{-1}A_{12} \end{bmatrix} \begin{bmatrix} \delta_1 \\ \delta_2 \end{bmatrix} = \begin{bmatrix} \beta_1 \\ -A_{21}A_{11}^{-1}\beta_1 + \beta_2 \end{bmatrix} = \begin{bmatrix} \gamma_1 \\ \gamma_2 \end{bmatrix} \tag{5.15}$$

However, the matrix $A_{22} - A_{21}A_{11}^{-1}A_{12}$ must be a null matrix since any nonzero value when combined with A_{11} would produce a matrix of rank $r + 1$, and by definition this is not possible. Therefore,

$$\begin{bmatrix} A_{11} & A_{12} \\ 0 & 0 \end{bmatrix} \begin{bmatrix} \delta_1 \\ \delta_2 \end{bmatrix} = \begin{bmatrix} \gamma_1 \\ \gamma_2 \end{bmatrix} \tag{5.16}$$

These equations may be solved for δ_1 if, and only if, $\gamma_2 = 0$, that is,

$$-A_{21}A_{11}^{-1}\beta_1 + \beta_2 = 0 \tag{5.17}$$

If $\gamma_2 \neq 0$, the solution of Eq. (5.16) is not possible, and the equations are said to be inconsistent.

If $\gamma_2 = 0$, the solution of Eq. (5.16) is possible and the equations are consistent.

Then from (5.16),

$$\begin{aligned} A_{11}\delta_1 + A_{12}\delta_2 &= \beta_1 \\ \delta_1 &= -A_{11}^{-1}(A_{12}\delta_2 + \beta_1) \end{aligned} \tag{5.18}$$

The unknowns δ_2 may be chosen arbitrarily.

These conditions may be expressed in terms of matrix rank by comparing the matrices

$$\begin{bmatrix} A_{11} & A_{12} \\ 0 & 0 \end{bmatrix} \quad \text{and} \quad \begin{bmatrix} A_{11} & A_{12} & \gamma_1 \\ 0 & 0 & \gamma_2 \end{bmatrix} \tag{5.19}$$

These two matrices are of the same rank if, and only if, $\gamma_2 = 0$. Premultiply both of these matrices by P^{-1}, a process that does not alter their ranks (see texts containing theory of elementary transformations), and return them to the form

$$\begin{bmatrix} A_{11} & A_{12} \\ A_{21} & A_{22} \end{bmatrix} \quad \begin{bmatrix} A_{11} & A_{12} & \beta_1 \\ A_{21} & A_{22} & \beta_2 \end{bmatrix} \tag{5.20}$$

Hence, the m nonhomogeneous linear equations in n unknowns are consistent and solvable in terms of $n - r$ parameters if, and only if, the rank of the augmented matrix $[A : \beta]$ is equal to the rank r of A.

5.2 DEFLECTION-INFLUENCE COEFFICIENTS; DEFLECTIONS; FOUNDATION SETTLEMENT; MAXWELL-MOHR SUPERPOSITION EQUATIONS, LOADS ALONG MEMBERS

It has been demonstrated in Sec. 5.1 that for the stable but statically indeterminate structure, the equations of joint equilibrium are insufficient for the calculation of all member forces and structure reactions. The indeterminacy (or redundancy) of the structure may be classified as internal, that is, member forces are unknown (as in Fig. 5.9*a*), or as external (as in Fig. 5.9*b*), in which case the number of reaction components are in excess of those necessary for the equilibrium of the structure when considered in its entirety as a rigid body. Naturally when the structure is complex in form, it will generally contain both types of indeterminate forces (as in Fig. 5.9*c*).

If the redundant structure is to be analyzed for member forces and node deflections due to applied loads and variation of member temperature, resort must be had to additional equations involving in some way the unknown force quantities independent of those already obtained from equilibrium conditions. It should be noticed that there will generally be several ways in which a given structure may be made determinate by cutting members or removing reactions. In Fig. 5.9*a*, for example, either members (1) and (2) may be chosen with F_1 and F_2, their forces, as the unknown quantities, or alternatively members A and B may be taken with F_A and F_B as the unknown forces.

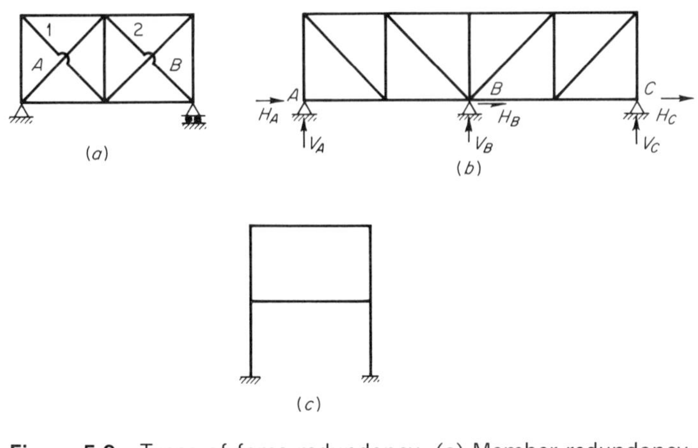

Figure 5.9 Types of force redundancy. (*a*) Member redundancy, number of redundants = 2; (*b*) reaction redundancy, number of redundants = 3; (c) member force and reaction redundancy, number of redundants = 6.

The analysis of the structure may therefore be conceived in two stages. In the first stage, the structure is cut so that all the redundancies are removed; the structure remaining must of course be made stable as well as determinate. A convenient way of doing this for a rigid frame structure is to choose a tree structure as the determinate base system. For this determinate structure, deflections are calculated as functions of the external load system and the stress resultants (or forces) at the cuts that have been selected. From Chap. 3, the b force transformation matrices may now be generated for each of these force systems by using the joint equilibrium equations. In addition, from Chap. 4, it is possible to calculate deflections throughout the structure for unit values of both force groups.

In the second stage of the analysis, it is realized that the continuity of deflections must be maintained between the various parts of the structure at the cut sections. For each condition of continuity, one equation must be formed involving the external (applied) loads and the unknown forces. That is, sufficient equations are generated to calculate the unknown forces. These equations are known historically as the *Maxwell-Mohr superposition equations* (see Ref. 28).

As an example, consider the continuous beam loaded as shown in Fig. 5.10 that is made determinate by placing hinges at the two internal supports. The deflected shape for the determinate structure exhibits discontinuities of slopes at these two points (Fig. 5.10*b*), which may be removed by applying the unknown support moments X_1 and X_2. Notice that this type of slope discontinuity may be expressed in terms of the normals to member tangents and that when continuity is restored, these normals coincide or at least resume their original relative dispositions.

Before the general matrix theory for the calculation of the unknown forces is developed, a simple example is considered with only a single unknown.

Example 5.4 The beam shown in Fig. 5.11 is a propped cantilever. That is, it is built in (or fixed) at end i and has a roller support at end j. Since the beam has only three equations of static equilibrium in the plane ($\Sigma F_x = 0$, $\Sigma F_y = 0$, and $\Sigma M = 0$ about any point)

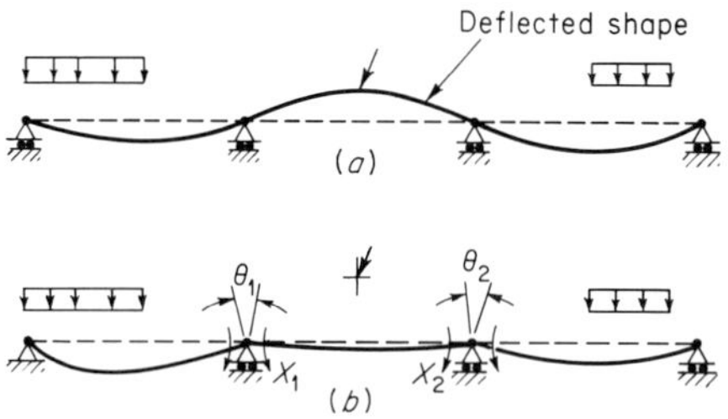

Figure 5.10 Statically indeterminate structure. (*a*) Continuous structure deflections; (*b*) discontinuity of normal slopes in cut structure.

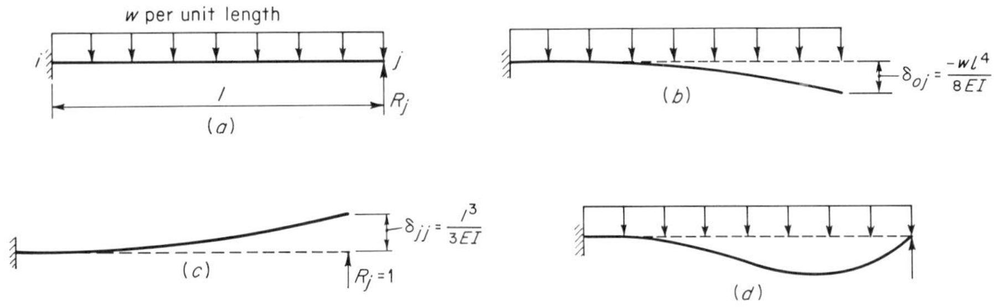

Figure 5.11 Analysis of propped cantilever. (*a*) Indeterminate structure; (*b*) determinate structure; (*c*) deflection due to unit value of redundant; (*d*) actual deflected shape.

but there are four reactions, three at i and one at j, it is statically indeterminate with one redundancy. The redundant force may be taken as the reaction at end j. It is required to calculate the reaction j for a distributed load w over the whole length l of the beam.

In Fig. 5.11*b* it is seen that when the redundant support R_j is removed, the vertical deflection at j is, assuming w positive as shown,

$$\delta_{oj} = \frac{-wl^4}{8EI} \qquad (i)$$

A unit value of R_j applied in its positive sense (Fig. 5.11*c*) produces the deflection at j,

$$\delta_{jj} = \frac{l^3}{3EI} \qquad (ii)$$

The actual value of R_j must be such that the net deflection at j is zero, since the support is assumed unyielding and the deflected shape is as shown in Fig. 5.11*d*. That is,

$$\delta_{oj} + R_j\delta_{jj} = 0 \qquad (iii)$$

Substitute for the values of δ_{oj} and δ_{jj} from Eqs. (*i*) and (*ii*). Then

$$\frac{-wl^4}{8EI} + R_j\frac{l^3}{3EI} = 0$$

$$R_j = \frac{3wl}{8}$$

The process given in Example 5.4 may be generalized by using the matrix notation introduced in Chap. 4 for the calculation of deflections. In the first discussion of sign conventions, a force applied to a structure is considered positive in the direction given to it, and the deflection is positive when in the direction of the force. With this convention the forces applied on either side of a cut in a redundant structure are all considered positive. This is in contrast with the sign convention in which forces and displacements in the directions of the positive coordinate axes are taken as positive. The connection between the sign conventions is developed later in this section and is useful if automatic programs are used in the analysis of joint equilibrium equations in the determinate structure.

Consider a statically indeterminate structure with degree of redundancy m and loaded with the force system $R^T = \langle R_1 \quad R_2 \quad \cdots \quad R_n \rangle$. It is required to calculate member forces and node deflections, in the first instance, in the directions $1, 2, \ldots, n$ of the forces R. The structure is cut in such a way that deflections for each component part of the structure may be calculated from the R forces acting on that part and also from the unknown forces $X^T = \langle X_1 \quad X_2 \quad \cdots \quad X_m \rangle$ acting to restore the continuity of the structure at the cuts. In general, the structure obtained by introducing the X forces will be statically determinate. This however is an unnecessary restriction since all that is required is that for each component part of the structure, b_o and b_1, the force transformation matrices due to R and X should be known. Then,

$$S = [b_o \quad b_1] \begin{bmatrix} R \\ X \end{bmatrix} \tag{5.21}$$

The deflections of the structure may be partitioned into r_o, those deflections at R and r_x, and the deflections at the cuts in directions X. The deflections are calculated from either of Eqs. (4.14) or (4.47) to be

$$\begin{bmatrix} r_o \\ r_x \end{bmatrix} = \begin{bmatrix} b_o{}^T \\ b_1{}^T \end{bmatrix} [f][b_o \quad b_1] \begin{bmatrix} R \\ X \end{bmatrix} \tag{5.22}$$

$$\begin{bmatrix} r_o \\ r_x \end{bmatrix} = \int_{\text{vol}} \begin{bmatrix} b_o{}^T \\ b_1{}^T \end{bmatrix} [f][b_o \quad b_1] \begin{bmatrix} R \\ X \end{bmatrix} \tag{5.23}$$

The first equation, (5.22), applies to structures with discrete elements and is the one usually employed in computer programs. A notable exception of this is for curved beams and arches. The second equation, (5.23), applies to structures in which stress resultants are specified in infinitesimal form. The theory is developed here from Eq. (5.22). When the multiplication of the b and f matrices on the right-hand side of Eq. (5.22) is carried out, it is seen that

$$\begin{bmatrix} r_o \\ r_x \end{bmatrix} = \begin{bmatrix} b_o{}^T f b_o & b_o{}^T f b_1 \\ b_1{}^T f b_o & b_1{}^T f b_1 \end{bmatrix} \begin{bmatrix} R \\ X \end{bmatrix} \tag{5.24}$$

The submatrices in Eq. (5.24) are simply the deflection-influence coefficients of the structure for the forces R and X. Since the X values represent a pair of equal and opposite forces applied to the structure, the terms of $b_1{}^T f b_1$ give the relative deflection coefficients at the cuts.

The shorthand notation is introduced

$$D_{00} = b_o{}^T f b_o \quad D_{01} = b_o f b_1$$
$$D_{10} = b_1{}^T f b_o \quad D_{11} = b_1{}^T f b_1$$

That is,

$$\begin{bmatrix} r_o \\ r_x \end{bmatrix} = \begin{bmatrix} D_{00} & D_{01} \\ D_{10} & D_{11} \end{bmatrix} \begin{bmatrix} R \\ X \end{bmatrix} \tag{5.25}$$

Thence,

$$r_o = D_{00}R + D_{01}X \tag{5.26a}$$

$$r_x = D_{10}R + D_{11}X \tag{5.26b}$$

However, it has been specified that continuity must be restored at the cuts, and for this to be accomplished

$$r_x \equiv 0 = D_{10}R + D_{11}X \tag{5.27}$$

On solving Eq. (5.27) for X,

$$X = -D_{11}{}^{-1}D_{10}R \tag{5.28}$$

When this value for X is substituted in Eq. (5.26a), the deflections r_o are calculated as

$$r_o = (D_{00} - D_{01}D_{11}{}^{-1}D_{10})R \tag{5.29}$$

Member forces are calculated from Eq. (5.21),

$$S = b_oR + b_1X \tag{5.30}$$

Again, substitution for X from Eq. (5.28) gives

$$S = (b_o - b_1D_{11}{}^{-1}D_{10})R \tag{5.31}$$

The matrices in Eqs. (5.29) and (5.31) may now be interpreted as deflection- and force-influence coefficient matrices for the indeterminate structure. That is,

$$F = D_{00} - D_{01}D_{11}{}^{-1}D_{10} \tag{5.32}$$

$$b = b_o - b_1D_{11}{}^{-1}D_{10} \tag{5.33}$$

It should be emphasized that Eqs. (5.26a) and (5.30) are the usual equations used in deflection and stress calculation. Equations (5.32) and (5.33) are used if the transformation matrices for the complete structure are required. For example, F may be used in the calculation of natural frequencies of the structure. That is, the F matrix is the flexibility matrix of the structure for the R forces.

The theory developed so far may be used satisfactorily with generalized flexibility matrices and generalized forces for members loaded between structure nodes. Alternatively, the methods developed in Sec. 4.8 may be applied. That is, from method 1, Sec. 4.8, the structure deflections for the loads R may be expressed in terms of the member forces and the loads along the member. Thus from Eq. (4.142),

$$\begin{bmatrix} r_o \\ r_x \end{bmatrix} = \begin{bmatrix} D_{00} & D_{01} \\ D_{10} & D_{11} \end{bmatrix} \begin{bmatrix} R \\ X \end{bmatrix} + \begin{bmatrix} b_o{}^T \\ b_1{}^T \end{bmatrix} [\delta_o] R \tag{5.34}$$

Notice in Eq. (5.34) that the X forces do not contribute to the δ_o terms. That is, R may contain distributed forces, but X does not.

From Eq. (5.34), for continuity at the cuts,

$$r_x \equiv 0 = D_{10}R + D_{11}X + b_1{}^T\delta_o R \tag{5.35}$$

Thence,

$$X = -D_{11}{}^{-1}(D_{10} + b_1{}^T\delta_o)R \tag{5.36}$$

Then,

$$r_o = D_{00}R + D_{01}X + b_o{}^T\delta_o R \tag{5.37}$$

Example 5.5 The portal frame shown in Fig. 5.12*a* has fixed supports. In such a case there are six reaction components and only three equations of static equilibrium. The portal frame is statically indeterminate with three unknown forces. These will be taken to be X_1, the moment at node 1; X_2, the moment at node 4; and X_3, the horizontal force at node 4. The bending-moment diagrams for unit values of each of these redundant forces are given in Fig. 5.12*b*.

The portal frame is loaded with a distributed force of 1 kip/ft run vertically downward on the top member (2).

It is required to calculate bending moments in the frame (at member ends only) where bending distortions only are considered in deflection calculations. For this structure, Eq. (5.36) will be used to calculate the redundant forces. From Fig. 5.12*b*, the

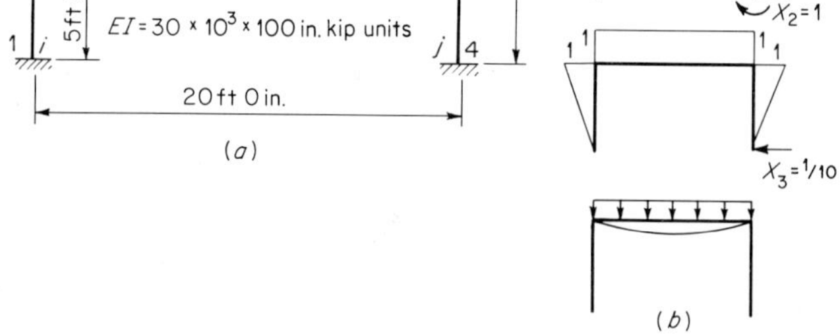

Figure 5.12 Analysis of portal frame. (*a*) Portal-frame dimensions and loading; (*b*) moment diagrams for unit values of load and redundants.

force transformation matrix $[b_o \quad b_1]$ is generated, connecting (M_i,M_j) for each member to (w,X_1,X_2,X_3). Notice that w produces no such member moments. That is,

$$[b_o \quad b_1] = \left[\begin{array}{c|ccc} 0 & -1 & 0 & 0 \\ 0 & -1 & 0 & -1 \\ 0 & -1 & 0 & -1 \\ 0 & 0 & -1 & -1 \\ 0 & 0 & -1 & -1 \\ 0 & 0 & -1 & 0 \end{array}\right]$$

The force $R = wl = 1 \times 20 = 20$ kips, so that

$$\delta_o R = \frac{50}{6EI}\begin{bmatrix} 0 \\ 0 \\ 1 \\ 1 \\ 0 \\ 0 \end{bmatrix} R$$

Note

$$\phi_i = \phi_j = \frac{wl^3}{24EI_2} = \frac{l^2}{24E(2I)} R$$

The matrix of member flexibilities f is given by

$$f = \frac{10}{6EI}\begin{bmatrix} 2 & 1 & 0 & 0 & 0 & 0 \\ 1 & 2 & 0 & 0 & 0 & 0 \\ 0 & 0 & 2 & 1 & 0 & 0 \\ 0 & 0 & 1 & 2 & 0 & 0 \\ 0 & 0 & 0 & 0 & 2 & 1 \\ 0 & 0 & 0 & 0 & 1 & 2 \end{bmatrix}$$

On calculation,

$$D_{11} = b_1{}^T f b_1 = \frac{10}{6EI}\begin{bmatrix} 8 & 1 & 6 \\ 1 & 8 & 6 \\ 6 & 6 & 10 \end{bmatrix}$$

On inversion,

$$D_{11}^{-1} = \frac{6EI}{1260}\begin{bmatrix} 44 & 26 & -42 \\ 26 & 44 & -42 \\ -42 & -42 & 63 \end{bmatrix}$$

Also,

$$b_1{}^T\delta_o = \frac{50}{6EI}\begin{bmatrix} -1 \\ -1 \\ -2 \end{bmatrix}$$

Then, since $b_1{}^T f b_o = D_{10} \equiv 0$ because $b_o \equiv 0$,

$$X = -D_{11}{}^{-1}(D_{10} + b_1{}^T\delta_o)R$$

$$X = \frac{-5}{126}\begin{bmatrix} 44 & 26 & -42 \\ 26 & 44 & -42 \\ -42 & -42 & 63 \end{bmatrix}\begin{bmatrix} -1 \\ -1 \\ -2 \end{bmatrix} 20 = \begin{bmatrix} -11.1 \\ -11.1 \\ 33.3 \end{bmatrix}$$

Then,

$$S = b_oR + b_1X = \begin{bmatrix} -1 & 0 & 0 \\ -1 & 0 & -1 \\ -1 & 0 & -1 \\ 0 & -1 & -1 \\ 0 & -1 & -1 \\ 0 & -1 & 0 \end{bmatrix}\begin{bmatrix} -11.1 \\ -11.1 \\ 33.3 \end{bmatrix} = \begin{bmatrix} 11.1 \\ -22.2 \\ -22.2 \\ -22.2 \\ -22.2 \\ 11.1 \end{bmatrix} \text{ft-kips}$$

These bending moments have been plotted on the tension side of members in Fig. 5.13. The parabolic diagram with center ordinate $wl^2/8$ has been added to the top member.

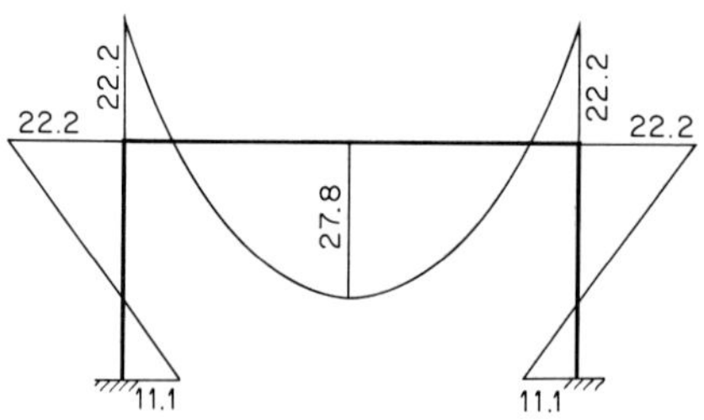

Figure 5.13 Bending-moment diagram in foot-kip units.

Example 5.6 In Sec. 4.8, method 2, it has been shown that deflections may be calculated for the statically determinate structure by first fixing all joints. Then, for this fixed-joint condition, fixed-end forces are calculated for all members with loads between nodes. When these fixed-end forces with reversed sign are applied to the nodes, the effect achieved is that of releasing the joints. The portal frame in Fig. 5.12 is reanalyzed with the use of release forces. The condition of all joints fixed is shown in Fig. 5.14*a*. For the frame shown in Fig. 5.14*b*, the node forces producing bending deflections are the two node moments, so that $R^T = \langle 33.33 \quad 33.33 \rangle$.

Then in the equation $S = [b_o \quad b_1]\begin{bmatrix} R \\ X \end{bmatrix}$, by using the same release system as in Fig. 5.12*b*, the matrix $[b_o \quad b_1]$ is given as

$$[b_o \quad b_1] = \left[\begin{array}{cc|ccc} 0 & 0 & -1 & 0 & 0 \\ 0 & 0 & -1 & 0 & -1 \\ 1 & 0 & -1 & 0 & -1 \\ 0 & 1 & 0 & -1 & -1 \\ 0 & 0 & 0 & -1 & -1 \\ \underbrace{0 \quad 0}_{b_o} & & \underbrace{0 \quad -1 \quad 0}_{b_1} & & \end{array}\right]$$

The member-flexibility matrix f and the b_1 matrix are the same as in Example 5.5, so that D_{11}^{-1} is given as before,

$$D_{11}^{-1} = \frac{6EI}{1260}\begin{bmatrix} 44 & 26 & -42 \\ 26 & 44 & -42 \\ -42 & -42 & 63 \end{bmatrix}$$

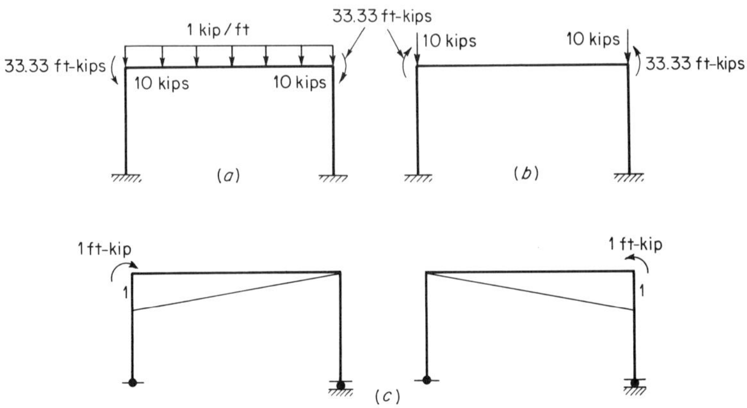

Figure 5.14 Analysis of portal-frame node loads. (*a*) Fixed-joint condition; (*b*) joint-release forces; (*c*) analysis of portal frame—node loads.

Then,

$$D_{10} = b_1{}^T f b_o = \frac{10}{6EI}\begin{bmatrix} -2 & -1 \\ -1 & -2 \\ -3 & -3 \end{bmatrix}$$

Then, from Eq. (5.28), $X = -D_{11}{}^{-1}D_{10}R$, so that

$$X = \frac{1}{126}\begin{bmatrix} 44 & 26 & -42 \\ 26 & 44 & -42 \\ -42 & -42 & 63 \end{bmatrix}\begin{bmatrix} 2 & 1 \\ 1 & 2 \\ 3 & 3 \end{bmatrix}\begin{bmatrix} 33.33 \\ 33.33 \end{bmatrix} = \begin{bmatrix} -11.11 \\ -11.11 \\ 33.33 \end{bmatrix}$$

For the node-force solution,

$S_N = b_o R + b_1 X$

That is,

$$\begin{bmatrix} M_{1i} \\ M_{1j} \\ M_{2i} \\ M_{2j} \\ M_{3i} \\ M_{3j} \end{bmatrix} = \begin{bmatrix} 0 & 0 \\ 0 & 0 \\ 1 & 0 \\ 0 & 1 \\ 0 & 0 \\ 0 & 0 \end{bmatrix}\begin{bmatrix} 33.33 \\ 33.33 \end{bmatrix} + \begin{bmatrix} -1 & 0 & 0 \\ -1 & 0 & -1 \\ -1 & 0 & -1 \\ 0 & -1 & -1 \\ 0 & -1 & -1 \\ 0 & -1 & 0 \end{bmatrix}\begin{bmatrix} -11.11 \\ -11.11 \\ 33.33 \end{bmatrix} = \begin{bmatrix} 11.11 \\ -22.22 \\ 11.11 \\ 11.11 \\ -22.22 \\ 11.11 \end{bmatrix}$$

To these moments must be added those fixed-end moments that exist before the nodes are released. These fixed-end moments may be written

$$S_F = \begin{bmatrix} M_{1i} \\ M_{1j} \\ M_{2i} \\ M_{2j} \\ M_{3i} \\ M_{3j} \end{bmatrix}_F = \begin{bmatrix} 0 \\ 0 \\ -33.33 \\ -33.33 \\ 0 \\ 0 \end{bmatrix}$$

Finally,

$$S = S_N + S_F = \begin{bmatrix} 11.11 \\ -22.22 \\ 11.11 \\ 11.11 \\ -22.22 \\ 11.11 \end{bmatrix} + \begin{bmatrix} 0 \\ 0 \\ -33.33 \\ -33.33 \\ 0 \\ 0 \end{bmatrix} = \begin{bmatrix} 11.11 \\ -22.22 \\ -22.22 \\ -22.22 \\ -22.22 \\ 11.11 \end{bmatrix} \text{ft-kips}$$

These moments are identical with those calculated in Example 5.5 and shown in Fig. 5.13.

Foundation Settlement in Statically Indeterminate Structures

There are several ways in which the deformation properties of flexible foundations may be specified. Thus, in the first instance, foundation settlement may be given by specific amounts. For example, a support in a structure may be given a unit displacement and the effects on forces in the structure examined. If, on the other hand, some knowledge of the foundation material is known, it may be treated as a spring whose flexibility is known. If only elastic analysis is being undertaken, this flexibility must be assumed to be linear.

Thus, in the first case, settlement of foundation by a known amount can be treated by considering the reaction a member whose deformation is known. Since the reaction on the structure is a node force, there will be a negative sign introduced into the displacement transformation matrix.

When the foundation is such that its load deflection characteristic is known (and assumed linear), it may be used to introduce a fictitious member whose properties reproduce the foundation behavior. These two types of settlement problem are illustrated in Fig. 5.15*a* and *b*, in which support *B* deflects when load is applied in *a* and, in *b*, undergoes a deflection of an amount δ. Because type *a* is so readily accommodated by the addition of the fictitious member, only type *b* is considered here.

Then, let b_{OF} and b_{XF} be the force transformations for the reactions due to R and X respectively in the statically determinate structure. Let the foundation settlement be given by the vector v_F, so that by applying the contragradient law from Eq. (4.14),

$$\tilde{r}_o = -b_{OF}{}^T v_F \tag{5.38a}$$

$$\tilde{r}_x = -b_{XF}{}^T v_F \tag{5.38b}$$

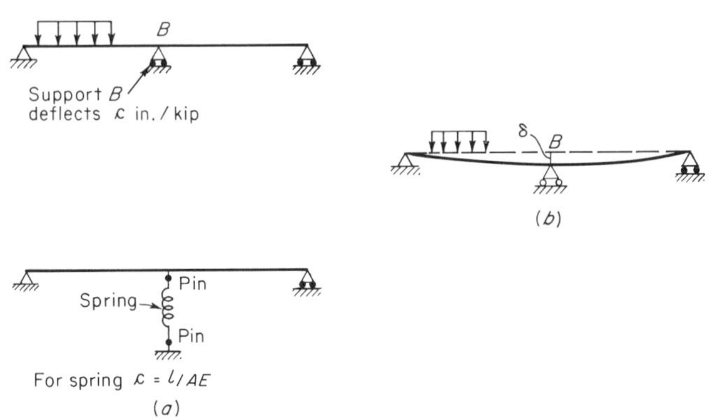

Figure 5.15 Foundation settlement. (*a*) Elastic support; (*b*) support *B* settles δ in.

Equation (5.25) may now be rewritten to include these deflection terms due to foundation settlement.

$$\begin{bmatrix} r_o \\ r_x \end{bmatrix} = \begin{bmatrix} D_{00} & D_{01} \\ D_{10} & D_{11} \end{bmatrix} \begin{bmatrix} R \\ X \end{bmatrix} - \begin{bmatrix} b_{OF}{}^T \\ b_{XF}{}^T \end{bmatrix} v_F \tag{5.39}$$

Continuity of the structure requires that $r_x \equiv 0$, as previously. Equations (5.39) are then solved for X, so that

$$X = -D_{11}{}^{-1}(D_{10}R - b_{XF}{}^T v_F) \tag{5.40}$$

Member forces are calculated by Eq. (5.30), and deflections at the load points are obtained from Eq. (5.39). That is,

$$r_o = D_{00}R + D_{10}X - b_{OF}{}^T v_F \tag{5.41}$$

Example 5.7 The above theory is used to analyze the two-span continuous beam, loaded as shown in Fig. 5.16*a*. In addition to the load $R = 10$ kips in the center of the span BC, the support B settles ¼ in. The unit-load diagrams $R = 1$ and $X = 1$ are shown in Fig. 5.16*b*, and from these the b_{OF} and b_{XF} matrices are obtained. Only reaction B need be considered since it alone suffers displacement. That is, from Fig. 5.16*b*, I,

$$b_{OF} = 0.5$$

and from Fig. 5.16*b*, II,

$$b_{XF} = -\frac{2}{l} = -\frac{1}{50}$$

The D_{11} and D_{10} matrices are single-integer values (number of redundants = 1) and are calculated from simple-beam theory to be

$$D_{11} = 2\frac{l}{3EI} \qquad \text{and} \qquad D_{10} = \frac{l^2}{16EI}$$

Substitute these values in Eq. (5.40), and hence obtain the moment X at B.

$$X = -\frac{3EI}{2l}\left[\frac{l^2}{16EI}R - \left(-\frac{1}{50}\right)\left(-\frac{1}{4}\right)\right]$$

$$= -\frac{3l}{32}R + \frac{3EI}{400l}$$

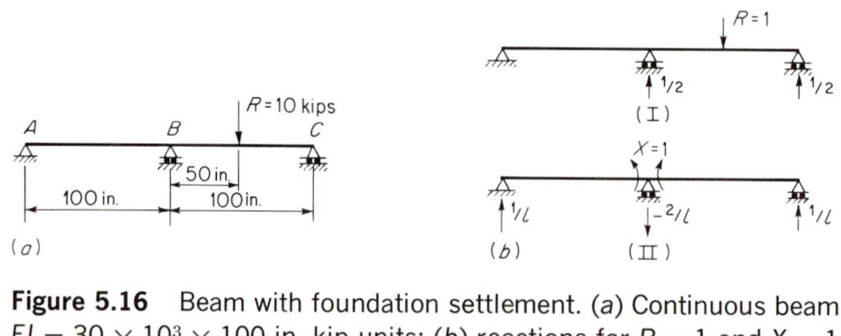

Figure 5.16 Beam with foundation settlement. (*a*) Continuous beam, $EI = 30 \times 10^3 \times 100$ in.-kip units; (*b*) reactions for $R = 1$ and $X = 1$.

Substitute the values given in Fig. 5.16 for R, E, I, and l. Then,

$$X = -\frac{3 \times 100}{32} \times 10 + \frac{3 \times 30 \times 10^3 \times 100}{400 \times 100}$$
$$= -93.75 + 225.0 = 131.25 \text{ in.-kips}$$

The bending-moment diagram is shown in Fig. 5.17. It is interesting to note that a settlement of only ¼ in. completely reverses the sign of the moment at B from -93.75 in.-kip to $+131.25$ in.-kips.

Generation of the Maxwell-Mohr Equations with Use of Equilibrium Sign Convention for Member Forces

In the theory developed for the analysis of the joint equilibrium equations for determinate structures in Chap. 3, it was found convenient to use the sign convention of forces and displacements' being positive in the directions of the positive coordinate directions. With this sign convention, when a structure is cut (as shown in Fig. 5.18a), the forces on one side of the cut are positive and on the other negative.

From Fig. 5.18a, it is seen that although the forces on the two sides of the cut have opposite signs, there are still only three unknown quantities. These are taken to be $\langle X_1 \quad X_2 \quad X_3 \rangle$, and they agree in sign with $\langle R_1 \quad R_2 \quad R_3 \rangle$. The forces given in Fig. 5.18b are related to $\langle X_1 \quad X_2 \quad X_3 \rangle$ by Eq. (5.42).

$$\begin{bmatrix} R_3 \\ R_4 \\ R_5 \\ R_6 \\ R_7 \\ R_8 \end{bmatrix} = \begin{bmatrix} 1 & 0 & 0 \\ 0 & 1 & 0 \\ 0 & 0 & 1 \\ -1 & 0 & 0 \\ 0 & -1 & 0 \\ 0 & 0 & -1 \end{bmatrix} \begin{bmatrix} X_1 \\ X_2 \\ X_3 \end{bmatrix} \tag{5.42}$$

This equation will be written

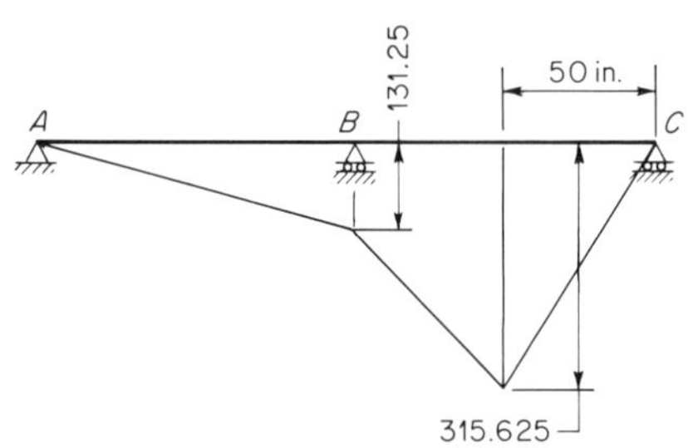

Figure 5.17 Bending-moment diagram.

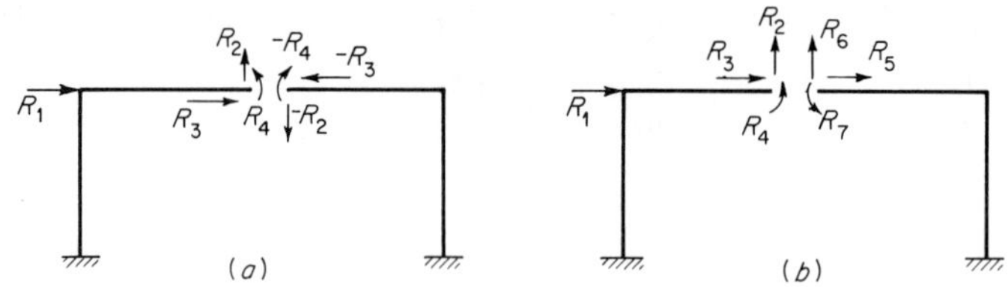

Figure 5.18 Forces at cut in joint sign convention. (a) Forces on sides of cut; (b) positive forces on structure.

$$R^* = BX \tag{5.43}$$

When the structure is analyzed (as in Chap. 3) for unit values of the joint forces R and the forces on the cuts R^*, all assumed to be in the positive sense, the force transformation matrices b_o and b_1 are obtained so that member forces are given by

$$S = b_o R + b_1 R^* \tag{5.44}$$

The substitution for R^* may now be made from Eq. (5.43). Then,

$$S = b_o R + b_1 B X \tag{5.45}$$

That is,

$$S = b_o R + b_x X \tag{5.46}$$

where b_x is defined by

$$b_x = b_1 B \tag{5.47}$$

The analysis proceeds from Eqs. (5.46) and (5.47) in the usual manner following Eqs. (5.36)ff.

Example 5.8 The joint equilibrium-force sign convention is used in the analysis of the continuous beam shown in Fig. 5.19a. The beam has external forces (R_1,R_2) as shown, and by

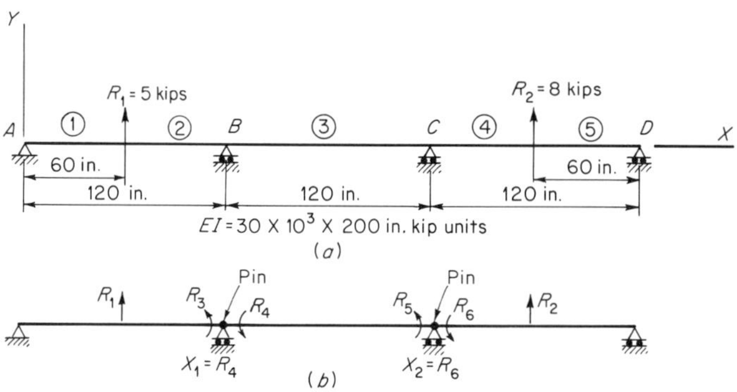

Figure 5.19 Beam analysis in joint sign convention. (a) Beam dimensions and loads; (b) cut structure with positive forces.

placing pins at the supports B and C, the four forces (R_3, R_4, R_5, R_6) are introduced (shown in Fig. 5.19*b*) to simulate the unknown moments (X_1, X_2).

The member bending-moment members (1), (2), . . . , (5) are related to the forces $\langle R_1 \quad R_2 \quad \cdots \quad R_6 \rangle$ by Eq. (*i*).

$$\begin{bmatrix} M_{1i} \\ M_{1j} \\ M_{2i} \\ M_{2j} \\ M_{3i} \\ M_{3j} \\ M_{4i} \\ M_{4j} \\ M_{5i} \\ M_{5j} \end{bmatrix} = \Bigg[\underbrace{\begin{matrix} 0 & 0 \\ -30 & 0 \\ 30 & 0 \\ 0 & 0 \\ 0 & 0 \\ 0 & 0 \\ 0 & 0 \\ 0 & -30 \\ 0 & 30 \\ 0 & 0 \end{matrix}}_{b_o} \Bigg| \underbrace{\begin{matrix} 0 & 0 & 0 & 0 \\ 0.5 & 0 & 0 & 0 \\ -0.5 & 0 & 0 & 0 \\ 1.0 & 0 & 0 & 0 \\ 0 & 1.0 & 0 & 0 \\ 0 & 0 & 1.0 & 0 \\ 0 & 0 & 0 & 1.0 \\ 0 & 0 & 0 & -0.5 \\ 0 & 0 & 0 & 0.5 \\ 0 & 0 & 0 & 0 \end{matrix}}_{b_1}\Bigg] \begin{bmatrix} R_1 \\ R_2 \\ R_3 \\ R_4 \\ R_5 \\ R_6 \end{bmatrix} \qquad (i)$$

From Fig. 5.19*b*, it is seen that the B matrix is given by

$$B = \begin{bmatrix} -1 & 0 \\ 1 & 0 \\ 0 & -1 \\ 0 & 1 \end{bmatrix} \qquad (ii)$$

Then,

$$b_x = b_1 B = \begin{bmatrix} 0 & 0 \\ -0.5 & 0 \\ 0.5 & 0 \\ -1.0 & 0 \\ 1.0 & 0 \\ 0 & -1.0 \\ 0 & 1.0 \\ 0 & -0.5 \\ 0 & 0.5 \\ 0 & 0 \end{bmatrix} \qquad (iii)$$

The member-flexibility matrices are calculated from

$$f_i = \frac{l_i}{6EI_i}\begin{bmatrix} 2 & -1 \\ -1 & 2 \end{bmatrix}$$

Then,

$$f_1 = f_2 = f_4 = f_5 = \frac{60}{6EI}\begin{bmatrix} 2 & -1 \\ -1 & 2 \end{bmatrix}$$

$$f_3 = \frac{120}{6EI}\begin{bmatrix} 2 & -1 \\ -1 & 2 \end{bmatrix}$$

Then,

$$f = \frac{60}{6EI}\begin{bmatrix} 2 & -1 & & & & & & & & \\ -1 & 2 & & & & & & & \mathbf{0} & \\ & & 2 & -1 & & & & & & \\ & & -1 & 2 & & & & & & \\ & & & & 4 & -2 & & & & \\ & & & & -2 & 4 & & & & \\ & & & & & & 2 & -1 & & \\ & & & & & & -1 & 2 & & \\ & \mathbf{0} & & & & & & & 2 & -1 \\ & & & & & & & & -1 & 2 \end{bmatrix} \qquad (iv)$$

The matrices D_{11} and D_{10} are calculated from b_o, b_x, and f.

$$D_{11} = b_x^T f b_x = \frac{10}{EI}\begin{bmatrix} 8.0 & 2.0 \\ 2.0 & 8.0 \end{bmatrix} \qquad (v)$$

$$D_{10} = b_x^T f b_o = \frac{10}{EI}\begin{bmatrix} 90.0 & 0 \\ 0 & 90.0 \end{bmatrix} \qquad (vi)$$

Then,

$$X = -D_{11}^{-1}D_{10}R = \frac{-1}{60}\begin{bmatrix} 8.0 & -2.0 \\ -2.0 & 8.0 \end{bmatrix}\begin{bmatrix} 90.0 & 0 \\ 0 & 90.0 \end{bmatrix}\begin{bmatrix} 5 \\ 8 \end{bmatrix} \qquad (vii)$$

$$X = -\begin{bmatrix} 36 \\ 81 \end{bmatrix} \qquad (viii)$$

Member forces may now be calculated with Eq. (5.46), $S = b_oR + b_xX$.

$$\begin{bmatrix} M_{1i} \\ M_{1j} \\ M_{2i} \\ M_{2j} \\ M_{3i} \\ M_{3j} \\ M_{4i} \\ M_{4j} \\ M_{5i} \\ M_{5j} \end{bmatrix} = \begin{bmatrix} 0 & 0 \\ -30 & 0 \\ 30 & 0 \\ 0 & 0 \\ 0 & 0 \\ 0 & 0 \\ 0 & 0 \\ 0 & -30 \\ 0 & 30 \\ 0 & 0 \end{bmatrix} \begin{bmatrix} 5 \\ 8 \end{bmatrix} + \begin{bmatrix} 0 & 0 \\ -0.5 & 0 \\ 0.5 & 0 \\ -1.0 & 0 \\ 1.0 & 0 \\ 0 & -1.0 \\ 0 & 1.0 \\ 0 & -0.5 \\ 0 & 0.5 \\ 0 & 0 \end{bmatrix} \begin{bmatrix} -36 \\ -81 \end{bmatrix} = \begin{bmatrix} 0 \\ -132 \\ 132 \\ 36 \\ -36 \\ 81 \\ -81 \\ -199.5 \\ 199.5 \\ 0 \end{bmatrix} \qquad (ix)$$

The bending moments in Eq. (ix) have been plotted in Fig. 5.20 on the tension side of the members. It should be remembered that the sign convention for moments in Eq. (ix) is that anticlockwise moment on the member is positive.

5.3 TEMPERATURE STRAINS IN STATICALLY INDETERMINATE STRUCTURES; INTERNAL FORCES AND NODE DEFLECTIONS

In Sec. 4.8 it is shown that temperature strains in members produce node deflections of the statically determinate structure. This means that in the cut structure, temperature strains will produce discontinuity of deflections at the cuts so that redundant forces must be introduced to restore continuity. Since the external loads are zero, the reactions due to temperature effects are self-equilibrating.

Let $\bar{v}_i$ be the member distortion due to the temperature variation in the member. The temperature distribution in the member is assumed to be such as to produce distortion without internal stresses. That is, the member itself is assumed determinate with respect to the temperature distribution. This means that for the bending element a linear variation of temperature across the member section is possible.

The temperature strains $\bar{v}_i$ may be considered together with those v_i due to the member force S_i. For all members,

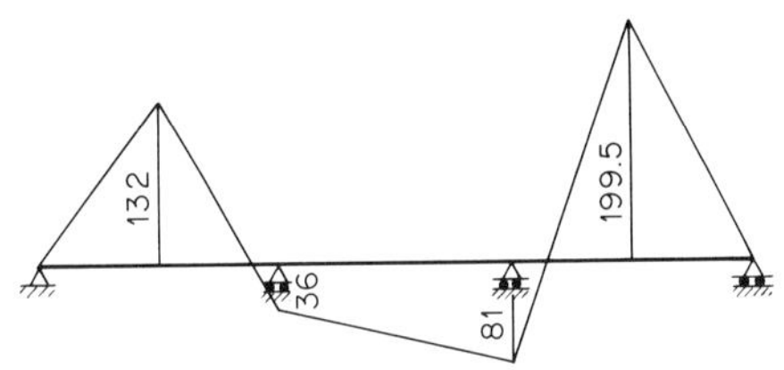

Figure 5.20 Member bending moments in inch-kips.

$$v_t = v + \bar{v} = fS + \bar{v} \tag{5.48}$$

In Eq. (5.48), $v_t^T = \langle v_{1t} \quad v_{2t} \quad \cdots \quad v_{nt} \rangle$, $v^T = \langle v_1 \quad v_2 \quad \cdots \quad v_n \rangle$, and $\bar{v}^T = \langle \bar{v}_1 \quad \bar{v}_2 \quad \cdots \quad \bar{v}_n \rangle$. Node deflections are calculated,

$$r = b_o^T(fS + \bar{v}) \tag{5.49}$$

From Eq. (5.25), the member forces S are expressed in terms of the external forces R and the redundant forces X. That is,

$$\begin{bmatrix} r_o \\ r_x \end{bmatrix} = \begin{bmatrix} D_{00} & D_{01} \\ D_{10} & D_{11} \end{bmatrix} \begin{bmatrix} R \\ X_t \end{bmatrix} + \begin{bmatrix} b_o^T \\ b_1^T \end{bmatrix} \bar{v} \tag{5.50}$$

In Eq. (5.50), X_t gives the redundant forces including temperature forces. It is still necessary for continuity of the structure to be restored at the cuts. Hence,

$$r_x \equiv 0 = D_{10}R + D_{11}X_t + b_1^T\bar{v} \tag{5.51}$$

Then

$$X_t = -D_{11}^{-1}(D_{10}R + b_1^T\bar{v}) \tag{5.52}$$

or

$$X_t = X - D_{11}^{-1}b_1^T\bar{v} \tag{5.53}$$

In Eq. (5.53), X is the value for redundants excluding temperature effects. Expressions for $\bar{v}$ are given for prismatic frame members in Sec. 4.8. That is, for the member given in Fig. 4.27,

$$\begin{bmatrix} \bar{\phi}_i \\ \bar{\phi}_j \\ \delta \end{bmatrix} = \frac{l\alpha}{2} \begin{bmatrix} -\Delta & +\Delta \\ -\Delta & +\Delta \\ 1 & 1 \end{bmatrix} \begin{bmatrix} t_1 \\ t_2 \end{bmatrix} \qquad \Delta = \frac{1}{c_1 - c_2} \tag{5.54}$$

From Eq. (5.50), node deflections are calculated to be

$$r_o = D_{00}R + D_{01}X_t + b_o^T\bar{v} \tag{5.55}$$

Calculation of Node Deflections in Statically Indeterminate Structures

In Eq. (5.31), it has been shown that member forces in the redundant structure are given as

$$S = (b_o - b_1 D_{11}^{-1} D_{10})R$$

and that

$$b = b_o - D_{11}^{-1}D_{10}$$

is the force transformation matrix for the redundant structure. With this notation, node deflections are calculated from

$$r = b_o^T f b R \tag{5.56}$$

In Eq. (5.56), it is seen that the member distortions due to member forces S are

$$v = f b R \tag{5.57}$$

These member distortions v are such that they satisfy the continuity requirements to return from the cut structure to the continuous structure. This is true for any cut structure, so that in Eq. (5.56) the transformation matrix b_o^T used to calculate node deflections from member distortions may be obtained from any cut structure and not necessarily that used in the calculation of X. To generalize Eq. (5.56), let $\tilde{b}_o^T$ be the distortion transformation matrix so that now

$$r = \tilde{b}_o^T f b R = \tilde{b}_o^T f (b_o R + b_1 X) = \tilde{b}_o^T f S \tag{5.58}$$

Example 5.9 The use of the force transformation matrix $\tilde{b}_o$ is illustrated in the calculation of the horizontal deflection of point A, the midpoint of member (1) for the portal frame in Fig. 5.12a.

From Fig. 5.12b it is seen that the vector of the redundants, X, has been calculated by using a moment release at each of the supports and a shear release at the right-hand support.

In order to calculate the horizontal deflection at A, it is necessary to obtain a $\tilde{b}_o$ matrix for a horizontal force at A. To do this, cut the structure so that it is a cantilever from the left-hand support, as shown in Fig. 5.21a. From Fig. 5.21a,

$$\tilde{b}_o^T = [-60 \quad 0]$$

$$(bR)^T = [11.1 \quad -5.55]$$

For the member AB,

$$f = \frac{60}{6 \times 30 \times 10^3 \times 100} \begin{bmatrix} 2 & 1 \\ 1 & 2 \end{bmatrix}$$

Thence,

$$v = \frac{1}{30 \times 10^4} \begin{bmatrix} 2 & 1 \\ 1 & 2 \end{bmatrix} \begin{bmatrix} 11.1 \\ -5.55 \end{bmatrix} = \frac{1}{30 \times 10^4} \begin{bmatrix} 16.65 \\ 0 \end{bmatrix} \text{rad}$$

and then

$$u_A = \frac{1}{30 \times 10^4} [-60 \quad 0] \begin{bmatrix} 16.65 \\ 0 \end{bmatrix} = 0.00333 \text{ in.}$$

Example 5.10 In this example the matrix interpretative program is used to analyze the structure shown in Fig. 5.22a. This two-bay portal frame has loads $\langle w_1 \quad w_2 \quad R_{7X} \rangle$, as shown,

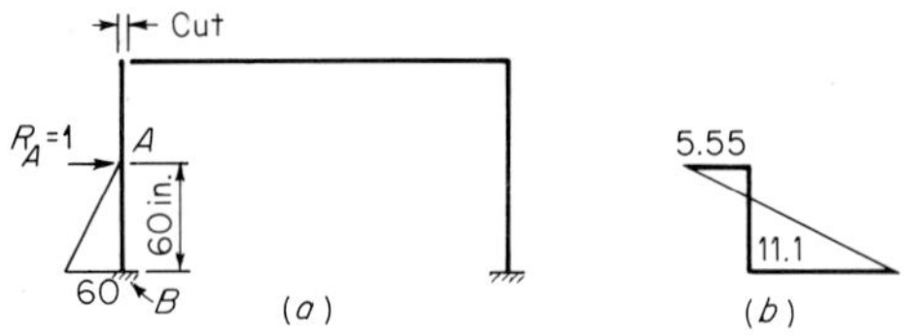

Figure 5.21 Cut structure for $\tilde{b}_o$.

acting on members (1) and (2) and on node 7 (in the X direction) respectively. In addition to these forces, members (2) and (5) have a temperature rise, constant along the length, but varying across the section linearly, as shown in Fig. 5.22*b*.

It is required to calculate node deflections and member bending moments for both external loads and temperature effects, both separately and combined. From the temperature distribution given in Fig. 5.22*b* for members (2) and (5), it is seen that these members suffer axial deformations as well as rotational deformations due to the temperature strains. Then, for members (2) and (5), axial forces and distortions must be considered in the deflection calculations. From Eq. (5.54) and the distribution in Fig. 5.22*b*, it is seen in this case that both members suffer the same unrestrained temperature distortions given in Eq. (i).

$$\begin{bmatrix} \bar{\phi}_i \\ \bar{\phi}_j \\ \delta \end{bmatrix}_{2\text{or}5} = \frac{33.541 \times 12 \times \alpha}{2} \begin{bmatrix} \frac{-1}{12} & \frac{1}{12} \\ \frac{-1}{12} & \frac{1}{12} \\ 1 & 1 \end{bmatrix} \begin{bmatrix} 110 \\ 80 \end{bmatrix} = 33.541 \times 12 \times \alpha \begin{bmatrix} -15 \\ -15 \\ 95 \end{bmatrix} \qquad (i)$$

The member flexibilities are given below:

Members 1, 4, and 7:

$$f = \frac{240}{6EI} \begin{bmatrix} 2 & 1 \\ 1 & 2 \end{bmatrix}$$

Members 3 and 6:

$$f = \frac{33.54 \times 12}{6EI} \begin{bmatrix} 2 & 1 \\ 1 & 2 \end{bmatrix} = \frac{240}{6EI} \begin{bmatrix} 3.354 & 1.677 \\ 1.677 & 3.354 \end{bmatrix}$$

Members 2 and 3:

$$f = \frac{33.54 \times 12}{6EI} \begin{bmatrix} 2 & 1 & 0 \\ 1 & 2 & 0 \\ 0 & 0 & 60 \end{bmatrix} = \frac{240}{6EI} \begin{bmatrix} 3.354 & 1.677 & 0 \\ 1.677 & 3.354 & 0 \\ 0 & 0 & 100.62 \end{bmatrix}$$

The end rotations for members (1) and (3) due to the member loads w_1 and w_2 are given by

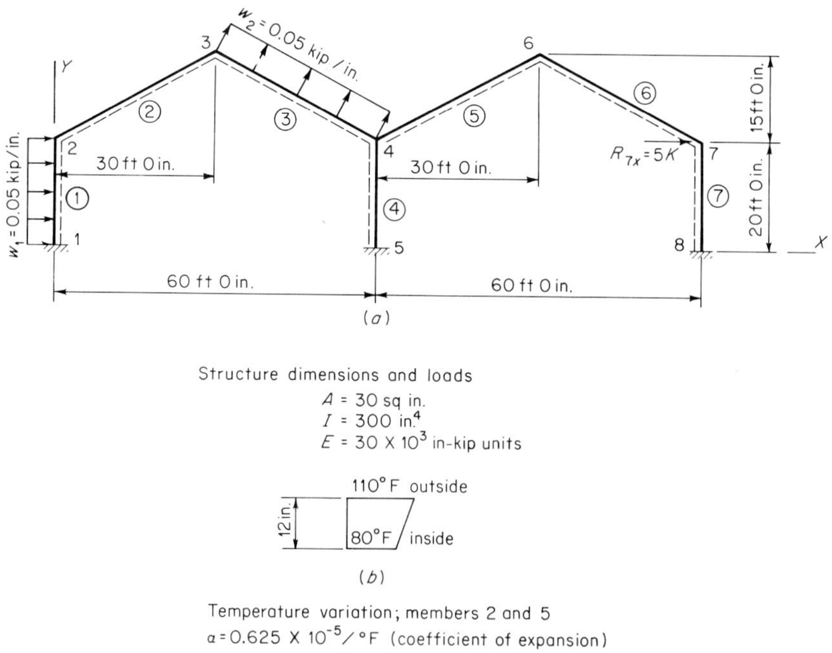

Figure 5.22 Analysis of two-bay portal frame.

$$\begin{bmatrix} \phi_{1i} \\ \phi_{1j} \end{bmatrix}_{w_1} = \frac{(20 \times 12)^3}{24EI} \begin{bmatrix} 1 \\ 1 \end{bmatrix} w_1$$

$$\begin{bmatrix} \phi_{3i} \\ \phi_{3j} \end{bmatrix}_{w_2} = -\frac{(33.541 \times 12)^3}{24EI} \begin{bmatrix} 1 \\ 1 \end{bmatrix} w_2 \qquad (ii)$$

The deflection equations in the cut structure are given by combining Eqs. (5.34) and (5.50).

$$\begin{bmatrix} r_o \\ r_x \end{bmatrix} = \begin{bmatrix} D_{00} & D_{01} \\ D_{10} & D_{11} \end{bmatrix} \begin{bmatrix} R \\ X \end{bmatrix} + \begin{bmatrix} b_o{}^T \\ b_1{}^T \end{bmatrix} [\delta_o] R + \begin{bmatrix} b_o{}^T \\ b_1{}^T \end{bmatrix} [\delta_t]$$

For structure continuity at the cuts,

$$r_x \equiv 0 = D_{10}R + D_{11}X + b_1{}^T\delta_o R + b_1{}^T\delta_t$$

Hence,

$$X = -D_{11}{}^{-1}(D_{10}R + b_1{}^T\delta_o R + b_1{}^T\delta_t)$$

The matrices δ_o and δ_t are generated from Eqs. (*i*) and (*ii*) and are found to be

$$\delta_o = \frac{(20 \times 12)^3}{24EI}\begin{bmatrix} 1 & 0 & 0 \\ 1 & 0 & 0 \\ 0 & 0 & 0 \\ 0 & 0 & 0 \\ 0 & -4.575 & 0 \\ 0 & -4.575 & 0 \\ 0 & 0 & 0 \\ 0 & 0 & 0 \\ 0 & 0 & 0 \\ 0 & 0 & 0 \\ 0 & 0 & 0 \\ 0 & 0 & 0 \\ 0 & 0 & 0 \\ 0 & 0 & 0 \\ 0 & 0 & 0 \\ 0 & 0 & 0 \end{bmatrix} \qquad \delta_t = 33.541 \times 12 \times \alpha \begin{bmatrix} 0 \\ 0 \\ -15 \\ -15 \\ 95 \\ 0 \\ 0 \\ 0 \\ 0 \\ -15 \\ -15 \\ 95 \\ 0 \\ 0 \\ 0 \\ 0 \end{bmatrix} \qquad (iii)$$

Calculation of Flexibility Matrices D_{11} and D_{10}

From Eqs. (5.24) and (5.25),

$$D_{10} = b_o{}^T f b_1$$

$$D_{11} = b_1{}^T f b_1$$

The structure is cut at nodes 3 and 5, and the redundant forces

$$X^T = \langle X_1 \quad X_2 \quad X_3 \quad X_4 \quad X_5 \quad X_6 \rangle$$

are shown in Fig. 5.23. Note that positive member moments cause tension on the sides of the members marked by the dotted line in Fig. 5.22*a*. From Fig. 5.23,

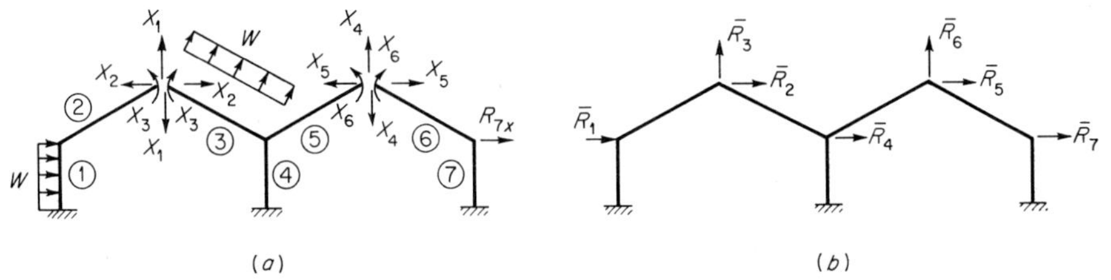

Figure 5.23 Cut-structure-load redundant forces.

$$
\begin{bmatrix} M_{1i} \\ M_{1j} \\ M_{2i} \\ M_{2j} \\ F_2 \\ M_{3i} \\ M_{3j} \\ M_{4i} \\ M_{4j} \\ M_{5i} \\ M_{5j} \\ F_5 \\ M_{6i} \\ M_{6j} \\ M_{7i} \\ M_{7j} \end{bmatrix}
=
\underbrace{\begin{bmatrix}
-28800 & 0 & 0 \\
0 & 0 & 0 \\
0 & 0 & 0 \\
0 & 0 & 0 \\
0 & 0 & 0 \\
0 & 81000 & 0 \\
0 & 81000 & 0 \\
0 & 1.24200 & 0 \\
0 & 0 & 0 \\
0 & 0 & 0 \\
0 & 0 & 0 \\
0 & 0 & 0 \\
0 & 0 & 0 \\
0 & 0 & 0 \\
0 & 0 & 0 \\
0 & 0 & 240
\end{bmatrix}}_{b_o}
\underbrace{\begin{bmatrix}
360 & 420 & 1.0 & 0 & 0 & 0 \\
360 & 180 & 1.0 & 0 & 0 & 0 \\
360 & 180 & 1.0 & 0 & 0 & 0 \\
0 & 0 & 1.0 & 0 & 0 & 0 \\
0.4472 & -0.8944 & 0 & 0 & 0 & 0 \\
0 & 0 & 1.0 & 0 & 0 & 0 \\
-360 & 180 & 1.0 & 0 & 0 & 0 \\
-360 & 180 & 1.0 & -360 & -180 & -1.0 \\
-360 & 420 & 1.0 & -360 & -420 & -1.0 \\
0 & 0 & 0 & 360 & 180 & 1.0 \\
0 & 0 & 0 & 0 & 0 & 1.0 \\
0 & 0 & 0 & 0.4472 & -0.8944 & 0 \\
0 & 0 & 0 & 0 & 0 & 1.0 \\
0 & 0 & 0 & -360 & 180 & 1.0 \\
0 & 0 & 0 & -360 & 180 & 1.0 \\
0 & 0 & 0 & -360 & 420 & 1.0
\end{bmatrix}}_{b_1}
\begin{bmatrix} w_1 \\ w_2 \\ R_{7x} \\ X_1 \\ X_2 \\ X_3 \\ X_4 \\ X_5 \\ X_6 \end{bmatrix}
\qquad (iv)
$$

Deflections will be calculated in the seven directions shown in Fig. 5.23*b*, that is, horizontal deflections at nodes 2, 4, and 7 and both horizontal and vertical deflections at nodes 3 and 6. For forces in these directions, the b_o matrix is given by

$$
b_o =
\begin{bmatrix}
-240 & -420 & 360 & 0 & 0 & 0 & 0 \\
0 & -180 & 360 & 0 & 0 & 0 & 0 \\
0 & -180 & 360 & 0 & 0 & 0 & 0 \\
0 & 0 & 0 & 0 & 0 & 0 & 0 \\
0 & 0.4472 & 0.8944 & 0 & 0 & 0 & 0 \\
0 & 0 & 0 & 0 & 0 & 0 & 0 \\
0 & 0 & 0 & 0 & 0 & 0 & 0 \\
0 & 0 & 0 & 0 & 180 & -360 & 0 \\
0 & 0 & 0 & 240 & 420 & -360 & 0 \\
0 & 0 & 0 & 0 & -180 & 360 & 0 \\
0 & 0 & 0 & 0 & 0 & 0 & 0 \\
0 & 0 & 0 & 0 & 0.4472 & 0.8944 & 0 \\
0 & 0 & 0 & 0 & 0 & 0 & 0 \\
0 & 0 & 0 & 0 & 0 & 0 & 0 \\
0 & 0 & 0 & 0 & 0 & 0 & 0 \\
0 & 0 & 0 & 0 & 0 & 0 & 240
\end{bmatrix}
\qquad (v)
$$

From Eq. (v), it is seen that

$$r_o = \bar{b}_o^T[(fb_oR + fb_1X) + \delta_oR + \delta_t]$$

Hence

$$r_o = \bar{b}_o^T[f(b_oR + b_1X) + \delta_oR + \delta_t]$$

In the program on pages 260 to 269 member forces and deflections are calculated first with no temperature strains and secondly with them included.

5.4 REMOVAL OF A MEMBER; CHANGE IN MEMBER FLEXIBILITY; GENERAL THEORY ON THE INTRODUCTION OF RELEASES INTO A STRUCTURE

In certain structures it may be convenient first to analyze a regular system and then to ascertain the effects of introducing certain cutouts. For example, in an aircraft fuselage or bus body, the regular structure may be first examined and then doors and openings introduced. In addition, it may be of value in complex structures to be able to calculate the effect of modifying a member size (and hence its flexibility) without reanalyzing the whole structural system. One might notice in passing that as the speed and size of the modern computer has increased, the size of problem for which reanalysis for change of member properties is feasible has also increased considerably. Finally, release systems are discussed in relation to the redundant forces in the structure. It is sometimes convenient to analyze a structure without releases and then later to introduce them. For example, in the analysis

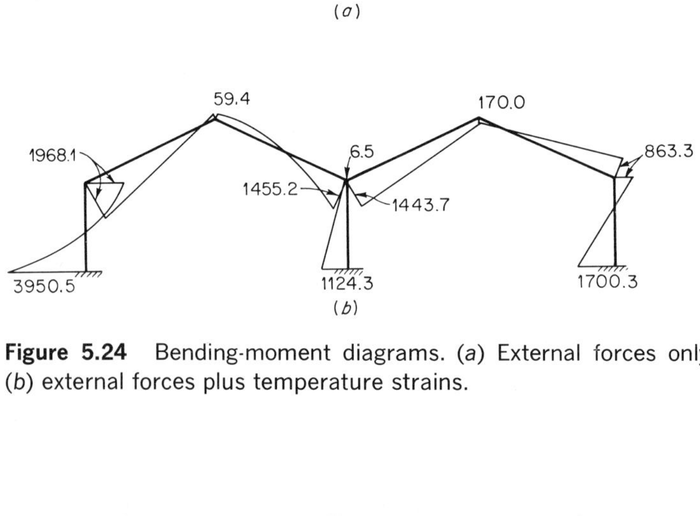

Figure 5.24 Bending-moment diagrams. (*a*) External forces only; (*b*) external forces plus temperature strains.

of structures by using a tree as the determinate system, releases are not permissible in tree members since they would cause collapse. Similarly, in the arch analysis presented in Sec. 5.5, moment releases are introduced after the fixed-end arch has been analyzed. Naturally, one will encounter difficulties if the releases introduced (or members removed) leave the remaining structure unstable. Such a condition will be recognized by a singular D_{11} matrix [Eq. (5.28)].

The simplest member modification is complete member removal. This process can be achieved by introducing initial strains (and hence forces) in the member concerned so that when the external loads are applied, the total force in the member will be zero. If the force in the member is zero, it necessarily has no part in the structural action of the system and may be removed.

For a structure with initial strains v_{oi} in members i, the deflections r_o and r_x are given [from Eq. (5.50)] in Eq. (5.59).

$$\begin{bmatrix} r_o \\ r_x \end{bmatrix} = \begin{bmatrix} D_{00} & D_{01} \\ D_{10} & D_{11} \end{bmatrix} \begin{bmatrix} R \\ X \end{bmatrix} + \begin{bmatrix} b_{oi}{}^T \\ b_{1i}{}^T \end{bmatrix} v_{oi} \tag{5.59}$$

For continuity of the structure at the cuts, $r_x \equiv 0$, so that

$$X = -D_{11}{}^{-1}(D_{10}R + b_{1i}{}^T v_{oi}) \tag{5.60}$$

Hence member forces are given as

$$S = b_o R + b_1 X$$

and in particular for those members i,

$$S_i = b_{oi} R + b_{1i} X \tag{5.61}$$

By substitution from Eq. (5.60),

$$S_i = b_{oi} R - b_{1i} D_{11}{}^{-1}(D_{10}R + b_{1i}{}^T v_{oi}) \tag{5.62}$$

The forces S_i have been produced by the initial strains and the external loads, and if these members are to be removed, they must be equal to zero. That is,

$$S_i = 0 = (b_{oi} - b_{1i} D_{11}{}^{-1} D_{10}) R - b_{1i} D_{11}{}^{-1} b_{1i}{}^T v_{oi} \tag{5.63}$$

The required initial strains may be calculated from Eq. (5.63), so that

$$v_{oi} = (b_{1i} D_{11}{}^{-1} b_{1i}{}^T)^{-1} (b_{oi} - b_{1i} D_{11}{}^{-1} D_{10}) R \tag{5.64}$$

The redundant forces are obtained from Eq. (5.60), which may now be rewritten

$$X_{\text{new}} = X_{\text{original}} - D_{11}{}^{-1} b_{1i}{}^T v_{oi} \tag{5.65}$$

```
START
  SOLUTION OF TWO BAY PORTAL FRAME FOR MEMBER LOADS
     AND NODE DEFLECTIONS
     EXTERNAL LOADS AND TEMPERATURE


LOAD BI 16 6
   16 ROWS,      6 COLUMNS

PRINT BI * FORCE TRANSFORMATION MATRIX FOR REDUNDANTS

              1              2              3              4              5
   1     360.00000      420.00000        1.00000        0.00000        0.00000
   2     360.00000      180.00000        1.00000        0.00000        0.00000
   3     360.00000      180.00000        1.00000        0.00000        0.00000
   4       0.00000        0.00000        1.00000        0.00000        0.00000
   5       0.44720       -0.89440        0.00000        0.00000        0.00000
   6       0.00000        0.00000        1.00000        0.00000        0.00000
   7    =360.00000      180.00000        1.00000        0.00000        0.00000
   8    =360.00000      180.00000        1.00000     -360.00000     -180.00000
   9    =360.00000      420.00000        1.00000     -360.00000     -420.00000
  10       0.00000        0.00000        0.00000      360.00000      180.00000
  11       0.00000        0.00000        0.00000        0.00000        0.00000
  12       0.00000        0.00000        0.00000        0.44720       -0.89440
  13       0.00000        0.00000        0.00000        0.00000        0.00000
  14       0.00000        0.00000        0.00000     -360.00000      180.00000
  15       0.00000        0.00000        0.00000     -360.00000      180.00000
  16       0.00000        0.00000        0.00000     -360.00000      420.00000

LOAD BO 16 3
   16 ROWS,      3 COLUMNS

PRINT BO   * FORCE TRANSFORMATION MATRIX FOR LOADS

              1              2              3
   1  =28800.00002        0.00000        0.00000
   2       0.00000        0.00000        0.00000
   3       0.00000        0.00000        0.00000
   4       0.00000        0.00000        0.00000
   5       0.00000        0.00000        0.00000
   6       0.00000        0.00000        0.00000
   7       0.00000    81000.00000        0.00000
   8       0.00000    81000.00000        0.00000
   9       0.00000   124200.00046        0.00000
  10       0.00000        0.00000        0.00000
  11       0.00000        0.00000        0.00000
  12       0.00000        0.00000        0.00000
  13       0.00000        0.00000        0.00000
  14       0.00000        0.00000        0.00000
  15       0.00000        0.00000        0.00000
  16       0.00000        0.00000      240.00000

LOAD F 16 16
   16 ROWS,     16 COLUMNS

PRINT F   * MEMBER FLEXIBILITY MATRIX AXIAL CPTS MEMBERS 2 AND 5

              1              2              3              4              5
   1       2.00000        1.00000        0.00000        0.00000        0.00000
   2       1.00000        2.00000        0.00000        0.00000        0.00000
   3       0.00000        0.00000        3.35400        1.67700        0.00000
   4       0.00000        0.00000        1.67700        3.35400        0.00000
   5       0.00000        0.00000        0.00000        0.00000      100.62400
   6       0.00000        0.00000        0.00000        0.00000        0.00000
   7       0.00000        0.00000        0.00000        0.00000        0.00000
   8       0.00000        0.00000        0.00000        0.00000        0.00000
   9       0.00000        0.00000        0.00000        0.00000        0.00000
  10       0.00000        0.00000        0.00000        0.00000        0.00000
  11       0.00000        0.00000        0.00000        0.00000        0.00000
  12       0.00000        0.00000        0.00000        0.00000        0.00000
  13       0.00000        0.00000        0.00000        0.00000        0.00000
  14       0.00000        0.00000        0.00000        0.00000        0.00000
  15       0.00000        0.00000        0.00000        0.00000        0.00000
  16       0.00000        0.00000        0.00000        0.00000        0.00000
```

0.00000
0.00000
0.00000
0.00000
0.00000
0.00000
0.00000
-1.00000
-1.00000
1.00000
1.00000
0.00000
1.00000
1.00000
1.00000
1.00000

	7	8
0.00000	0.00000	0.00000
0.00000	0.00000	0.00000
0.00000	0.00000	0.00000
0.00000	0.00000	0.00000
0.00000	0.00000	0.00000
3.35400	1.67700	0.00000
1.67700	3.35400	0.00000
0.00000	0.00000	2.00000
0.00000	0.00000	1.00000
0.00000	0.00000	0.00000
0.00000	0.00000	0.00000
0.00000	0.00000	0.00000
0.00000	0.00000	0.00000
0.00000	0.00000	0.00000
0.00000	0.00000	0.00000
0.00000	0.00000	0.00000

```
               9            10            11            12            13
     1      0.00000      0.00000      0.00000      0.00000      0.00000
     2      0.00000      0.00000      0.00000      0.00000      0.00000
     3      0.00000      0.00000      0.00000      0.00000      0.00000
     4      0.00000      0.00000      0.00000      0.00000      0.00000
     5      0.00000      0.00000      0.00000      0.00000      0.00000
     6      0.00000      0.00000      0.00000      0.00000      0.00000
     7      0.00000      0.00000      0.00000      0.00000      0.00000
     8      1.00000      0.00000      0.00000      0.00000      0.00000
     9      2.00000      0.00000      0.00000      0.00000      0.00000
    10      0.00000      3.35400      1.67700      0.00000      0.00000
    11      0.00000      1.67700      3.35400      0.00000      0.00000
    12      0.00000      0.00000      0.00000    100.62000      0.00000
    13      0.00000      0.00000      0.00000      0.00000      3.35400
    14      0.00000      0.00000      0.00000      0.00000      1.67700
    15      0.00000      0.00000      0.00000      0.00000      0.00000
    16      0.00000      0.00000      0.00000      0.00000      0.00000

SCALE F 0.04444
 SCALAR=  0.4444000E-01

SCALE F 0.0001
 SCALAR=  0.1000000E-03

TRANS BI BIT

MULT F BO B

MULT BIT B FIO

MULT F BI A

MULT BIT A FII

INVERT FII

LOAD R 3 1
    3 ROWS,      1 COLUMNS

PRINT R    *LOADS

               1
     1      0.10000
     2      0.10000
     3     10.00000

MULT FIO R C

LOAD DELO 16 3
   16 ROWS,      3 COLUMNS

PRINT DELO    *CONTRIBUTION OF LOADS ON MEMBERS TO DEFLECTIONS

               1             2             3
     1      1.00000       0.00000       0.00000
     2      1.00000       0.00000       0.00000
     3      0.00000       0.00000       0.00000
     4      0.00000       0.00000       0.00000
     5      0.00000       0.00000       0.00000
     6      0.00000      -4.57500       0.00000
     7      0.00000      -4.57500       0.00000
     8      0.00000       0.00000       0.00000
     9      0.00000       0.00000       0.00000
    10      0.00000       0.00000       0.00000
    11      0.00000       0.00000       0.00000
    12      0.00000       0.00000       0.00000
    13      0.00000       0.00000       0.00000
    14      0.00000       0.00000       0.00000
    15      0.00000       0.00000       0.00000
    16      0.00000       0.00000       0.00000

SCALE DELO 0.064
 SCALAR=  0.6400000E-01

MULT DELO R V
```

	15	16
0.00000	0.00000	0.00000
0.00000	0.00000	0.00000
0.00000	0.00000	0.00000
0.00000	0.00000	0.00000
0.00000	0.00000	0.00000
0.00000	0.00000	0.00000
0.00000	0.00000	0.00000
0.00000	0.00000	0.00000
0.00000	0.00000	0.00000
0.00000	0.00000	0.00000
0.00000	0.00000	0.00000
0.00000	0.00000	0.00000
1.67700	0.00000	0.00000
3.35400	0.00000	0.00000
0.00000	2.00000	1.00000
0.00000	1.00000	2.00000

```
MULT BIT V VA

ADD VA C

MULT FII C X

SCALE X -1.0
 SCALAR= -0.1000000E+01

PRINT X  * REDUNDANTS  NO TEMPERATURE

              1
    1        11.85475
    2       -12.35248
    3      -263.87623
    4         2.87102
    5         1.18169
    6       -45.39601

DELETE B

DELETE A
     CALCULATE DEFLECTIONS  AND  MEMBER FORCES NO TEMPERATURE

MULT BO R SO

MULT BI X SI

ADD SO SI

PRINT SI  * MEMBER FORCES NO TEMPERATURE

              1
    1     -4064.20594
    2      1780.38867
    3      1780.38867
    4      -263.87623
    5        16.34950
    6      -263.87623
    7      1344.96695
    8       144.09055
    9      1215.89000
   10      1200.87639
   11       -45.39601
   12         0.22702
   13       -45.39601
   14      -866.25953
   15      -866.25953
   16      1817.34639

MULT F SI VB

ADD V VB

LOAD BOA 16 7
   16 ROWS,      7 COLUMNS

PRINT BOA  * STATICS MATRIX

              1             2             3            4           5
    1    -240.00000    -420.00000    360.00000      0.00000      0.0000
    2       0.00000    -180.00000    360.00000      0.00000      0.0000
    3       0.00000    -180.00000    360.00000      0.00000      0.0000
    4       0.00000       0.00000      0.00000      0.00000      0.0000
    5       0.00000       0.44720      0.89440      0.00000      0.0000
    6       0.00000       0.00000      0.00000      0.00000      0.0000
    7       0.00000       0.00000      0.00000      0.00000      0.0000
    8       0.00000       0.00000      0.00000      0.00000    180.0000
    9       0.00000       0.00000      0.00000    240.00000    420.0000
   10       0.00000       0.00000      0.00000      0.00000   -180.0000
   11       0.00000       0.00000      0.00000      0.00000      0.0000
   12       0.00000       0.00000      0.00000      0.00000      0.8944
   13       0.00000       0.00000      0.00000      0.00000      0.0000
   14       0.00000       0.00000      0.00000      0.00000      0.0000
   15       0.00000       0.00000      0.00000      0.00000      0.0000
   16       0.00000       0.00000      0.00000      0.00000      0.0000
```

6	7
0.00000	0.00000
0.00000	0.00000
0.00000	0.00000
0.00000	0.00000
0.00000	0.00000
0.00000	0.00000
0.00000	0.00000
60.00000	0.00000
60.00000	0.00000
60.00000	0.00000
0.00000	0.00000
0.44720	0.00000
0.00000	0.00000
0.00000	0.00000
0.00000	0.00000
0.00000	240.00000

```
TRANS BOA BOB

MULT BOB VB DEF

DELETE VB

PRINT DEF  *DEFLECTIONS NO TEMPERATURE

              1
    1          5.23455
    2          3.99175
    3          2.49867
    4          2.74732
    5          2.85007
    6         -0.20526
    7          2.95270
  TEMPERATURE  EFFECTS

LOAD DELI 16 1
   16 ROWS,       1 COLUMNS

PRINT DELI  * TEMPERATURE STRAIN MATRIX

              1
    1          0.00000
    2          0.00000
    3         -3.33300
    4         -3.33300
    5         50.00000
    6          0.00000
    7          0.00000
    8          0.00000
    9          0.00000
   10         -3.33300
   11         -3.33300
   12         50.00000
   13          0.00000
   14          0.00000
   15          0.00000
   16          0.00000

SCALE DELI 0.0025156
 SCALAR=  0.2515600E-02

MULT BIT DELI D

DELETE X

DELETE SI

DELETE BO

DELETE BIT

DELETE DEF

ADD D C

MULT FII C XT

SCALE XT -1.0
 SCALAR= -0.1000000E+01

PRINT XT  *  REDUNDANTS TEMPERATURE INCLUDED

              1
    1         11.96235
    2        -12.66102
    3        -59.36156
    4          3.21116
    5          0.68167
    6        169.96915

MULT BI XT SI
```

```
ADD SO SI

PRINT SI   *  MEMBER FORCES TEMPERATURE INCLUDED

                  1
     1     -3950.54497
     2      1968.10044
     3      1968.10044
     4       -59.36156
     5        16.67358
     6       -59.36156
     7      1455.20830
     8         6.52232
     9      1124.27606
    10      1448.68596
    11       169.96915
    12         0.82634
    13       169.96915
    14      -863.34645
    15      -863.34645
    16      1700.25436

MULT F SI VB

ADD V VB

ADD DELI VB

MULT BOB VB DEF

PRINT DEF   *  NODE DEFLECTIONS INCLUDING TEMPERATURE STRAINS

                  1
     1         4.79189
     2         3.61343
     3         2.59526
     4         2.40517
     5         2.52612
     6        -0.15983
     7         2.70604
```

Example 5.11 The pin-jointed truss structure shown in Fig. 5.25 has one redundant, taken as the force in member (6), and is loaded by the horizontal force of 10 kips, as shown. All members are assumed to have unit flexibility.

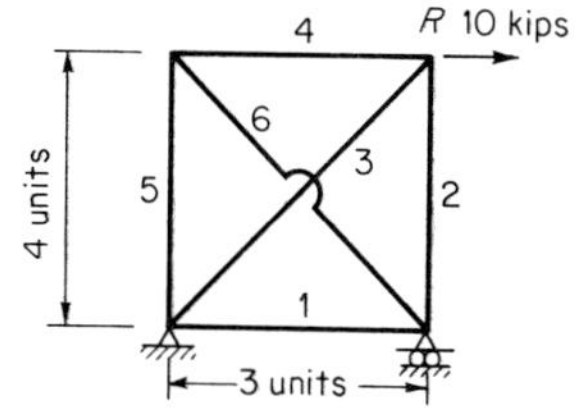

Figure 5.25 Member removal from truss.

The effect of removing member (4) is examined. Of course this is a trivial case leaving the structure determinate with forces in members (1), (2), and (3) only.

1 Member forces in terms of R and X $(= F_6)$, that is, b_o and b_1:

$$\begin{bmatrix} S_1 \\ S_2 \\ S_3 \\ S_4 \\ S_5 \\ S_6 \end{bmatrix} = \begin{bmatrix} 0 & -3/5 \\ -4/3 & -4/5 \\ 5/3 & 1 \\ 0 & -3/5 \\ 0 & -4/5 \\ 0 & 1 \end{bmatrix} \begin{bmatrix} R \\ X \end{bmatrix}$$

2 Member flexibilities:

$$f = \begin{bmatrix} 1 & & & & & \mathbf{0} \\ & 1 & & & & \\ & & 1 & & & \\ & & & 1 & & \\ & & & & 1 & \\ \mathbf{0} & & & & & 1 \end{bmatrix}$$

3 Then,

$$b^T f b = \begin{bmatrix} 41/9 & 41/15 \\ 41/15 & 100/25 \end{bmatrix}$$

4 To remove member (4), calculate $(b_{1i} D_{11}^{-1} b_{1i}^T)^{-1}$. That is,

$$b_{14} D_{11}^{-1} b_{14}^T = (-3/5)(25/100)(-3/5) = 9/100$$

$$(b_{14} D_{11}^{-1} b_{14}^T)^{-1} = 100/9$$

The matrix $b_{04} = 0$ and

$$-b_{14} D_{11}^{-1} D_{10} = -(3/5 \; 25/100 \; 41/15) = 41/100$$

5 To remove member (4),

$$v_{04} = 100/9 \; 41/100 R = 41/9 R$$

Thence,

$$X = -D_{11}^{-1}(D_{10} R + b_{14}^T v_{04})$$

$$= -D_{11}^{-1}(41/15 - 3/5 \; 41/9) R = 0$$

It will be seen that this is a necessary result since with member (4) removed, member (6) carries zero force. The member forces are given by b_o alone, the required result for this structure.

Example 5.12 Hinge (or Moment) release in a member

The beam shown in Fig. 5.26, continuous over two spans, is one degree indeterminate. A load R is placed at the center of the right-hand span as shown, and to make this a node force, the beam is considered composed of the three members as shown.

The effect of the release $M_{2j} = 0$, that is, a hinge at the midpoint of the right-hand span, will be investigated. As in Example 5.11, the effect of the hinge is known before the analysis is commenced since it renders the structure determinate.

The moment at the center support is chosen as the unknown force quantity so that the b matrix is given as

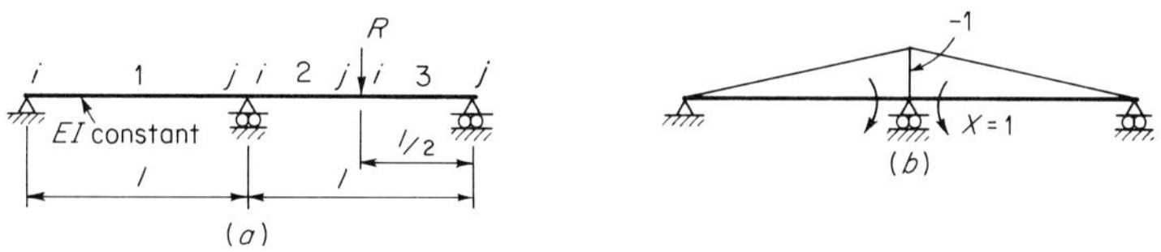

Figure 5.26 Beam-structure releases. (*a*) $6EI = 1$; (*b*) redundant system.

$$b = [b_o \quad b_1] = \begin{bmatrix} 0 & 0 \\ 0 & -1 \\ 0 & -1 \\ \frac{l}{4} & -\frac{1}{2} \\ \frac{l}{4} & -\frac{1}{2} \\ \underbrace{0}_{b_o} & \underbrace{0}_{b_1} \end{bmatrix}$$

The member-flexibility matrix is given by

$$f = l\begin{bmatrix} 2 & 1 & 0 & 0 & 0 & 0 \\ 1 & 2 & 0 & 0 & 0 & 0 \\ 0 & 0 & 1 & \frac{1}{2} & 0 & 0 \\ 0 & 0 & \frac{1}{2} & 1 & 0 & 0 \\ 0 & 0 & 0 & 0 & 1 & \frac{1}{2} \\ 0 & 0 & 0 & 0 & \frac{1}{2} & 1 \end{bmatrix}$$

Thence,

$$b^T f b = \begin{bmatrix} \frac{l^3}{8} & -\frac{3l^2}{8} \\ -\frac{3l^2}{8} & \frac{16l}{4} \end{bmatrix}$$

With no release,

$$X = -D_{11}{}^{-1}D_{10}R = \frac{3l}{32}R$$

For a hinge release at the load point, $S_{2j} = 0$. Thence,

$$v_{02j} = \left[-\frac{1}{2}\frac{4}{16l}\left(-\frac{1}{2}\right)\right]^{-1}\left[\frac{l}{4} - \left(-\frac{1}{2}\right)\left(\frac{4}{16l}\right)\left(-\frac{3l^2}{8}\right)\right]R$$

$$= l^2\frac{13}{4}R$$

Then from Eq. (5.65),

$$X_{\text{new}} = \frac{3l}{32}R - \frac{4}{16l}\left(-\frac{1}{2}\right)l^2\frac{13}{4}R = \frac{Rl}{2}$$

It is seen that this is the required cantilever bending moment produced at the center support when a hinge is placed at the load point.

The Stress Transformation Matrix and Flexibility Matrix for Structure with Members Removed

So far it has been shown, Eq. (5.65), how the removal of a member may be accommodated by modification of the redundant forces in the system. These new redundant values may now be used to generate the b matrix for the modified structure and, in addition, to calculate the flexibility matrix $(F)_{\text{modified}}$.

From Eq. (5.33), it is known that

$$S = bR,$$

where

$$b = b_o - b_1 D_{11}^{-1} D_{10}$$

for the redundant structure.

Partition b into b_h, that transformation matrix for the unaltered members, and b_i for members removed (or those in which releases occur). Then from Eq. (5.33),

$$b_h = b_{oh} - b_{1h} D_{11}^{-1} D_{10} \tag{5.66}$$

$$b_i = b_{oi} - b_{1i} D_{11}^{-1} D_{10} \tag{5.67}$$

From Eq. (5.64), [by using Eq. (5.67)] it is seen that

$$v_{oi} = (b_{1i} - D_{11}^{-1} b_{1i}^{T})^{-1} b_i R \tag{5.68}$$

Thence,

$$X_{\text{new}} = X_{\text{original}} - D_{11}^{-1} b_{1i}^{T} (b_{1i} D_{11}^{-1} b_{1i}^{T})^{-1} b_i R \tag{5.69}$$

Then,

$$S_i \equiv 0$$

$$S_h = b_{oh} R + b_{1h} X$$

From Eq. (5.69),

$$S_h = (b_{oh} - b_{1h} D_{11}^{-1} D_{10}) R - b_{1h} D_{11}^{-1} b_{1i}^{T} (b_{1i} D_{11}^{-1} b_{1i}^{T})^{-1} b_i R$$

That is,

$$S_h = [b_h - b_{1h}D_{11}^{-1}b_{1i}^T(b_{1i}D_{11}^{-1}b_{1i}^T)^{-1}b_i]R \tag{5.70}$$

The stress transformation in the modified structure is

$$b_{h(\text{new})} = b_h - b_{1h}D_{11}^{-1}b_{1i}^T(b_{1i}D_{11}^{-1}b_{1i}^T)^{-1}b_i \tag{5.71}$$

Modified Structure-Flexibility Matrix

From Eq. (5.59), it is seen that the deflections at the load points are given by

$$r_o = D_{00}R + D_{01}X + b_{oi}^T v_{oi} \tag{5.72}$$

From Eq. (5.60),

$$r_o = (D_{00} - D_{01}D_{11}^{-1}D_{10})R + (b_{oi}^T - D_{01}D_{11}^{-1}b_{1i}^T)v_{oi} \tag{5.73}$$

That is,

$$r_o = FR + (b_{oi}^T - D_{01}D_{11}^{-1}b_{1i}^T)v_{oi} \tag{5.74}$$

The flexibility matrix F is again that for the original unaltered structure.

However, since $D_{01} = D_{10}^T$, from Eq. (5.67) it is seen that

$$r_o = FR + b_i^T v_{oi} \tag{5.75}$$

Then from Eq. (5.68),

$$r_o = FR + b_i^T(b_{1i}D_{11}^{-1}b_{1i}^T)^{-1}b_iR \tag{5.76}$$

Define

$$M_i = (b_{1i}D_{11}^{-1}b_{1i}^T)^{-1} \tag{5.77}$$

Then,

$$r_o = FR + b_i^T M_i b_i R \tag{5.78}$$

The modified flexibility matrix is then,

$$F_{\text{modified}} = F + b_i^T M_i b_i \tag{5.79}$$

Change in Member Flexibility

The removal of a member or the introduction of a release is only a special case of the alteration of a member or member-force flexibility. In such a case the flexibility is set to zero. Evidently this theory is useful when part of the structure is cut out, as for openings in an aircraft fuselage. In addition it is useful to know how member forces and node deflections will be altered when a member undergoes a change in flexibility that results in other than zero final value.

As previously, suppose that the structure is partitioned into members i, whose flexibilities are altered, and h, those remaining unchanged. Partition the diagonal member-flexibility matrix accordingly.

$$f = \begin{bmatrix} f_h & 0 \\ 0 & f_i \end{bmatrix} \qquad \text{unaltered}$$

$$f_m = \begin{bmatrix} f_h & 0 \\ 0 & f_i + \Delta f_i \end{bmatrix} \qquad \text{modified} \tag{5.80}$$

By a suitable choice of initial strains v_{oi} in the members i of the unmodified structure, the forces S_i in this structure will be the same as those S_i^* in the modified structure, and in addition the deformations of the two sets will be the same.

These two requirements give

$$S_i^* = S_i = b_i R - b_{1i} D_{11}^{-1} b_{1i}^T v_{oi} \tag{5.81}$$

$$v_i^* = v_i \tag{5.82}$$

In the unmodified structure,

$$v_i = f_i S_i + v_{oi} \tag{5.83}$$

Then, Eq. (5.82) becomes

$$(f_i + \Delta f_i) S_i^* = f_i S_i + v_{oi} \tag{5.84}$$

That is,

$$v_{oi} = \Delta f_i S_i^* \tag{5.85}$$

Substitution from Eq. (5.81) gives

$$v_{oi} = \Delta f_i (b_i R - b_{1i} D_{11}^{-1} b_{1i}^T v_{oi}) \tag{5.86}$$

This equation may now be solved for the initial strains v_{oi}.

$$v_{oi} = (b_{1i} D_{11}^{-1} b_{1i}^T + \Delta f_i^{-1})^{-1} b_i R \tag{5.87}$$

The internal forces in both modified and original structures are identical, so that

$$S = S^* = [b - b_1 D_{11}^{-1} b_{1i}^T (b_{1i} D_{11}^{-1} b_{1i}^T + \Delta f_i^{-1})^{-1} b_i] R \tag{5.88}$$

or

$$S^* = (b + \Delta b) R \tag{5.89}$$

Flexibility Matrix for Modified Structure

From Eq. (5.75),

$$r_o = FR + b_i^T v_{oi}$$

In this case, substitution for v_{oi} gives

$$r_o = FR + b_i^T (b_{1i} D_{11}^{-1} b_{1i}^T + \Delta f_i^{-1})^{-1} b_i R \tag{5.90}$$

That is,

$$r = (F + \Delta F)R = [F + b_i^T(b_{1i}D_{11}^{-1}b_{1i}^T + \Delta f_i^{-1})^{-1}b_i]R \tag{5.91}$$

Then,

$$\Delta F = b_i^T(b_{1i}D_{11}^{-1}b_{1i}^T + \Delta f_i^{-1})^{-1}b_i \tag{5.92}$$

It is now possible to examine two limiting cases for Δf_i. The first occurs when $f_i + \Delta f_i \to \infty$, that is, when the member is removed. For this, the theory must give identical results with the previous expressions. The second limiting case arises when $f_i + \Delta f_i = 0$, that is, when a member is made rigid or, in other words, a flexibility is removed from the system.

Case 1 Suppose the member is removed.

Then the flexibility is infinite so that $\Delta f_i \to \infty$, and hence the stiffness $\Delta f_i^{-1} \to o$. With $\Delta f_i^{-1} = 0$, Eq. (5.88) becomes

$$S = [b - b_1D_{11}^{-1}b_{1i}^T(b_{1i}D_{11}^{-1}b_{1i}^T)^{-1}b_i]R \tag{5.93}$$

For the unaltered members h, this becomes

$$S_h = [b_h - b_{1h}D_{11}^{-1}b_{1i}^T(b_{1i}D_{11}^{-1}b_{1i}^T)^{-1}b_i]R \tag{5.94}$$

This is identical with Eq. (5.70). Also,

$$S_i = [b_i - b_{1i}D_{11}^{-1}b_{1i}^T(b_{1i}D_{11}^{-1}b_{1i}^T)^{-1}b_i]R \equiv 0$$

Case 2 Suppose the member is made rigid. The member then has infinite stiffness, and

$$f_i + \Delta f_i = 0$$

Then,

$$\Delta f_i = -f_i \tag{5.95}$$

Let $k_i = f_i^{-1}$ be the member stiffness. Then, Eq. (5.88) becomes

$$S = [b - b_1D_{11}^{-1}b_{1i}^T(b_{1i}D_{11}^{-1}b_{1i}^T - k_i)^{-1}b_i]R \tag{5.96}$$

And

$$\Delta F = b_i^T(b_{1i}D_{11}^{-1}b_{1i}^T - k_i)^{-1}b_i \tag{5.97}$$

Introduction of Releases by Removal of Redundant Forces

When a release is introduced into a structure, the degree of indeterminacy is automatically reduced by 1. It is assumed of course that the structure is stable and indeterminate. In the first instance, assume that the releases to be removed correspond to certain of the redundants selected in the analysis.

The structure is assumed to have been analyzed for the redundants X with no releases so that from Eq. (5.28),

$$X = -D_{11}{}^{-1}D_{10}R \tag{5.98}$$

Now suppose that X is partitioned into X_l and X_m, where $X_l \equiv 0$ and $X_m \neq 0$. That is, X_l are the redundants to be released. Write Eq. (5.98) in partitioned form, using the notation

$$\begin{bmatrix} \alpha_l \\ \alpha_m \end{bmatrix} = D_{10}R \tag{5.99}$$

Then,

$$\begin{bmatrix} r_{xl} \\ r_{xm} \end{bmatrix} = \begin{bmatrix} \alpha_l \\ \alpha_m \end{bmatrix} + \begin{bmatrix} D_{ll} & D_{lm} \\ D_{ml} & D_{mm} \end{bmatrix} \begin{bmatrix} X_l \\ X_m \end{bmatrix} \tag{5.100}$$

It is seen that whereas $X_l = 0$, $r_{xl} \neq 0$ and that $r_{xm} = 0$ for continuity of the structure. Thence,

$$X_m = -D_{mm}{}^{-1}\alpha_m \tag{5.101}$$

The theory is readily extended as follows to the structure where the releases are not at the positions of the X forces.

Let the stress resultants corresponding to the releases be S_l, and relate S_l to the set X through the usual type of force transformation. That is,

$$S_l = b_l X \tag{5.102}$$

Arrange the first l columns of b_1 so that each contains at least one nonzero value. This will entail a rearrangement of the order of the redundants X. In this new order, label the first l redundants X_l and the remainder X_m. Then $\langle S_l \quad X_m \rangle^T$ forms a complete, independent redundant set for the structure. It is possible to relate this new set, designated $\overline{X}$, to X through the reversible transformation

$$\begin{bmatrix} S_l \\ X_m \end{bmatrix} = \begin{bmatrix} b_l \\ 0000100 \\ 0000010 \\ 0000001 \end{bmatrix} \begin{bmatrix} X_l \\ X_m \end{bmatrix} \tag{5.103}$$

That is,

$$\overline{X} = BX \tag{5.104}$$

Then the contragradient law gives the displacement relationship

$$r_x = B^T \bar{r}_x \tag{5.105}$$

From Eq. (5.100), it is known that

$$r_x = \alpha + D_{11}X \tag{5.106}$$

so that

$$B^T\bar{r}_x = \alpha + D_{11}B^{-1}\bar{X} \tag{5.107}$$

and thence,

$$\bar{r}_x = (B^T)^{-1}\alpha + (B^T)^{-1}D_{11}B^{-1}\bar{X} \tag{5.108}$$

This equation is written

$$\bar{r}_x = \bar{\alpha} + \bar{D}_{11}\bar{X} \tag{5.109}$$

Now partition $\bar{\alpha}$, $\bar{D}_{11}$, etc., in the same way as α and D_{11}. Then when $\bar{X}_l \equiv 0$,

$$\bar{X}_m = -\bar{D}_{mm}{}^{-1}\bar{\alpha}_m \tag{5.110}$$

To return to X,

$$X = B^{-1}\bar{X} = B^{-1}\begin{bmatrix} S_l \equiv 0 \\ \bar{X}_m \end{bmatrix} \tag{5.111}$$

Example 5.13 Moment releases in a portal frame

The portal frame shown in Fig. 5.12 has been analyzed for the fixed-base condition (Example 5.5). In the present example, two release cases are considered, (1) moment releases at the supports, nodes 1 and 4 and (2) moment releases 5 ft 0 in. from the base of each column.

Both these cases are shown in Fig. 5.27. From Example 5.5 it is seen that

$$D_{11} = \frac{10}{6EI}\begin{bmatrix} 8 & 1 & 6 \\ 1 & 8 & 6 \\ 6 & 6 & 10 \end{bmatrix}$$

and

$$\alpha = b_1{}^T\delta_o R = \frac{50}{6EI}\begin{bmatrix} -1 \\ -1 \\ -2 \end{bmatrix} R \qquad R = wl = \text{load on span}$$

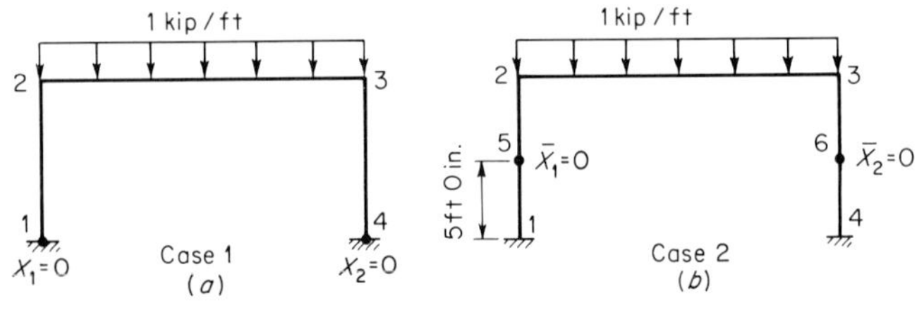

Figure 5.27 Structure with releases.

Case 1 When $X_1 = X_2 = 0$, the Eq. (5.101) may be used since the release forces are already included in the redundant set. That is,

$$\frac{10}{6EI}\begin{bmatrix} 8 & 1 & 6 \\ 1 & 8 & 6 \\ 6 & 6 & 10 \end{bmatrix}\begin{bmatrix} 0 \\ 0 \\ X_3 \end{bmatrix} + \frac{50}{6EI}\begin{bmatrix} -1 \\ -1 \\ -2 \end{bmatrix} R = \begin{bmatrix} r_{x1} \\ r_{x2} \\ 0 \end{bmatrix}$$

Then,

$$X_3 = R = wl = 20 \text{ kips} \qquad \text{for } w = 1 \text{ kip/ft}$$
$$X_1 = 0$$
$$X_2 = 0$$

Member forces are given by

$$S = b_oR + b_1X = \begin{bmatrix} -1 & 0 & 0 \\ -1 & 0 & -1 \\ -1 & 0 & -1 \\ 0 & -1 & -1 \\ 0 & -1 & -1 \\ 0 & -1 & 0 \end{bmatrix}\begin{bmatrix} 0 \\ 0 \\ 20 \end{bmatrix} = \begin{bmatrix} 0 \\ -20 \\ -20 \\ -20 \\ -20 \\ 0 \end{bmatrix}$$

The bending-moment diagram is shown in Fig. 5.28*a*.

Case 2 For case 2 (Fig. 5.27*b*) it is first necessary to transform

$$\langle X_1 \quad X_2 \quad X_3 \rangle \qquad \text{to} \qquad \langle \overline{X}_1 \quad \overline{X}_2 \quad \overline{X}_3 \rangle$$

before setting redundant values to zero [Eq. (5.108)]. As the new redundants, take $\overline{X}_1$ as the moment at point 5 and $\overline{X}_2$ as the moment at point 6. Then,

$$\begin{bmatrix} \overline{X}_1 \\ \overline{X}_2 \\ \overline{X}_3 \end{bmatrix} = \underbrace{\begin{bmatrix} -1 & 0 & -\frac{1}{2} \\ 0 & -1 & -\frac{1}{2} \\ 0 & 0 & 1 \end{bmatrix}}_{B}\begin{bmatrix} X_1 \\ X_2 \\ X_3 \end{bmatrix}$$

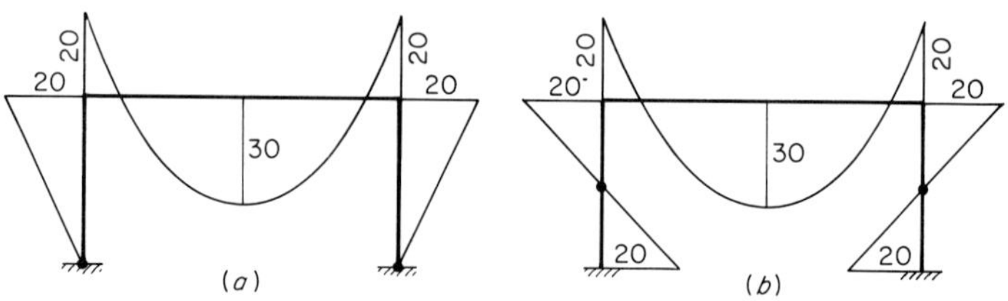

Figure 5.28

Then

$$B^{-1} = \begin{bmatrix} -1 & 0 & -\tfrac{1}{2} \\ 0 & -1 & -\tfrac{1}{2} \\ 0 & 0 & 1 \end{bmatrix}$$

Then Eq. (5.108) becomes

$$\frac{50}{6EI}\begin{bmatrix} -1 & 0 & 0 \\ 0 & -1 & 0 \\ -\tfrac{1}{2} & -\tfrac{1}{2} & 1 \end{bmatrix}\begin{bmatrix} -1 \\ -1 \\ -2 \end{bmatrix} R + \frac{10}{6EI}\begin{bmatrix} -1 & 0 & 0 \\ 0 & -1 & 0 \\ -\tfrac{1}{2} & -\tfrac{1}{2} & 1 \end{bmatrix}\begin{bmatrix} 8 & 1 & 6 \\ 1 & 8 & 6 \\ 6 & 6 & 10 \end{bmatrix}\begin{bmatrix} -1 & 0 & -\tfrac{1}{2} \\ 0 & -1 & -\tfrac{1}{2} \\ 0 & 0 & 1 \end{bmatrix}\begin{bmatrix} 0 \\ 0 \\ \overline{X}_3 \end{bmatrix} = \begin{bmatrix} \bar{r}_{x1} \\ \bar{r}_{x2} \\ 0 \end{bmatrix}$$

That is,

$$\frac{50}{6EI}\begin{bmatrix} 1 \\ 1 \\ -1 \end{bmatrix} R + \frac{10}{6EI}\begin{bmatrix} 8 & 1 & -\tfrac{3}{2} \\ 1 & 8 & -\tfrac{3}{2} \\ -\tfrac{3}{2} & -\tfrac{3}{2} & \tfrac{5}{2} \end{bmatrix}\begin{bmatrix} 0 \\ 0 \\ \overline{X}_3 \end{bmatrix} = \begin{bmatrix} \bar{r}_{x1} \\ \bar{r}_{x2} \\ 0 \end{bmatrix}$$

Thence,

$$\overline{X}_3 = 2R$$

Then

$$\begin{bmatrix} X_1 \\ X_2 \\ X_3 \end{bmatrix} = \begin{bmatrix} -1 & 0 & -\tfrac{1}{2} \\ 0 & -1 & -\tfrac{1}{2} \\ 0 & 0 & 1 \end{bmatrix}\begin{bmatrix} 0 \\ 0 \\ 2R \end{bmatrix} = \begin{bmatrix} -R \\ -R \\ 2R \end{bmatrix}$$

Member forces are given by

$$S = b_o R + b_1 X$$

$$S = \begin{bmatrix} -1 & 0 & 0 \\ -1 & 0 & -1 \\ -1 & 0 & -1 \\ 0 & -1 & -1 \\ 0 & -1 & -1 \\ 0 & -1 & 0 \end{bmatrix}\begin{bmatrix} -R \\ -R \\ 2R \end{bmatrix} = \begin{bmatrix} 1 \\ -1 \\ -1 \\ -1 \\ -1 \\ 1 \end{bmatrix} R = \begin{bmatrix} 20 \\ -20 \\ -20 \\ -20 \\ -20 \\ 20 \end{bmatrix} \text{ in.-kips}$$

The bending-moment diagram is shown in Fig. 5.28*b*.

5.5 ANALYSIS OF STATICALLY INDETERMINATE ARCHES

It has been shown in Sec. 4.9 that arch and curved beam structures may be approximated in their structural action by an equivalent structure composed of straight chord members. However, it was shown in that section that when the arch shape is given by an analytic function (i.e., circular, parabolic, etc.), it is possible to use numerical integration to calculate deflection functions. In this section the theory developed in Sec. 4.9 is extended to cover statically indeterminate arches. The theory for deflection calculations having been completely developed in Sec. 4.9, all that is now required is to be able to distinguish between applied forces and redundant forces so that the continuity condition $r_x = 0$ may be applied. Fixed-ended arches are considered first and the Eqs. (5.104)ff applied so that moment releases may be introduced. It is important to realize that the use of Eq. (5.104) allows the integrations to be performed before the releases are considered. Then the positions of the hinges have no effect on the intervals of subdivision used in the numerical quadrature.

Concentrated Forces on the Arch

The arch shown in Fig. 5.29 is loaded with forces at its left-hand end, taken as the origin of coordinates, and at the general point P on the arch. The redundant forces X will now be specified as

$$X^T = \langle X_1 \quad X_2 \quad X_3 \rangle = \langle R_{xo} \quad R_{yo} \quad R_{zo} \rangle$$

Deflections at O due to unit values of these forces produces the flexibility matrix D_{11}. From Eq. (4.153),

$$D_{11} \equiv F_O = \int_o^l b_1{}^T f b_1 \, dx \tag{5.112}$$

The matrix b_1 transforms X to axial force, shear and moment, at a point in the arch. In the numerical integration of Eq. (5.112), b_1 will be evaluated along the arch for equal intervals of x.

In the cantilever beam, fixed at Q, the forces at P produce forces in the structure only in the part PQ. The force at P will be designated

$$R^T = \langle R_{px} \quad R_{py} \quad M_{pz} \rangle$$

and the matrix b_o transforms R to internal forces in the portion PQ.

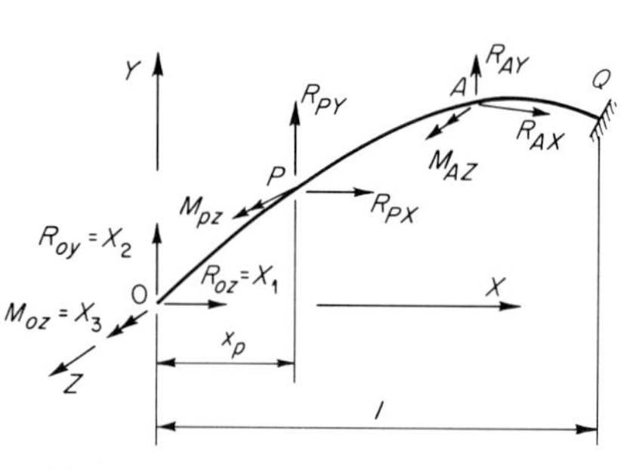

Figure 5.29 Fixed-ended arch.

The deflection at O due to R at P is given by [Eq. (4.155)]

$$r_{op} = \int_{x_p}^{l} b_1{}^T f b_o \, dx \, R \equiv D_{10} R \tag{5.113}$$

The deflection at P due to R will be

$$r_{pp} = \int_{x_p}^{l} b_o{}^T f b_o \, dx \, R \equiv D_{00} R \tag{5.114}$$

The values of the forces X to restore continuity at O may be calculated in the usual way, so that

$$X = -D_{11}{}^{-1} D_{10} R \tag{5.115}$$

Forces at various points in the arch are given by

$$S = b_1 X + b_o R \tag{5.116}$$

Deflections at P are obtained from R and X by using the relationship

$$D_{01} = (D_{10})^T$$

That is,

$$r_p = r_{pp} + r_{px} = D_{00} R + (D_{10})^T X \tag{5.117}$$

If deflections are required at the point A of the arch (Fig. 5.29), the force transformation matrix b for loads at A is generated. Then,

$$r_A = \int_A^l b_A{}^T f S \, dx = \int_A^l b_A{}^T f b_1 \, dx \, X + \int_A^l b_A{}^T f b_o \, dx \, R \tag{5.118}$$

Distributed Forces on the Arch

For the cantilever beam supported at Q, Eq. (4.159) gives the recurrence relationship that will generate forces throughout the arch for distributed loads alone.

Deflections at O are given [Eq. (4.156)] by

$$r_{ow} = \int_o^l b_1{}^T f R'_w \, dx \tag{5.119}$$

For zero deflection at O,

$$X = -D_{11}{}^{-1} r_{ow} \tag{5.120}$$

Forces throughout the arch are calculated by

$$S = S_w + b_1 X = R'_w + b_1 X \tag{5.121}$$

If deflections are to be calculated at point A,

$$r_A = \int_A^l b_A{}^T f S \, dx$$

Hence,

$$r_A = \int_A^l b_A{}^T f b_1 \, dx \, X + \int_A^l b_A{}^T f R_w' \, dx \tag{5.122}$$

Introduction of Hinges into the Arch Rib

Moment releases such as those shown at C and D in the arch of Fig. 5.30 may be accommodated as a special case of Eqs. (5.104) and (5.111). For the arch with hinges C and D, the redundants $X^T = \langle R_{ox} \quad R_{oy} \quad M_{oz} \rangle$ are first transformed to the equivalent set $\langle M_C \quad M_D \quad M_{oz} \rangle$ and then M_C and M_D are set equal to zero. From statics, it is seen that,

$$\begin{bmatrix} M_C \\ M_D \\ M_{oz} \end{bmatrix} = \begin{bmatrix} y_C & -x_C & 1 \\ y_D & -x_D & 1 \\ 0 & 0 & 1 \end{bmatrix} \begin{bmatrix} R_{ox} \\ R_{oy} \\ M_{oz} \end{bmatrix} \tag{5.123}$$

This is the equation

$$\overline{X} = BX$$

The matrix B^{-1} may be obtained by inversion of B, so that

$$B^{-1} = \frac{1}{x_C y_D - x_D y_C} \begin{bmatrix} -x_D & x_C & x_D - x_C \\ -y_D & y_C & y_D - y_C \\ 0 & 0 & x_D y_D - x_D y_C \end{bmatrix} \tag{5.124}$$

The results follow directly from Eqs. (5.108), (5.110), and (5.111).

5.6 ALGORITHMS IN THE FORCE METHOD SUITABLE FOR COMPUTER PROGRAMMING

In Secs. 5.1 to 5.5 it has been shown how the analysis of statically indeterminate structures may be considered separated into two basic problems. The first of these is the analysis of statically determinate structures as outlined in Chap. 4. This process has been complicated by the necessity for generating both the b_o and the b_1 matrices. For the determinate analysis, it is required to know the structure geometry, the member-node connectivity (topology), the support conditions, and the position of the redundant forces.

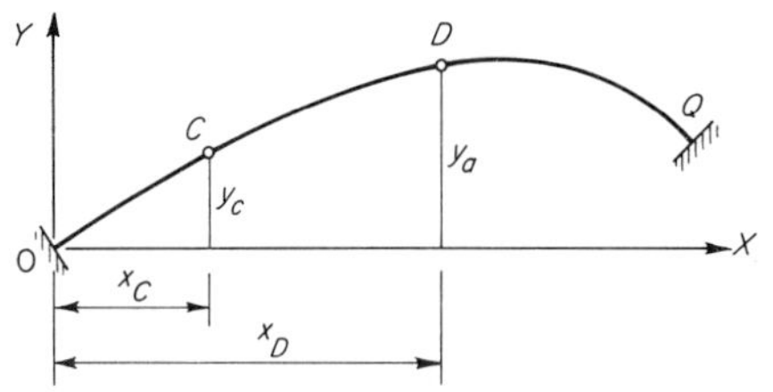

Figure 5.30 Arch with hinges.

The second phase requires a deflection analysis for the structure for both the external and redundant forces, the actual values of these later forces being such as to force compatibility. This process involves the knowledge of member flexibilities and follows the methods given in Chap. 4.

Possibly the most difficult problem in the analysis lies in the distinguishing between external forces and redundant forces. Several ways have been devised so that this may be carried out automatically. The first of these given in this section considers a structure with rigid joints and supports, for which a tree structure may be either selected by inspection or generated by a computer program. Considerable effort is required if this program is to select a tree that provides a well-conditioned set of redundants. This type of program suffers the disadvantage that releases must be introduced when the primary analysis is completed.

The second method herein sets up all the joint equilibrium equations in terms of the member forces and reactions and proceeds to select a nonsingular matrix from as many columns as there are equations. It has the advantage that member releases can be incorporated in the member-type specification and that it is not necessary then to select a primary tree structure.

Method 1 Tree Structure Analysis

In Chap. 3, alternative methods are given in Eqs. (3.52) and (3.83) for expressing forces from a member acting on its nodes, even when the member forces are the same fundamental set. It will be seen that the difference is only in the order in which the transformations are performed.

In Eq. (3.52) for the plane frame, the forces from the joints on the member are

$$\underset{6\times 1}{\begin{bmatrix} F_i \\ F_j \end{bmatrix}} = \underset{6\times 6}{\begin{bmatrix} L_1 & 0 \\ 0 & L_1 \end{bmatrix}} \underset{6\times 3}{T_1} \underset{3\times 1}{S} = L_D T_1 S \tag{5.125}$$

In this expression,

$$L_1 = \begin{bmatrix} \cos\alpha & -\sin\alpha & 0 \\ \sin\alpha & \cos\alpha & 0 \\ 0 & 0 & 1 \end{bmatrix}$$

and in the three-dimensional frame would be composed of two 3 × 3 direction-cosine blocks. Then L_D is a 12 × 12 matrix with four diagonal blocks.

In Eq. (5.125) the member force S, $S^T = \langle F \quad S_i \quad M_i \rangle$, is transformed first to six components, three at each member end in the local coordinates, and then these are all transformed to the global system. All the force components are shown again for reference in Fig. 5.31.

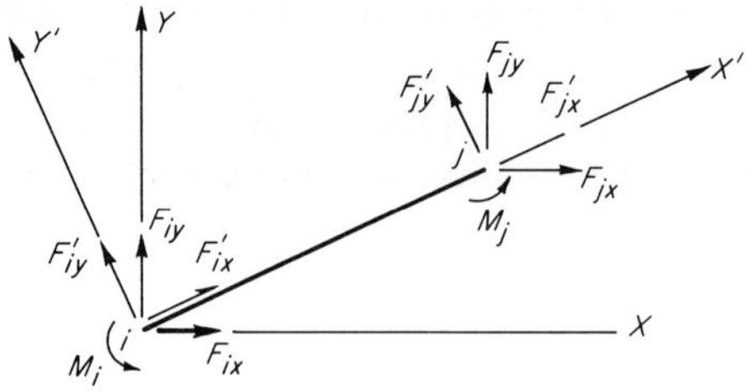

Figure 5.31 Plane frame member-force components.

In Eq. (3.83), the order of transformations is reversed. That is, S is transformed first to global components and then to forces at each end of the member. From Eq. (3.83), it is seen that

$$\underset{6\times 1}{\begin{bmatrix} F_i \\ F_j \end{bmatrix}} = \underset{6\times 3}{\begin{bmatrix} I \\ -U^{-1} \end{bmatrix}} \underset{3\times 3}{L_D} \underset{3\times 1}{S} \tag{5.126}$$

In this expression L_D is a 3×3 matrix only and for the three-dimensional frame expands to a 6×6 matrix composed of two direction-cosine blocks. It is seen here that in the two methods there is a difference in the definition of L_D, which should not cause any confusion, however, since the equations will occur independently. In this section, Eq. (5.126) is used in the first instance. From Eq. (5.126) all expressions may be calculated by using member forces defined in global coordinates.

Thus, define

$$S_G = L_D S \tag{5.127}$$

Then S_G are the global components of S. The inverse relationship is,

$$S = L_D{}^{-1} S_G = L_D{}^T S_G \tag{5.128}$$

The flexibility matrix for end i in global coordinates has been given in Eq. (4.206) and in the present notation is written

$$f_G = L_D f' L_D{}^T \tag{5.129}$$

In Eq. (5.126) write

$$V = -U^{-1} \tag{5.130}$$

Then,

$$\begin{bmatrix} F_i \\ F_j \end{bmatrix} = \begin{bmatrix} I \\ V \end{bmatrix} S_G \tag{5.131}$$

From Eq. (3.73), V may be calculated:

For the three-dimensional member,

$$V = \begin{bmatrix} -1 & 0 & 0 & 0 & 0 & 0 \\ 0 & -1 & 0 & 0 & 0 & 0 \\ 0 & 0 & -1 & 0 & 0 & 0 \\ 0 & -z_{12} & y_{12} & -1 & 0 & 0 \\ z_{12} & 0 & -x_{12} & 0 & -1 & 0 \\ -y_{12} & x_{12} & 0 & 0 & 0 & -1 \end{bmatrix} \tag{5.132}$$

For the plane frame,

$$V = \begin{bmatrix} -1 & 0 & 0 \\ 0 & -1 & 0 \\ -y_{12} & x_{12} & -1 \end{bmatrix} \tag{5.133}$$

The joint equilibrium equations are given from Eq. (3.38) as

$$R = a^T S \tag{5.134}$$

As shown in Chap. 3, matrix a^T may be built up from Eq. (5.126).

The matrix a^T may be defined as follows by the matrix Q. A member is considered positively or negatively incident on a node as shown in Fig. 5.32. Then Q is composed of blocks Q_{NM} as submatrices, such that

$$Q_{NM} = \begin{cases} 0 & \text{if } M\text{th branch is not incident on node } N \\ I & \text{if } M \text{ is positively incident on } N \\ V & \text{if } M \text{ is negatively incident on } N \end{cases}$$

The member-force and -displacement vectors in global components are defined by

$$\begin{aligned} S_G{}^T &= \langle S_1 \quad S_2 \quad \cdots \quad S_M \quad \cdots \quad S_K \rangle_G \\ \delta_G{}^T &= \langle \delta_1 \quad \delta_2 \quad \cdots \quad \delta_M \quad \cdots \quad \delta_K \rangle_G \end{aligned} \tag{5.135}$$

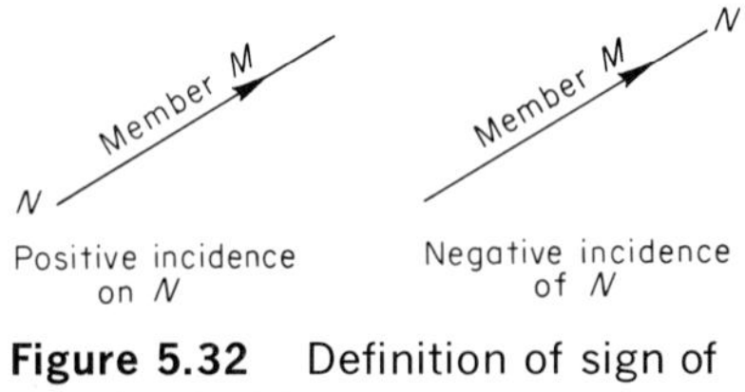

Figure 5.32 Definition of sign of member incidence.

For member M, the contragradient law gives [from Eq. (5.131)] the relationship between the member distortions δ_G and the member-end displacements r_i, r_j to be

$$\delta_G = r_i + V^T r_j \tag{5.136}$$

From the total structure, a tree structure is selected, connecting all joints. Then S_G and δ_G may be partitioned into groups containing values of tree and link members respectively. That is,

$$S_G{}^T = \langle S_T{}^T \quad S_L{}^T \rangle_G \tag{5.137}$$

$$\delta_G{}^T = \langle \delta_T{}^T \quad \delta_L{}^T \rangle_G \tag{5.138}$$

The equilibrium equations are written for all the joints, excluding supports, as

$$R = QS_G \tag{5.139}$$

Partition Q, according to Eq. (5.137), into Q_T and Q_L, that is, into contributions to the joint equilibrium equations from the tree and link members. Then,

$$R = [Q_T \quad Q_L] \begin{bmatrix} S_{TG} \\ S_{LG} \end{bmatrix} \tag{5.140}$$

For the tree structure, it is known that Q_T is a nonsingular matrix [see Eq. (3.81)], so that the forces in the tree members may be expressed in terms of R and S_{LG} such that

$$S_{TG} = Q_T{}^{-1}(R - Q_L S_{LG}) \tag{5.141}$$

The member distortions δ_G have been related to the node distortions r for member M, in Eq. (5.136). If all such equations are combined, it is seen that

$$\delta_G = Q^T r \tag{5.142}$$

Again partition according to link and tree members so that,

$$\begin{bmatrix} \delta_T \\ \delta_L \end{bmatrix}_G = \begin{bmatrix} Q_T{}^T \\ Q_L{}^T \end{bmatrix} r \tag{5.143}$$

Thence,

$$r = (Q_T{}^{-1})^T \delta_{TG} \tag{5.144}$$

This is the almost self-evident result that the node deflections can be calculated from the distortions of the tree members alone.

The link distortions are given from Eqs. (5.143) and (5.144),

$$\delta_{LG} = Q_L{}^T (Q_T{}^{-1})^T \delta_{TG} \tag{5.145}$$

Since the member-flexibility matrices have been given in Eq. (5.129), the member distortions may be expressed in terms of member forces, again partitioned into tree and link members,

$$\delta_{TG} = f_{TG}S_{TG} \tag{5.146}$$

$$\delta_{LG} = f_{LG}S_{LG} \tag{5.147}$$

The matrices f_{TG} and f_{LG} for tree and link members have the f_G matrices as diagonal blocks. From Eqs. (5.145) and (5.146), the expression obtained for δ_{LG} is

$$\delta_{LG} = Q_L{}^T(Q_T{}^{-1})^T f_{TG}S_{TG} \tag{5.148}$$

Then, also, from Eqs. (5.141) and (5.147),

$$\delta_{LG} = f_{LG}S_{LG} = Q_L{}^T(Q_T{}^{-1})^T f_{TG}Q_T{}^{-1}(R - Q_LS_{LG}) \tag{5.149}$$

The Eq. (5.149) may be solved for the unknown link forces S_{LG}. First collect terms containing S_{LG}. That is,

$$[f_{LG} + Q_L{}^T(Q_T{}^{-1})^T f_{TG}Q_T{}^{-1}Q_L]S_{LG} = Q_L{}^T(Q_T{}^{-1})^T f_{TG}Q_T{}^{-1}R \tag{5.150}$$

In fact, it is now found that Eq. (5.150) is very similar to the expression obtained in Eq. (5.27). Define the matrices b_o, b_{1T}, and b_1 as follows:

$$b_o = Q_T{}^{-1} \tag{5.151}$$

$$b_{1T} = -Q_T{}^{-1}Q_L \tag{5.152}$$

$$b_1 = \begin{bmatrix} I \\ b_{1T} \end{bmatrix} \tag{5.153}$$

Then Eq. (5.150) is written

$$b_1{}^T\begin{bmatrix} f_{LG} & \cdot \\ \cdot & f_{TG} \end{bmatrix} b_1S_{LG} = -b_{1T}{}^T f_{TG}b_oR \tag{5.154}$$

Finally,

$$S_{LG} = -\left\{b_1{}^T\begin{bmatrix} f_{LG} & \cdot \\ \cdot & f_{TG} \end{bmatrix} b_1\right\}^{-1} b_{1T}{}^T f_{TG}b_oR \tag{5.155}$$

When the link forces have been calculated, all the tree forces and the node deflections may be evaluated, since from Eq. (5.141),

$$S_{TG} = Q^{T-1}(R - Q_LS_{LG}) \tag{5.156}$$

Member forces will be required in local coordinates, so that

$$S = L_D{}^T S_G \tag{5.157}$$

The forces at end j of the member in global coordinates are given by

$$F_j = VS_G \tag{5.158}$$

Then in member coordinates,

$$F_j' = L_D{}^T F_j = L_D{}^T V S_G \tag{5.159}$$

Node deflections are given by Eq. (5.144) and in the b notation by

$$r = b_o{}^T \delta_{TG} \tag{5.160}$$

From the preceding theory it can be seen that the procedure may be completely automated once the tree structure and the link members have been defined. This may be done by inspection and the results used as data in a suitable program, or a program may be written for this purpose. As mentioned, a good deal of care must be taken if the tree selected is to be such as to produce a well-conditioned matrix for inversion in Eq. (5.155). The tree chosen may be many branched, but if the matrix Q^T is to be a lower-triangular matrix so that advantage may be taken of the inversion process in Eq. (3.81), the first member, (1), of the tree should be incident on node 1, and this should be an end point of the tree as shown in Fig. 5.33. It also seems evident that in an efficient tree structure, all support points will be in the tree.

For the tree structure and links of the structure shown in Fig. 5.33a and b, the Q_T and Q_L matrices are

$$\begin{array}{cc} & \begin{array}{ccc} 1 & 2 & 3 \end{array} \leftarrow \textit{Member} \\ Q_T = \begin{array}{c} 1 \\ 2 \\ 3 \\ \uparrow \\ \textit{Node} \end{array} & \begin{bmatrix} V_1 & 0 & 0 \\ I & V_2 & 0 \\ 0 & I & V_3 \end{bmatrix} \end{array}$$

$$\begin{array}{cc} & \begin{array}{ccc} 1 & 2 & 3 \end{array} \leftarrow \textit{Link} \\ Q_L = \begin{array}{c} 1 \\ 2 \\ 3 \\ \uparrow \\ \textit{Node} \end{array} & \begin{bmatrix} 0 & 0 & I \\ 0 & I & 0 \\ I & 0 & 0 \end{bmatrix} \end{array}$$

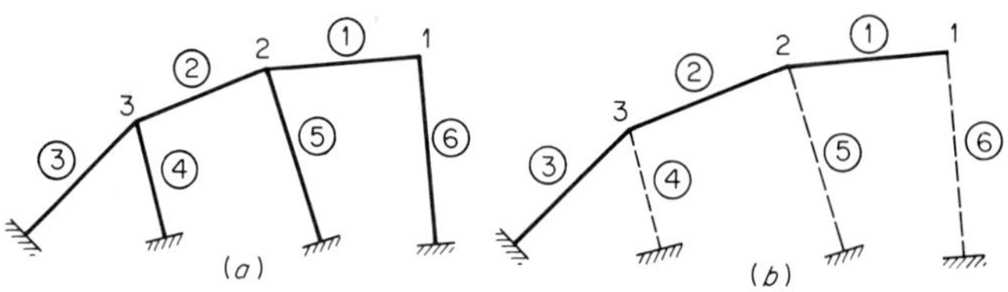

Figure 5.33 Tree structure with node and member numbering. (a) Continuous structure; (b) tree structure and links.

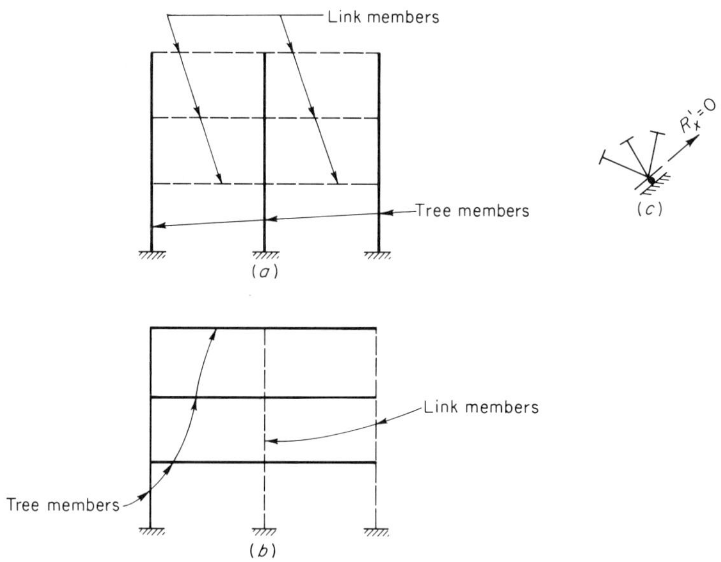

Figure 5.34 Tree structures. (*a*) Good choice of tree; (*b*) poor choice of tree; (*c*) release $R_x' = 0$.

In Fig. 5.34 a structure is shown with a good and bad choice of tree structure. It should be seen that for the tree in Fig. 5.34*b*, the D_{11} matrix will have relatively large off-diagonal flexibility terms.

Releases

Releases may be accommodated by using the theory developed in Sec. 5.4, Eqs. (5.98)ff. It will be apparent that use of such a program to analyze a structure with many releases (for example, a pin-jointed truss) would be most uneconomical. Considerable ingenuity would be required also to program a release of the type shown in Fig. 5.34*c*, where several members frame into an inclined roller support. This would require calculation of reactions, and then the reaction R_x' in the direction of the roller would be set equal to zero.

Nonuniform Members; Loads between Node Points

Both these features may be readily accommodated in the tree-structure algorithm by using the theory outlined in Sec. 4.8, method 2, and also given in Example 5.6. That is, fixed-end forces are calculated for each member and these summed at every node. These totals are negated and applied as release forces. When a member is nonuniform, the flexibility matrix must be calculated (see Sec. 4.10, subsec. 2) and transformed to that for forces at end *i*. The program may be considered subdivided into two stages. Stage 1 is a load generator program that takes all member forces, converts to node forces, and generates flexibility matrices. Stage 2 analyzes the structure, introduces releases, and finally produces node deflections and member forces.

Alternative Tree-structure Algorithm

The solution to the tree-structure problem is now set up by using Eq. (5.125). That is, although the equilibrium equations for the joints are solved in the global coordinate system, the continuity requirements between the links and the tree structure are expressed in the local coordinates of each link.

The member forces are defined in the identical manner with that used previously, so that $S^T = \langle F \quad S_i \quad M_i \rangle$. Then, from Eq. (5.125), end forces in frame components are given as

$$\begin{bmatrix} F_i \\ F_j \end{bmatrix} = L_D T_1 S$$

From Eq. (3.50), it is seen that for the plane frame member,

$$T_1 = \begin{bmatrix} & I_3 & \\ -1 & 0 & 0 \\ 0 & -1 & 0 \\ 0 & l & -1 \end{bmatrix} = \begin{bmatrix} I_3 \\ W \end{bmatrix} \tag{5.161}$$

For the three-dimensional element,

$$T_1 = \begin{bmatrix} I_6 \\ W \end{bmatrix} \tag{5.162}$$

where

$$W = \begin{bmatrix} -1 & \cdot & \cdot & \cdot & \cdot & \cdot \\ \cdot & -1 & \cdot & \cdot & \cdot & \cdot \\ \cdot & \cdot & -1 & \cdot & \cdot & \cdot \\ \cdot & \cdot & & -1 & \cdot & \cdot \\ \cdot & \cdot & -l & \cdot & -1 & \cdot \\ \cdot & l & \cdot & \cdot & \cdot & -1 \end{bmatrix} \tag{5.163}$$

Then from Eq. (5.161) (or 5.162), the forces on the joints from the member in global coordinates are given by

$$F_i = L_1 S \qquad F_j = L_1 W S \tag{5.164}$$

The Q matrix is now composed of blocks Q_{NM} such that by using the notation of Fig. 5.31,

$$Q_{NM} = \begin{cases} 0 & \text{if the Mth branch is not incident on node } N \\ L_1 & \text{if } M \text{ is positively incident on } N \\ L_1 W & \text{if } M \text{ is negatively incident on } N \end{cases}$$

The vectors composed of member forces and member displacements are defined by

$$\begin{aligned} S^T &= \langle S_1 \quad S_2 \quad \cdots \quad S_K \rangle \\ \delta^T &= \langle \delta_1 \quad \delta_2 \quad \cdots \quad \delta_K \rangle \end{aligned} \tag{5.165}$$

For member M, the contragradient law gives the relationship between member distortions and the displacements of the member nodes i and j to be

$$\delta = T_1{}^T L_D{}^T \begin{bmatrix} r_i \\ r_j \end{bmatrix} = [I_3 \quad W^T] \begin{bmatrix} L_1{}^T & 0 \\ 0 & L_1{}^T \end{bmatrix} \begin{bmatrix} r_i \\ r_j \end{bmatrix}$$

That is,

$$\delta = L_1{}^T r_i + L_1{}^T W^T r_j \tag{5.166}$$

Then, if δ is partitioned into δ_T and δ_L,

$$\begin{bmatrix} \delta_T \\ \delta_L \end{bmatrix} = \begin{bmatrix} Q_T{}^T \\ Q_L{}^T \end{bmatrix} r \tag{5.167}$$

The expressions obtained to calculate S_L are now identical in form with Eq. (5.155) in which the flexibility matrices f_L and f_T are given in terms of the member forces S.

Method 2 Analysis by Using the Structure-cutter Concept[1]

As already mentioned at the beginning of this section, the structure-cutter approach starts the analysis by setting up all the equations of joint equilibrium. Both member forces and external reactions are included, and instead of the resulting coefficient matrix's being square and nonsingular as for the determinate structures in Chap. 3, this matrix will now have more columns than rows, that is, more unknown forces than equations. These additional columns are the expression of the degree of indeterminacy of the system. Each member will contribute forces to the joint equilibrium equations at its extremities, the type and number of independent member forces being dependent on the number of releases it contains. Since the arbitrary addition of releases in members may result in null joint equilibrium equations, a general program should check all joints for such conditions and delete

[1] See Ref. 6.

all such equations. For example, in a plane frame, if all members meeting at a joint have a moment release there, the equation $\Sigma M = 0$ can never be satisfied for arbitrary applied external moment. In this case the equation should be deleted and only the forces R_x and R_y applied at the joint.

It should be apparent that this method has an advantage over the tree-structure approach in that releases are considered as the equilibrium equations are being developed. The primary structure is selected automatically by the computer and will not necessarily be a tree structure. It should be selected in some way that ensures that it is well conditioned. The equations of joint equilibrium, including reaction components R_1, are written

$$R + AR_1 - a^T S = 0$$

That is, grouping R_1 and S, the unknown quantities,

$$R - [a^T \quad -A]\begin{bmatrix} S \\ R_1 \end{bmatrix} = 0 \tag{5.168}$$

Suppose now that a determinate system is chosen, so that its forces S_o may be expressed in terms of R and unknown quantities S_x. These forces S_o and S_x are extracted from the vector $\begin{bmatrix} S \\ R_1 \end{bmatrix}$ and in fact constitute this vector in re-arranged form. By using this notation, Eq. (5.168) is now written

$$R - [a_o{}^T \quad a_1{}^T]\begin{bmatrix} S_o \\ S_x \end{bmatrix} = 0 \tag{5.169}$$

Thence,

$$S_o = (a_o{}^T)^{-1}R - (a_o{}^T)^{-1}a_1{}^T S_x \tag{5.170}$$

The inverse $(a_o{}^T)^{-1}$ exists if the S_o forces constitute a stable determinate system.

It is seen that $(a_o{}^T)^{-1} = \bar{b}_o$, the matrix that transforms the R forces to member forces S_o in the determinate structure. The product $-(a_o{}^T)^{-1}a_1{}^T$ will be defined as $\bar{b}_1$ and transforms S_x to forces in the determinate structure. However, if the S_x values are member forces, they also produce forces at the indeterminate sections, and thence,

$$b_1 = \begin{bmatrix} \bar{b}_1 \\ I \end{bmatrix} \tag{5.171}$$

If these same sections are included in the b_o transformation, then

$$b_o = \begin{bmatrix} \bar{b}_o \\ 0 \end{bmatrix} \tag{5.172}$$

Since b_1 and b_o have been determined, the analysis now proceeds in the same way as in Eqs. (5.24)ff. That is,

$$S_x = -(b_1^T f b_1)^{-1}(b_1^T f b_o) \tag{5.173}$$

Example 5.14 The continuous beam shown in Fig. 5.35 is to be analyzed for moments R_1, R_2, R_3 applied at the joints as shown. This is a particular case in which the reactions do not enter into the equilibrium equations $\Sigma M_z = 0$ and so need not be considered. The moment equilibrium equations at the joints are written

$$\begin{bmatrix} 1 & 0 & 0 & 0 \\ 0 & 1 & 1 & 0 \\ 0 & 0 & 0 & 1 \end{bmatrix} \begin{bmatrix} M_{1i} \\ M_{1j} \\ M_{2i} \\ M_{2j} \end{bmatrix} = \begin{bmatrix} R_1 \\ R_2 \\ R_3 \end{bmatrix} \tag{i}$$

These equations show the interesting point that in the matrix of coefficients obtained, it is not sufficient merely to select the first square submatrix. If this is attempted in Eq. (i), it is clearly seen to be singular. Instead, interchange columns 3 and 4 in the coefficient matrix. This will be equivalent to choosing M_{2i} as the single unknown force of the system. Then, Eq. (5.169) is written,

$$\begin{bmatrix} 1 & \cdot & \cdot \\ \cdot & 1 & \cdot \\ \cdot & \cdot & 1 \end{bmatrix} \begin{bmatrix} M_{1i} \\ M_{1j} \\ M_{2j} \end{bmatrix} + \begin{bmatrix} 0 \\ 1 \\ 0 \end{bmatrix} (M_{2i}) = \begin{bmatrix} R_1 \\ R_2 \\ R_3 \end{bmatrix} \tag{ii}$$

Then,

$$\begin{bmatrix} M_{1i} \\ M_{1j} \\ M_{2j} \end{bmatrix} = I \begin{bmatrix} R_1 \\ R_2 \\ R_3 \end{bmatrix} - I \begin{bmatrix} 0 \\ 1 \\ 0 \end{bmatrix} M_{2i} \tag{iii}$$

The member forces may be given in a form compatible with the b_1 and b_o matrices in Eqs. (5.171) and (5.172), but it is found more convenient to use the order given in Eq. (i). This allows f to be formed in the usual way. That is,

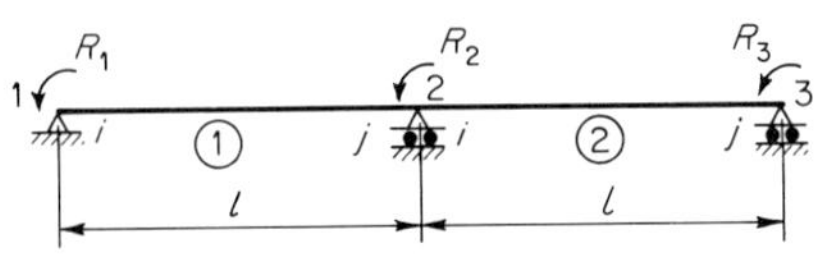

Figure 5.35 Two-span beam with applied joint moments.

$$\begin{bmatrix} M_{1i} \\ M_{1j} \\ M_{2i} \\ M_{2j} \end{bmatrix} = \underbrace{\begin{bmatrix} 1 & 0 & 0 \\ 0 & 1 & 0 \\ 0 & 0 & 0 \\ 0 & 0 & 1 \end{bmatrix}}_{b_o} \begin{bmatrix} R_1 \\ R_2 \\ R_3 \end{bmatrix} + \underbrace{\begin{bmatrix} 0 \\ -1 \\ 1 \\ 0 \end{bmatrix}}_{b_1} M_{2i} \tag{iv}$$

In this case,

$$f = \frac{l}{6EI} \begin{bmatrix} 2 & -1 & 0 & 0 \\ -1 & 2 & 0 & 0 \\ 0 & 0 & 2 & -1 \\ 0 & 0 & -1 & 2 \end{bmatrix}$$

so that

$$b_1{}^T f b_1 = \frac{4l}{6EI}$$

$$b_1{}^T f b_o = \frac{l}{6EI} [1 \quad -2 \quad -1]$$

Thence,

$$S_x \equiv M_{2i} = -\frac{1}{4} [1 \quad -2 \quad -1] \begin{bmatrix} R_1 \\ R_2 \\ R_3 \end{bmatrix}$$

Let $R_1 = 20$ and $R_2 = R_3 = 0$. Then $S_x = -5.0$ and

$$\begin{bmatrix} M_{1i} \\ M_{1j} \\ M_{2i} \\ M_{2j} \end{bmatrix} = \begin{bmatrix} 1 & 0 & 0 \\ 0 & 1 & 0 \\ 0 & 0 & 0 \\ 0 & 0 & 1 \end{bmatrix} \begin{bmatrix} 20 \\ 0 \\ 0 \end{bmatrix} + \begin{bmatrix} 0 \\ -1 \\ 1 \\ 0 \end{bmatrix} (-5.0) = \begin{bmatrix} 20.0 \\ 5.0 \\ -5.0 \\ 0 \end{bmatrix}$$

The member-force diagram is shown in Fig. 5.36.

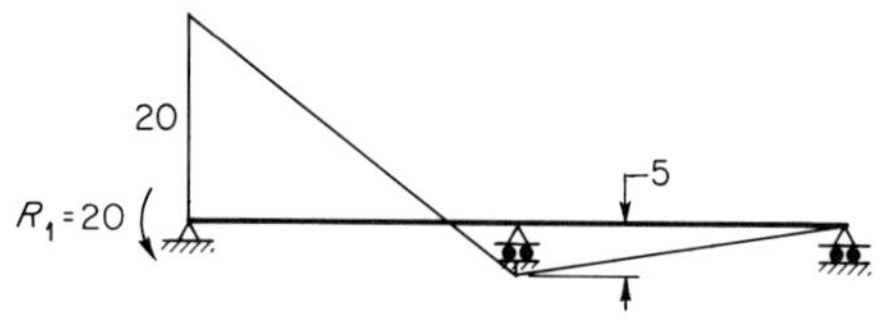

Figure 5.36

Example 5.15 In the previous example it is seen that one of the requirements of a structure-cutter program is that it must be able to select a set of member forces that have a nonzero matrix for their coefficients in the equilibrium equations. Just how this may be achieved is discussed later. If such a condition cannot be found, the structure must be unstable.

In this example it is shown that in addition the structure-cutter program must be able to extract member forces from the aggregate vector containing member forces and reactions and that the redundant quantities may be reactions and not member forces, in which case the unit matrix I in Eq. (5.171) will not be present.

The propped cantilever beam shown in Fig. 5.37 will be analyzed for the joint forces $\langle F_{1y} \quad M_{1z} \quad F_{2y} \quad M_{2z}\rangle$. Since there are no forces in the X direction, the equations for $\Sigma F_x = 0$ will be deleted. The member forces are $\langle M_{1i} \quad M_{1j}\rangle$ and the reactions $\langle R_1 \quad R_2 \quad R_3\rangle$. The equations of equilibrium [Eq. (5.168)] are written

$$\underbrace{\left[\begin{array}{cccc|c} \frac{1}{l} & \frac{1}{l} & -1 & 0 & 0 \\ 1 & 0 & 0 & -1 & 0 \\ -\frac{1}{l} & -\frac{1}{l} & 0 & 0 & -1 \\ 0 & 1 & 0 & 0 & 0 \end{array}\right]}_{a_o^T \qquad\qquad a_1^T} \begin{bmatrix} M_{1i} \\ M_{1j} \\ R_1 \\ R_2 \\ R_3 \end{bmatrix} = \begin{bmatrix} F_{1y} \\ M_{1z} \\ F_{2y} \\ M_{2z} \end{bmatrix} \tag{i}$$

In this case it is found that a_o is nonsingular and that

$$(a_o^T)^{-1} = \begin{bmatrix} 0 & 0 & -l & -1 \\ 0 & 0 & 0 & 1 \\ -1 & 0 & -1 & 0 \\ 0 & -1 & -l & -1 \end{bmatrix} \tag{ii}$$

It should be evident from Fig. 5.37 that this must be so since the choice of R_3 as unknown implies that the determinate structure is the cantilever beam fixed at node 1. Then, from Eqs. (ii) and (5.170),

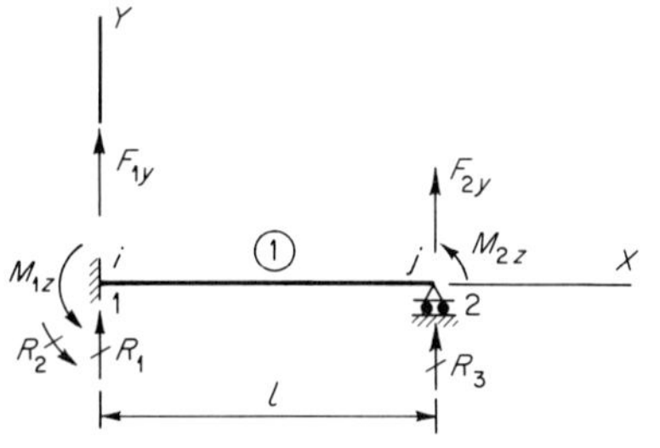

Figure 5.37 Propped cantilever beam.

$$\begin{bmatrix} M_{1i} \\ M_{1j} \\ R_1 \\ R_2 \end{bmatrix} = \begin{bmatrix} 0 & 0 & -l & -1 \\ 0 & 0 & 0 & 1 \\ -1 & 0 & -1 & 0 \\ 0 & -1 & -l & -1 \end{bmatrix} \left\{ \begin{bmatrix} F_{1y} \\ M_{1z} \\ F_{2y} \\ M_{2z} \end{bmatrix} - \begin{bmatrix} 0 \\ 0 \\ -1 \\ 0 \end{bmatrix} R_3 \right\} \tag{iii}$$

Then,

$$b_o = \begin{bmatrix} 0 & 0 & -l & -1 \\ 0 & 0 & 0 & 1 \end{bmatrix} \tag{iv}$$

$$b_1 = \begin{bmatrix} -l \\ 0 \end{bmatrix} \tag{v}$$

It follows that,

$$b_1{}^T f b_1 = \frac{l^3}{3EI} \tag{vi}$$

$$b_1{}^T f b_o = \frac{l^2}{6EI} [0 \quad 0 \quad 2l \quad 3] \tag{vii}$$

Thence

$$X = R_3 = -\frac{1}{2l} [0 \quad 0 \quad 2l \quad 3] \begin{bmatrix} F_{1y} \\ M_{1z} \\ F_{2y} \\ M_{2z} \end{bmatrix} \tag{viii}$$

If, for example,

$$\langle F_{1y} \quad M_{1z} \quad F_{2y} \quad M_{2z} \rangle = \langle 0 \quad 0 \quad 0 \quad 10 \rangle$$

then

$$X = -\frac{15}{l}$$

$$\begin{bmatrix} M_{1i} \\ M_{1j} \\ R_1 \\ R_2 \end{bmatrix} = \begin{bmatrix} -10 \\ 10 \\ 0 \\ -10 \end{bmatrix} + \begin{bmatrix} 15 \\ 0 \\ \frac{15}{l} \\ 15 \end{bmatrix} = \begin{bmatrix} 5 \\ 10 \\ \frac{15}{l} \\ 5 \end{bmatrix}$$

Determination of $\bar{b}_o$ and $\bar{b}_1$

From Eqs. (5.171) and (5.172), it is seen that the fundamental force transformation matrices to be determined for the structure are $\bar{b}_o$ and $\bar{b}_1$. When these are known, the actual procedure for determining the redundant forces may be formulated in several ways. If, for example, Eqs. (5.171) and (5.172) are used in Eq. (5.173), then reaction forces must be given zero flexibilities, and also some reordering in S_o may be needed if f is to be a diagonal matrix. For example, in Example 5.14, Eq. (*iv*), the zero row corresponding to the null submatrix in Eq. (5.172), appears as the third row and not the last.

It is convenient to use gaussian elimination in the solution, Eq. (5.170), rather than to use the formal inversion of a_o^T. Then a^T is reduced to a form so that the first square submatrix is an upper triangular matrix with unit diagonal terms. At each stage of the reduction, the remaining matrix to be reduced is searched for its largest term and the rows and columns interchanged so that this value may be the next pivot term. If instead of a load vector R, a series of unit loads is used such that $[R] = [I]$, the back-substitution process produces $\bar{b}_o$. Also, if the effects of unit values of S_x are evaluated so that $[S_x] = [I]$, the back-substitution process operating on the now reduced $[-a_1^T]$ gives $\bar{b}_1$.

If the elimination process is to produce a well-conditioned set of redundants, then the least-flexible members should be chosen for their forces to be included in the determinate set. To do this, columns in a^T should be weighted inversely to the member flexibility for the corresponding member force. In this way the largest terms will occur for members of greatest stiffness. If reactions are to be included, they should be weighted by a factor greater than the maximum value for a member force.

A program has been written to illustrate the structure-cutter technique for producing the b_o and b_1 matrices. This program accepts the joint equilibrium equations as input. It then generates a unit matrix of dimensions equal to the number of equations and writes this by columns on tape. Then the coefficient matrix for member forces is reduced so that the first square submatrix of maximum size is an upper triangular matrix. Back-substitution as outlined produces $\bar{b}_o$ and $\bar{b}_1$. These are then reordered so that the original member-force sequence is preserved and the appropriate rows are added to produce b_o and b_1. The program is given in Appendix E and has been used in Example 5.16 below.

Example 5.16 Use of a structure-cutter program to generate b_o and b_1 force transformation matrices

The portal frame to be analyzed is shown in Fig. 5.38. There are four joints and three members, and thence with six reaction components, the frame has five degrees of indeterminacy. The vector of member forces and reactions will be S^T where

$$S^T = \langle M_{1i} \quad M_{1j} \quad F_1, \quad M_{2i} \quad M_{2j} \quad F_2, \quad M_{3i} \quad M_{3j} \quad F_3, \quad R_{e1} \quad R_{e2} \quad R_{e3}, \quad R_{e4} \quad R_{e5} \quad R_{e6}\rangle$$

The reaction forces $\langle R_{e1} \quad R_{e2} \quad R_{e3}\rangle$ act at joint 1, and $\langle R_{e4} \quad R_{e5} \quad R_{e6}\rangle$ at joint 4. The Eq. (5.169) is given in Eq. (*i*). Note that in Eq. (*i*) the coefficients for member forces are obtained by evaluating $L_D T$ for members 1, 2, and 3 [see Eq. (3.49)].

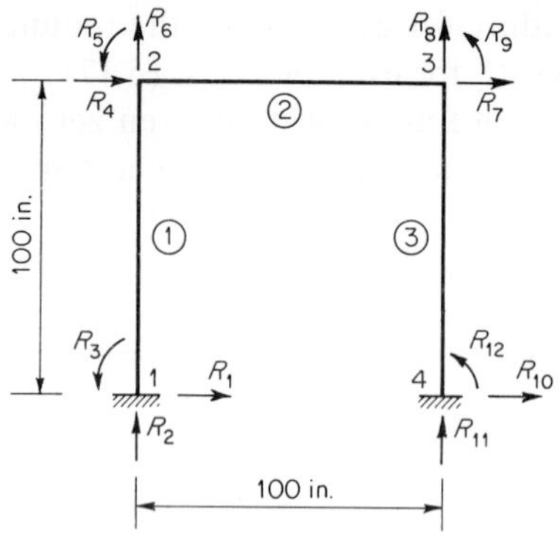

Figure 5.38 Portal-frame dimensions and joint loads.

$$
\begin{bmatrix}
-.01 & -.01 & \cdot & \cdot & \cdot & \cdot & \cdot & \cdot & \cdot & -1 & \cdot & \cdot & \cdot & \cdot & \cdot \\
\cdot & \cdot & -1 & \cdot & \cdot & \cdot & \cdot & \cdot & \cdot & \cdot & -1 & \cdot & \cdot & \cdot & \cdot \\
1 & \cdot & \cdot & \cdot & \cdot & \cdot & \cdot & \cdot & \cdot & \cdot & \cdot & -1 & \cdot & \cdot & \cdot \\
.01 & .01 & \cdot & \cdot & \cdot & -1 & \cdot & \cdot & \cdot & \cdot & \cdot & \cdot & \cdot & \cdot & \cdot \\
\cdot & \cdot & 1 & .01 & .01 & \cdot & \cdot & \cdot & \cdot & \cdot & \cdot & \cdot & \cdot & \cdot & \cdot \\
\cdot & 1 & \cdot & 1 & \cdot & \cdot & \cdot & \cdot & \cdot & \cdot & \cdot & \cdot & \cdot & \cdot & \cdot \\
\cdot & \cdot & \cdot & \cdot & \cdot & 1 & .01 & .01 & \cdot & \cdot & \cdot & \cdot & \cdot & \cdot & \cdot \\
\cdot & \cdot & \cdot & -0.1 & -.01 & \cdot & \cdot & \cdot & 1 & \cdot & \cdot & \cdot & \cdot & \cdot & \cdot \\
\cdot & \cdot & \cdot & \cdot & 1 & \cdot & 1 & \cdot & \cdot & \cdot & \cdot & \cdot & \cdot & \cdot & \cdot \\
\cdot & \cdot & \cdot & \cdot & \cdot & \cdot & -.01 & -.01 & \cdot & \cdot & \cdot & \cdot & -1 & \cdot & \cdot \\
\cdot & \cdot & \cdot & \cdot & \cdot & \cdot & \cdot & \cdot & -1 & \cdot & \cdot & \cdot & \cdot & -1 & \cdot \\
\cdot & \cdot & \cdot & \cdot & \cdot & \cdot & \cdot & 1 & \cdot & \cdot & \cdot & \cdot & \cdot & \cdot & -1
\end{bmatrix}
\begin{bmatrix}
M_{1i} \\ M_{1j} \\ F_1 \\ M_{2i} \\ M_{2j} \\ F_2 \\ M_{3i} \\ M_{3j} \\ F_2 \\ R_{e1} \\ R_{e2} \\ R_{e3} \\ R_{e4} \\ R_{e5} \\ R_{e6}
\end{bmatrix}
= I_{12} \qquad (i)
$$

Equations (i) have been written with I_{12} on the right-hand side. Then when a solution is obtained, it will be for unit loads applied at all nodes. It will be noticed that only free nodes really need be considered in the load vector. In this case columns of I_{12} corresponding to reactions can be omitted.

When the coefficient matrix in Eq. (i) is reduced so that the first 12×12 submatrix is an upper triangular matrix, it is found that the columns have been rearranged in the sequence

10, 3, 1, 6, 11, 4, 9, 7, 13, 14, 8, 12, 5, 2, 15

The last three columns correspond to the redundant forces S_x. It is found from the member force vector in Eq. (i) that M_{1j}, M_{2j}, and R_{e6} are the selected redundant quantities. The cut structure is shown in Fig. 5.39a.

The reduced equations are given in Eq. (ii), and the b_o and b_1 matrices in Eqs. (iii) and (iv). The force diagrams from columns of b_o and b_1 have been plotted in Fig. 5.39b.

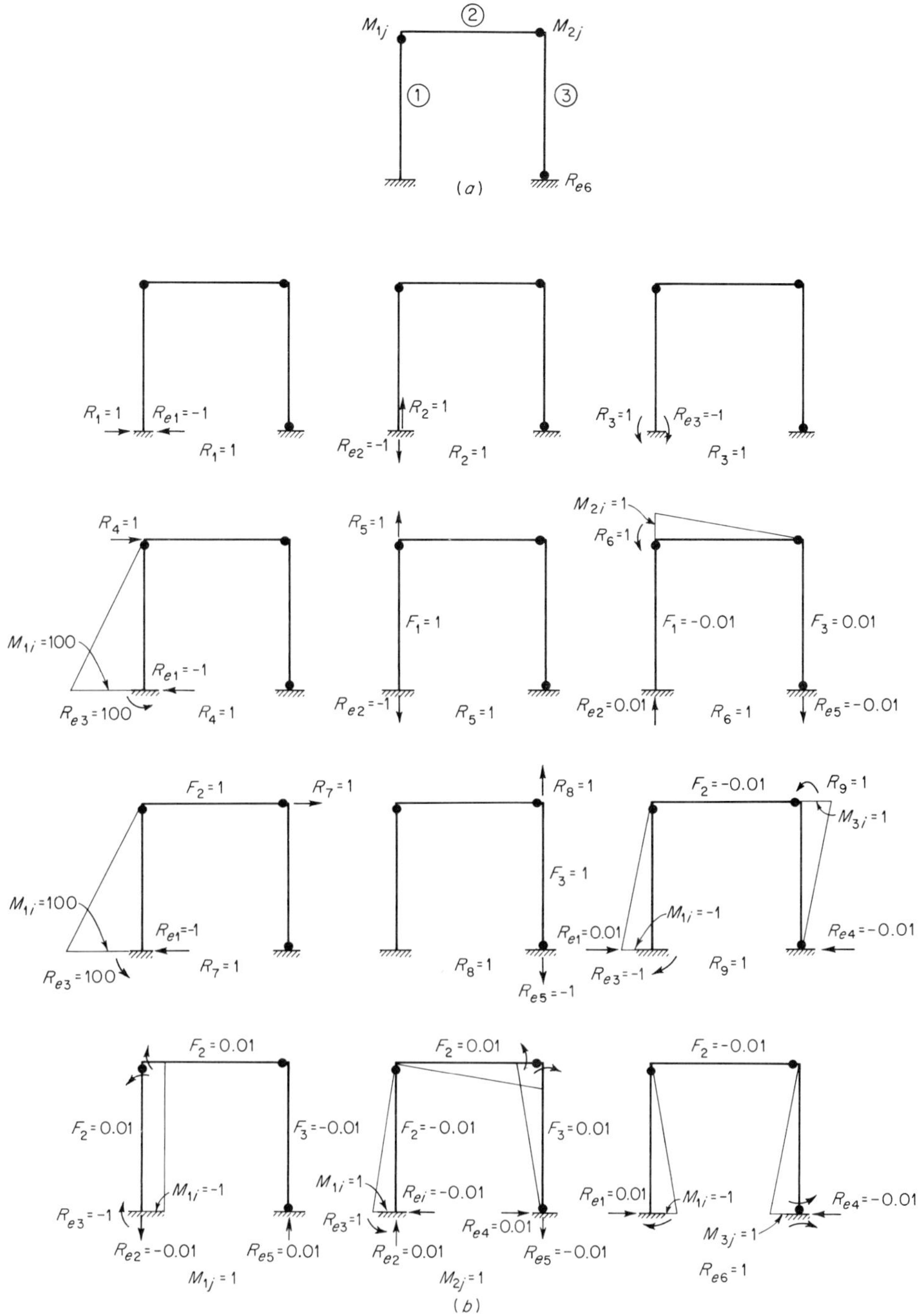

Figure 5.39 Cut-structure analysis. (*a*) Cut structure and releases selected by computer program; (*b*) member forces for unit applied forces from Eq. (*iii*).

$$
\begin{array}{r}
1\\2\\3\\4\\5\\6\\8\\9\\10\\11\\12\\7
\end{array}
\left[\begin{array}{cccccccccccc:ccc}
1 & \cdot & .01 & \cdot & \cdot & \cdot & \cdot & \cdot & \cdot & \cdot & \cdot & \cdot & \cdot & .01 & \cdot \\
 & 1 & \cdot & \cdot & 1 & \cdot & \cdot & \cdot & \cdot & \cdot & \cdot & \cdot & \cdot & \cdot & \cdot \\
 & & 1 & \cdot & \cdot & \cdot & \cdot & \cdot & \cdot & \cdot & \cdot & -1 & \cdot & \cdot & \cdot \\
 & & & 1 & \cdot & \cdot & \cdot & \cdot & \cdot & \cdot & \cdot & -.01 & \cdot & -.01 & \cdot \\
 & & & & 1 & -.01 & \cdot & \cdot & \cdot & \cdot & \cdot & \cdot & -.01 & \cdot & \cdot \\
 & & & & & 1 & \cdot & \cdot & \cdot & \cdot & \cdot & \cdot & \cdot & 1 & \cdot \\
 & & & & & & 1 & \cdot & \cdot & \cdot & \cdot & \cdot & -.01 & .01 & \cdot \\
 & & & & & & & 1 & \cdot & \cdot & \cdot & \cdot & 1 & \cdot & \cdot \\
 & & & & & & & & 1 & \cdot & -.01 & \cdot & -.01 & \cdot & \cdot \\
 & & & & & & & & & 1 & \cdot & \cdot & .01 & -.01 & \cdot \\
 & & & & & & & & & & 1 & \cdot & \cdot & \cdot & -1 \\
\mathbf{0} & & & & & & & & & & & 1 & -1 & 1 & 1
\end{array}\right] \qquad (ii)
$$

$$
\uparrow \quad R_{e1} \; F_1 \; M_{1i} \; F_2 \; R_{e2} \; M_{2i} \; F_3 \; M_{3j} \; R_{e4} \; R_{e5} \; M_{3j} \; R_{e3} \; M_{2j} \; M_{1j} \; R_{e6} \leftarrow \textit{Member force or reaction}
$$

No. of joint equation

On solving the reduced equations,

$$
b_o = \begin{bmatrix}
\cdot & \cdot & \cdot & 100 & \cdot & \cdot & 100 & \cdot & -1 & \cdot & \cdot & -1 \\
\cdot & \cdot & \cdot & \cdot & \cdot & \cdot & \cdot & \cdot & \cdot & \cdot & \cdot & \cdot \\
\cdot & \cdot & \cdot & \cdot & 1 & -.01 & \cdot & \cdot & \cdot & \cdot & \cdot & \cdot \\
\cdot & \cdot & \cdot & \cdot & \cdot & 1 & \cdot & \cdot & \cdot & \cdot & \cdot & \cdot \\
\cdot & \cdot & \cdot & \cdot & \cdot & \cdot & \cdot & \cdot & \cdot & \cdot & \cdot & \cdot \\
\cdot & \cdot & \cdot & \cdot & \cdot & \cdot & 1 & \cdot & -.01 & \cdot & \cdot & -.01 \\
\cdot & \cdot & \cdot & \cdot & \cdot & \cdot & \cdot & \cdot & 1 & \cdot & \cdot & \cdot \\
\cdot & \cdot & \cdot & \cdot & \cdot & \cdot & \cdot & \cdot & \cdot & \cdot & \cdot & 1 \\
\cdot & \cdot & \cdot & \cdot & \cdot & .01 & \cdot & 1 & \cdot & \cdot & \cdot & \cdot \\
-1 & \cdot & \cdot & -1 & \cdot & \cdot & -1 & \cdot & .01 & \cdot & \cdot & .01 \\
\cdot & -1 & \cdot & \cdot & -1 & .01 & \cdot & \cdot & \cdot & \cdot & \cdot & \cdot \\
\cdot & \cdot & -1 & 100 & \cdot & \cdot & 100 & \cdot & 1 & \cdot & \cdot & -1 \\
\cdot & \cdot & \cdot & \cdot & \cdot & \cdot & \cdot & \cdot & -.01 & -1 & \cdot & -.01 \\
\cdot & \cdot & \cdot & \cdot & \cdot & -.01 & \cdot & -1 & \cdot & \cdot & -1 & \cdot \\
\cdot & \cdot & \cdot & \cdot & \cdot & \cdot & \cdot & \cdot & \cdot & \cdot & \cdot & \cdot
\end{bmatrix} \qquad (iii)
$$

$$b_1 = \begin{bmatrix} 1 & -1 & -1 \\ & 1 & \cdot \\ -.01 & .01 & \cdot \\ \cdot & -1 & \cdot \\ 1 & \cdot & \cdot \\ .01 & \cdot & -.01 \\ -1 & \cdot & \cdot \\ \cdot & \cdot & 1 \\ .01 & -.01 & \cdot \\ -.01 & \cdot & .01 \\ .01 & -.01 & \cdot \\ 1 & -1 & -1 \\ .01 & \cdot & -.01 \\ -.01 & .01 & \cdot \\ \cdot & \cdot & 1 \end{bmatrix} \qquad (iv)$$

5.7 METHODS OF ANALYSIS OF STRUCTURES BY TEARING INTO SEPARATE SUBSTRUCTURES

It may happen that the complexity of a structure is such that the computer capacity is exceeded for a single-pass program. Then it may be possible to break down (*tear*) the complete structure into a set of substructures, each of a size that may be handled satisfactorily. Finally, all these parts must be reassembled so that compatibility is restored at the connection points. It will be found that the procedure followed is rather similar to that introduced for generalized forces in Chap. 4. In Sec. 4.7, it has been shown that forces on members between nodes may be treated as generalized forces rather than node forces. In the same way a set of forces acting on a structure of many members may be treated as a generalized force for a member that is now composed of the structure in its entirety. If this unit is only a portion of a whole system consisting of several substructures, unknown force values will exist between the parts when continuity is enforced. These forces may be evaluated in the usual way with Eq. (5.28), etc. In this section the arrangement of these calculations will be considered.

A particularly simple case of tearing occurs when the separate structures are such that forces on an individual part cause deformations of that part alone. That is, in the statically determinate state no forces act between substructures. One such structure torn into two parts is shown in Fig. 5.40. In this case the substructures are themselves indeterminate. Two examples are given, showing how the complete structure is built up from the substructures.

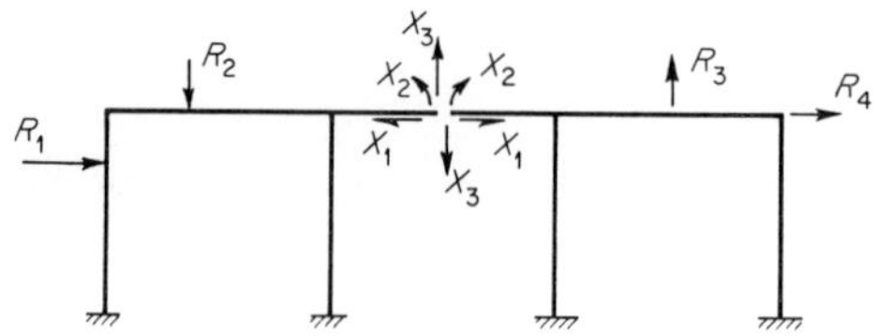

Figure 5.40 Structure torn into two subsystems.

Example 5.17 The two-span continuous beam shown in Fig. 5.41 is considered composed of the structures I and II (in this case each substructure is a single member), and the tearing has been achieved by placing a hinge at the center support 2. Then the influences of the forces R_1 and R_2 are contained in parts I and II respectively in the determinate structure. For each of the parts the generalized forces will be $\langle M_i \quad R \quad M_j \rangle$, and the generalized flexibility matrix from Eq. (4.134) for the central loads R_1 and R_2 is for each member,

$$f = \frac{l}{6EI}\begin{bmatrix} 2 & \frac{3l}{8} & 1 \\ \frac{3l}{8} & \frac{l^2}{8} & \frac{3l}{8} \\ 1 & \frac{3l}{8} & 2 \end{bmatrix} \qquad (i)$$

The force transformation matrices b_o and b_1 are then given with unit or zero values. That is,

$$b = \left[\begin{array}{cc|c} 0 & 0 & 0 \\ 1 & 0 & 0 \\ 0 & 0 & 1 \\ 0 & 0 & 1 \\ 0 & 1 & 0 \\ 0 & 0 & 0 \end{array}\right] \qquad (ii)$$
$$\qquad \underbrace{\quad}_{b_o} \quad \underbrace{\quad}_{b_1}$$

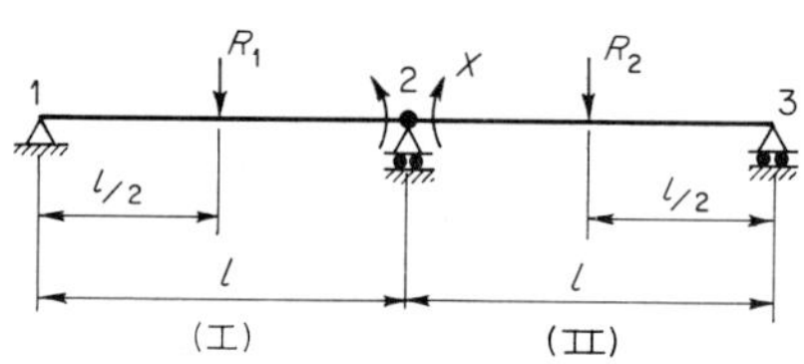

Figure 5.41

Thence,

$$b_1^T f b_1 = D_{11} = \frac{4l}{6EI} \tag{iii}$$

$$D_{10} = \frac{-l}{6EI}\begin{bmatrix}\frac{3l}{8} & \frac{3l}{8}\end{bmatrix} \tag{iv}$$

The unknown force X is given by

$$X = -D_{11}^{-1}D_{10}R = \frac{-1}{4}\begin{bmatrix}\frac{3l}{8} & \frac{3l}{8}\end{bmatrix}\begin{bmatrix}R_1 \\ R_2\end{bmatrix} \tag{v}$$

If $R_1 = R_2 = 1$, then $X = 3l/16$.

Although the member-flexibility matrix in Eq. (*i*) may be written down by inspection, the idea of tearing is embodied in Eq. (*ii*). There, the unit values mean that flexibility connection between the substructures is obtained by direct addition rather than by the formal multiplication in Eq. (*iii*). That is,

$$b_1^T f b_1 = \frac{2l}{6EI} + \frac{2l}{6EI} = f_{\mathrm{I}(M_j)} + f_{\mathrm{II}(M_i)}$$

Example 5.18 The structure shown in Fig. 5.42 has been torn into two structures designated A and B, that in the torn state act independently of one another under the action of the force systems R and $\bar{R}$ respectively. The connecting forces X are common to both systems. The sign convention used in Eqs. (5.42)ff (see Fig. 5.18) will be used here. With the forces shown in Fig. 5.42, the member forces for the two parts will be S and $\bar{S}$ such that

$$\begin{aligned} S^T &= \langle S_1 \quad S_2 \quad S_3 \quad S_4 \quad S_5 \quad S_6\rangle = \langle R_1 \quad R_2 \quad R_3 \quad -X_1 \quad -X_2 \quad -X_3\rangle \\ \bar{S}^T &= \langle \bar{S}_1 \quad \bar{S}_2 \quad \bar{S}_3 \quad \bar{S}_4 \quad \bar{S}_5 \quad \bar{S}_6 \quad \bar{S}_7\rangle = \langle \bar{R}_1 \quad \bar{R}_2 \quad \bar{R}_3 \quad \bar{R}_4 \quad X_1 \quad X_2 \quad X_3\rangle \end{aligned} \tag{i}$$

The flexibility matrices and force transformation matrices must be calculated for each part of the structure by using Eqs. (5.32) and (5.33) respectively. From Eq. (5.32),

$$f = D_{00} - D_{01}D_{11}^{-1}D_{10}$$

and from Eq. (5.33), (*ii*)

$$b = b_o - b_1 D_{11}^{-1}D_{10}$$

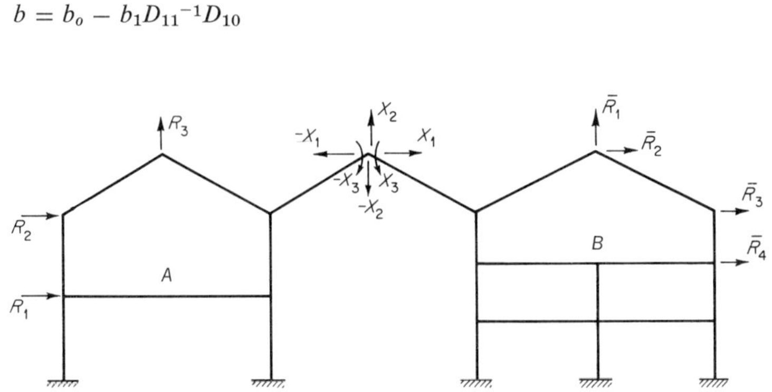

Figure 5.42 Torn structure.

The matrix $\lceil f \rfloor$ of member flexibilities for the whole structure is given by

$$\lceil f \rfloor = \begin{bmatrix} f_A & 0 \\ 0 & f_B \end{bmatrix} \tag{iii}$$

The force transformation matrices for the parts A and B will be given by the upper-case notation $B = [B_o \quad B_1]$. These are obtained from Eq. (i). Thence,

$$\begin{bmatrix} S_1 \\ S_2 \\ S_3 \\ S_4 \\ S_5 \\ S_6 \\ \bar{S}_1 \\ \bar{S}_2 \\ \bar{S}_3 \\ \bar{S}_4 \\ \bar{S}_5 \\ \bar{S}_6 \\ \bar{S}_7 \end{bmatrix} = \Bigg[\underbrace{\begin{matrix} 1 & 0 & 0 & 0 & 0 & 0 & 0 \\ 0 & 1 & 0 & 0 & 0 & 0 & 0 \\ 0 & 0 & 1 & 0 & 0 & 0 & 0 \\ 0 & 0 & 0 & 0 & 0 & 0 & 0 \\ 0 & 0 & 0 & 0 & 0 & 0 & 0 \\ 0 & 0 & 0 & 0 & 0 & 0 & 0 \\ 0 & 0 & 0 & 1 & 0 & 0 & 0 \\ 0 & 0 & 0 & 0 & 1 & 0 & 0 \\ 0 & 0 & 0 & 0 & 0 & 1 & 0 \\ 0 & 0 & 0 & 0 & 0 & 0 & 1 \\ 0 & 0 & 0 & 0 & 0 & 0 & 0 \\ 0 & 0 & 0 & 0 & 0 & 0 & 0 \\ 0 & 0 & 0 & 0 & 0 & 0 & 0 \end{matrix}}_{B_o} \Bigg| \underbrace{\begin{matrix} 0 & 0 & 0 \\ 0 & 0 & 0 \\ 0 & 0 & 0 \\ -1 & 0 & 0 \\ 0 & -1 & 0 \\ 0 & 0 & -1 \\ 0 & 0 & 0 \\ 0 & 0 & 0 \\ 0 & 0 & 0 \\ 0 & 0 & 0 \\ 1 & 0 & 0 \\ 0 & 1 & 0 \\ 0 & 0 & 1 \end{matrix}}_{B_1}\Bigg] \begin{bmatrix} R_1 \\ R_2 \\ R_3 \\ \bar{R}_1 \\ \bar{R}_2 \\ \bar{R}_3 \\ \bar{R}_4 \\ X_1 \\ X_2 \\ X_3 \end{bmatrix} \tag{iv}$$

That is,

$$S = B_o R + B_1 X \tag{v}$$

Equation (iv) may be written with B_o and B_1 in partitioned form.

$$\begin{bmatrix} S \\ \bar{S} \end{bmatrix} = \left[\begin{array}{cc|c} I_3 & 0 & 0 \\ 0 & 0 & -I_3 \\ \hline 0 & I_4 & 0 \\ 0 & 0 & I_3 \end{array}\right] \begin{bmatrix} R \\ \bar{R} \\ X \end{bmatrix} \tag{vi}$$

$$\qquad\qquad B_o \qquad\quad B_1$$

In Eq. (vi) the zero terms are matrices whose number of rows and columns are given by the I matrices that lie in these same rows and columns.

Because of the simplicity in the forms of B_o and B_1, the formal multiplications $B_o^T f B_o$, $B_o^T f B_1$, and $B_1^T f B_1$ need not be carried out and, instead, the appropriate terms in f_A and f_B may be added.

Partition f_A and f_B according to the R and X force groups. That is, let

$$f_A = \begin{bmatrix} f_{A11} & f_{A12} \\ f_{A21} & f_{A22} \end{bmatrix}$$

$$f_B = \begin{bmatrix} f_{B11} & f_{B12} \\ f_{B21} & f_{B22} \end{bmatrix}$$

Then from Eq. (*vi*),

$$B_1{}^T f B_1 = [0 \quad -I_3 \quad 0 \quad I_3] \begin{bmatrix} f_{A11} & f_{A12} & \cdot & \cdot \\ f_{A21} & f_{A22} & \cdot & \cdot \\ \cdot & \cdot & f_{B11} & f_{B12} \\ \cdot & \cdot & f_{B21} & f_{B22} \end{bmatrix} \begin{bmatrix} 0 \\ -I_3 \\ 0 \\ I_3 \end{bmatrix} = [f_{A22} + f_{B22}]$$

$$B_1{}^T[f]B_o = [0 \quad -I_3 \quad 0 \quad I_3][f] \begin{bmatrix} I_3 & 0 \\ 0 & 0 \\ 0 & I_4 \\ 0 & 0 \end{bmatrix} = [-f_{A21} \quad f_{B21}]$$

$$B_o{}^T f B_o = \begin{bmatrix} I_3 & 0 & 0 & 0 \\ 0 & 0 & I_4 & 0 \end{bmatrix} [f] \begin{bmatrix} I_3 & 0 \\ 0 & 0 \\ 0 & I_4 \\ 0 & 0 \end{bmatrix} = \begin{bmatrix} f_{A11} & 0 \\ 0 & f_{B11} \end{bmatrix}$$

Then,

$$X = -\ [f_{A22} + f_{B22}]^{-1}[-f_{A21} \quad f_{B21}] \begin{bmatrix} R \\ \bar{R} \end{bmatrix}$$

The flexibility matrix for the whole structure is

$$F = \begin{bmatrix} f_{A11} & 0 \\ 0 & f_{B22} \end{bmatrix} - \begin{bmatrix} f_{A21}^T \\ f_{B21}^T \end{bmatrix} [f_{A22} + f_{B22}]^{-1}[f_{A21} \quad f_{B21}]$$

Hence,

$$F_{AA} = f_{A11} - f_{A21}^T[f_{A22} + f_{B22}]^{-1} f_{A21}$$
$$F_{BB} = f_{B22} - f_{B21}^T[f_{A22} + f_{B22}]^{-1} f_{B21}$$
$$F_{AB} = F_{BA}{}^T = -f_{A21}^T[f_{A22} + f_{B22}]^{-1} f_{B21}$$

General Tearing Procedure

The process of tearing may be extended to cover cases for which forces from one substructure in the torn condition have reactions on other parts of the structure. For example, the plane frame shown in Fig. 5.43 has been torn into the two parts A and B. It is evident that part B of the system must supply reactions for A when a force such as R is applied to A. A method for specifying the various types of coupling and external forces is given here from Ref. 23.

For a given component structure, let external forces acting be R_h, and redundant forces X_q. Then this structure is analyzed for X_q in the usual manner. That is,

$$\begin{bmatrix} r_h \\ r_q \end{bmatrix} = \begin{bmatrix} D_{hh} & D_{hq} \\ D_{qh} & D_{qq} \end{bmatrix} \begin{bmatrix} R_h \\ X_q \end{bmatrix} \tag{5.174}$$

The forces X_q are internal forces in the component structure and not coupling redundants.

For continuity within the component,

$$X_q = -D_{qq}{}^{-1} D_{qh} R \tag{5.175}$$

Deflections at the load points are

$$\Delta_h = (D_{hh} - D_{hq} D_{qq}{}^{-1} D_{qh}) R_h \tag{5.176}$$

That is,

$$\Delta_h = F_h R_h \tag{5.177}$$

Then F_h is the flexibility matrix for the component structure for the as yet undefined forces R_h. Member forces are given [Eq. (5.33)] by

$$S_h = (b_{oh} - b_{1h} D_{qq}{}^{-1} D_{qh}) R_h \tag{5.178}$$

or

$$S_h = b_h R_h \tag{5.179}$$

and

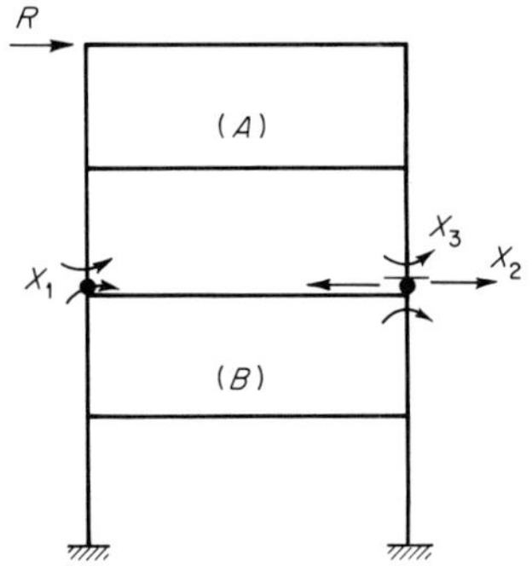

Figure 5.43 Dependent torn systems.

$$b_h = b_{oh} - b_{1h} D_{qq}{}^{-1} D_{qh} \tag{5.180}$$

That is, b_h is the force transformation matrix in the redundant component structure. The matrices F_h and b_h are generated for each component structure.

The Complete Structure

A complete structure composed of substructures I, II, and III is shown in Fig. 5.44. These have been separated in such a manner that the whole system is statically determinate with respect to the reactions acting on I, II, and III.

Then the force vector R_h acting on a substructure is given by

$$R_h = [a_{hj} \quad b_{ht}] \begin{bmatrix} R_j \\ X_t \end{bmatrix} \tag{5.181}$$

The force vector R_j represents the complete external force system on the structure, and the X_t are the coupling redundants. To avoid confusion, a careful delineation must be made of the forces acting on a component. On a particular substructure, there will be three distinguishable types of load points. That is, there will be three R_h types. These are designated types i, b, and t as follows:

Type i is a point directly loaded by external forces.

Type b is a point on a statically determinate connection.

Type t is a point where coupling redundant forces act.

For the complete structure, define the matrices A_{hj} and B_{ht} so that

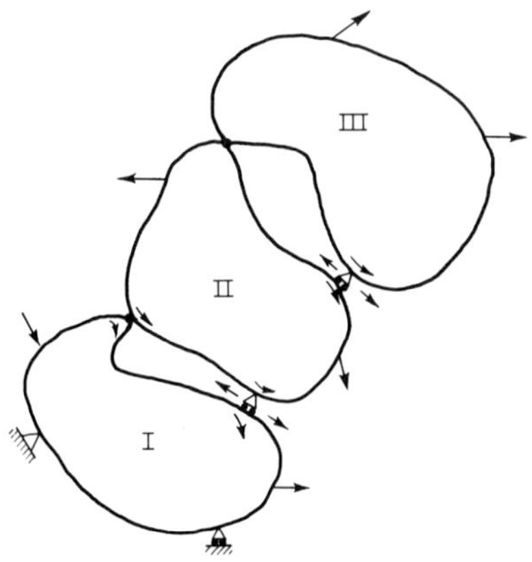

Figure 5.44 Structure subsystems.

$$A_{hj} = \begin{bmatrix} (a_{hj})_I \\ (a_{hj})_{II} \\ (a_{hj})_{III} \end{bmatrix} \qquad B_{ht} = \begin{bmatrix} (b_{ht})_I \\ (b_{ht})_{II} \\ (b_{ht})_{III} \end{bmatrix} \tag{5.182}$$

Then,

$$\begin{bmatrix} R_{h\text{I}} \\ R_{h\text{II}} \\ R_{h\text{III}} \end{bmatrix} = [A_{hj} \quad B_{ht}] \begin{bmatrix} R_j \\ X_t \end{bmatrix} \tag{5.183}$$

The R_h forces may now be partitioned according to the three types i, b, and t above. Then

$$R_h = \begin{bmatrix} R_{hi} \\ R_{hb} \\ R_{ht} \end{bmatrix} = \begin{bmatrix} a_{hij} & 0 \\ a_{hbj} & b_{hbt} \\ a_{htj} & b_{htt} \end{bmatrix} \begin{bmatrix} R_j \\ X_t \end{bmatrix} \tag{5.184}$$

It is seen that whereas X_t may produce type b forces, they do not produce type i forces. Note that

$$[a_{hj}]_\text{I} = \begin{bmatrix} a_{hij} \\ a_{hbj} \\ a_{htj} \end{bmatrix}_\text{I}$$

$$[b_{hj}]_\text{I} = \begin{bmatrix} 0 \\ b_{hbt} \\ b_{htt} \end{bmatrix}_\text{I}$$

For R_j and X_t the flexibility matrix may be calculated from Eqs. (5.177) and (5.183).

$$\begin{bmatrix} \delta_{jj} & \delta_{jt} \\ \delta_{tj} & \delta_{tt} \end{bmatrix} = \begin{bmatrix} A_{hj}{}^T \\ B_{ht}{}^T \end{bmatrix} \begin{bmatrix} F_{h\text{I}} & \cdot & \cdot \\ \cdot & F_{h\text{II}} & \cdot \\ \cdot & \cdot & F_{h\text{III}} \end{bmatrix} [A_{hj} \quad B_{ht}] \tag{5.185}$$

By using the definitions of A_{hj} and B_{hj}, it is seen that Eq. (5.185) may be expanded to give

$$\begin{bmatrix}\delta_{jj} & \delta_{jt}\\ \delta_{tj} & \delta_{tt}\end{bmatrix}=\begin{bmatrix}(a_{hj})_{\mathrm{I}}^T & (a_{hj})_{\mathrm{II}}^T & (a_{hj})_{\mathrm{III}}^T\\ (b_{ht})_{\mathrm{I}}^T & (b_{ht})_{\mathrm{II}}^T & (b_{ht})_{\mathrm{III}}^T\end{bmatrix}\begin{bmatrix}F_{h\mathrm{I}} & \cdot & \cdot\\ \cdot & F_{h\mathrm{II}} & \cdot\\ \cdot & \cdot & F_{h\mathrm{III}}\end{bmatrix}\begin{bmatrix}(a_{hj})_{\mathrm{I}} & (b_{ht})_{\mathrm{I}}\\ (a_{hj})_{\mathrm{II}} & (b_{ht})_{\mathrm{II}}\\ (a_{hj})_{\mathrm{III}} & (b_{ht})_{\mathrm{III}}\end{bmatrix} \tag{5.186}$$

The deflections are calculated

$$\begin{bmatrix}r_i\\ r_t\end{bmatrix}=\begin{bmatrix}\delta_{jj} & \delta_{jt}\\ \delta_{tt} & \delta_{tt}\end{bmatrix}\begin{bmatrix}R_i\\ X_t\end{bmatrix} \tag{5.187}$$

For continuity across the torn sections connecting the substructures,

$$r_t \equiv 0$$

Hence,

$$X_t = -\delta_{tt}{}^{-1}\delta_{tj}R_j \tag{5.188}$$

Deflections of the complete structure are given by

$$r_j = (\delta_{jj} - \delta_{jt}\delta_{tt}{}^{-1}\delta_{tj})R_j = \delta_{jj}R_j + \delta_{jt}X_t \tag{5.189}$$

These calculations may be systematized by defining the matrices x_{tj} and f_{hj} as follows:

$$x_{tj} = -\delta_{tt}{}^{-1}\delta_{tj} \tag{5.190}$$

$$f_{hj} = a_{hj} - b_{ht}\delta_{tt}{}^{-1}\delta_{tj} = a_{hj} + b_{ht}x_{tj} \tag{5.191}$$

Then

$$R_h = f_{hj}R_j \tag{5.192}$$

$$S_h = b_h R_h = b_h f_{hj} R_j \tag{5.193}$$

Example 5.19 The four-story plane frame shown in Fig. 5.45*a* is torn into two identical parts I and II, shown in Fig. 5.45*b*. It is to be analyzed for the horizontal loads $\langle R_1 \quad R_2 \quad R_3 \quad R_4\rangle = \langle 10 \quad 10 \quad 10 \quad 10\rangle$ kips. When parts I and II are considered the members of the system, their member forces will be composed of the R and X forces and their reactions as given in Eq. (5.183)ff. These member forces for each of the parts are shown in Fig. 5.46. It will be noticed that for this, structure member II is loaded by forces acting on part I. Since the reverse is not true, zero forces are omitted in part I to simplify the analysis. Since, however, the two parts are identical, only part II need be analyzed. From this analysis the appropriate parts are extracted for part I. The matrices a_{hj} and b_{ht} for both parts are given below in Eqs. (*i*) and (*ii*).

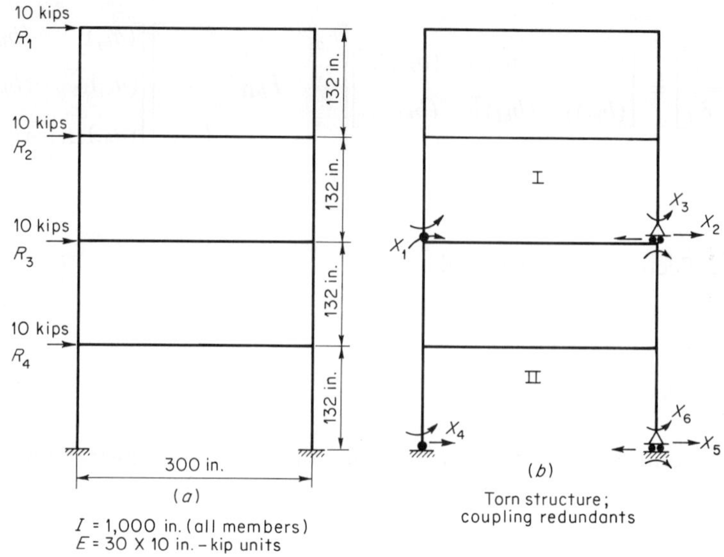

Figure 5.45 Analysis by tearing.

Force Transformations

Part I

$$
\begin{matrix} i\left\{\begin{matrix} \\ \\ \end{matrix}\right. \\ t\left\{\begin{matrix} \\ \\ \\ \end{matrix}\right. \end{matrix}
\begin{bmatrix} S_1 \\ S_2 \\ S_3 \\ S_4 \\ S_5 \end{bmatrix} =
\left[\begin{array}{cccc|cccccc}
0 & 1 & 0 & 0 & 0 & 0 & 0 & 0 & 0 & 0 \\
1 & 0 & 0 & 0 & 0 & 0 & 0 & 0 & 0 & 0 \\
0 & 0 & 0 & 0 & 1 & 0 & 0 & 0 & 0 & 0 \\
0 & 0 & 0 & 0 & 0 & 0 & 1 & 0 & 0 & 0 \\
\underbrace{0 \quad 0 \quad 0 \quad 0}_{[a_{hj}]_{\text{I}}} & & & & \underbrace{0 \quad 1 \quad 0 \quad 0 \quad 0 \quad 0}_{[b_{ht}]_{\text{I}}} & & & & &
\end{array}\right]
\begin{bmatrix} R_1 \\ R_2 \\ R_3 \\ R_4 \\ \hline X_1 \\ X_2 \\ X_3 \\ X_4 \\ X_5 \\ X_6 \end{bmatrix} \qquad (i)
$$

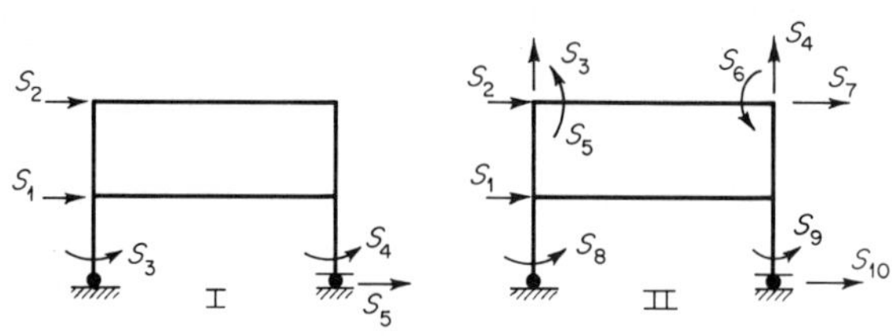

Figure 5.46 Member forces.

Part II

$$
\begin{matrix} i \\ \\ b \\ \\ \\ \\ t \\ \\ \\ \\ \end{matrix}
\begin{Bmatrix} S_1 \\ S_2 \\ S_3 \\ S_4 \\ S_5 \\ S_6 \\ S_7 \\ S_8 \\ S_9 \\ S_{10} \end{Bmatrix}
=
\underbrace{\left[\begin{array}{cccc|cccccc}
0 & 0 & 0 & 1 & 0 & 0 & 0 & 0 & 0 & 0 \\
1 & 1 & 1 & 0 & 0 & 1 & 0 & 0 & 0 & 0 \\
-0.88 & -0.44 & 0 & 0 & -0.00333 & 0 & -0.00333 & 0 & 0 & 0 \\
0.88 & 0.44 & 0 & 0 & 0.00333 & 0 & 0.00333 & 0 & 0 & 0 \\
0 & 0 & 0 & 0 & -1 & 0 & 0 & 0 & 0 & 0 \\
0 & 0 & 0 & 0 & 0 & 0 & -1 & 0 & 0 & 0 \\
0 & 0 & 0 & 0 & 0 & -1 & 0 & 0 & 0 & 0 \\
0 & 0 & 0 & 0 & 0 & 0 & 0 & 1 & 0 & 0 \\
0 & 0 & 0 & 0 & 0 & 0 & 0 & 0 & 0 & 1 \\
0 & 0 & 0 & 0 & 0 & 0 & 0 & 0 & 1 & 0
\end{array}\right]}_{[a_{hj}]_{II} \;\; [b_{ht}]_{II}}
\begin{Bmatrix} R_1 \\ R_2 \\ R_3 \\ R_4 \\ X_1 \\ X_2 \\ X_3 \\ X_4 \\ X_5 \\ X_6 \end{Bmatrix}
\qquad (ii)
$$

alysis of Subsystem I *or* II The subsystem analyzed is actually part II with its member forces $\langle S_1 \quad S_2 \quad \cdots \quad S_{10} \rangle$. In addition there will be the internal redundants $\langle M_1 \quad M_2 \quad M_3 \rangle$ taken to be the stress resultants at the junction of members (2) and (6), as shown in Fig. 5.47. Positive member moments cause tension on the side of the members shown by the dotted line in Fig. 5.47. The force transformation matrices b_o and b_1 for this frame are given in Eq. (*iii*).

$$
\begin{Bmatrix} M_{1i} \\ M_{1j} \\ M_{2i} \\ M_{2j} \\ M_{3i} \\ M_{3j} \\ M_{4i} \\ M_{4j} \\ M_{5i} \\ M_{5j} \\ M_{6i} \\ M_{6j} \\ M_{7i} \\ M_{7j} \end{Bmatrix}
=
\left[\begin{array}{cccccccccc|ccc}
0 & 0 & 0 & 0 & 0 & 0 & 0 & -1 & 0 & 0 & 0 & 0 & 0 \\
132 & 132 & 0 & 0 & 0 & 0 & 132 & -1 & 0 & 132 & 0 & 0 & 0 \\
0 & 0 & 0 & 0 & 0 & 0 & 0 & 0 & 0 & 0 & 0 & 150 & 1 \\
0 & 0 & 0 & 0 & 0 & 0 & 0 & 0 & 0 & 0 & 0 & 0 & 1 \\
132 & 132 & 0 & 0 & 0 & 0 & 132 & -1 & 0 & 132 & 0 & -150 & -1 \\
132 & 264 & 0 & 0 & 0 & 0 & 264 & -1 & 0 & 264 & 132 & -150 & -1 \\
132 & 264 & 0 & 0 & -1 & 0 & 264 & -1 & 0 & 264 & 132 & -150 & -1 \\
0 & 0 & 0 & 0 & 0 & 1 & 0 & 0 & 1 & 264 & 132 & 150 & -1 \\
0 & 0 & 0 & 0 & 0 & 0 & 0 & 0 & 1 & 264 & 132 & 150 & -1 \\
0 & 0 & 0 & 0 & 0 & 0 & 0 & 0 & 1 & 132 & 0 & 150 & -1 \\
0 & 0 & 0 & 0 & 0 & 0 & 0 & 0 & 0 & 0 & 0 & 0 & 1 \\
0 & 0 & 0 & 0 & 0 & 0 & 0 & 0 & 0 & 0 & 0 & -150 & 1 \\
0 & 0 & 0 & 0 & 0 & 0 & 0 & 0 & 1 & 132 & 0 & 0 & 0 \\
0 & 0 & 0 & 0 & 0 & 0 & 0 & 0 & 1 & 0 & 0 & 0 & 0
\end{array}\right]
\begin{Bmatrix} S_1 \\ S_2 \\ S_3 \\ S_4 \\ S_5 \\ S_6 \\ S_7 \\ S_8 \\ S_9 \\ S_{10} \\ M_1 \\ M_2 \\ M_3 \end{Bmatrix}
\qquad (iii)
$$

$$\underbrace{\qquad\qquad}_{b_o} \qquad \underbrace{\qquad}_{b_1}$$

The matrix of member flexibilities is composed of the following member flexibilities as diagonal blocks:

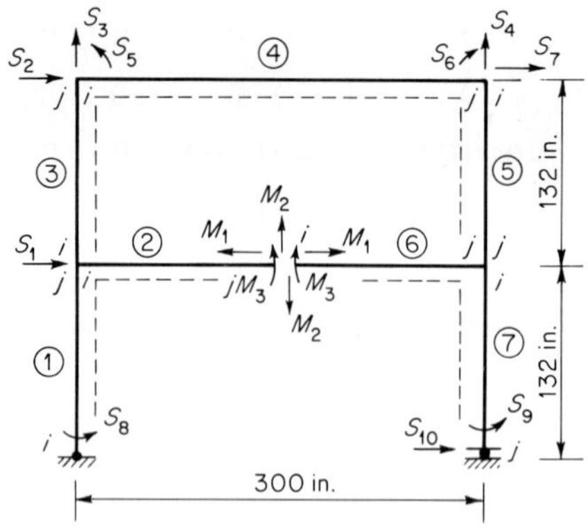

Figure 5.47 Subsystem analysis.

$$f_1 = f_3 = f_5 = f_7 = \frac{132}{6EI}\begin{bmatrix} 2 & 1 \\ 1 & 2 \end{bmatrix}$$

$$f_2 = f_6 = \frac{132}{6EI}\begin{bmatrix} 2.27272 & 1.13636 \\ 1.13636 & 2.27272 \end{bmatrix}$$

$$f_4 = \frac{132}{6EI}\begin{bmatrix} 4.54544 & 2.27272 \\ 2.27272 & 4.54544 \end{bmatrix}$$

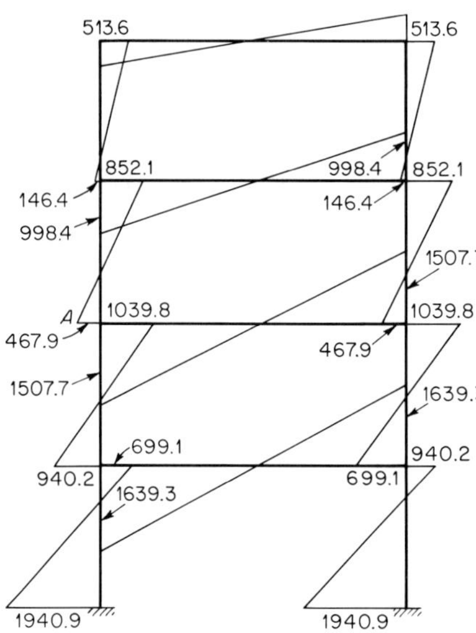

Figure 5.48 Bending-moment diagram of complete structure.

The calculation of the matrices $F_{h\text{II}}$ and $b_{h\text{II}}$ has been carried out by using SMIS in the accompanying coding. From these, $F_{h\text{I}}$ and $b_{h\text{I}}$ have been extracted and are given below. These are then punched on cards and used as input in the complete run given here. (See pages 314 to 327.)

$$F_{h\text{I}} \times 100 = \begin{bmatrix} 4.63866 & 5.36464 & -0.04482 & -0.00148 & 4.44368 \\ 5.36464 & 7.13609 & -0.05032 & -0.00698 & 4.44368 \\ -0.04482 & -0.05032 & 0.00056 & 0.00001 & -0.04334 \\ -0.00148 & -0.00698 & 0.00001 & 0.00056 & 0.04334 \\ 4.44368 & 4.44368 & -0.04334 & 0.04334 & 8.88736 \end{bmatrix}$$

$$b_{h\text{I}} = \begin{bmatrix} 0.0 & 0.0 & -1.0 & 0.0 & 0.0 \\ 132.0 & 132.0 & -1.0 & 0.0 & 132.0 \\ 66.08017 & 99.08013 & -0.50061 & -0.28388 & 28.60859 \\ 14.3043 & 14.30429 & -0.10837 & 0.10837 & 28.60859 \\ 65.91983 & 32.91987 & -0.49939 & 0.28388 & 103.39141 \\ 8.36169 & 41.36169 & -0.06335 & -0.15217 & -11.72483 \\ 8.36169 & 41.36169 & -0.06335 & -0.15217 & -11.72483 \\ -20.08657 & -53.08664 & 0.15217 & 0.06335 & -11.72483 \\ -20.08657 & -53.08664 & 0.15217 & 0.06335 & -11.72483 \\ 37.47157 & 70.47155 & -0.28388 & 0.49939 & 103.39139 \\ 14.3043 & 14.30429 & -0.10837 & 0.10837 & 28.6086 \\ -37.47157 & -70.47155 & 0.28388 & 0.50061 & 28.6086 \\ 0.0 & 0.0 & 0.0 & 1.0 & 132.0 \\ 0.0 & 0.0 & 0.0 & 1.0 & 0.0 \end{bmatrix}$$

The matrices δ_{jj}, δ_{jt}, and δ_{tt} are calculated from Eq. (5.186), by using Eqs. (*i*) and (*ii*), to be

$$\delta_{jj} = \sum_{\text{all substructures}} a_{hj}{}^T F_h a_{hj}$$

$$\delta_{jt} = \sum_{\text{all substructures}} a_{hj}{}^T F_h b_{ht}$$

$$\delta_{tt} = \sum_{\text{all substructures}} b_{ht}{}^T F_h b_{ht}$$

The coupling redundants X_t are calculated in the SMIS program as the vector E. Deflections are then calculated by using Eq. (5.189), and member forces in each part by using Eqs. (*i*) and (*ii*) together with the b_h matrices.

The bending-moment diagram is plotted in Fig. 5.48. Note that the effect of coupling should be checked. For example, the moments at the joints where coupling occurs should sum to zero. At joint A,

```
START

LOAD B02 14 10
   14 ROWS,     10 COLUMNS

PRINT B02  * FORCE TRANSFORMATION PART 2

             1              2              3              4              5
   1      0.00000        0.00000        0.00000        0.00000        0.00000
   2    132.00000      132.00000        0.00000        0.00000        0.00000
   3      0.00000        0.00000        0.00000        0.00000        0.00000
   4      0.00000        0.00000        0.00000        0.00000        0.00000
   5    132.00000      132.00000        0.00000        0.00000        0.00000
   6    132.00000      264.00000        0.00000        0.00000        0.00000
   7    132.00000      264.00000        0.00000        0.00000       -1.00000
   8      0.00000        0.00000        0.00000        0.00000        0.00000
   9      0.00000        0.00000        0.00000        0.00000        0.00000
  10      0.00000        0.00000        0.00000        0.00000        0.00000
  11      0.00000        0.00000        0.00000        0.00000        0.00000
  12      0.00000        0.00000        0.00000        0.00000        0.00000
  13      0.00000        0.00000        0.00000        0.00000        0.00000
  14      0.00000        0.00000        0.00000        0.00000        0.00000

             9             10
   1      0.00000        0.00000
   2      0.00000      132.00000
   3      0.00000        0.00000
   4      0.00000        0.00000
   5      0.00000      132.00000
   6      0.00000      264.00000
   7      0.00000      264.00000
   8      1.00000      264.00000
   9      1.00000      264.00000
  10      1.00000      132.00000
  11      0.00000        0.00000
  12      0.00000        0.00000
  13      1.00000      132.00000
  14      1.00000        0.00000

LOAD B12 14 3
   14 ROWS,      3 COLUMNS

PRINT B12  * FORCE TRANSFORMATION PART 2(M FORCES)

             1              2              3
   1      0.00000        0.00000        0.00000
   2      0.00000        0.00000        0.00000
   3      0.00000      150.00000        1.00000
   4      0.00000        0.00000        1.00000
   5      0.00000     -150.00000       -1.00000
   6    132.00000     -150.00000       -1.00000
   7    132.00000     -150.00000       -1.00000
   8    132.00000      150.00000       -1.00000
   9    132.00000      150.00000       -1.00000
  10      0.00000      150.00000       -1.00000
  11      0.00000        0.00000        1.00000
  12      0.00000     -150.00000        1.00000
  13      0.00000        0.00000        0.00000
  14      0.00000        0.00000        0.00000

ZERO F2 14 14
   14 ROWS,     14 COLUMNS

LOAD A 2 2
    2 ROWS,      2 COLUMNS

PRINT A * SUBMATRICES OF MEMBER FLEXIBILITY MATRIX

             1              2
   1      2.00000        1.00000
   2      1.00000        2.00000
```

6	7	8
0.00000	0.00000	-1.00000
0.00000	132.00000	-1.00000
0.00000	0.00000	0.00000
0.00000	0.00000	0.00000
0.00000	132.00000	-1.00000
0.00000	264.00000	-1.00000
0.00000	264.00000	-1.00000
1.00000	0.00000	0.00000
0.00000	0.00000	0.00000
0.00000	0.00000	0.00000
0.00000	0.00000	0.00000
0.00000	0.00000	0.00000
0.00000	0.00000	0.00000
0.00000	0.00000	0.00000

```
ADDSMI A F2 1 1

LOAD A 2 2
   2 ROWS,      2 COLUMNS

PRINT A * SUBMATRICES OF MEMBER FLEXIBILITY MATRIX

                1               2
   1          2.27272         1.13636
   2          1.13636         2.27272

ADDSMI A F2  3 3

LOAD A 2 2
   2 ROWS,      2 COLUMNS

PRINT A * SUBMATRICES OF MEMBER FLEXIBILITY MATRIX

                1               2
   1          2.00000         1.00000
   2          1.00000         2.00000

ADDSMI A F2  5 5

LOAD A 2 2
   2 ROWS,      2 COLUMNS

PRINT A * SUBMATRICES OF MEMBER FLEXIBILITY MATRIX

                1               2
   1          4.54544         2.27272
   2          2.27272         4.54542

ADDSMI A F2  7 7

LOAD A 2 2
   2 ROWS,      2 COLUMNS

PRINT A * SUBMATRICES OF MEMBER FLEXIBILITY MATRIX

                1               2
   1          2.00000         1.00000
   2          1.00000         2.00000

ADDSMI A F2  9 9

LOAD A 2 2
   2 ROWS,      2 COLUMNS

ADDSMI A F2  11 11

LOAD A 2 2
   2 ROWS,      2 COLUMNS

PRINT A * SUBMATRICES OF MEMBER FLEXIBILITY MATRIX

                1               2
   1          2.00000         1.00000
   2          1.00000         2.00000

ADDSMI A F2  13 13

SCALE F2 7.33333
 SCALAR=  0.7333330E+01

SCALE F2 0.0000001
 SCALAR=  0.1000000E-06

MULT F2 B02 A

TRANS B02 B02T

MULT B02T A D00

MULT F2 B12 B
```

```
TRANS BI2 BI2T

MULT BI2T B DII

MULT BI2T A DIO

TRANS DIO DOI

INVERT DII

MULT DII DIO C

MULT DOI C D

SUB DI DOO

SCALE DOO 100.0
 SCALAR=   0.1000000E+03

PRINT DOO   *FLEXIBILITY MATRIX PART2     SCALE FACTOR OF 100.0

             1            2            3            4            5
    1     4.63866      5.36466      0.00000      0.00000      0.0005
    2     5.36466      7.13609      0.00000      0.00000     -0.0049
    3     0.00000      0.00000      0.00000      0.00000      0.0000
    4     0.00000      0.00000      0.00000      0.00000      0.0000
    5    -0.00056     -0.00494      0.00000      0.00000      0.0001
    6    -0.00530     -0.01080      0.00000      0.00000      0.0000
    7     5.36466      7.13609      0.00000      0.00000     -0.0049
    8    -0.04482     -0.05032      0.00000      0.00000     -0.0000
    9    -0.00148     -0.00698      0.00000      0.00000      0.0000
   10     4.44368      4.44368      0.00000      0.00000      0.0058

             9           10
    1    -0.00148      4.44368
    2    -0.00698      4.44368
    3     0.00000      0.00000
    4     0.00000      0.00000
    5     0.00004      0.00586
    6    -0.00000     -0.00586
    7    -0.00698      4.44368
    8     0.00001     -0.04334
    9     0.00056      0.04334
   10     0.04334      8.88736

SCALE DOO .01
 SCALAR=   0.1000000E-01

MULT BI2 C D

SUB DI BO2

PRINT BO2        * FORCE TRANSFORMATION MATRIX IN PART 2

             1            2            3            4            5
    1     0.00000      0.00000      0.00000      0.00000      0.000
    2   132.00000    132.00000      0.00000      0.00000      0.000
    3    66.08017     99.08013      0.00000      0.00000      0.065
    4    14.30430     14.30429      0.00000      0.00000      0.044
    5    65.91983     32.91987      0.00000      0.00000      0.065
    6     8.36169     41.36169      0.00000      0.00000      0.499
    7     8.36169     41.36169      0.00000      0.00000     -0.500
    8   -20.08657    -53.08664      0.00000      0.00000      0.283
    9   -20.08657    -53.08664      0.00000      0.00000      0.283
   10    37.47157     70.47155      0.00000      0.00000     -0.159
   11    14.30430     14.30429      0.00000      0.00000      0.044
   12   -37.47157    -70.47155      0.00000      0.00000      0.159
   13     0.00000      0.00000      0.00000      0.00000      0.00
   14     0.00000      0.00000      0.00000      0.00000      0.00
```

	7	8
0.00530	5.36466	-0.04482
0.01080	7.13609	-0.05032
0.00000	0.00000	0.00000
0.00000	0.00000	0.00000
0.00001	-0.00494	-0.00000
0.00012	-0.01080	0.00004
0.01080	7.13609	-0.05032
0.00004	-0.05032	0.00056
0.00000	-0.00698	0.00001
0.00586	4.44368	-0.04334

	7	8
0.00000	0.00000	-1.00000
0.00000	132.00000	-1.00000
0.15217	99.08013	-0.50061
0.04441	14.30429	-0.10837
0.15217	32.91987	-0.49939
0.28388	41.36169	-0.06335
0.28388	41.36169	-0.06335
0.50061	-53.08664	0.15217
0.49939	-53.08664	0.15217
.06335	70.47155	-0.28388
.04441	14.30429	-0.10837
.06335	-70.47155	0.28388
.00000	0.00000	0.00000
.00000	0.00000	0.00000

```
               9             10
  1      0.00000        0.00000
  2      0.00000      132.00000
  3     -0.28388       28.60859
  4      0.10837       28.60860
  5      0.28388      103.39141
  6     -0.15217      -11.72483
  7     -0.15217      -11.72483
  8      0.06335      -11.72485
  9      0.06335      -11.72485
 10      0.49939      103.39139
 11      0.10837       28.60860
 12      0.50061       28.60861
 13      1.00000      132.00000
 14      1.00000        0.00000

DELETE A

DELETE B

DELETE C

DELETE D

DELETE BI2

DELETE DOI

DELETE DII

DELETE F2

DELETE BI2T

DELETE BO2T
 MATRICES DOO AND BO2 CONTAIN INFORMATION FOR PART 2
 PART 1 IS SIMILAR AND THE FLEXIBILITY IS TAKEN FROM COLUMNS AND ROWS
 1,2,8,9,10 OF DOO SIMILARLY FOR BO2
 THESE HAVE BEEN EXTRACTED ON A PREVIOUS RUN AND ARE READ IN AS REQUIRED

LOAD F1 5 5
    5 ROWS,      5 COLUMNS

PRINT F1  *DOO FOR PART 1

               1             2             3             4             5
  1      4.63866       5.36464      -0.04482      -0.00148       4.44368
  2      5.36464       7.13609      -0.05032      -0.00698       4.44368
  3     -0.04482      -0.05032       0.00056       0.00001      -0.04334
  4     -0.00148      -0.00698       0.00001       0.00056       0.04334
  5      4.44368       4.44368      -0.04334       0.04334       8.88736

SCALE F1 .01
 SCALAR=  0.1000000E-01

LOAD AHJ1 5 4
    5 ROWS,      4 COLUMNS

PRINT AHJ1  *LOAD MATRIX PART 1

               1             2             3             4
  1      0.00000       1.00000       0.00000       0.00000
  2      1.00000       0.00000       0.00000       0.00000
  3      0.00000       0.00000       0.00000       0.00000
  4      0.00000       0.00000       0.00000       0.00000
  5      0.00000       0.00000       0.00000       0.00000

LOAD BHT1 5 6
    5 ROWS,      6 COLUMNS

PRINT BHT1  *  REDUNDANT MATRIX PART 1
```

```
                 1              2              3              4              5
      1      0.00000        0.00000        0.00000        0.00000        0.00000
      2      0.00000        0.00000        0.00000        0.00000        0.00000
      3      1.00000        0.00000        0.00000        0.00000        0.00000
      4      0.00000        0.00000        1.00000        0.00000        0.00000
      5      0.00000        1.00000        0.00000        0.00000        0.00000

MULT F1 AHJ1 A

MULT F1 BHT1 B

TRANS AHJ1 AHJ1T

TRANS BHT1 BHT1T

MULT BHT1T B DII1

MULT AHJ1T B DOI1

MULT AHJ1T A DOO1

DELETE AHJ1T

DELETE BHT1T

LOAD AHJ2 10 4
   10 ROWS,      4 COLUMNS

PRINT AHJ2  * LOAD MATRIX PART 2

                 1              2              3              4
      1      0.00000        0.00000        0.00000        1.00000
      2      1.00000        1.00000        1.00000        0.00000
      3     -0.88000       -0.44000        0.00000        0.00000
      4      0.88000        0.44000        0.00000        0.00000
      5      0.00000        0.00000        0.00000        0.00000
      6      0.00000        0.00000        0.00000        0.00000
      7      0.00000        0.00000        0.00000        0.00000
      8      0.00000        0.00000        0.00000        0.00000
      9      0.00000        0.00000        0.00000        0.00000
     10      0.00000        0.00000        0.00000        0.00000

LOAD BHT2 10 6
   10 ROWS,      6 COLUMNS

PRINT BHT2  * REDUNDANT MATRIX PART 2

                 1              2              3              4              5
      1      0.00000        0.00000        0.00000        0.00000        0.00000
      2      0.00000        1.00000        0.00000        0.00000        0.00000
      3     -0.00333        0.00000       -0.00333        0.00000        0.00000
      4      0.00333        0.00000        0.00333        0.00000        0.00000
      5     -1.00000        0.00000        0.00000        0.00000        0.00000
      6      0.00000        0.00000       -1.00000        0.00000        0.00000
      7      0.00000       -1.00000        0.00000        0.00000        0.00000
      8      0.00000        0.00000        0.00000        1.00000        0.00000
      9      0.00000        0.00000        0.00000        0.00000        0.00000
     10      0.00000        0.00000        0.00000        0.00000        1.00000

MULT DOO AHJ2 A

MULT DOO BHT2 B

TRANS AHJ2 AHJ2T

TRANS BHT2 BHT2T

MULT BHT2T B C

MULT AHJ2T A DOO2

ADD DOO2 DOO1

ADD DII1 C
```

```
6
 0.00000
 0.00000
 0.00000
 0.00000
 0.00000
```

```
0.00000
0.00000
0.00000
0.00000
.00000
.00000
.00000
.00000
.00000
.00000
```

```
INVERT C

DELETE BHT2T

MULT AHU2T B D

DELETE AHJ2T

ADD DOI1 D

TRANS D DT

MULT C DT E

SCALE E -1.0
 SCALAR= -0.1000000E+01

MULT  D E G

ADD 3 D001

LOAD R 4 1  *LOADS
    4 ROWS,      1 COLUMNS

MULT D001 R DEF

PRINT DEF  * NODE DEFLECTIONS

              1
    1         1.28768
    2         1.08941
    3         0.75009
    4         0.30809

DELETE A

DELETE B

DELETE C

MULT BHT1 E A

ADD AHJ1 A

MULT A R R1

LOAD B01 14 5
   14 ROWS,      5 COLUMNS

PRINT B01 * FORCE TRANSFORMATION PART 1

             1             2             3            4            5
    1      0.00000       0.00000      -1.00000      0.00000      0.00000
    2    132.00000     132.00000      -1.00000      0.00000    132.00000
    3     66.08017      99.08013      -0.50061     -0.28388     28.60859
    4     14.30430      14.30429      -0.10837      0.10837     28.60859
    5     65.91983      32.91987      -0.49939      0.28388    103.39141
    6      8.36169      41.36169      -0.06335     -0.15217    -11.72483
    7      8.36169      41.36169      -0.06335     -0.15217    -11.72483
    8    -20.08657     -53.08664       0.15217      0.06335    -11.72483
    9    -20.08657     -53.08664       0.15217      0.06335    -11.72483
   10     37.47167      70.47155      -0.28388      0.49939    103.39139
   11     14.30430      14.30429      -0.10837      0.10837     28.60860
   12    -37.47157     -70.47155       0.28388      0.50061     28.60860
   13      0.00000       0.00000       0.00000      1.00000    132.00000
   14      0.00000       0.00000       0.00000      1.00000      0.00000

MULT B01 R1 FORCE1

PRINT FORCE1  * MEMBER FORCES PART 1
```

```
               1
  1      -467.91604
  2       852.08281
  3       998.44088
  4        -0.00005
  5      -146.35808
  6       513.63666
  7       513.63666
  8      -513.63832
  9      -513.63832
 10       146.35888
 11        -0.00015
 12      -998.44110
 13      -852.08332
 14       467.91783

MULT BHT2 E B

ADD AHJ2 B

MULT B R R2

MULT BO2 R2 FORCE2

PRINT FORCE2  *MEMBER FORCES PART  2

               1
  1     -1940.92758
  2       699.07305
  3      1639.25326
  4        -0.00004
  5      -940.18002
  6      1039.81828
  7      1507.73431
  8     -1507.73827
  9     -1039.82044
 10       940.18104
 11        -0.00004
 12     -1639.25334
 13      -699.07259
 14      1940.92707

ALL CARDS PROCESSED  END OF JOB
```

$$\Sigma M = 467.9 + 1039.8 - 1507.7 = 0$$

Also the sum of shears across the base columns should equal 40 kips.

$$\text{Base shear} = \frac{2(699.1) + 2(1940.9)}{132} = \frac{5280}{132} = 40$$

PROBLEMS

5.1 Examine the conditions of the determinacy, indeterminacy, and stability of the structures shown in Fig. P5.1.

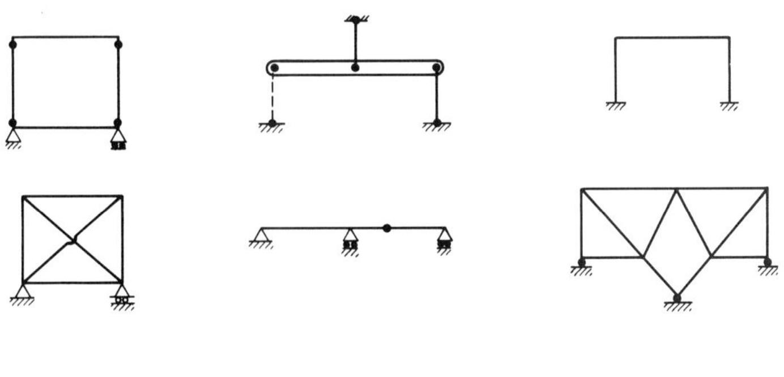

Figure P5.1

5.2 Use the force method to analyze the indeterminate structures shown in Fig. P5.2. Calculate member forces and deflections in terms of EI at the load points.

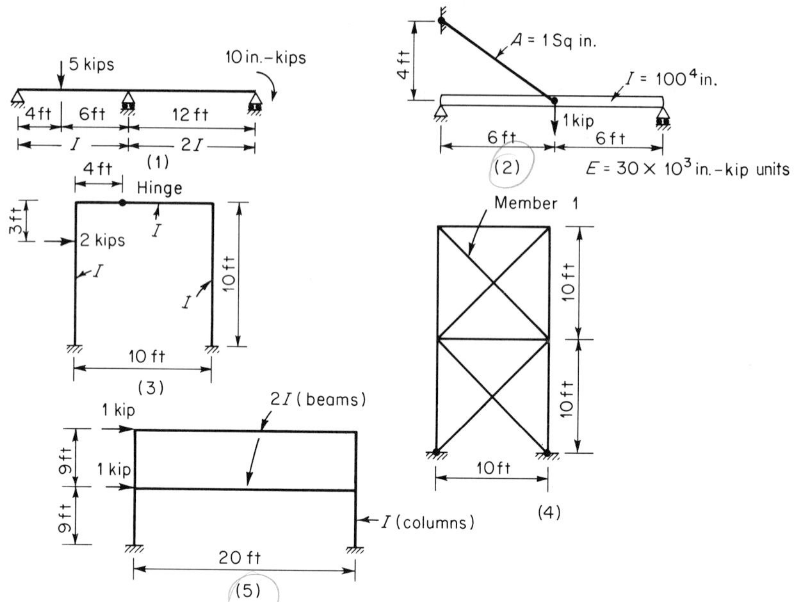

Figure P5.2

5.3 The structures shown in Fig. P5.3 all have distributed loading. Calculate member forces and node deflections by the following methods:

a Initial strains
b Generalized member forces
c Analysis from a fixed-joint position

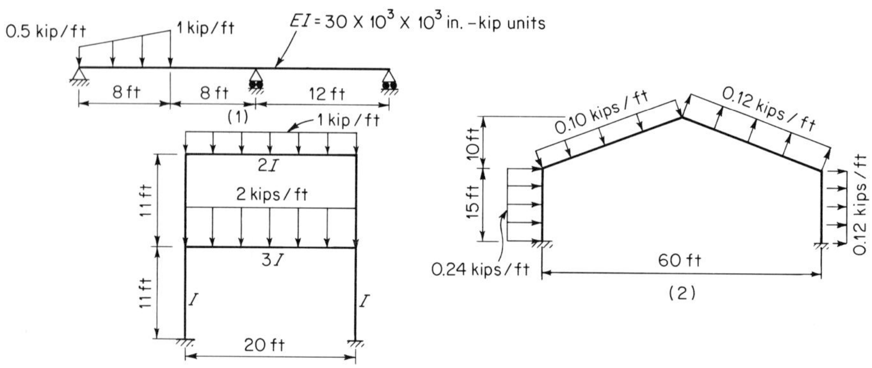

Figure P5.3

5.4 Calculate the bending moments in the frame shown in Fig. P5.3(1) if the support moments and A and B are released.

5.5 Calculate the bending moments induced in the frame shown in Fig. P5.3(2) if support B settles 1.0 in.

5.6 Analyze the frames shown in Fig. P5.4 by tearing into suitable substructures as shown by the section AA.

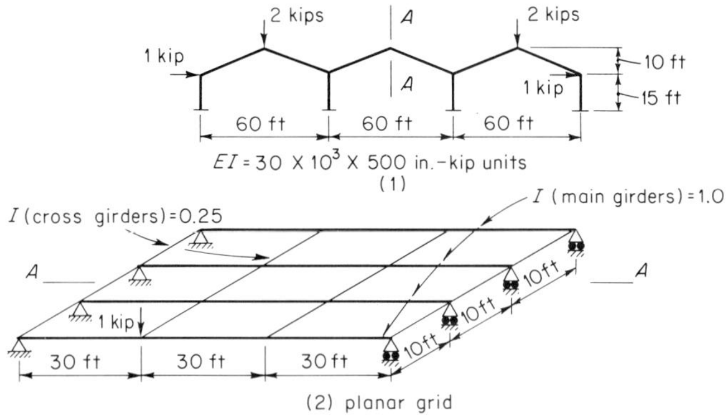

Figure P5.4

5.7 Analyze structure (4) in Fig. P5.2 making allowance for the removal of member 1.

6 THE DISPLACEMENT METHOD OF ANALYSIS

6.1 GENERAL CONCEPTS OF STRUCTURE STIFFNESS AND STRAIN ENERGY EXPRESSED IN TERMS OF DISPLACEMENTS; GENERATION OF STIFFNESS MATRICES FROM DISPLACEMENT TRANSFORMATION MATRICES; KINEMATICALLY INDETERMINATE STRUCTURES

In Eqs. (5.29) and (5.32), the flexibility relationship [Eq. (6.1)] has been obtained for a given linear elastic structure

$$r = FR \tag{6.1}$$

The reverse process to that in Eq. (6.1) is always possible; that is, given a set of node displacements r, the node forces R may be expressed as linear functions of these displacements.

Let the inverse of the flexibility matrix be K defined as a stiffness matrix. Then

$$K = F^{-1} \tag{6.2}$$

This K matrix must be specified as the stiffness matrix of the structure for the displacements r. It should be noted that whereas the flexibility coefficients in F are independent of the number of loads chosen, those of K will alter when the number of displacement points is changed. This will become apparent when the force-deflection relationship is examined in detail.

The forces R expressed as linear functions of the displacements r are given by Eq. (6.3).

$$R = Kr \tag{6.3}$$

Expand the ith equation of this set, expressing the force R_i in terms of the displacements $\langle r_1 \quad r_2 \quad \cdots \quad r_n \rangle$.

$$R_i = K_{i1}r_1 + K_{i2}r_2 + \cdots + K_{ii}r_i + \cdots + K_{in}r_n \tag{6.4}$$

The coefficient K_{ij} is the value of the force R_i for a displacement $r_j = 1$ when all other displacements are held equal to zero. When the number of points n is changed, the coefficients K_{ij} will change.

Example 6.1 The effect of constraints on stiffness terms

The cantilever beam shown in Fig. 6.1 has uniform inertia EI and length l. The stiffness at end 1 may be calculated for a displacement $r_y = 1$, with the displacement θ_Z either not constrained or constrained to be equal to zero. These two displacement patterns are shown in Fig. 6.1*a* and *b* respectively.

The stiffness terms are

$$K_Y = \frac{3EI}{l^3}$$

$$K_Y = \frac{12EI}{l^3}$$

The added constraint $\theta_z = 0$ evidently has a very marked effect on the stiffness K_Y.

Elastic Strain Energy

It has been shown in Eq. (4.80) that the elastic strain energy U for a structure may be expressed as a quadratic function of the node loads R with flexibility terms as coefficients. That is,

$$U = \tfrac{1}{2}R^TFR \tag{6.5}$$

When Eq. (6.3) is substituted in Eq. (6.5), the expression for strain energy is obtained as a quadratic function of the node displacements r with stiffness terms as coefficients. From Eq. (6.5),

$$U = \tfrac{1}{2}r^TK^TFKr$$

Hence

$$U = \tfrac{1}{2}r^TKr \tag{6.6}$$

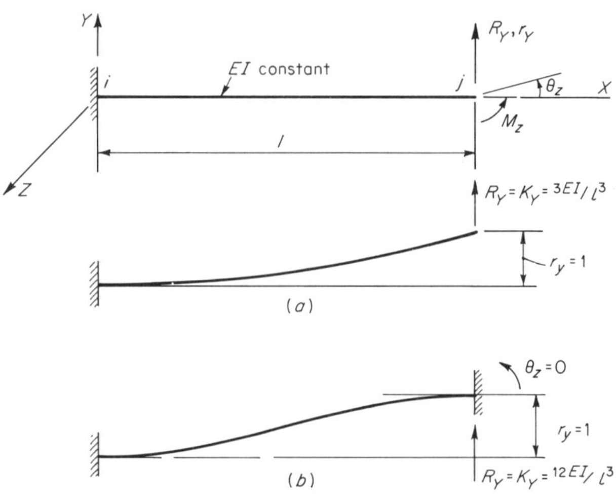

Figure 6.1 Variation of stiffness with introduction of constraints.

It is noted that since F is symmetrical so is its inverse, and thence K is symmetrical. Then

$$K^T F = KF = I$$

Stiffness Matrices for Members and for the Whole Structure

In Eq. (6.4) it has been shown that the K_{ij}th term of the stiffness matrix K of a structure is the value of the force R_i for a displacement $r_j = 1$, all other specified displacements in the set r being zero. This could provide a means of generation of terms of K. Alternatively they could be obtained by firstly calculating the flexibility matrix F and then inverting to produce K. It will be shown later that a third approach will be to differentiate the strain-energy expression Eq. (6.6) with respect to displacements. For the structure composed of many members, it is most convenient to build up K in a manner similar to that in which F may be generated from the individual member flexibilities through the transformation

$$F = b^T f b \tag{6.7}$$

Displacement Transformation Matrices; Influence Coefficients for Member Displacements

It has been shown in Chap. 3 that when member distortions are expressed in terms of node displacements through the equation

$$v = ar \tag{6.8}$$

the equations of joint equilibrium may be expressed

$$R - a^T S = 0 \tag{6.9}$$

In the particular case of static determinacy,

$$S = (a^T)^{-1} R = bR$$

In this way a method was devised for generation of the b matrix, and displacement transformations were calculated (see Sec. 3.5) for member distortions in terms of node displacements. In the displacement method of analysis, Eqs. (6.8) and (6.9) are applied to the general structure for which determinacy has not been assumed.

Equation (6.9) has been obtained by application of the contragradient law that states [see Eqs. (3.18)ff] that when (6.8) is true, then

$$R = a^T S \tag{6.10}$$

From Eqs. (6.8) and (6.10) it is evident that if a stiffness relationship relates member forces S to member distortions v, then R may be related to r. That is, let

$$S = kv \tag{6.11}$$

Then

$$R = a^T kv = a^T kar \tag{6.12}$$

Before use is made of Eq. (6.12) in the generation of the structure-stiffness matrix, further discussion is given to the displacement transformation matrix a and the matrix of member stiffnesses k.

Just as the force transformation matrix b may be generated by successive applications of a unit load at various nodes in a structure, so the displacement transformation matrix a may be produced by unit values' being given in turn to each node displacement. This process was used in Chap. 3, in which it was shown that for a statically determinate structure,

$$b = (a^T)^{-1}$$

For example, in Fig. 6.2, the displacements of node 1 will be $r^T = \langle r_x \quad r_y \quad \theta \rangle$. When unit values are given in turn to r_x, r_y, and θ, the distortions of members appear as in Fig. 6.2b. If the member distortions are taken as $v^T = \langle \theta_i \quad \theta_j \quad \delta \rangle$, the a matrix is given in Eq. (i), being generated by inspection of Fig. 6.2b.

$$\begin{bmatrix} \begin{pmatrix} \theta_i \\ \theta_j \\ \delta \end{pmatrix}_1 \\ \begin{pmatrix} \theta_i \\ \theta_j \\ \delta \end{pmatrix}_2 \end{bmatrix} = \begin{bmatrix} 0 & -\dfrac{1}{l_1} & 0 \\ 0 & -\dfrac{1}{l_1} & 1 \\ 1 & 0 & 0 \\ 0 & \dfrac{1}{l_2} & 1 \\ 0 & \dfrac{1}{l_2} & 0 \\ -1 & 0 & 0 \end{bmatrix} \begin{bmatrix} r_x \\ r_y \\ \theta \end{bmatrix} \tag{i}$$

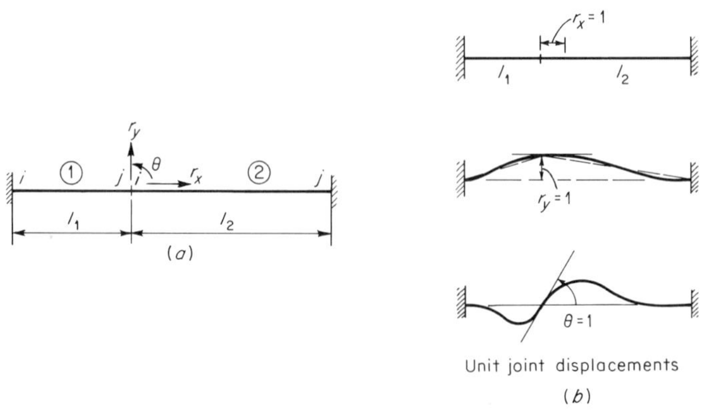

Figure 6.2 Node and member displacements.

In Sec. 3.2, the relationships between member distortions and node displacements have been developed in considerable detail. Therein it is shown how node displacements at the member ends in global coordinates are transformed to those in local coordinates for the member and thence to member distortions.

The transformations are given in the form,

$$v = T^T L_D^T r \tag{6.13}$$

It will be found later, in Chap. 7, that these T and L matrices will be used to transform member-stiffness matrices in the basic force deformation quantities to those for the forces and displacements of member ends in global coordinates. These stiffness matrices are used in the method of analysis generally designated the *direct stiffness method.* Transformation equations such as Eq. (6.13) will be found useful also in certain structures such as the rigid headstock on flexible piles.

In this chapter the main concern will be in the general formulation of the displacement method, and in the examples given, the a matrices will be developed from elementary displacement patterns. These will sometimes be complicated by the assumption of no change in axial length of members.

Consider the general structure, and let $a_{i1}, a_{i2}, \ldots, a_{ik}$ be the values of the member distortion v_i due to $r_1 = 1, r_2 = 1, \ldots, r_k = 1$, respectively. That is, for all displacements $r^T = \langle r_1 \quad r_2 \quad \cdots \quad r_k \rangle$, the particular distortion v_i is given by the linear relationship

$$v_i = a_{i1}r_1 + a_{i2}r_2 + \cdots + a_{ik}r_k \tag{6.14}$$

For all such member distortions and joint displacements,

$$\begin{bmatrix} v_1 \\ v_2 \\ \vdots \\ v_i \\ \vdots \\ v_n \end{bmatrix} = \begin{bmatrix} a_{11} & a_{12} & \cdots & a_{1k} \\ a_{21} & a_{22} & \cdots & a_{2k} \\ \cdots & \cdots & \cdots & \cdots \\ a_{i1} & a_{i2} & \cdots & a_{ik} \\ \cdots & \cdots & \cdots & \cdots \\ a_{n1} & a_{n2} & \cdots & a_{nk} \end{bmatrix} \begin{bmatrix} r_1 \\ r_2 \\ \vdots \\ r_k \end{bmatrix} \tag{6.15}$$

Equation (6.15) is of course simply

$$v = ar \tag{6.16}$$

For a typical member a of the structure with member forces S_a and member distortions v_a, a stiffness relationship will exist that relates S_a to v_a. This member-stiffness matrix may be obtained from force or displacement patterns that have been assumed to exist in the member. It may be calculated from these directly or from inversion of the flexibility matrix f_a if v_a includes only basic member distortions. It may be shown that

$$k_a = f_a^{-1}$$

if v_a and S_a include only basic distortions, or

$$k_a = Tf_a^{-1}T^T$$

where rigid-body displacements are included in v_a. Notice that f_a always refers to a basic set of forces since the equations of equilibrium must be satisfied when unit forces are applied to a member. In either case,

$$S_a = k_a v_a \tag{6.17}$$

All such equations for members a, b, c, etc., may be combined, with the member-stiffness matrices forming diagonal submatrices.

$$\begin{bmatrix} S_a \\ S_b \\ S_c \\ \vdots \end{bmatrix} = \begin{bmatrix} k_a & & & \mathbf{0} \\ & k_b & & \\ & & k_c & \\ \mathbf{0} & & & \ddots \end{bmatrix} \begin{bmatrix} v_a \\ v_b \\ v_c \\ \vdots \end{bmatrix} \tag{6.18}$$

This is the equation

$$S = kv \tag{6.19}$$

Then from Eq. (6.16),

$$S = kar \tag{6.20}$$

Equation (6.20) provides a convenient method for determining member forces when the node displacements are known. The relationship,

$$R = a^T S$$

may be obtained, as in Eq. (6.10), as a direct application of the contragradient law, or alternatively from the principle of virtual displacements. That is, the R force system acting on the structure together with the member-node forces forms a system in equilibrium, and hence if the structure is given a set of compatible displacements,

$$\Sigma\, W_i = \Sigma\, W_e$$

As a suitable virtual displacement, choose $\bar{r}_i = \Delta$ (a small quantity) with all other displacements zero. Then

$$\Sigma\, W_e = R_i \times \bar{r}_i = R_i\,\Delta$$

To calculate $\Sigma\, W_i$, consider the member distortions $(\bar{v}_{1i}, \bar{v}_{2i}, \ldots, \bar{v}_{ni})$ associated with $\bar{r}_i = \Delta$.

From Eq. (6.15) it is seen that these distortions are simply Δ multiplied by the ith column of a. That is,

$$\begin{bmatrix} \bar{v}_{1i} \\ \bar{v}_{2i} \\ \bar{v}_{3i} \\ \vdots \\ \bar{v}_{ni} \end{bmatrix} = \begin{bmatrix} \bar{a}_{1i} \\ \bar{a}_{2i} \\ \bar{a}_{3i} \\ \vdots \\ \bar{a}_{ni} \end{bmatrix} \Delta$$

The work done by the member forces is

$$\Sigma\, W_i = (\bar{a}_{1i}S_1 + \bar{a}_{2i}S_2 + \cdots + \bar{a}_{ni}S_n)\,\Delta$$

$$= [\bar{a}_{1i} \quad \bar{a}_{2i} \quad \cdots \quad \bar{a}_{ni}] \begin{bmatrix} S_1 \\ S_2 \\ \vdots \\ S_n \end{bmatrix} \Delta$$

That is, $\Sigma\, W_i = \bar{a}_i{}^T S\,\Delta$, where $\bar{a}_i$ is the ith column of a. Hence

$$R_i = \bar{a}_i{}^T S \tag{6.21}$$

When the above procedure is repeated in turn for each of the loads R_i, then Eq. (6.21) is only one row of the complete set

$$R = \bar{a}^T S \tag{6.22}$$

In the above, the bar notation has been used to indicate that $\bar{r}_i = \Delta$ need not be applied to the same constrained structure as the r deflections. However, it is generally convenient to make $\bar{a} \equiv a$.

Finally

$$R = a^T k a r \tag{6.23}$$

The stiffness matrix for the displacements r has been found to be

$$K = a^T k a \tag{6.24}$$

The actual application of the equation $K = a^T k a$ to the calculation of the structure-stiffness matrix takes many forms, and it will be largely this topic that occupies the remainder of this chapter.

Generally, of course, the node forces R are known, and the displacements r are the required unknowns. The solution of Eq. (6.23) may be written formally as

$$r = K^{-1} R \tag{6.25}$$

Some adjustment may be necessary to K if it embodies the rigid-body displacements of the structure before a solution may be obtained to Eq. (6.25). Details of this will be given later.

Kinematically Indeterminate Structures

A structure is said to be kinematically determinate when all node displacements are known. Then node forces may be calculated directly from inversion of Eq. (6.23). The condition of all joints fixed is a kinematically determinate state since then $r \equiv 0$. As has been demonstrated in Chaps. 3 and 4, this state is a convenient starting point in the analysis of structures for which loads appear on members between node points.

In the more general case, suppose that only some of the nodes of an elastic structure are given specified displacements. Those at the remaining nodes will be unknown but may be calculated from the known displacements as follows.

In the equation,

$$R = a^T k a r$$

the vector r will be partitioned into r_o, the known, and r_x, the unknown, joint displacements. Partition R similarly into R_o and R_x, and also a into a_o and a_1, so that

$$\begin{bmatrix} R_o \\ R_x \end{bmatrix} = \begin{bmatrix} a_o{}^T \\ a_1{}^T \end{bmatrix} k \begin{bmatrix} a_o & a_1 \end{bmatrix} \begin{bmatrix} r_o \\ r_x \end{bmatrix} \tag{6.26}$$

On multiplication of the partitioned matrices,

$$\begin{bmatrix} R_o \\ R_x \end{bmatrix} = \begin{bmatrix} a_o{}^T k a_o & a_o{}^T k a_1 \\ a_1{}^T k a_o & a_1{}^T k a_1 \end{bmatrix} \begin{bmatrix} r_o \\ r_x \end{bmatrix} \tag{6.27}$$

The notation is introduced:

$$C_{00} = a_o{}^T k a_o$$

$$C_{01} = C_{10}{}^T = a_o{}^T k a_1$$

$$C_{11} = a_1{}^T k a_1$$

With this notation, Eq. (6.27) is written

$$\begin{bmatrix} R_o \\ R_x \end{bmatrix} = \begin{bmatrix} C_{00} & C_{01} \\ C_{10} & C_{11} \end{bmatrix} \begin{bmatrix} r_o \\ r_x \end{bmatrix} \tag{6.28}$$

In Eq. (6.28), r_o are the known displacements. The nodes corresponding to r_x may be unloaded, in which case $R_x \equiv 0$ or have known forces applied. It is seen that the unknown quantities in Eq. (6.28) are the forces R_o and the displacements r_x.

On expansion of Eq. (6.28), two equations are obtained, and from the second of these the r_x displacements are obtained in terms of r_o and R_x. That is,

$$r_x = C_{11}^{-1}(R_x - C_{10}r_o) \tag{6.29}$$

when $R_x \equiv 0$,

$$r_x = -C_{11}^{-1}C_{10}r_o \tag{6.30}$$

The R_o forces are obtained from Eq. (6.28),

$$R_o = C_{00}r_o + C_{01}r_x$$

Substitution for r_x from Eq. (6.28) gives

$$R_o = (C_{00} - C_{01}C_{11}^{-1}C_{10})r_o + C_{01}C_{11}^{-1}R_x \tag{6.31}$$

Again when $R_x \equiv 0$,

$$R_o = (C_{00} - C_{01}C_{11}^{-1}C_{10})r_o \tag{6.32}$$

The similarity between kinematic and static indeterminacy is shown when Eqs. (6.30) and (6.32) are compared with Eqs. (5.28) and (5.29).

Member forces may be calculated from the expression

$$S = kar$$

In the partitioned form of Eq. (6.26),

$$S = k[a_o \quad a_1]\begin{bmatrix} r_o \\ r_x \end{bmatrix}$$

that is,

$$S = k[(a_o - a_1C_{11}^{-1}C_{10})r_o + a_1C_{11}^{-1}R_x] \tag{6.33}$$

An interesting class of problems arises in which both r_o and r_x are unknown but $R_x \equiv 0$. That is, how are the displacements at the unloaded joints related to those of the load points? As in Eq. (6.30),

$$r_x = -C_{11}^{-1}C_{10}r_o$$

and from Eq. (6.32),

$$r_o = (C_{00} - C_{01}C_{11}^{-1}C_{10})^{-1}R_o \tag{6.34}$$

Consider now Eq. (6.28) in the form

$$\begin{bmatrix} R_o \\ 0 \end{bmatrix} = K\begin{bmatrix} r_o \\ r_x \end{bmatrix} \tag{6.35}$$

The flexibility matrix for the structure may be obtained from direct inversion of K, and in partitioned form,

$$F = \begin{bmatrix} F_{11} & F_{12} \\ F_{21} & F_{22} \end{bmatrix} \tag{6.36}$$

Then, deflections are given by

$$\begin{bmatrix} r_o \\ r_x \end{bmatrix} = \begin{bmatrix} F_{11} & F_{12} \\ F_{21} & F_{22} \end{bmatrix} \begin{bmatrix} R_o \\ 0 \end{bmatrix} \tag{6.37}$$

That is,

$$\begin{aligned} r_o &= F_{11} R_o \\ r_x &= F_{21} R_o \end{aligned} \tag{6.38}$$

A comparison of Eqs. (6.34) and (6.38) shows that

$$F_{11} = (C_{00} - C_{01} C_{11}{}^{-1} C_{10})^{-1} \tag{6.39}$$

Equation (6.39) is a useful result since it means that the stiffness matrix for the displacements r_o has been found with the displacements r_x unconstrained (see Ref. 34). That is,

$$K_{00} = F_{11}{}^{-1} \tag{6.40}$$

This procedure may be used in the analysis of a frame structure in which K_{00} is required in an analysis for dynamic loading. The frame shown in Fig. 6.3 is analyzed for the unit loads shown at each floor level. Deflections are calculated for these unit forces and when partitioned correctly, give F_{11} and F_{21}.

Some Member-stiffness Matrices

1 Prismatic member of length l and area A with Young's modulus of elasticity E: The stiffness is the inverse of the flexibility.

$$k = \frac{EA}{l} \tag{6.41}$$

2 Beam element that is prismatic, of length l, and of bending stiffness EI: The sign convention for end moments on the member is for anticlockwise

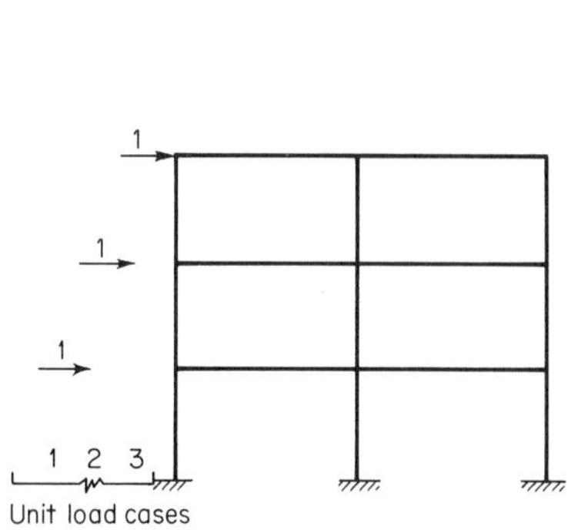

Figure 6.3 Analysis to produce flexibility and stiffness matrices.

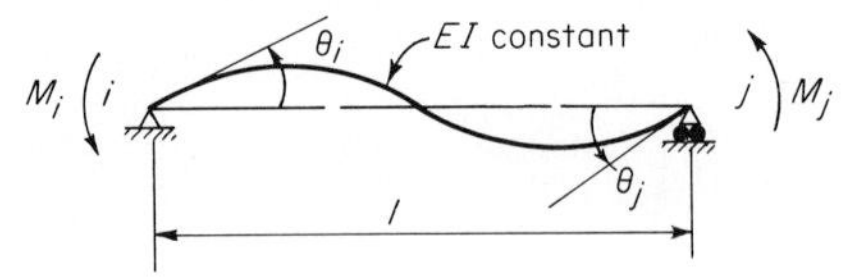

Figure 6.4 Beam element with sign convention for end rotations and moments.

moments positive. This corresponds to M_Z positive in a clockwise sense and is the convention used in the equilibrium equations in Chap. 3 and again in computer solutions in Chaps. 4 and 5. The beam with end rotations ϕ_i and ϕ_j is shown in Fig. 6.4.

In this case the stiffness matrix is obtained as the inverse of the flexibility matrix given in Eq. (4.96). That is,

$$k = \frac{2EI}{l}\begin{bmatrix} 2 & 1 \\ 1 & 2 \end{bmatrix} \tag{6.42}$$

The stiffness matrix for the member under distortions $\langle \phi_i \quad \phi_j \quad \delta \rangle$ is given in Eq. (6.43). It should be noted that when the distortions of the member are small, bending and axial stiffnesses are uncoupled. As was demonstrated for flexibilities in Sec. 4.10, this is not the case for a member in which the axial load is significant compared with the Euler load for the member.

$$\begin{bmatrix} M_i \\ M_j \\ F \end{bmatrix} = \frac{2EI}{l}\begin{bmatrix} 2 & 1 & 0 \\ 1 & 2 & 0 \\ 0 & 0 & \frac{A}{2I} \end{bmatrix}\begin{bmatrix} \phi_i \\ \phi_j \\ \delta \end{bmatrix} \tag{6.43}$$

Example 6.2 Analysis of a frame structure by the displacement method

The portal frame is shown in Fig. 6.5 and is analyzed for the 10-kip horizontal force on node 1. Axial distortions are neglected, and the inertia values for the members are shown in the figure. The analysis follows the sequence of operations:

1 $v = ar$

Ascertain the relevant node displacements, and calculate the displacement transformation matrix.

2 $S = kv = kar$

From member dimensions and properties, form the k matrix of member stiffnesses.

3 $R = a^T kar = Kr$

Generate the structure-stiffness matrix for the displacements r.

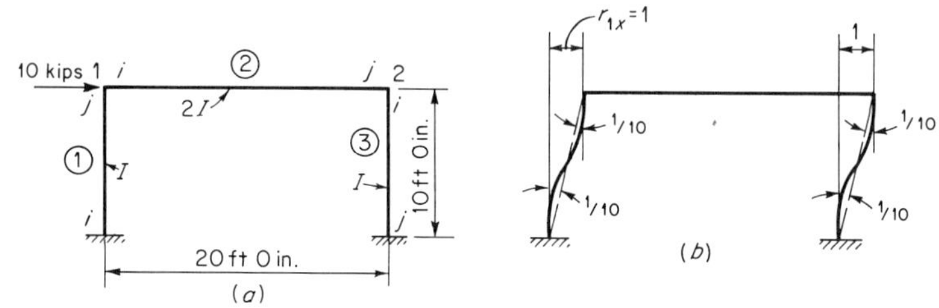

Figure 6.5 Portal-frame dimensions, loads, and sway displacement.

4 Calculate displacements

$$r = K^{-1}R$$

5 Calculate member forces

$$S = kar$$

In this example Eqs. (6.30) and (6.32) will be used to reduce the amount of hand calculation. There are three possible displacements of the two free nodes 1 and 2. These are the two rotations θ_1 and θ_2 and the displacement r_{1x}. Notice that because the members are assumed inextensible,

$$r_{2x} = r_{1x}$$

$$r_{1y} = r_{2y} = 0$$

Thence,

$$v = \begin{bmatrix} \phi_{1i} \\ \phi_{1j} \\ \phi_{2i} \\ \phi_{2j} \\ \phi_{3i} \\ \phi_{3j} \end{bmatrix} = \underbrace{\begin{bmatrix} \cdot & \cdot & 0.1 \\ 1 & \cdot & 0.1 \\ 1 & \cdot & \cdot \\ \cdot & 1 & \cdot \\ \cdot & 1 & 0.1 \\ \cdot & \cdot & 0.1 \end{bmatrix}}_{a} \begin{bmatrix} \theta_1 \\ \theta_2 \\ r_{1x} \end{bmatrix}$$

The matrix of member stiffnesses is

$$k = \frac{2EI}{10} \begin{bmatrix} 2 & 1 & \cdot & \cdot & \cdot & \cdot \\ 1 & 2 & \cdot & \cdot & \cdot & \cdot \\ \cdot & \cdot & 2 & 1 & \cdot & \cdot \\ \cdot & \cdot & 1 & 2 & \cdot & \cdot \\ \cdot & \cdot & \cdot & \cdot & 2 & 1 \\ \cdot & \cdot & \cdot & \cdot & 1 & 2 \end{bmatrix}$$

Then

$$ka = \frac{2EI}{10}\begin{bmatrix} 1 & 0 & 0.3 \\ 2 & 0 & 0.3 \\ 2 & 1 & 0 \\ 1 & 2 & 0 \\ 0 & 2 & 0.3 \\ 0 & 1 & 0.3 \end{bmatrix} \qquad (i)$$

$$K = a^T ka = \frac{2EI}{10}\begin{bmatrix} 4 & 1 & 0.3 \\ 1 & 4 & 0.3 \\ 0.3 & 0.3 & 0.12 \end{bmatrix}$$

Then,

$$\begin{bmatrix} 0 \\ 0 \\ 10 \end{bmatrix} = \frac{2EI}{10}\begin{bmatrix} 4 & 1 & 0.3 \\ 1 & 4 & 0.3 \\ 0.3 & 0.3 & 0.12 \end{bmatrix}\begin{bmatrix} \theta_1 \\ \theta_2 \\ r_{1x} \end{bmatrix} \qquad (ii)$$

In this equation note that C_{11} appears as the first submatrix. Then,

$$0 = \begin{bmatrix} 4 & 1 \\ 1 & 4 \end{bmatrix}\begin{bmatrix} \theta_1 \\ \theta_2 \end{bmatrix} + \begin{bmatrix} 0.3 \\ 0.3 \end{bmatrix} r_{1x}$$

Solve for θ_1 and θ_2 in terms of r_{1x}:

$$\begin{bmatrix} \theta_1 \\ \theta_2 \end{bmatrix} = -\frac{3}{50}\begin{bmatrix} 1 \\ 1 \end{bmatrix} r_{1x} \qquad (iii)$$

The second equation in (*ii*) is written

$$10 = \frac{2EI}{10}\left\{[0.3 \quad 0.3]\begin{bmatrix} \theta_1 \\ \theta_2 \end{bmatrix} + 0.12 r_{1x}\right\}$$

Substitute for $\langle \theta_1 \quad \theta_2 \rangle$ from Eq. (*iii*)

$$10 = \frac{2EI}{10}\left\{-[0.3 \quad 0.3]\begin{bmatrix} 0.06 \\ 0.06 \end{bmatrix} + 0.12 r_{1x}\right\}$$

$$= \frac{2EI}{10}(0.084) r_{1x}$$

$$r_{1x} = \frac{50{,}000}{84EI}$$

From Eq. (*iii*),

$$\begin{bmatrix}\theta_1\\ \theta_2\end{bmatrix} = -\frac{3}{50}\begin{bmatrix}1\\1\end{bmatrix}\frac{50{,}000}{84EI} = \frac{-1{,}000}{28EI}\begin{bmatrix}1\\1\end{bmatrix}$$

Member moments are calculated from Eq. (*i*):

$$\begin{bmatrix}M_{1i}\\ M_{1j}\\ M_{2i}\\ M_{2j}\\ M_{3i}\\ M_{3j}\end{bmatrix} = \frac{2EI}{10}\begin{bmatrix}1 & 0 & 0.3\\ 2 & 0 & 0.3\\ 2 & 1 & 0\\ 1 & 2 & 0\\ 0 & 2 & 0.3\\ 0 & 1 & 0.3\end{bmatrix}\frac{1{,}000}{EI}\begin{bmatrix}-\frac{1}{28}\\ -\frac{1}{28}\\ \frac{50}{84}\end{bmatrix} = \begin{bmatrix}28.5\\ 21.4\\ -21.4\\ -21.4\\ 21.4\\ 28.5\end{bmatrix}$$

The bending-moment diagram is drawn in Fig. 6.6.

Analysis of Structures by the Displacement Method When Loads Occur on Members between Node Points

It has been shown in Chaps. 3, 4, and 5, that in the force method, loaded members may be treated by first considering the structure to have all nodes fixed in space. For this condition fixed-end forces are calculated for each member, and when all these are applied as release forces at the nodes, the total-structure behavior is the sum of the two analyses. The actual calculation of fixed-end forces has been left to the present discussion because fixed-end moments have traditionally been associated with displacement analyses (for example, in slope deflection and moment distribution). It is noted that the joint-fixed condition is the kinematically determinate solution $r \equiv 0$. The definitions of member forces in global and local coordinates are given in Sec. 4.8. It is repeated here that it is most convenient to transform member forces into local coordinate components for the calculation of fixed-end moments and shears. When these forces are used in fully automated computer programs, it will be found necessary to transform back to global coordinates, just as in general force programs.

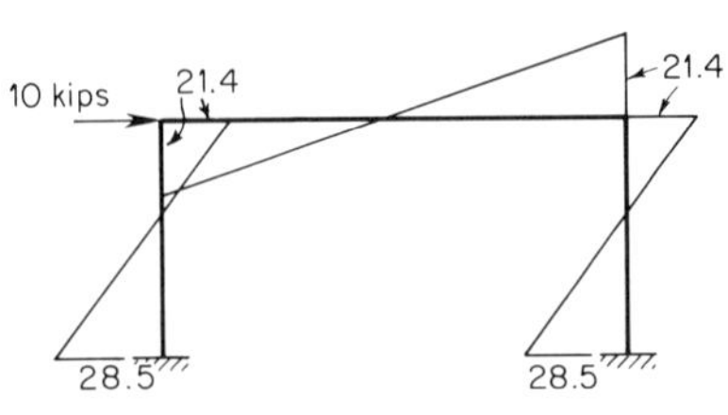

Figure 6.6 Bending-moment diagram in foot-kip units.

Example 6.3 In this example the end deflection and rotation of the cantilever beam with uniformly distributed load are calculated by the displacement method. The analysis should be compared with the same problem solved by the force method in Example 4.16.

The loaded beam is shown in Fig. 6.7*a*. The kinematically determinate situation $r \equiv 0$ is shown in Fig. 6.7*b*, and from this, release forces are determined for node 1 in Fig. 6.7*c*.

The displacement transformation for node 1 to member (1) is given by

$$\begin{bmatrix} \phi_i \\ \phi_j \end{bmatrix} = \begin{bmatrix} -\tfrac{1}{16} & 0 \\ -\tfrac{1}{16} & 1 \end{bmatrix} \begin{bmatrix} r_y \\ \theta_z \end{bmatrix}$$

Member stiffness is

$$k = \frac{2EI}{16} \begin{bmatrix} 2 & 1 \\ 1 & 2 \end{bmatrix}$$

Thence,

$$ka = \frac{2EI}{16} \begin{bmatrix} -\tfrac{3}{16} & 1 \\ -\tfrac{3}{16} & 2 \end{bmatrix}$$

$$a^T ka = \frac{2EI}{256} \begin{bmatrix} \tfrac{6}{16} & -3 \\ -3 & 32 \end{bmatrix}$$

In the equation $R = a^T k a r_1$, the forces R are given by $R^T = \langle -8 \quad 21.33 \rangle$. Solving for the displacements,

$$\begin{bmatrix} r_y \\ \theta_z \end{bmatrix} = \frac{256}{2EI} \frac{1}{3} \begin{bmatrix} 32 & 3 \\ 3 & \tfrac{3}{8} \end{bmatrix} \begin{bmatrix} -8 \\ 21.33 \end{bmatrix}$$

$$= \frac{128}{3EI} \begin{bmatrix} -192 \\ -16 \end{bmatrix}$$

Member forces are given as $S = kar$. That is,

$$\begin{bmatrix} M_i \\ M_j \end{bmatrix} = \frac{2EI}{16} \begin{bmatrix} -\tfrac{3}{16} & 1 \\ -\tfrac{3}{16} & 2 \end{bmatrix} \frac{128}{3EI} \begin{bmatrix} -192 \\ -16 \end{bmatrix} = \frac{64}{3} \begin{bmatrix} 5 \\ 1 \end{bmatrix}$$

The complete solution must include the fixed-end moments shown in Fig. 6.7*b*, so that

$$\begin{bmatrix} M_i \\ M_j \end{bmatrix}_{\text{total}} = \frac{64}{3} \begin{bmatrix} 5 \\ 1 \end{bmatrix} + \frac{64}{3} \begin{bmatrix} 1 \\ -1 \end{bmatrix} = \begin{bmatrix} 128 \\ 0 \end{bmatrix}$$

This result is to be expected since for the statically determinate cantilever, it is seen by inspection that $M_i = wl^2/2$.

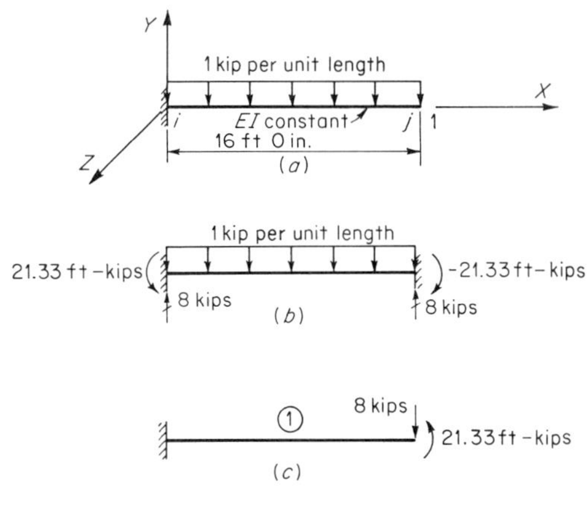

Figure 6.7 Analysis for member loads. (*a*) Cantilever beam; (*b*) fixed-joint solution; (*c*) release forces.

It is seen that in the displacement analysis, the equation $R = Kr$ must be solved even when the structure is statically determinate. Of course, if the a matrix had been checked, it would have been found to be square and nonsingular, and the solution for member forces would follow directly from $R = a^T S$.

It was found in Chaps. 3 and 5 that in order to produce automated force-method programs (or to solve joint equilibrium equations), it is necessary that axial forces be considered in all members. In an analogous way it is found that, generally, axial strains must be considered in frame members in the displacement method when automatic analysis procedures are developed (see Chap. 7). The complexity that arises in producing displacement patterns and equivalent node forces when bending distortions only are considered is illustrated in the analysis of the gable frame in Example 6.4.

Example 6.4 The gable frame is shown in Fig. 6.8*a* and is loaded on member (2) only, with a distributed load of 2 kips/ft of horizontal projection. The flexural properties EI are the same for all members. The fixed-end forces for member (2) are shown in Fig. 6.8*b*.

Independent node displacements must be calculated, and the inextensibility of the members taken into consideration. Evidently there are three rotation displacements $(\theta_{1z}, \theta_{2z}, \theta_{3z})$. There are however only two independent sidesway mechanisms since if node 1 is to have a horizontal displacement, it is necessary for node 2 also to be displaced.

The two independent displacement mechanisms $U_{1x} = 1$ and $U_{3x} = 1$ are shown in Fig. 6.9. It should be noted that any two independent patterns would suffice. For example, symmetrical and antisymmetrical patterns could be used.

From Fig. 6.9 and the rotation displacements, the displacement transformation $v = ar$ is found to be

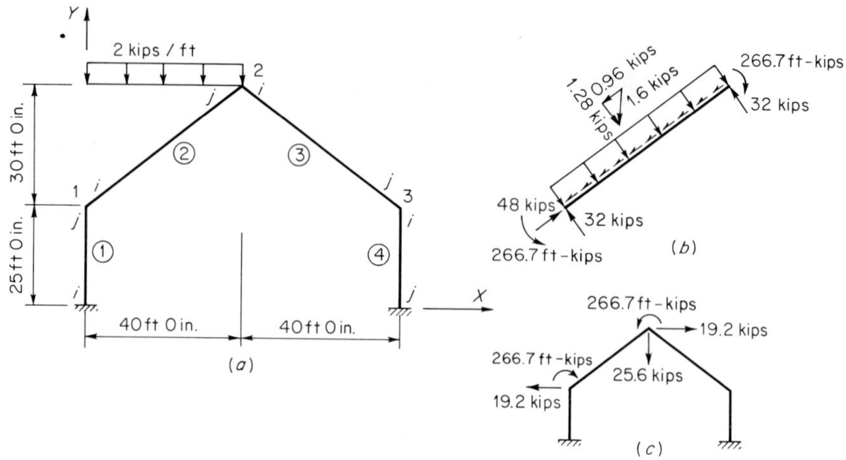

Figure 6.8 Analysis of gable frame. (*a*) Gable-frame dimensions and loading; (*b*) fixed-end forces for member (2); (*c*) release forces.

$$
\begin{bmatrix} \phi_{1i} \\ \phi_{1j} \\ \phi_{2i} \\ \phi_{2j} \\ \phi_{3i} \\ \phi_{3j} \\ \phi_{4i} \\ \phi_{4j} \end{bmatrix}
=
\begin{bmatrix}
\cdot & \cdot & \cdot & \frac{1}{25} & \cdot \\
1 & \cdot & \cdot & \frac{1}{25} & \cdot \\
1 & \cdot & \cdot & -\frac{0.833}{50} & \frac{0.833}{50} \\
\cdot & 1 & \cdot & -\frac{0.833}{50} & \frac{0.833}{50} \\
\cdot & 1 & \cdot & -\frac{0.833}{50} & -\frac{0.833}{50} \\
\cdot & \cdot & 1 & -\frac{0.833}{50} & -\frac{0.833}{50} \\
\cdot & \cdot & 1 & \cdot & \frac{1}{25} \\
\cdot & \cdot & \cdot & \cdot & \frac{1}{25}
\end{bmatrix}
\begin{bmatrix} \theta_{1z} \\ \theta_{2z} \\ \theta_{3z} \\ r_{1x} \\ r_{3x} \end{bmatrix}
\qquad (i)
$$

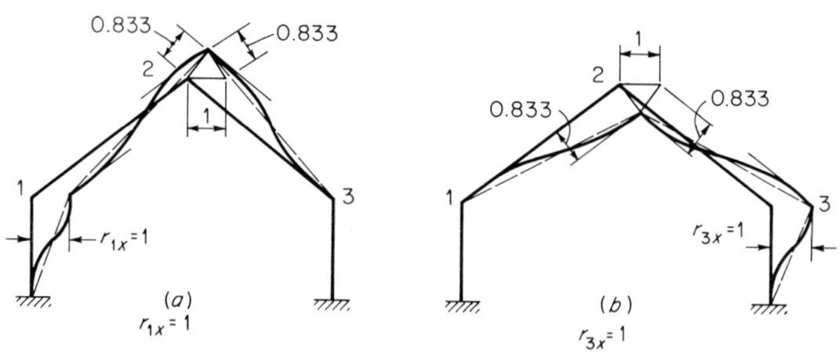

Figure 6.9 Displacement patterns.

The matrix of member stiffnesses, is the 8×8 matrix k.

$$k = \frac{2EI}{50}\begin{bmatrix} 4 & 2 & \cdot & \cdot & \cdot & \cdot & \cdot & \cdot \\ 2 & 4 & \cdot & \cdot & \cdot & \cdot & \cdot & \cdot \\ \cdot & \cdot & 2 & 1 & \cdot & \cdot & \cdot & \cdot \\ \cdot & \cdot & 1 & 2 & \cdot & \cdot & \cdot & \cdot \\ \cdot & \cdot & \cdot & \cdot & 2 & 1 & \cdot & \cdot \\ \cdot & \cdot & \cdot & \cdot & 1 & 2 & \cdot & \cdot \\ \cdot & \cdot & \cdot & \cdot & \cdot & \cdot & 4 & 2 \\ \cdot & \cdot & \cdot & \cdot & \cdot & \cdot & 2 & 4 \end{bmatrix} \qquad (ii)$$

From Eqs. (*i*) and (*ii*) the product ka is found to be

$$ka = \frac{2EI}{50}\begin{bmatrix} 2 & \cdot & \cdot & \frac{6}{25} & \cdot \\ 4 & \cdot & \cdot & \frac{6}{25} & \cdot \\ 2 & 1 & \cdot & -\frac{2.5}{50} & \frac{2.5}{50} \\ 1 & 2 & \cdot & -\frac{2.5}{50} & \frac{2.5}{50} \\ \cdot & 2 & 1 & \frac{2.5}{50} & -\frac{2.5}{50} \\ \cdot & 1 & 2 & \frac{2.5}{50} & -\frac{2.5}{50} \\ \cdot & \cdot & 4 & \cdot & \frac{6}{25} \\ \cdot & \cdot & 2 & \cdot & \frac{6}{25} \end{bmatrix} \qquad (iii)$$

Finally,

$$a^T ka = \begin{bmatrix} 6 & 1 & 0 & 0.015833 & 0.004167 \\ & 4 & 1 & 0 & 0 \\ & & 6 & 0.004167 & 0.015833 \\ & \text{symmetrical} & & 0.0001564 & -0.0000231 \\ & & & & 0.0001564 \end{bmatrix}$$

Before the equations $R = a^T kar$ may be solved, the consequence of using the displacement patterns in Fig. 6.9 must be evaluated in terms of the load vector R. The generalized force Q_1 corresponding to the displacement $r_{1x} = 1$ must be such that for a virtual displacement δr_{1x}, the work done,

$\delta r_{1x} Q_1 =$ work done by node forces

It is seen that the displacement patterns are in fact producing sidesway type of equilibrium equations. That is, in the equations $R - a^T S$, the particular rows in $a^T S$ corresponding to r_{1x} are applicable to several node forces. In this case, from Figs. 6.8*c* and 6.9,

$$Q_1 = (-19.20 \times 1 + 19.20 \times 0.5 - 25.6 \times 0.667) = -26.67 \text{ kips}$$

$$Q_2 = 0.5 \times 19.2 + 25.6 \times 0.667 = 26.67 \text{ kips}$$

Thence

$$R^T = \langle -266.7 \quad 266.7 \quad 0 \quad -26.67 \quad 26.67 \rangle.$$

Solving the equations $r = K^{-1}R$ gives deflection values, to be multiplied by $300/EI$:

$$\theta_{1z} = -934.9 \text{ rad}$$

$$\theta_{2z} = 1280.3 \text{ rad}$$

$$\theta_{3z} = -985.6 \text{ rad}$$

$$r_{1x} = 6234.6 \text{ in.}$$

$$r_{3x} = 294{,}318.0 \text{ in.}$$

The bending moments in inch-kips are (from $S = kar$)

$$\begin{bmatrix} M_{1i} \\ M_{1j} \\ M_{2i} \\ M_{2j} \\ M_{3i} \\ M_{3j} \\ M_{4i} \\ M_{4j} \end{bmatrix} = \begin{bmatrix} -1984.5 \\ -3864.3 \\ 662.8 \\ 2878.0 \\ 322.7 \\ -1943.2 \\ 1944.0 \\ 3915.2 \end{bmatrix} \qquad \begin{aligned} M_{2i,\text{total}} &= 662.8 + 3200.4 = 3863.2 \\ M_{2j,\text{total}} &= 2878.0 - 3200.4 = -322.4 \end{aligned}$$

The bending-moment diagram is plotted in Fig. 6.10.

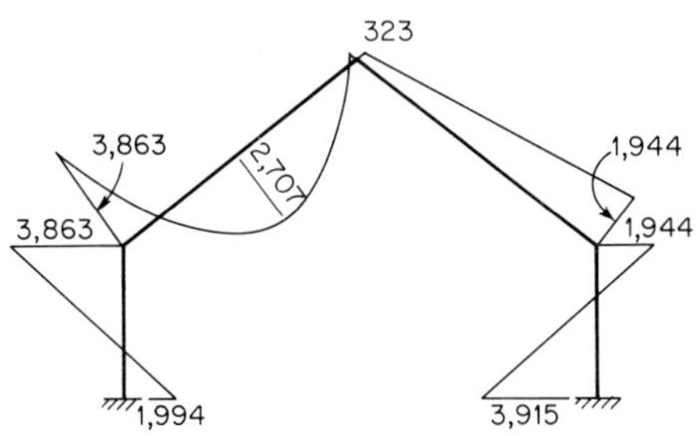

Figure 6.10 Bending-moment diagram in inch-kip units.

6.2 ELASTIC STRAIN ENERGY IN TERMS OF DISPLACEMENTS

It has been shown in Eq. (6.6) that when the flexibility matrix F for the structure with loads R and displacements r is known, the stiffness matrix $K = F^{-1}$, and the expression for strain energy in terms of displacements is the quadratic function

$$U = \tfrac{1}{2}r^T K r$$

This expression may be used to calculate the strain energy for a member as well as for the whole structure, and thus, in a manner analogous to that in Sec. 4.4, strain energy for the whole structure may be evaluated as the sum of that of the individual parts. Just as this gave a method for calculating F, it gives an expression for K in the displacement analysis.

For member i with displacements v_i and stiffness matrix k_i, the Eq. (6.6) gives the strain energy for the member as

$$U_i = \tfrac{1}{2}v_i^T k_i v_i \tag{6.44}$$

For all members of the structure, labeled $a, b, c, \ldots,$ the total strain energy is given by

$$U_{\text{total}} = U_a + U_b + U_c + \cdots = \tfrac{1}{2}\langle v_a^T \quad v_b^T \quad v_c^T \quad \cdots\rangle \begin{bmatrix} k_a & & & \mathbf{0} \\ & k_b & & \\ & & k_c & \\ \mathbf{0} & & & \ddots \end{bmatrix} \begin{bmatrix} v_a \\ v_b \\ v_c \\ \vdots \end{bmatrix} \tag{6.45}$$

However, in terms of the definitions of v and k in Eqs. (6.18)ff, this is simply the expression

$$U_{\text{total}} = \tfrac{1}{2}v^T k v \tag{6.46}$$

When, now, the displacement transformation equation (6.16) is used to express member distortions in terms of node displacements, an alternative expression to Eq. (6.6) is obtained for U_{total}. Substitute $v = ar$ in Eq. (6.46). Then,

$$U_{\text{total}} = \tfrac{1}{2}r^T a^T k a r \tag{6.47}$$

It follows that

$$K = a^T k a \tag{6.48}$$

Derivatives of Strain Energy with Respect to Displacements

When the strain energy is differentiated with respect to the displacement r_i, it is found that the value obtained is the corresponding force R_i. This result is known as *Castigliano's first theorem.*

Differentiate Eq. (6.46) with respect to r_i.

$$\frac{dU}{dr_i} = \tfrac{1}{2}\frac{dr^T}{dr_i}Kr + \tfrac{1}{2}r^TK\frac{dr}{dr_i}$$

$$= \tfrac{1}{2}[0 \quad 0 \quad \cdots \quad \underset{\substack{\uparrow \\ i\text{th column}}}{1} \quad \cdots \quad 0]\begin{bmatrix} \cdots & \cdots & \cdots & \cdots \\ k_{i1} & k_{i2} & \cdots & k_{in} \\ \cdots & \cdots & \cdots & \cdots \end{bmatrix} r + \tfrac{1}{2}r^T \begin{bmatrix} \cdots & k_{1i} & \cdots \\ \cdots & k_{2i} & \cdots \\ \cdots & \cdots & \cdots \\ \cdots & k_{ni} & \cdots \end{bmatrix}\begin{bmatrix} 0 \\ 0 \\ \vdots \\ 1 \\ \vdots \\ 0 \end{bmatrix} \leftarrow i\text{th row}$$

That is, using the symmetry property of K, in which $K_{ij} = K_{ji}$,

$$\frac{dU}{dr_i} = K_{i1}r_1 + K_{i2}r_2 + \cdots + K_{in}r_n = R_i$$

that is,

$$\frac{dU}{dr_i} = R_i \tag{6.49}$$

In Sec. 4.4 it was found that energy considerations gave rise to an expression [Eq. (4.88)], known as Betti's theorem, connecting two force groups applied to a structure and their mutual displacements. A similar expression arises when two sets of displacements are applied to a structure.

Consider two sets of node displacements r_1 and r_2 such that the stiffness relationship in partitioned form may be written

$$\begin{bmatrix} R_1 \\ R_2 \end{bmatrix} = \begin{bmatrix} K_{11} & K_{22} \\ K_{21} & K_{22} \end{bmatrix}\begin{bmatrix} r_1 \\ r_2 \end{bmatrix} \tag{6.50}$$

The displacements r_1 and r_2 will be applied to the structure in this and the reverse order, and the expressions for U_{total} calculated.

1 Apply r_1, holding $r_2 = 0$. Then forces will be developed at node 2:

$R_{21} = K_{21}r_1$

The strain energy will be

$U_1 = \tfrac{1}{2}r_1{}^TK_{11}r_1$

2 Now apply r_2, holding r_1 constant. The increase in strain energy will be

$U_2 = \tfrac{1}{2}r_2{}^TK_{22}r_2 + R_{21}{}^Tr_2$

Thence

$$U_{\text{total}} = U_1 + U_2 = \tfrac{1}{2}r_1{}^T K_{11} r_1 + \tfrac{1}{2}r_2{}^T K_{22} r_2 + R_{21}{}^T r_2 \qquad (i)$$

Now reverse the order of displacement application when the r_2 group is applied; with $r_1 = 0$, the forces generated at 1 will be R_{12}. Now,

$$U_1 = \tfrac{1}{2}r_2{}^T K_{22} r_2$$

When r_1 is applied with r_2 held constant,

$$U_2 = \tfrac{1}{2}r_1{}^T K_{11} r_1 + R_{12}{}^T r_1$$

Then,

$$U_{\text{total}} = \tfrac{1}{2}r_1{}^T K_{11} r_1 + \tfrac{1}{2}r_2{}^T K_{22} r_2 + R_{12}{}^T r_1 \qquad (ii)$$

A comparison of Eqs. (i) and (ii) leads to the conclusion that for conservation of energy,

$$R_{12}{}^T r_1 = R_{21}{}^T r_2 \qquad (6.51)$$

6.3 INFINITESIMAL SYSTEMS; EXPRESSIONS FOR STIFFNESS MATRICES AND STRAIN ENERGY AS INTEGRALS OF STRAIN FUNCTIONS

In Sec. 4.3, it was shown that for a structure or member in which the stress distribution was known, the displacements can be calculated by the equation

$$r = \int_{\text{vol}} b^T f b \, dV \, R$$

In the same way the stiffness relationship may be expressed by Eq. (6.52) when the strain-displacement transformation function a is known.

$$R = \int_{\text{vol}} a^T k a \, dV \, r \qquad (6.52)$$

Define the matrices T_1 and T_2 as in Sec. 4.3. That is,

$$T_1 = \begin{bmatrix} dy\,dz & \cdot & \cdot & \cdot & \cdot & \cdot \\ \cdot & dz\,dx & \cdot & \cdot & \cdot & \cdot \\ \cdot & \cdot & dx\,dy & \cdot & \cdot & \cdot \\ \cdot & \cdot & \cdot & dy\,dz & \cdot & \cdot \\ \cdot & \cdot & \cdot & \cdot & dz\,dx & \cdot \\ \cdot & \cdot & \cdot & \cdot & \cdot & dx\,dy \end{bmatrix}$$

$$T_2 = \begin{bmatrix} dx & \cdot & \cdot & \cdot & \cdot & \cdot \\ \cdot & dy & \cdot & \cdot & \cdot & \cdot \\ \cdot & \cdot & dz & \cdot & \cdot & \cdot \\ \cdot & \cdot & \cdot & dx & \cdot & \cdot \\ \cdot & \cdot & \cdot & \cdot & dy & \cdot \\ \cdot & \cdot & \cdot & \cdot & \cdot & dz \end{bmatrix}$$

For the elemental cube dx, dy, dz, Hooke's law in terms of strains can be expressed

$$\sigma = f^{-1}\varepsilon = k\varepsilon \tag{6.53}$$

The matrix f is given in Eq. (4.30), and on inversion, k is found to be

$$k = 2G \begin{bmatrix} \rho + 1 & \rho & \rho & 0 & 0 & 0 \\ \rho & \rho + 1 & \rho & 0 & 0 & 0 \\ \rho & \rho & \rho + 1 & 0 & 0 & 0 \\ 0 & 0 & 0 & \frac{1}{2} & 0 & 0 \\ 0 & 0 & 0 & 0 & \frac{1}{2} & 0 \\ 0 & 0 & 0 & 0 & 0 & \frac{1}{2} \end{bmatrix} \tag{6.54}$$

In Eq. (6.54),

$$\rho = \frac{\nu E}{2G(1 + \nu)(1 - 2\nu)} \tag{6.55}$$

It should be noted that for plane stress and strain, Eq. (6.54) reduces to

$$k = E_1 \begin{bmatrix} \Gamma & \nu & 0 \\ \nu & \Gamma & 0 \\ 0 & 0 & \lambda \end{bmatrix} \tag{6.56}$$

For plane stress:

$$\Gamma = 1 \qquad \lambda = \frac{1 - \nu}{2} \qquad E_1 = \frac{E}{1 - \nu^2}$$

For plane strain:

$$\Gamma = 1 - \nu \qquad \lambda = 0.5 - \nu \qquad E_1 = \frac{E}{(1 + \nu)(1 - 2\nu)}$$

For the elementary cube, the stress and strain quantities have been defined in Eqs. (4.27) and (4.28). That is,

$$\sigma^T = \langle \sigma_x \quad \sigma_y \quad \sigma_z \quad \tau_{xy} \quad \tau_{yz} \quad \tau_{zx} \rangle$$

$$\varepsilon^T = \langle \varepsilon_x \quad \varepsilon_y \quad \varepsilon_z \quad \gamma_{xy} \quad \gamma_{yz} \quad \gamma_{zx} \rangle$$

The stress resultants on the element are

$$S = T_1 \sigma$$

and the deformations of the elementary cube are

$$v = T_2 \varepsilon$$

Then,

$$v = T_2 f T_1{}^{-1} S \tag{6.57}$$

Solve Eq. (6.57) for stress resultants in terms of deformations.

$$S = (T_2 f T_1{}^{-1})^{-1} v = T_1 k T_2{}^{-1} v \tag{6.58}$$

This is the stiffness relationship between S and v and is written

$$S = \bar{k} v \tag{6.59}$$

In the present case the strains at the point x, y, z are assumed to be related to the node displacements r by the transformation matrix a_{xyz}. That is,

$$\varepsilon = a_{xyz} r \tag{6.60}$$

The deformations of the element are given by

$$v = T_2 a_{xyz} r \tag{6.61}$$

Write Eq. (6.61) as

$$v = \bar{a}_{xyz} r \tag{6.62}$$

The contribution to the node forces R from the stress resultant S will be designated δR, and from the contragradient law these forces are known to be

$$\delta R = \bar{a}_{xyz}^T S \tag{6.63}$$

Combining Eqs. (6.59), (6.62), and (6.63),

$$\delta R = \bar{a}_{xyz}^T \bar{k} \bar{a}_{xyz} r \tag{6.64}$$

When $\bar{a}_{xyz}$ and $\bar{k}$ are expressed in terms of the stress-strain relationship and the T_1 and T_2 matrices,

$$\delta R = a_{xyz}^T T_2{}^T T_1 k T_2{}^{-1} T_2 a_{xyz} r \tag{6.65}$$

[in Sec. 4.2, Eq. (4.45)]. As before,

$T_2{}^T T_1 = I\, dV$, so that

$$\delta R = a_{xyz}^T k a_{xyz}\, dv\, r \tag{6.66}$$

When contributions from all such elements dV are summed over the structure, it is seen that the integral expression is obtained

$$R = \int_{\text{vol}} a_{xyz}^T k a_{xyz} \, dV \, r \tag{6.67}$$

Equation (6.67) is found to be useful in two- or three-dimensional elements for which displacement (and hence strain) patterns are specified. These displacement patterns form a method for the development of finite-element models.

Example 6.5 In Fig. 6.11, the cantilever beam has a displacement r_x of its free end that produces the strain field

$$\varepsilon_x = \frac{r_x}{l}$$

It is required to calculate the stiffness relationship between R_x and r_x. From Eq. (6.67),

$$R_x = \int_{\text{vol}} \frac{1}{l} E \frac{1}{l} \, dV \, r_x$$

$$= \frac{EA}{l^2} \int_0^l dx \, r_x = \frac{EA}{l} r_x$$

6.4 INITIAL STRAINS IN MEMBERS; REMOVAL OF A MEMBER; MODIFICATION OF A MEMBER STIFFNESS

Initial strains may arise from a variety of causes, such as loads between nodes of the member, temperature change, lack of fit due to manufacture (either by chance or by intention), and nonlinear plastic strains. The last type of course will involve more detailed consideration at a later stage. All these strains may be included in the analysis by starting from the kinematically determinate state $r \equiv 0$. Since $v = ar$, this implies that all member distortions must be eliminated by applying fixed-end forces. The structure analysis (as distinct from the member analysis) consists in determining the effects of the release forces in the structure. For any member, the total forces will be the sum of the fixed-end and release forces, that is, the forces in the state $r \equiv 0$ together with those producing the deflections r.

Suppose then that member i with stiffness k_i has a set of initial strains θ_i. For compatibility in the condition of all joints fixed, the displacements $-\theta_i$ must be applied to the member and these will require forces

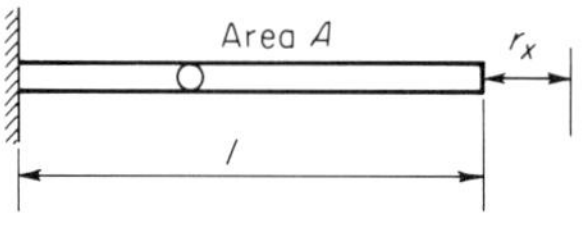

Figure 6.11 Cantilever-beam unit displacement.

$$J_i = -k_i\theta_i \tag{6.68}$$

To produce J_i, it will be necessary to apply joint forces

$$R_i^F = a_i^T J_i \tag{6.69}$$

The release forces are

$$R_i = -R_i^F = a_i^T k_i \theta_i \tag{6.70}$$

If R is the vector of externally applied node loads, the structure is analyzed for the loads,

$$R^* = R + R_i = R + a_i^T k_i \theta_i \tag{6.71}$$

The member forces S^* calculated from R^* will be

$$\begin{aligned} S^* &= kar^* = kaK^{-1}R^* \\ S^* &= kaK^{-1}(R + R_i) \end{aligned} \tag{6.72}$$

For the member i,

$$S_i = S_i^* + J_i = S_i^* - k_i\theta_i \tag{6.73}$$

Hence,

$$S_i = k_i a_i K^{-1}(R + a_i^T k_i \theta_i) - k_i\theta_i$$

Finally,

$$S_i = k_i a_i K^{-1}R + k_i(a_i K^{-1} a_i^T k_i - I)\theta_i \tag{6.74}$$

The above theory may be used to calculate the effects of the removal of a member on the rest of the structure, since by suitable choice of initial strains, the forces in the member may be set to zero. Evidently if the member has zero forces, it is not participating in the action of the structure and may be removed.

For the whole structure with loads R, the force in member i will be

$$S_i = k_i a_i r$$

Apply initial strains θ_i to the member i so that the new force in i, $\overline{S}_i$, is equal to zero. That is,

$$\overline{S}_i = 0 = S_i^* + J_i$$

That is,

$$k_i a_i r + (-k_i + k_i a_i K^{-1} a_i^T k_i)\theta_i = 0$$

Therefore,

$$\theta_i = -(-k_i + k_i a_i K^{-1} a_i^T k_i)^{-1} k_i a_i r \tag{6.75}$$

The joint forces applied are

$$R^* = R + R_i = R + a_i^T k_i \theta_i$$

The total node deflections with the member removed, will be

$$r_{\text{total}} = K^{-1}(R + a_i{}^T k_i \theta_i) \tag{6.76}$$

Substitution for θ_i from (6.75), gives

$$r_{\text{total}} = K^{-1}[R - a_i{}^T k_i(-k_i + k_i a_i K^{-1} a_i{}^T k_i)^{-1} k_i a_i r] \tag{6.77}$$

Since $r = K^{-1}R$ gives the deflections before modification,

$$r_{\text{total}} = r + \bar{r} = r - K^{-1} a_i{}^T k_i(-k_i + k_i a_i K^{-1} a_i{}^T k_i)^{-1} k_i a_i r$$

or

$$\bar{r} = -K^{-1} a_i{}^T k_i(-k_i + k_i a_i K^{-1} a_i{}^T k_i)^{-1} k_i a_i K^{-1} R \tag{6.78}$$

From this it is seen that

$$r_{\text{total}} = K^{-1}[I - a_i{}^T k_i(-k_i + k_i a_i K^{-1} a_i{}^T k_i)^{-1} k_i a_i K^{-1}]R \tag{6.79}$$

That is, the modified flexibility matrix for the structure is

$$F_{\text{mod}} = K^{-1}[I - a_i{}^T(-k_i{}^{-1} + a_i K^{-1} a_i{}^T)^{-1} a_i K^{-1}] \tag{6.80}$$

Member forces are calculated

$$S_{\text{total}} = k a r_{\text{total}} = ka(r + \bar{r}) \tag{6.81}$$

For member i,

$$S_{\text{total},i} = S + \bar{S} - J_i \theta_i = 0$$

Example 6.6 The pin-jointed truss member shown in Fig. 6.12 has an initial extension of a units. Calculate the fixed-end forces to restore the member to its original shape.

The stiffness relationship between end displacements $\langle r'_{ix} \quad r'_{jx} \rangle$ in local coordinates and the forces $\langle F'_{ix} \quad F'_{jx} \rangle$ is easily shown to be

$$\begin{bmatrix} F'_{ix} \\ F'_{jx} \end{bmatrix} = \frac{EA}{l} \begin{bmatrix} 1 & -1 \\ -1 & 1 \end{bmatrix} \begin{bmatrix} r'_{ix} \\ r'_{jx} \end{bmatrix}$$

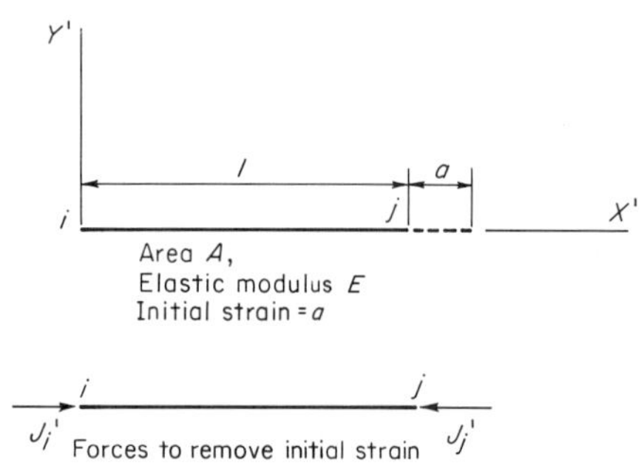

Figure 6.12 Truss member with initial strain.

For initial strains, let $r'_{ix} = 0$, and then $r'_{jx} = a$. Applying Eq. (6.68),

$$\begin{bmatrix} J'_i \\ J'_j \end{bmatrix} = \frac{EA}{l}\begin{bmatrix} 1 & -1 \\ -1 & 1 \end{bmatrix}\begin{bmatrix} 0 \\ a \end{bmatrix} = \frac{EA}{l}\begin{bmatrix} a \\ -a \end{bmatrix}$$

Example 6.7 The rigid frame shown in Fig. 6.13a has a uniform load of ½ kip/ft on the beam member, which also suffers a temperature variation of 100°F on the outside and 50°F on the inside.

The frame is composed of a steel I section, 24 × 9 in., with $I = 2{,}364$ in.^{4} and $A = 24.71$ sq. in. The coefficient of thermal expansion is taken to be 0.625×10^{-5} in./(in.)(°F), and Young's modulus, $E = 30 \times 10^3$ in.-kip units.

The end rotations due to the uniform load will be a measure of the lack of fit introduced from this source. The temperature variation will also cause end rotations and, in addition, an axial extension.

1 *Rotations due to uniform load*

$$\begin{bmatrix} \phi_{2i} \\ \phi_{2j} \end{bmatrix} = \frac{wl^3}{24EI}\begin{bmatrix} -1 \\ 1 \end{bmatrix} = \frac{60^3}{96EI}\begin{bmatrix} -1 \\ 1 \end{bmatrix} = 0.009137\begin{bmatrix} -1 \\ 1 \end{bmatrix}$$

2 *Rotations due to temperature strains* From Eq. (4.137), noting the sign change in rotation ϕ_i due to the different sign convention,

$$\begin{bmatrix} \phi_i \\ \phi_j \end{bmatrix}_t = \frac{\alpha l}{2(c_1 - c_2)}\begin{bmatrix} 1 & -1 \\ -1 & 1 \end{bmatrix}\begin{bmatrix} t_1 \\ t_2 \end{bmatrix}$$

Then,

$$\begin{bmatrix} \phi_{2i} \\ \phi_{2j} \end{bmatrix}_t = \frac{0.625 \times 10^{-5} \times 60 \times 12}{2 \times 24}\begin{bmatrix} 1 & -1 \\ -1 & 1 \end{bmatrix}\begin{bmatrix} 100 \\ 50 \end{bmatrix}$$

$$= \frac{9.375 \times 10^{-3}}{2}\begin{bmatrix} 1 \\ -1 \end{bmatrix}$$

Axial extension is given by

$$\Delta l_{2t} = 0.625 \times 10^{-5} \times 60 \times 12 \times 75$$

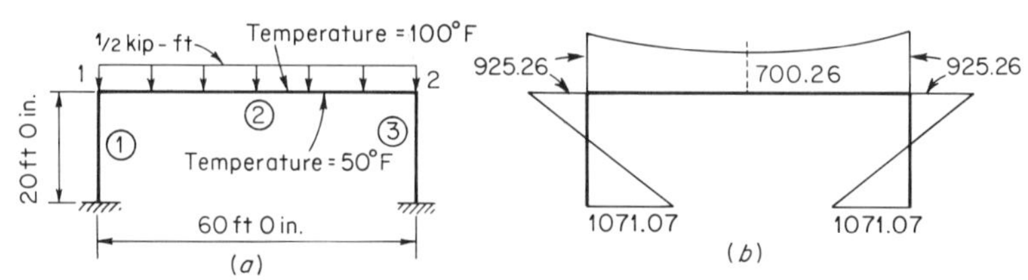

Figure 6.13 Frame analysis for distributed load and temperature strain.

The displacement transformation equation

$$v = ar$$

is here given by

$$\begin{bmatrix} \phi_{1i} \\ \phi_{1j} \\ \phi_{2i} \\ \phi_{2j} \\ \delta_2 \\ \phi_{3i} \\ \phi_{3j} \end{bmatrix} = \begin{bmatrix} 0 & 0 & 1/240 & 0 \\ 1 & 0 & 1/240 & 0 \\ 1 & 0 & 0 & 0 \\ 0 & 1 & 0 & 0 \\ 0 & 0 & -1 & 1 \\ 0 & 1 & 0 & 1/240 \\ 0 & 0 & 0 & 1/240 \end{bmatrix} \begin{bmatrix} \theta_1 \\ \theta_2 \\ U_1 \\ U_2 \end{bmatrix}$$

The member stiffnesses are

$$k_1 = k_3 = \frac{2 \times 30 \times 10^3 \times 2{,}364}{20 \times 12} \begin{bmatrix} 2 & 1 \\ 1 & 2 \end{bmatrix} = 591 \times 10^3 \begin{bmatrix} 2 & 1 \\ 1 & 2 \end{bmatrix}$$

$$k_2 = \frac{2 \times 30 \times 10^3 \times 2{,}364}{60 \times 12} \begin{bmatrix} 2 & 1 & 0 \\ 1 & 2 & 0 \\ 0 & 0 & \dfrac{24.71}{2 \times 2{,}364} \end{bmatrix} = 197 \times 10^3 \begin{bmatrix} 2 & 1 & 0 \\ 1 & 2 & 0 \\ 0 & 0 & 0.005226 \end{bmatrix}$$

The SMIS coding for the above matrices is given below. The analysis produces the bending-moment diagram shown in Fig. 6.13*b*.

```
START
 DISPLACEMENT ANALYSIS INCLUDING TEMPERATURE STRAINS

LOAD A 7 4
     7 ROWS,      4 COLUMNS

PRINT A   * DISPLACEMENT TRANSFORMATION MATRIX

                    1                 2                 3                 4
       1        0.00000           0.00000           0.00417           0.00000
       2        1.00000           0.00000           0.00417           0.00000
       3        1.00000           0.00000           0.00000           0.00000
       4        0.00000           1.00000           0.00000           0.00000
       5        0.00000           0.00000          -1.00000           1.00000
       6        0.00000           1.00000           0.00000           0.00417
       7        0.00000           0.00000           0.00000           0.00417

ZERO KM 7 7
     7 ROWS,      7 COLUMNS

LOAD KI 2 2
     2 ROWS,      2 COLUMNS

PRINT KI   * MEMBER 1 AND 3 STIFFNESS
```

```
              1                2
    1         2.00000          1.00000
    2         1.00000          2.00000

SCALE KI  591000.
 SCALAR=  0.5910000E+06

ADDSM KI KM 1 1

ADDSM KI KM 6 6

LOAD KJ 3 3
    3 ROWS,      3 COLUMNS

PRINT KJ  * MEMBER 5 STIFFNESS

              1                2                3
    1         2.00000          1.00000          0.00000
    2         1.00000          2.00000          0.00000
    3         0.00000          0.00000          0.00523

SCALE KJ  197000.
 SCALAR=  0.1970000E+06

ADDSM KJ KM 3 3

MULT KM A AA

TRANS A AT

MULT AT AA K

INVERT K
  CALCULATE FIXED END EFFECTS

LOAD THETA 7 1
    7 ROWS,      1 COLUMNS

PRINT THETA  * MEMBER DISTORTIONS FROM ROTA

              1
    1         0.00000
    2         0.00000
    3        -0.00445
    4         0.00445
    5         0.33750
    6         0.00000
    7         0.00000

MULT KM THETA TEMP

DUPL TEMP FIXED

SCALE FIXED -1.0
 SCALAR= -0.1000000E+01

MULT AT TEMP FORCE

MULT K FORCE DISP

PRINT  DISP  *  NODE DISPLACEMENTS

              1
    1         0.00025
    2        -0.00025
    3        -0.16471
    4         0.16471
 CALCULATE MEMBER FORCES

MULT AA DISP S

ADD FIXED S
```

```
PRINT S  * MEMBER FORCES TOTALS

                 1
  1       -1071.07416
  2        -925.25628
  3         925.25591
  4        -925.25605
  5          -8.31871
  6         925.25567
  7        1071.07368

ALL CARDS PROCESSED  END OF JOB
```

Modified Member Stiffness

The removal of a member is simply a special case in which the member stiffness is altered by an amount Δk_i. The variation of K, the structure stiffness, will be designated ΔK, and

$$K + \Delta K = K + a_i^T \, \Delta k_i \, a_i \tag{6.82}$$

The Householder algorithm for the modified inverse of a matrix is

$$(B - USV^T)^{-1} = B^{-1} + B^{-1}U(S^{-1} - V^TB^{-1}U)^{-1}V^TB^{-1} \tag{6.83}$$

In Eq. (6.83), B is an $n \times n$ matrix, S is an $r \times r$ matrix, and U and V are $n \times r$ matrices. In the present case,

$$S = -\Delta k_i$$

$$U = a_i^T \qquad \text{or} \qquad U^T = a_i$$

$$V^T = a_i \qquad \text{and} \qquad V = a_i^T$$

Hence,

$$(K - a_i^T[-\Delta k_i \, a_i])^{-1} = K^{-1} + K^{-1}a_i^T[(-\Delta k_i)^{-1} - a_iK^{-1}a_i^T]^{-1}a_iK^{-1} \tag{6.84}$$

Notice that this expression is identical with Eq. (6.80) if $\Delta k_i = -k_i$.

Example 6.8 The portal frame shown in Fig. 6.5 will be modified so that member (3) has zero bending stiffness. That is,

$$\Delta k_3 = -k_3$$

Notice that member (3) is still assumed to have infinite axial stiffness so that the modified structure will be equivalent to that shown in Fig. 6.14.

The matrices K, a_i, and k_i are given in Example 6.2 and are listed for reference below. The constant term EI is omitted from the calculations.

$$K = \begin{bmatrix} 0.4 & 0.1 & 0.03 \\ 0.1 & 0.4 & 0.03 \\ 0.03 & 0.03 & 0.012 \end{bmatrix}$$

$$a_3 = \begin{bmatrix} 0 & 1 & 0.1 \\ 0 & 0 & 0.1 \end{bmatrix}$$

$$-\Delta k_3{}^{-1} = k_3{}^{-1} = \frac{10}{6}\begin{bmatrix} 2 & -1 \\ -1 & 2 \end{bmatrix}$$

The calculations in Eq. (6.84) have been coded in SMIS and the processing of these with the results are listed below.

```
START
   MODIFICATION OF MEMBER STIFFNESS

LOAD K 3 3
     3 ROWS,        3 COLUMNS

PRINT K   * STIFFNESS MATRIX

                  1                  2                  3
     1         0.40000            0.10000            0.03000
     2         0.10000            0.40000            0.03000
     3         0.03000            0.03000            0.01200

SCALE K 2.0
 SCALAR=   0.2000000E+01

INVERT K

PRINT K   * FLEXIBILITY MATRIX UNALTERED

                  1                  2                  3
     1         1.54762           -0.11905           -3.57143
     2        -0.11905            1.54762           -3.57143
     3        -3.57143           -3.57143           59.52381

LOAD AI 2 3
     2 ROWS,        3 COLUMNS

PRINT AI    * DISPLACEMENT TRANSFORMATION MATRIX

                  1                  2                  3
     1         0.00000            1.00000            0.10000
     2         0.00000            0.00000            0.10000

LOAD DKI 2 2
     2 ROWS,        2 COLUMNS

PRINT DKI   * MEMBER FLEXIBILITY MATRIX

                  1                  2
     1         2.00000           -1.00000
     2        -1.00000            2.00000

SCALE DKI 1.666667
 SCALAR=   0.1666667E+01

MULT AI K AA

TRANS AI AIT

MULT AA AIT AB

SUB AB DKI

INVERT DKI
```

```
MULT DKI AA AC

MULT AIT AC AD

MULT K AD AE

ADD AE K

PRINT K * MODIFIED FLEXIBILITY MATRIX

                 1            2            3
     1      2.50000     -1.25000    -12.49999
     2     -1.25000      3.12500      6.24999
     3    -12.49999      6.24999    145.83326
```

6.5 MEMBER–STIFFNESS MATRICES AND FIXED–END FORCES

So far only two types of member stiffnesses have been used, namely, axial stiffness and bending stiffness of prismatic members. For the displacement method to be generally applicable, a variety of member-stiffness matrices must be developed. In addition, means of calculating fixed-end forces must now be introduced since these exist in the kinematically determinate solution $r \equiv 0$.

Member-stiffness Matrices

1 Prismatic Member in Torsion

a Circular cross section: $k = GI_P/l$

b Rectangular cross section:

$$k = k_1 \frac{G(2a)^3 b}{l} \qquad b > a$$

See Sec. 4.5 for values of k_1.

c For the general prismatic member of closed cross section, the St. Venant's torsion constant may be calculated by finite-element analysis (see Chap. 8) from the stress function ϕ, since

$$M_{\text{torque}} = 2 \times \text{volume under } \phi \text{ surface}$$

2 Prismatic Members with Inclusion of Shear Distortions in the Bending-stiffness Matrix

It has been shown in Eq. (4.183) that the flexibility matrix for the prismatic member with end moments (M_i, M_j) and including shear distortions is

$$f = \frac{l}{6EI}\begin{bmatrix} 2+\alpha & 1-\alpha \\ 1-\alpha & 2+\alpha \end{bmatrix} \qquad \alpha = \frac{6EI}{GA_{\text{eff}}l^2}$$

In this expression the sign used for moments is that tension on one side is positive. When the convention that anticlockwise moment on the member

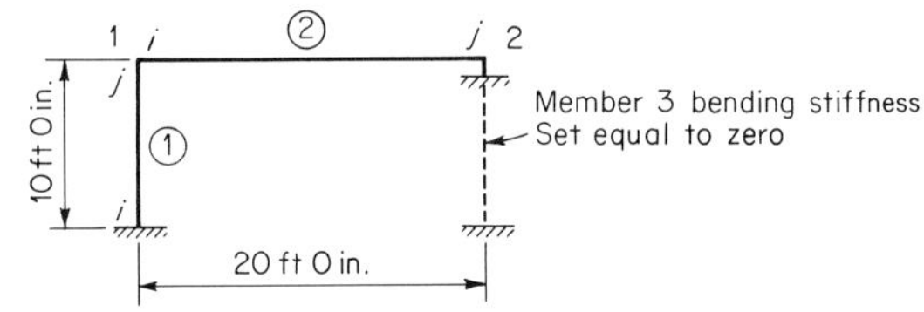

Figure 6.14 Member modification.

ends is positive is used, the signs of the off-diagonal terms must be changed. To obtain the stiffness matrix, change the signs of these terms and invert. Thence,

$$k = \frac{2EI}{l(1+2\alpha)}\begin{bmatrix} 2+\alpha & 1-\alpha \\ 1-\alpha & 2+\alpha \end{bmatrix}$$

3 Bending Stiffness of Members with Infinite Inertias in the End Sections

The flexibility matrix for this member type was developed in Eq. (4.176). The stiffness matrix could be obtained by inversion of this expression. However, it is an interesting and useful kinematic analysis to set up k directly from Eq. (6.24). Let $\langle \phi_i \quad \phi_j \rangle$ be the rotations of the tangents relative to the chord over the length l. It should be noted that the end rotation $\theta_i = 1$ shown in Fig. 6.15 is possible by virtue of the flexibility of the center section. From the Fig. 6.15 it is seen that

$$\begin{bmatrix} \phi_i \\ \phi_j \end{bmatrix} = \begin{bmatrix} 1+\dfrac{A}{l} & \dfrac{B}{l} \\ \dfrac{A}{l} & 1+\dfrac{B}{l} \end{bmatrix}\begin{bmatrix} \theta_i \\ \theta_j \end{bmatrix}$$

The stiffness matrix for portion l of the member is of the form

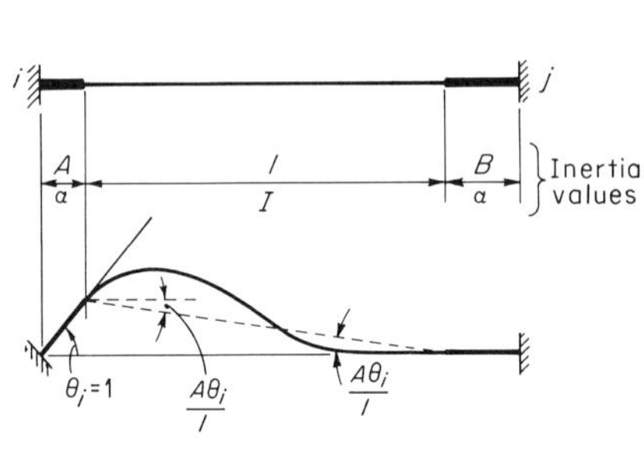

Figure 6.15 Kinematic analysis of beam with infinite end inertias.

$$k = \begin{bmatrix} k_{11} & k_{12} \\ k_{21} & k_{22} \end{bmatrix}$$

For the portion l, which is prismatic,

$$k = \frac{2EI}{l}\begin{bmatrix} 2 & 1 \\ 1 & 2 \end{bmatrix}$$

Then, for $\langle \theta_i \quad \theta_j \rangle$ the stiffness matrix is given by

$$K = a^T k a = \begin{bmatrix} 1 + \dfrac{A}{l} & \dfrac{A}{l} \\ \dfrac{B}{l} & a + \dfrac{B}{l} \end{bmatrix} \begin{bmatrix} k_{11} & k_{12} \\ k_{21} & k_{22} \end{bmatrix} \begin{bmatrix} 1 + \dfrac{A}{l} & \dfrac{B}{l} \\ \dfrac{A}{l} & 1 + \dfrac{B}{l} \end{bmatrix} \tag{6.85}$$

By multiplication of matrices in Eq. (6.85), explicit values can be obtained for the terms of K. However, in computer programming it is probably more convenient to use Eq. (6.85).

4 Bending Stiffness of Members with an Intermediate Hinge (Moment Release)

In this case it is assumed that the stiffness matrix k for end rotations $\langle \phi_i \quad \phi_j \rangle$ and moments $\langle M_i \quad M_j \rangle$ of the member is known. It is required to ascertain the effect on k of a hinge placed in the span as shown in Fig. 6.16*a*. In addition the modification to fixed-end forces must be calculated.

The usefulness of this type of transformation is in that for nonprismatic members no special provisions need be made in the description of member properties. That is, the member with the hinge is treated as a normal member and then modified.

The member may be assumed to be nonprismatic and the stiffness matrix k for the member calculated as in Eqs. (6.94) and (6.95). The stiffness matrix will have the form

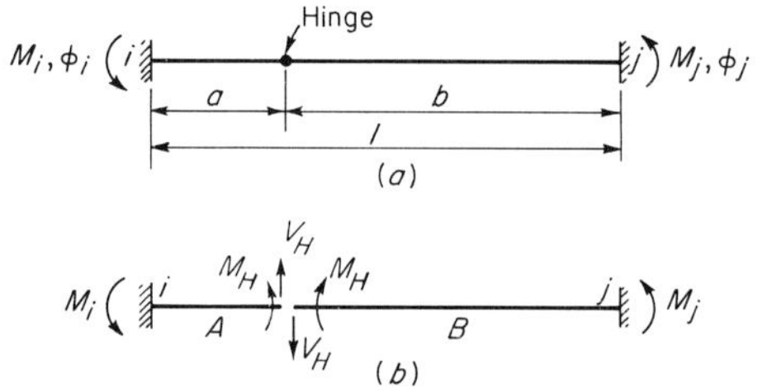

Figure 6.16 Member with hinge.

$$k = \begin{bmatrix} k_{ii} & k_{ij} \\ k_{ji} & k_{jj} \end{bmatrix} \qquad \text{and} \qquad k = \frac{2EI}{l}\begin{bmatrix} 2 & 1 \\ 1 & 2 \end{bmatrix} \qquad \text{for prismatic members}$$

The choice of displacements $\langle \phi_i \quad \phi_j \rangle$ and moments $\langle M_i \quad M_j \rangle$ to express the stiffness relationship of the member is an arbitrary one. Any two independent quantities may be used. In this case it will be convenient to use $\langle M_H \quad V_H \rangle$, the moment and shear at the release point. As yet, of course, the release has not been introduced. The transformation from end moments to the forces at the hinge is obtained by considering the equilibrium of forces on the portions A and B of the beam shown in Fig. 6.16*b*. Then,

$$\begin{bmatrix} M_i \\ M_j \end{bmatrix} = \begin{bmatrix} -1 & -a \\ 1 & -b \end{bmatrix} \begin{bmatrix} M_H \\ V_H \end{bmatrix} \qquad (i)$$

This equation will be written

$$S_{ij} = TS_H \qquad (ii)$$

Solving for S_H gives

$$S_H = T^{-1}S_{ij} \qquad (iii)$$

The stiffness relationship for end forces and displacements will be

$$S_{ij} = k\phi_{ij} \qquad (iv)$$

From the contragradient law, the expression for displacements at the releases in terms of $\langle \phi_i \quad \phi_j \rangle$ will be

$$\delta_H = T^T\phi_{ij} \qquad (v)$$

Combining Eqs. (*iii*), (*iv*), and the inverse of (*v*), the stiffness expressed in terms of δ_H and S_H is given by

$$k_H = T^{-1}k(T^{-1})^T \qquad (6.86)$$

The reverse transformation also holds that

$$k = Tk_HT^T \qquad (6.87)$$

Let k_H be of the form

$$k_H = \begin{bmatrix} h_{ii} & h_{ij} \\ h_{ji} & h_{jj} \end{bmatrix}$$

At the hinge, $M_H = 0$; hence,

$$\begin{bmatrix} 0 \\ V_H \end{bmatrix} = \begin{bmatrix} h_{ii} & h_{ij} \\ h_{ji} & h_{jj} \end{bmatrix} \begin{bmatrix} \theta_H \\ v_H \end{bmatrix} \qquad (vi)$$

From Eq. (*vi*),

$$\theta_H = -\frac{h_{ij}}{h_{ii}} v_H$$

and hence

$$V_H = \left(h_{jj} - \frac{h_{ij}h_{ji}}{h_{ii}}\right) v_H = h'_{jj} v_H \tag{6.88}$$

The reduced stiffness matrix with the hinge introduced is

$$k'_H = \begin{bmatrix} 0 & 0 \\ 0 & h'_{jj} \end{bmatrix} \tag{6.89}$$

then

$$k' = T k'_H T^T \tag{6.90}$$

Example 6.9 Suppose that for a prismatic member, the hinge is at the midpoint of the member so that $a = b = l/2$. Then

$$T = \begin{bmatrix} -1 & -\frac{l}{2} \\ 1 & -\frac{l}{2} \end{bmatrix}$$

$$T^{-1} = \frac{1}{l}\begin{bmatrix} -\frac{l}{2} & \frac{l}{2} \\ -1 & -1 \end{bmatrix}$$

From Eq. (6.86),

$$k_H = T^{-1}k(T^{-1})^T = \frac{2EI}{l^3}\begin{bmatrix} -\frac{l}{2} & \frac{l}{2} \\ -1 & -1 \end{bmatrix}\begin{bmatrix} 2 & 1 \\ 1 & 2 \end{bmatrix}\begin{bmatrix} -\frac{l}{2} & -1 \\ \frac{l}{2} & -1 \end{bmatrix} = \frac{2EI}{l^2}\begin{bmatrix} \frac{l^2}{2} & 0 \\ 0 & 6 \end{bmatrix}$$

The reduced stiffness is simply

$$k'_H = \frac{2EI}{l^3}\begin{bmatrix} 0 & 0 \\ 0 & 6 \end{bmatrix}$$

For the end forces and rotations

$$k' = T k'_H T^T = \frac{3EI}{l}\begin{bmatrix} 1 & 1 \\ 1 & 1 \end{bmatrix}$$

Notice that now k' is singular. This implies that a linear relationship exists between M_i and M_j. This must be the case since the bending-moment diagram must always pass through the hinge. This condition was also examined in Chap. 4, where a flexibility matrix was established in which only one of the end moments M_i or M_j was used.

The Effect of the Hinge on Fixed-end Moments Suppose that the fixed-end moments $\langle M_{iF} \quad M_{jF} \rangle$ have been calculated for the member without the hinge. The moment and shear at the hinge point will be a combination of these together with the simply supported values $\langle M_o \quad V_o \rangle$. Then,

$$\begin{bmatrix} M_H \\ V_H \end{bmatrix} = T^{-1} \begin{bmatrix} M_{iF} \\ M_{jF} \end{bmatrix} + \begin{bmatrix} M_o \\ V_o \end{bmatrix}$$

To introduce the hinge, apply $-M_H$ to the member, allowing distortions θ_H, but maintaining $v_H = 0$. Then,

$$\begin{bmatrix} -M_H \\ V_H \end{bmatrix} = \begin{bmatrix} h_{ii} & h_{ij} \\ h_{ji} & h_{jj} \end{bmatrix} \begin{bmatrix} \theta_H \\ 0 \end{bmatrix}$$

Hence

$$\theta_H = \frac{-M_H}{h_{ii}}$$

$$V_H = \frac{-h_{ji} M_H}{h_{ii}}$$

The effect of the release $-M_H$ on $\langle M_i \quad M_j \rangle$ is given in Eq. (6.91).

$$\begin{bmatrix} M_i \\ M_j \end{bmatrix}' = \begin{bmatrix} M_i \\ M_j \end{bmatrix} + T \begin{bmatrix} -1 \\ -\dfrac{h_{ji}}{h_{ii}} \end{bmatrix} M_H \tag{6.91}$$

Example 6.10 The beam shown in Fig. 6.17*a* has a hinge at the center, $a = b = l/2$, and is loaded there with a force P. The effect of the hinge on the original fixed-end moments $\langle M_i \quad M_j \rangle = (Pl/8)\langle 1 \quad -1 \rangle$ will be calculated by using Eq. (6.91). The final condition is shown in Fig. 6.17*b*. From Example 6.9,

$$k_H = \frac{2EI}{l^3} \begin{bmatrix} \dfrac{l^2}{2} & 0 \\ 0 & 6 \end{bmatrix}$$

$$T = \begin{bmatrix} -1 & -\dfrac{l}{2} \\ 1 & -\dfrac{l}{2} \end{bmatrix} \qquad T^{-1} = \frac{1}{l} \begin{bmatrix} -\dfrac{l}{2} & \dfrac{l}{2} \\ -1 & -1 \end{bmatrix}$$

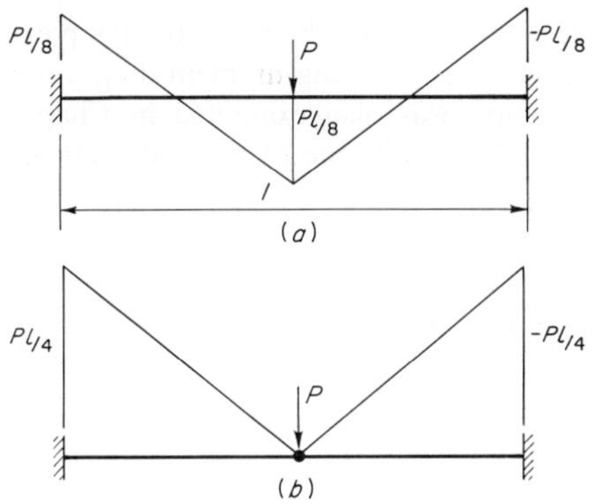

Figure 6.17 Fixed-end moments. (*a*) No hinge; (*b*) hinge at center.

Then

$$\begin{bmatrix} M_H \\ V_H \end{bmatrix} = \frac{1}{l}\begin{bmatrix} -\frac{l}{2} & \frac{l}{2} \\ -1 & -1 \end{bmatrix}\begin{bmatrix} 1 \\ -1 \end{bmatrix}\frac{Pl}{8} + \begin{bmatrix} 2 \\ 0 \end{bmatrix}\frac{Pl}{8} = \begin{bmatrix} \frac{Pl}{8} \\ 0 \end{bmatrix}$$

In this case, $h_{ji} = 0$, so that

$$\begin{bmatrix} M_i \\ M_j \end{bmatrix}' = \begin{bmatrix} 1 \\ -1 \end{bmatrix}\frac{Pl}{8} + \begin{bmatrix} -1 & -\frac{l}{2} \\ 1 & -\frac{l}{2} \end{bmatrix}\begin{bmatrix} -1 \\ 0 \end{bmatrix}\frac{Pl}{8} = \begin{bmatrix} 1 \\ -1 \end{bmatrix}\frac{Pl}{8} + \begin{bmatrix} 1 \\ -1 \end{bmatrix}\frac{Pl}{8} = \begin{bmatrix} 1 \\ -1 \end{bmatrix}\frac{Pl}{4}$$

These moments are shown in Fig. 6.17*b*.

5 Axial and Bending Stiffnesses for Nonuniform Members

The same assumption is made here as was used in the calculation of flexibility matrices for nonuniform members. That is, the centroidal axis of the member is straight, and the material in the member is distributed symmetrically about the axis.

The bending-stiffness matrix could be obtained by inversion of the flexibility matrix. Alternatively the area-moment theorems may be used, and this approach is developed below. The member with its area and inertia properties is shown in Fig. 6.18. These properties are given for four equal subdivisions of the member, and Simpson's rule will be used for the numerical quadrature.

From Eq. (4.177), with axial stiffness $k_{\text{axial}} = f^{-1}{}_{\text{axial}}$,

$$f_{\text{axial}} = \frac{h}{3EA_1} W^T\alpha$$

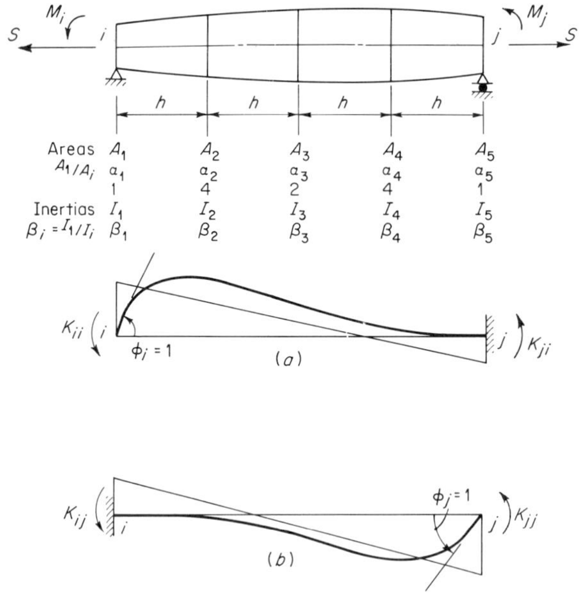

Figure 6.18 Member properties with unit rotations of ends of member.

where

$$W^T = \langle 1 \quad 4 \quad 2 \quad 4 \quad 1 \rangle \qquad \text{and} \qquad \alpha^T = \langle \alpha_1 \quad \alpha_2 \quad \alpha_3 \quad \alpha_4 \quad \alpha_5 \rangle$$

Bending Stiffnesses If a unit rotation is applied at end i while end j is held clamped, the force at end i will be k_{ii}, and that at end j will be k_{ji}. Similarly when a unit rotation is given to end j, the force at end i will be k_{ij}, and that at j will be k_{jj}.

The first area-moment theorem used states that the rotation of the tangent at end i relative to that at end j is equal to the area of the M/EI diagram between i and j.

From Fig. 6.18a, by Simpson's rule,

$$\phi_i = 1 = \frac{h}{3EI_1}\{k_{ii}[\beta_1 + 4\beta_2(3/4) + 2\beta_3(2/4) + 4\beta_4(1/4)] - k_{ji}[4\beta_2(1/4) + 2\beta_3(2/4) + 4\beta_4(3/4) + \beta_5]\}$$

Write this equation in the form

$$1 = \frac{h}{3EI_1}(k_{ii}A_1 - k_{ji}A_2) \tag{6.92}$$

As in Sec. 4.10, let

$$\zeta_1{}^T = \frac{x}{l} = (0,\tfrac{1}{4},\tfrac{2}{4},\tfrac{3}{4},1)$$

$$\zeta_2{}^T = 1 - \frac{x}{l} = (1,\tfrac{3}{4},\tfrac{2}{4},\tfrac{1}{4},0)$$

Also

$$[\beta\rfloor = \text{diag}\,\langle \beta_1 \quad \beta_2 \quad \beta_3 \quad \beta_4 \quad \beta_5 \rangle$$

In this notation,

$$A_1 = W^T[\beta\rfloor\zeta_2$$
$$A_2 = W^T[\beta\rfloor\zeta_1$$

The second area-moment theorem says that the deflection at i relative to the tangent at end j is equal to the moment of the M/EI diagram beween i and j taken about i. Since i is fixed in position and the tangent at j remains horizontal, the moment of the M/EI diagram about i is equal to zero. In this case,

$$0 = k_{ii}B_1 - k_{ji}B_2 \tag{6.93}$$

where

$$B_1 = 4\beta_2(\tfrac{3}{4})(\tfrac{1}{4}) + 2\beta_3(\tfrac{2}{4})(\tfrac{2}{4}) + 4\beta_4(\tfrac{1}{4})(\tfrac{3}{4})$$
$$B_2 = 4\beta_2(\tfrac{1}{4})(\tfrac{1}{4}) + 2\beta_3(\tfrac{2}{4})(\tfrac{2}{4}) + 4\beta_4(\tfrac{3}{4})(\tfrac{3}{4}) + \beta_5$$

That is,

$$B_1 = W^T[\beta\rfloor\zeta_1 * \zeta_2$$
$$B_2 = W^T[\beta\rfloor\zeta_1 * \zeta_1$$

It is seen from Eq. (6.93) that

$$k_{ji} = \frac{B_1 k_{ii}}{B_2} \tag{6.94}$$

Substituting for k_{ji} in Eq. (6.92) gives

$$k_{ii} = \frac{3EI_1 B_2}{h(A_1B_2 - A_2B_1)} \tag{6.95}$$

The process may be repeated to obtain k_{jj}.

Calculation of Fixed-end Moments and Shears Consider the case in which the load w_x is a continuous function of x producing moments m_x.

End rotations are calculated in Eq. (4.181). In the present case the sign of ϕ_i is changed so that

$$\begin{bmatrix} \phi_i \\ \phi_i \end{bmatrix} = \frac{h}{3EI_1}\begin{bmatrix} -W^T & \cdot \\ \cdot & W^T \end{bmatrix}\begin{bmatrix} [\beta\rfloor & \cdot \\ \cdot & [\beta\rfloor \end{bmatrix}\begin{bmatrix} \zeta_2 * m \\ \zeta_1 * m \end{bmatrix} \tag{6.96}$$

It will be assumed that w_x has been calculated at the same intervals as m_x as shown in Fig. 6.19. The reactions are given by

$$R_i = \frac{h}{3}\,[w_1 + 4w_2\,(3/4) + 2w_3(2/4) + 4w_4(1/4)]$$

that is,

$$R_i = \frac{h}{3}\,W^T \lceil w \rfloor \zeta_2$$
$$R_j = \frac{h}{3}\,W^T \lceil w \rfloor \zeta_1 \tag{6.97}$$

The matrix

$$\lceil w \rfloor = \text{diag}\,\langle w_1 \quad w_2 \quad w_3 \quad w_4 \quad w_5 \rangle$$

From Eq. (6.96) the fixed-end moments will be such that the rotations $\langle \phi_i \quad \phi_j \rangle$ are brought back to zero. Then

$$\begin{bmatrix} M_i \\ M_j \end{bmatrix}_F = -k \begin{bmatrix} \phi_i \\ \phi_j \end{bmatrix} \tag{6.98}$$

These moments produce the reactions

$$\begin{bmatrix} R_i \\ R_j \end{bmatrix}_F = \frac{1}{l} \begin{bmatrix} M_i + M_j \\ -(M_i + M_j) \end{bmatrix}_F = \frac{1}{l} \begin{bmatrix} 1 & 1 \\ -1 & -1 \end{bmatrix} \begin{bmatrix} M_i \\ M_j \end{bmatrix}_F \tag{6.99}$$

6 The Effect of Axial Loads on the Bending Stiffness of Prismatic Members

When the bending moments produced by the interaction between the axial force in the member and the deflected shape are taken into account, the behavior is nonlinear and requires some type of iteration procedure to arrive at a solution. That is, the bending stiffness depends on the axial load in the member, which in turn is calculated by using the bending stiffness. These problems are discussed further in Chap. 9. The classical expressions for flexibility matrices were derived for straight prismatic members in Sec. 4.10. If the same theory is used to develop the stiffness matrix, it is found that

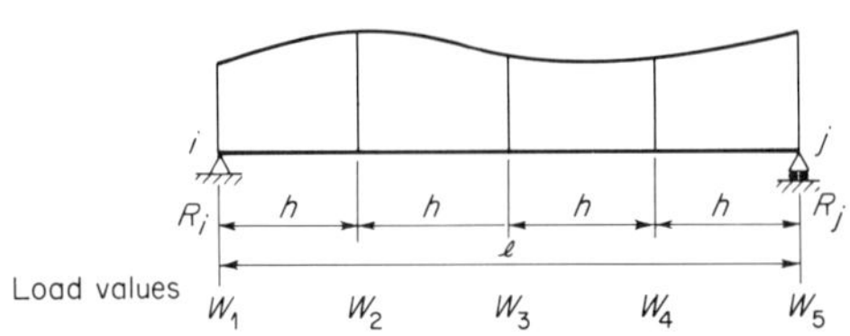

Figure 6.19 Variable load intensity.

$$\begin{bmatrix} M_i \\ M_j \end{bmatrix} = \frac{2EI}{l} \begin{bmatrix} 2\phi_3 & \phi_4 \\ \phi_4 & 2\phi_3 \end{bmatrix} \begin{bmatrix} \phi_i \\ \phi_j \end{bmatrix} \tag{6.100}$$

For axial compression in the member

$$\phi_1 = \beta \cot \beta \qquad \phi_2 = \frac{\beta^2}{3(1 - \beta \cot \beta)}$$

then

$$\phi_3 = \tfrac{1}{4}(\phi_1 + 3\phi_2) \qquad \phi_4 = \tfrac{1}{2}(3\phi_2 - \phi_1)$$

For axial tension in the member,

$$\phi_1 = \beta \coth \beta \qquad \phi_2 = \frac{\beta^2}{3(\beta \coth \beta - 1)}$$

$$\phi_3 = \tfrac{1}{4}(\phi_1 + 3\phi_2) \qquad \phi_4 = \tfrac{1}{2}(\phi_1 - 3\phi_2)$$

$$\beta = \frac{l}{2}\sqrt{\frac{P}{EI}}$$

Notice that in this approach, no account is made for the effect of the bending moments on axial stiffness. That is,

$$k_{\text{axial}} = \frac{EA}{l}$$

A useful alternative idea for the incorporation of the axial-force effects in the bending-stiffness matrix assumes that the deflected shape is a given polynomial function (see Ref. 18). For example, if a cubic function is assumed, these effects are readily determined. The error in the method lies in the fact that the axial force will modify the assumed shape. This should not be a large source of error if the axial force is significantly less than the critical load. In Fig. 6.20 the member is shown with end rotations $\langle \phi_i \quad \phi_j \rangle$ and moments $\langle M_i \quad M_j \rangle$. Instead of using a single coordinate x to describe the position of a point on the beam, the two nondimensional (natural) coordinates $\langle \zeta_1 \quad \zeta_2 \rangle$ will be used.

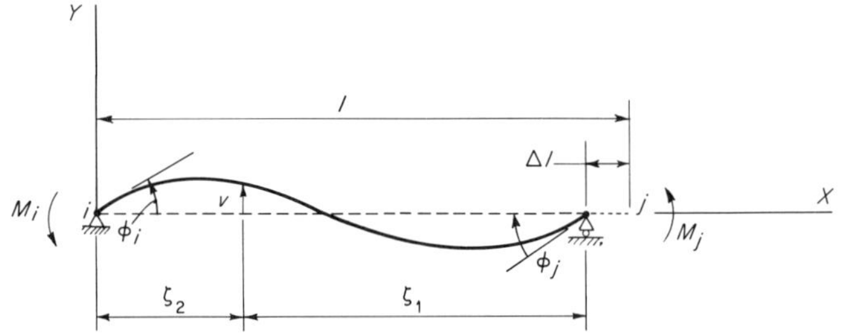

Figure 6.20 Assumed deflected shape $v = f(\zeta_1, \zeta_2)$.

These are defined

$$\zeta_1 = 1 - \zeta_2 = 1 - \frac{x}{l}$$

$$\zeta_2 = \frac{x}{l}$$

The deflected shape v may be expressed as a cubic function of $\langle \zeta_1 \quad \zeta_2 \rangle$, so that

$$v = r^T \begin{bmatrix} l\zeta_1{}^2\zeta_2 \\ -l\zeta_1\zeta_2{}^2 \end{bmatrix} \tag{6.101}$$

In Eq. (6.101),

$$r^T = \langle \phi_i \quad \phi_j \rangle$$

The vector containing the ζ functions in Eq. (6.101) will be termed an *interpolation polynomial* and will be designated Φ. It is seen that

$$v = r^T\Phi = \Phi^T r \tag{6.102}$$

The slope of the curve v will be designated v_x, and

$$v_x = r^T \begin{bmatrix} \zeta_1(\zeta_1 - 2\zeta_2) \\ \zeta_2(\zeta_2 - 2\zeta_1) \end{bmatrix} = r^T\Phi_x = \Phi_x{}^T r \tag{6.103}$$

In Eq. (6.103), write Φ_x in terms of ζ_2 only, using the relationship

$$\zeta_1 = 1 - \zeta_2$$

Then,

$$\Phi_x{}^T = \langle (1 - \zeta_2)(1 - 3\zeta_2) \quad \zeta_2(-2 + 3\zeta_2) \rangle$$

Also,

$$v_x{}^2 = (\Phi_x{}^T r)^2 = r^T\Phi_x\Phi_x{}^T r = r^T \begin{bmatrix} (1 - \zeta_2)(1 - 3\zeta_2) \\ \zeta_2(-2 + 3\zeta_2) \end{bmatrix} [(1 - \zeta_2)(1 - 3\zeta_2) \quad \zeta_2(-2 + 3\zeta_2)] r$$

The length l of the member is calculated by

$$l = \int_o^{l-\Delta l} (1 + v_x{}^2)^{1/2}\, dx \approx l - \Delta l + \tfrac{1}{2} \int_o^l v_x{}^2\, dx$$

Then the shortening due to bowing is

$$\Delta l = \tfrac{1}{2} \int_o^l v_x{}^2\, dx$$

However,

$$\int_o^l v_x^2\, dx = r^T \int_o^l \Phi_x \Phi_x^T\, dx\, r$$

$$= lr^T \int_o^l \Phi_x \Phi_x^T\, d\zeta_2\, r$$

$$= \frac{lr^T}{30}\begin{bmatrix} 4 & -1 \\ -1 & 4 \end{bmatrix} r$$

That is,

$$\int_o^l v_x^2\, dx = \frac{l}{30}(4\phi_i^2 - 2\phi_i\phi_j + 4\phi_j^2)$$

Then

$$\Delta l = \tfrac{1}{2}\int_o^l v_x^2\, dx = \frac{l}{30}(2\phi_i^2 - \phi_i\phi_j + 2\phi_j^2) \tag{6.104}$$

The small-deflection flexibility matrix is given as

$$\begin{bmatrix} \phi_i \\ \phi_j \end{bmatrix} = \frac{l}{6EI}\begin{bmatrix} 2 & -1 \\ -1 & 2 \end{bmatrix}\begin{bmatrix} M_i \\ M_j \end{bmatrix}$$

With the deflected shape v given in terms of $\langle \phi_i \quad \phi_j \rangle$, the moments due to the axial force P (tension positive and hogging moment positive) are given (see Fig. 6.21) by

$$M_P = -Pv = -Pr^T\begin{bmatrix} l\zeta_1^2\zeta_2 \\ -l\zeta_1\zeta_2^2 \end{bmatrix}$$

By using the virtual force $M_i = 1$ in Fig. 6.21, the rotation at end i is calculated by

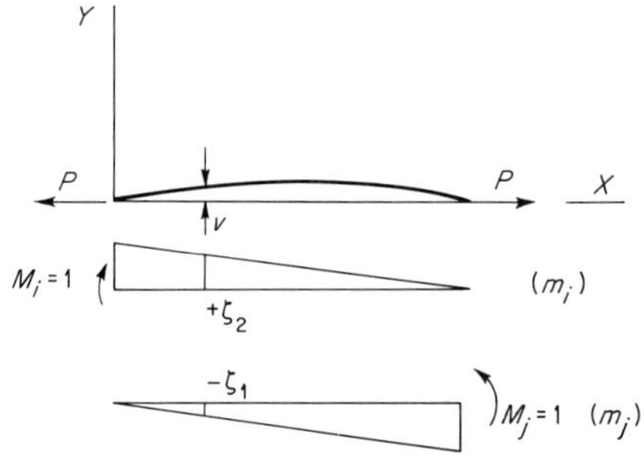

Figure 6.21 Calculations at end rotations due to P.

$$\phi_{iP} = \int_o^l \frac{M_P m_i}{EI}\,dx = \int_o^l \frac{M_P m_i l}{EI}\,d\zeta_2$$

That is,

$$\phi_{iP} = -\frac{P r^T l^2}{EI}\int_o^l \begin{bmatrix} \zeta_1{}^2\zeta_2 \\ -\zeta_1\zeta_2{}^2 \end{bmatrix} \zeta_1\,d\zeta_2$$

On integration of the ζ functions,

$$\phi_{iP} = -\frac{l^2 P r^T}{60EI}\begin{bmatrix} 3 \\ -2 \end{bmatrix}$$

$$\phi_{jP} = -\frac{l^2 P r^T}{60EI}\begin{bmatrix} -2 \\ 3 \end{bmatrix}$$

In matrix form,

$$\phi_P{}^T = \langle \phi_{iP} \quad \phi_{jP} \rangle = \frac{-l^2 P r^T}{60EI}\begin{bmatrix} 3 & -2 \\ -2 & 3 \end{bmatrix} \tag{6.105}$$

That is, given the deflected shape defined by $\langle \phi_i \quad \phi_j \rangle$, the axial force produced rotations $\phi_P{}^T = \langle \phi_{iP} \quad \phi_{jP} \rangle$.

For rotations to remain at $\langle \phi_i \quad \phi_j \rangle$, additional moments must be applied to produce rotations equal to $-\phi_P$. That is,

$$\begin{bmatrix} M_i \\ M_j \end{bmatrix} = \frac{2EI}{l}\begin{bmatrix} 2 & 1 \\ 1 & 2 \end{bmatrix}\left\{\begin{bmatrix} \phi_i \\ \phi_j \end{bmatrix} + \frac{l^2 P}{60EI}\begin{bmatrix} 3 & -2 \\ -2 & 3 \end{bmatrix}\begin{bmatrix} \phi_i \\ \phi_j \end{bmatrix}\right\}$$

Hence

$$\begin{bmatrix} M_i \\ M_j \end{bmatrix} = \frac{2EI}{l}\begin{bmatrix} 2 & 1 \\ 1 & 2 \end{bmatrix}\begin{bmatrix} \phi_i \\ \phi_j \end{bmatrix} + \frac{Pl}{30}\begin{bmatrix} 4 & -1 \\ -1 & 4 \end{bmatrix}\begin{bmatrix} \phi_i \\ \phi_j \end{bmatrix} \tag{6.106}$$

It is seen that the first term on the right-hand side of Eq. (6.106) is the normal, small-deflection stiffness matrix and that the second term gives the effect of axial force.

The expressions (6.104) and (6.106) will be considered further in Chap. 9 on nonlinear analysis.

Problems

6.1 Calculate node deflections and member forces for the trusses shown in Fig. P6.1, using Eqs. (6.20), (6.23), and (6.25).

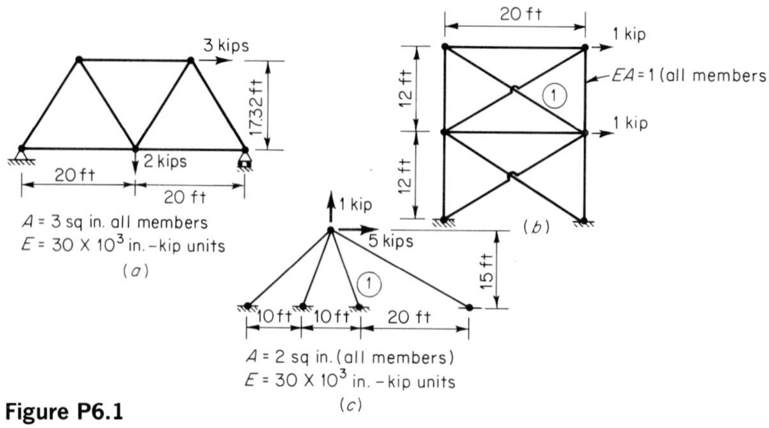

Figure P6.1

6.2 Calculate the effect to node deflections and member forces if member (1) is removed from the structures b and c in Fig. P6.1.

6.3 Calculate the node deflections and member forces (bending moments only) for the frame structures shown in Figs. P6.2 and P6.3.

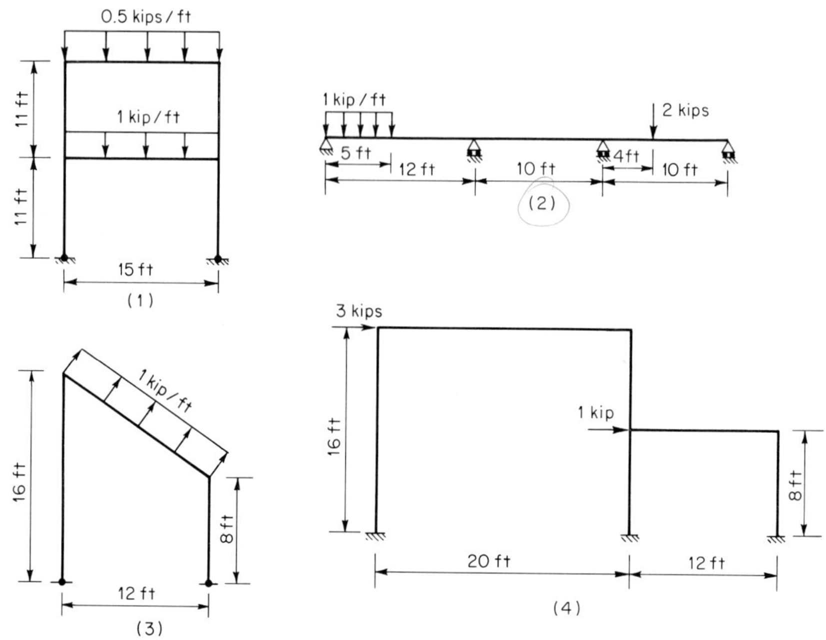

Figure P6.2

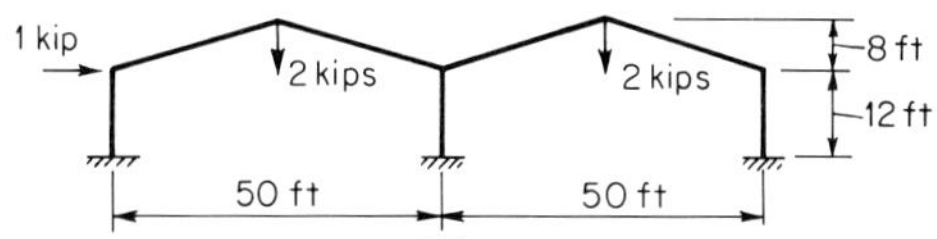

(5)

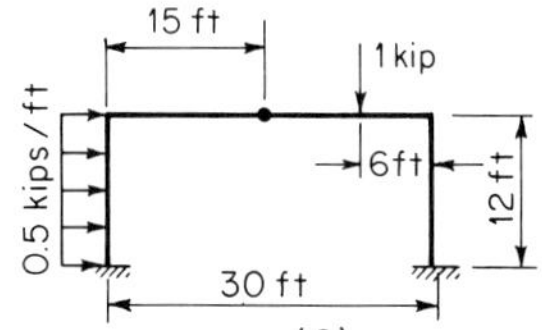

(6)

$EI = 10^7$ in.–kip units, all members

Figure P6.3

7 THE DIRECT STIFFNESS METHOD. SOLUTION TECHNIQUES

7.1 DISPLACEMENT TRANSFORMATION MATRICES

It has been shown in Chap. 6 that the calculation of member forces and node deflections by the displacement method involves the solution of the equations

$$R = Kr \tag{7.1}$$

The matrix K is the stiffness matrix of the structure for the deflections r and may be generated by the transformation

$$K = a^T k a \tag{7.2}$$

In Chap. 6, it was noted that the matrix a is composed of displacement-influence functions expressing member distortions in terms of node displacements. As yet, no indication has been given as to how a may be generated, except by the graphic method of drawing the deflected shape for unit values of each displacement in turn. Evidently such an approach will be difficult to program, and recourse must be made to the displacement transformations developed in Chaps. 2 and 3. The form of the a matrix will depend on how k is formulated for members. Thus, for example, in the direct stiffness variation of the displacement method, a contains only unit values and k may then be loaded into K without actually performing the formal multiplication of Eq. (7.2).

It will be remembered that displacement transformations are used to generate the equilibrium equations in the form

$$R - a^T S = 0 \tag{7.3}$$

In Chap. 3, it was shown that to generate a^T, transformation matrices can be developed that relate basic member distortions to node displacements, firstly in the local coordinate system for the member and then in the global coordinate system for the whole structure. If the node under consideration is

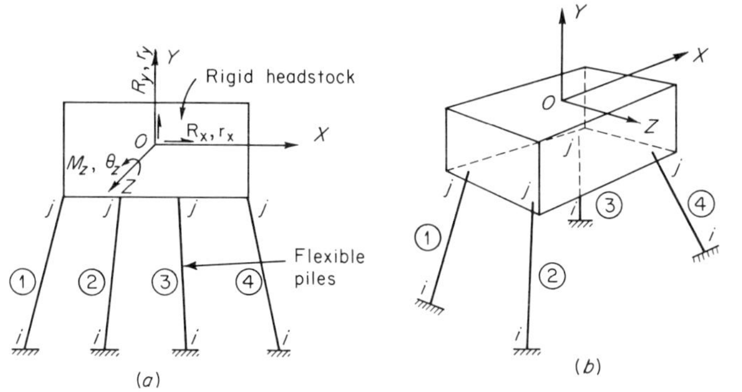

Figure 7.1 Pile structure idealization. (*a*) Planar structure; (*b*) three-dimensional structure.

a rigid body with finite dimensions, rather than an idealized point, then the transformation in Eq. (2.62) must first be used to transfer the given node displacements to the point where the member is connected. These types of transformation are useful in a special class of problem, generally referred to as the *rigid headstock on flexible piles*. This structure may be idealized in some way to a planar system, as shown in Fig. 7.1*a*, or treated in more accurate fashion as a three-dimensional structure, as in Fig. 7.1*b*. The limitation on the assumption of headstock stiffness will not be considered here.

Consider firstly the planar structure, and develop the transformations that relate the displacements at the end j of each pile in the local coordinates for that pile. In Fig. 7.2, the point O is taken as the reference origin for forces and displacements in the headstock. The displacements of O are $\langle r_x \quad r_y \quad \theta_z \rangle$. From Eq. (2.62), the displacements at P, in xy coordinates are

$$\begin{bmatrix} r_{xP} \\ r_{yP} \\ \theta_{zP} \end{bmatrix} = \begin{bmatrix} 1 & 0 & -y \\ 0 & 1 & x \\ 0 & 0 & 1 \end{bmatrix} \begin{bmatrix} r_x \\ r_y \\ \theta_z \end{bmatrix} \tag{7.4}$$

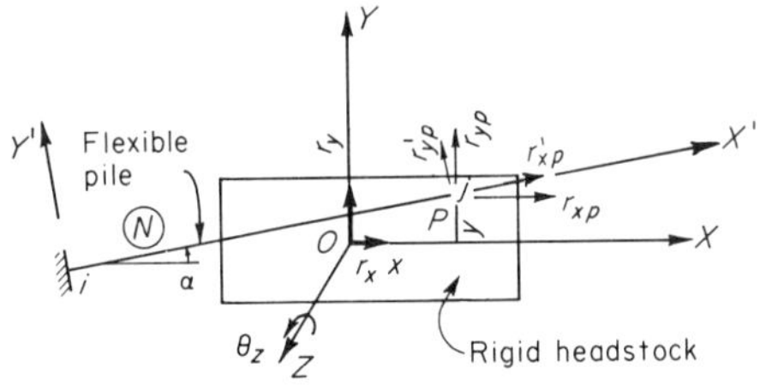

Figure 7.2 Displacements at origin O and at pile connection *P*.

That is,

$$r_P = T^T r_o \tag{7.5}$$

The components in the $x'y'$ coordinate system for pile N will be

$$r'_P = L^T r_P \tag{7.6}$$

In this case

$$L^T = \begin{bmatrix} \cos\alpha & \sin\alpha & 0 \\ -\sin\alpha & \cos\alpha & 0 \\ 0 & 0 & 1 \end{bmatrix} \tag{7.7}$$

Combining Eqs. (7.5) and (7.6),

$$r'_P = L^T T^T r_o \tag{7.8}$$

By Eq. (7.8), the displacements at the top of each pile may be expressed in terms of r_o.

Let the basic distortions of a member be $\langle \phi_i \quad \phi_j \quad \delta \rangle$. How these are related to r'_P will depend on the support condition at end i. This may be considered either pinned or fixed. Both conditions are shown in Fig. 7.3. In condition a, the terms ϕ_i and M_i may be neglected, and

$$\begin{bmatrix} \phi_j \\ \delta \end{bmatrix} = \begin{bmatrix} 0 & -\dfrac{1}{l} & 1 \\ 1 & 0 & 0 \end{bmatrix} \begin{bmatrix} r_{xP} \\ r_{yP} \\ \theta_z \end{bmatrix}' \tag{7.9}$$

and in b,

$$\begin{bmatrix} \phi_i \\ \phi_j \\ \delta \end{bmatrix} = \begin{bmatrix} 0 & -\dfrac{1}{l} & 0 \\ 0 & -\dfrac{1}{l} & 1 \\ 1 & 0 & 0 \end{bmatrix} \begin{bmatrix} r_{xP} \\ r_{yP} \\ \theta_z \end{bmatrix}'$$

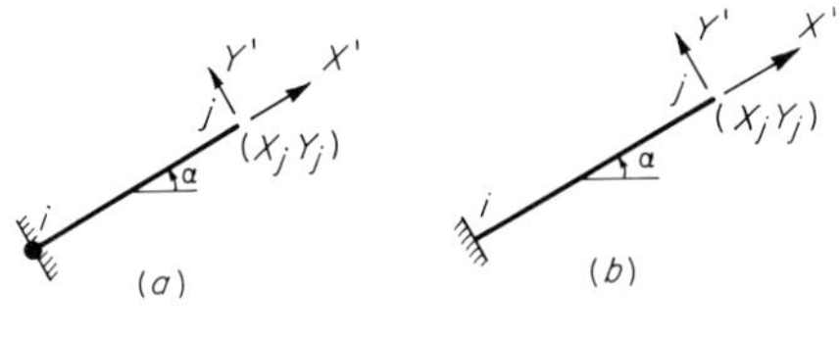

Figure 7.3 End support conditions. (*a*) End *i* pinned; (*b*) end *j* fixed.

The stiffness matrices for the types are

$$k_{\text{pinned}} = \frac{3EI}{l}\begin{bmatrix} 1 & 0 \\ 0 & \dfrac{A}{3I} \end{bmatrix} \qquad k_{\text{fixed}} = \frac{2EI}{l}\begin{bmatrix} 2 & 1 & 0 \\ 1 & 2 & 0 \\ 0 & 0 & \dfrac{A}{2I} \end{bmatrix} \tag{7.10}$$

Equation (7.9) will be written

$$v_N = Ur'_P \tag{7.11a}$$

Then finally,

$$v_N = UL^T T^T r_o \tag{7.11b}$$

This is the equation

$$v_N = a_N r_o \tag{7.12}$$

For the whole structure of piles 1, 2, . . . , M,

$$\begin{bmatrix} v_1 \\ v_2 \\ \vdots \\ v_N \\ \vdots \\ v_M \end{bmatrix} = \begin{bmatrix} a_1 \\ a_2 \\ \vdots \\ a_N \\ \vdots \\ a_M \end{bmatrix} r_o \tag{7.13}$$

Then

$$K = a^T k a = \sum_{1-M} a_N{}^T k_N a_N \tag{7.14}$$

The final stiffness matrix K for the headstock is then a 3×3 matrix. When forces $\langle R_X \quad R_Y \quad M_Z \rangle$ are applied at O, the displacements are calculated by

$$r_o = K^{-1} R \tag{7.15}$$

Member forces are calculated by

$$S_N = k_N a_N r_o \tag{7.16}$$

The Three-dimensional Pile Headstock

The rigid headstock supported on flexible piles is shown in Fig. 7.1*b*. In this structure two major problems arise. The first is similar to that already solved in the two-dimensional problem. That is, given displacements at O, transform these to displacements at the end of a pile. Secondly, the principal axes of the pile must be related through direction cosines to the global coordinate system. For circular piles and perhaps also hexagonal, where the inertia terms are constant, or nearly constant, for rotation of axes, the actual positions of the axes are not important. However for *H* piles these may have a considerable influence on the performance of the structure.

In the present analysis the principal axes of the pile cross section are specified so that one principal axis lies in the plane that projects the pile onto the *X* plane, and the other being perpendicular to this plane is parallel to the *XZ* plane. These axes together with positive force and displacement quantities of *O* are shown in Fig. 7.4.

Member length is

$$l = \sqrt{(x_j - x_i)^2 + (y_j - y_i)^2 + (z_j - z_i)^2}$$

From Fig. 7.4*a*, the direction cosines of jX', jY', and jZ' may be calculated.

For jX':

$$l_{11} = \frac{x_j - x_i}{l} \qquad l_{12} = \frac{y_j - y_i}{l} \qquad l_{13} = \frac{z_j - z_i}{l}$$

For jZ':

$$l_{31} = -\frac{(z_j - z_i)}{l_1} \qquad l_{32} = 0 \qquad l_{33} = \frac{x_j - x_i}{l_1}$$

The projected length l_1 is

$$l_1 = \sqrt{(x_j - x_i)^2 + (z_j - z_i)^2}$$

The axis jY' is obtained by taking the cross product of jZ' with jX'. Thence

$$l_{21} = \frac{-(y_j - y_i)(x_j - x_i)}{ll_1} \qquad l_{22} = \frac{(z_j - z_i)^2 + (x_j - x_i)^2}{ll_1}$$

$$l_{23} = \frac{-(z_j - z_i)(y_j - y_i)}{ll_1}$$

The matrix

$$L = \begin{bmatrix} L_1 & 0 \\ 0 & L_1 \end{bmatrix}$$

is a 6×6 matrix, in which

$$L_1{}^T = \begin{bmatrix} l_{11} & l_{12} & l_{13} \\ l_{21} & l_{22} & l_{23} \\ l_{31} & l_{32} & l_{33} \end{bmatrix}$$

The development of the 6×6 stiffness matrix for the displacements $r_o{}^T = \langle r_x \quad r_y \quad r_z \quad \theta_x \quad \theta_y \quad \theta_z \rangle$ at the origin of coordinates, *O*, follows the

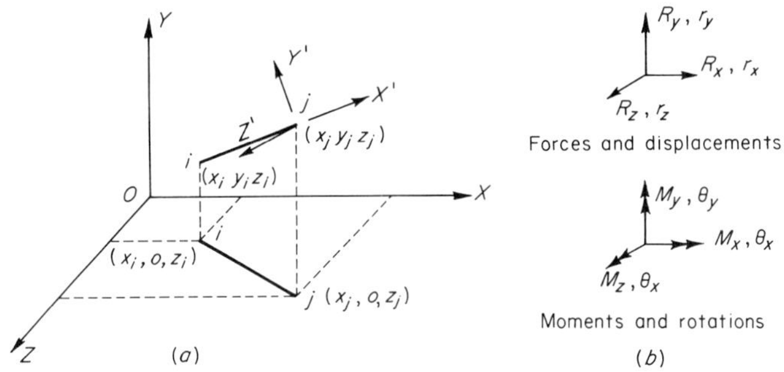

Figure 7.4 Pile axes and coordinates.

same process as for the planar problem. The transformation for displacements at O to those at P, the point of connection of a pile, is given by

$$r_P = T^T r_o$$

In this case,

$$T^T = \begin{bmatrix} 1 & \cdot & \cdot & \cdot & z & -y \\ \cdot & 1 & \cdot & -z & \cdot & x \\ \cdot & \cdot & 1 & y & -x & \cdot \\ \cdot & \cdot & \cdot & 1 & \cdot & \cdot \\ \cdot & \cdot & \cdot & \cdot & 1 & \cdot \\ \cdot & \cdot & \cdot & \cdot & \cdot & 1 \end{bmatrix}$$

By use of the matrix L, the r_P vector is now transformed to components in the member axes, r'_P

$$r'_P = L^T r_P = L^T T^T r_o \tag{7.17}$$

From r'_P the basic member distortions are calculated, since

$$\begin{bmatrix} \phi_{iz} \\ \phi_{jz} \\ \phi_{iy} \\ \phi_{jy} \\ \theta_x \\ \delta \end{bmatrix}_N = \begin{bmatrix} \cdot & -\dfrac{1}{l} & \cdot & \cdot & \cdot & \cdot \\ \cdot & -\dfrac{1}{l} & \cdot & \cdot & \cdot & 1 \\ \cdot & \cdot & \dfrac{1}{l} & \cdot & \cdot & \cdot \\ \cdot & \cdot & \dfrac{1}{l} & \cdot & 1 & \cdot \\ \cdot & \cdot & \cdot & 1 & \cdot & \cdot \\ 1 & \cdot & \cdot & \cdot & \cdot & \cdot \end{bmatrix} \begin{bmatrix} r_x \\ r_y \\ r_z \\ \theta_x \\ \theta_y \\ \theta_z \end{bmatrix}'_P$$

That is,

$$v_N = UL^T T^T r_o$$

or

$$v_N = a_N r_o$$

The stiffness matrix k_N for member N is given by the combination of the bending stiffnesses in each of the principal planes together with the torsion and axial stiffness. That is,

$$k_N = \begin{bmatrix} \frac{4EI_1}{l} & \frac{2EI_1}{l} & & & & \mathbf{0} \\ \frac{2EI_1}{l} & \frac{4EI_1}{l} & & & & \\ & & \frac{4EI_2}{l} & \frac{2EI_2}{l} & & \\ & & \frac{2EI_2}{l} & \frac{4EI_2}{l} & & \\ & & & & k_T & \\ \mathbf{0} & & & & & \frac{EA}{l} \end{bmatrix}$$

The moments of inertia are

$$I_1 = \int_{\text{area}} y^2\, dA \qquad \text{and} \qquad I_2 = \int_{\text{area}} z^2\, dA$$

The torsion-stiffness constant k_T will depend on the cross-section shape of the pile (see the tabulation of member stiffnesses in Chap. 6). The stiffness matrix at O is calculated by

$$K = \sum_{N=1,M} a_N{}^T k_N a_N$$

Thence

$$r_o = K^{-1} R_o$$

$$S_N = k_N a_N r_o$$

7.2 TRANSFORMATION OF STIFFNESS MATRICES

In Sec. 7.1 it has been shown how displacement transformation matrices may be developed so that the equation

$$K = a^T k a \tag{7.18}$$

can be used to generate the structure-stiffness matrix.

A subtle variation to this occurs if k is not the stiffness matrix for the fundamental member distortions but has been expanded to be the stiffness matrix for the member in the global coordinate system. In this case rows of a contain either unit or zero values. This implies that instead of the formal multiplication in Eq. (7.18), the member-stiffness terms may be added directly to K.

Before this method can be used, it is necessary to be able to obtain member-stiffness matrices in global components. This requires a knowledge of how stiffness matrices transform.

For example, consider the joint A of the plane structure shown in Fig. 7.5, and let the stiffness relationship between $\langle r_X \quad r_Y \rangle$ and $\langle R_X \quad R_Y \rangle$ be

$$R = Kr \tag{7.19}$$

It is required to calculate the stiffness matrix K' when the forces and displacements are expressed in the $X'Y'$ coordinate axes obtained by a positive rotation α from XY. The relationships between (R,R') and (r,r') have been established to be

$$R = LR'$$
$$r = Lr'$$

where

$$L = \begin{bmatrix} \cos\alpha & -\sin\alpha \\ \sin\alpha & \cos\alpha \end{bmatrix}$$

If these values are substituted in Eq. (7.19), and it is noted that $L^{-1} = L^T$, then

$$R' = L^T K L r' \tag{7.20}$$

That is,

$$K' = L^T K L \tag{7.21}$$

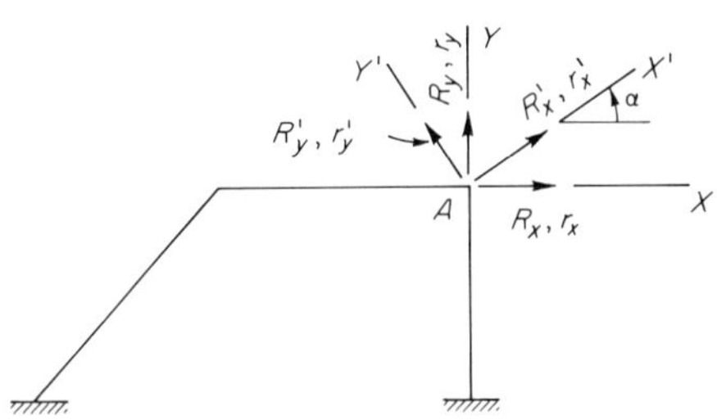

Figure 7.5 Rotation of coordinate axes.

The Stiffness Matrix for Plane Frame Members in Global Coordinates

The planar line element i, j is shown in Fig. 7.6*a*, with displacement components $\langle r_x \quad r_y \quad \theta_z \rangle$ and force components $\langle R_X \quad R_Y \quad M_Z \rangle$ at its ends. The fundamental member-force quantities $\langle M_i \quad M_j \quad F \rangle$ and displacement quantities $\langle \phi_i \quad \phi_j \quad \delta \rangle$ are defined in Fig. 7.6*b*. The fundamental stiffness relationship is of the form

$$\begin{bmatrix} M_i \\ M_j \\ F \end{bmatrix} = \begin{bmatrix} K_{11} & K_{12} & 0 \\ K_{21} & K_{22} & 0 \\ 0 & 0 & K_{33} \end{bmatrix} \begin{bmatrix} \phi_i \\ \phi_j \\ \delta \end{bmatrix} \tag{7.22}$$

Equation (7.22) is written

$$S = k'v \tag{7.23}$$

For prismatic members without shear distortion,

$$k' = \frac{2EI}{l} \begin{bmatrix} 2 & 1 & 0 \\ 1 & 2 & 0 \\ 0 & 0 & \dfrac{A}{2I} \end{bmatrix}$$

The member distortions v are obtained by the transformation Eq. (3.68), that is,

$$\begin{bmatrix} \phi_i \\ \phi_j \\ \delta \end{bmatrix} = \begin{bmatrix} 0 & \dfrac{1}{l} & 1 & 0 & -\dfrac{1}{l} & 0 \\ 0 & \dfrac{1}{l} & 0 & 0 & -\dfrac{1}{l} & 1 \\ -1 & 0 & 0 & 1 & 0 & 0 \end{bmatrix} \begin{bmatrix} r'_{ix} \\ r'_{iy} \\ \theta_i \\ r'_{jx} \\ r'_{jy} \\ \theta_j \end{bmatrix} \tag{7.24}$$

This is the equation

$$v = T^T r' \tag{7.25}$$

The transformation from r to r' has been developed in Chap. 2, Eq. (2.40*a*) such that

$$r' = L_D{}^T r \tag{7.26}$$

The matrix L_D is the superdiagonal matrix

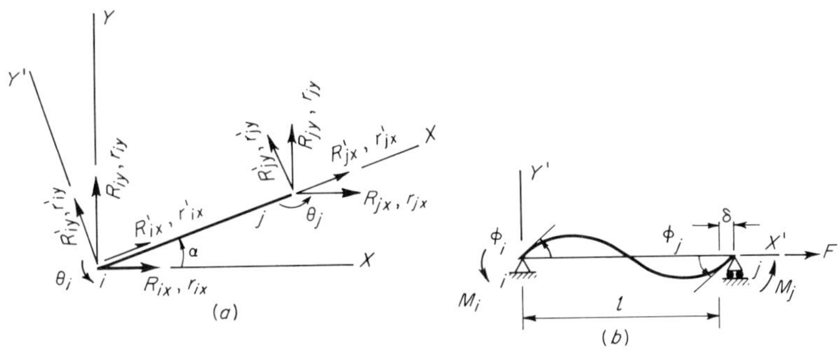

Figure 7.6 Member forces and displacements.

$$L_D = \begin{bmatrix} L & 0 \\ 0 & L \end{bmatrix}$$

in which

$$L = \begin{bmatrix} \cos\alpha & -\sin\alpha & 0 \\ \sin\alpha & \cos\alpha & 0 \\ 0 & 0 & 1 \end{bmatrix} \tag{7.27}$$

Combining Eqs. (7.25) and (7.26),

$$v = T^T L_D{}^T r \tag{7.28}$$

The relationship between R and S is given by the contragradient law to be

$$R = L_D TS \tag{7.29}$$

Hence, by combining Eqs. (7.23), (7.28), and (7.29), the expression is obtained for the member-stiffness matrix in global force components. That is,

$$R = L_D Tk' T^T L_D{}^T r \tag{7.30}$$

or

$$R = kr$$

where

$$k = L_D Tk' T^T L_D{}^T \tag{7.31}$$

Since the member forces S are of interest when the effects of stresses on the member material properties are being examined, they may be obtained from r through the equation

$$S = k' T^T L_D{}^T r \tag{7.32}$$

The Stiffness Matrix for Grid Members in Global Components

In this particular case the transformation matrices will be developed for components in the coordinate axes shown in Fig. 7.7*a*. The z axis will then coincide with the positive gravity-force axis, and the grid structure will lie in the XY plane.

For the grid member the basic member forces are $\langle M_{iy} \quad M_{jy} \quad M_T \rangle$ and the displacements are $\langle \phi_{iy} \quad \phi_{jy} \quad \theta_T \rangle$. The force M_T is the torque in the member, and θ_T is the relative rotation between the ends i and j. The stiffness relationship is then given by

$$\begin{bmatrix} M_{iy} \\ M_{jy} \\ M_T \end{bmatrix} = \begin{bmatrix} K_{11} & K_{12} & 0 \\ K_{21} & K_{22} & 0 \\ 0 & 0 & K_{33} \end{bmatrix} \begin{bmatrix} \phi_{iy} \\ \phi_{jy} \\ \theta_T \end{bmatrix} \tag{7.33}$$

The term K_{33} is the torsion stiffness of the member. Equation (7.33) will be written as for the plane frame member,

$$S = k'v \tag{7.34}$$

In this case, the member distortions are related to the displacements $\langle \theta_x \quad \theta_y \quad r_z \rangle$ of ends i and j of the member by Eq. (7.35). That is,

$$\begin{bmatrix} \phi_{iy} \\ \phi_{jy} \\ \theta_T \end{bmatrix} = \begin{bmatrix} 0 & 1 & -\frac{1}{l} & 0 & 0 & \frac{1}{l} \\ 0 & 0 & -\frac{1}{l} & 0 & 1 & \frac{1}{l} \\ -1 & 0 & 0 & 1 & 0 & 0 \end{bmatrix} \begin{bmatrix} \theta_{ix} \\ \theta_{iy} \\ r_{iz} \\ \theta_{jx} \\ \theta_{jy} \\ r_{jz} \end{bmatrix}' \tag{7.35}$$

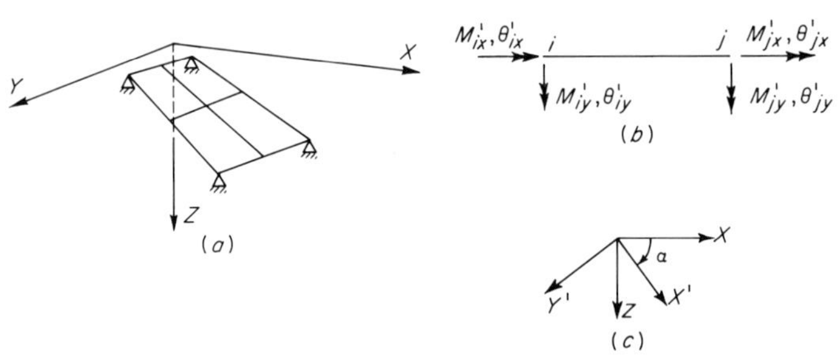

Figure 7.7 Grid structure. (*a*) Grid structure coordinates; (*b*) rotations and moments at member ends (local coordinates); (*c*) rotation of axes.

This is the equation

$$v = T^T r' \tag{7.36}$$

The transformation from global coordinate components to member coordinate components is identical with that in Eq. (7.26) if the angle α measures the rotation of the $X'Y'$ set of axes from XY about the Z axis (see Fig. 7.7c). Then

$$v = T^T L_D{}^T r' \tag{7.37}$$

The relationship between the forces R, for which

$$R^T = \langle M_{iX} \quad M_{iY} \quad R_{iZ} \quad M_{jX} \quad M_{jY} \quad R_{jZ} \rangle$$

on the member ends and the member forces S is given by the contragradient law to be

$$R = L_D T S \tag{7.38}$$

Hence, finally,

$$R = L_D T k' T^T L_D{}^T r \tag{7.39}$$

That is, the member-stiffness matrix in global components is

$$k = L_D T k' T^T L_D{}^T \tag{7.40}$$

It is seen that the transformations for the planar frame and grid members are identical in form and that an analogy exists between the two, relating twisting moment and twist rotation, on the one hand, to axial force and axial extension, on the other. The analogy is easily extended to the forces and displacements on the member ends but is not pursued further here.

The Stiffness Matrix for the Prismatic Member in Three-dimensional Space

The procedures for frame and grid structures may be extended to cover the general three-dimensional element. It is found that a certain amount of difficulty arises in defining the coordinate axes of the member relative to the global system.

The basic member distortions will be taken as:

$\delta =$ axial change in length

$\theta_T =$ relative twist of two ends of the member about the OX' axis of the member

$\left.\begin{matrix}\phi_{iz}\\ \phi_{iy}\end{matrix}\right\} =$ relative rotations of tangents at end i relative to chord

$\left.\begin{matrix}\phi_{jz}\\ \phi_{jy}\end{matrix}\right\} =$ relative rotations of tangents at end j relative to chord

The displacement vectors at ends i and j in the local coordinates will be

$$r_i'^T = \langle r_{ix}' \quad r_{iy}' \quad r_{iz}' \quad \theta_{ix}' \quad \theta_{iy}' \quad \theta_{iz}' \rangle$$
$$r_j'^T = \langle r_{jx}' \quad r_{jy}' \quad r_{jz}' \quad \theta_{jx}' \quad \theta_{jy}' \quad \theta_{jz}' \rangle$$

Then,

$$\begin{bmatrix} \delta \\ \phi_{iz} \\ \phi_{iy} \\ \theta_T \\ \phi_{jz} \\ \phi_{jy} \end{bmatrix} = \begin{bmatrix} -1 & 0 & 0 & 0 & 0 & 0 & 1 & 0 & 0 & 0 & 0 & 0 \\ 0 & \frac{1}{l} & 0 & 0 & 0 & 1 & 0 & -\frac{1}{l} & 0 & 0 & 0 & 0 \\ 0 & 0 & -\frac{1}{l} & 0 & 1 & 0 & 0 & 0 & \frac{1}{l} & 0 & 0 & 0 \\ 0 & 0 & 0 & -1 & 0 & 0 & 0 & 0 & 0 & 1 & 0 & 0 \\ 0 & \frac{1}{l} & 0 & 0 & 0 & 0 & 0 & -\frac{1}{l} & 0 & 0 & 0 & 1 \\ 0 & 0 & -\frac{1}{l} & 0 & 0 & 0 & 0 & 0 & \frac{1}{l} & 0 & 1 & 0 \end{bmatrix} \begin{bmatrix} r_{ix}' \\ r_{iy}' \\ r_{iz}' \\ \theta_{ix}' \\ \theta_{iy}' \\ \theta_{iz}' \\ r_{jx}' \\ r_{jy}' \\ r_{jz}' \\ \theta_{jx}' \\ \theta_{jy}' \\ \theta_{jz}' \end{bmatrix} \tag{7.41}$$

This is the equation

$$v = T^T r' \tag{7.42}$$

The relationship between member forces S, for which

$$S^T = \langle F \quad M_{iz} \quad M_{iy} \quad M_T \quad M_{jz} \quad M_{jy} \rangle,$$

and the member distortions v are given

$$S = k'v \tag{7.43}$$

where

$$k' = \begin{bmatrix} \frac{AE}{l} & 0 & 0 & 0 & 0 & 0 \\ 0 & \frac{4EI_z}{l} & 0 & 0 & \frac{2EI_z}{l} & 0 \\ 0 & 0 & \frac{4EI_y}{l} & 0 & 0 & \frac{2EI_y}{l} \\ 0 & 0 & 0 & \frac{GI_P}{l} & 0 & 0 \\ 0 & \frac{2EI_z}{l} & 0 & 0 & \frac{4EI_z}{l} & 0 \\ 0 & 0 & \frac{2EI_y}{l} & 0 & 0 & \frac{4EI_y}{l} \end{bmatrix} \tag{7.44}$$

The member with its coordinate axes is shown in Fig. 7.8, and it is necessary (to transform the stiffness matrix to global coordinates) to construct the direction-cosine matrix

$$L^T = \begin{bmatrix} l_{11} & l_{12} & l_{13} \\ l_{21} & l_{22} & l_{23} \\ l_{31} & l_{32} & l_{33} \end{bmatrix}$$

The member ij is of course simply the general directed-line element defined in Chap. 2. From the coordinates (x_i,y_i,z_i) and (x_j,y_j,z_j), the projections of ij, namely, $(\Delta x,\Delta y,\Delta z)$, on the coordinate axes are obtained. That is,

$$\Delta x = x_j - x_i$$
$$\Delta y = y_j - y_i$$
$$\Delta z = z_j - z_i$$

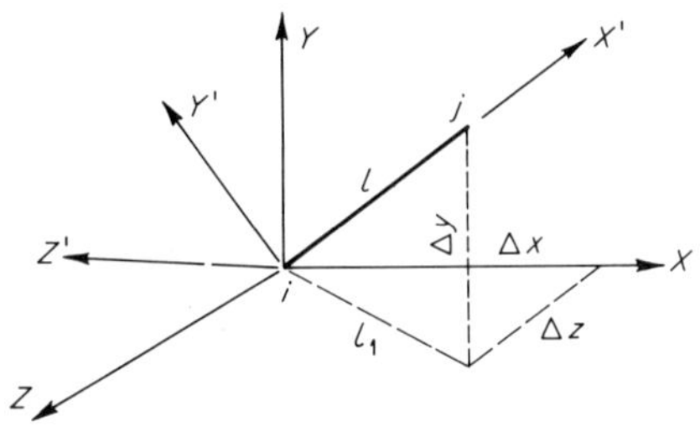

Figure 7.8

The member length is

$$l = \sqrt{(\Delta x)^2 + (\Delta y)^2 + (\Delta z)^2}$$

The projected length on the XZ plane is

$$l_1 = \sqrt{(\Delta x)^2 + (\Delta z)^2}$$

Hence

$$l = \sqrt{l_1{}^2 + (\Delta y)^2}$$

The L matrix may be generated as the product of three transformation matrices that are defined by rotations about the axes OX', OY', and OZ'. These three rotations are given by the Euler angles of the $X'Y'Z'$ axes. Only one of these three angles need be specified as data in addition to the coordinates of the ends i and j of the member.

It is seen in Fig. 7.9 that the $X'Y'Z'$ axes are completely specified when the X' direction is known, together with the angle θ_3 measured from the XZ plane to the axis iZ'. Notice, however, in Fig. 7.9 that a θ_3 angle can specify two positions for the $Z'Y'$ axes.

In order to use a consistent transformation, only one of these sets of coordinate positions will be permissible. Thus θ_3 is defined as the angle for which the projection of Y' on the Y axis is positive when the $Z'Y'$ axes are rotated about iX' through θ_3 so that Z' lies in the XZ plane. The axes Z_1', Y_1' are of course quite acceptable but are defined by the rotation $(180° + \theta_3)$.

Notice also in this derivation of L that θ_3 is measured positive in the positive sense about iX' from the XY plane to iZ'.

For the first transformation (Fig. 7.10), let

$$c_3 = \cos\theta_3$$
$$s_3 = \sin\theta_3$$

Define

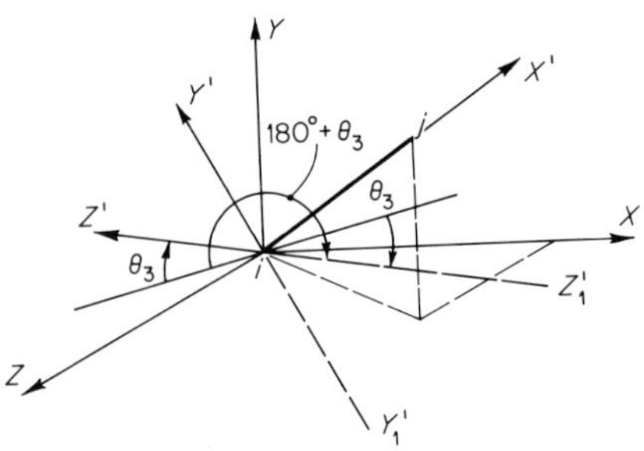

Figure 7.9 Alternate positions of coordinate axes.

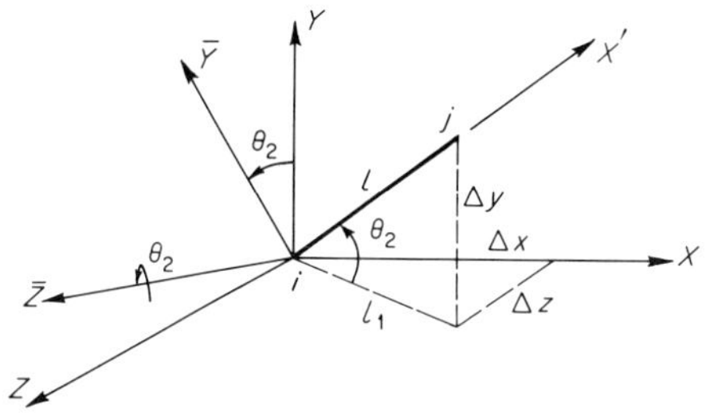

Figure 7.10 Position of axes after first rotation.

$$L_3 = \begin{bmatrix} 1 & 0 & 0 \\ 0 & c_3 & -s_3 \\ 0 & s_3 & c_3 \end{bmatrix} \tag{7.45}$$

When Y', Z' have been rotated so that Z' lies in the XZ plane, let these new axes be designated $\overline{Z}$, $\overline{Y}$. The transformation of components of vectors in the $X'\overline{Y}\overline{Z}$ system to those in $X'Y'Z'$ is given by

$$U' = L_3{}^T\overline{U} \tag{7.46}$$

The second rotation of the coordinate axes will be for $X'\overline{Y}$ about the $\overline{Z}$ axis so that $\overline{Y}$ coincides with the iY axis. These axes will be designated $\overline{Z}Y\overline{X}$ and are shown in Fig. 7.11. For this transformation, it is seen from Fig. 7.10 that

$$\cos\theta_2 = c_2 = \frac{l_1}{l}$$

$$\sin\theta_2 = s_2 = \frac{\Delta y}{l}$$

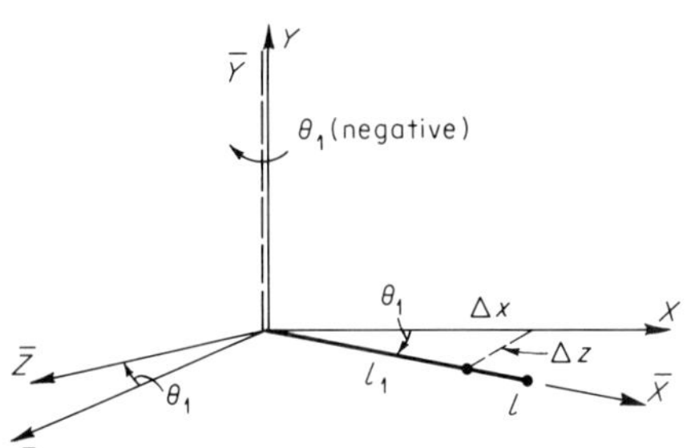

Figure 7.11 Position of axes after second rotation.

Define the matrix

$$L_2 = \begin{bmatrix} c_2 & -s_2 & 0 \\ s_2 & c_2 & 0 \\ 0 & 0 & 0 \end{bmatrix} \tag{7.47}$$

The transformation of components of vectors in the $\overline{X}Y\overline{Z}$ system to those in the $X'\overline{Y}\overline{Z}$ system is given by

$$\overline{U} = L_2{}^T\overline{\overline{U}} \tag{7.48}$$

The third rotation of the axes (see Fig. 7.11) will be for $\overline{X}\overline{Z}$ about the iY axis defined by the angle θ_1 such that

$$\cos\theta_1 = c_1 = \frac{\Delta x}{l_1}$$

$$\sin\theta_1 = s_1 = \frac{-\Delta z}{l_1}$$

Then,

$$L_1 = \begin{bmatrix} c_1 & 0 & -s_1 \\ 0 & 1 & 0 \\ s_1 & 0 & c_1 \end{bmatrix} \tag{7.49}$$

The transformation of the components of the vector U in the XYZ coordinate system to those of $\overline{\overline{U}}$ in the $\overline{X}Y\overline{Z}$ axes is given by

$$\overline{\overline{U}} = L_1{}^TU \tag{7.50}$$

It is seen from Eqs. (7.46), (7.48), and (7.50) the transformation from U to U' is given since

$$U' = L_3{}^TL_2{}^TL_1{}^TU \tag{7.51}$$

However it is known that

$$U' = L^TU \tag{7.52}$$

where L is the matrix of direction cosines of the XYZ axes with respect to XYZ. Hence,

$$L^T = L_3{}^TL_2{}^TL_1{}^T \tag{7.53}$$

or

$$L = L_1L_2L_3 \tag{7.54}$$

If the multiplication in Eq. (7.54) is carried out, the explicit expression obtained for L is

$$L = \begin{bmatrix} c_1c_2 & -c_1c_3s_2 + s_1s_3 & c_1s_2s_3 + s_1c_3 \\ s_2 & c_2c_3 & -c_2s_3 \\ -s_1c_2 & s_1s_2s_3 + c_1s_3 & -s_1s_2s_3 + c_3c_1 \end{bmatrix} \tag{7.55}$$

The matrix L_D is defined as the 12 × 12 superdiagonal matrix with L as submatrix diagonal blocks. That is,

$$L_D = \begin{bmatrix} L & & & \mathbf{0} \\ & L & & \\ & & L & \\ \mathbf{0} & & & L \end{bmatrix} \tag{7.56}$$

The displacement vector r' may now be obtained in terms of the global displacement vector r.

$$r' = L_D{}^T r \tag{7.57}$$

Hence

$$v = T^T L_D{}^T r \tag{7.58}$$

The contragradient law gives the force relationship

$$R = L_D T S \tag{7.59}$$

Finally, the force displacement relationship in global components is

$$R = L_D T k' T^T L_D{}^T r \tag{7.60}$$

Hence

$$k = L_D T k' T^T L_D{}^T \tag{7.61}$$

Again, if node displacements are known, member forces are

$$S = k' T^T L_D{}^T r \tag{7.62}$$

On examination of Eqs. (7.31), (7.40), and (7.61), it is noticed that their forms are identical. They differ only in the internal construction of the matrices T and L_D. The plane frame and grid structure matrices are only condensed special cases of the three-dimensional matrices and could be extracted from these.

The stiffness matrices k, obtained above, are singular since they have been expanded so that rigid-body displacements are contained in the vector r. Write the stiffness equation

$$R = kr$$

in partitioned form so that the vectors $\langle R \quad r \rangle$ are subdivided into the components at each end of the member. Then

$$\begin{bmatrix} R_i \\ R_j \end{bmatrix} = \begin{bmatrix} k_{ii} & k_{ij} \\ k_{ji} & k_{jj} \end{bmatrix} \begin{bmatrix} r_i \\ r_j \end{bmatrix} \tag{7.63}$$

It is now shown that the submatrices k_{ii}, k_{ij}, and k_{jj} are not independent, but that given any one, the others may be determined. This gives a useful alternative method for calculating k that may be used for curved members, for which it is convenient to calculate the effects of forces at one end.

The forces R_i and R_j acting on the member are in equilibrium. If then R_i is transformed to end j, the components are equal and opposite to R_j. Note that U here is identical with T in Eq. (2.45) but is used here to save confusion with T given in Eq. (7.42). That is,

$$R_j = -U^{-1}R_i \tag{7.64}$$

The matrix U^{-1} has been defined in Eq. (2.48) for the three-dimension case.

$$U^{-1} = \begin{bmatrix} 1 & \cdot & \cdot & \cdot & \cdot & \cdot \\ \cdot & 1 & \cdot & \cdot & \cdot & \cdot \\ \cdot & \cdot & 1 & \cdot & \cdot & \cdot \\ \cdot & \Delta z & -\Delta y & 1 & \cdot & \cdot \\ -\Delta z & \cdot & -\Delta x & \cdot & 1 & \cdot \\ \Delta y & \Delta x & \cdot & \cdot & \cdot & 1 \end{bmatrix} \tag{7.65}$$

The U^{-1} matrix for the plane frame and grid members is obtained by deletion of the appropriate rows and columns in Eq. (7.65). Substitute for R_j in Eq. (7.63) so that

$$\begin{bmatrix} R_i \\ -U^{-1}R_i \end{bmatrix} = \begin{bmatrix} k_{ii} & k_{ij} \\ k_{ji} & k_{jj} \end{bmatrix} \begin{bmatrix} r_i \\ r_j \end{bmatrix} \tag{7.66}$$

On expansion of Eq. (7.66),

$$R_i = k_{ii}r_i + k_{ij}r_j$$
$$R_i = -Uk_{ji}r_i - Uk_{jj}r_j$$

Comparing the coefficients of r_i, it is seen that

$$k_{ji} = -U^{-1}k_{ii} \tag{7.67}$$

Hence, from symmetry,

$$k_{ij} = -k_{ii}(U^{-1})^T \tag{7.68}$$

Comparing coefficients of r_j,

$$k_{jj} = -U^{-1}k_{ij} = U^{-1}k_{ii}(U^{-1})^T \tag{7.69}$$

It is seen that basic member force and displacement quantities could be chosen as $\langle R_i \quad r_i \rangle$. It will be convenient to use the definitions in Eqs. (7.22), (7.33), etc., however, since these use the rotations $\langle \phi_i \quad \phi_j \rangle$ relative to the chord and these have been used extensively in the calculation of fixed-end forces when loads occur along the member or temperature strains are present. If, for example, the plane frame member shown in Fig. 7.12 has a distributed load ω'_y over its whole length, the end rotations are

$$\phi_{iz} = -\phi_{jz} = \frac{\omega'_y l^3}{24EI}$$

The displacement vector for the member in the global system is thus given by

$$r^T = \langle r_{ix} \quad r_{iy} \quad \theta_{iz} \quad r_{jx} \quad r_{jy} \quad \theta_{jz} \rangle = \langle 0 \quad 0 \quad \frac{\omega'_y l^3}{24EI} \quad 0 \quad 0 \quad -\frac{\omega'_y l^3}{24EI} \rangle$$

The multiplication

$$R = -kr$$

then gives the fixed-end forces acting to restrain the member, and

$$R = kr$$

gives the joint release forces.

7.3 CONDENSATION OF STIFFNESS MATRICES

Consider first the stiffness matrix for the frame member given in Eq. (7.63). It has been shown that R_i and R_j are dependent forces and that consequently k can be generated from any one of its four submatrices. It is also found that k is a singular matrix and that the submatrix of largest size that is nonsingular is six for the three-dimensional member and three for the frame or grid member. This property can be demonstrated by calculating the stiffness at end j when end i has been released. This stiffness must necessarily be zero

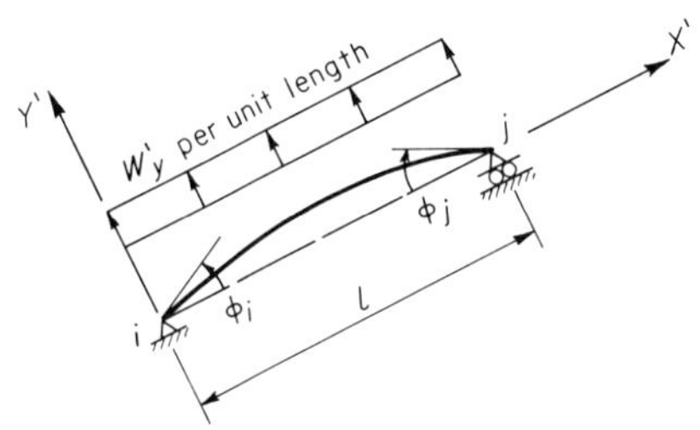

Figure 7.12 Distributed load on plane frame member.

since any displacement at j, the member being unrestrained at i, will simply cause rigid-body motion.

Thus in Eq. (7.63), to release end i, set $R_i \equiv 0$, so that

$$0 = k_{ii}r_i + k_{ij}r_j$$

Hence

$$r_i = -k_{ii}^{-1}k_{ij}r_j$$

Substitute now for r_i in the second equation in Eq. (7.63); then

$$R_j = (k_{jj} - k_{ji}k_{ii}^{-1}k_{ij})r_j \tag{7.70}$$

The reduced matrix in this case is a null matrix. This is proved by substituting for k_{jj}, k_{ji}, and k_{ij} in terms of k_{ii} and the U matrix from Eq. (7.67). That is,

$$(k_{jj} - k_{ji}k_{ii}^{-1}k_{ij}) = U^{-1}k_{ii}(U^{-1})^T - (-U^{-1}k_{ii})k_{ii}^{-1}[-k_{ii}(U^{-1})^T] = 0$$

Condensation of stiffness matrices can be of importance when a structure is composed of substructures that may be separated, so that internal subsets can be isolated in such a way that they do not affect the connection of the substructures. This is shown diagrammatically in Fig. 7.13.

In Fig. 7.13 there are three substructures I, II, and III. The subsets A, B, and C do not contribute stiffness at connection points between I, II, and III, and consequently their nodes may be released before the analysis proceeds from the kinematically determinate position $r \equiv 0$. In this case, r will be a subset of the total number of nodes that does not contain the released nodes. As the nodes in A, B, and C are released, their forces will produce modified fixed-end forces on the remaining nodes, and in addition within subsets, deflections will occur. Finally, when the deflections r are found, the extra deflections in A, B, and C must be calculated.

The problem is first discussed from a partitioning viewpoint and then from the expediency of practical calculations.

Consider one of the substructures, I, with its subset A. The node forces and deflections may be partitioned into the groups 1 and 2, according to whether they are in A or in the set connecting I to the rest of the system. Then,

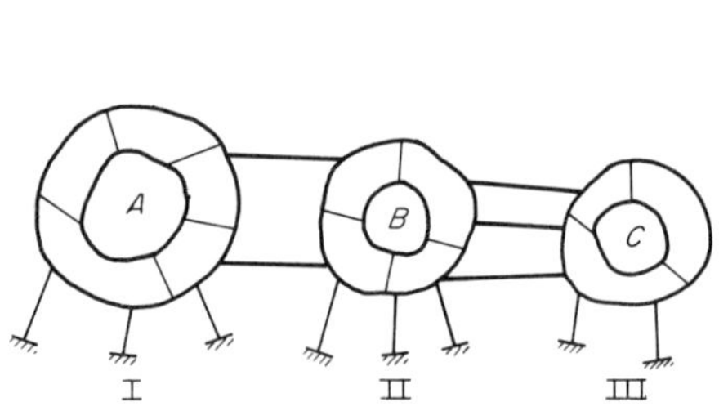

Figure 7.13 Substructures and subsystems.

$$\begin{bmatrix} R_1 \\ R_2 \end{bmatrix} = \begin{bmatrix} k_{11} & k_{12} \\ k_{21} & k_{22} \end{bmatrix} \begin{bmatrix} r_1 \\ r_2 \end{bmatrix} \tag{7.71}$$

1 First determine deflections r_{10} that occur when the forces R_1 act on A with r_2 held at zero. Evidently from (7.71),

$$r_{10} = k_{11}{}^{-1}R_1 \tag{7.72}$$

2 The forces R_1 now cause additional fixed-end forces at nodes r_2. With deflections r_1,

$$R_{2F} = k_{21}r_1 = k_{21}k_{11}{}^{-1}R_1 \tag{7.73}$$

The release forces are the negative of R_{2F}. That is,

$$\overline{R}_2 = R_2 - k_{21}k_{11}{}^{-1}R_1 \tag{7.74}$$

3 The reduced stiffness matrix $\overline{k}_{22}$ is obtained by setting $R_1 = 0$ in (7.71). Then

$$0 = k_{11}r_1 + k_{12}r_2$$

$$r_1 = -k_{11}{}^{-1}k_{12}r_2 \tag{7.75}$$

This equation gives the deflections r_1 when r_2 occur, and the r_1 nodes are free to deflect. Substitution for r_1 in the second equation (7.71), gives

$$\overline{R}_2 = (k_{22} - k_{21}k_{11}{}^{-1}k_{12})r_2 \tag{7.76}$$

or

$$\overline{R}_2 = \overline{k}_{22}r_2 \tag{7.77}$$

The above process is conveniently carried out without resort to the formal inversion of the matrix k_{11} by reducing the first submatrix k_{11} to an upper triangular matrix by gaussian reduction. That is, reduce Eq. (7.71) to

$$\begin{bmatrix} \overline{R}_1 \\ \overline{R}_2 \end{bmatrix} = \begin{bmatrix} k_{11(u)} & \overline{k}_{12} \\ 0 & \overline{k}_{22} \end{bmatrix} \begin{bmatrix} r_1 \\ r_2 \end{bmatrix} \tag{7.78}$$

Then the deflections r_{10} are obtained by back-substitution in the equation

$$\overline{R}_1 = k_{11(u)}r_{10} \tag{7.79}$$

The forces $\overline{R}_2$ are the required release forces. That is, the $\overline{R}_2$ in Eq. (7.78) are the same as those in Eq. (7.74). Finally, when r_2 are known, then r_1, the additional deflections of the released nodes, are calculated by back-substitution in the equation

$$k_{11(u)}r_1 = -\overline{k}_{12}r_2 \tag{7.80}$$

This is actually back-substitution applied to the first r_1 terms in the system

$$[k_{11(u)} \quad \bar{k}_{12}]\begin{bmatrix} r_1 \\ r_2 \end{bmatrix} = \begin{bmatrix} 0 \\ 0 \end{bmatrix} \tag{7.81}$$

To examine the gaussian reduction process in more detail, write the stiffness equation in full, instead of partitioned, form.

$$\begin{bmatrix} R_1 \\ R_2 \\ R_3 \\ \vdots \\ R_n \end{bmatrix} = \left[\begin{array}{c|ccccc} k_{11} & k_{12} & k_{13} & \cdots & k_{1n} \\ \hline k_{21} & k_{22} & k_{23} & \cdots & k_{2n} \\ k_{31} & k_{32} & k_{33} & \cdots & k_{3n} \\ \cdots & \cdots & \cdots & \cdots & \cdots \\ k_{n1} & k_{n2} & k_{n3} & \cdots & k_{nn} \end{array}\right] \begin{bmatrix} r_1 \\ r_2 \\ r_3 \\ \vdots \\ r_n \end{bmatrix} \tag{7.82}$$

The first step in the reduction divides row 1 by k_{11}, that is, by $k_{11}{}^{-1} = 1/k_{11}$ for a single term. The first row becomes

$$\frac{R_1}{k_{11}} = \begin{bmatrix} 1 & \dfrac{k_{12}}{k_{11}} & \dfrac{k_{13}}{k_{11}} & \cdots & \dfrac{k_{1n}}{k_{11}} \end{bmatrix} \begin{bmatrix} r_1 \\ r_2 \\ r_3 \\ \vdots \\ r_n \end{bmatrix} \tag{7.83}$$

It is seen that if all node deflections $r_2, \ldots, r_n$ are held equal to zero,

$$\frac{R_1}{k_{11}} = R_1 f_{11} = \text{the deflection of } r_1 \equiv \bar{r}_{10}$$

Note that $\bar{r}_{10}$ is modified to produce r_{10} by the other deflections of nodes released. In addition, if $r_2, r_3, \ldots, r_n$ are known, then from Eq. (7.83),

$$r_1 = -\begin{bmatrix} \dfrac{k_{12}}{k_{11}} & \dfrac{k_{13}}{k_{11}} & \cdots & \dfrac{k_{1n}}{k_{11}} \end{bmatrix} \begin{bmatrix} r_2 \\ r_3 \\ \vdots \\ r_n \end{bmatrix} \tag{7.84}$$

This is the equation $r_1 = -k_{11}{}^{-1}k_{12}r_2$ and is simply the back-substitution process referred to in Eq. (7.75).

Continuing the reduction process, for $i > 1$,

$$\bar{R}_i = R_i - R_1 \frac{k_{i1}}{k_{11}} \tag{7.85}$$

Once again this is seen to be

$$\bar{R}_2 = R_2 - k_{21}k_{11}^{-1}R_1$$

For the $(n-1) \times (n-1)$ submatrix with k_{22} as its leading terms,

$$k_{ij} = k_{ij} - \frac{k_{i1}k_{1i}}{k_{11}} \qquad \begin{array}{l} i = 2, n \\ j = 2, n \end{array} \tag{7.86}$$

This is the equation

$$\bar{k}_{22} = k_{22} - k_{21}k_{11}^{-1}k_{12}$$

When joint 1 has been released, then the process can proceed to joint 2, which is in turn released. Notice that the only effect that this and subsequent releases have on r_1 is to produce additional deflection that is accounted for in Eq. (7.84).

7.4 THE DIRECT STIFFNESS METHOD FOR THE ANALYSIS OF ELASTIC SYSTEMS

The direct stiffness method is one of the most useful tools of analysis to emerge from the current development of computer techniques. It has the advantage of being applicable to many areas of analysis such as finite-element idealization of continua, as well as to framed structures.

It has been shown in Chap. 6 and in the beginning of the present chapter that the solution to the structures problem by the displacement method can be considered in the formal steps:

1	$v = ar$	displacement transformation
2	$S = kv$	member forces in terms of member displacements
3	$R = a^T S$	joint equilibrium equations
4	$r = (a^T k a)^{-1} R$	solution for joint displacements
5	$S = kar$	member forces

In Sec. 7.2 expressions have been developed for the member-stiffness matrix k, that relates force and displacement components of the member, S, v, expressed in their global components. In such cases the displacement transformation matrix a in equation 1 above will contain only unit or zero values. Moreover, if the member displacements are given in the same order as the node displacements (for example, in the plane frame, r_x, r_y, θ_z), then the unit values form unit submatrices in a.

Consider a particular member N connecting the nodes P and Q, and write down that part of $a^T ka$ concerned with member N. Partition r, and a in blocks so that the transformation $v_N = a_n r$ is as in Eq. (7.87).

$$v_N = \begin{bmatrix} r_p \\ r_q \end{bmatrix} = \begin{bmatrix} 0 & 0 & 0 & 0 & I & 0 & 0 & 0 & 0 \\ 0 & 0 & 0 & 0 & \underset{\substack{\uparrow \\ p\text{th} \\ \text{column}}}{0} & 0 & 0 & \underset{\substack{\uparrow \\ q\text{th} \\ \text{column}}}{I} & 0 \end{bmatrix} \begin{bmatrix} r_1 \\ r_2 \\ \vdots \\ r_p \\ \vdots \\ r_q \\ \vdots \end{bmatrix} \tag{7.87}$$

The member-stiffness matrix is partitioned as in Eq. (7.63) so that the contribution to

$K = a^T k a$

for member N is given in Eq. (7.88) by K_N.

$$K_N = \begin{bmatrix} 0 & 0 \\ 0 & 0 \\ 0 & 0 \\ 0 & 0 \\ I & 0 \\ 0 & 0 \\ 0 & 0 \\ 0 & I \\ 0 & 0 \end{bmatrix} \begin{matrix} \\ \\ \\ \\ \leftarrow p\text{th row} \\ \\ \\ \leftarrow q\text{th row} \\ \\ \end{matrix} \begin{bmatrix} k_{ii} & k_{ij} \\ k_{ji} & k_{jj} \end{bmatrix} \begin{bmatrix} 0 & 0 & 0 & 0 & \overset{p\text{th column}}{\overset{\downarrow}{I}} & 0 & 0 & 0 & 0 \\ 0 & 0 & 0 & 0 & 0 & 0 & 0 & \underset{\substack{\uparrow \\ q\text{th} \\ \text{column}}}{I} & 0 \end{bmatrix} = \begin{bmatrix} 0 & 0 \\ 0 & 0 \\ 0 & 0 \\ 0 & 0 \\ I & 0 \\ 0 & 0 \\ 0 & 0 \\ 0 & I \\ 0 & 0 \end{bmatrix} \begin{bmatrix} 0 & 0 & 0 & 0 & k_{ii} & 0 & 0 & k_{ij} & 0 \\ 0 & 0 & 0 & 0 & k_{ji} & 0 & 0 & k_{jj} & 0 \end{bmatrix} \tag{7.88}$$

Thence

$$K_N = \begin{matrix} \\ \\ \\ \\ p\text{th row} \rightarrow \\ \\ \\ q\text{th row} \rightarrow \\ \\ \end{matrix} \begin{bmatrix} 0 & 0 & 0 & 0 & \overset{\substack{p\text{th} \\ \text{column} \\ \downarrow}}{0} & 0 & 0 & \overset{\substack{q\text{th} \\ \text{column} \\ \downarrow}}{0} & 0 \\ 0 & 0 & 0 & 0 & 0 & 0 & 0 & 0 & 0 \\ 0 & 0 & 0 & 0 & 0 & 0 & 0 & 0 & 0 \\ 0 & 0 & 0 & 0 & 0 & 0 & 0 & 0 & 0 \\ 0 & 0 & 0 & 0 & k_{ii} & 0 & 0 & k_{ij} & 0 \\ 0 & 0 & 0 & 0 & 0 & 0 & 0 & 0 & 0 \\ 0 & 0 & 0 & 0 & 0 & 0 & 0 & 0 & 0 \\ 0 & 0 & 0 & 0 & k_{ji} & 0 & 0 & k_{jj} & 0 \\ 0 & 0 & 0 & 0 & 0 & 0 & 0 & 0 & 0 \end{bmatrix} \tag{7.89}$$

In Eq. (7.89) each term is of course a submatrix whose size will depend on the type of structure being analyzed. For example, for truss structures, the size is 2×2, whereas for plane frames and grids, it is 3×3, and for three-dimensional frames it is 6×6.

From Eq. (7.89) it is seen that it is not necessary to carry out the formation of the a matrix in Eq. (7.88). The blocks k_{ii}, k_{ij}, etc., can be added directly into K by using the node numbers P and Q to locate the appropriate submatrix.

If the order of k_{ii}, etc., is $n \times n$, then the row and column locations in K for the first terms are as follows:

		Row No.	*Column No.*
1	$(k_{ii})_1$	$n(p-1)+1$	$n(p-1)+1$
2	$(k_{ij})_1$	$n(p-1)+1$	$n(q-1)+1$
3	$(k_{ji})_1$	$n(q-1)+1$	$n(p-1)+1$
4	$(k_{jj})_1$	$n(q-1)+1$	$n(q-1)+1$

Suitable computer program coding will then add in all the additional terms of the k submatrices.

Example 7.1 Structure-stiffness matrix

The frame shown in Fig. 7.14 has six members and four free joints. Since the three support joints are fixed, they may be omitted from the stiffness calculation $R = Kr$. In Fig. 7.14, two different node numberings are shown, and the effect of this on the structure of K will be examined.

For the planar frame, the member stiffness k_N is a 6×6 matrix composed for four 3×3 blocks. From Fig. 7.14a the table of member-node numbering is given with support nodes zero:

Member	*Node i*	*Node j*
1	0	1
2	1	2
3	2	3
4	3	0
5	3	4
6	4	0

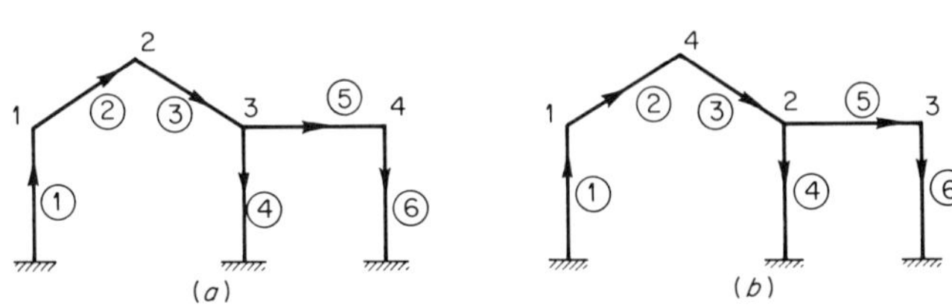

Figure 7.14

The structure-stiffness matrix for the free joints 1, 2, 3, and 4 is obtained by adding in the contributions from the member stiffnesses k_N. In this case K will be a 12×12 matrix composed of four 3×3 blocks. It is seen that all diagonal blocks will have stiffness terms, whereas off-diagonal blocks are nonzero only if the row-column number combination represents a physical connection of two nodes by a member. The row and column numbers then give the member connectivity.

With the node numbering given in Fig. 7.14*a*, the K matrix will be as in Eq. (7.90).

$$K = \begin{bmatrix} K_{11} & K_{12} & 0 & 0 \\ K_{21} & K_{22} & K_{23} & 0 \\ 0 & K_{32} & K_{33} & K_{34} \\ 0 & 0 & K_{43} & K_{44} \end{bmatrix} \tag{7.90}$$

In this case nonzero terms are packed closely on either side of the diagonal and the matrix is banded.

The same stiffness matrix is shown in Eq. (7.91) for the different node numbering given in Fig. 7.14*b*. Note that with this numbering the nonzero off-diagonal terms are dispersed through the matrix.

$$K = \begin{bmatrix} K_{11} & 0 & 0 & K_{14} \\ 0 & K_{22} & K_{23} & K_{24} \\ 0 & K_{32} & K_{33} & 0 \\ K_{41} & K_{42} & 0 & K_{44} \end{bmatrix} \tag{7.91}$$

It will be a general feature of the direct stiffness method that the K matrix can be banded as in Eq. (7.90) by a suitable node-numbering sequence. Since the stiffness matrix is symmetrical, only the upper half of the band may be stored if the equation solution is carried out with an efficient band solver. If the equation solution is carried out by iteration methods, then the band property need not be exploited. For efficient storage, however, only nonzero blocks should be stored so that advantage can be taken of the general sparsity of terms in the stiffness matrix.

Calculation of Member Forces

For member N, node numbers p and q, the member displacements at ends i and j, are located in the n rows of r following the locations $n(p-1)+1$ for i and $n(q-1)+1$ for j. When these are loaded into the vector r_N, basic member forces are calculated with equations such as (7.32) and (7.62). That is,

$$S = k' T^T L_D{}^T r_N \tag{7.92}$$

Support Conditions

A support condition may be allowed for in the inversion (or solution) of the structure-stiffness matrix by deleting the row and column corresponding to the support displacement. In this way free-node deflections are calculated.

However the support deflections must be set equal to zero since in the formation of the vector r_N, the support values may be involved. This procedure may be incorporated in the solution of the equations by zeroing the row and column corresponding to the support condition and placing unity on the diagonal. When the force in this direction is set equal to zero, the result is that a zero displacement is generated.

If reactions are to be calculated for the structure, they may be found at the time of back-substitution in the solution of the equations. If when the equations are being solved (by gaussian elimination), the pivot point corresponding to each support condition is skipped, then this particular equation becomes a reduced-stiffness equation and when the back-substitution takes place, the support reaction appears in the deflection vector.

The calculation of reactions in this manner follows closely the idea of kinematic determinacy and indeterminacy discussed in Sec. 6.1, Eqs. (6.26)ff. For illustration, suppose that the nodes are all free or all fixed, as for the frame in Fig. 7.14. Partition R and r into $\langle R_F \quad R_S \rangle$ and $\langle r_F \quad r_S \rangle$. The suffixes F and S refer to free and fixed nodes respectively. Then it is assumed that r_S is known, generally equal to zero. Partition the complete stiffness matrix according to free and fixed nodes so that

$$\begin{bmatrix} R_F \\ R_S \end{bmatrix} = \begin{bmatrix} K_{FF} & K_{FS} \\ K_{SF} & K_{SS} \end{bmatrix} \begin{bmatrix} r_F \\ r_S \end{bmatrix} \tag{7.93}$$

Since r_S is known from the first equation in (7.93),

$$r_F = K_{FF}{}^{-1}(R_F - K_{FS} r_S) \tag{7.94}$$

The reactions are given by

$$R_S = K_{SF} r_F + K_{SS} r_S \tag{7.95}$$

Support Condition in a Direction other than that of the Coordinate Axes

A support condition such as a roller on an inclined plane (see Fig. 7.15) is not allowed for in the previous discussion, in which it has been assumed that

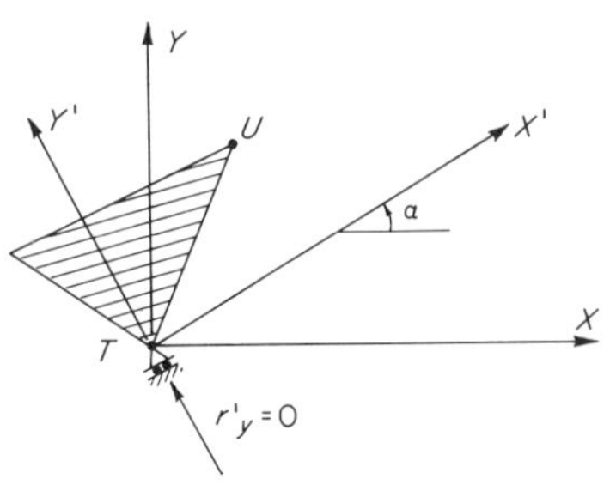

Figure 7.15 Inclined roller support.

the constraint condition $r_S = 0$ applies to displacements in the global coordinates. Suppose that for the planar support shown in Fig. 7.15, $r'_y \equiv 0$. Consider the stiffness equations at joints T and U in detail. Joint U is a neighboring joint to T, and has a connection to T. Then

$$R_T = K_{T1}r_1 + K_{T2}r_2 + \cdots + K_{TT}r_T + K_{TU}r_U + \cdots \tag{7.96}$$

Transform R_T and r_T to the oblique coordinate axes, using Eqs. (2.29*a*) and (2.29*b*).

$$R'_T = L^T R_T \qquad \text{or} \qquad R_T = LR'_T$$

$$r'_T = L^T r_T \qquad \text{or} \qquad r_T = Lr'_T$$

Substitute in Eq. (7.96) so that

$$LR'_T = K_{T1}r_1 + K_{T2}r_2 + \cdots + K_{TT}Lr'_T + \cdots K_{TU}r_U$$

Finally, multiplying both sides of this equation by L^T, and substituting $K' = L^T K_{TT} L$ [see Eq. (7.21)],

$$R'_T = L^T K_{T1}r_1 + L^T K_{T2}r_2 + \cdots + K'_{TT}r'_T + \cdots L^T K_{TU}r_U \tag{7.97}$$

It is seen that all terms in the Tth row must be premultiplied by L^T, and in addition K'_{TT} must be formed with the extra postmultiplication by L. Once the transformation has been made to Eq. (7.97) for joint T, the support condition $r'_y = 0$ may be treated in any of the previously mentioned ways.

Modification to all stiffness equations for joints such as U that are connected to T must be made. Write the equation for node U in full.

$$R_U = K_{U1}r_1 + K_{U2}r_2 + \cdots + K_{UU}r_U + \cdots + K_{UT}r_T + \cdots$$

Substitution for r_T gives

$$R_U = K_{U1}r_1 + K_{U2}r_2 + \cdots + K_{UU}r_U + \cdots + K_{UT}Lr'_T + \cdots \tag{7.98}$$

All terms in the Tth column must therefore be postmultiplied by the rotation matrix L.

Example 7.2 Analysis of a plane truss by the direct stiffness method

The stiffness matrix for force and displacement components of the member ij (see Fig. 7.16) with axial force as the only basic force (that is, the member has pins at each end) is readily deduced in the same way as that in Eq. (7.31). From Eq. (7.24), the transformation from displacements of i and j in member coordinates to member extension δ, is given by

$$\delta = [-1 \quad 0 \quad 1 \quad 0] \begin{bmatrix} r'_{ix} \\ r'_{iy} \\ r'_{jx} \\ r'_{jy} \end{bmatrix} \tag{i}$$

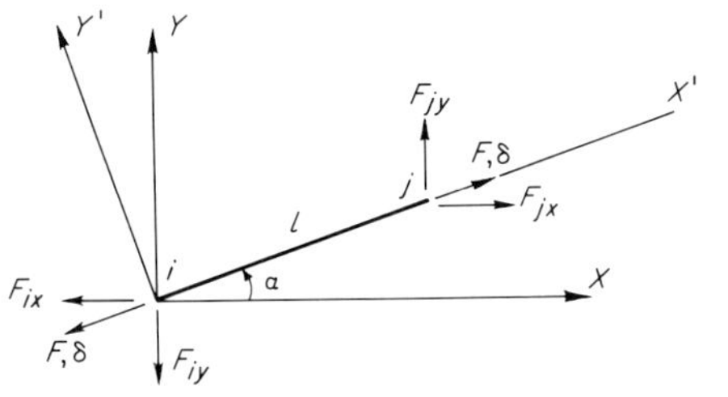

Figure 7.16 Pin-jointed truss member.

The transformation from global to local components of displacements is

$$\begin{bmatrix} r'_{ix} \\ r'_{iy} \\ r'_{jx} \\ r'_{jy} \end{bmatrix} = \begin{bmatrix} \cos\alpha & \sin\alpha & 0 & 0 \\ -\sin\alpha & \cos\alpha & 0 & 0 \\ 0 & 0 & \cos\alpha & \sin\alpha \\ 0 & 0 & -\sin\alpha & \cos\alpha \end{bmatrix} \begin{bmatrix} r_{ix} \\ r_{iy} \\ r_{jx} \\ r_{jy} \end{bmatrix} \qquad (ii)$$

From (i) and (ii),

$$\delta = [-\cos\alpha \quad -\sin\alpha \quad \cos\alpha \quad \sin\alpha] \begin{bmatrix} r_{ix} \\ r_{iy} \\ r_{jx} \\ r_{jy} \end{bmatrix} \qquad (iii)$$

The member stiffness is

$$k' = \frac{EA}{l}$$

Hence

$$k = \begin{bmatrix} -\cos\alpha \\ -\sin\alpha \\ \cos\alpha \\ \sin\alpha \end{bmatrix} \frac{EA}{l} [-\cos\alpha \quad -\sin\alpha \quad \cos\alpha \quad \sin\alpha] \qquad (iv)$$

or

$$k = \frac{EA}{l} \begin{bmatrix} \cos^2\alpha & \cos\alpha\ \sin\alpha & -\cos^2\alpha & -\cos\alpha\sin\alpha \\ \cos\alpha\sin\alpha & \sin^2\alpha & -\cos\alpha\sin\alpha & -\sin^2\alpha \\ -\cos^2\alpha & -\cos\alpha\sin\alpha & \cos^2\alpha & \cos\alpha\sin\alpha \\ -\cos\alpha\sin\alpha & -\sin^2\alpha & \cos\alpha\sin\alpha & \sin^2\alpha \end{bmatrix} \qquad (v)$$

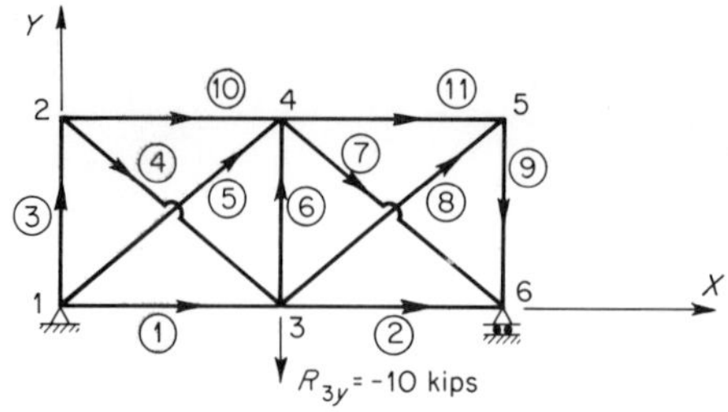

Figure 7.17 Truss analysis.

The truss to be analyzed is shown in Fig. 7.17. Only node deflections and member forces will be calculated, so that equations related to the support conditions will be eliminated. Node and member numbering is given in Fig. 7.17. The area of all members will be taken to be 1 sq in., an $E = 30 \times 10^3$ in.-kip units. Hence

$$\frac{EA}{l} = 250 \qquad \text{for members 10 ft 0 in. long}$$

$$\frac{EA}{l} = 186.75 \qquad \text{for diagonal members}$$

The positive-member senses are shown in Fig. 7.17. Notice they have been chosen so that node number at j is always greater than node number at i. This is a useful convention that allows k to be formed in the arrangement that is added into the stiffness matrix K. In Eq. (v) it is seen that the four submatrices are numerically the same. The off-diagonal blocks simply have their signs changed.

TABLE 7.1

Member number	Node number i	j	cos α	sin α
1	1	3	1.0	0.0
2	3	6	1.0	0.0
3	1	2	0.0	1.0
4	2	3	0.707	−0.707
5	1	4	0.707	0.707
6	3	4	0.0	1.0
7	4	6	0.707	−0.707
8	3	5	0.707	0.707
9	5	6	0.0	−1.0
10	2	4	1.0	0.0
11	4	5	1.0	0.0

From Eq. (v) and Table 7.1 of member properties, the stiffness matrices have been calculated for the members and are listed below:

Member number	*Stiffness matrix*
1, 2, 10, 11	$250\left[\begin{array}{cc\|cc} 1 & 0 & -1 & 0 \\ 0 & 0 & 0 & 0 \\ \hline -1 & 0 & 1 & 0 \\ 0 & 0 & 0 & 0 \end{array}\right]$
3, 6, 9	$250\left[\begin{array}{cc\|cc} 0 & 0 & 0 & 0 \\ 0 & 1 & 0 & -1 \\ \hline 0 & 0 & 0 & 0 \\ 0 & -1 & 0 & 1 \end{array}\right]$
5, 8	$250\left[\begin{array}{cc\|cc} 0.3735 & 0.3735 & -0.3735 & -0.3735 \\ 0.3735 & 0.3735 & -0.3735 & -0.3735 \\ \hline -0.3735 & -0.3735 & 0.3735 & 0.3735 \\ -0.3735 & -0.3735 & 0.3735 & 0.3735 \end{array}\right]$
4, 7	$250\left[\begin{array}{cc\|cc} 0.3735 & -0.3735 & -0.3735 & 0.3735 \\ -0.3735 & 0.3735 & 0.3735 & -0.3735 \\ \hline -0.3735 & 0.3735 & 0.3735 & -0.3735 \\ 0.3735 & -0.3735 & -0.3735 & 0.3735 \end{array}\right]$

The total-stiffness matrix [Eq. (vi)] for the structure is constructed by the addition of the 2×2 submatrices composing the member-stiffness matrices above into the locations indicated by the node numbers given in Table 7.1. (That is, Table 7.1 gives the topology of the structure.)

The stiffness matrix K for the free nodes is obtained from K_T by deleting those rows and columns for which displacements are zero. That is, for $r_{1x} = r_{1y} = r_{6y} = 0$. These give the row and column values, 1, 2, and 12 to be deleted.

The load vector R is given by

$$R^T = \langle 0 \quad 0 \quad 0 \quad -10 \quad 0 \quad 0 \quad 0 \quad 0 \quad 0 \rangle$$

The node displacements are calculated by

$$r = K^{-1}R$$

Then, given in inches,

$$K_T = 250\left[\begin{array}{cc|cc|cc|cc|cc|cc}
1.3735 & 0.3735 & 0 & 0 & -1 & 0 & -0.3735 & -0.3735 & 0 & 0 & 0 & 0 \\
0.3735 & 1.3735 & 0 & -1 & 0 & 0 & -0.3735 & -0.3735 & 0 & 0 & 0 & 0 \\
\hline
0 & 0 & 1.3735 & -0.3735 & -0.3735 & 0.3735 & -1 & 0 & 0 & 0 & 0 & 0 \\
0 & -1 & -0.3735 & 1.3735 & 0.3735 & -0.3735 & 0 & 0 & 0 & 0 & 0 & 0 \\
\hline
-1 & 0 & -0.3735 & 0.3735 & 2.747 & 0 & 0 & 0 & -0.3735 & -0.3735 & -1 & 0 \\
0 & 0 & 0.3735 & -0.3735 & 0 & 1.747 & 0 & -1 & -0.3735 & -0.3735 & 0 & 0 \\
\hline
-0.3735 & -0.3735 & -1 & 0 & 0 & 0 & 2.747 & 0 & -1 & 0 & -0.3735 & 0.3735 \\
-0.3735 & -0.3735 & 0 & 0 & 0 & -1 & 0 & 1.747 & 0 & 0 & 0.3735 & -0.3735 \\
\hline
0 & 0 & 0 & 0 & -0.3735 & -0.3735 & -1 & 0 & 1.3735 & 0.3735 & 0 & 0 \\
0 & 0 & 0 & 0 & -0.3735 & -0.3735 & 0 & 0 & 0.3735 & 1.3735 & 0 & -1 \\
\hline
0 & 0 & 0 & 0 & -1 & 0 & -0.3735 & 0.3735 & 0 & 0 & 1.3735 & -0.3735 \\
0 & 0 & 0 & 0 & 0 & 0 & 0.3735 & -0.3735 & 0 & -1 & -0.3735 & 1.3735
\end{array}\right] \tag{vi}$$

$$r = \begin{bmatrix} 0.02000 \\ -0.01097 \\ 0.00903 \\ -0.05129 \\ 0.00903 \\ -0.03322 \\ -0.00193 \\ -0.01097 \\ 0.01807 \end{bmatrix} \begin{matrix} \left.\begin{matrix} \\ \\ \end{matrix}\right\} 2 \\ \left.\begin{matrix} \\ \\ \end{matrix}\right\} 3 \\ \left.\begin{matrix} \\ \\ \end{matrix}\right\} 4 \\ \left.\begin{matrix} \\ \\ \end{matrix}\right\} 5 \\ \} 6 \end{matrix} \quad \text{(node)}$$

Member forces may now be calculated with Eq. (*iii*) by extracting the appropriate deflections given above, together with the zero values at the supports. When this has been done,

$$S = \begin{bmatrix} 2.2575 \\ 2.26000 \\ -2.74250 \\ 3.87515 \\ -3.19386 \\ 4.51750 \\ -3.19254 \\ 3.87647 \\ -2.74250 \\ -2.74250 \\ -2.74000 \end{bmatrix} \begin{matrix} 1 \\ 2 \\ 3 \\ 4 \\ 5 \\ 6 \\ 7 \\ 8 \\ 9 \\ 10 \\ 11 \end{matrix} \quad \text{(member)}$$

Notice that some round of error has occurred in these member forces, possibly because of truncation of the r displacements. The above problem has been worked with the use of SMIS. It will be found that direct stiffness problems are not so conveniently worked by using the matrix interpretative program because they require a good deal of indexing of material within the matrices and are better served by general programs.

7.5 METHODS FOR THE SOLUTION OF THE STIFFNESS EQUATIONS

The same problem of efficient utilization of the high-speed store of the computer arises here in the direct stiffness method as has already been encountered in the structure-cutter and tree-structure algorithms for the force method. It is found that except for small problems, the storing of the whole K matrix in core soon exhausts available memory. Fortunately in the direct stiffness method it is relatively simple either to take account of zero submatrices by omitting them, or alternatively to arrange node numbering so that the matrix is narrowly banded. Furthermore in structures that exhibit a regular pattern such as high-rise building frames, both iterative and tridiagonalization procedures can be most effective. In the direct solution methods it may be found necessary to use double precision to iterate on residuals.

Various equation-solution techniques are discussed in this section. Analysis of high-rise structures is left to a separate section, 7.6.

1 Gauss-Seidel Iteration

Gauss-Seidel iteration at first sight has considerable appeal since it can be arranged so that nonzero submatrices are omitted, and hence band width is of no consequence in allocation of available storage. However it is found that convergence is not necessarily ensured for all classes of problems. Various types of overrelaxation factors are found to improve the convergence rate.

The stiffness equations will be written for the displacements and forces at each node of the structure. For example, for a planar structure, R_i and r_i for node i will each be 3×1 vectors, and K then is partitioned into 3×3 blocks. Write the equations then

$$\begin{aligned} K_{11}r_1 + K_{12}r_2 + K_{13}r_3 &= R_1 \\ K_{21}r_1 + K_{22}r_2 + K_{23}r_3 &= R_2 \\ K_{31}r_1 + K_{32}r_2 + K_{33}r_3 &= R_3 \end{aligned} \tag{7.99}$$

In the iteration procedure the equations are written so that each of the vectors $r_1, r_2, \ldots, r_n$ is given in terms of the other values. That is,

$$\begin{aligned} r_1 &= K_{11}^{-1}(R_1 - K_{12}r_2 - K_{13}r_3) \\ r_2 &= K_{22}^{-1}(R_2 - K_{21}r_1 - K_{23}r_3) \\ r_3 &= K_{33}^{-1}(R_3 - K_{31}r_1 - K_{32}r_2) \end{aligned} \tag{7.100}$$

The general expression is

$$r_i = K_{ii}^{-1}(R_i - \sum_{l=1,i-1} K_{il}r_l - \sum_{l=i+1,N} K_{il}r_l) \tag{7.101}$$

It is seen that $K_{ii}^{-1} = F_{ii}$, the flexibility matrix for node i with all other joints fixed. The term inside the parentheses in Eq. (7.101) is the summation of the total joint force at i when all other joints have been given their prescribed displacements. If then all the displacements $r_1, r_2, \ldots, r_{i-1}, r_{i+1}, \ldots, r_N$ have been given to the structure, with joint i held fixed, the summation in the parentheses gives the force necessary to produce r_i.

The iteration process starts with values' being assumed (usually zero) for $r_1, r_2, \ldots, r_N$. In the Gauss-Seidel iteration, the value of r_i at the $s + 1$ iteration cycle is given by

$$r_i^{s+1} = F_{ii}(R_i - \sum_{l=1,i-1} K_{il}r_l^{s+1} - \sum_{l=i+1,N} K_{il}r_l^s) \tag{7.102}$$

Notice that in the calculation of r_i^{s+1}, advantage is taken of the improved values of $r_1, r_2, \ldots, r_{i-1}$ calculated in this cycle. It is more convenient to work with the increment to r_i, namely, Δr_i, during each cycle of iteration. Let

$$\Delta r_i^s = r_i^{s+1} - r_i^s \tag{7.103}$$

Then from Eq. (7.102),

$$\Delta r_i^s = F_{ii}(R_i - \sum_{l=1,i-1} K_{il}r_l^{s+1} - \sum_{l=i,N} K_{il}r_l^s) \tag{7.104}$$

The new r_i value is then given by

$$r_i^{s+1} = r_i^s + \Delta r_i^s \tag{7.105}$$

This form is convenient since it allows for an overrelaxation factor to be applied to Δr_i^s.

It is found that convergence is generally accelerated if a factor β is applied, $1 < \beta < 2$, so that

$$r_i^{s+1} = r_i^s + \beta\,\Delta r_i^s \tag{7.106}$$

The value of β must be obtained from experience by considering the structural type. For example, in finite-element problems, values of $\beta = 1.8$ to 1.9 have been found to be near optimum.

When an iteration cycle is complete, substitution of the displacement vector in Eqs. (7.99) will give the difference between the forces applied to the joints and the externally applied loads. Evidently this difference should be as small as possible. For example, if the sum of the absolute values of these differences is less than 10^{-3} times the sum of the applied loads, the system should have converged.

For many structures the Gauss-Seidel process does not converge. For example, if small values of certain displacements cause large values elsewhere, then there is a distinct possibility that convergence will be poor. Such is the case for the cantilever beam shown in Fig. 7.18, where small rotations at sections near (1) cause large deflections at the end of the cantilever. It is

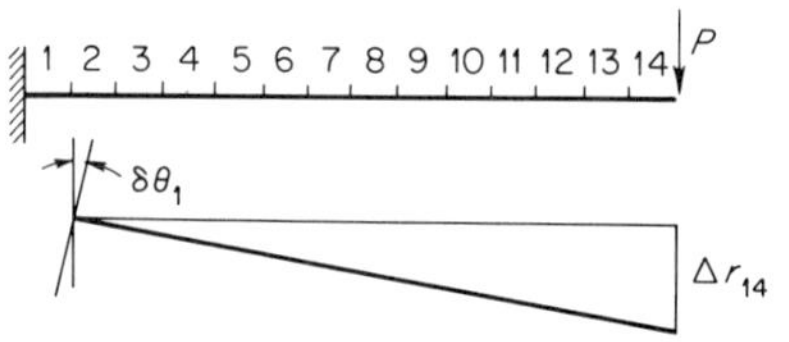

Figure 7.18 Poorly coupled displacements.

found that this problem also has round-off error difficulties when gaussian elimination is used.

This type of convergence problem may be overcome by using an energy balance on the structure. After a given number of cycles, let r_1 denote the calculated displacements, and suppose that the true values r are related to r_1 by a factor δ such that

$$r = \delta r_1 \tag{7.107}$$

The total strain energy may be calculated from the load vector R and also as a quadratic function of displacements [see Eq. (6.6)]. That is,

$$U = \tfrac{1}{2}R^T r = \tfrac{1}{2}r^T K r \tag{7.108}$$

Since r_1 is known, this affords a means by which to calculate δ. Substitute for r in Eq. (7.108).

$$R^T \delta r_1 = \delta r_1{}^T K r_1$$

Then

$$\delta = \frac{R^T r_1}{r_1{}^T K r_1} \tag{7.109}$$

The factor δ may be used every 10 to 20 cycles. The physical action of δ is that it induces the structure to take up rigid-body type of displacements such as those shown in Fig. 7.18, where the displacement $\delta\theta$ has caused little strain energy but a considerable amount of external work.

In the iteration process it is necessary to store only nonzero terms and in addition to make use of the symmetry of K by storing only those terms on or above the diagonal. This may be done in the computer program by generating an array that gives the column positions of successive nonzero matrices in any row. Such an array must also give the starting location of each row.

Example 7.3 Storage arrangement suitable for Gauss-Seidel iteration

Suppose that the structure has two displacements per joint so that the matrices in Eqs. (7.99) and (7.100) will be 2×2. In the computer program, K will be dimensioned as a three-dimensional array, that is, $K(500,2,2)$. A K matrix with zero blocks but no band structure is shown in Eq. (i).

$$K = \begin{bmatrix} K_{11} & 0 & K_{13} & 0 & K_{15} \\ & K_{22} & K_{23} & 0 & K_{25} \\ & & K_{33} & K_{34} & 0 \\ \text{symmetrical} & & & K_{44} & K_{45} \\ & & & & K_{55} \end{bmatrix} \tag{i}$$

The storage of the upper triangle in $K(500,2,2)$ will be as in Fig. 7.19. Note that the diagonal blocks have been inverted to give F_{11}, etc., modified if necessary for the support conditions.

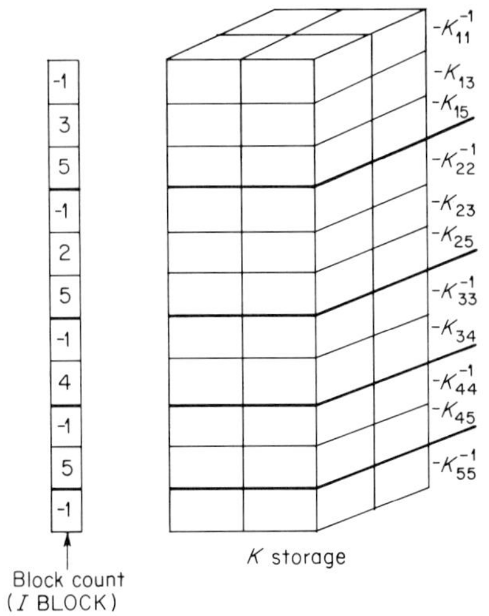

Figure 7.19

The row-count column in Fig. 7.19 gives the column number in K where the stored values are located. For example, the third value in IBLOCK is 5 and indicates that K_{15} is in the four locations at $K(3,1,1)$, $K(3,1,2)$, $K(3,2,1)$, and $K(3,2,2)$. In IBLOCK, the start of a row of K is indicated by -1.

The coding for the iteration is given below:

```
*       NUMNP = NUMBER OF NODE POINTS
   100  SUM=0.0
        MC=1
        MR=1
        PUT LOADS INTO VECTORS TO OPERATE EQ. 7.104
        DO 101 1=1, NUMNP
        TX(I)=XLOAD(I)
   101  TY(I)=YLOAD(I)
        DO 200 N=1, NUMNP
   150  MC = MC+1
        J=IBLOCK(MC)
        IF(J.LE.O) GO TO 160
        TX(J)=TX(J)-K(MC,1,1)*DISX(J)-K(MC,1,2)*DISY(J)
        TY(J)=TY(J)-K(MC,2,1)*DISX(J)-K(MC,2,2)*DISY(J)
        GO TO 150
*       CALCULATE INCREMENT TO NODE DISPLACEMENTS
   160  DX=K(MR,1,1)*TX(J)+K(MR,2,1)*TY(J)-DISX(J)
        DY=K(MR,1,2)*TX(J)+K(MR,2,2)*TY(J)-DISY(J)
        DISX(J) = DISX(J) + RELAX*DX
        DISY(J) = DISY(J) + RELAX*DY
        IF (IBLOCK(MR).EQ.-1) GO TO 170
```

```
*         CALCULATE JOINT FORCE UNBALANCE
          DET=K(MR,1,1)*K(MR,2,2)-K(MR,1,2)**2
          XUNBAL=(K(MR,2,2)*DX-K(MR,2,1)*DY)/DET
          YUNBAL=(-K(MR,1,2)*DX+K(MR,1,1)*DY)/DET
*         FIND VECTOR RESULTANT OF FORCE COMPONENTS
          SUM=SUM+SQRT (XUNBAL**2+YUNBAL**2)
    170   MR=MR+1
          IF(MR.EQ.MC) GO TO 200
          L=IBLOCK(MR)
          TX(L)=TX(L)-K(MR,1,1)*DISX(J)-K(MR,2,1)*DISY(J)
          TY(L)=TY(L)-K(MR,1,2)*DISX(J)-K(MR,2,2)*DISY(J)
          GO TO 170
    200   CONTINUE
          IF(SUM.LE.TOLER.OR.NCYCM.LE.NCYCLE) GO TO 300
          NCYCLE=NCYCLE+1
          GO TO 100
    300   CONTINUE
```

2 Cholesky Decomposition

The Cholesky method for the decomposition of a symmetrical matrix into an upper triangular matrix and its transpose may be adopted for use with band matrices. It is found to be particularly useful for problems of medium size, where all the equations can be retained in core. In addition, the decomposition may be divided into a forward and a later process so that if many solutions are required, only the later part need be repeated. Again this is useful when iteration on residuals is necessary for extra precision in the solution.

The stiffness equations here are written

$$Kr = R \tag{7.110}$$

Then K is a banded matrix that may be decomposed

$$K = L^T L \tag{7.111}$$

The matrix L is also banded and nonzero values lie entirely within the upper half-band width of K. Then,

$$L^T L r = R \tag{7.112}$$

Let

$$Lr = y \tag{7.113}$$

so that

$$L^T y = R \tag{7.114}$$

Because of the nature of L and L^T, Eqs. (7.113) and (7.114) may be solved by back-substitution, firstly to obtain y and finally r. The process of obtaining L is given in subroutine DECOM in Appendix F, and that for obtaining r, in the subroutine RESOLV. These subroutines require only the upper half of the band to be stored, so that if K is given as a 7×7 matrix with three nonzero band terms off the diagonal, the resulting storage allocation will be as in Eq. (7.115).

$$K_{\text{band}} = \left[\begin{array}{cccc} K_{11} & K_{12} & K_{13} & K_{14} \\ K_{22} & K_{23} & K_{24} & K_{25} \\ K_{33} & K_{34} & K_{35} & K_{36} \\ K_{44} & K_{45} & K_{46} & K_{47} \\ K_{55} & K_{56} & K_{57} & 0 \\ K_{66} & K_{67} & 0 & 0 \\ K_{77} & 0 & 0 & 0 \end{array}\right] \begin{array}{l} \swarrow \text{5th column} \\ \\ \\ \\ \leftarrow \text{5th row} \\ \\ \\ \end{array} \tag{7.115}$$

Notice that in this storage pattern, rows remain as such but columns now lie diagonally in the matrix as shown in Eq. (7.115). This necessitates special care in indexing operations both in the solution process and in the loading of the member-stiffness matrices into K_{band}.

For example, suppose that the submatrix blocks K_{11}, K_{12}, etc., in Eq. (7.115) are 3×3 matrices and that a member connects nodes 3 to 5 of the structure. Then K_N for the member (3-5) is a 6×6 matrix with 3×3 submatrix blocks, so that, symbolically,

$$K_N = \begin{bmatrix} A & B \\ B^T & C \end{bmatrix}$$

In Eq. (7.115) the matrices A, B, and C will appear in the locations given below:

Node									
3	A_{11}	A_{12}	A_{13}				B_{11}	B_{12}	B_{13}
	A_{22}	A_{23}				B_{21}	B_{22}	B_{23}	
	A_{33}				B_{31}	B_{32}	B_{33}		
4									
5	C_{11}	C_{12}	C_{13}						
	C_{22}	C_{23}							
	C_{33}								

It is also convenient to number members IJ or IJK according to an ascending order of nodes. That is, $K > J > I$. Since this is an inconvenient limitation to be placed on input data, a sorting process may easily be undertaken in the computer program.

Iteration on Residuals

The equations

$$Kr = R$$

are solved to yield only approximate values r_1, owing to round-off errors accentuated perhaps by poor conditioning of the K matrix. A better approximation to the true solution r may be obtained by determining small increments r' such that

$$K(r_1 + r') = R \tag{7.116}$$

or

$$Kr' = R - Kr_1 = R'_1 \tag{7.117}$$

The R'_n are the residuals corresponding to the approximate solution vector r_n. The arithmetic operations involved in obtaining these residuals, that is,

$$R'_n = R - Kr_n$$

must be carried out in double-precision arithmetic because of the smallness of the quantities involved. The refinement possible to the solution appears to be dependent on the degree of precision in which the arithmetic operations involved in the calculation of residuals from the approximate solution are carried out.

The equation

$$Kr'_1 = R'_1 \tag{7.118}$$

is solved, and an improved solution

$$r_2 = r_1 + r'_1$$

is produced. Notice that once K has been decomposed as in Eq. (7.111), the solution in Eq. (7.118) simply involves the RESOLV subroutine. This process may be repeated until increments r'_1 are smaller than a prescribed level.

Example 7.4 Use of iteration on residuals and the precision obtained

The equations to be solved are derived from the cantilever beam shown in Fig. 7.20*a*, subdivided into a number of segments and loaded with a single concentrated force on the free end. With $EI = ½$ and the subdivision interval $l = 1$, the form of the equations is

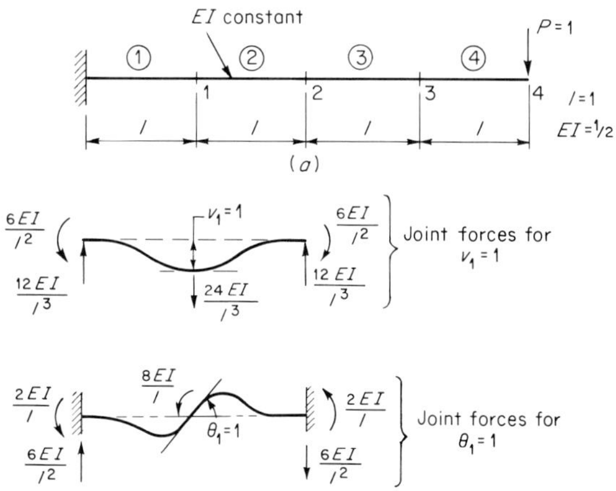

Figure 7.20 Development of stiffness matrix for cantilever beam.

$$
\begin{bmatrix}
12 & 0 & -6 & -3 \\
4 & 3 & 1 & 0 \\
12 & 0 & -6 & -3 \\
4 & 3 & 1 & 0 \\
\cdots & \cdots & \cdots & \cdots \\
12 & 0 & -6 & -3 \\
4 & 3 & 1 & 0 \\
6 & 3 & 0 & 0 \\
2 & 0 & 0 & 0
\end{bmatrix}
[r] =
\begin{bmatrix}
0 \\ 0 \\ 0 \\ 0 \\ \vdots \\ 0 \\ 0 \\ 1 \\ 0
\end{bmatrix}
$$

At each joint, two displacements v and θ exist, and it is known that the equations are not well conditioned since residual errors in θ values near the fixed end have an exaggerated influence on deflections at the free end. The error in the initial-solution vector becomes more pronounced as the number of subdivisions is increased. The solutions have been carried out for 100 and 200 equations corresponding to the cantilevers with 50 and 100 subdivisions respectively. The magnitude of the sum of the squares of the residuals $\Sigma(R')^2$ is plotted against the number of cycles of iteration in Fig. 7.21. It is seen that the shape of the two curves is substantially the same and that in each case $\Sigma\,(R')^2$ is reduced by a factor of 10 or more. This corresponds to a much greater increase in accuracy of the solution vector. Notice also that the increase in accuracy is only marginal after two cycles.

For example, in the case with 200 equations, the deflection at the free end after successive cycles is

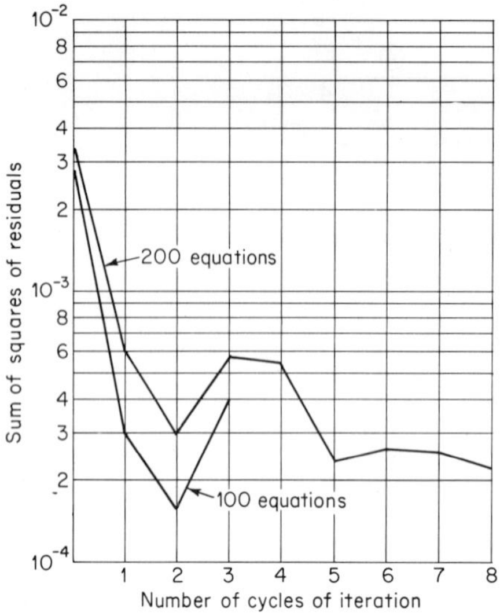

Figure 7.21

6.433×10^5	*Initial solution*
6.55877×10^5	1
6.66629×10^5	2
6.666557×10^5	3
6.66656700×10^5	4
6.66656731×10^5	5
6.66656732×10^5	6
6.66656732×10^5	7
6.66656733×10^5	8

The exact solution is

$$\frac{(100)^3}{3EI} = \frac{(100)^3}{3 \times \frac{1}{2}} = 6.6 \times 10^5$$

3 Large-capacity Band Solvers

When the structure to be analyzed has so many unknown joint displacements that core storage is exhausted, resort must be had to the use of high-speed peripheral storage such as drum or disk, or some other algorithm such as forward partitioning (see the next subsection of this chapter) must be used.

It is seen from Fig. 7.22 that when a set of banded equations is being reduced by gaussian elimination, only the triangular portion of the band shaded is influenced by the operation on the row marked i, and hence this is the only portion needed in core at that time. A similar remark applies to the force vector also shown in Fig. 7.22.

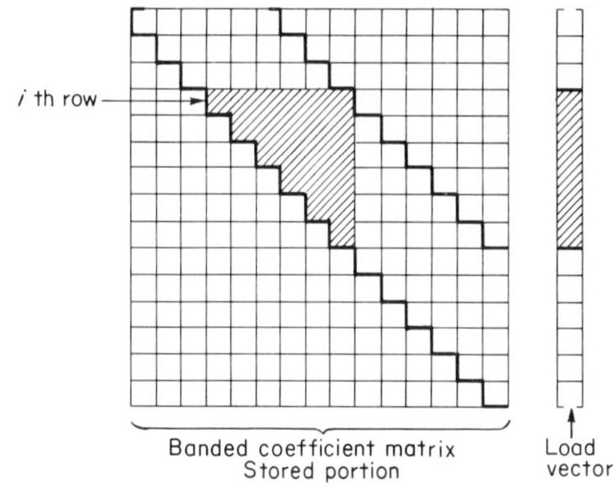

Figure 7.22

A suitable procedure to follow in the solution of a system by the direct stiffness method is outlined in the following steps:

1 From storage available and the band width of the K matrix, determine how many blocks are necessary to fit the band of K into the core. These blocks are shown in Fig. 7.23.

2 From the member-connectivity arrays, determine which members contribute stiffness to each block. Notice that some members will contribute to both blocks I and II or II and III. It should be clear that the number of rows to a block should be a function of the number of unknowns per joint, so that member k's can be loaded by submatrices according to their nodes.

3 The blocks I, II, and III are formed in turn from member-stiffness matrices. These may have been calculated previously and stored on random axis drum or disk. Modify as necessary for boundary conditions. When a given block is complete, write it on tape, column by column as shown systematically in Fig. 7.23.

4 When all blocks have been completed, reduce K by a gaussian process that recalls the columns in step 3 to memory.

Solution of Equations by the Method of Forward Partitioning

In this scheme the stiffness equations are partitioned according to the number of displacement unknowns per joint. For example, in a planar, pin-jointed truss this will be two, whereas for a three-dimension, rigid-jointed structure it will be six. These partitioned and banded equations are shown in Fig. 7.24. It will be noticed that the step size in the band is in fact equal to the partition size.

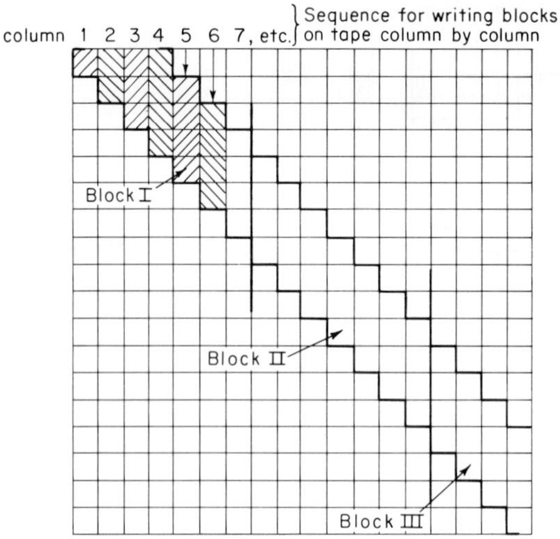

Figure 7.23

When the reduction process commences, the partitioning for joint 1 will be given in Eq. (7.119).

$$\begin{bmatrix} K_{11} & K_{1m} \\ K_{m1} & K_{mm} \end{bmatrix} \begin{bmatrix} r_1 \\ r_m \end{bmatrix} = \begin{bmatrix} R_1 \\ R_m \end{bmatrix} \tag{7.119}$$

It should be noted that $K_{1m}{}^T = K_{m1}$ and that K_{11} and K_{mm} are symmetrical. It is now possible to solve for the displacements r_1 in terms of the remaining r_m displacements and the forces R_1 on node 1. Thus,

$$r_1 = K_{11}{}^{-1}R_1 - K_{11}{}^{-1}K_{1m}r_m \tag{7.120}$$

When this value for r_1 is substituted in the remaining equations, they become independent of r_1. That is,

$$(K_{mm} - K_{1m}{}^T K_{11}{}^{-1}K_{1m})r_m = R_m - K_{1m}{}^T K_{11}{}^{-1}R_1 \tag{7.121}$$

The expressions (7.120) and (7.121) provide a pair of recurrence relationships that if applied successively to the equations (joint by joint), supply the means to obtain the displacements r. When the last joint is reached, the solution of Eq. (7.121) gives its displacements, and then successive applications of (7.120) give the remainder. Of course Eqs. (7.120) and (7.121) apply whether or not the banded nature of the K matrix as shown in Fig. 7.24 is taken into consideration. Computer storage is saved if the matrices K_{1m} and K_{mm} are partitioned to separate nonzero parts from the zeros according to the half-band width p (see Fig. 7.24). The partitioned forms of K_{1m} and K_{mm} and of r_m and R_m are given in Eqs. (7.122).

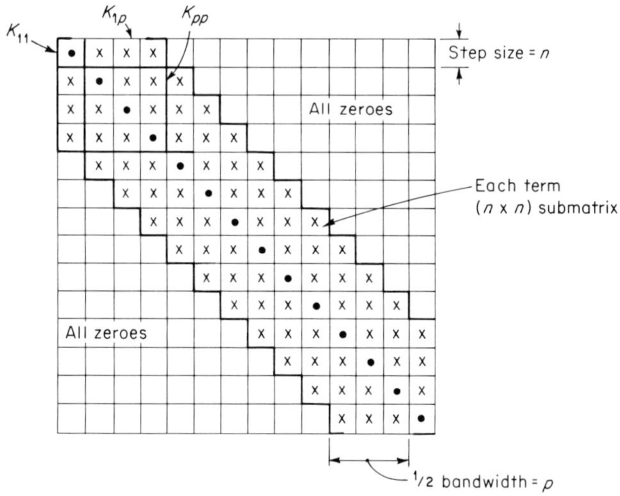

Figure 7.24 Banded matrix form partitioned for structure with n displacements per joint.

$$K_{1m} = [K_{ip} \quad 0] \qquad K_{m1} = K_{1m}{}^T = \begin{bmatrix} K_{1p}{}^T \\ 0 \end{bmatrix} \qquad K_{mm} = \begin{bmatrix} K_{pp} & K_{p,m\text{-}p} \\ K_p{}^T{}_{,m\text{-}p} & K_{m\text{-}p,m\text{-}p} \end{bmatrix}$$

$$R_m = \begin{bmatrix} r_p \\ r_{m\text{-}p} \end{bmatrix} \qquad r_m = \begin{bmatrix} r_p \\ r_{m\text{-}p} \end{bmatrix} \tag{7.122}$$

With this new notation it is possible to rewrite both sides of Eqs. (7.120) and (7.121) and to observe the effects of the null submatrices. From Eq. (7.120),

$$K_{11}{}^{-1}K_{1m}r_m = K_{11}{}^{-1}[K_{ip} \quad 0]\begin{bmatrix} r_p \\ r_{m\text{-}p} \end{bmatrix} = K_{11}{}^{-1}K_{1p}r_p \tag{7.123}$$

From the left-hand side of Eq. (7.121),

$$K_{mm} - K_{1m}{}^TK_{11}{}^{-1}K_{1m} = \begin{bmatrix} K_{pp} & K_{p,m\text{-}p} \\ K_p{}^T{}_{,m\text{-}p} & K_{m\text{-}p,m\text{-}p} \end{bmatrix} - \begin{bmatrix} K_{1p}{}^TK_{11}{}^{-1}K_{1p} & 0 \\ 0 & 0 \end{bmatrix} \tag{7.124}$$

From the right-hand side of Eq. (7.121),

$$R_m - K_{1m}{}^TK_{11}{}^{-1}R_1 = \begin{bmatrix} R_p \\ R_{m\text{-}p} \end{bmatrix} - \begin{bmatrix} K_{1p}{}^T \\ 0 \end{bmatrix} K_{11}{}^{-1}R_1$$

Thence,

$$R_m - K_{1m}{}^T K_{11}{}^{-1} R_1 = \begin{bmatrix} R_p - K_{1p}{}^T K_{11}{}^{-1} R_1 \\ R_{m\text{-}p} \end{bmatrix} \tag{7.125}$$

From Eqs. (7.123), (7.124), and (7.125), it is seen that in the operations involved in Eqs. (7.120) and (7.121), the only portion of K_{mm} affected will be K_{pp}, and similarly with R_m, only R_p will be affected. Also in the calculation of r_1 from r_m (back-substitution), only R_1 and the next r_p quantities are required. The matrices K_{1p} and K_{pp} are shown by the heavy lines in Fig. 7.24. The saving of computer storage is even more marked if only the upper triangle of K_{pp} is stored in core as a single array. Of course, as in all compact schemes, such refinements increase the complexity of the indexing in the program. To use the method, the structure analysis takes the following general steps:

1 Read in all node coordinate and member arrays (recording node connectivity, member material properties, and boundary conditions).

2 Set up a two-dimensional array with as many columns as there are nodes. In these columns, the elements that contribute stiffness to each node are recorded. For example, an element IJK ($I < J < K$) connecting nodes I, J, and K is recorded in the Ith column. When the upper triangle of the element's stiffness matrix is being loaded into K (when node I is being reduced), those portions pertaining to J and K are loaded into K_{pp}, whereas those pertaining to I, J, and K are loaded into K_{ip}.

3 The procedure outlined in Eqs. (7.120) and (7.121)ff is then carried out, the pivot matrix K_{11} being modified for boundary conditions before inversion.

4 When the displacements r have been determined, member forces are generated from the equation (in the general form)

$S = kT^T L_D{}^T r$

The essential coding for the above algorithm is set out below. The scheme is dimensioned here to handle up to 10 load cases at one time.

Program sketch for solution of equations by forward partitioning: written of a three-dimensional frame structure

```
      DIMENSION IONARY (9,100), NARRAY (400), AKK (6000),
     1 R1(100,10), R2(100,10), R3(6,10), AK1(6,100), AK2(6,100),
     2 ND(100), A11(6,6), NPI(100), NPJ(100), ESTIF(6,6)
     3 DIS(100,6)
      EQUIVALENCE (IONARY, NARRAY)
C     Number of nodes = NUMNP
C     Half-band width = NDIF
```

```
C          Maximum size of AKK = 6 * NDIF
C          Number of load cases = NLD
C          Read in first-joint load values.
           READ 101, ((R(I,M), I=1, 6), M=1, NLD)
      101  FORMAT (6F10.3)
           PRINT 101, ((R(I,M), I=1, 6), M=1, NLD)
           NSIZ = NDIF * 6
C          Set up the array ND so that ND(N) gives the
C          number that when added to NSIZ, gives the first
C          term location in AKK of the Nth row.
           DO 200 N=1, NSIZ
           ND(N) = NSIZ*(N-1) - (N*(N-1))/2
C          In IONARY columns the connectivity of members to
C          joints has been determined. That is, in IONARY (K,N),
C          K=1, 9, those members (I,J), J > I and I=N are
C          recorded. This matrix is condensed into a single
C          array NARRAY to conserve storage. The start
C          of a new joint is indicated by a negative sign.
           NARRAY(1) = -IONARY(1,1)
           L=1
           DO204 I=1,NUMNP
           NARRAY(L) = -IONARY(1,I)
           L=L+1
           DO 202 M=2,9
           IF (IONARY(M,I).LE.O) GO TO 204
           NARRAY(L) = IONARY(M,I)
      202  L=L+1
      204  CONTINUE
C          Initialize by clearing arrays AK1, R1,
C          AKK, A11, and R3
           DO 208 N=1, NSIZ
           DO 206 M=1,6
      206  AK1(M,N)=0.0
           DO 208 M=1,NLD
      208  R1(N,M)=0.0
           NSK = (NSIZ * (NSIZ+1))/2
           DO 210 M=1,NSK
      210  AKK(M)=0.0
           DO 221 I=1,6
           DO 220 J=1,6
      220  A11(I,J) = 0.0
           DO 221 J=1,NLD
      221  R3(I,J) = 0.0
           NLC=1
           NC=1
```

```
          L=0
    245   L=L+1
C         Calculate actual size of AKK for node
C         under consideration.
          NSIZ = MINO(NPDIF,(NUMNP-L)) * 6
C         Check if last joint is being processed.
          IF (NSIZ.LE.O) NSIZ=6
C         Calculate stiffness for member. This is done
C         in the subroutine STIFF and the result is
C         stored in ESTIF(6,6)
          DO 300NK = 1,9
          N = NARRAY(NC)
          IF(N) 302,301,250
C         N = negative next joint
C         N = 0 no more members
C         N = positive. Proceed with current joint.
    250   NC = NC+1
          CALL STIFF (N,ESTIF)
          K = NPJ(N)
C         Set up ESTIF submatrices in A11, AK1, AKK
          N = (K-L-1) * 6
          DO 280 I=1, 6
          II = N+I
          DO 275 J=1, 6
          JJ = N+J
          A11(I,J) = A11(I,J)+ESTIF(I,J)
    275   AK1(I,JJ) = AK1(I,JJ)+ESTIF(I,J+6)
          DO 280 J=1,6
          JJ = N+J
          M = ND(II)+JJ
    280   AKK(M) = AKK(M) + ESTIF(I+6,J+6)
C         Note on Indexing in AKK
C         N = number of rows to block under consideration
C         in AKK.
C         II = N+I = the actual row
C         Therefore ND(II) + N + I = start position of row in AKK
C         Hence ND(II) + N + J, J=I, 6 gives all locations of terms
C         in row N + I.
    301   NC = NC+1
          GO TO 303
    302   NARRAY(NC) = -NARRAY(NC)
    303   CONTINUE
          DO 310 I=1,6
          DO 310 J=1,NLD
    310   R3(I,J) = R(I,J) + R3(I,J)
```

```
C          If the last joint is being treated, bypass
C          the read-in of joint loads since this would
C          be joint (NUMNP+1).
           IF(L.GE.NUMNP) GO TO 322
           READ 101, ((R(I,M), I=1,6), M=1,NLD)
C          Modify for boundary conditions.
C          NFIX(L) contains six digits each of
C          which signifies a fixed boundary.
C          If NFIX(L) = 0, the point is free.
C          By successive division of NFIX(L) ≠ 0
C          by 10, the various boundary fixities
C          are picked up.
     322   IF(NFIX(L).EQ.O) GO TO 400
           I = 100000
           DO 380 J = 1,6
           IF(NFIX(L).LT.I) GO TO 380
           DO 340 K = 1,6
           A11(J,K) = 0.0
     340   A11(K,J) = 0.0
           A11(J,J) = 1.0
           DO 345 K = 1,NLD
     345   R3(J,K) = 0.0
           DO 350 K = 1,NSIZ
     350   AK1(J,K) = 0.0
           NFIX(L) = NFIX(L) - I
     380   I = I/10
C          Invert diagonal term using the
C          Symmetric matrix invert routine SYMINV.
     400   CALL SYMINV (A11,6,6)
C          If this is last term, skip the alteration
C          of A11, AKK, and AK1.
           IF(L.GE.NUMNP) GO TO 600
C          Modify load and off-diagonal for back-substitution.
           DO 420 I = 1,6
           DO 410 K = 1,NLD
           R2(I,K) = 0.0
           DO 410 J = 1,6
     410   R2(I,K) = R2(I,K) + A11(I,J) * R3(J,K)
           DO 420 J = 1,NSIZ
           AK2(I,J) = 0.0
           DO 420 K = 1,6
     420   AK2(I,J) = AK2(I,J) + A11(I,K) * AK1(K,J)
C          Write modified matrices on tape.
           WRITE(1) ((R2(K,J),J=1,NLD), (AK2(K,J),J=1,NSIZ), I=1,6)
C          Modify terms below for next joint
```

```
      DO 440 I = 1,NSIZ
      DO 440 K = 1,6
      DO 430 J = I,NSIZ
      M = ND(I) + J
  430 AKK(M) = AKK(M) - AK1(K,I) * AK2(K,J)
      DO 440 J = 1,NLD
  440 R1(I,J) = R1(I,J) - AK1(K,I) * R2(K,J)
C     Shift up portion of AKK into A11, and
C     into AK1, and R1 into R3 for the next
C     joint.
      DO 460 I = 1,6
      DO 455 J = 1,NLD
  455 R3(K,J) = R1(I,J)
      DO 460 J = I,6
      M = ND(I) + J
      A11(I,J) = AKK(M)
  460 A11(J,I) = A11(I,J)
C     Check if last joint. If so, no further
C     shifting of AKK is needed.
      IF (NSIZ.LE.6) GO TO 485
      DO 470 I = 1,6
      DO 470 J = 7,NSIZ
      M = ND(I) + J
  470 AK1(I,J-6) = AKK(M)
      DO 480 I = 7,NSIZ
      DO 475 J = 1,NLD
  475 R1(I-6,J) = R1(I,J)
      DO 480 J = I,NSIZ
      M = ND(I) + J
      M1 = ND(I-6) + J-6
  480 AKK(M1) = AKK(M)
  485 CONTINUE
      K = NSIZ - 5
      K1 = K - 1
C     The bottom and side portions of AKK and AK1
C     must now be cleared because they contain
C     material appropriate for previous joint.
      DO 487 I = 1,K1
      DO 487 J = K,NSIZ
      M = ND(I) + J
  487 AKK(M) = 0.0
      DO 490 I = K,NSIZ
      DO 490 J = I,NSIZ
      M = ND(I) + J
  490 AKK(M) = 0.0
```

```
      DO 500 J = K,NSIZ
      DO 495 I = 1,NLD
  495 R1(J,I) = 0.0
      DO 500 I = 1,6
  500 AK1(I,J) = 0.0
C     Now process next point by returning
C     to statement 245 and repeating cycle.
      GO TO 245
C     At 600 all nodes have been processed
C     and program proceeds on to back-substitution.
  600 BACKSPACE 1
      REWIND 2
C     Displacements of node NUMNP
      DO 610 I = 1,6
      DO 610 K = 1,NLD
      R2(I,K) = 0.0
      DO 610 J = 1,6
  610 R2(I,K) = R2(I,K) + A11(I,J) * R3(J,K)
C     Put one set of displacements in the
C     displacement vector and the rest on tape.
      DO 615 J = 1,6
  615 DIS(NUMNP,J) = R2(J,1)
      L = NUMNP
      IF (NLD.EQ.1) GO TO 620
  616 WRITE (2) ((R2(I,J),J = 2,NLD), I = 1,6)
C     Back-substitution process
  620 L = L - 1
      NSIZ = MINO (NDPIF, (NUMNP-L)) * 6
      READ (1) ((R1(I,J), J = 1,NLD), (AK2(I,J),J = 1,NSIZ),I = 1,6)
      BACKSPACE 1
      DO 640 I = 1,6
      DO 640 J = 1,NSIZ
      DO 640 K = 1,NLD
  640 R1(I,K) = R1(I,K) - AK2(I,J) * R2(J,K)
      DO 642 J = 1,6
  642 DIS(L,J) = R1(J,1)
      K = NSIZ-6
      IF (NSIZ.LT.6 * NPDIF) K = NSIZ
C     Displacements are rearranged for next point
      DO 670 I = 1,K
      II = K - I + 1
      DO 670 J = 1,NLD
  670 R2(II + 6,J) = R2(II,J)
      BACKSPACE 1
      DO 680 I = 1,6
```

```
          DO 680 J = 1,NLD
      680 R2(I,J) = R1(I,J)
          IF (NLD.EQ.1) GO TO 688
          WRITE (2), ((R2(I,J),J=2,NLD), I=1,6)
      688 IF (L.GT.1) GO TO 620
C         If L = 1, last point has been processed
C         in the back-substitution.
```

7.6 ANALYSIS OF HIGH–RISE FRAMES WITH RIGID JOINTS

The algorithms developed so far in the direct stiffness method require a complete description of the node geometry together with the node connectivity of every member. These data are required so that the programs may be as general as possible.

The structure type to be considered now is such that it may be subdivided into a number of similar substructures. These substructures have the same internal topology but may have different member properties and node loads. The description of the structure connectivity now has only to contain that for the substructures and their assemblage to form the complete structure. For the high-rise frame, the first of these is given by the number of bays in the frame, and the second by the number of stories.

The Tridiagonalization Solution Technique

In the first solution technique discussed, the tridiagonal method, the linear graph obtained by replacing each substructure by a single member must be a tree. The most familiar class of structure to which the method may be applied is the high-rise rigid-frame building. Such a frame is shown in Fig. 7.25*a*. The substructures are shown in Fig. 7.25*b*, and each of these is

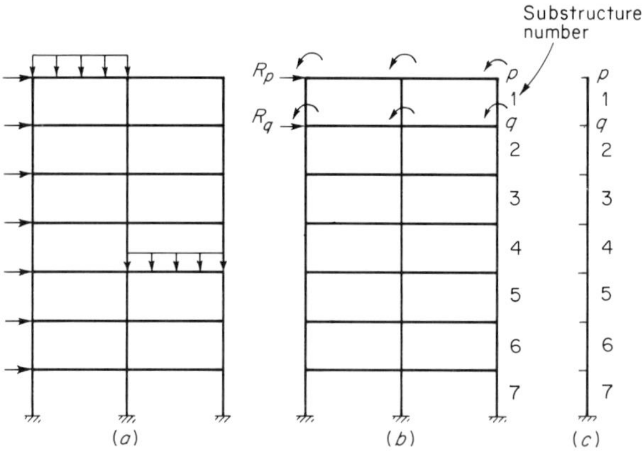

Figure 7.25 (*a*) High-rise frame; (*b*) substructures; (*c*) linear graph of substructures.

composed of the beams at one floor level together with the columns from the story below. The linear graph (tree structure) produced by replacing each substructure with a single member is shown in Fig. 7.25*c*.

The substructure is considered a single unit of the tree structure, so that the first step at each cycle of the reduction must be the calculation of the stiffness matrix of the substructure from its individual members. This matrix will be partitioned according to the joint groups at its extremities (in the tree configuration, see *pq* in Fig. 7.25*b* and *c*). In the high-rise frame, this will be according to the joints at the beam level and those at the foot of the columns. It is evident that each of these groups has the same number of displacements, so that $K_{\text{substructure}}$ will be partitioned into four square blocks of the same dimension. The term *tridiagonal* applies to the total-stiffness matrix which when partitioned in the above manner, consists of three diagonal bands lying on and immediately above and below the principal diagonal. The method is very similar to that already described as the method by forward partitioning. However, it is simplified considerably since the band structure is known and constant. The tridiagonalization process may be discussed as a purely mathematical technique for solving this class of equations. Here it will be developed from the physical description of what happens in the structure as the solution progresses.

The solution starts with member (1) in Fig. 7.25*b*, and its stiffness matrix is arranged in the groups *p* and *q* according to the upper and lower floor levels respectively. That is,

$$\begin{bmatrix} R_p \\ R_q \end{bmatrix} = \begin{bmatrix} K_{pp} & K_{pq} \\ K_{qp} & K_{qq} \end{bmatrix} \begin{bmatrix} r_p \\ r_q \end{bmatrix} \tag{7.126}$$

Consider the initial condition of the structure to be all joints locked against rotation and translation, and proceed down from the top, releasing one floor group at a time. That is, imagine the level *p* to be released with *q* held fixed. Then displacements will occur at *p* and forces will be transferred to level *q*. Then,

$$\bar{r}_p = K_{pp}{}^{-1}R_p \qquad \bar{r}_q \equiv 0 \tag{7.127}$$

The deflections $\bar{r}_p$ will be the in-story deflections for the substructure. The reactions caused at the level *q* will be

$$K_{qp}K_{pp}{}^{-1}R_p$$

Hence the new active joint forces at *q* will be

$$\bar{R}_q = R_q - K_{qp}K_{pp}{}^{-1}R_p \tag{7.128}$$

The stiffness at *q* will be different from the K_{qq} which was calculated with the *p* level held fixed.

To calculate the new *q* joint stiffness, suppose the *p* level is free, that is, $R_p = 0$. Then from Eq. (7.126),

$$0 = K_{pp}r_p + K_{pq}r_q$$
$$R_q = K_{qp}r_p + K_{qq}r_q$$

Thence

$$\bar{\bar{r}}_p = -K_{pp}^{-1}K_{pq}r_q \tag{7.129}$$

and also,

$$R_q = (K_{qq} - K_{qp}K_{pp}^{-1}K_{pq})r_q \tag{7.130}$$

or

$$R_q = \overline{K}_{qq}r_q \tag{7.131}$$

Equation (7.129) gives the displacements at p from those at q when the q level is displaced. The analysis now moves to the next substructure, $\overline{K}_{qq}$ being added to the K_{pp} for this new level. The process is set out in the flow chart below. In this flow chart, matrix names are written with the indices from Eqs. (7.126) to (7.131). Thus, for example,

R_p becomes *RP*

K_{pp} becomes *KPP*

Deflection terms are prefixed with a D; for example, r_p becomes DRP.

It is seen that the tridiagonalization method leads to very simple programming with a minimum of input data. It will however suffer the same round-off errors mentioned in Sec. 7.5 and will be prone to this in tall frames with large numbers of stories. This could be overcome by iteration on residuals in double-precision arithmetic or the use of double precision in the calculation of K_{pp}^{-1} and the associated transfer matrices.

Frame Analysis by Iteration Methods

The tridiagonal solution technique just discussed may be used for any structural type whose stiffness matrix may be arranged to have the required band form.

The iteration methods developed in this section have their special application for high-rise frames and take cognizance of the various types of joint (or structure) displacements and equilibrium equations that occur in these structures. The first of the methods discussed is known as *Gasper Kani's Iteration method* (see Ref. 12). The second method has appeared in several papers by Clough, Wilson, and King (see Ref. 5). Both methods provide algorithms that are suitable for computer programming. The Kani method iterates on member forces as the unknown quantities, whereas the Clough method uses iteration on the stiffness equations to produce joint and story displacements.

Flow chart for computer program for high-rise frame analysis

Set KMOD = 0
Read in RP, the loads at level p

↓

Loop on the number of stories

↓

Read in column and beam properties ◄ (loop return)

↓

Read in RQ, the loads at level q

↓

Calculate KPP, KPQ, KQQ for the substructure
Do this member by member in the form a^Tka,
writing ka on tape 1 for each member in a
prescribed sequence, so that member forces may
be calculated, S = kar.

↓

KMOD = KPP + KMOD   Stiffness at level p

↓

DRP1 = $KMOD^{-1}$ RP { In story deflections; Store DRP1 on tape 2 }

↓

RQ = RQ − $KPQ^T KMOD^{-1}$ RP   New forces at level q

↓

Store −$KMOD^{-1}$ KPQ on tape 3

↓

KMOD = KQQ − KPQ^T $KMOD^{-1}$ KPQ   New reduced stiffness at level q

↓

Transfer RQ back to RP

↓

End of story loop ──► (return to "Read in column and beam properties")

↓

The last value of KMOD obtained will be at the level
of the supports. Modify in the usual way for any
foundation releases.

↓

Write end of file on tape 3
Write end of file on tape 2

↓

BACK-SUBSTITUTION

↓

At base deflections DRQ = $KMOD^{-1}$ RQ.
If all bases are fixed, DRQ ≡ 0

↓

Store DRQ on tape 4

↓

Loop on the number of stories.

↓

Backspace tape 3, 2 records ◀

Read one record from tape 3 (that is $-\text{KMOD}^{-1}$ KPQ)

↓

Backspace tape 2, 2 records.

Read one record from tape 2 (this is DRP1)

↓

Then DRP = DRP1 − KMOD^{-1} KPQ DRQ

↓

Print DRP

↓

Store DRP on tape 4

↓

End of story loop for deflection calculations ▶

Write end of file on tape 4

Member force calculation

Rewind tape 1

↓

Loop on the number of stories

↓

Backspace tape 4, 3 records ◀

↓

Read 2 records from tape 4 (these will be DRP and DRQ for top story in the first instance)

↓

Loop on the number of members in the substructure ◀

↓

Read ka from tape 1

↓

Calculate member force S = kar

↓

Print S

↓

End of member loop ▶

↓

End of story loop ▶

Gasper Kani's Method of Moment and Shear Iteration

In this method, members of the frame are considered inextensible so that the only distortions that arise are from bending moments. The equilibrium equations to be solved will be those for moment at each joint and for the shear in members across each column line equal to the sum of the horizontal forces above that line. These two functions will be treated separately, so that an artificial uncoupling of the equations will have been produced. The distorted position of member ij is shown in Fig. 7.26. For any member, the moment M_{ij} at end i may be considered composed of four contributions:

1 The moment existing in the all-joints-fixed condition (fixed-end moment) will be $\overline{M}_{ij}$.

2 The moment due to the rotation θ_i of end i will be $K_{ii}\theta_i$.

3 The moment due to rotation θ_j of end j. If end i is fixed and rotation θ_j is given to end j, the moment at i will be $K_{ij}\theta_j$. Alternatively, if C_{ji} is the carryover factor j to i such that $C_{ji} = K_{ij}/K_{jj}$, then the moment at i will be $C_{ji}K_{jj}\theta_j$.

4 The moment due to the relative displacement δ_{ij} of i and j. If this is positive as shown in Fig. 7.26, the moment induced at end i will be

$(K_{ii} + C_{ji}K_{jj})(\delta_{ij}/l_{ij})$

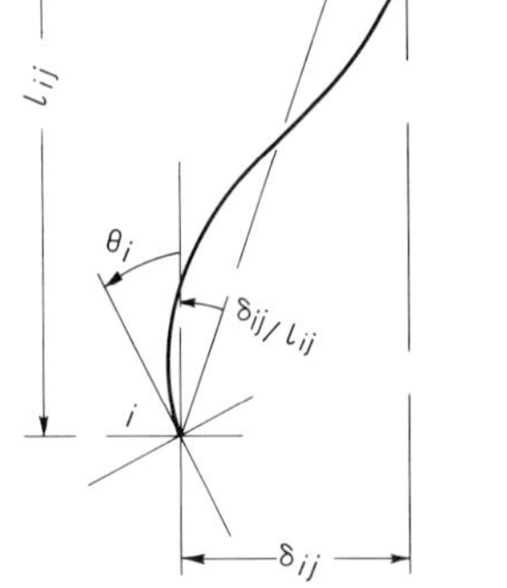

Figure 7.26 Member distortions.

Then, from all contributions,

$$M_{ij} = \overline{M}_{ij} + K_{ii}\theta_i + C_{ji}K_{jj}\theta_j + (K_{ii} + C_{ji}K_{jj})\frac{\delta_{ij}}{l_{ij}} \tag{7.132}$$

By definition, let

$$M'_{ij} = C_{ij}K_{ii}\theta_i \tag{7.133a}$$

$$M'_{ji} = C_{ji}K_{jj}\theta_j \tag{7.133b}$$

$$M''_{ji} = 3C_{ji}K_{jj}\frac{\delta_{ij}}{l_{ij}} = 3C_{ij}K_{ii}\frac{\delta_{ij}}{l_{ij}} = M''_{ij} \tag{7.133c}$$

Notice in Eq. (7.133*c*) that

$$C_{ji}K_{jj} = K_{ij} = K_{ji} = C_{ij}K_{ii}$$

Substitute these expressions in Eq. (7.132) to give

$$M_{ij} = \overline{M}_{ij} + \frac{1}{C_{ij}}M'_{ij} + M'_{ji} + B_{ij}M''_{ij} \tag{7.134}$$

In Eq. (7.134) it is seen that

$$B_{ij} = \frac{1 + C_{ij}}{3C_{ij}} \tag{7.135}$$

Computation of M'_{ij}, M'_{ji}, and M''_{ij}

It will be seen that M'_{ij} and M'_{ji} are associated with the joint rotations, whereas M''_{ij} is associated only with sidesway displacements and will be present, then, only for the column members. At any point i,

$$\Sigma M_{ij} = 0$$

which is the sum of all ij members with i the given joint number. Then,

$$0 = \sum_j \overline{M}_{ij} + \sum_j \frac{1}{C_{ij}}M'_{ij} + \sum_j M'_{ji} + \sum_j B_{ij}M''_{ij} \tag{7.136}$$

Assume that the values of M'_{ji} and M''_{ij} are known, and write

$$\overline{M}_i = \sum_j \overline{M}_{ij}$$

This term is a constant for each load case since it is the sum of all the fixed-end moments at the joint i. From Eq. (7.136),

$$\sum_j \frac{1}{C_{ij}}M'_{ij} = -\left(\overline{M}_i + \sum_j M'_{ji} + \sum_j B_{ij}M''_{ij}\right) \tag{7.137}$$

The value of M'_{ij} for member ij can be calculated now from the sum on the right-hand side of Eq. (7.137) since all members meeting at joint i have the same joint rotation θ_i and hence moments will be proportioned inversely to the member stiffness. That is,

$$\frac{1}{C_{ij}} M'_{ij} = K_{ii}\theta_i$$

and hence

$$\sum_j \frac{1}{C_{ij}} M'_{ij} = \theta_i \sum_j K_{ii} \tag{7.138}$$

The sum on K_{ii} is for all members meeting at joint i. The value of θ_i from Eq. (7.138) is

$$\theta_i = \frac{\sum_j (1/C_{ij}) M'_{ij}}{\sum_j K_{ii}} \tag{7.139}$$

Then,

$$\frac{1}{C_{ij}} M'_{ij} = \frac{K_{ii}}{\sum_j K_{ii}} \left(\sum_j \frac{1}{C_{ij}} M'_{ij} \right) \tag{7.140}$$

Let

$$\mu_{ij} = \frac{-C_{ij}K_{ii}}{\sum_j K_{ii}} \tag{7.141}$$

Then from Eq. (7.137),

$$M'_{ij} = \mu_{ij}(\overline{M}_i + \sum_j M'_{ji} + \sum_j B_{ij}M''_{ij}) \tag{7.142}$$

It should be evident that if $M''_{ij} \equiv 0$ for all members, that is, the frame has no sidesway, then Eq. (7.142) provides a recurrence relationship for calculating firstly M'_{ij} for every member (and of course M'_{ji}), and hence from Eq. (7.134), M_{ij}. It can be shown that this is simply the Gauss-Seidel iteration method applied to the calculation of moments produced by joint rotations.

Computations for M''_{ij}

When the frame has sidesway displacements it will be necessary to calculate M''_{ij} for all column members. Consider the rth story (of height h_r) as shown in Fig. 7.27. For this story the equation of equilibrium for horizontal forces must be

$$\sum_{l=1\text{ to }r} P_l + \sum_r H_{ij} = 0 \tag{7.143}$$

The sum $\sum_{l=1,r} P_l$ is the sum of all the external horizontal loads acting above the top of the rth story. The sum $\sum_r H_{ij}$ is the sum of the column shears in

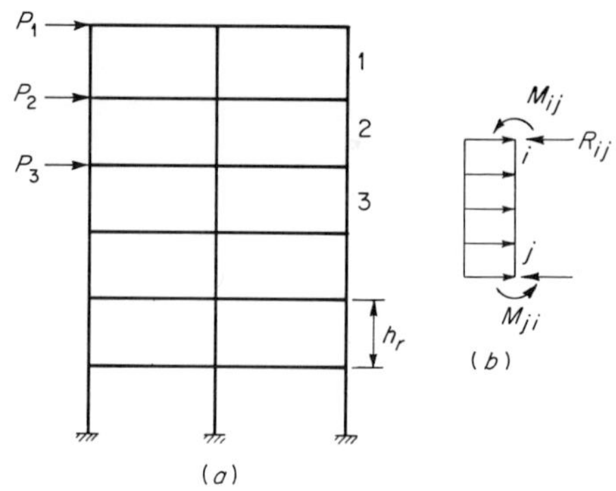

Figure 7.27 Horizontal forces.

the rth story. This may include distributed load on the column as shown in Fig. 7.27b, so that

$$H_{ij} = R_{ij} - \frac{M_{ij} + M_{ji}}{h_r}$$

Thence, in Eq. (7.143) for each story,

$$h_r \Sigma P_l + h_r \Sigma R_{ij} - \Sigma (M_{ij} + M_{ji}) = 0$$

Substitute now for M_{ij} and M_{ji} by using Eq. (7.134) to give

$$h_r \Sigma P_l + h_r \Sigma R_{ij} - [\Sigma (\overline{M}_{ij} + \overline{M}_{ji}) + 3 \Sigma B_{ij}M'_{ij} + 3 \Sigma B_{ji}M'_{ji} + \Sigma (B_{ij} + B_{ji})M''_{ij}] = 0 \tag{7.144}$$

This sum may be apportioned to each column in the rth story according to stiffnesses, since all columns undergo the same relative displacement δ_{ij}.

Then for each column,

$$M''_{ij} = 3C_{ij}K_{ij}\frac{\delta_{ij}}{l_{ij}} \tag{7.145}$$

Therefore,

$$\Sigma (B_{ij} + B_{ji})M''_{ij} = 3\frac{\delta_{ij}}{l_{ij}}\Sigma (B_{ij} + B_{ji})C_{ij}K_{ij}$$

The deflection δ_{ij} is calculated by

$$\frac{3\delta_{ij}}{l_{ij}} = \frac{\Sigma (B_{ij} + B_{ji})M''_{ij}}{\Sigma (B_{ij} + B_{ji})C_{ij}K_{ij}} \tag{7.146}$$

Substitution in Eq. (7.145) gives

$$M''_{ij} = \frac{C_{ij}K_{ij}}{\Sigma (B_{ij} + B_{ji})C_{ij}K_{ij}}[\Sigma (B_{ij} + B_{ji})M''_{ij}] \tag{7.147}$$

Let

$$\nu_{ij} = \frac{3C_{ij}K_{ij}}{\Sigma\,(B_{ij} + B_{ji})C_{ij}K_{ij}} \qquad \text{a constant for each column}$$

$$\overline{M}_R = \frac{h_r}{3}(\Sigma\,P + \Sigma\,S_{ij}) \qquad \text{a constant for each load case}$$

where

$$\Sigma\,S_{ij} = \Sigma\,R_{ij} - \frac{\Sigma\,(\overline{M}_{ij} + \overline{M}_{ji})}{h_r}$$

Then

$$M''_{ij} = \nu_{ij}(\overline{M}_R - \Sigma\,B_{ij}M'_{ij} - \Sigma\,B_{ji}M'_{ji}) \tag{7.148}$$

The iteration process uses Eqs. (7.142) and (7.148) in turn. To start, the moment values M''_{ij}, M'_{ij}, and M'_{ji} are set equal to zero for all members. Equation (7.148) might then be used to obtain M''_{ij} values for the columns and then these values used in Eq. (7.142) and so on. Experience may show that one of the above equations should be applied more frequently than the other depending on the relative convergence rates. When M' and M'' values have been obtained to the desired tolerance, convergence is considered complete. Then member moments are obtained from Eq. (7.134). Kani's method is in some ways similar to the displacement iteration method developed by Clough, Wilson, and King, except that in the latter work axial distortions of the columns are taken into consideration. Kani's method is very easy to program and should be highly convergent and hence should give programs of very fast execution time. Its disadvantage is that in the primary analysis deflections are not calculated. These could be obtained from Eq. (7.146) and accumulated for successive stories. Also, because axial distortions in the columns are neglected, frame moments may be in error.

Frame Analysis by Iteration on Displacements; the Clough, Wilson, and King Algorithm

The notation used to identify frame members is shown in Fig. 7.28. In the frame, stories are labeled 1, 2, 3, . . . , N and bays 1, 2, 3, . . . , M. A joint is referred to by the indices (n,m) according to story and bay members. Girder (n,m) and column (n,m) are similarly indexed, as shown in Fig. 7.28. In the analysis it is recognized that axial strains in the columns are much higher than those in the beams and that for this reason the latter will be neglected, whereas column strains will be included.

Since displacements are to be calculated, various stiffnesses for joint translation and story translation must be evaluated.

Consider first the column force-distortion relationship. The column (n,m) is shown in its distorted position in Fig. 7.29. The member forces considered will be $\langle M_C{}^T \quad M_C{}^B \quad Q_C \quad P_C \rangle$, that is, the end moments, the transverse

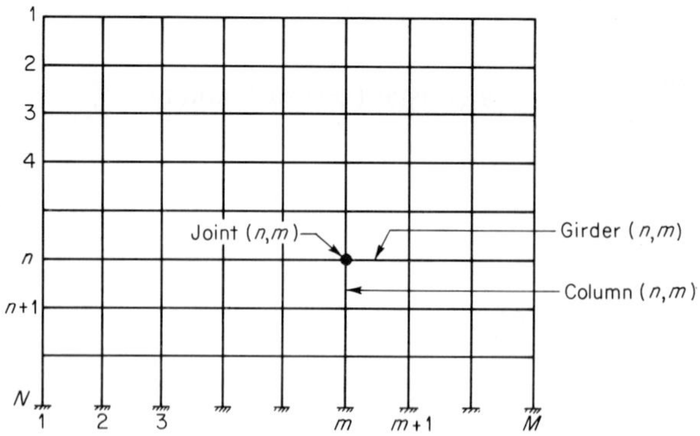

Figure 7.28 Frame indexing of nodes, girders, and columns.

shear, and the axial force. Let $k_B = 4EI/l$ and $k_A = EA/l$ be the bending and axial stiffness terms for the member. Consider the displacements of column (n,m), shown in Fig. 7.29 to be the joint rotations $\langle \theta_{n,m} \quad \theta_{n+1,m} \rangle$ and relative displacement u_n, together with the vertical displacements $\langle v_{n,m} \; v_{n+1,m} \rangle$. Then,

$$\begin{bmatrix} \phi_i \\ \phi_j \end{bmatrix} = \begin{bmatrix} 1 & 0 & \dfrac{1}{l} \\ 0 & 1 & \dfrac{1}{l} \end{bmatrix} \begin{bmatrix} \theta_{n,m} \\ \theta_{n+1,m} \\ u_n \end{bmatrix}$$

Hence

$$a^T k a = \begin{bmatrix} 1 & 0 \\ 0 & 1 \\ \dfrac{1}{l} & \dfrac{1}{l} \end{bmatrix} \begin{bmatrix} k_B & \tfrac{1}{2}k_B \\ \tfrac{1}{2}k_B & k_B \end{bmatrix} \begin{bmatrix} 1 & 0 & \dfrac{1}{l} \\ 0 & 1 & \dfrac{1}{l} \end{bmatrix}$$

$$= k_B \begin{bmatrix} 1 & \tfrac{1}{2} & \dfrac{3}{2l} \\ \tfrac{1}{2} & 1 & \dfrac{3}{2l} \\ \dfrac{3}{2l} & \dfrac{3}{2l} & \dfrac{3}{l^2} \end{bmatrix}$$

Finally,

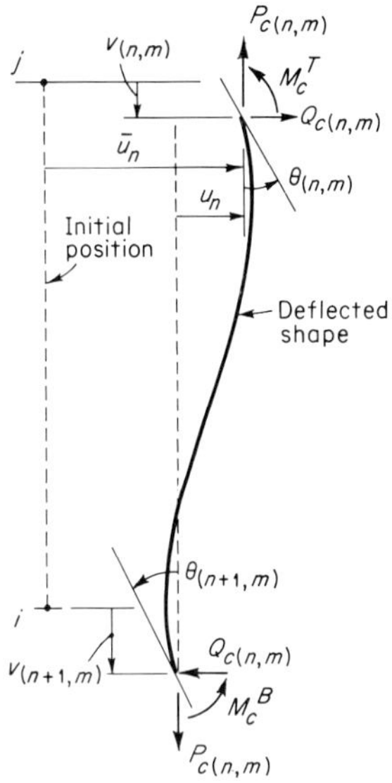

Figure 7.29 Column forces and deformations.

$$\begin{bmatrix} M_C{}^T \\ M_C{}^B \\ Q_C \\ P_C \end{bmatrix}_{(n,m)} = \begin{bmatrix} k_B & \dfrac{k_B}{2} & \dfrac{3k_B}{2l} & 0 & 0 \\ \dfrac{k_B}{2} & k_B & \dfrac{3k_B}{2l} & 0 & 0 \\ \dfrac{3k_B}{2l} & \dfrac{3k_B}{2l} & \dfrac{3k_B}{l^2} & 0 & 0 \\ 0 & 0 & 0 & -k_A & k_A \end{bmatrix} \begin{bmatrix} \theta_{n,m} \\ \theta_{n+1,m} \\ u_n \\ v_{n,m} \\ v_{n+1,m} \end{bmatrix} \tag{7.149}$$

Consider now the beam element (n,m) shown in Fig. 7.30, whose stiffness matrix is to be expressed in terms of the displacements $\langle \theta_{n,m} \quad \theta_{n,m+1} \quad v_{n,m} \quad v_{n,m+1} \rangle$.

In the equation $k = a^T k a$ used to obtain the stiffness matrix,

$$a = \begin{bmatrix} 1 & 0 & -\dfrac{1}{l} & \dfrac{1}{l} \\ 0 & 1 & -\dfrac{1}{l} & \dfrac{1}{l} \end{bmatrix} \qquad k = \begin{bmatrix} k_B & \tfrac{1}{2}k_B \\ \tfrac{1}{2}k_B & k_B \end{bmatrix}$$

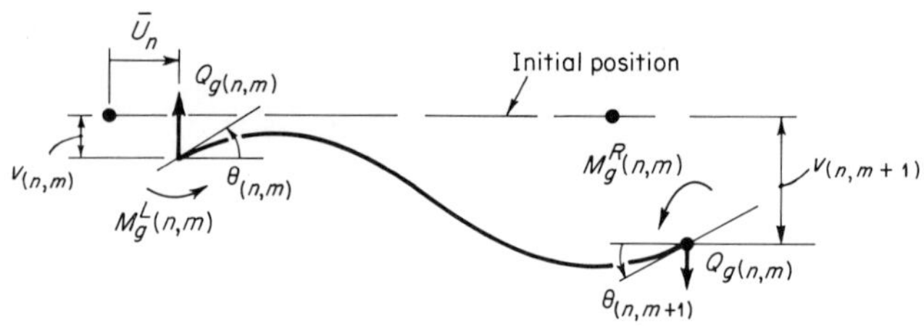

Figure 7.30 Beam member.

Hence

$$K = a^T k a = k_B \begin{bmatrix} 1 & \frac{1}{2} & -\frac{3}{2l} & \frac{3}{2l} \\ \frac{1}{2} & 1 & -\frac{3}{2l} & \frac{3}{2l} \\ -\frac{3}{2l} & -\frac{3}{2l} & \frac{3}{l^2} & -\frac{3}{l^2} \\ \frac{3}{2l} & \frac{3}{2l} & -\frac{3}{l^2} & \frac{3}{l^2} \end{bmatrix}$$

That is,

$$\begin{bmatrix} M_g{}^L \\ M_g{}^R \\ F_V{}^L \\ F_V{}^R \end{bmatrix} = k_B \begin{bmatrix} 1 & \frac{1}{2} & -\frac{3}{2l} & \frac{3}{2l} \\ \frac{1}{2} & 1 & -\frac{3}{2l} & \frac{3}{2l} \\ -\frac{3}{2l} & -\frac{3}{2l} & \frac{3}{l^2} & -\frac{3}{l^2} \\ \frac{3}{2l} & \frac{3}{2l} & -\frac{3}{l^2} & \frac{3}{l^2} \end{bmatrix} \begin{bmatrix} \theta_{n,m} \\ \theta_{n,m+1} \\ v_{n,m} \\ v_{n,m+1} \end{bmatrix}$$

However, for equilibrium of the beam member,

$F_V{}^R = -F_V{}^L = Q_g$ the girder shear

Then,

$$\begin{bmatrix} M_g{}^L \\ M_g{}^R \\ Q_g \end{bmatrix} = k_B \begin{bmatrix} 1 & \frac{1}{2} & -\frac{3}{2l} & \frac{3}{2l} \\ \frac{1}{2} & 1 & -\frac{3}{2l} & \frac{3}{2l} \\ \frac{3}{2l} & \frac{3}{2l} & -\frac{3}{l^2} & \frac{3}{l^2} \end{bmatrix} \begin{bmatrix} \theta_{n,m} \\ \theta_{n,m+1} \\ v_{n,m} \\ v_{n,m+1} \end{bmatrix} \tag{7.150}$$

From the column and beam member-stiffness relationships in Eqs. (7.149) and (7.150), the various stiffnesses for displacements in the frame can be constructed. The three basic stiffness types will be:

1 Stiffness for joint rotation

2 Stiffness for shear distortion across a story

3 Vertical shear stiffness for a column line

These stiffnesses will be characterized by the three displacements $\theta_{n,m}$, u_n, and $v_{n,m}$. These displacements are shown in Fig. 7.31*a*, *b*, and *c* respectively. Notice that both displacements u_n and $v_{n,m}$ take place without joint rotations.

The stiffnesses for each of the displacements shown in Fig. 7.31 may be built up from Eqs. (7.149) and (7.150) as follows:

1 Joint-rotation stiffness for joint (n,m)

$$k_{\theta(n,m)} = k_{BC(n,m)} + k_{BC(n-1,m)} + k_{Bg(n,m)} + k_{Bg(n,m-1)} \tag{7.151}$$

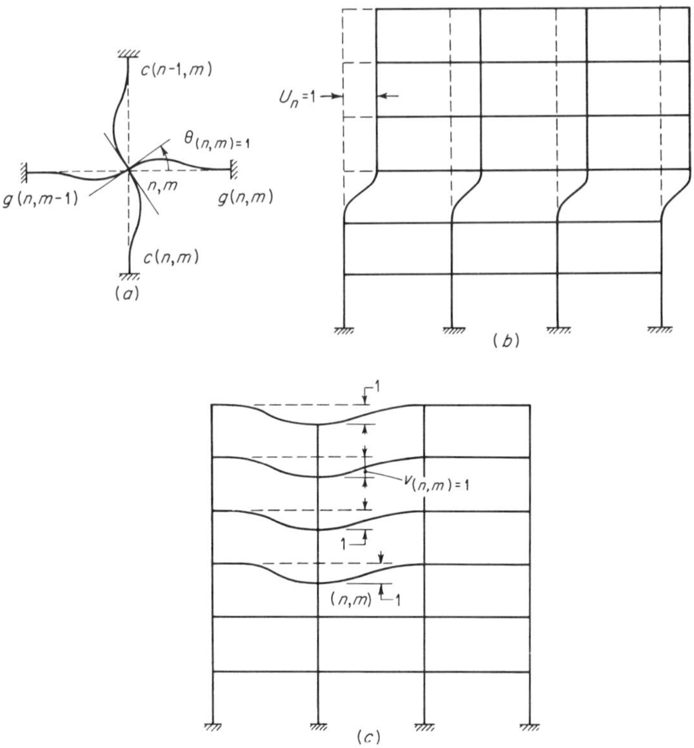

Figure 7.31 Vertical deflection. (*a*) Joint rotation; (*b*) story translation.

2 Lateral stiffness for each story level (nth story)

$$k_{u(n)} = \sum_{m=1,M} \frac{3k_{BC(n,m)}}{l_n^2} \tag{7.152}$$

3 Vertical stiffness for displacement of joint (n,m)

$$k_{v(n,m)} = k_{A(n,m)} + \sum_{j=1,n} \frac{3k_{Bg(j,m)}}{l_m^2} + \sum_{j=1,n} \frac{3k_{Bg(j,m-1)}}{l_{m-1}^2} \tag{7.153}$$

These expressions may be used now in the iteration process, starting from the joint forces calculated in the all-joints-fixed condition. The release forces are calculated in the usual way for loads along the members, as the negative of the fixed-end forces summed at each joint.

The actual force F corresponding to a given displacement is computed from these applied loads and from the joint displacements that exist just prior to this relaxation.

For a joint, the joint constraint is relaxed and the joint allowed to rotate. The new rotational displacement $\theta_{(n,m)}$ of the joint will be given by

$$\theta_{(n,m)} = \theta^*_{(n,m)} + \frac{F_{\theta(n,m)}}{k_{\theta(n,m)}} \tag{7.154}$$

The rotation $\theta^*_{(n,m)}$ exists just prior to the current relaxation. The force $F_{\theta(n,m)}$ is obtained from the applied load and by the use of the appropriate rows in Eqs. (7.149) and (7.150).

For the story translation, the unbalanced shear force $F_{u(n)}$ acting at any level n is the difference between the total external lateral force applied above that level and the sum of the shear forces acting in all of the columns at the nth level. That is,

$$F_{u(n)} = \sum_{j=1,n} R_j - \sum_{m=1,M} Q_{c(n,m)} \tag{7.155}$$

It is seen that when the nth story suffers lateral displacement, the whole structure above is given a rigid-body movement. Thus the story translation process corresponding to the release of the constraint to $F_{u(n)}$ should permit the entire structure above this level to displace as a rigid unit.

The new story displacement is given by

$$u_{(n)} = u^*_{(n)} + \beta \frac{F_{u(n)}}{K_{u(n)}} \tag{7.156}$$

In Eq. (7.156), the star indicates that value existing just prior to the current relaxation. The overrelaxation factor β is similar to that used in the Gauss-Seidel iteration. The authors in Ref. 5 recommend a β value of 1.7.

The last type of iteration, that for the vertical joint displacements, is the most complex of the three. The unbalanced vertical force $F_{v(n,m)}$ at any joint (n,m) may be expressed as the sum of the forces V applied at all the joints

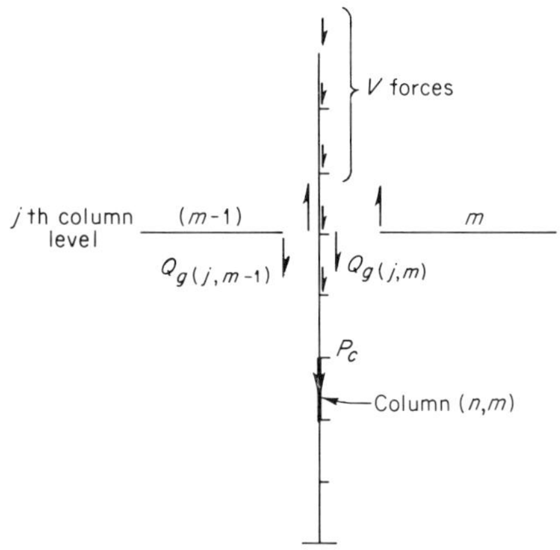

Figure 7.32 Positive forces on column.

above that level. These consist of release shears from fixed-end forces for members loaded between nodes as well as externally applied loads.

The mth-column line is shown in Fig. 7.32 with the V forces together with the girder shears and reactions at the beam level j. It is seen that $F_{v(n,m)}$ is given by

$$F_{v(n,m)} = \sum_{j=1,n} V_{(j,m)} + P_{c(n,m)} - \sum_{j=1,n} Q_{g(j,m-1)} + \sum_{j=1,n} Q_{g(j,m)} \tag{7.157}$$

The vertical-joint-relaxation process consists in evaluating the constraint force F_v and thence the resulting change in the vertical-deflection component Δv of this joint and all joints above it, as follows:

$$\Delta v_{(j,m)} = \frac{F_{v(n,m)}}{k_{v(n,m)}} \qquad j = 1, n \tag{7.158}$$

Then

$$v_{(j,m)} = v^*_{(j,m)} + \Delta v_{(j,m)} \qquad j = 1, n \tag{7.159}$$

In Ref. 5 this type of relaxation was found to be highly convergent and was applied at every 10 cycles of joint rotation and sidesway relaxation. This is probably to be expected since the column strains must necessarily produce displacements that are small relative to those from sidesway and joint rotation.

Convergence of the whole iteration process may be judged by the ratio of the current unbalance of forces to the initial unbalance. If this is less than a factor of, for example, 10^{-4}, the system will be considered to have converged. An alternative criterion could be based on the ratio of the deflection increment to the current value. Again if this is less than 10^{-4} for all increments, the system will have converged.

When the displacements have been calculated, column and beam forces are calculated from Eqs. (7.149) and (7.150) respectively.

8
FINITE–ELEMENT THEORY

8.1 INTRODUCTION TO THE FINITE–ELEMENT CONCEPT

The members discussed so far have all been line elements. That is, although member stiffnesses have been defined in terms of area and inertia properties, the member dimensions that contribute to these are not involved in the structural model insofar as the compatibility of strains at member ends is defined by point contact or coincidence of normal member slopes. Alternatively, force compatibility may be specified as the equality of stress resultants at member ends. This type of structural model is well suited to the analysis of common structural types such as frames, grids, or trusses since in these, transverse member dimensions are usually small compared with member lengths. One may enquire whether the structural theorems developed so far can be extended to cover two- and three-dimensional elements. Perhaps the first attempt at such a structural replacement was the Hrennikoff model in which a panel of material is replaced with a truss panel with double-diagonal members. Member areas are adjusted to give the E and ν values for the material. This type of structural model has difficulties (apart from interpreting the resulting member forces), since it is virtually impossible to simulate anisotropic conditions and also flow conditions such as von Mises-Hencky's in the elastoplastic range. An alternative approach (the finite-element method) has been to subdivide the domain under consideration into a number of polygonal elements whose properties are the same as that of the original structure. In each of the domains the strain (or stress) patterns are assumed to be given functions of certain values at specified points (nodes), and compatibility or equilibrium conditions are satisfied at these points for the whole assemblage of elements. In Fig. 8.1*a*, a plate is loaded in its plane so that it may be considered in a state of plane stress.

In Fig. 8.1*b* and *c*, two alternative element types are shown. It is evident from Fig. 8.1*b* that the rectangular elements will have difficulty in conforming to the actual boundary of the plate, whereas the triangular elements will be able to approximate the irregular boundary more closely. A combination of both element types may be possible. A great variety of element types are now available. Since much of the basic theory is common to all types, only the triangular elements are discussed herein. It must be emphasized that the

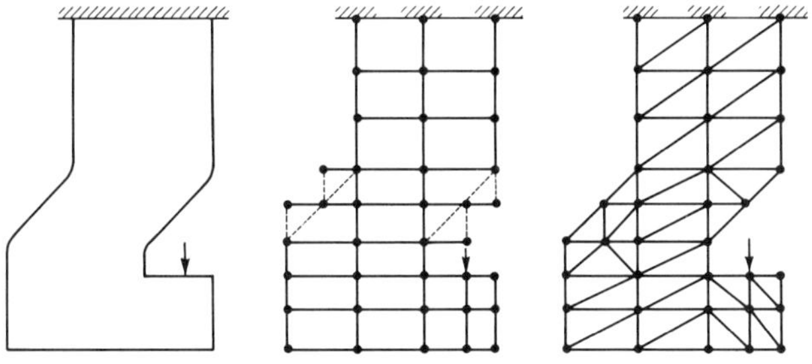

Figure 8.1 (*a*) Member in state of plane stress; (*b*) rectangular elements; (c) triangular elements.

models shown in Fig. 8.1*b* and *c* are essentially mathematical in concept rather than physical. That is, if a structure were actually constructed of triangular pieces as in Fig. 8.1*c*, these being connected at their apices by pins, the structure would be found to be far too flexible since very high stress and strain fields would be produced in these regions. It would be necessary to apply (unknown) forces along the sides of all elements to constrain their stress (strain) fields to the specified functions.

In this chapter, plane stress and strain, plate bending, axisymmetrical stress distribution, and field problems are discussed. At the present time two methods of analysis of the finite-element structures are available. In the first method, element-stiffness matrices are developed from the element dimensions, material properties, and given displacement (force) fields that have been prescribed for the element. The analysis then follows the identical procedure with that for framed structures in the direct stiffness method, and in practice a structure may be composed of a combination of both finite and line elements.

The second method seeks strain energy functions that may be expressed in terms of the nodal-displacement values. The solution becomes analogous to a Rayleigh-Ritz approach when the strain energy function is minimized with respect to the nodal parameters.

The first method generally will be used in the analyses developed herein, although the latter method will be used for field problems.

In the line element it has been usual (although not obligatory) to specify member behavior by stress-resultant patterns rather than by displacement patterns. For example, a beam element with axial force F has the force field specified simply by F that equals a constant over the member. Alternatively it could be stated that strains in the member are constant or that displacements in the member vary linearly along its length. These equivalent patterns are shown in Fig. 8.2*a* and *b*.

The stress field is given simply

$$S = F = \text{constant}$$

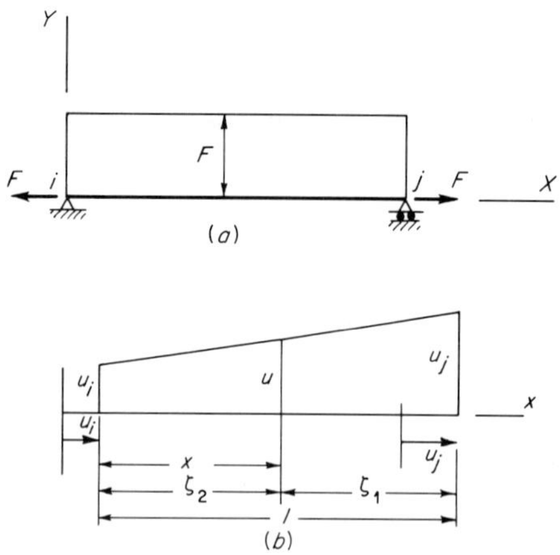

Figure 8.2 Member force and displacement fields. (*a*) Member force $S = F =$ constant; (*b*) linear displacement field.

The displacement field $u(x)$ is given as

$$u(x) = A + Bx$$

The constants A and B may be determined from the end displacements

$$u(o) = U_i, \text{ and}$$
$$u(l) = U_j$$

Then

$$U_i = A$$
$$U_j = A + Bl$$

so that

$$B = \frac{1}{l}(U_j - U_i)$$

Finally,

$$u(x) = U_i + \frac{1}{l}(U_j - U_i)x$$

Alternatively, the nondimensional natural coordinates $\langle \zeta_1 \quad \zeta_2 \rangle$ may be used (see Fig. 8.2*b*) so that

$$u(x) = u(\zeta) = A\zeta_1 + B\zeta_2$$

The boundary values are now given by

$$u(\zeta) = U_i \text{ when } \zeta_1 = 1 \qquad \zeta_2 = 0$$

and

$u(\zeta) = U_j$ when $\zeta_1 = 0 \qquad \zeta_2 = 1$

From these,

$A = U_i \qquad$ and $\qquad B = U_j$

Then

$$u(\zeta) = U_i\zeta_1 + U_j\zeta_2 \tag{8.1}$$

It will be found that these natural nondimensional coordinates are particularly useful in the description of force and displacement fields. Their advantages become evident when they are used to express polynomial functions over triangular domains.

Equation (8.1) may be regarded as an interpolation formula for the function $u(\zeta)$ expressed as a linear function between the nodal values at i and j. Equation (8.1) may be written in matrix form

$$u(\zeta) = \langle \zeta_1 \quad \zeta_2 \rangle \begin{bmatrix} U_i \\ U_j \end{bmatrix} = \zeta^T U = U^T \zeta \tag{8.2}$$

By definition, in Eq. (8.2),

$$\begin{aligned} \zeta^T &= \langle \zeta_1 \quad \zeta_2 \rangle \\ U^T &= \langle U_i \quad U_j \rangle \end{aligned} \tag{8.3}$$

In the next section, the general problem of interpolation will be discussed in more detail.

As a second example, consider the prismatic element ij shown in Fig. 8.3, subjected to the terminal moments M_i and M_j such that $m(\zeta)$ varies linearly over the element. For elastic distortions due to m, the curvature of the ele-

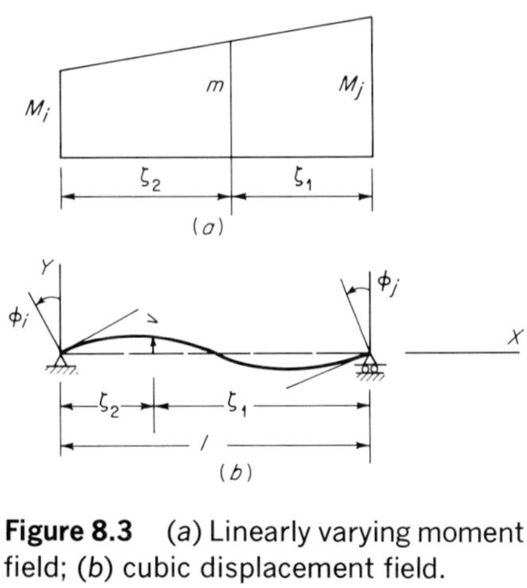

Figure 8.3 (*a*) Linearly varying moment field; (*b*) cubic displacement field.

ment $v'' = M/EI$ will also vary linearly. Double integration of the curvature will give the transverse displacement $v(\zeta)$ that must be a cubic function of either x, the distance along the member, or of the natural coordinates $\langle \zeta_1 \quad \zeta_2 \rangle$. In this case,

$$m(\zeta) = \zeta^T M \tag{8.4}$$

where $M^T = \langle M_i \quad M_j \rangle$

The displacement field, relative to the chord ij, is a cubic function of either x or $\langle \zeta_1 \quad \zeta_2 \rangle$. That is,

$$v(x) = Ax^3 + Bx^2 + Cx + D$$

The constant coefficients may be determined from the boundary conditions

$$v_i = 0$$
$$v_j = 0$$
$$v_i' = \phi_i$$
$$v_j' = \phi_j$$

Alternatively the deflection may be expressed as a function $v(\zeta)$ such that

$$v(\zeta) = A\zeta_1{}^3 + B\zeta_2{}^3 + C\zeta_1{}^2\zeta_2 + D\zeta_1\zeta_2{}^2 \tag{8.5}$$

Notice that all terms are cubic functions since if they were not, they could always be multiplied by the requisite power of $\zeta_1 + \zeta_2 = 1$.

The boundary conditions for Eq. (8.5) are,

1 $v_i = 0$ when $\zeta_1 = 1$ and $\zeta_2 = 0$; hence $A = 0$.

2 $v_j = 0$ when $\zeta_1 = 0$ and $\zeta_2 = 1$; hence $B = 0$.

Equation (8.5) is reduced to

$$v(\zeta) = C\zeta_1{}^2\zeta_2 + D\zeta_1\zeta_2{}^2 \tag{8.6}$$

The constants C and D may be found from the slope conditions ϕ_i and ϕ_j at the ends i and j. To use these conditions, differentiate $v(\zeta)$ with respect to x. Then,

$$\frac{\partial v}{\partial x} \equiv v' = \frac{\partial v}{\partial \zeta_1}\frac{\partial \zeta_1}{\partial x} + \frac{\partial v}{\partial \zeta_2}\frac{\partial \zeta_2}{\partial x} \tag{8.7}$$

Now,

$$\zeta_1 = \frac{l - x}{l} \qquad \text{so that} \qquad \frac{\partial \zeta_1}{\partial x} = -\frac{1}{l}$$

$$\zeta_2 = \frac{x}{l} \qquad \text{so that} \qquad \frac{\partial \zeta_2}{\partial x} = \frac{1}{l}$$

Then from Eq. (8.7),

$$v' = \frac{1}{l}\left(-\frac{\partial v}{\partial \zeta_1} + \frac{\partial v}{\partial \zeta_2}\right) = \frac{1}{l}(-2C\zeta_1\zeta_2 - D\zeta_2^2 + C\zeta_1^2 + 2D\zeta_1\zeta_2) \tag{8.8}$$

At the ends i and j:

1 End i: $v' = \phi_i = \dfrac{1}{l}C$; hence $C = l\phi_i$.

2 End j: $v' = \phi_j = \dfrac{1}{l}(-D)$; hence $D = -l\phi_j$.

The interpolation formula Eq. (8.6) becomes

$$v(\zeta) = l\phi_i\zeta_1^2\zeta_2 - l\phi_j\zeta_1\zeta_2^2 \tag{8.9}$$

In matrix form,

$$v(\zeta) = l\langle \zeta_1^2\zeta_2 \quad -\zeta_1\zeta_2^2\rangle \begin{bmatrix} \phi_i \\ \phi_j \end{bmatrix} \tag{8.10}$$

That is,

$$v(\zeta) = \zeta_v{}^T r \tag{8.11}$$

where

$$r^T = \langle \phi_i \quad \phi_j \rangle$$

and by definition in Eq. (8.11),

$$\zeta_v{}^T = l\langle \zeta_1^2\zeta_2 - \zeta_1\zeta_2^2\rangle \tag{8.12}$$

It will be shown later how Eqs. (8.4) and (8.12) may be used in the derivation of the stiffness matrix of the prismatic beam element.

8.2 COMPATIBLE FINITE ELEMENTS; POLYNOMIAL DISPLACEMENT FUNCTIONS; GENERAL CONCEPT OF POLYNOMIAL INTERPOLATION

The behavior of the actual structure is approximated by that of a discretized structure formed by an assemblage of bounded, simply connected domains called *finite elements*. Since for computational expediency the geometry of the elements must be simple, the geometry of the actual structure is in general only approximated by the model (cf. the rectangular and triangular elements in Fig. 8.1).

Every element displacement field is constrained to belong to a finite class (for example, linear, quadratic, or cubic polynomials) or space of functions continuous within the finite elements and satisfying deformation continuity across element interfaces.

The coordinate functions $\{\varphi_i\}$ component of a basis for the element function space [see Eq. (8.10)] will be called *displacement modes,* and their amplitudes are the generalized coordinates. These generalized coordinates will usually be the displacements or slopes at given points of the element, for example, at the apices and midpoints of triangular elements.

The element-stiffness matrix expresses the relation between the generalized coordinates and the associated generalized forces. It is a function of the geometric and material properties of the element [see, for example, Eq. (6.43)].

For application of the direct stiffness method, the generalized forces and displacements of all elements must be expressed in the global coordinate system. For the combination of individual stiffnesses to be possible, the displacement modes must be defined by their values at the nodal points usually located on the element interfaces. The generalized coordinates become the nodal displacements, and the components of the associated orthonormal basis will be the interpolating functions for the element.

Interelement compatibility requires that the displacements of any surface S_{MN} common to two elements M and N be completely specified in terms of nodal displacements on S_{MN}.

Types of Compatibility Requirements

For general three-dimensional elements, only the displacement field needs to be continuous across element interfaces. This will be referred to as *type C-o compatibility*.

A problem may be reduced to two dimensions if a displacement law of order q is assumed to hold in the third direction. Then the satisfaction of three-dimensional compatibility requires that normal gradients of transverse components up to the qth order should be continuous at the interelement boundaries. This will be *type C-q compatibility*.

The plane stress elements discussed herein are type *C-o* elements. Compatible plate bending elements are type *C*-1 since they require continuity of normal slopes on the boundaries.

Polynomial Interpolation

The space P_n of all polynomials of degree $\leq n$ in the euclidean space E_s contains

$$m = \frac{(n+s)!}{n!s!}$$

independent functions. For example, let

$$s = 1 \qquad n = 1 \qquad m = 2! = 2$$

That is, for a line element

$$f(x) = A + Bx$$

let $s = 2$ and $n = 2$; that is, a quadratic function in two-dimension space

$$m = \frac{4!}{2!2!} = 6$$

Thence

$$f(x,y) = A + Bx + Cy + Dx^2 + Exy + Fy^2$$

Call P_n a complete polynomial space of order n, and then P_n contains all P_r $(r \leq n)$ as linear subspaces.

Let $X = (x_1, x_2, \ldots, x_s)$ denote a point in a bounded, simply connected domain D of euclidean space E_s. Consider the m-dimensional vector space Γ of functions $f(X)$ defined and differentiable in D and spanned by a basis $\varphi^T = \langle \varphi_1(X) \quad \varphi_2(X) \quad \cdots \quad \varphi_m(X) \rangle$.
By definition, any $f(X)$ contained in Γ can be expressed as a linear function of φ, as follows:

$$f(X) = a_1\varphi_1 + a_2\varphi_2 + \cdots + a_m\varphi_m$$

or

$$f(X) = a^T\varphi(X) = \varphi(X)^T a \tag{8.13}$$

In Eq. (8.13), $a^T = \langle a_1 \quad a_2 \quad \cdots \quad a_m \rangle$ are the generalized coordinates of $f(X)$ in the basis φ.

Consider now a configuration (N) of m distinct points of D;

$$(N): (X_1, X_2, \cdots, X_m)$$

at which $f(X)$ takes the values $f_i = f(X_i)$. Then let

$$f^T = \langle f_1 \quad f_2 \quad \cdots \quad f_m \rangle \tag{8.14}$$

The value f_i is obtained by evaluating $f(X)$ at the node value X_i. Then from Eq. (8.13),

$$\begin{bmatrix} f_1 \\ f_2 \\ \vdots \\ f_i \\ \vdots \\ f_m \end{bmatrix} = \begin{bmatrix} \varphi_1(X_1) & \varphi_2(X_1) & \cdots & \varphi_m(X_1) \\ \varphi_1(X_2) & \varphi_2(X_2) & \cdots & \varphi_m(X_2) \\ \cdots & \cdots & \cdots & \cdots \\ \varphi_1(X_i) & \varphi_2(X_i) & \cdots & \varphi_m(X_i) \\ \cdots & \cdots & \cdots & \cdots \\ \varphi_1(X_m) & \varphi_2(X_m) & \cdots & \varphi_m(X_m) \end{bmatrix} \begin{bmatrix} a_1 \\ a_2 \\ \vdots \\ a_m \end{bmatrix} \tag{8.15}$$

The coefficient matrix on the right-hand side of Eq. (8.15) will be defined

$$A = [\varphi_r(X_s)]$$

constructed with columns φ_r. Then,

$$f = Aa \tag{8.16}$$

If the determinant $A \neq 0$, it is possible to obtain the generalized coordinates a, since then

$$a = A^{-1}f \tag{8.17}$$

An orthogonal basis $\phi^T = \langle \phi_1(X) \quad \phi_2(X) \quad \cdots \quad \phi_m(X) \rangle$ is defined as one for which the nodal parameters f are also the generalized coordinates. That is,

$$f(X) = \varphi(X)^T a = \phi(X)^T f \tag{8.18}$$

Let $\bar{A}$ be the matrix constructed from $\phi(X)$, in the same way as A, by substitution for X_s in ϕ_r. Then from Eq. (8.18),

$$f = \bar{A}f$$

Hence $\bar{A} = I$, the unit matrix.

It is required now to construct the basis ϕ if that φ is given. Substitute for a from Eq. (8.17) in Eq. (8.18), so that

$$\varphi(X)^T A^{-1} f = \phi(X)^T f$$

Hence,

$$\phi(X) = (A^{-1})^T \varphi(X) \tag{8.19}$$

Example 8.1 Construction of an orthogonal bases system

Consider the quadratic function in one dimension for which values f are known at $x = 0$, ½, and 1. In this case

$$m = \frac{(n+s)!}{n!s!} = \frac{3!}{2!} = 3$$

Thence

$$\varphi^T = \langle 1 \quad x \quad x^2 \rangle$$

$$f(x) = a^T \begin{bmatrix} 1 \\ x \\ x^2 \end{bmatrix}$$

The matrix A is generated by substituting for $x = 0$, ½, and 1 in Eq. (8.13) for φ. Then

$$A^T = \begin{bmatrix} 1 & 1 & 1 \\ 0 & \tfrac{1}{2} & 1 \\ 0 & \tfrac{1}{4} & 1 \end{bmatrix}$$

$$A^{-1} = \begin{bmatrix} 1 & 0 & 0 \\ -3 & 4 & -1 \\ 2 & -4 & 2 \end{bmatrix}$$

From Eq. (8.19),

$$\phi(X) = \begin{bmatrix} 1 & -3 & 2 \\ 0 & 4 & -4 \\ 0 & -1 & 2 \end{bmatrix} \begin{bmatrix} 1 \\ x \\ x^2 \end{bmatrix} = \begin{bmatrix} 1 - 3x + 2x^2 \\ 4x(1 - x) \\ x(2x - 1) \end{bmatrix}$$

Example 8.2 In this example the quadratic function is required for the region $x = 0, 1$, with the boundary values being at

$$x = 0, f_o, \left.\frac{\partial f}{\partial x}\right|_o$$

$$x = 1, f_1$$

Choose the polynomial $\varphi^T = \langle 1 \quad x \quad x^2 \rangle$ as in Example 8.1. Then

$$f(X) = a^T \begin{bmatrix} 1 \\ x \\ x^2 \end{bmatrix}$$

$$\frac{\partial f}{\partial x} = f'(X) = a^T \begin{bmatrix} 0 \\ 1 \\ 2x \end{bmatrix}$$

Evaluating f and f' at 0 and 1 gives

$$\begin{bmatrix} f_o \\ f'_o \\ f_1 \end{bmatrix} = a^T \begin{bmatrix} 1 & 0 & 1 \\ 0 & 1 & 1 \\ 0 & 0 & 1 \end{bmatrix}$$

Hence

$$A = \begin{bmatrix} 1 & 0 & 0 \\ 0 & 1 & 0 \\ 1 & 1 & 1 \end{bmatrix}$$

$$A^{-1} = \begin{bmatrix} 1 & 0 & 0 \\ 0 & 1 & 0 \\ -1 & -1 & 1 \end{bmatrix}$$

From Eq. (8.19),

$$\phi(X) = \begin{bmatrix} 1 & 0 & -1 \\ 0 & 1 & -1 \\ 0 & 0 & 1 \end{bmatrix} \begin{bmatrix} 1 \\ x \\ x^2 \end{bmatrix} = \begin{bmatrix} 1 - x^2 \\ x(1 - x) \\ x^2 \end{bmatrix}$$

8.3 COORDINATE SYSTEMS FOR PLANE TRIANGLES

There are three types of coordinate systems used in the description of triangular-element properties, stiffness-matrix generation, and stress and strain (displacement) identification. These are:

1 The global coordinates of the whole system
Data are input to the computer program in this system, and all calculations for node displacements and forces are in the global system.

2 The local coordinate systems associated with each side
There are three such sets. These coordinates are useful in the expressions of derivatives of functions normal to the triangle sides.

3 The area or natural coordinates
These are useful for general polynomial interpolation over the triangle. Integrals of these functions are easily calculated. They will be used generally in the derivation of triangle-stiffness matrices.

1 Global Coordinates

A plane triangle 1-2-3 (or ijk) lies on the xy plane of the global system xyz. The coordinates of the corners are (x_i, y_i), $i = 1, 2, 3$. The global dimensions of the triangle are taken parallel to the x and y axes and are denoted by a_i and b_i. Thus (a_1, b_1) are the global dimensions of side 1, side 1 being the side opposite to vertex 1 (see Fig. 8.4). From Fig. 8.4 it is seen that

$$
\begin{aligned}
a_1 &= x_3 - x_2 \\
a_2 &= x_1 - x_3 \\
a_3 &= x_2 - x_1 \\
b_1 &= y_2 - y_3 \\
b_2 &= y_3 - y_1 \\
b_3 &= y_1 - y_2
\end{aligned}
\tag{8.20}
$$

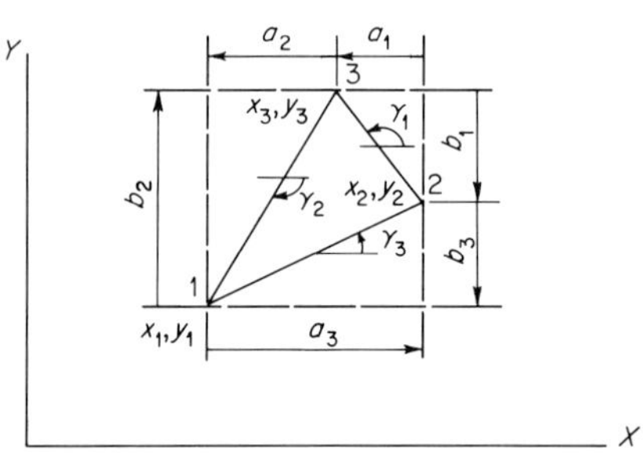

Figure 8.4 Global dimensions of triangle.

It is seen from Eq. (8.20) that

$$\sum_{i=1,3} a_i = \sum_{i=1,3} b_i = 0 \tag{8.21}$$

Let A be the area of the triangle; then

$$2A = a_3 b_2 - a_2 b_3 \tag{8.22}$$

Cyclic permutation of the indices in Eq. (8.22) gives two further expressions for the area.

2 Local Coordinates

The three local coordinate systems (s_i, n_i), $i = 1, 2, 3$ (shown in Fig. 8.5) are defined with the s_i axes along the sides in a counterclockwise cyclic sense and with the origins at the apices. The n_i axes are taken as the inward normal of the corresponding sides.

The coordinates of the corners of the triangles are expressed in terms of the side lengths l_i, the heights h_i, and the opposite-corner projections d_i. The subscript is always related to the opposite-corner number. The values (l_i, h_i, d_i) are called the *intrinsic dimensions* of the triangle. They are shown in Fig. 8.6. From Fig. 8.6 it is seen that in local coordinates the expression for the area of the triangle is

$$2A = l_i h_i \qquad i = 1, 2, 3 \tag{8.23}$$

3 Area or Natural Coordinates

Let A_1, A_2, and A_3 be the areas of the three subtriangles subtended by P (see Fig. 8.7) and the corners, the index designating the opposite-corner number. That is, area A_1, for example, has l_1 as its base. The three triangular coordinates of P are

$$\zeta_1 = \frac{A_1}{A} \qquad \zeta_2 = \frac{A_2}{A} \qquad \zeta_3 = \frac{A_3}{A} \tag{8.24}$$

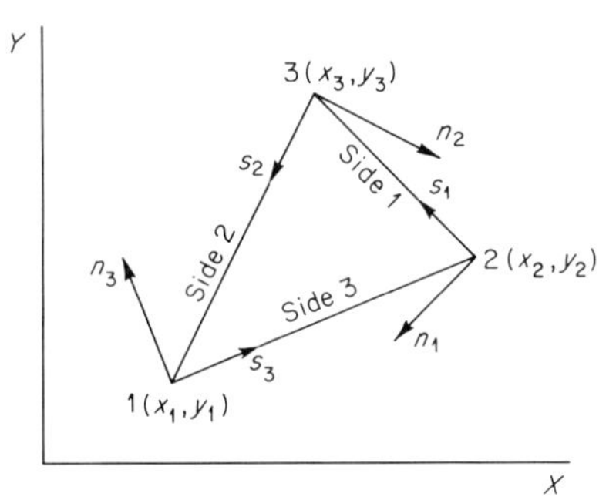

Figure 8.5 Local coordinate systems.

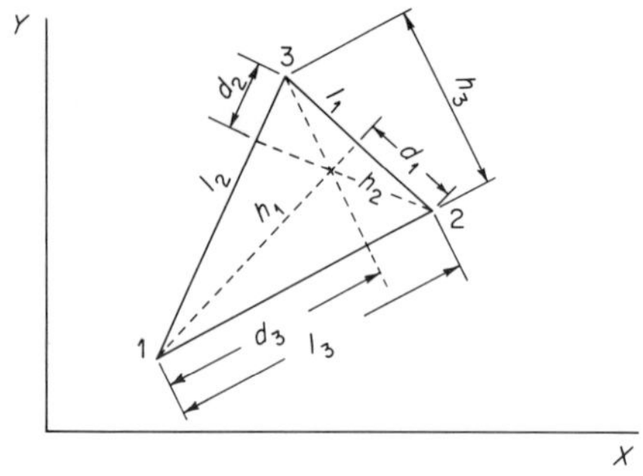

Figure 8.6 Intrinsic dimensions.

It is seen that

$$\zeta_1 + \zeta_2 + \zeta_3 = 1 \tag{8.25}$$

The point P will thus have coordinates

$$\zeta^T = \langle \zeta_1 \quad \zeta_2 \quad \zeta_3 \rangle \tag{8.26}$$

Notice that the apices of the triangle now have the nondimensional coordinates

Apex 1: (1,0,0)
Apex 2: (0,1,0)
Apex 3: (0,0,1)

Also, side 1 has $\zeta_1 = 0$, etc. It is seen that the line ζ_i = constant is parallel with the side i, $i = 1, 2, 3$.

The natural coordinates will be used to construct the interpolation formulas for functions whose nodal values are assumed known. Since strains are given as derivatives of displacements, it is necessary to derive these expressions in the various coordinate systems and to obtain the transformation equations relating the coordinate systems.

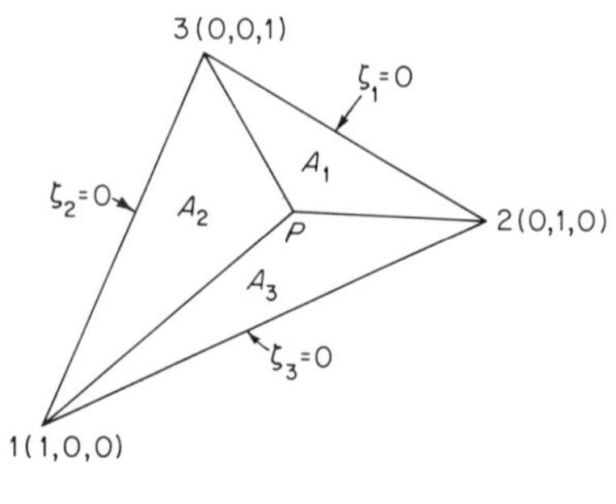

Figure 8.7 Area or natural coordinates.

Thus partial derivatives of the function $f(\zeta)$ will be referred to as:

a *Global derivatives* when with respect to x and y axes

b *Side derivatives* when with respect to s_i axes

c *Normal derivatives* when with respect to n_i axes

The Relationships between the Coordinate Systems

1 Between Global and Local Coordinate Systems

These are given in trigonometric form by using the γ_i angles shown in Fig. 8.4, or alternatively by using the global dimensions. That is,

$$\begin{bmatrix} s_i \\ n_i \end{bmatrix} = \begin{bmatrix} \cos\gamma_i & \sin\gamma_i \\ -\sin\gamma_i & \cos\gamma_i \end{bmatrix} \begin{bmatrix} x - x_j \\ y - y_j \end{bmatrix} \tag{8.27}$$

or

$$\begin{bmatrix} s_i \\ n_i \end{bmatrix} = \frac{1}{l_i} \begin{bmatrix} a_i & -b_i \\ b_i & a_i \end{bmatrix} \begin{bmatrix} x - x_j \\ y - y_j \end{bmatrix} \tag{8.28}$$

In Eqs. (8.27) and (8.28) the indices i and j have the permutations

i	j	k
1	2	3
2	3	1
3	1	2

The transformation from local to global coordinates is obtained by solving Eq. (8.28) for x and y. That is,

$$\begin{bmatrix} x \\ y \end{bmatrix} = \frac{1}{l_i} \begin{bmatrix} a_i & b_i \\ -b_i & a_i \end{bmatrix} \begin{bmatrix} s_i \\ n_i \end{bmatrix} + \begin{bmatrix} x_j \\ y_j \end{bmatrix} \tag{8.29}$$

2 Between Global and Natural Coordinates

From Figs. 8.4 and 8.7, it is seen that the xy coordinates are obtained from ζ_1, ζ_2, and ζ_3 by the expression

$$\begin{bmatrix} 1 \\ x \\ y \end{bmatrix} = \begin{bmatrix} 1 & 1 & 1 \\ x_1 & x_2 & x_3 \\ y_1 & y_2 & y_3 \end{bmatrix} \begin{bmatrix} \zeta_1 \\ \zeta_2 \\ \zeta_3 \end{bmatrix} \tag{8.30}$$

On solving Eq. (8.30) for ζ_1, ζ_1, and ζ_3, it is found

$$\begin{bmatrix} \zeta_1 \\ \zeta_2 \\ \zeta_3 \end{bmatrix} = \frac{1}{2A} \begin{bmatrix} x_2y_3 - x_3y_2 & y_2 - y_3 & x_3 - x_2 \\ x_3y_1 - x_1y_3 & y_3 - y_1 & x_1 - x_3 \\ x_1y_2 - x_2y_1 & y_1 - y_2 & x_2 - x_1 \end{bmatrix} \begin{bmatrix} 1 \\ x \\ y \end{bmatrix} \tag{8.31}$$

Equation (8.31) is simplified if the terms of the transformation matrix are expressed in terms of the global dimensions a_i and b_i. Thence

$$\begin{bmatrix} \zeta_1 \\ \zeta_2 \\ \zeta_3 \end{bmatrix} = \frac{1}{2A} \begin{bmatrix} 2A_{23} & b_1 & a_1 \\ 2A_{31} & b_2 & a_2 \\ 2A_{12} & b_3 & a_3 \end{bmatrix} \begin{bmatrix} 1 \\ x \\ y \end{bmatrix} \tag{8.32}$$

The coefficient A_{ij} is given as the area subtended by the corners i, j and the origin of coordinates. In terms of the permutation table given in relationship 1 (page 459), the area of the triangle is expressed

$$2A = a_jb_i - a_ib_j$$

Since $\zeta_i =$ constant is the equation for a line parallel to side i, so also is

$$2A_{jk} + b_ix + a_iy = \text{constant}$$

In particular, the equation for side i is

$$2A_{jk} + b_ix + a_iy = 0$$

3 Between Local and Natural Coordinates

In this case,

$$\begin{bmatrix} 1 \\ s_i \\ n_i \end{bmatrix} = \begin{bmatrix} 1 & 1 & 1 \\ 0 & l_i & d_i \\ 0 & 0 & h_i \end{bmatrix} \begin{bmatrix} \zeta_j \\ \zeta_k \\ \zeta_i \end{bmatrix} \tag{8.33}$$

On solving Eq. (8.33) for ζ values, it is seen that

$$\begin{bmatrix} \zeta_j \\ \zeta_k \\ \zeta_i \end{bmatrix} = \frac{1}{2A} \begin{bmatrix} 2A & -h_i & d_i - l_i \\ 0 & h_i & -d_i \\ 0 & 0 & l_i \end{bmatrix} \begin{bmatrix} 1 \\ s_i \\ n_i \end{bmatrix} \tag{8.34}$$

Derivatives of ζ functions From 1 and 2 above, it is seen that

$$\frac{\partial \zeta_i}{\partial x} = \frac{b_i}{2A} \quad \text{and} \quad \frac{\partial \zeta_i}{\partial y} = \frac{a_i}{2A} \tag{8.35}$$

Alternatively,

$$\frac{\partial x}{\partial \zeta_i} = x_i \qquad \text{and} \qquad \frac{\partial y}{\partial \zeta_i} = y_i \tag{8.36}$$

For derivatives with respect to s_i, n_i,

$$\begin{bmatrix} \dfrac{\partial \zeta_1}{\partial s_1} & \dfrac{\partial \zeta_1}{\partial s_2} & \dfrac{\partial \zeta_1}{\partial s_3} \\ \dfrac{\partial \zeta_2}{\partial s_1} & \dfrac{\partial \zeta_2}{\partial s_2} & \dfrac{\partial \zeta_2}{\partial s_3} \\ \dfrac{\partial \zeta_3}{\partial s_1} & \dfrac{\partial \zeta_3}{\partial s_2} & \dfrac{\partial \zeta_3}{\partial s_3} \end{bmatrix} = \begin{bmatrix} 0 & \dfrac{1}{l^2} & \dfrac{-1}{l^3} \\ \dfrac{-1}{l_1} & 0 & \dfrac{1}{l_3} \\ \dfrac{1}{l_1} & \dfrac{-1}{l_2} & 0 \end{bmatrix} \tag{8.37}$$

$$\begin{bmatrix} \dfrac{\partial \zeta_1}{\partial n_1} & \dfrac{\partial \zeta_1}{\partial n_2} & \dfrac{\partial \zeta_1}{\partial n_3} \\ \dfrac{\partial \zeta_2}{\partial n_1} & \dfrac{\partial \zeta_2}{\partial n_2} & \dfrac{\partial \zeta_2}{\partial n_3} \\ \dfrac{\partial \zeta_3}{\partial n_1} & \dfrac{\partial \zeta_3}{\partial n_2} & \dfrac{\partial \zeta_3}{\partial n_3} \end{bmatrix} = \frac{1}{2A} \begin{bmatrix} l_1 & -d_2 & d_3 - l_3 \\ d_1 - l_1 & l_2 & -d_3 \\ -d_1 & d_2 - l_2 & l_3 \end{bmatrix} \tag{8.38}$$

From the above values, derivatives of a function $f(\zeta_1,\zeta_2,\zeta_3)$ with respect to local or global cartesian axes can be obtained by the chain rule. Thus

$$\frac{\partial f}{\partial s_i} = \frac{1}{l_i}\left(\frac{\partial f}{\partial \zeta_k} - \frac{\partial f}{\partial \zeta_j}\right) \tag{8.39a}$$

$$\frac{\partial f}{\partial n_i} = \frac{1}{2A}\left[\frac{\partial f}{\partial \zeta_i} l_i + \frac{\partial f}{\partial \zeta_j}(d_i - l_i) - \frac{\partial f}{\partial \zeta_k} d_i\right] \tag{8.39b}$$

$$\frac{\partial f}{\partial x} = \frac{1}{2A}\left(\frac{\partial f}{\partial \zeta_1} b_1 + \frac{\partial f}{\partial \zeta_2} b_2 + \frac{\partial f}{\partial \zeta_3} b_3\right) \tag{8.40a}$$

$$\frac{\partial f}{\partial y} = \frac{1}{2A}\left(\frac{\partial f}{\partial \zeta_1} a_1 + \frac{\partial f}{\partial \zeta_2} a_2 + \frac{\partial f}{\partial \zeta_3} a_3\right) \tag{8.40b}$$

8.4 POLYNOMIAL INTERPOLATION FOR PLANE TRIANGLES; INTEGRATION OF FUNCTIONS OVER THE AREA OF THE TRIANGLE

A complete polynomial space P_n in two dimensions is spanned by coordinate functions

$$\{\varphi_i\} = \langle 1 \quad x \quad y \quad x^2 \quad xy \quad y^2 \quad \cdots \quad y^n \rangle$$

and has dimension

$$m = \frac{(n+2)(n+1)}{2}$$

Any polynomial $P_n(x,y)$ of order n may also be expressed as a polynomial $P_r(\zeta)$, $r \geq n$, in triangular coordinates. Such a representation is not unique since each term may contain arbitrary factors $(\zeta_1 + \zeta_2 + \zeta_3)^q \equiv 1$. When all such factors have been removed, $P_r(\zeta) \equiv P_n(\zeta)$ has degree n and is said to be *irreducible*.

The number of polynomial terms that can be formed in P_n from the three variables ζ_1, ζ_2, and ζ_3 is given by

$$r = \frac{(n+3)!}{n!3!} = \frac{1}{6}(n+3)(n+2)(n+1)$$

However in the euclidean space $s = 2$, that is, the xy plane, the number of independent functions

$$m = \frac{(n+2)!}{n!2!} = \frac{1}{2}(n+2)(n+1)$$

Hence the number of dependent functions is

$$r - m = \tfrac{1}{6}(n+2)(n+1)n$$

It is found in the triangle that for $n \geq 3$, a complete irreducible polynomial $P_n(\zeta)$ cannot be specified uniquely by boundary values. This is because terms of the type of $\zeta_1{}^{i_1} \quad \zeta_2{}^{i_2} \quad \zeta_3{}^{i_3}$ with $i_i \geq 1$ may be added, and these belong to a subspace of polynomials of order $\leq n$ that have zero values on all sides. Similarly for $n \geq 6$ a complete irreducible polynomial $P_n(\zeta)$ cannot be uniquely determined for values of P_n and the normal gradient $\partial P_n/\partial n$ on the boundary. In this case arbitrary terms $\zeta_1{}^{i_1} \quad \zeta_2{}^{i_2} \quad \zeta_3{}^{i_3}$ with $i_i \geq 2$ belonging to the subspace of all polynomials of order $\leq n$ may be added, whose values and normal derivatives are zero on the sides.

Consider now the three important cases of linear, quadratic, and cubic interpolation of a function over the triangle. The orthogonal interpolation functions will be obtained together with the requisite number of nodal values.

Case 1 *Linear Function* In this case $n = 1$ and $s = 2$, so that

$$m = \frac{(1+2)!}{1!2!} = 3$$

That is, three nodes are required. A suitable choice for these will be the vertices (i,j,k), Fig. 8.8a.

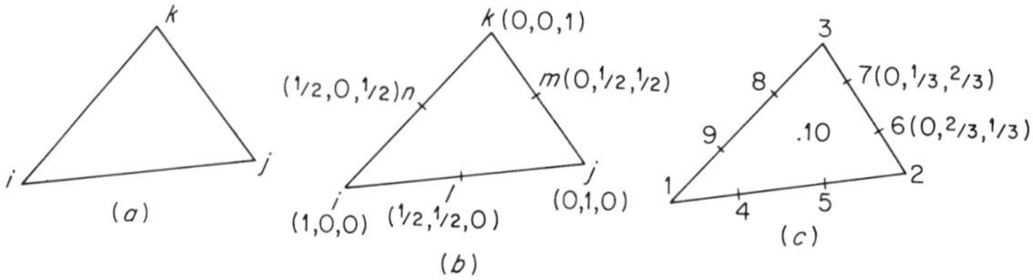

Figure 8.8 Triangle nodes. (*a*) Linear; (*b*) quadratic; (*c*) cubic.

The function f may be expressed

$$\tilde{f}(x,y) = \alpha_1 + \alpha_2 x + \alpha_3 y \tag{8.41}$$

The generalized α coordinates may be obtained by substituting for x and y at the vertices i, j, and k. Then,

$$\begin{bmatrix} f_i \\ f_j \\ f_k \end{bmatrix} = \begin{bmatrix} 1 & x_i & y_i \\ 1 & x_j & y_j \\ 1 & x_k & y_k \end{bmatrix} \begin{bmatrix} \alpha_1 \\ \alpha_2 \\ \alpha_3 \end{bmatrix} \tag{8.42}$$

However, the orthogonal interpolation functions have been introduced expressly to avoid the inversion needed in Eq. (8.42), even though in this case the inverse may be obtained by inspection.

If $\tilde{f}$ is expressed as a function of the ζ coordinates, then

$$\tilde{f} = \alpha_1\zeta_1 + \alpha_2\zeta_2 + \alpha_3\zeta_3 \tag{8.43}$$

When nodal coordinates are substituted in Eq. (8.43), it is found that

$$\begin{bmatrix} f_i \\ f_j \\ f_k \end{bmatrix} = \begin{bmatrix} 1 & 0 & 0 \\ 0 & 1 & 0 \\ 0 & 0 & 1 \end{bmatrix} \begin{bmatrix} \alpha_1 \\ \alpha_2 \\ \alpha_3 \end{bmatrix} \tag{8.44}$$

The interpolation polynomial $\zeta^T = \langle \zeta_1 \quad \zeta_2 \quad \zeta_3 \rangle$ is the orthogonal linear interpolation set. Then,

$$\tilde{f} = [\zeta_1 \quad \zeta_2 \quad \zeta_3] \begin{bmatrix} f_i \\ f_j \\ f_k \end{bmatrix} = \phi_1{}^T f = f^T \phi_1 \tag{8.45}$$

In Eq. (8.45),

$$\begin{aligned} f^T &= \langle f_i \quad f_j \quad f_k \rangle \\ \phi_1{}^T &= \langle \zeta_1 \quad \zeta_2 \quad \zeta_3 \rangle \end{aligned} \tag{8.46}$$

Case 2 *Quadratic Interpolation Formula* In this case, $n = 2$ and $s = 2$; hence,

$$m = \frac{(2+2)!}{2!2!} = 6$$

It is seen that six nodal values are required for the complete polynomial, and in the xy coordinates the function $\tilde{f}$ is expressed by

$$\tilde{f}(x,y) = \alpha_1 + \alpha_2 x + \alpha_3 y + \alpha_4 x^2 + \alpha_5 xy + \alpha_6 y^2 \tag{8.47}$$

Again the α values may be obtained by substituting x, y values at the six nodes. It will be convenient to use the triangle vertices and the midpoints of the sides as the six nodes (see Fig. 8.8*b*).

The quadratic polynomial expressed in terms of the coordinates is written [see Eq. (8.13)]

$$\varphi^T = \langle \zeta_1^2 \; \zeta_2^2 \; \zeta_3^2 \; \zeta_1\zeta_2 \; \zeta_2\zeta_3 \; \zeta_3\zeta_1 \rangle \tag{8.48}$$

From this, the orthogonal polynomial is constructed to produce an expression similar to Eq. (8.45). It will be assumed that values of the function $\tilde{f}$ are given at $\langle i \quad j \quad k \quad l \quad m \quad n \rangle$.

The orthogonal basis is constructed from Eq. (8.19). From the nodal coordinates given in Fig. 8.8*b*, the A matrix is found to be

$$A^T = \left[\begin{array}{ccc|ccc} 1 & 0 & 0 & \frac{1}{4} & 0 & \frac{1}{4} \\ 0 & 1 & 0 & \frac{1}{4} & \frac{1}{4} & 0 \\ 0 & 0 & 1 & 0 & \frac{1}{4} & \frac{1}{4} \\ \hline 0 & 0 & 0 & \frac{1}{4} & 0 & 0 \\ 0 & 0 & 0 & 0 & \frac{1}{4} & 0 \\ 0 & 0 & 0 & 0 & 0 & \frac{1}{4} \end{array}\right]$$

This matrix may be inverted by partitioning into four submatrices shown by the dashed lines. Thence,

$$(A^{-1})^T = \begin{bmatrix} 1 & 0 & 0 & -1 & 0 & -1 \\ 0 & 1 & 0 & -1 & -1 & 0 \\ 0 & 0 & 1 & 0 & -1 & -1 \\ 0 & 0 & 0 & 4 & 0 & 0 \\ 0 & 0 & 0 & 0 & 4 & 0 \\ 0 & 0 & 0 & 0 & 0 & 4 \end{bmatrix}$$

Then,

$$(A^{-1})^T\varphi = \phi_2 = \begin{bmatrix} \zeta_1^2 - \zeta_1\zeta_2 - \zeta_3\zeta_1 \\ \zeta_2^2 - \zeta_1\zeta_2 - \zeta_2\zeta_3 \\ \zeta_3^2 - \zeta_2\zeta_3 - \zeta_3\zeta_1 \\ 4\zeta_1\zeta_2 \\ 4\zeta_2\zeta_3 \\ 4\zeta_3\zeta_1 \end{bmatrix} \tag{8.49}$$

The first three terms in Eq. (8.49) may be simplified if they are written respectively in terms of ζ_1 only, ζ_2 only, and ζ_3 only. For example, the first term may be written

$$\zeta_1^2 - \zeta_1(\zeta_2 + \zeta_3) = \zeta_1^2 - \zeta_1(1 - \zeta_1) = \zeta_1(2\zeta_1 - 1)$$

Making this and similar substitutions for the second and third terms,

$$\phi_2 = \begin{bmatrix} \zeta_1(2\zeta_1 - 1) \\ \zeta_2(2\zeta_2 - 1) \\ \zeta_3(2\zeta_3 - 1) \\ 4\zeta_1\zeta_2 \\ 4\zeta_2\zeta_3 \\ 4\zeta_3\zeta_1 \end{bmatrix} \tag{8.50}$$

Case 3 *Cubic Interpolation Formula* For $n = 3$ and $s = 2$, the number of independent terms is

$$m = \frac{(3+2)!}{3!2!} = 10$$

It has been shown that it is not possible to define the complete third-degree ζ polynomial in terms of values on the boundary because the product $\zeta_1\zeta_2\zeta_3 = 0$ on all sides. To define this term, a point (10 in Fig. 8.8*c*) must be used. It is convenient to use the center point, coordinates (⅓,⅓,⅓).

It should be noted that in a derivation of stiffness matrices for such a triangle, the theory in Sec. 7.3 applies and the center point may be relaxed before the analysis of nodal deflections is undertaken. The cubic interpolation formula for the function whose values are given at the nodes 1 to 10 in Fig. 8.8*c* is given in Eq. (8.51). Its derivation follows the same lines as for the quadratic function.

$$\phi_3 = \frac{1}{2} \begin{bmatrix} \zeta_1(3\zeta_1 - 1)(3\zeta_1 - 2) \\ \zeta_2(3\zeta_2 - 1)(3\zeta_2 - 2) \\ \zeta_3(3\zeta_3 - 1)(3\zeta_3 - 2) \\ 9\zeta_1\zeta_2(3\zeta_1 - 1) \\ 9\zeta_1\zeta_2(3\zeta_2 - 1) \\ 9\zeta_2\zeta_3(3\zeta_2 - 1) \\ 9\zeta_2\zeta_3(3\zeta_3 - 1) \\ 9\zeta_3\zeta_1(3\zeta_3 - 1) \\ 9\zeta_3\zeta_1(3\zeta_1 - 1) \\ 54\zeta_1\zeta_2\zeta_3 \end{bmatrix} \tag{8.51}$$

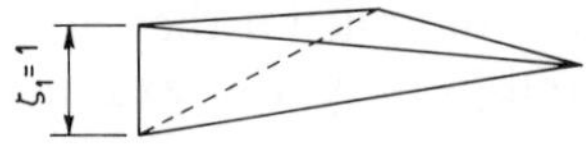

Figure 8.9 $f(\zeta) = \zeta_1$

Integration of ζ Functions over the Area of the Triangle It has been shown (Ref. 27) that the integral

$$I = \int_{\text{area}} \zeta_1{}^i \zeta_2{}^j \zeta_3{}^k \, dA$$

in which i, j, and k are integers is given by the factorial expression

$$I = \frac{(i!j!k!)2A}{(n+2)!} \tag{8.52}$$

where A is the area of the triangle, and n is the order of the polynomial, that is, $n = i + j + k$. For example, for $n = 1$, $i = 1$, and $j = k = 0$,

$$I = \frac{2A}{3!} = \frac{1}{3}A = \int_{\text{area}} \zeta_1 \, dA \tag{8.53}$$

The function $f(\zeta) = \zeta_1$ is shown in Fig. 8.9, and the value $I = \frac{1}{3}A$ is obtained by inspection. The expression for I in Eq. (8.52) has been verified in Ref. 27 for values of n through to the sixth order. Values for $n = 1, 2, 3$ are given in Table 8.1. When function $f(\zeta_1\zeta_2\zeta_3)$ is not a simple polynomial expression as is the case in the axisymmetric finite element in which terms containing $1/r$ occur, resort may be had to gaussian quadrature which gives an approximate value to the integral. Then,

$$\int_{\text{area}} f(\zeta_1,\zeta_2,\zeta_3) \, dA \simeq A \sum_{k=1}^{N} w_k f(\zeta_k)$$

The values ζ_k are the triangular coordinates at the integration points P_k $(k = 1, 2, \ldots, N)$. The values of the ordinates P_k, and weighting factors for the seven-point quadrature formula taken from Ref. 10 are given below.

Order	Figure	Points	Coordinates			Weights
Quintic	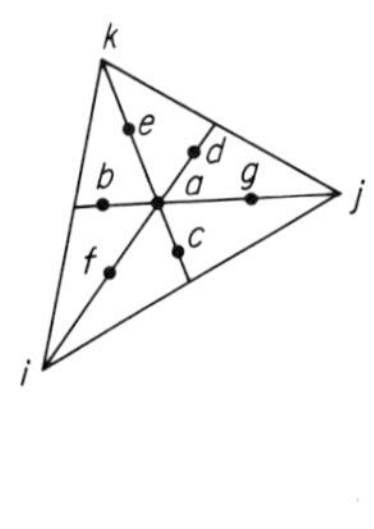	a	⅓	⅓	⅓	0.225
		b	β_1	α_1	β_1	
		c	β_1	β_1	α_1	0.13239415
		d	α_1	β_1	β_1	
		e	β_2	β_2	α_2	
		f	α_2	β_2	β_2	0.12593918
		g	β_2	α_2	β_2	

$$\alpha_1 = 0.05971587$$
$$\beta_1 = 0.47014206$$
$$\alpha_2 = 0.79742699$$
$$\beta_2 = 0.10128651$$

TABLE 8.1 Area integrals

$I = \alpha A$

Order $n = i + j + k$	i	j	k	$M\alpha$	M
1	1	0	0	1	3
2	2	0	0	2	12
	1	1	0	1	
3	3	0	0	6	60
	2	1	0	2	
	1	1	1	1	

8.5 CALCULATION OF ELEMENT–STIFFNESS MATRICES; PLANE STRESS AND STRAIN

The formation of the stiffness matrix for a triangular element as given here follows closely that given in Sec. 6.3 for the continuum for which strains have been expressed as known functions of the nodal displacements.

From Eq. (6.67),

$$R = \int_{\text{vol}} a_{xyz}^T k a_{xyz}\, dV\, r$$

the matrix a_{xyz} has terms that are functions of the coordinate system. Since there is a relationship [Eq. (8.32)] between the xy coordinates in the plane triangle and the natural coordinates $\zeta_1, \zeta_2, \zeta_3$, the expression for the element-stiffness matrix may be written

$$k = \int_{\text{vol}} a_\zeta{}^T k a_\zeta\, dV \tag{8.54}$$

The elemental volume dV may be a function of ζ if the thickness varies over the triangle. The matrix a_ζ expresses the relationship between the strains ε and the nodal displacements r. The stiffness matrix k depends on the constitutive equations for the material [see Eqs. (6.54) and (6.56)]. These may vary from point to point through the material, but in the theory developed herein, k is assumed constant within an element.

The element may have surface or body forces and be subjected to a temperature change. All these effects are considered later, and at this stage only node forces are applied.

Equation (8.54) will be used in the following general steps:

Firstly, the interpolation polynomial is chosen relating triangle displacements to the node values. This is expressed

$$\tilde{r}(\zeta) = \phi^T r \tag{8.55}$$

The tilde notation $\tilde{r}$ will be used for variables as distinct from node values.

Differentiation of Eq. (8.55) with respect to x and y gives the equation relating strains $\tilde{\varepsilon}$ to the nodal displacements r. This is of the form

$$\tilde{\varepsilon} = a_\zeta r \tag{8.56}$$

However, given nodal strains ε_C, the strain $\tilde{\varepsilon}$ may also be expressed by an interpolation polynomial of a lower order than ϕ, as

$$\tilde{\varepsilon} = \phi_\varepsilon{}^T \varepsilon_C \tag{8.57}$$

The nodal strains ε_C may be obtained from Eq. (8.56) by evaluating a_ζ at the node points. Then,

$$\varepsilon_C = \Phi_C r \tag{8.58}$$

where $\Phi_C \equiv a_\zeta$ evaluated at the strain node points. The node points required to define $\tilde{\varepsilon}$ will be less than that for $\tilde{r}$ since the differentiation to obtain strains reduces the order of the interpolation polynomial. For example, if ϕ in Eq. (8.55) is a quadratic function and $\tilde{\varepsilon}$ is obtained from $\tilde{r}$ by a single differentiation, then ϕ_ε contains linear functions. From Eqs. (8.57) and (8.58),

$$\tilde{\varepsilon} = \phi_\varepsilon{}^T \Phi_C r \tag{8.59}$$

It is seen that

$$a_\zeta = \phi_\varepsilon{}^T \Phi_C$$

The matrix Φ_C is a constant matrix, so that substitution in Eq. (8.54) gives

$$R = \Phi_C{}^T \int_{\text{vol}} \phi_\varepsilon k \phi_\varepsilon{}^T \, dV \, \Phi_C r \tag{8.60}$$

That is, the element-stiffness matrix k is

$$K = \Phi_C{}^T \int_{\text{vol}} \phi_\varepsilon k \phi_\varepsilon{}^T \, dV \, \Phi_C \tag{8.61}$$

Example 8.3 In this example the stiffness matrix for the beam element shown in Fig. 8.3 is calculated by using Eq. (8.61). It has been shown in Eq. (8.10) that the cubic equation to the deflected shape is

$$v(\zeta) = l[\zeta_1{}^2\zeta_2 \quad -\zeta_1\zeta_2{}^2]\begin{bmatrix}\phi_i \\ \phi_j\end{bmatrix}$$

Let $\tilde{C}$ be the curvature at any point in the beam; then

$$\tilde{C} = \frac{\partial^2 v}{\partial x^2}$$

By using the chain rule for differentiation, the curvature is found to be

$$\tilde{C} = \frac{2}{l}[(\zeta_2 - 2\zeta_1) \quad -(\zeta_1 - 2\zeta_2)]\begin{bmatrix}\phi_i \\ \phi_j\end{bmatrix} \tag{i}$$

The values of $\tilde{C}$ at the ends i and j will be designated C_i and C_j and are obtained from Eq. (i) by substituting $\langle \zeta_1 \quad \zeta_2 \rangle = (1,0)$ and $(0,1)$ respectively. That is,

$$\begin{bmatrix} C_i \\ C_j \end{bmatrix} = \frac{2}{l} \begin{bmatrix} -2 & -1 \\ 1 & 2 \end{bmatrix} \begin{bmatrix} \phi_i \\ \phi_j \end{bmatrix} \tag{ii}$$

Equation (ii) is the expression of Eq. (8.58) for the beam element.

The curvature $\tilde{C}$ varies linearly along the element, so that Eq. (8.57) is given by

$$\tilde{C} = [\zeta_1 \quad \zeta_2] \begin{bmatrix} C_i \\ C_j \end{bmatrix} \tag{iii}$$

The stiffness for an element of the beam length, dx, is

$$k = EI\,dx$$

Thence

$$\phi_\varepsilon k \phi_\varepsilon^T = \begin{bmatrix} \zeta_1 \\ \zeta_2 \end{bmatrix} [\zeta_1 \quad \zeta_2] EI\,dx = EI \begin{bmatrix} \zeta_1^2 & \zeta_1\zeta_2 \\ \zeta_1\zeta_2 & \zeta_2^2 \end{bmatrix} dx$$

Integration over the length of the beam O to l gives

$$\int_{\text{vol}} \phi_\varepsilon k \phi_\varepsilon^T\,dV = \int_o^l EI \begin{bmatrix} \zeta_1^2 & \zeta_1\zeta_2 \\ \zeta_1\zeta_2 & \zeta_2^2 \end{bmatrix} dx = l \int_o^1 EI \begin{bmatrix} \zeta_1^2 & \zeta_1\zeta_2 \\ \zeta_1\zeta_2 & \zeta_2^2 \end{bmatrix} d\zeta_2$$

Finally, for EI constant,

$$\int_{\text{vol}} \phi_\varepsilon k \phi_\varepsilon^T\,dV = \frac{EIl}{6} \begin{bmatrix} 2 & 1 \\ 1 & 2 \end{bmatrix} = k_C$$

Then,

$$K = \Phi_C^T k_C \Phi_C = \frac{EIl}{6} \frac{4}{l^2} \begin{bmatrix} -2 & 1 \\ -1 & 2 \end{bmatrix} \begin{bmatrix} 2 & 1 \\ 1 & 2 \end{bmatrix} \begin{bmatrix} -2 & -1 \\ 1 & 2 \end{bmatrix} = \frac{2EI}{l} \begin{bmatrix} 2 & 1 \\ 1 & 2 \end{bmatrix}$$

The result is of course that obtained previously in Chap. 6, indicating that the cubic displacement function chosen is in fact that produced by the linear moment field. Although the derivation here appears somewhat complex, the power of Eq. (8.61) will become apparent when the plane stress and strain element-stiffness matrices are discussed in the following section.

Plane Stress and Strain Elements

The element may vary in thickness, but this must be such that the general two-dimensional behavior is not affected. The complete displacement field is specified by the two displacement components $\tilde{u}(x,y)$ and $\tilde{v}(x,y)$ given in the global coordinate system. Equally well, each of these displacement components may be specified in triangular coordinates by the interpolation formulas,

$$\tilde{u}(\zeta) = \phi_u{}^T U \tag{8.62}$$

$$\tilde{v}(\zeta) = \phi_v{}^T V \tag{8.63}$$

The interpolation formula for $\tilde{u}$ and for $\tilde{v}$ will be the same, so that $\phi_u = \phi_v = \phi$. In what follows, both linear and quadratic functions are discussed and their stiffness matrices developed.

Equations (8.62) and (8.63) may be combined to give the interpolation equation for the vector $\tilde{r}^T = \langle \tilde{u}, \tilde{v} \rangle$:

$$\tilde{r} \equiv \begin{bmatrix} \tilde{u}(\zeta) \\ \tilde{v}(\zeta) \end{bmatrix} = \begin{bmatrix} \phi^T & 0 \\ 0 & \phi^T \end{bmatrix} \begin{bmatrix} U \\ V \end{bmatrix} = \psi r \tag{8.64}$$

For the linear displacement-field triangle (that is, constant strain), the nodes i, j, and k at the triangle corners will be used, so that

$$\phi^T = \phi_1{}^T = \langle \zeta_1 \quad \zeta_2 \quad \zeta_3 \rangle \tag{8.65}$$

$$U^T = \langle u_i \quad u_j \quad u_k \rangle \tag{8.66a}$$

$$V^T = \langle v_i \quad v_j \quad v_k \rangle \tag{8.66b}$$

For the quadratic displacement-field triangle (that is, linear strain), nodes i, j, and k together with those l, m, and n at the midpoints of the sides are used. Then from Eq. (8.50),

$$\phi^T = \phi_2{}^T = \langle \zeta_1(2\zeta_1 - 1) \quad \zeta_2(2\zeta_2 - 1) \quad \zeta_3(2\zeta_3 - 1) \quad 4\zeta_1\zeta_2 \quad 4\zeta_2\zeta_3 \quad 4\zeta_3\zeta_1 \rangle \tag{8.67}$$

The nodal-displacement vectors are

$$U^T = \langle u_i \quad u_j \quad u_k \quad u_l \quad u_m \quad u_n \rangle \tag{8.68a}$$

$$V^T = \langle v_i \quad v_j \quad v_k \quad v_l \quad v_m \quad v_n \rangle \tag{8.68b}$$

The strain functions may be evaluated from the derivatives of the displacement functions such that

$$\tilde{\varepsilon}_x(\zeta) = \frac{\partial \tilde{u}}{\partial x} = \phi_x{}^T U \tag{8.69}$$

$$\tilde{\varepsilon}_y(\zeta) = \frac{\partial \tilde{v}}{\partial x} = \phi_y{}^T V \tag{8.70}$$

$$\tilde{\gamma}_{xy}(\zeta) = \frac{\partial \tilde{v}}{\partial x} + \frac{\partial \tilde{u}}{\partial y} = \phi_y{}^T U + \phi_x{}^T V \tag{8.71}$$

In Eqs. (8.69) to (8.71) the x and y subscripts on ϕ^T indicate partial differentiation.

Then, Eqs. (8.69) to (8.71) in matrix form become

$$\tilde{\varepsilon}(\zeta) = \begin{bmatrix} \tilde{\varepsilon}_x(\zeta) \\ \tilde{\varepsilon}_y(\zeta) \\ \tilde{\gamma}_{xy}(\zeta) \end{bmatrix} = \begin{bmatrix} \phi_x{}^T & 0 \\ 0 & \phi_y{}^T \\ \phi_y{}^T & \phi_x{}^T \end{bmatrix} \begin{bmatrix} U \\ V \end{bmatrix} = a_\zeta r \tag{8.72}$$

For the constant-strain triangle,

$$a_\zeta = \frac{1}{2A} \begin{bmatrix} b_1 & b_2 & b_3 & 0 & 0 & 0 \\ 0 & 0 & 0 & a_1 & a_2 & a_3 \\ a_1 & a_2 & a_3 & b_1 & b_2 & b_3 \end{bmatrix} \tag{8.73}$$

$$a_\zeta = \frac{1}{2A} \left[\begin{array}{cccccc|cccccc} (4\zeta_1-1)b_1 & (4\zeta_2-1)b_2 & (4\zeta_3-1)b_3 & 4(\zeta_2 b_1+\zeta_1 b_2) & 4(\zeta_2 b_3+\zeta_3 b_2) & 4(\zeta_3 b_1+\zeta_1 b_3) & 0 & 0 & 0 & 0 & 0 & 0 \\ \hline 0 & 0 & 0 & 0 & 0 & 0 & (4\zeta_1-1)a_1 & (4\zeta_2-1)a_2 & (4\zeta_3-1)a_3 & 4(\zeta_2 a_1+\zeta_1 a_2) & 4(\zeta_2 a_3+\zeta_3 a_2) & 4(\zeta_3 a_1+\zeta_1 a_3) \\ \hline (4\zeta_1-1)a_1 & (4\zeta_2-1)a_2 & (4\zeta_3-1)a_3 & 4(\zeta_2 a_1+\zeta_1 a_2) & 4(\zeta_2 a_3+\zeta_3 a_2) & 4(\zeta_3 a_1+\zeta_1 a_3) & (4\zeta_1-1)b_1 & (4\zeta_2-1)b_2 & (4\zeta_3-1)b_3 & 4(\zeta_2 b_1+\zeta_1 b_2) & 4(\zeta_2 b_3+\zeta_3 b_2) & 4(\zeta_3 b_1+\zeta_1 b_3) \end{array} \right] \tag{8.74}$$

The stress at any point (or more correctly the infinitesimal-element-stiffness matrix) is given by the constitutive law for the material (see Sec. 6.3), that is,

$$\begin{bmatrix} \sigma_x \\ \sigma_y \\ \gamma_{xy} \end{bmatrix} = \begin{bmatrix} C_{11} & C_{12} & C_{13} \\ C_{21} & C_{22} & C_{23} \\ C_{31} & C_{32} & C_{33} \end{bmatrix} \begin{bmatrix} \varepsilon_x \\ \varepsilon_y \\ \gamma_{xy} \end{bmatrix}$$

That is,

$$\tilde{\sigma} = k\tilde{\varepsilon} \tag{8.75}$$

For isotropic materials,

$$k = E_r \begin{bmatrix} \tau & \nu & 0 \\ \nu & \tau & 0 \\ 0 & 0 & \lambda \end{bmatrix}$$

where for plane stress,

$$\tau = 1 \qquad \lambda = \frac{1-\nu}{2} \qquad E_r = \frac{E}{1-\nu^2}$$

and for plane strain,

$$\tau = 1 - \nu \qquad \lambda = 0.5 - \nu \qquad E_r = \frac{E}{(1+\nu)(1-2\nu)}$$

The strain field $\tilde{\varepsilon}$ is constant for the linear displacement-field triangle [see Eq. (8.73)] and varies linearly [see Eq. (8.74)] for the quadratic displacement-field triangle. Thence, for the constant-strain triangle,

$$\Phi_C = a_\zeta = \frac{1}{2A}\begin{bmatrix} b_1 & b_2 & b_3 & 0 & 0 & 0 \\ 0 & 0 & 0 & a_1 & a_2 & a_3 \\ a_1 & a_2 & a_3 & b_1 & b_2 & b_3 \end{bmatrix} \tag{8.76}$$

For the linear-strain triangle substitution of the coordinate values for $i(1,0,0)$, $j(0,1,0)$, and $k(0,0,1)$ in Eq. (8.74),

$$\begin{bmatrix} \varepsilon_{xi} \\ \varepsilon_{xj} \\ \varepsilon_{xk} \\ \varepsilon_{yi} \\ \varepsilon_{yj} \\ \varepsilon_{yk} \\ \gamma_{xyi} \\ \gamma_{xyj} \\ \gamma_{xyk} \end{bmatrix} = \frac{1}{2A}\left[\begin{array}{cccccc|cccccc} 3b_1 & -b_2 & -b_3 & 4b_2 & 0 & 4b_3 & 0 & 0 & 0 & 0 & 0 & 0 \\ -b_1 & 3b_2 & -b_3 & 4b_1 & 4b_3 & 0 & 0 & 0 & 0 & 0 & 0 & 0 \\ -b_1 & -b_2 & 3b_3 & 0 & 4b_2 & 4b_1 & 0 & 0 & 0 & 0 & 0 & 0 \\ \hline 0 & 0 & 0 & 0 & 0 & 0 & 3a_1 & -a_2 & -a_3 & 4a_2 & 0 & 4a_3 \\ 0 & 0 & 0 & 0 & 0 & 0 & -a_1 & 3a_2 & -a_3 & 4a_1 & 4a_3 & 0 \\ 0 & 0 & 0 & 0 & 0 & 0 & -a_1 & -a_2 & 3a_3 & 0 & 4a_2 & 4a_1 \\ \hline 3a_1 & -a_2 & -a_3 & 4a_2 & 0 & 4a_3 & 3b_1 & -b_2 & -b_3 & 4b_2 & 0 & 4b_3 \\ -a_1 & 3a_2 & -a_3 & 4a_1 & 4a_3 & 0 & -b_1 & 3b_2 & -b_3 & 4b_1 & 4b_3 & 0 \\ -a_1 & -a_2 & 3a_3 & 0 & 4a_2 & 4a_1 & -b_1 & -b_2 & 3b_3 & 0 & 4b_2 & 4b_1 \end{array}\right]\begin{bmatrix} u_i \\ u_j \\ u_k \\ u_l \\ u_m \\ u_n \\ v_i \\ v_j \\ v_k \\ v_l \\ v_m \\ v_n \end{bmatrix} \tag{8.77}$$

This is the equation

$$\varepsilon_C = \Phi_C r \tag{8.78}$$

It only remains now to write down the interpolation polynomials ϕ_ε and calculate the integral

$$k_C = \int \phi_\varepsilon k \phi_\varepsilon{}^T \, dV$$

Consider firstly the constant-thickness triangle, thickness h. Then for the constant-strain triangle, this integral is simply equal to Ah multiplied by k, so that

$$k_C = AhE_r\begin{bmatrix} \tau & \nu & 0 \\ \nu & \tau & 0 \\ 0 & 0 & \lambda \end{bmatrix} \tag{8.79}$$

For the linear-strain triangle, the strain interpolation is given by

$$\begin{bmatrix} \tilde{\varepsilon}_x \\ \tilde{\varepsilon}_y \\ \tilde{\gamma}_{xy} \end{bmatrix} = \begin{bmatrix} \zeta_1 & \zeta_2 & \zeta_3 & 0 & 0 & 0 & 0 & 0 & 0 \\ 0 & 0 & 0 & \zeta_1 & \zeta_2 & \zeta_3 & 0 & 0 & 0 \\ 0 & 0 & 0 & 0 & 0 & 0 & \zeta_1 & \zeta_2 & \zeta_3 \end{bmatrix} \begin{bmatrix} \varepsilon_{xi} \\ \varepsilon_{xj} \\ \varepsilon_{xk} \\ \varepsilon_{yi} \\ \varepsilon_{yj} \\ \varepsilon_{yk} \\ \gamma_{xyi} \\ \gamma_{xyj} \\ \gamma_{xyk} \end{bmatrix} \tag{8.80}$$

Let $\phi_1{}^T = \langle \zeta_1 \quad \zeta_2 \quad \zeta_3 \rangle$, and write Eq. (8.80) as

$$\tilde{\varepsilon} = \begin{bmatrix} \phi_1{}^T & 0 & 0 \\ 0 & \phi_1{}^T & 0 \\ 0 & 0 & \phi_1{}^T \end{bmatrix} \varepsilon_C \tag{8.81}$$

That is,

$$\phi_\varepsilon{}^T = \begin{bmatrix} \phi_1{}^T & 0 & 0 \\ 0 & \phi_1{}^T & 0 \\ 0 & 0 & \phi_1{}^T \end{bmatrix} \tag{8.82}$$

Then,

$$\phi_\varepsilon k \phi_\varepsilon{}^T = \begin{bmatrix} \phi_1 & 0 & 0 \\ 0 & \phi_1 & 0 \\ 0 & 0 & \phi_1 \end{bmatrix} E_r \begin{bmatrix} \tau & \nu & 0 \\ \nu & \tau & 0 \\ 0 & 0 & \lambda \end{bmatrix} \begin{bmatrix} \phi_1{}^T & 0 & 0 \\ 0 & \phi_1{}^T & 0 \\ 0 & 0 & \phi_1{}^T \end{bmatrix} = E_r \begin{bmatrix} \phi_1\phi_1{}^T\tau & \phi_1\phi_1{}^T\nu & 0 \\ \phi_1\phi_1{}^T\nu & \phi_1\phi_1{}^T\tau & 0 \\ 0 & 0 & \phi_1\phi_1{}^T\lambda \end{bmatrix}$$

The product

$$\phi_1\phi_1{}^T = \begin{bmatrix} \zeta_1{}^2 & \zeta_1\zeta_2 & \zeta_1\zeta_3 \\ \zeta_1\zeta_2 & \zeta_2{}^2 & \zeta_2\zeta_3 \\ \zeta_1\zeta_3 & \zeta_2\zeta_3 & \zeta_3{}^2 \end{bmatrix}$$

Integrating $\phi_1\phi_1{}^T$ over the area of the triangle (see Table 8.1) and defining Λ

such that

$$\Lambda = \begin{bmatrix} 2 & 1 & 1 \\ 1 & 2 & 1 \\ 1 & 1 & 2 \end{bmatrix}$$

Then

$$\int_{\text{vol}} \phi_\varepsilon k \phi_\varepsilon^T \, dV = \frac{Ah}{12} E_r \begin{bmatrix} \Lambda\tau & \Lambda\nu & 0 \\ \Lambda\nu & \Lambda\tau & 0 \\ 0 & 0 & \Lambda\lambda \end{bmatrix} = k_C \tag{8.83}$$

The matrix k_C is of order 9×9 composed of the 3×3 submatrices shown in Eq. (8.83). The stiffness matrix K obtained by using Eq. (8.61) is 12×12.

Element of Varying Thickness

Let h_r be a reference thickness for the element, and define the relative thickness,

$$\tilde{\xi}(\zeta) = \frac{\tilde{h}(\zeta)}{h_r}$$

Then

$$\tilde{h}(\zeta) = \tilde{\xi}(\zeta) h_r$$

The interpolation polynomial for $\tilde{\xi}(\zeta)$ will be defined ϕ_ξ, so that in terms of nodal thicknesses,

$$\begin{aligned} \tilde{\xi}(\zeta) &= \phi_\xi^T \xi \\ \tilde{h}(\zeta) &= h_r \phi_\xi^T \xi \end{aligned} \tag{8.84}$$

For an element of linearly varying thickness, the ξ values may be defined at the corners so that

$$\xi^T = \langle \xi_i \quad \xi_j \quad \xi_k \rangle$$
$$\phi_\xi^T = \langle \zeta_1 \quad \zeta_2 \quad \zeta_3 \rangle$$

The elementary volume dV is now expressed by

$$dV = \tilde{h} \, dA = h_r \phi_\xi^T \xi \, dA \tag{8.85}$$

The stiffness matrix k_C is calculated from

$$k_C = h_r \int_{\text{area}} \phi_\varepsilon \phi_\xi^T \xi k \phi_\varepsilon^T \, dA \tag{8.86}$$

Notice that the thickness term $\phi_\xi^T \xi$ enters the integral as a constant multiplier of each term of the product $\phi_\varepsilon k \phi_\varepsilon^T$. If, for example, the constant-strain triangle is being used with a linear thickness variation, $\phi_\varepsilon k \phi_\varepsilon^T$ is constant, and then the multiplier is given by

$$\beta = \int_{\text{area}} \phi_\xi^T \xi \, dA = \int_{\text{area}} (\xi_i \zeta_1 + \xi_j \zeta_2 + \xi_k \zeta_3) \, dA$$

Hence

$$\beta = \frac{A}{3} (\xi_i + \xi_j + \xi_k) = A \times \xi_{\text{average}}$$

Consider the linear-strain triangle with a linear thickness variation. In this case,

$$\phi_\xi^T \xi = [\zeta_1 \quad \zeta_2 \quad \zeta_3] \begin{bmatrix} \xi_i \\ \xi_j \\ \xi_k \end{bmatrix}$$

Let

$$\pi = \phi_\xi^T \xi$$

Then

$$\phi_1^T \phi_\xi^T \xi \phi_1 = \begin{bmatrix} \zeta_1^2 \pi & \zeta_1 \zeta_2 \pi & \zeta_1 \zeta_3 \pi \\ & \zeta_2^2 \pi & \zeta_2 \zeta_3 \pi \\ \text{symmetrical} & & \zeta_3^2 \pi \end{bmatrix}$$

and thence,

$$\int_{\text{area}} \phi_1^T \phi_\xi^T \xi \phi_1 \, dA = \frac{A}{60} \begin{bmatrix} 6\xi_i + 2\xi_j + 2\xi_k & 2\xi_i + 2\xi_j + \xi_k & 2\xi_i + \xi_j + 2\xi_k \\ 2\xi_i + 2\xi_j + \xi_k & 2\xi_i + 6\xi_j + 2\xi_k & \xi_i + 2\xi_j + 2\xi_k \\ 2\xi_i + \xi_j + 2\xi_k & \xi_i + 2\xi_j + 2\xi_k & 2\xi_i + 2\xi_j + 6\xi_k \end{bmatrix}$$

Calculation of Nodal Stresses

The generalized forces corresponding to ε_c will be denoted S_c. From Eq. (8.57) it is seen that the generalized force ΔS_c corresponding to the strain ε is

$$\Delta S_c = \phi_\varepsilon \tilde{\sigma} = \phi_\varepsilon k \tilde{\varepsilon} = \phi_\varepsilon k \phi_\varepsilon^T \varepsilon_c$$

Integration over the area of the triangle gives

$$S_c = \int \phi_\varepsilon k \phi_\varepsilon^T \, dV \, \varepsilon_c = k_c \varepsilon_c$$

The generalized forces S_c are not the nodal stresses. The strains ε_c and the generalized forces S_c are similar in many ways to member distortions and forces associated with line elements. At any point in the element the stresses may be calculated from

$$\tilde{\sigma} = k\tilde{\varepsilon} = k\phi_{\varepsilon}{}^{T}\varepsilon_c$$

Nodal stresses are calculated by substitution of the appropriate coordinate values in $\phi_{\varepsilon}{}^{T}$.

For the constant-strain triangle, the stresses are constant so that the nodal stress vector σ_c is a 3×1 column matrix. In the linear-strain triangle, however, the stresses vary linearly so that σ_c is a 9×1 matrix. In this case,

$$\sigma_c = [k]\varepsilon_c$$

The matrix $[k]$ will depend on the ordering of the stresses σ_c, and if these are as in Eq. (8.77), then

$$\begin{bmatrix} \sigma_{xi} \\ \sigma_{xj} \\ \sigma_{xk} \\ \sigma_{yi} \\ \sigma_{yj} \\ \sigma_{yk} \\ \tau_{xyi} \\ \tau_{xyj} \\ \tau_{xyk} \end{bmatrix} = E_r \begin{bmatrix} \tau & 0 & 0 & \nu & 0 & 0 & 0 & 0 & 0 \\ 0 & \tau & 0 & 0 & \nu & 0 & 0 & 0 & 0 \\ 0 & 0 & \tau & 0 & 0 & \nu & 0 & 0 & 0 \\ \nu & 0 & 0 & \tau & 0 & 0 & 0 & 0 & 0 \\ 0 & \nu & 0 & 0 & \tau & 0 & 0 & 0 & 0 \\ 0 & 0 & \nu & 0 & 0 & \tau & 0 & 0 & 0 \\ 0 & 0 & 0 & 0 & 0 & 0 & \lambda & 0 & 0 \\ 0 & 0 & 0 & 0 & 0 & 0 & 0 & \lambda & 0 \\ 0 & 0 & 0 & 0 & 0 & 0 & 0 & 0 & \lambda \end{bmatrix} \begin{bmatrix} \varepsilon_{xi} \\ \varepsilon_{xj} \\ \varepsilon_{xk} \\ \varepsilon_{yi} \\ \varepsilon_{yj} \\ \varepsilon_{yk} \\ \gamma_{xyi} \\ \tau_{xyj} \\ \gamma_{xyk} \end{bmatrix}$$

8.6 SURFACE AND BODY FORCES; TEMPERATURE STRAINS; STRATIFIED MATERIAL

Thus far, the stiffness matrices for two types of plane stress or strain elements have been developed. These are expressed in terms of nodal forces and displacements in the global coordinate system so that the direct stiffness method of analysis may be applied to the assemblage of elements. The complete stiffness matrix is singular until the boundary support conditions are included. The same procedures are followed here as given for frame structures in Chap. 7.

In general, the types of forces applied to a continuum will be surface forces distributed over some portion of the body, or body forces such as gravity applied as a force per unit volume. If in addition various elements undergo different temperature variations, there will be stresses induced from this source. Each of these topics will now be developed in detail so that the equivalent node forces may be used in the direct stiffness program.

The treatments of surface and body forces are similar in that node forces are produced that are statically equivalent to the given distributed forces.

Surface Forces

The in-plane surface loading $\tilde{p}(\zeta_B)$ acting on a given boundary B has been integrated through the thickness h, so that it is a force per unit length. Let R_p be the node forces that are statically equivalent to $\tilde{p}(\zeta_B)$. Then when virtual node displacements r_B are produced, the work done by these node forces must be equal to that done by the surface forces $\tilde{p}(\zeta_B)$. That is,

$$R_p{}^T r_B = \int_B \tilde{p}^T \tilde{r}_B \, ds \tag{8.87}$$

Let the distributed force $\tilde{p}(\zeta_B)$ vary as an n_pth-order polynomial. The number of node values needed to specify $\tilde{p}(\zeta_B)$ will be $m_p = (n_p + 1)!/n_p! = n_p + 1$, since along the boundary the variable $\tilde{p}(\zeta_B)$ is a function of a single parameter. Then

$$\tilde{p}(\zeta_B) = \begin{bmatrix} p_x(\zeta_B) \\ p_y(\zeta_B) \end{bmatrix} = \begin{bmatrix} \Phi_p{}^T & 0 \\ 0 & \Phi_p{}^T \end{bmatrix} \begin{bmatrix} p_x \\ p_y \end{bmatrix} = \psi_p{}^T p \tag{8.88}$$

In Eq. (8.88), the terms $\tilde{p}_x(\zeta_B)$ and $\tilde{p}_y(\zeta_B)$ are the X and Y components of the distributed force, and p_x and p_y are the values specified at the $n_p + 1$ node points. For convenience, the order n_p must be restricted for the constant-strain and linear-strain triangles. For the constant-strain triangle, $\tilde{p}(\zeta_B)$ can vary linearly along a side without having to define extra node points, and for the linear-strain triangle it may vary parabolically.

The displacement of the boundary B depends only on the nodal points located on the element interface that contains B; hence

$$\tilde{r}_B(\xi_B) = \begin{bmatrix} \tilde{u}(\zeta_B) \\ \tilde{v}(\zeta_B) \end{bmatrix} = \begin{bmatrix} \phi_B{}^T & 0 \\ 0 & \phi_B{}^T \end{bmatrix} \begin{bmatrix} u_B \\ v_B \end{bmatrix} = \psi_B{}^T r_B \tag{8.89}$$

The vector $r_B{}^T = \langle u_B{}^T \quad v_B{}^T \rangle$ includes only those components of the nodal displacements r that determine $\tilde{r}_B(\zeta_B)$ of B.

The interpolating function ϕ_B for side B is obtained from ϕ by extracting interpolating functions corresponding to $\tilde{u}_B(\zeta_B)$ and $\tilde{v}_B(\zeta_B)$ and evaluating them on B where they become functions of a single triangular coordinate. From Eqs. (8.87)ff,

$$R_p{}^T r_B = p^T \int_B \psi_p \psi_B{}^T \, ds \, r_B$$

That is,

$$R_p = \int_B \psi_B \psi_p{}^T \, ds \, p \tag{8.90}$$

Hence

$$\begin{bmatrix} R_{px} \\ R_{py} \end{bmatrix} = \int_B \begin{bmatrix} \phi_B \phi_p{}^T & 0 \\ 0 & \phi_B \phi_p{}^T \end{bmatrix} ds \begin{bmatrix} p_x \\ p_y \end{bmatrix}$$

$$R_{px} = \int_B \phi_B \phi_p{}^T \, ds \, p_x \tag{8.91a}$$

$$R_{py} = \int_B \phi_B \phi_p{}^T \, ds \, p_y \tag{8.91b}$$

Evaluation of Eqs. (8.91) for Constant-strain and Linear-strain Triangles

1 *Constant-strain Triangle* The linear variation of $\tilde{p}(\zeta_B)$ on side 1 is shown for the constant-strain triangle in Fig. 8.10. The equivalent node forces will act on nodes j and k. On side 1, the term $\zeta_1 = 0$, and hence

$$\phi_B = \begin{bmatrix} \zeta_2 \\ \zeta_3 \end{bmatrix} = \begin{bmatrix} \zeta_2 \\ 1 - \zeta_2 \end{bmatrix}$$

For a linear variation of $\tilde{p}$, the interpolation polynomial $\phi_p \equiv \phi_B$, and thence

$$\phi_B \phi_p{}^T = \begin{bmatrix} \zeta_2 \\ 1 - \zeta_2 \end{bmatrix} [\zeta_2 \quad 1 - \zeta_2] = \begin{bmatrix} \zeta_2{}^2 & \zeta_2(1 - \zeta_2) \\ \zeta_2(1 - \zeta_2) & (1 - \zeta_2)^2 \end{bmatrix}$$

Expressed in terms of the local coordinates,

$$\zeta_2 = \frac{l_1 - s_1}{l_1}$$

hence

$$ds_1 = -l_1 \, d\zeta_2$$

Then,

$$R_{px} = \int_0^s \phi_B \Phi_p{}^T \, ds_1 \, p_x = l_1 \int_0^1 \phi_B \phi_p{}^T \, d\zeta_2 \, p_x$$

$$\int_0^1 \phi_B \phi_p{}^T \, d\zeta_2 = \int_0^1 \begin{bmatrix} \zeta_2{}^2 & \zeta_2(1 - \zeta_2) \\ \zeta_2(1 - \zeta_2) & (1 - \zeta_2)^2 \end{bmatrix} d\zeta_2 = \tfrac{1}{6} \begin{bmatrix} 2 & 1 \\ 1 & 2 \end{bmatrix}$$

Finally,

$$\begin{bmatrix} R_{pxj} \\ R_{pxk} \end{bmatrix} = \frac{l_1}{6} \begin{bmatrix} 2 & 1 \\ 1 & 2 \end{bmatrix} \begin{bmatrix} p_{xj} \\ p_{xk} \end{bmatrix} \tag{8.92}$$

A similar expression is obtained for the y components, and the node forces

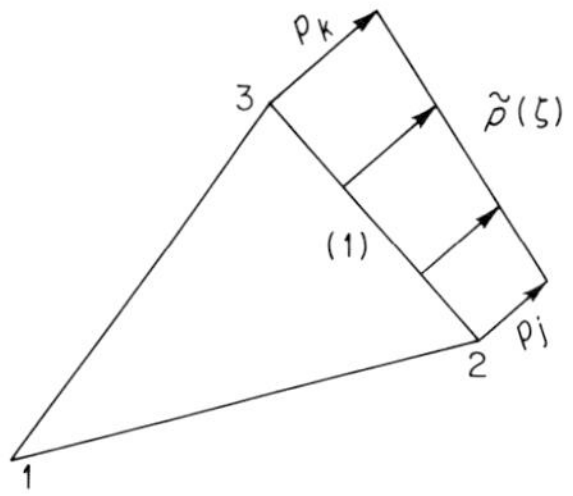

Figure 8.10

for distributed forces on the other two sides are obtained by cyclic permutation of the indices *i*, *j*, and *k*.

2 *Linear-strain Triangle* The quadratic variation of $\tilde{p}(\zeta_B)$ on side 1 is shown in Fig. 8.11. The values are specified at the nodes *j*, *k*, and *m*. In this case, ϕ_B is obtained from Eq. (8.67) by substituting $\zeta_1 = 0$, so that

$$\phi_B = \begin{bmatrix} \zeta_2(2\zeta_2 - 1) \\ \zeta_3(2\zeta_3 - 1) \\ 4\zeta_2\zeta_3 \end{bmatrix} = \begin{bmatrix} \zeta_2(2\zeta_2 - 1) \\ (1 - \zeta_2)(1 - 2\zeta_2) \\ 4\zeta_2(1 - \zeta_2) \end{bmatrix}$$

The load interpolation polynomial Φ_p is again the same as ϕ_B since they are both quadratic. On integration it is found that

$$\int_B \phi_B \phi_p{}^T \, ds_1 = l_1 \int_o^1 \phi_B \phi_p{}^T \, d\zeta_2 = \frac{l_1}{30} \begin{bmatrix} 4 & -1 & 2 \\ -1 & 4 & 2 \\ 2 & 2 & 16 \end{bmatrix}$$

Thence

$$\begin{bmatrix} R_{xj} \\ R_{xm} \\ R_{xk} \end{bmatrix} = \frac{l_1}{30} \begin{bmatrix} 4 & 2 & -1 \\ 2 & 16 & 2 \\ -1 & 2 & 4 \end{bmatrix} \begin{bmatrix} p_{xj} \\ p_{xm} \\ p_{xk} \end{bmatrix} \tag{8.93}$$

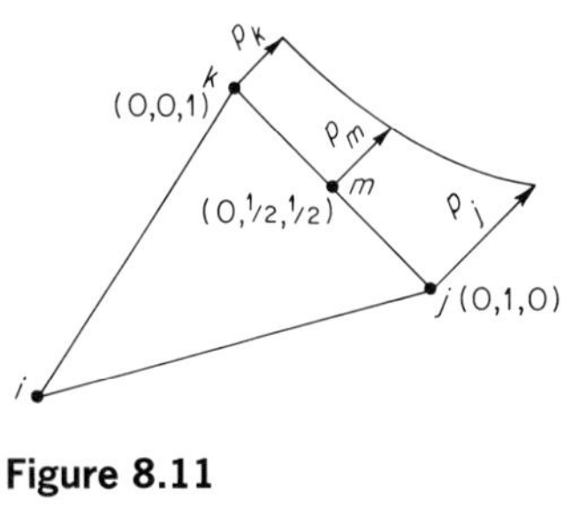

Figure 8.11

A similar expression is obtained for the y components of force, and permutation of the i, j, k, l, m, n indices gives the forces for the other two sides.

A comparison must be made between Eq. (8.93) and Eq. (4.133) in which the generalized flexibility matrix is given for a member with quadratic variation of curvatures M/EI. It is seen that the expressions for the forces here and the rotations given in Eq. (4.133) are identical.

Body Forces

The body force per unit area will be designated $\tilde{f}(\zeta)$ and may be a function (that is, constant, linear, or quadratic) of the natural coordinates. The most common body force encountered is the gravity force that is constant for a given material of constant density.

The vector $\tilde{f}(\zeta)$ has components such that

$$\tilde{f}(\zeta) = \begin{bmatrix} \tilde{f}_x(\zeta) \\ \tilde{f}_y(\zeta) \end{bmatrix} \tag{8.94}$$

A nodal system N_f is necessary to define $\tilde{f}(\zeta)$, and it will be convenient to use only that variation which allows the use of the same nodes in N_f as for the stiffness-matrix allocation.

Let the nodal body forces in force per unit volume be

$$f^T = \langle f_x^T \quad f_y^T \rangle \tag{8.95}$$

If then ϕ_f is the interpolation polynomial for body forces,

$$\tilde{f}_x(\zeta) = \tilde{h}\phi_f^T f_x \tag{8.96a}$$

$$\tilde{f}_y(\zeta) = \tilde{h}\phi_f^T f_y \tag{8.96b}$$

For static equivalence of the node forces R_f replacing the body forces,

$$R_f^T r = \int_{\text{area}} \tilde{f}^T \tilde{r}\, dA \tag{8.97}$$

where

$$\tilde{f} = \tilde{h}\begin{bmatrix} \phi_f^T & 0 \\ 0 & \phi_f^T \end{bmatrix}\begin{bmatrix} f_x \\ f_y \end{bmatrix} \tag{8.98}$$

or

$$\tilde{f} = \tilde{h}\psi_f^T f \tag{8.99}$$

Then, substitution in Eq. (8.97) gives

$$R_f^T r = f^T \int_{\text{area}} \psi_f \tilde{h} \psi \, dA \, r$$

or

$$R_f = \int_{\text{area}} \psi^T \tilde{h} \psi_f^T \, dA \, f \tag{8.100}$$

From the definitions of ψ and ψ_f, it follows that

$$R_{xf} = \int_{\text{area}} \phi \tilde{h} \phi_f^T \, dA \, f_x$$

$$R_{yf} = \int_{\text{area}} \phi \tilde{h} \phi_f^T \, dA \, f_y$$

From Eq. (8.84),

$$\tilde{h} = h_r \phi_\xi^T \xi$$

so that

$$R_{xf} = h_r \int_{\text{area}} \phi \phi_\xi^T \xi \phi_f^T \, dA \, f_x \tag{8.101a}$$

$$R_{yf} = h_r \int_{\text{area}} \phi \phi_\xi^T \xi \phi_f^T \, dA \, f_y \tag{8.101b}$$

Constant Body Force; Gravity Axis

The gravity axis is shown in Fig. 8.12, inclined at the angle α with the x axis, so that

$$\tilde{f}_x = \rho \cos \alpha = f_x$$
$$\tilde{f}_y = \rho \sin \alpha = f_y$$

Furthermore, let the element have a constant thickness. Then,

$$h_r \int_{\text{area}} \phi \phi_\xi^T \xi \phi_f^T \, dA = h \int_{\text{area}} \phi \, dA$$

Constant-strain triangle:

$$h \int_{\text{area}} \phi \, dA = \frac{hA}{3} \begin{bmatrix} 1 \\ 1 \\ 1 \end{bmatrix} \tag{8.102}$$

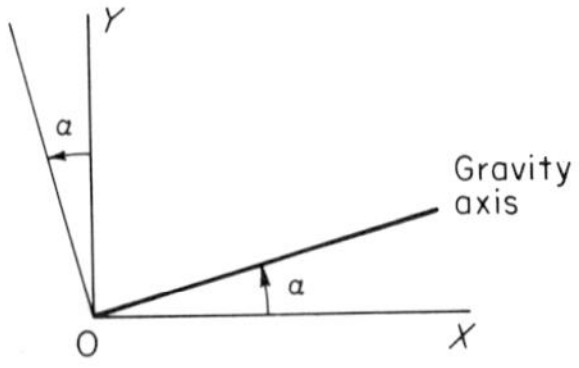

Figure 8.12

Linear-strain triangle:

$$h\int_{\text{area}} \phi\, dA = \frac{hA}{3}\begin{bmatrix}0\\0\\0\\1\\1\\1\end{bmatrix} \tag{8.103}$$

It is seen that for the constant-strain triangle, one-third of the body force is applied at each of the nodes *i, j,* and *k*, whereas for the linear-strain triangle, the same fraction is applied to the midpoints of sides at *l, m,* and *n.*

Temperature Strains

The method employed in Sec. 7.4 for framed structures is used for the finite-element assemblage. That is, every node of the structure is considered completely restrained. In this situation the temperature variations within the elements will cause node forces. The analysis then involves the application of equal and opposite release forces to the nodes. In each element the final stresses will be the combination of those in the fixed-node condition and those produced in the release of the nodes.

Thus, let $\tilde{\alpha}(\zeta)$ be the coefficient of thermal expansion; $\tilde{\theta}(\zeta)$, the temperature variation; and $\tilde{\sigma}_t(\zeta)$, the restraint stresses. The strains due to the temperature variation are

$$\begin{bmatrix}\tilde{\varepsilon}_x\\ \tilde{\varepsilon}_y\\ \tilde{\gamma}_{xy}\end{bmatrix}_t = -\tilde{\alpha}(\zeta)\tilde{\theta}(\zeta)\begin{bmatrix}1\\1\\0\end{bmatrix}$$

From the constitutive law $\tilde{C}$ for the element, the restraint stresses $\tilde{\sigma}$ are

$$\begin{bmatrix}\tilde{\sigma}_x\\ \tilde{\sigma}_y\\ \tilde{\tau}_{xy}\end{bmatrix}_t = -\tilde{C}\tilde{\alpha}(\zeta)\tilde{\theta}(\zeta)\begin{bmatrix}1\\1\\0\end{bmatrix} \tag{8.104}$$

Let N_θ be the nodal system used to define the element temperature field. For convenience, $N_\theta = N_\varepsilon$, and then the temperature variation is given by the interpolation expression

$$\tilde{\theta}(\zeta) = \phi_\theta{}^T\theta \tag{8.105}$$

It will be reasonable to assume that the coefficient of thermal expansion does not vary within an element, so that it may be supposed

$\tilde{\alpha}(t) = \alpha \times \text{a constant}$

Equation (8.104) is now written

$$\begin{bmatrix} \tilde{\sigma}_x \\ \tilde{\sigma}_y \\ \tilde{\tau}_{xy} \end{bmatrix}_t = -\tilde{C}\alpha \begin{bmatrix} 1 \\ 1 \\ 0 \end{bmatrix} \phi_\theta{}^T\theta \tag{8.106}$$

The nodal stresses σ_t at N_θ will also define the stress field $\tilde{\sigma}_t(\zeta)$, so that since $C\alpha$ is a constant, it is possible to take $\phi_\sigma \equiv \phi_\theta$. Then,

$$\tilde{\sigma}_{xt} = \phi_\theta{}^T\sigma_{xt}$$
$$\tilde{\sigma}_{yt} = \phi_\theta{}^T\sigma_{yt}$$

In matrix form,

$$\begin{bmatrix} \tilde{\sigma}_{xt} \\ \tilde{\sigma}_{yt} \end{bmatrix} = \begin{bmatrix} \phi_\theta{}^T & 0 \\ 0 & \phi_\theta{}^T \end{bmatrix} \begin{bmatrix} \sigma_{xt} \\ \sigma_{yt} \end{bmatrix}$$

or

$$\tilde{\sigma}_t = \psi_\theta\sigma_t \tag{8.107}$$

The nodal forces required to produce the $\tilde{\sigma}_t$ field will be such that if the element is given a set of virtual strains, the internal work will be equal to the external work. The virtual strains must conform to the specified pattern for the element, given in Eq. (8.82). That is,

$$\tilde{\varepsilon} = \phi_\varepsilon{}^T\varepsilon_C = \phi_\varepsilon{}^T\Phi_C r$$

Let R_t be the node restraint forces. Then, for constant thickness h,

$$r^TR_t = \int_{\text{vol}} \tilde{\varepsilon}^T\tilde{\sigma}\,dV = r^T\Phi_C{}^Th\int_{\text{area}} \phi_\varepsilon\psi_\theta\,dA\,\sigma_t$$

Thence,

$$R_t = \Phi_C{}^Th\int_{\text{area}} \phi_\varepsilon\psi_\theta\,dA\,\sigma_t$$

When h is not constant, but given by the interpolation formula, Eq. (8.84), then

$$r^TR_t = r^T\Phi_C{}^Th_r\int_{\text{area}} \phi_\varepsilon\phi_\xi{}^T\xi\psi_\theta\,dA\,\sigma_t$$

In this case,

$$R_t = \Phi_C{}^T h_r \int_{\text{area}} \phi_\varepsilon \phi_\xi{}^T \xi \psi_\theta \, dA \, \sigma_t \tag{8.108}$$

Example 8.4 Consider the constant-strain triangle of constant thickess in plane stress under a constant temperature variation. In this case,

$$\sigma_t = \begin{bmatrix} \sigma_x \\ \sigma_y \end{bmatrix} = \frac{E}{1-\nu^2}\begin{bmatrix} 1 & \nu \\ \nu & 1 \end{bmatrix}\begin{bmatrix} \alpha\theta \\ \alpha\theta \end{bmatrix} = \frac{E\alpha\theta}{1-\nu}\begin{bmatrix} 1 \\ 1 \end{bmatrix}$$

$$\Phi_C{}^T = \frac{1}{2A}\begin{bmatrix} b_1 & 0 \\ b_2 & 0 \\ b_3 & 0 \\ 0 & a_1 \\ 0 & a_2 \\ 0 & a_3 \end{bmatrix}$$

Also in this case,

$$\phi_\varepsilon = 1 \qquad \phi_\xi{}^T \xi = 1 \qquad h_r = h \qquad \psi_\theta = \begin{bmatrix} 1 & 0 \\ 0 & 1 \end{bmatrix}$$

The node forces are given by

$$\begin{bmatrix} R_{xi} \\ R_{xj} \\ R_{xk} \\ R_{yi} \\ R_{yj} \\ R_{yk} \end{bmatrix} = \frac{h}{2A}\begin{bmatrix} b_1 & 0 \\ b_2 & 0 \\ b_3 & 0 \\ 0 & a_1 \\ 0 & a_2 \\ 0 & a_3 \end{bmatrix} A \begin{bmatrix} 1 & 0 \\ 0 & 1 \end{bmatrix}\begin{bmatrix} 1 \\ 1 \end{bmatrix}\frac{E\alpha\theta}{1-\nu} = \frac{E\alpha\theta h}{2(1-\nu)}\begin{bmatrix} b_1 \\ b_2 \\ b_3 \\ a_1 \\ a_2 \\ a_3 \end{bmatrix}$$

Example 8.5 Consider in this example the linear-strain triangle of constant thickness, subject to a linear temperature variation. In this case,

$$\phi_\varepsilon{}^T = \begin{bmatrix} \zeta_1 & \zeta_2 & \zeta_3 & 0 & 0 & 0 \\ 0 & 0 & 0 & \zeta_1 & \zeta_2 & \zeta_3 \end{bmatrix}$$

Let the node temperatures θ be $\theta^T = \langle \theta_1 \quad \theta_2 \quad \theta_3 \rangle$. Then, σ_t is given by

$$\begin{bmatrix} \sigma_{xi} \\ \sigma_{xj} \\ \sigma_{xk} \\ \sigma_{yi} \\ \sigma_{yj} \\ \sigma_{yk} \end{bmatrix} = \frac{E\alpha}{1-\nu}\begin{bmatrix} \theta_1 \\ \theta_2 \\ \theta_3 \\ \theta_1 \\ \theta_2 \\ \theta_3 \end{bmatrix}$$

Because the temperature varies linearly, $\psi_\theta = \phi_\varepsilon$, so that

$$\int \phi_\varepsilon \psi_\theta \, dA = \int \begin{bmatrix} \zeta_1 & 0 \\ \zeta_2 & 0 \\ \zeta_3 & 0 \\ 0 & \zeta_1 \\ 0 & \zeta_2 \\ 0 & \zeta_3 \end{bmatrix} \begin{bmatrix} \zeta_1 & \zeta_2 & \zeta_3 & 0 & 0 & 0 \\ 0 & 0 & 0 & \zeta_1 & \zeta_2 & \zeta_3 \end{bmatrix} dA = \frac{A}{12} \begin{bmatrix} 2 & 1 & 1 & 0 & 0 & 0 \\ 1 & 2 & 1 & 0 & 0 & 0 \\ 1 & 1 & 2 & 0 & 0 & 0 \\ 0 & 0 & 0 & 2 & 1 & 1 \\ 0 & 0 & 0 & 1 & 2 & 1 \\ 0 & 0 & 0 & 1 & 1 & 2 \end{bmatrix}$$

Finally, by using Φ_C from Eq. (8.78) and remembering to omit the shear terms, the node forces are given by

$$\begin{bmatrix} R_{xi} \\ R_{xj} \\ R_{xk} \\ R_{xl} \\ R_{xm} \\ R_{xn} \\ R_{yi} \\ R_{yj} \\ R_{yk} \\ R_{yl} \\ R_{ym} \\ R_{yn} \end{bmatrix} = \frac{E\alpha h}{6(1-\nu)} \begin{bmatrix} b_1 & 0 & 0 \\ 0 & b_2 & 0 \\ 0 & 0 & b_3 \\ 2b_2 + b_1 & b_2 + 2b_1 & b_2 + b_1 \\ b_3 + b_2 & 2b_3 + b_2 & b_3 + 2b_2 \\ 2b_3 + b_1 & b_3 + b_1 & b_3 + 2b_1 \\ a_1 & 0 & 0 \\ 0 & a_2 & 0 \\ 0 & 0 & a_3 \\ 2a_2 + a_1 & a_2 + 2a_1 & a_2 + a_1 \\ a_3 + a_2 & 2a_3 + a_2 & a_3 + 2a_2 \\ 2a_3 + a_1 & a_3 + a_1 & a_3 + 2a_1 \end{bmatrix} \begin{bmatrix} \theta_1 \\ \theta_2 \\ \theta_3 \end{bmatrix}$$

Stratified-material Constitutive Law

In certain materials (for example, sedimentary rock and timber), the moduli of elasticity in two directions at right angles may be significantly different. Furthermore, the axes in which the moduli are defined may not coincide with the global coordinate axes. Such a material is shown in Fig. 8.13.

In the particular case shown, the axis OX' perpendicular to the plane of stratification is obtained by a rotation through the angle θ about the OZ axis. It should be noticed that such a layered system may exhibit other properties, such as low tensile strength in the X' direction (see Ref. 39).

In Fig. 8.13, the $Y'Z'$ plane is taken as the plane of stratification so that properties in the directions of the Y' and Z' axes are the same, whereas those in the X' axis direction are different from the other two. The X' axis bears the same relationship (in material properties) to both the Y' and Z' axes.

Thus, let E_1 be Young's modulus of elasticity in the direction OX', and E_2

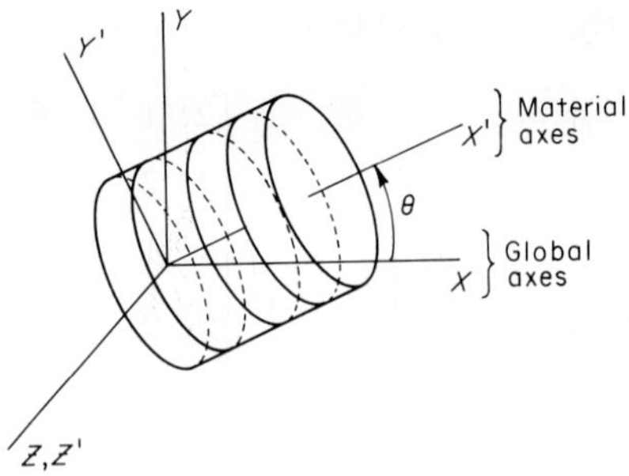

Figure 8.13 Stratified material.

be that for the OY' and OZ' directions. Furthermore, let Poisson's ratio for the $Y'Z'$ plane be ν_2, and $\nu_{12} \neq \nu_{21}$ be those values connecting the strains in the $Y'Z'$ plane to those in the X' direction, and vice versa.

Then, the prime indicating the material axes, the direct strains are given by

$$\begin{bmatrix} \varepsilon_x \\ \varepsilon_y \\ \varepsilon_z \end{bmatrix}' = \begin{bmatrix} \dfrac{1}{E_1} & -\dfrac{\nu_{12}}{E_2} & -\dfrac{\nu_{12}}{E_2} \\ -\dfrac{\nu_{21}}{E_1} & \dfrac{1}{E_2} & -\dfrac{\nu_2}{E_2} \\ -\dfrac{\nu_{21}}{E_1} & -\dfrac{\nu_2}{E_2} & \dfrac{1}{E_2} \end{bmatrix} \begin{bmatrix} \sigma_x \\ \sigma_y \\ \sigma_z \end{bmatrix}' \tag{8.109}$$

Now, from symmetry considerations (Maxwell's reciprocal theorem),

$$\frac{\nu_{12}}{E_2} = \frac{\nu_{21}}{E_2}$$

That is,

$$\nu_{21} = \frac{E_1}{E_2}\nu_{12}$$

Define the modular ratio n such that

$$n = \frac{E_1}{E_2}$$

It is convenient to write $\nu_{12} = \nu_1$, so that

$$\nu_{21} = n\nu_{12} = n\nu_1$$

Substitution in Eq. (8.109) gives

$$\begin{bmatrix} \varepsilon_x \\ \varepsilon_y \\ \varepsilon_z \end{bmatrix}' = \frac{1}{E_1} \begin{bmatrix} 1 & -n\nu_1 & -n\nu_1 \\ -n\nu_1 & n & -n\nu_2 \\ -n\nu_1 & -n\nu_2 & n \end{bmatrix} \begin{bmatrix} \sigma_x \\ \sigma_y \\ \sigma_z \end{bmatrix}' \tag{8.110}$$

Of particular interest will be the plane strain situation for which $\varepsilon_z \equiv 0$. From Eq. (8.110), it is found in this case that

$$\sigma_z' = [\nu_1 \quad \nu_2] \begin{bmatrix} \sigma_x \\ \sigma_y \end{bmatrix}'$$

Substitution for σ_z' in Eq. (8.110) gives the equivalent two-dimensional constitutive equation to be

$$\begin{bmatrix} \varepsilon_x \\ \varepsilon_y \end{bmatrix}' = \frac{1}{E_1} \begin{bmatrix} 1 - n\nu_1{}^2 & -n\nu_1(1 + \nu_2) \\ -n\nu_1(1 + \nu_2) & n(1 - \nu_2{}^2) \end{bmatrix} \begin{bmatrix} \sigma_x \\ \sigma_y \end{bmatrix}'$$

The inverse stiffness relationship is given then by

$$\begin{bmatrix} \sigma_x \\ \sigma_y \end{bmatrix}' = \frac{E_1}{(1 + \nu_2)(1 - \nu_2 - 2n\nu_1{}^2)} \begin{bmatrix} 1 - \nu_2{}^2 & \nu_1(1 + \nu_2) \\ \nu_1(1 + \nu_2) & \dfrac{1}{n} - \nu_1{}^2 \end{bmatrix} \begin{bmatrix} \varepsilon_x \\ \varepsilon_y \end{bmatrix}'$$

Now the shear stresses in the $X'Y'$ plane will be given by

$$\tau_{xy} = G_1 \gamma_{xy}$$

The shear modulus G_1 will be taken as

$$G_1 = \frac{E_1}{2(1 + \nu_1)}$$

Finally, Eq. (8.75) for the local member axes is given with k' as

$$k' = \frac{E_1}{(1 + \nu_2)(1 - \nu_2 - 2n\nu_1{}^2)} \begin{bmatrix} 1 - \nu_2{}^2 & \nu_1(1 + \nu_2) & 0 \\ \nu_1(1 + \nu_2) & \dfrac{1}{n} - \nu_1{}^2 & 0 \\ 0 & 0 & \Omega \end{bmatrix} \tag{8.111}$$

where

$$\Omega = \frac{(1 + \nu_2)(1 - \nu_2 - 2n\nu_1{}^2)}{2(1 - \nu_1)}$$

Consider the situation shown in Fig. 8.13 in which the $X'Y'$ axes are obtained from the global system XY by the rotation θ. The transformations for the stress and strain tensors have been obtained in Chap. 2 [see Eqs. (2.115) and (2.168)]. That is,

$$\begin{aligned} \sigma' &= L^T\sigma L \\ \varepsilon' &= L^T\varepsilon L \end{aligned} \qquad (i)$$

The inverse transformations are

$$\begin{aligned} \sigma &= L\sigma' L^T \\ \varepsilon &= L\varepsilon L^T \end{aligned} \qquad (ii)$$

In these equations,

$$L = \begin{bmatrix} \cos\theta & -\sin\theta \\ \sin\theta & \cos\theta \end{bmatrix}$$

$$\sigma = \begin{bmatrix} \sigma_x & \tau_{xy} \\ \tau_{yx} & \sigma_y \end{bmatrix}$$

$$\varepsilon = \begin{bmatrix} \varepsilon_x & \varepsilon_{xy} \\ \varepsilon_{yx} & \varepsilon_y \end{bmatrix}$$

When Eq. (*ii*) is expanded, the relationship is obtained between

$$S^T = \langle \sigma_x \quad \sigma_y \quad \tau_{xy} \rangle \qquad \text{and} \qquad (S')^T = \langle \sigma_x' \quad \sigma_y' \quad \tau_{xy}' \rangle$$

That is,

$$\begin{bmatrix} \sigma_x \\ \sigma_y \\ \tau_{xy} \end{bmatrix} = \begin{bmatrix} c^2 & s^2 & -2cs \\ s^2 & c^2 & 2cs \\ cs & -cs & c^2 - s^2 \end{bmatrix} \begin{bmatrix} \sigma_x \\ \sigma_y \\ \tau_{xy} \end{bmatrix}' \qquad (8.112)$$

This equation is written

$$S = TS' \qquad (iii)$$

If Eq. (*i*) is expanded, it is found that

$$v' = T^T v$$

It should be seen that this result could be obtained directly from (*iii*) by application of the contragradient law. Substitution in the equation $S' = k'v'$ gives

$$S = Tk'T^T v$$

That is, the stiffness matrix in global coordinates is given by

$$k = Tk'T^T$$

Stresses may be calculated in global coordinates, or alternatively in the stratified directions. Thus,

$$S' = T^{-1}S$$

or

$$S' = k'T^Tv$$

8.7 AXISYMMETRICAL STRUCTURES WITH AXISYMMETRICAL LOADING

There are many situations for which both the structure and its loading are symmetrical about a given axis, for example, a pressure vessel under internal pressure or a spherical dome under its own dead load. In the theory of elasticity, the problem is encountered in a variety of analytic solutions, such as the Lamé solution for thick cylinders and the Boussinesq solution for the half-plane with a point load normal to the surface. Because of the symmetry, the essentially three-dimensional problem is reduced to a two-dimensional one. The basic finite element is taken as a toroid of triangular cross section, as shown in Fig. 8.14. All the interpolation formulas of the previous section apply here also. There is the added complication however, that hoop strains and stresses must be taken into consideration. The global coordinates (x_i, y_i), etc., of the corners of the triangular cross section are shown in Fig. 8.14*b*.

The displacement interpolation is given by Eq. (8.64), that is,

$$\tilde{r} = \psi r$$

The strain vector $\tilde{\varepsilon}(\zeta)$ is now different from that for the planar element since the $\tilde{\varepsilon}_\theta$ term is included.

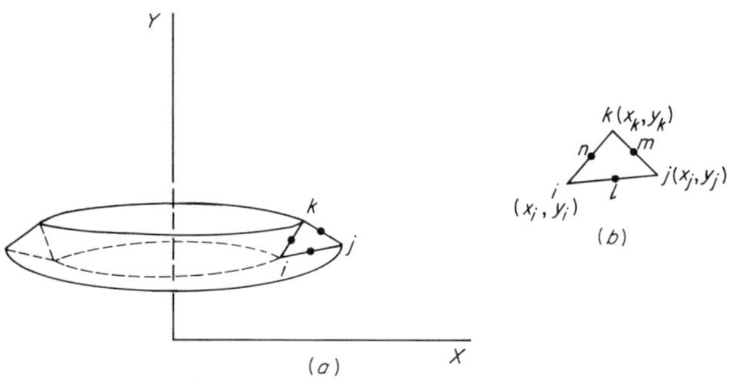

Figure 8.14 Axisymmetric element.

$$\tilde{\varepsilon}(\zeta) = \begin{bmatrix} \varepsilon_x \\ \varepsilon_y \\ \varepsilon_\theta \\ \gamma_{xy} \end{bmatrix} = \begin{bmatrix} \dfrac{\partial u}{\partial x} \\ \dfrac{\partial v}{\partial y} \\ \dfrac{u}{x} \\ \dfrac{\partial u}{\partial y} + \dfrac{\partial v}{\partial x} \end{bmatrix} \tag{8.113}$$

It is seen that the hoop strain ε_θ is a necessary result of the displacement u and is equal to u/x since

$$\varepsilon_\theta = \frac{2\pi(x+u) - 2\pi x}{2\pi x} = \frac{u}{x}$$

The constitutive equation for strains in terms of the stresses is written

$$\begin{bmatrix} \varepsilon_x \\ \varepsilon_y \\ \varepsilon_\theta \\ \gamma_{xy} \end{bmatrix} = \frac{1}{E} \begin{bmatrix} 1 & -\nu & -\nu & 0 \\ -\nu & 1 & -\nu & 0 \\ -\nu & -\nu & 1 & 0 \\ 0 & 0 & 0 & 2(1+\nu) \end{bmatrix} \begin{bmatrix} \sigma_x \\ \sigma_y \\ \sigma_\theta \\ \tau_{xy} \end{bmatrix} \tag{8.114}$$

This is the equation

$$\tilde{\varepsilon} = f\tilde{\sigma}$$

On inversion of Eq. (8.114), the stiffness relationship is found to be

$$\begin{bmatrix} \sigma_x \\ \sigma_y \\ \sigma_\theta \\ \tau_{xy} \end{bmatrix} = 2G \begin{bmatrix} \rho+1 & \rho & \rho & 0 \\ \rho & \rho+1 & \rho & 0 \\ \rho & \rho & \rho+1 & 0 \\ 0 & 0 & 0 & 0.5 \end{bmatrix} \begin{bmatrix} \varepsilon_x \\ \varepsilon_y \\ \varepsilon_\theta \\ \gamma_{xy} \end{bmatrix} \tag{8.115}$$

That is,

$$\tilde{\sigma} = k\tilde{\varepsilon}$$

In Eq. (8.115),

$$2G = \frac{E}{(1+\nu)(1-2\nu)} \tag{8.116}$$

The stiffness matrix for the toroidal element is conveniently calculated from Eq. (8.54) as

$$K = \int_{\text{vol}} a_\zeta{}^T k a_\zeta \, dV = 2\pi \int_{\text{area}} x a_\zeta{}^T k a_\zeta \, dA \tag{8.117}$$

The evaluation of the integral $\int_{\text{area}} x a_\zeta{}^T k a_\zeta \, dA$ is considerably more complex than that for the planar triangle, and consequently the stiffness matrices for the constant-strain and linear-strain triangle toroids will be developed independently.

1 Constant-strain Triangle; Stiffness Matrix

In this case,

$$\tilde{u}(\zeta) = \phi_1{}^T u$$
$$\tilde{v}(\zeta) = \phi_1{}^T v$$

From Eq. (8.113),

$$\begin{bmatrix} \varepsilon_x \\ \varepsilon_y \\ \varepsilon_\theta \\ \gamma_{xy} \end{bmatrix} = \frac{1}{2A} \begin{bmatrix} b_1 & b_2 & b_3 & 0 & 0 & 0 \\ 0 & 0 & 0 & a_1 & a_2 & a_3 \\ \beta_1 & \beta_2 & \beta_3 & 0 & 0 & 0 \\ a_1 & a_2 & a_3 & b_1 & b_2 & b_3 \end{bmatrix} \begin{bmatrix} u_i \\ u_j \\ u_k \\ v_i \\ v_j \\ v_k \end{bmatrix}$$

where $\beta_i = 2A\zeta_i/x$. Hence

$$a_\zeta = \frac{1}{2A} \begin{bmatrix} b_1 & b_2 & b_3 & 0 & 0 & 0 \\ 0 & 0 & 0 & a_1 & a_2 & a_3 \\ \beta_1 & \beta_2 & \beta_3 & 0 & 0 & 0 \\ a_1 & a_2 & a_3 & b_1 & b_2 & b_3 \end{bmatrix} \tag{8.118}$$

Let

$$\alpha = [b_1 \quad b_2 \quad b_3]$$
$$\gamma = [a_1 \quad a_2 \quad a_3]$$
$$\theta = [\beta_1 \quad \beta_2 \quad \beta_3]$$

Then

$$a_\zeta = \frac{1}{2A} \begin{bmatrix} \alpha & 0 \\ 0 & \gamma \\ \theta & 0 \\ \gamma & \alpha \end{bmatrix} \tag{8.119}$$

The form of a_ζ in Eq. (8.119) is used both here and for the linear-strain triangle and is useful in defining the various types of integrals to be calculated in Eq. (8.117). On multiplication,

$$xa^Tka = \frac{xG}{2A^2}\left[\begin{array}{c|c} (\rho+1)\alpha^T\alpha + \rho\alpha^T\theta + \rho\theta^T\alpha + (\rho+1)\theta^T\theta + \tfrac{1}{2}\gamma^T\gamma & \rho\alpha^T\alpha + \rho\theta^T\gamma + \tfrac{1}{2}\gamma^T\alpha \\ \hline \text{symmetrical} & (\rho+1)\gamma^T\gamma + \tfrac{1}{2}\alpha^T\alpha \end{array}\right] \tag{8.120}$$

Note that

$$x(\zeta) = [\zeta_1 \quad \zeta_2 \quad \zeta_3]\begin{bmatrix} x_i \\ x_j \\ x_k \end{bmatrix}$$

On examination of Eq. (8.120), it is found that there are three distinct types of terms to be integrated. These are:

Group 1:

$x\alpha^T\alpha \qquad x\alpha^T\gamma \qquad x\gamma^T\alpha \qquad x\gamma^T\gamma$

Group 2:

$x\alpha^T\theta \qquad x\alpha^T\theta \qquad x\theta^T\alpha \qquad x\theta^T\gamma$

Group 3:

$x\theta^T\theta$

For groups 1 and 2, explicit expressions for the integrals over the triangle area are easily found. For group 3, the $1/x$ term is present, and for this integration, the numerical quadrature may be used (see page 466). The quintic formula given there is particularly applicable because then the $1/x$ singularity when x_i or x_j or $x_k = 0$ does not occur since all points lie within the triangle.

When the integration of the matrices of groups 1 and 2 is carried out, the resulting contribution to the stiffness matrix k will be designated X. Furthermore, let $e^T = \langle 1 \quad 1 \quad 1\rangle$. Then,

$$X = \frac{A(x_i + x_j + x_k)}{3}\left[\begin{array}{c|c} (\rho+1)\alpha\alpha^T + \tfrac{1}{2}\gamma\gamma^T & \rho\alpha\gamma^T + \tfrac{1}{2}\gamma\gamma^T \\ \hline \text{symmetrical} & (\rho+1)\gamma\gamma^T + \tfrac{1}{2}\gamma\gamma^T \end{array}\right] + \frac{2A\rho}{3}\left[\begin{array}{c|c} e\alpha^T + \alpha e^T & e\gamma^T \\ \hline \text{symmetrical} & 0 \end{array}\right] \tag{8.121}$$

From Eq. (8.120) it is seen that X must be multiplied by the scalar $G/2A^2$. The group 3 integrals may be obtained numerically. In the term $\int x\theta^T\theta \, dA$ the integrals, $\int(\zeta_i^2/x)\, dA$ and $\int(\zeta_i\zeta_j/x)\, dA$ occur.

Body Forces for the Constant-strain Triangle

It will be assumed that the body force acting is a gravity force in the direction of the y axis. The only other axisymmetrical body force possible is a centrifugal force that would be produced in a disk or cylinder rotating about the y axis. In the present case the body force vector is

$$\tilde{f}(\zeta) = \begin{bmatrix} 0 \\ \tilde{f}_y(\zeta) \end{bmatrix}$$

The equivalent node forces are calculated by producing a virtual displacement of the nodes and equating internal and external work. For a virtual displacement δr, the displacement at any point of the triangle will be $\delta\tilde{r}$ such that

$$\delta\tilde{r} = \delta\begin{bmatrix} \tilde{u} \\ \tilde{v} \end{bmatrix} = \begin{bmatrix} \zeta_1 & \zeta_2 & \zeta_3 & 0 & 0 & 0 \\ 0 & 0 & 0 & \zeta_1 & \zeta_2 & \zeta_3 \end{bmatrix} \delta \begin{bmatrix} u \\ v \end{bmatrix}$$

Choose the virtual displacement so that $\delta u = 0$, $\delta v \neq 0$. Then

$$\delta\tilde{v} = [\zeta_1 \quad \zeta_2 \quad \zeta_3]\,\delta v$$

The internal work done by the body forces, W_i, is calculated from

$$\delta W_i = \tilde{f}_y\,dV\,\delta\tilde{v}$$

For constant gravity

$$\tilde{f}_y = f_y$$

Hence

$$W_i = f_y \int_{\text{vol}} [\zeta_1 \quad \zeta_2 \quad \zeta_3]\,dV\,\delta v$$

and then

$$W_i = f_y 2\pi \int_{\text{area}} x[\zeta_1 \quad \zeta_2 \quad \zeta_3]\,dA\,\delta v$$

On integration,

$$W_i = f_y \frac{\pi A}{6}[2r_i + r_j + r_k \quad r_i + 2r_j + r_k \quad r_i + r_j + 2r_k]\,\delta v$$

The external work W_e is given by

$$W_e = R_{By}{}^T\,\delta v$$

Hence for $W_i = W_e$,

$$R_{By}{}^T = f_y \frac{\pi A}{6}[2r_i + r_j + r_k \quad r_i + 2r_j + r_k \quad r_i + r_j + 2r_k] \tag{8.122}$$

Notice if $\tilde{f}_y$ is variable over the triangle, a nodal system N_f and interpolation polynomial ϕ_f must be defined such that

$$f(\zeta) = \phi_f{}^T f$$

Then

$$W_i = f_y{}^T 2\pi \int_{\text{area}} x\phi_f \phi_1{}^T \, dA \; \delta v$$

Hence

$$R_{By}{}^T = f_y{}^T 2\pi \int_{\text{area}} x\phi_f \phi_1{}^T \, dA$$

Surface Forces for the Constant-strain Triangle

For the constant-strain triangle, only two special cases of surface forces will be developed (shown in Fig. 8.15) for normal force on transverse planes and circumferential surfaces.

The statically equivalent node forces $\langle R_{pyi} \quad R_{pyj}\rangle$ and $\langle R_{pxi} \quad R_{pxj}\rangle$ are calculated in the same way as that given for the planar triangle. From Eq. (8.91),

$$R_{px} = \int_B \phi_B \phi_p{}^T \, ds \, p_x$$

$$R_{py} = \int_B \phi_B \phi_p{}^T \, ds \, p_y$$

For R_{px}, from Fig. 8.15*b*, for p_x = constant

$$R_{px} = 2\pi \int_{y_i}^{y_j} x_i \phi_B \, dy \, p_x$$

In this case

$$\phi_B = \begin{bmatrix} \zeta_1 \\ 1 - \zeta_1 \end{bmatrix}$$

$$\int_{y_i}^{y_j} x_i \begin{bmatrix} \zeta_1 \\ 1 - \zeta_1 \end{bmatrix} dy = (y_i - y_j) x_i \int_o^1 \begin{bmatrix} \zeta_1 \\ 1 - \zeta_1 \end{bmatrix} d\zeta_1$$

$$= x_i (y_i - y_j) \begin{bmatrix} \frac{1}{2} \\ \frac{1}{2} \end{bmatrix}$$

Thence,

$$R_{px} = \pi x_i (y_i - y_j) \begin{bmatrix} 1 \\ 1 \end{bmatrix} p_x \tag{8.123}$$

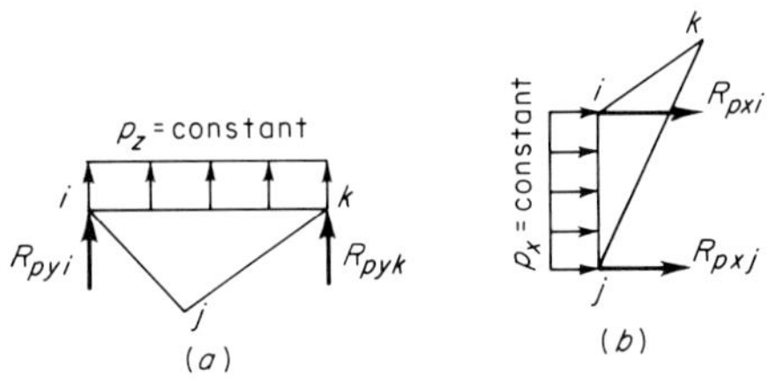

Figure 8.15 Surface forces.

By similar reasoning, for side 3 of the triangle shown in Fig. 8.15*a*,

$$\begin{bmatrix} R_{pyi} \\ R_{pyk} \end{bmatrix} = \frac{\pi p_z}{3}(x_k - x_i)\begin{bmatrix} 2 & 1 \\ 1 & 2 \end{bmatrix}\begin{bmatrix} r_i \\ r_k \end{bmatrix} \tag{8.124}$$

2 Linear-strain Triangle; Stiffness Matrix

The procedure followed here is identical with that for the constant-strain triangle. The actual computation involved in evaluating the quadratic function product $a_\zeta{}^T k a_\zeta$ has increased enormously. For the linear-strain triangle,

$$\begin{bmatrix} \tilde{u} \\ \tilde{v} \end{bmatrix} = \begin{bmatrix} \phi_2{}^T & 0 \\ 0 & \phi_2{}^T \end{bmatrix}\begin{bmatrix} u \\ v \end{bmatrix} \tag{8.125}$$

As defined previously,

$$\begin{aligned} \phi_2{}^T &= \langle \zeta_1(2\zeta_1 - 1) \quad \zeta_2(2\zeta_2 - 1) \quad \zeta_3(2\zeta_3 - 1) \quad 4\zeta_1\zeta_2 \quad 4\zeta_2\zeta_3 \quad 4\zeta_3\zeta_1 \rangle \\ u^T &= \langle u_i \quad u_j \quad u_k \quad u_l \quad u_m \quad u_n \rangle \\ v^T &= \langle v_i \quad v_j \quad v_k \quad v_l \quad v_m \quad v_n \rangle \end{aligned} \tag{8.126}$$

The strains, given in Eq. (8.113), are calculated from Eq. (8.126) to be

$$\begin{bmatrix} \varepsilon_x \\ \varepsilon_y \\ \varepsilon_\theta \\ \gamma_{xy} \end{bmatrix} = \frac{1}{2A}\begin{bmatrix} \alpha & 0 \\ 0 & \gamma \\ \theta & 0 \\ \gamma & \alpha \end{bmatrix}\begin{bmatrix} u \\ v \end{bmatrix} \tag{8.127}$$

The α, γ, and θ matrices are redefined to be

$$\begin{aligned} \alpha^T = \langle b_1(4\zeta_1 - 1) \quad b_2(4\zeta_2 - 1) \quad b_3(4\zeta_3 - 1) \quad 4(b_1\zeta_2 + b_2\zeta_1) \\ 4(b_2\zeta_3 + b_3\zeta_2) \quad 4(b_1\zeta_3 + b_3\zeta_1)\rangle \end{aligned} \tag{8.128}$$

$$\begin{aligned} \gamma^T = \langle a_1(4\zeta_1 - 1) \quad a_2(4\zeta_2 - 1) \quad a_3(4\zeta_3 - 1) \quad 4(a_1\zeta_2 + a_2\zeta_1) \\ 4(a_2\zeta_3 + a_3\zeta_2) + 4(a_1\zeta_3 + a_3\zeta_1)\rangle \end{aligned} \tag{8.129}$$

$$\theta^T = \frac{2A\phi_2{}^T}{x} \tag{8.130}$$

The stiffness matrix is evaluated by using Eq. (8.117) and on multiplication of $a_\xi{}^T k a_\xi$ [by using Eq. (8.127)], the same basic integral groups 1 to 3 are obtained.

Group 1:

$x\alpha^T\alpha \quad x\alpha^T\gamma \quad x\gamma^T\alpha \quad x\gamma^T\gamma$

Group 2:

$x\alpha^T\theta \quad x\theta^T\alpha \quad x\theta^T\gamma$

Group 3:

$x\theta^T\theta$

For group 1, the general integration expression is

$I = \int x t^T u \, dA$

where t and u are defined by Eq. (8.128), if substituted there for b. The integrals of group 1 are then generated by substituting $t = b, u = b, t = b, u = a$, etc., in I. The integral I is a 6×6 matrix that is conveniently subdivided into four 3×3 submatrices that are generated independently. Thus let

$$I = \frac{A}{15}\begin{bmatrix} I_{11} & I_{12} \\ I_{12}{}^T & I_{22} \end{bmatrix} \tag{8.131}$$

To generate I_{11}, define

$$Q1 = TU^T \qquad T^T = \langle t_1 \quad t_2 \quad t_3 \rangle \qquad U^T = \langle u_1 \quad u_2 \quad u_3 \rangle \tag{8.132}$$

and

$$A1 = \begin{bmatrix} 9 & -2 & -2 \\ -2 & 3 & -1 \\ -2 & -1 & 3 \end{bmatrix} \tag{8.133}$$

The row or column interchange matrices are defined by

$$W1 = \begin{bmatrix} 0 & 0 & 1 \\ 0 & 1 & 0 \\ 1 & 0 & 0 \end{bmatrix} \qquad W2 = \begin{bmatrix} 0 & 1 & 0 \\ 1 & 0 & 0 \\ 0 & 0 & 1 \end{bmatrix} \qquad W3 = \begin{bmatrix} 1 & 0 & 0 \\ 0 & 0 & 1 \\ 0 & 1 & 0 \end{bmatrix} \tag{8.134}$$

It may be shown then that

$$I_{11} = (A1x_i + W2A1W2x_j + W1A1W1x_k) * Q1 \tag{8.135}$$

The * multiplication has been defined previously in Chap. 4 and is similar to matrix addition. That is, if $C = A * B$, then $C_{ij} = a_{ij}b_{ij}$ (no sum on i or j).

Let

$$B2 = (A1x_i + W2A1W2x_j + W1A1W1x_k)$$

so that

$$I_{11} = B2 * Q1 \tag{8.136}$$

For the generation of I_{12}, define $C1$:

$$C1 = \begin{bmatrix} 3 & 3 & 14 \\ 3 & -1 & -2 \\ -1 & 3 & -2 \end{bmatrix}$$

and

$$B1 = (C1x_i + W2C1W1x_j + W1C1W3x_k)$$

Let

$$Q2 = Q1W2W3$$

Then

$$I_{12} = B1 * Q1 + B2 * Q2 \tag{8.137}$$

Finally the lower right-hand block of I, that is, I_{22}, must be calculated. Generate $Q3$ and $Q4$ such that

$$Q3 = W3W2Q1$$
$$Q4 = Q3W2W3$$

Let $P = x_i + x_j + x_k$, and define

$$D1 = \begin{bmatrix} P + 2x_j & P - \tfrac{1}{2}x_i & P - \tfrac{1}{2}x_k \\ & P + 2x_k & P - \tfrac{1}{2}x_j \\ \text{symmetric} & & P + 2x_i \end{bmatrix}$$

Then

$$D2 = D1W2W1$$
$$D3 = W3W2D1$$
$$D4 = W2W1D2$$

Then

$$I_{22} = 8(D1 * Q1 + D2 * Q2 + D3 * Q3 + D4 * Q4) \tag{8.138}$$

The integrals of group 2 are given by the general expression in Eq. (8.139) with the various substitutions $t = a$ and $t = b$. Thus

$$\int \theta^T T = \frac{2A^2}{15} \begin{bmatrix} 2t_1 & -t_2 & -t_3 & 2t_2 - t_1 & -(t_2 + t_3) & -t_1 + 2t_3 \\ -t_1 & 2t_2 & -t_3 & 2t_1 - t_2 & -t_2 + 2t_3 & -(t_1 + t_3) \\ -t_1 & -t_2 & 2t_3 & -(t_1 + t_2) & 2t_2 - t_3 & -t_3 + 2t_1 \\ 3t_1 & 3t_2 & -t_3 & 8(t_1 + t_2) & 4(t_2 + 2t_3) & 4(t_1 + 2t_3) \\ -t_1 & 3t_2 & 3t_3 & 4(2t_1 + t_2) & 8(t_2 + t_3) & 4(2t_1 + t_3) \\ 3t_1 & -t_2 & 3t_3 & 4(t_1 + 2t_2) & 4(t_3 + 2t_2) & 8(t_3 + t_1) \end{bmatrix} \tag{8.139}$$

The group 3 integrals are again found by numerical quadrature. Altogether 21 separate integrals of the $f(\zeta)/x$ are required.

Surface Forces for the Linear-strain Triangle

Consider the linearly varying force, intensity $\tilde{p}(\zeta_B)$, acting on side 1 of the element. Force per unit length will be $2\pi x\tilde{p}(\zeta_B)$ (see Fig. 8.16). From Eq. (8.91*a*) for the X components of force,

$$R_{px} = \int_B \phi_B \phi_p{}^T \, dA \, p_x$$

Here,

$$dA = 2\pi x \, ds$$

so that

$$R_{px} = 2\pi \int_j^k x\phi_B\phi_p{}^T \, ds \, p_x$$

For the linear variation of $\tilde{p}(\zeta_B)$ on side 1,

$$\phi_p{}^T = [\zeta_2(1 - \zeta_2)]$$

$$\phi_B{}^T = [\zeta_2(2\zeta_2 - 1) \quad (1 - \zeta_2)(1 - 2\zeta_2) \quad 4\zeta_2(1 - \zeta_2)]$$

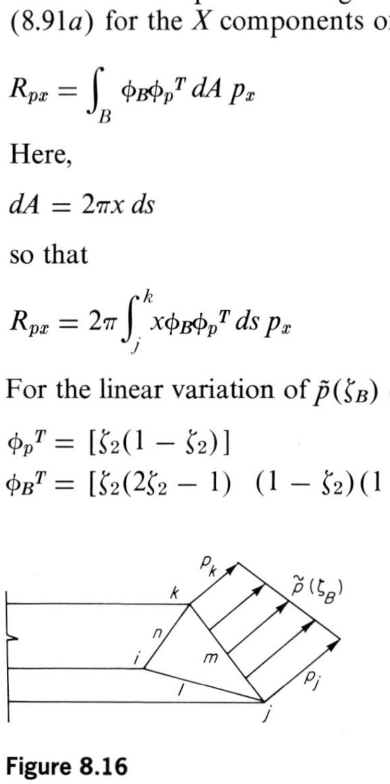

Figure 8.16

also

$$x = [\zeta_2(1-\zeta_2)]\begin{bmatrix} x_i \\ x_k \end{bmatrix}$$

and is a scalar multiplier of every term of $\phi_B\phi_p{}^T$. Hence

$$R_{px} = 2\pi l_1 \int_0^1 x \begin{bmatrix} \zeta_2(2\zeta_2 - 1) \\ (1-\zeta_2)(1-2\zeta_2) \\ 4\zeta_2(1-\zeta_2) \end{bmatrix} [\zeta_2(1-\zeta_2)]\, d\zeta_2\, p_x$$

When this expression is integrated and x is noted to be a function of ζ_2 and a multiplier of all terms, it is found that

$$R_{px} = \frac{\pi l_1}{30}\left\{ x_j \begin{bmatrix} 9 & 1 \\ -1 & 1 \\ 12 & 8 \end{bmatrix} + x_k \begin{bmatrix} 1 & -1 \\ 1 & 9 \\ 8 & 12 \end{bmatrix} \right\} \begin{bmatrix} p_{xj} \\ p_{xk} \end{bmatrix} \tag{8.140}$$

Suppose as an example that $x_j = x_k$ and $p_{xj} = p_{xk} = p_x$. This is a uniform pressure on a vertical face, and then

$$R_{px} = \begin{bmatrix} R_{xj} \\ R_{xk} \\ R_{xm} \end{bmatrix} = \frac{\pi l_1 x_j p_x}{30}\begin{bmatrix} 10 \\ 10 \\ 40 \end{bmatrix} = \frac{2\pi l_1 x_j p_x}{6}\begin{bmatrix} 1 \\ 1 \\ 4 \end{bmatrix}$$

For the planar triangle in Sec. 8.6, substitute for the uniform force p_x and unit thickness in Eq. (8.93); then

$$\begin{bmatrix} R_{xi} \\ R_{xk} \\ R_{xm} \end{bmatrix} = \frac{l_1}{30}\begin{bmatrix} 4 & -1 & 2 \\ -1 & 4 & 2 \\ 2 & 2 & 16 \end{bmatrix}\begin{bmatrix} p_x \\ p_x \\ p_x \end{bmatrix} = \frac{l_1 p_x}{6}\begin{bmatrix} 1 \\ 1 \\ 4 \end{bmatrix}$$

A comparison of the two expressions shows that in this case the axisymmetrical forces are simply those for the plane triangle multiplied by the circumference $2\pi x_j$.

Body Forces for the Linear-strain Triangle

The only body force considered will be parallel with the Y axis, so that, from the constant-strain triangle, the expression for node forces is

$$R_{By}{}^T = f_y{}^T 2\pi \int_{\text{area}} x\phi_f \phi_1{}^T \, dA$$

Consider the constant body force case. Then

$$R_{By} = f_y 2\pi \int_{\text{area}} x\phi_1 \, dA$$

On integrating $x\phi_1$,

$$x = [\zeta_1 \quad \zeta_2 \quad \zeta_3] \begin{bmatrix} x_i \\ x_j \\ x_k \end{bmatrix}$$

$$\begin{bmatrix} R_{Byi} \\ R_{Byj} \\ R_{Byk} \\ R_{Byl} \\ R_{Bym} \\ R_{Byn} \end{bmatrix} = \frac{\pi f_y A}{30} \begin{bmatrix} 2 & -1 & -1 \\ -1 & 2 & -1 \\ -1 & -1 & 2 \\ 8 & 4 & 8 \\ 8 & 8 & 4 \\ 4 & 8 & 8 \end{bmatrix} \begin{bmatrix} x_i \\ x_j \\ x_k \end{bmatrix} \tag{8.141}$$

8.8 COMBINATIONS OF DIFFERENT TYPES OF PLANE ELASTICITY ELEMENTS

It is found that both the constant-strain and linear-strain triangular elements discussed so far have problem types for whose solution they are particularly suited. For example, linear-strain triangles will be better suited to problems where strain (or stress) gradients are high. Again, in one problem there may be such zones in which the linear-strain triangle is best used and other zones for which coarse subdivision with constant-strain triangle elements is quite adequate. In this way a considerable economy may be achieved in the number of node points required for the satisfactory coverage of a given stress field. When both types of elements are used in one problem, some way must be found for matching elements along the interface between the zones of the two types. Such an interface is shown in Fig. 8.17*a*. In Fig. 8.17*b* it is seen that the matching of the two types of elements can occur if the five- and four-nodal-point elements 2 and 3 are developed. In these elements the displacements are described in different ways along each side. Thus in element 2 in Fig. 8.17*b*, one side of the element is constrained so that the displacements there are linear. Similarly, for element 3, the two sides are constrained in this manner. Such elements (that is, their stiffness matrices) are easily constructed if the displacement at the midpoint of a side is considered composed of a linear part plus the parabolic variation therefrom. If a side is to be constrained so that displacements in it are linear, the parabolic variation must be set equal

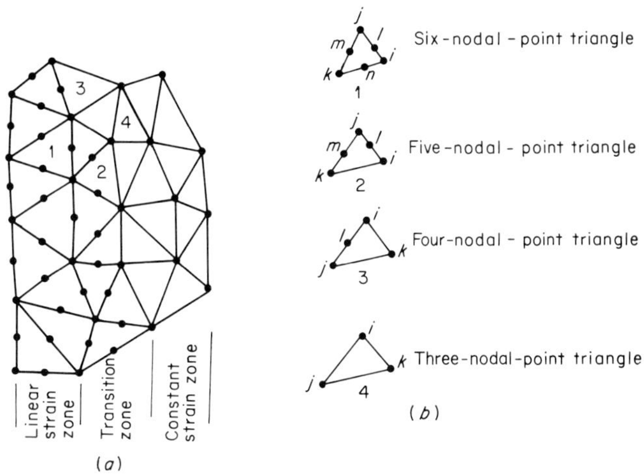

Figure 8.17 Matching finite elements.

to zero. Furthermore, if this variation is zero, it may henceforth be omitted from the calculations. The new (constrained) stiffness matrix is generated from the original linear-strain-triangle-stiffness matrix. Evidently, as one proceeds from element type 1 to element type 4 in Fig. 8.17*b*, the stiffness matrix is reduced from a 12×12 to a 6×6. It is seen that the final reduction must produce a constant-strain triangle element.

Let the original nodal-point displacements be

$$r^T = \langle u_i \quad v_i \quad u_j \quad v_j \quad u_k \quad v_k \quad u_l \quad v_l \quad u_m \quad v_m \quad u_n \quad v_n \rangle$$

and consider triangle 2 for which node n is to be omitted. First construct the stiffness matrix for the displacements r, including the node n, such that

$$R = kr$$

Now the displacements u_n and v_n are expressed as

$$u_n = \tfrac{1}{2}(u_k + u_i) + \alpha_n$$
$$v_n = \tfrac{1}{2}(v_k + v_i) + \beta_n$$

The variables α_n and β_n respectively express the variation from linearity of u and v in side 2, node n. That is, let

$$q^T = \langle u_i \quad v_i \quad u_j \quad v_j \quad u_k \quad v_k \quad u_l \quad v_l \quad u_m \quad v_m \quad \alpha_n \quad \beta_n \rangle$$

Then,

$$r = Tq \tag{8.142}$$

where in this case,

$$
T = \begin{bmatrix}
1 & \cdot & \cdot & \cdot & \cdot & \cdot & \cdot & \cdot & \cdot & \cdot & \cdot & \cdot \\
\cdot & 1 & \cdot & \cdot & \cdot & \cdot & \cdot & \cdot & \cdot & \cdot & \cdot & \cdot \\
\cdot & \cdot & 1 & \cdot & \cdot & \cdot & \cdot & \cdot & \cdot & \cdot & \cdot & \cdot \\
\cdot & \cdot & \cdot & 1 & \cdot & \cdot & \cdot & \cdot & \cdot & \cdot & \cdot & \cdot \\
\cdot & \cdot & \cdot & \cdot & 1 & \cdot & \cdot & \cdot & \cdot & \cdot & \cdot & \cdot \\
\cdot & \cdot & \cdot & \cdot & \cdot & 1 & \cdot & \cdot & \cdot & \cdot & \cdot & \cdot \\
\cdot & \cdot & \cdot & \cdot & \cdot & \cdot & 1 & \cdot & \cdot & \cdot & \cdot & \cdot \\
\cdot & \cdot & \cdot & \cdot & \cdot & \cdot & \cdot & 1 & \cdot & \cdot & \cdot & \cdot \\
\cdot & \cdot & \cdot & \cdot & \cdot & \cdot & \cdot & \cdot & 1 & \cdot & \cdot & \cdot \\
\cdot & \cdot & \cdot & \cdot & \cdot & \cdot & \cdot & \cdot & \cdot & 1 & \cdot & \cdot \\
0.5 & \cdot & \cdot & \cdot & 0.5 & \cdot & \cdot & \cdot & \cdot & \cdot & 1 & \cdot \\
\cdot & 0.5 & \cdot & \cdot & \cdot & 0.5 & \cdot & \cdot & \cdot & \cdot & \cdot & 1
\end{bmatrix} \tag{8.143a}
$$

For each additional node treated in this fashion, an extra two sets of 0.5 terms appear in the lower left-hand submatrix of T, so that when all have been thus modified,

$$
T = \begin{bmatrix}
1 & \cdot & \cdot & \cdot & \cdot & \cdot & \cdot & \cdot & \cdot & \cdot & \cdot & \cdot \\
\cdot & 1 & \cdot & \cdot & \cdot & \cdot & \cdot & \cdot & \cdot & \cdot & \cdot & \cdot \\
\cdot & \cdot & 1 & \cdot & \cdot & \cdot & \cdot & \cdot & \cdot & \cdot & \cdot & \cdot \\
\cdot & \cdot & \cdot & 1 & \cdot & \cdot & \cdot & \cdot & \cdot & \cdot & \cdot & \cdot \\
\cdot & \cdot & \cdot & \cdot & 1 & \cdot & \cdot & \cdot & \cdot & \cdot & \cdot & \cdot \\
\cdot & \cdot & \cdot & \cdot & \cdot & 1 & \cdot & \cdot & \cdot & \cdot & \cdot & \cdot \\
0.5 & \cdot & 0.5 & \cdot & \cdot & \cdot & 1 & \cdot & \cdot & \cdot & \cdot & \cdot \\
\cdot & 0.5 & \cdot & 0.5 & \cdot & \cdot & \cdot & 1 & \cdot & \cdot & \cdot & \cdot \\
\cdot & \cdot & 0.5 & \cdot & 0.5 & \cdot & \cdot & \cdot & 1 & \cdot & \cdot & \cdot \\
\cdot & \cdot & \cdot & 0.5 & \cdot & 0.5 & \cdot & \cdot & \cdot & 1 & \cdot & \cdot \\
0.5 & \cdot & \cdot & \cdot & 0.5 & \cdot & \cdot & \cdot & \cdot & \cdot & 1 & \cdot \\
\cdot & 0.5 & \cdot & \cdot & \cdot & 0.5 & \cdot & \cdot & \cdot & \cdot & \cdot & 1
\end{bmatrix} \tag{8.143b}
$$

Let Q be the generalized nodal forces corresponding to the displacements q. The contragradient law gives

$$Q = T^T R$$

Thence the stiffness matrix k_q is obtained from k through the transformation

$$k_q = T^T k T \tag{8.144}$$

The matrix T to be entered in Eq. (8.144) will depend on the type of triangle under consideration. Thus for the linear-strain triangle, it will be sufficient to take $T = I$, and for the five-node-point triangle, T is given in Eq. (8.143*a*) for the node n being considered. The required stiffness matrix is given by the first 10×10 submatrix of k_q. Finally, the constant-strain-triangle-stiffness matrix is obtained by using T as defined in Eq. (8.143*b*) and by extraction of the first 6×6 submatrix from k_q. It should be clear that the above procedure can be very easily programmed. All that will be required is some method for recognizing element types and branching accordingly.

8.9 A TRIANGULAR PLATE BENDING ELEMENT

The construction of compatible plate bending elements has proved more difficult than for the stress type of elements already discussed. The plate elements exhibit the C-1 type of compatibility. That is, along the boundary common to two elements, compatibility is achieved if both the deflections and normal slopes are defined by the node values within that boundary. For example, the deflected shape w of the element may be a cubic function, with normal slopes constrained to vary linearly along the sides. If then the apices i, j, and k are taken as nodes, it will be sufficient to define the deflection w and slopes θ_x and θ_y at each node.

However, a further requirement for the finite-element model to give solutions that converge to the differential equation solution must not be overlooked. This is that the uniform-strain states in the element must always be included in the displacement pattern. This was automatically satisfied in the constant-strain and linear-strain triangle elements for plane stress. With the plate bending element, the inclusion of constant curvature terms is not self-evident when the interpolation polynomial is constructed for the transverse deflections of the plate.

The complete cubic interpolation polynomial contains 10 terms and, as has been mentioned in Sec. 8.4, cannot be specified by nodal values on the triangle sides alone, since these must necessarily exclude the function $\zeta_1\zeta_2\zeta_3$ which is zero everywhere on the boundary.

It would be possible to construct an element with an internal node and eliminate this point by reduction of the 10×10 stiffness matrix to a 9×9 by including boundary displacements only. The element that is presented here is that given by Zienkiewiez (Ref. 38) and does not have compatible normal slopes between adjacent elements. It is designed however to include the constant strain terms to ensure convergence to differential equation solutions.

The coordinate system used to describe plate dimensions, deflections, and forces is shown in Fig. 8.18. In Fig. 8.18, the moments $\langle M_x \quad M_y \quad M_{xy} \rangle$ per

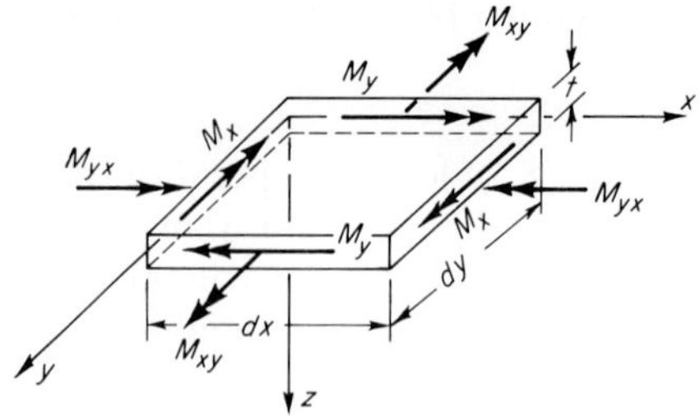

Figure 8.18 Infinitesimal plate element.

unit length are shown on the infinitesimal plate element dx, dy, of thickness t. The basic stress quantities for the plate element will be $\tilde{M}^T = \langle M_x \quad M_y \quad M_{xy}\rangle$, and these produce the strains or curvatures

$$\tilde{\varepsilon}^T = \left\langle -\frac{\partial^2\omega}{\partial x^2}, -\frac{\partial^2\omega}{\partial y^2}, \frac{2\partial^2\omega}{\partial x\,\partial y}\right\rangle$$

From the theory of plate bending it may be shown that

$$\begin{bmatrix} M_x \\ M_y \\ M_{xy} \end{bmatrix} = \frac{Et^3}{12(1-\nu^2)}\begin{bmatrix} 1 & \nu & 0 \\ \nu & 1 & 0 \\ 0 & 0 & \dfrac{1-\nu}{2} \end{bmatrix}\begin{bmatrix} -\dfrac{\partial^2\omega}{\partial x^2} \\ -\dfrac{\partial^2\omega}{\partial y^2} \\ \dfrac{2\partial^2\omega}{\partial x\,\partial y} \end{bmatrix} \tag{8.145}$$

This is the constitutive law for the plate element of isotropic material, whose plane sections remain plane in the deformed state. Equation (8.145) will be written

$$\tilde{M} = k\tilde{\varepsilon} \tag{8.146}$$

The deformation of the triangular element will be defined by nine nodal values. These will be the deflections $\langle\omega_i \quad \omega_j \quad \omega_k\rangle$ and the slopes

$$\langle\theta_{xi} \quad \theta_{yi} \quad \theta_{xj} \quad \theta_{yj} \quad \theta_{xk} \quad \theta_{yk}\rangle$$

These values are shown in Fig. 8.19.

It is convenient to measure the actual distortion of the plate w^* relative to the plane through i, j, and k. That is, the movement of the plate from the unloaded position i_o, j_o, k_o in Fig. 8.19 to i, j, k is a rigid-body motion only.

The deflection to the plane ijk will be denoted w^R, and the equation to w^R is

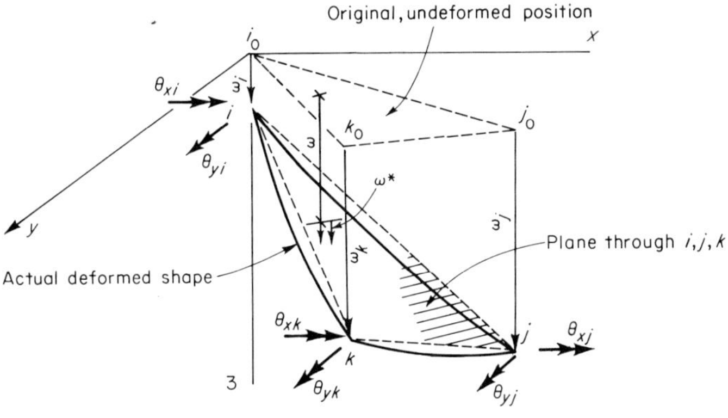

Figure 8.19 Plate element deflections.

$$w^R = \phi_1{}^T w_c = \langle \zeta_1 \quad \zeta_2 \quad \zeta_3 \rangle \begin{bmatrix} w_i \\ w_j \\ w_k \end{bmatrix} \tag{8.147}$$

The deflection of any point on the surface will be

$$w = w^R + w^* \tag{8.148}$$

For the relative deflections w^*, the nodal values are

$$w_i^* = w_j^* = w_k^* = 0$$

and hence only six nodal values $(\theta_x^*, \theta_y^*)_l$, $l = i, j, k$, are required to specify the displacement field assumed to be cubic in form but omitting for the moment the term $\zeta_1\zeta_2\zeta_3$.

The slopes θ_x^* and θ_y^* are defined by

$$\theta_x^* = \frac{\partial w^*}{\partial y} = \theta_x - \frac{\partial w^R}{\partial y} \tag{8.149}$$

$$\theta_y^* = -\frac{\partial w^*}{\partial x} = \theta_y + \frac{\partial w^R}{\partial x} \tag{8.150}$$

From Eqs. (8.40) and (8.147),

$$\begin{bmatrix} -\dfrac{\partial w^R}{\partial y} \\ \dfrac{\partial w^R}{\partial x} \end{bmatrix} = \frac{1}{2A} \begin{bmatrix} -a_1 & -a_2 & -a_3 \\ b_1 & b_2 & b_3 \end{bmatrix} \begin{bmatrix} w_i \\ w_j \\ w_k \end{bmatrix} \tag{8.151}$$

Define the nodal values of θ^* and r^* such that

$$r^{*T} = \langle \theta^*_{xi} \quad \theta^*_{yi} \quad \theta^*_{xj} \quad \theta^*_{yj} \quad \theta^*_{xk} \quad \theta^*_{yk} \rangle \tag{8.152}$$

The r^* values are obtained by evaluating θ^*_x and θ^*_y at i, j, and k by using Eqs. (8.149) and (8.151). Then,

$$r^* = \begin{bmatrix} \theta^*_{xi} \\ \theta^*_{yi} \\ \theta^*_{xj} \\ \theta^*_{yj} \\ \theta^*_{xk} \\ \theta^*_{yk} \end{bmatrix} = \frac{1}{2A} \begin{bmatrix} -a_1 & 2A & 0 & -a_2 & 0 & 0 & -a_3 & 0 & 0 \\ b_1 & 0 & 2A & b_2 & 0 & 0 & b_3 & 0 & 0 \\ -a_1 & 0 & 0 & -a_2 & 2A & 0 & -a_3 & 0 & 0 \\ b_1 & 0 & 0 & b_2 & 0 & 2A & b_3 & 0 & 0 \\ -a_1 & 0 & 0 & -a_2 & 0 & 0 & -a_3 & 2A & 0 \\ b_1 & 0 & 0 & b_2 & 0 & 0 & b_3 & 0 & 2A \end{bmatrix} \begin{bmatrix} w_i \\ \theta_{xi} \\ \theta_{yi} \\ w_j \\ \theta_{xj} \\ \theta_{yj} \\ w_k \\ \theta_{xk} \\ \theta_{yk} \end{bmatrix} \tag{8.153}$$

This is the equation

$$r^* = Tr \tag{8.154}$$

If R^* and R are the nodal forces corresponding to r^* and r respectively, the contragradient law gives

$$R = T^T R^* \tag{8.155}$$

It is evident that

$$R^{*T}_l = \langle M_{xl} \quad M_{yl} \rangle \qquad l = i, j, k$$
$$R_l{}^T = \langle F_{zl} \quad M_{xl} \quad M_{yl} \rangle \qquad l = i, j, k$$

If now the stiffness relationship connecting R^* and r^* is known, that is,

$$R^* = k^* r^* \tag{8.156}$$

then

$$R = T^T k^* T r$$

or

$$K = T^T k^* T \tag{8.157}$$

The stiffness matrix k^* must be constructed from the interpolation polynomial to w^* which has been arranged to include constant curvature terms.

The interpolation polynomial ϕ_3 for the relative shape function w^* will be such that

$$w^* = \phi_3{}^T r^* \tag{8.158}$$

Choose the polynomial φ such that

$$\varphi^T = \langle \zeta_1{}^2\zeta_2 \quad \zeta_1{}^2\zeta_3 \quad \zeta_2{}^2\zeta_1 \quad \zeta_2{}^2\zeta_3 \quad \zeta_3{}^2\zeta_1 \quad \zeta_3{}^2\zeta_2 \rangle \tag{8.159}$$

and obtain ϕ_3 by the orthogonalization process given in Sec. 8.2, Eq. (8.19). Thus,

$$w^* = \varphi^T \alpha$$

where α is the generalized coordinate vector. Nodal values are obtained from the slopes and from Eq. (8.159),

$$-2A\,\frac{\partial w^*}{\partial x} = -\begin{bmatrix} 2\zeta_1\zeta_2 b_1 + \zeta_1{}^2 b_2 \\ 2\zeta_1\zeta_3 b_1 + \zeta_1{}^2 b_3 \\ 2\zeta_2\zeta_1 b_2 + \zeta_2{}^2 b_1 \\ 2\zeta_2\zeta_3 b_2 + \zeta_2{}^2 b_3 \\ 2\zeta_3\zeta_1 b_3 + \zeta_3{}^2 b_1 \\ 2\zeta_3\zeta_2 b_3 + \zeta_3{}^2 b_2 \end{bmatrix}^T \alpha$$

$$2A\,\frac{\partial w^*}{\partial y} = \begin{bmatrix} 2\zeta_1\zeta_2 a_1 + \zeta_1{}^2 a_2 \\ 2\zeta_1\zeta_3 a_1 + \zeta_1{}^2 a_3 \\ 2\zeta_2\zeta_1 a_2 + \zeta_2{}^2 a_1 \\ 2\zeta_2\zeta_3 a_2 + \zeta_2{}^2 a_3 \\ 2\zeta_3\zeta_1 a_3 + \zeta_3{}^2 a_1 \\ 2\zeta_3\zeta_2 a_3 + \zeta_3{}^2 a_2 \end{bmatrix}^T \alpha$$

These slopes are evaluated at i, (1,0,0); j, (0,1,0), and k, (0,0,1).

Write Eq. (8.19) here as

$$\phi = (\Delta^{-1})^T \varphi \tag{8.160}$$

Then

$$\Delta^T = \frac{1}{2A}\begin{bmatrix} a_2 & -b_2 & 0 & 0 & 0 & 0 \\ a_3 & -b_3 & 0 & 0 & 0 & 0 \\ 0 & 0 & a_1 & -b_1 & 0 & 0 \\ 0 & 0 & a_3 & -b_3 & 0 & 0 \\ 0 & 0 & 0 & 0 & a_1 & -b_1 \\ 0 & 0 & 0 & 0 & a_2 & -b_2 \end{bmatrix}$$

On inversion of Δ,

$$(\Delta^T)^{-1} = (\Delta^{-1})^T = \begin{bmatrix} -b_3 & b_2 & 0 & 0 & 0 & 0 \\ -a_3 & a_2 & 0 & 0 & 0 & 0 \\ 0 & 0 & b_3 & -b_1 & 0 & 0 \\ 0 & 0 & a_3 & -a_1 & 0 & 0 \\ 0 & 0 & 0 & 0 & -b_2 & b_1 \\ 0 & 0 & 0 & 0 & -a_2 & a_1 \end{bmatrix}$$

Substitution in Eq. (8.160) gives

$$\phi = \begin{bmatrix} \zeta_1^2(-\zeta_2 b_3 + \zeta_3 b_2) \\ \zeta_1^2(-\zeta_2 a_3 + \zeta_3 a_2) \\ \zeta_2^2(\zeta_1 b_3 - \zeta_3 b_1) \\ \zeta_2^2(\zeta_1 a_3 - \zeta_3 a_1) \\ \zeta_3^2(-\zeta_1 b_2 + \zeta_2 b_1) \\ \zeta_3^2(-\zeta_1 a_2 + \zeta_2 a_1) \end{bmatrix} \tag{8.161}$$

Unfortunately the ϕ function thus obtained does not satisfy the constant-strain criterion necessary for convergence. It must be modified by adding the function $\beta\zeta_1\zeta_2\zeta_3$. It is found that $\beta = \frac{1}{2}$.

Let ϕ be modified as in Eq. (8.162), and it will then be necessary to calculate a numerical value for the constant β.

$$\phi_3 = \begin{bmatrix} -(\zeta_1^2\zeta_2 + \beta\zeta_1\zeta_2\zeta_3)b_3 + (\zeta_1^2\zeta_3 + \beta\zeta_1\zeta_2\zeta_3)b_2 \\ -(\zeta_1^2\zeta_2 + \beta\zeta_1\zeta_2\zeta_3)a_3 + (\zeta_1^2\zeta_3 + \beta\zeta_1\zeta_2\zeta_3)a_2 \\ (\zeta_2^2\zeta_1 + \beta\zeta_1\zeta_2\zeta_3)b_3 - (\zeta_2^2\zeta_3 + \beta\zeta_1\zeta_2\zeta_3)b_1 \\ (\zeta_2^2\zeta_1 + \beta\zeta_1\zeta_2\zeta_3)a_3 - (\zeta_2^2\zeta_3 + \beta\zeta_1\zeta_2\zeta_3)a_1 \\ -(\zeta_3^2\zeta_1 + \beta\zeta_1\zeta_2\zeta_3)b_2 + (\zeta_3^2\zeta_2 + \beta\zeta_1\zeta_2\zeta_3)b_1 \\ -(\zeta_3^2\zeta_1 + \beta\zeta_1\zeta_2\zeta_3)a_2 + (\zeta_3^2\zeta_2 + \beta\zeta_1\zeta_2\zeta_3)a_1 \end{bmatrix} \tag{8.162}$$

Double differentiation of $w*$, that is, ϕ_3 with respect to x and y gives the curvature functions, and if these are to be constant,

$$w^* = A_1\zeta_2\zeta_3 + A_2\zeta_3\zeta_1 + A_3\zeta_1\zeta_2 \tag{8.163}$$

If the two expressions for w^* from Eqs. (8.158) and (8.163) are to be identical, it is evident that A_1, A_2, and A_3 must contain the multiplier $\zeta_1 + \zeta_2 + \zeta_3 = 1$. Equation (8.163) is written

$$w^* = [\zeta_2\zeta_3 \quad \zeta_3\zeta_1 \quad \zeta_1\zeta_2]\begin{bmatrix} A_1 \\ A_2 \\ A_3 \end{bmatrix}$$

so that on differentiation with respect to x and y,

$$2A\begin{bmatrix} -\dfrac{\partial w^*}{\partial x} \\ \dfrac{\partial w^*}{\partial y} \end{bmatrix} = \begin{bmatrix} -(b_2\zeta_3 + b_3\zeta_2) & -(b_3\zeta_1 + b_1\zeta_3) & -(b_1\zeta_2 + b_2\zeta_1) \\ a_2\zeta_3 + a_3\zeta_2 & a_3\zeta_1 + a_1\zeta_3 & a_1\zeta_2 + a_2\zeta_1 \end{bmatrix}\begin{bmatrix} A_1 \\ A_2 \\ A_3 \end{bmatrix} \tag{8.164}$$

At the nodal points i, j, and k, Eq. (8.164) gives the alternative expression for r^* to be

$$2Ar^* = 2A\begin{bmatrix} \theta^*_{xi} \\ \theta^*_{yi} \\ \theta^*_{xj} \\ \theta^*_{yj} \\ \theta^*_{xk} \\ \theta^*_{yk} \end{bmatrix} = \begin{bmatrix} 0 & a_3 & a_2 \\ 0 & -b_3 & -b_2 \\ a_3 & 0 & a_1 \\ -b_3 & 0 & -b_1 \\ b_2 & a_1 & 0 \\ -b_2 & -b_1 & 0 \end{bmatrix}\begin{bmatrix} A_1 \\ A_2 \\ A_3 \end{bmatrix}$$

Substitute this value for r^* in Eq. (8.158). Thence

$$2Aw^* = \phi_3{}^T\begin{bmatrix} 0 & a_3 & a_2 \\ 0 & -b_3 & -b_2 \\ a_3 & 0 & a_1 \\ -b_3 & 0 & -b_1 \\ a_2 & a_1 & 0 \\ -b_2 & -b_1 & 0 \end{bmatrix}\begin{bmatrix} A_1 \\ A_2 \\ A_3 \end{bmatrix}$$

Write $\phi_3{}^T = \langle\phi_{31} \quad \phi_{32} \quad \phi_{33} \quad \phi_{34} \quad \phi_{35} \quad \phi_{36}\rangle$; then,

$$2Aw^* = \begin{bmatrix} \phi_{33}a_3 - \phi_{34}b_3 + \phi_{35}a_2 - \phi_{36}b_2 \\ \phi_{31}a_3 - \phi_{32}b_2 + \phi_{35}a_1 - \phi_{36}b_1 \\ \phi_{31}a_2 - \phi_{32}b_2 + \phi_{33}a_1 - \phi_{34}b_1 \end{bmatrix}^T\begin{bmatrix} A_1 \\ A_2 \\ A_3 \end{bmatrix}$$

Evaluate the first term $(\phi_{33}a_3 - \phi_{34}b_3 + \phi_{35}a_2 - \phi_{36}b_2)$ by using Eq. (8.162). That is,

$$\begin{aligned}\phi_{33}a_3 - \phi_{34}b_3 + \phi_{35}a_2 - \phi_{36}b_2 &= 2A(\zeta_3{}^2\zeta_2 + \beta\zeta_1\zeta_2\zeta_3) + 2A(\zeta_2{}^2\zeta_3 + \beta\zeta_1\zeta_2\zeta_3)\\ &= 2A\zeta_2\zeta_3(\zeta_3 + \zeta_2 + 2\beta\zeta_1)\end{aligned}$$

It is seen that if this is to reduce to the first term in Eq. (8.163), $\beta = ½$. When the remaining terms are evaluated, it is found again that $\beta = ½$.

Now that the interpolation polynomial ϕ_3 is known, the generation of k^* follows the same procedure as that for the planar element. In this case the basic strain quantities $\tilde{\varepsilon}$ are the curvatures and twist of the plate middle surface, and these will be related to nodal values ε_C. In turn, ε_C may be found by double differentiations of Eq. (8.158) and substitution of the nodal coordinates of i, j, and k in turn. From Eq. (8.158)

$$\begin{bmatrix} w^*_{,xx} \\ w^*_{,yy} \\ 2w^*_{,xy} \end{bmatrix} = \begin{bmatrix} \phi^T_{,xx} & & \mathbf{0} \\ & \phi^T_{,yy} & \\ \mathbf{0} & & 2\phi^T_{,xy} \end{bmatrix} r^* \tag{8.165}$$

For nodes i, j, and k,

$$\begin{bmatrix} -w^*_{,xxi} \\ -w^*_{,xxj} \\ -w^*_{,xxk} \\ -w^*_{,yyi} \\ -w^*_{,yyj} \\ -w^*_{,yyk} \\ 2w^*_{,xyi} \\ 2w^*_{,xyj} \\ 2w^*_{,xyk} \end{bmatrix} = \begin{bmatrix} -\phi^T_{,xxi} \\ -\phi^T_{,xxj} \\ -\phi^T_{,xxk} \\ -\phi^T_{,yyi} \\ -\phi^T_{,yyj} \\ -\phi^T_{,yyk} \\ 2\phi^T_{,xyi} \\ 2\phi^T_{,xyj} \\ 2\phi^T_{,xyk} \end{bmatrix} r^*$$

This is the equation

$$\varepsilon_c = \Phi_c r^* \tag{8.166}$$

When the differentiations of ϕ_3 have been performed and the coordinates of i, j, and k substituted, then in Eq. (8.165) it is found that Φ_c is as shown in Eq. (8.167).

Notice that Φ_c given in Eq. (8.167) has its columns arranged so that recurrence relationships can be devised for its generation. These columns should then be interchanged to give the original sequence.

$$
\Phi_c = \frac{1}{4A^2}
\begin{bmatrix}
b_2b_3(b_3-b_2) & b_2b_3(2b_1-b_2) & b_2b_3(b_3-2b_1) & b_2b_3(a_3-a_2)+8b_1A & b_2b_3(a_1-a_3)-2b_2{}^2a_3 & b_2b_3(a_2-a_1)+2b_3{}^2a_2 \\
b_3b_1(b_1-2b_2) & b_3b_1(b_1-b_3) & b_3b_1(2b_2-b_3) & b_3b_1(a_3-a_2)+2b_1{}^2a_3 & b_3b_1(a_1-a_3)+8b_2A & b_3b_1(a_2-a_1)-2b_3{}^2a_1 \\
b_1b_2(2b_3-b_1) & b_1b_2(b_2-2b_3) & b_1b_2(b_2-b_1) & b_1b_2(a_3-a_2)-2b_1{}^2a_2 & b_1b_2(a_1-a_3)+2b_2{}^2a_1 & b_1b_2(a_2-a_1)+8b_3A \\
a_2a_3(b_3-b_2)-8a_1A & a_2a_3(b_1-b_3)-2a_2{}^2b_3 & a_2a_3(b_2-b_1)+2a_3{}^2b_2 & a_2a_3(a_3-a_2) & a_2a_3(2a_1-a_2) & a_2a_3(a_3-2a_1) \\
a_3a_1(b_3-b_2)+2a_1{}^2b_3 & a_3a_1(b_1-b_3)-8a_2A & a_3a_1(b_2-b_1)-2a_3{}^2b_1 & a_3a_1(a_1-2a_2) & a_3a_1(a_1-a_3) & a_3a_1(2a_2-a_3) \\
a_1a_2(b_3-b_2)-2a_1{}^2b_2 & a_1a_2(b_1-b_3)+2a_2{}^2b_1 & a_1a_2(b_2-b_1)-8a_3A & a_1a_2(2a_3-a_1) & a_1a_2(a_2-2a_3) & a_1a_2(a_2-a_1) \\
(b_2-b_3)(a_3b_2+a_2b_3)+8b_1A & (b_3-b_1)(a_3b_2+a_2b_3)+4a_2b_2b_3 & (b_1b_2)(a_3b_2+a_2b_3)-4a_3b_2b_3 & (a_2-a_3)(a_3b_2+a_2b_3)-8a_1A & (a_3-a_1)(a_3b_2+a_2b_3)+4a_2a_3b_2 & (a_1-a_2)(a_3b_2+a_2b_3)-4a_2a_3b_3 \\
(b_2-b_3)(a_1b_3+a_3b_1)-4a_1b_1b_3 & (b_3-b_1)(a_1b_3+a_3b_1)+8b_2A & (b_1-b_2)(a_1b_3+a_3b_1)+4a_3b_1b_3 & (a_2-a_3)(a_1b_3+a_3b_1)-4a_1a_3b_1 & (a_3-a_1)(a_1b_3+a_3b_1)-8a_2A & (a_1-a_2)(a_1b_3+a_3b_1)+4a_1a_3b_3 \\
(b_2-b_3)(a_2b_1+a_1b_2)+4a_1b_1b_2 & (b_3-b_1)(a_2b_1+a_1b_2)-4a_2b_2b_1 & (b_1-b_2)(a_2b_1+a_1b_2)+8b_3A & (a_2-a_3)(a_2b_1+a_1b_2)+4a_1a_2b_1 & (a_3-a_1)(a_2b_1+a_1b_2)-4a_2a_1b_2 & (a_1-a_2)(a_2b_1+a_1b_2)-8a_3A
\end{bmatrix}
\begin{matrix}
\left.\begin{matrix} \\ \\ \\ \end{matrix}\right\} -w^*_{xx} \\
\left.\begin{matrix} \\ \\ \\ \end{matrix}\right\} -w^*_{yy} \\
\left.\begin{matrix} \\ \\ \\ \end{matrix}\right\} 2w^*_{xy}
\end{matrix}
$$

$$
\begin{matrix} \phi_1 & \phi_3 & \phi_5 & \phi_2 & \phi_4 & \phi_6 \end{matrix}
$$

Note the column order for ϕ's (1, 3, 5, 2, 4, 6) for ease of generation. (8.167)

The stiffness matrix k_c for the nodal curvatures is given by

$$k_c = \int \phi_\varepsilon k \phi_\varepsilon{}^T \, dA$$

$$k_c^* = \Phi_c{}^T k_c \Phi_c$$

Since curvatures vary linearly over the element, the same expressions apply as those derived in Eqs. (8.80)ff. That is,

$$\phi_\varepsilon{}^T = \begin{bmatrix} \phi_1{}^T & 0 & 0 \\ 0 & \phi_1{}^T & 0 \\ 0 & 0 & \phi_1{}^T \end{bmatrix} = \begin{bmatrix} \zeta_1 & \zeta_2 & \zeta_3 & 0 & 0 & 0 & 0 & 0 & 0 \\ 0 & 0 & 0 & \zeta_1 & \zeta_2 & \zeta_3 & 0 & 0 & 0 \\ 0 & 0 & 0 & 0 & 0 & 0 & \zeta_1 & \zeta_2 & \zeta_3 \end{bmatrix}$$

Then,

$$\phi_\varepsilon k \phi_\varepsilon{}^T = \frac{Et^3}{12(1-\nu^2)} \begin{bmatrix} \phi_1 & 0 & 0 \\ 0 & \phi_1 & 0 \\ 0 & 0 & \phi_1 \end{bmatrix} \begin{bmatrix} 1 & \nu & 0 \\ \nu & 1 & 0 \\ 0 & 0 & \dfrac{1-\nu}{2} \end{bmatrix} \begin{bmatrix} \phi_1{}^T & 0 & 0 \\ 0 & \phi_1{}^T & 0 \\ 0 & 0 & \phi_1{}^T \end{bmatrix}$$

$$= \frac{Et^3}{12(1-\nu^2)} \begin{bmatrix} \phi_1\phi_1{}^T & \nu\phi_1\phi_1{}^T & 0 \\ \nu\phi_1\phi_1{}^T & \phi_1\phi_1{}^T & 0 \\ 0 & 0 & \dfrac{1-\nu}{2}\phi_1\phi_1{}^T \end{bmatrix}$$

As before, let

$$\Lambda = \begin{bmatrix} 2 & 1 & 1 \\ 1 & 2 & 1 \\ 1 & 1 & 2 \end{bmatrix}$$

so that

$$k_c^* = \int_{\text{area}} \phi_\varepsilon k \phi_\varepsilon{}^T \, dA = \frac{Et^3 A}{144(1-\nu^2)} \begin{bmatrix} \Lambda & \nu\Lambda & 0 \\ \nu\Lambda & \Lambda & 0 \\ 0 & 0 & \dfrac{1-\nu}{2}\Lambda \end{bmatrix} \tag{8.168}$$

It is seen that k_c^* is a 9×9 matrix composed of the 3×3 submatrices in Eq. (8.168). From k_c^*,

$$k^* = \Phi_c{}^T k_c^* \Phi_c \tag{8.169}$$

and finally, from Eq. (8.157)

$$K = T^T k^* T \tag{8.170}$$

Surface Forces on the Plate Element

Surface forces perpendicular to the plane of the plate (that is, parallel with the z axis) are equivalent to the body forces of the stress elements. In these stress elements the body forces are replaced by equivalent node forces that are applied to the kinematically determinate structure $r \equiv 0$. In this condition the body forces will produce stresses within the elements that are usually overlooked. However in the present case, the transverse loads will produce fixed-end moments in the condition $r \equiv 0$, which, just as in the case of framed structures, must be added to those from the nodal-force analysis.

The expression for deflections

$$w = w^* + w^R$$

may be written in terms of the nodal displacement vector r as

$$w = (\phi_3{}^T T + U)r \tag{8.171}$$

In Eq. (8.171), ϕ_3 and T are as defined previously, and from Eq. (8.147),

$$U = [\zeta_1 \quad 0 \quad 0 \quad \zeta_2 \quad 0 \quad 0 \quad \zeta_3 \quad 0 \quad 0] \tag{8.172}$$

Let the distribution of load on the element be

$$\tilde{f}_z(\zeta) = \phi_f{}^T f_z \tag{8.173}$$

In Eq. (8.173), f_z is the vector of nodal values of $\tilde{f}(\zeta)$, and ϕ_f is the interpolation polynomial. Since only displacement nodes i, j, and k have been defined, $\tilde{f}_z(\zeta)$ may vary only linearly without defining extra nodes.

The equivalence of external and internal work for virtual displacements gives the nodal forces R_f to be

$$R_f{}^T r = \int_{\text{area}} \tilde{f}^T w \, dA$$

Substitution for w gives

$$R_f{}^T r = f_z{}^T \int_{\text{area}} \phi_f(\phi_3{}^T T + U) \, dA \, r$$

Thence

$$R_f = \int_{\text{area}} (T^T \phi_3 + U^T)\phi_f{}^T \, dA \, f_z$$

Since T is a constant matrix, this expression may be written

$$R_f = (T^T \int_{\text{area}} \phi_3 \phi_f{}^T \, dA + \int_{\text{area}} U^T \phi_f{}^T \, dA) f_z \tag{8.174}$$

Consider the special case of a uniformly distributed force f per unit area, and in this case $f_z = f$ (a scalar). Then ϕ_f is unity, so that

$$\int_{\text{area}} \phi_3 \, dA = \frac{A}{24} \begin{bmatrix} b_2 - b_3 \\ a_2 - a_3 \\ b_3 - b_1 \\ a_3 - a_1 \\ b_1 - b_2 \\ a_1 - a_2 \end{bmatrix} \tag{8.175}$$

$$\int_{\text{area}} U^T \, dA = \frac{A}{3} \begin{bmatrix} 1 \\ 0 \\ 0 \\ 1 \\ 0 \\ 0 \\ 1 \\ 0 \\ 0 \end{bmatrix}$$

The first expression gives the release moments applied to the nodes; the second, the statical forces from the applied load. Notice that the premultiplication by the T^T matrix in Eq. (8.174) may modify these reactions. The fixed-end moments are obtained as the negative of the release moments given in Eq. (8.175). (These fixed-end moments are not moments per unit length, and since there are nine undetermined values within the triangle, the effect of the fixed-end forces cannot be uniquely determined. They may be approximated, however, by saying that twisting moments are zero in the fixed-end case.) The joint forces will be the sum of the contributions from all loaded elements.

Example 8.6 Calculation of fixed-end forces for a prismatic beam element

The beam element has been considered in Sec. 8.1 and its cubic interpolation polynomial calculated. It is shown again in Fig. 8.20 in the same coordinate system as used for the plate element in Fig. 8.19. The uniformly distributed load f is applied over the whole length of the member. From Fig. 8.20,

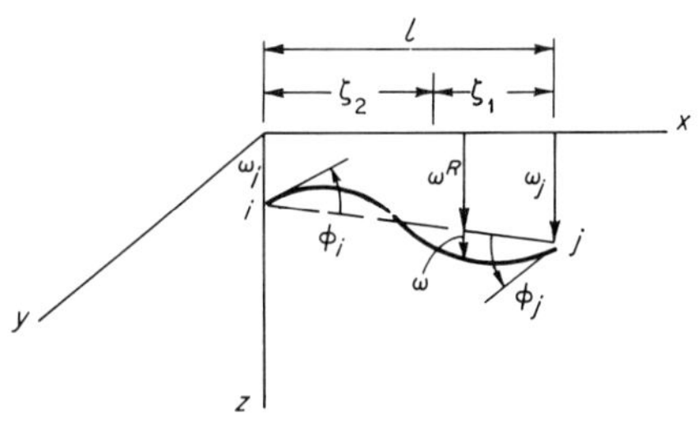

Figure 8.20

$$w^R = [\zeta_1 \quad 0 \quad \zeta_2 \quad 0]\begin{bmatrix} w_i \\ \theta_{iy} \\ w_j \\ \theta_{jy} \end{bmatrix} = Ur$$

From transformation matrices (see Chap. 3),

$$r^* = \begin{bmatrix} \phi_i \\ \phi_j \end{bmatrix} = \begin{bmatrix} -\dfrac{1}{l} & 1 & \dfrac{1}{l} & 0 \\ -\dfrac{1}{l} & 0 & \dfrac{1}{l} & 1 \end{bmatrix}\begin{bmatrix} w_i \\ \theta_{iy} \\ w_j \\ \theta_{jy} \end{bmatrix} = Tr$$

The interpolation polynomial for w^* has been given in Eq. (8.12). That is,

$$w^* = l[-\zeta_1{}^2\zeta_2 \quad \zeta_1\zeta_2{}^2]\begin{bmatrix} \phi_i \\ \phi_j \end{bmatrix} = \phi_3{}^T r^*$$

In this case, Eq. (8.174) becomes

$$R_f = (lT^T \int_o^1 \phi_3 \, d\zeta_2 + \int_o^1 U \, d\zeta_2) f$$

Now,

$$\int_o^1 \phi_3{}^T \, d\zeta_2 = \int_o^1 [-(1-\zeta_2)^2\zeta_2 \quad (1-\zeta_2)\zeta_2{}^2] \, d\zeta_2 = \frac{l}{12}[-1 \quad 1]$$

and

$$\int_o^1 U \, d\zeta_2 = ½\begin{bmatrix} 1 \\ 0 \\ 1 \\ 0 \end{bmatrix}$$

Finally,

$$R_f = \frac{l^2}{12}\begin{bmatrix} 0 \\ -1 \\ 0 \\ 1 \end{bmatrix} f + \frac{l}{2}\begin{bmatrix} 1 \\ 0 \\ 1 \\ 0 \end{bmatrix} f$$

It is seen that the first term gives the release moments at joints i and j, their positive sense being in the clockwise sense about the y axis. Reversed in sign, they will be the fixed-end moments.

8.10 FIELD PROBLEMS; SOLUTIONS TO LAPLACE AND POISSON EQUATIONS

Several physical problems, for example, small deflections of a stretched membrane under transverse load, torsion of closed prismatic members, and flow through porous media, may be represented mathematically by the second-order differential equations of the type of

$$\nabla^2\phi = 0 \qquad \text{Laplace} \tag{8.176}$$

$$\nabla^2\phi = p \qquad \text{Poisson} \tag{8.177}$$

When applied to the membrane problem, the first of these equations may be considered to represent the unloaded membrane with displacements applied to the boundary points, and the second to represent the membrane under transverse pressure p. In the present context of structural analysis, only the membrane and torsion problems are discussed. The membrane problem is interesting not only because of the useful analogy used in the torsion of bars, but also because it is an example of a structure with zero elastic stiffness but considerable geometric stiffness (see Chap. 9). It thus illustrates that elastic stiffness is not necessary for a linear load-deflection relationship for small deflections. It has been found useful to derive the joint equilibrium equations of the finite-element model of the membrane by minimization of the potential-energy function with respect to the node deflections, and this approach will be followed here.

It should be noted that this method of analysis has been used also for the various problems considered thus far (see Ref. 38). Its only difficulty lies in the selection of the correct energy function. This is particularly simple in the present case. Its further appeal lies in the application to field problems (torsion, heat transfer, etc.), in which the concepts of stiffness and flexibility have no real meaning and can only be used in loose analogy with the structures problem.

Differential Equation for the Deflected Membrane; Potential-energy Function

The thin membrane with uniform tension S per unit length and under transverse load p per unit area is shown in Fig. 8.21a. The differential equation

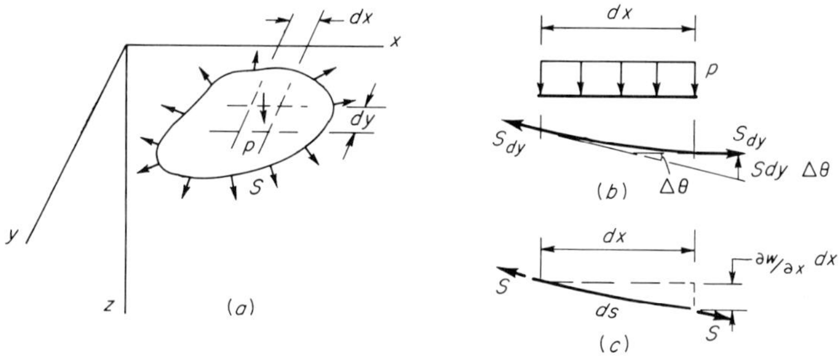

Figure 8.21 (*a*) Loaded membrane; (*b*) forces on the *X* faces of the element *dx dy*; (*c*) element deflected shape.

for the deflected surface *w* of the membrane may be calculated (for small deflections) from consideration of the equilibrium of the infinitesimal area *dx dy*. Forces acting on the *X* faces of the element are shown in Fig. 8.21*b*. It is evident that the vertical component of the inclined forces *S dy* is

$$S_{XV} = -S\,dy\,\Delta\theta$$

Note the negative sign since $\Delta\theta$ is positive and S_{XV} is in the negative *Z* direction. The small angle $\Delta\theta$ is approximately equal to the change in slope of the membrane surface in the *X* direction across the distance *dx*, so that

$$\Delta\theta \simeq \frac{\partial^2 w}{\partial x^2}\,dx$$

Then

$$S_{XV} = -S\frac{\partial^2 w}{\partial x^2}\,dx\,dy$$

In a similar manner the resultant vertical force S_{YV} from the tensions on the *Y* sides of the infinitesimal element is given

$$S_{YV} = -S\frac{\partial^2 w}{\partial y^2}\,dx\,dy$$

For equilibrium of forces in the *Z* direction,

$$-S\frac{\partial^2 w}{\partial x^2}\,dx\,dy - S\frac{\partial^2 w}{\partial y^2}\,dx\,dy + p\,dx\,dy = 0$$

Hence

$$S\frac{\partial^2 w}{\partial x^2} + S\frac{\partial^2 w}{\partial y^2} - p = 0$$

or

$$\nabla^2 w = \frac{p}{S} \tag{8.178}$$

Strain Energy Stored in Membrane

When the membrane is deflected under the load p, the element $dx\,dy$ is stretched so that the constant tension forces S must do work. From Fig. 8.21c, it is seen that the deformed length ds is given by

$$ds = dx\left[1 + \left(\frac{\partial w}{\partial x}\right)^2\right]^{1/2} \simeq dx\left[1 + \frac{1}{2}\left(\frac{\partial w}{\partial x}\right)^2\right]$$

for small deflections. The increase in length is then

$$ds - dx = dx\,\frac{1}{2}\left(\frac{\partial w}{\partial x}\right)^2$$

Hence the work done by the tension forces $S\,dy$ as the membrane is extended is

$$\Delta U_x = S\,dy\,dx\,\frac{1}{2}\left(\frac{\partial w}{\partial x}\right)^2$$

For both the forces $S\,dy$ and $S\,dx$, the total work done is

$$\Delta U = \frac{1}{2}\left[\left(\frac{\partial w}{\partial x}\right)^2 + \left(\frac{\partial w}{\partial y}\right)^2\right]S\,dx\,dy$$

For the whole membrane,

$$U = \frac{1}{2}\iint S\left[\left(\frac{\partial w}{\partial x}\right)^2 + \left(\frac{\partial w}{\partial y}\right)^2\right]dx\,dy$$

The potential energy of the load is

$$V = -\iint pw\,dx\,dy$$

It can be shown that the minimization of the potential-energy function $U + V$ leads to Eq. (8.178). Let

$$U + V = \chi = \frac{1}{2}\iint S\left[\left(\frac{\partial w}{\partial x}\right)^2 + \left(\frac{\partial w}{\partial y}\right)^2\right]dx\,dy - \iint pw\,dx\,dy \tag{8.179}$$

If now the membrane is subdivided into a number of domains (finite elements) for which the displacement w is approximated by a given function in terms of nodal parameters $(w,w, \ldots, w_N)$, χ is then expressed in terms of these parameters. Hence the minimization process reverts to that of differentiation with respect to these parameters, and solution of the resulting simultaneous equations is obtained from the condition

$$\frac{\partial \chi}{\partial w_i} = 0 \qquad i = 1, N \tag{8.180}$$

Notice the similarity between this and the Rayleigh-Ritz procedure in which the parameters used describe functions that are defined over the whole membrane. It should be evident that the finite-element version of the Rayleigh-Ritz procedure should give better approximation where irregular boundaries occur than that method which seeks functions over the whole domain.

St. Venant's Torsion of Closed Tubes

For the prismatic member (whose cross section is shown in Fig. 8.22) subject to a torque M_T producing a twist per unit length θ, it can be shown that a stress function ϕ may be chosen which satisfies Poisson's equation

$$\frac{\partial^2\phi}{\partial x^2} + \frac{\partial^2\phi}{\partial y^2} = -2G\theta \tag{8.181}$$

The stress function ϕ, which is zero on external boundaries, is defined such that the shear stresses in the cross section are

$$\tau_{zx} = \frac{\partial\phi}{\partial y} \qquad \tau_{zy} = -\frac{\partial\phi}{\partial x} \tag{8.182a}$$

It is seen that the analogy between the deflected membrane and the torsion problem is drawn by the equating of p/S to $-2G\theta$. If $p = -2\theta$, then $1/S = G$ gives the element elasticity. That is, if G is small (the material for the element being flexible), then S must be large in the analogous membrane. In the case of the torsion of a hollow shaft, the material of the hole may be considered of G very small, whereas in the membrane, the hole becomes a lid or flat surface for which S is very large.

Minimization Procedure

It has been shown that the function χ in Eq. (8.179) may be expressed in terms of the nodal values $w_1, w_2, \ldots, w_N$ of the finite-element configuration, when of course the displacements within each element are functions of these values.

The contributions to χ from the elements (and thence also to $\partial\chi/\partial x_i$) may

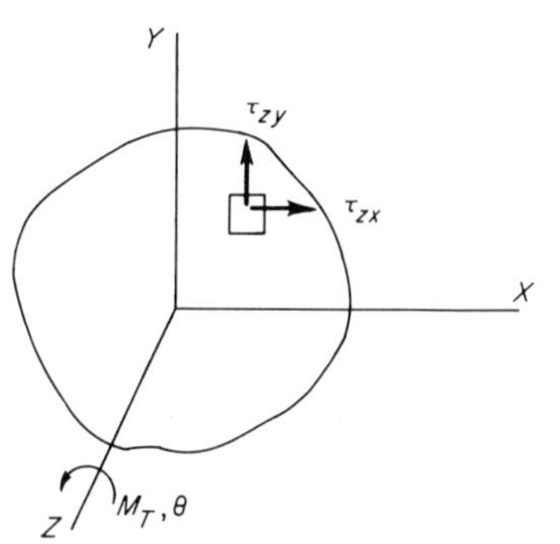

Figure 8.22 Torsion of prismatic member.

thus be calculated separately. It should be evident that for the rth element, the contribution to χ, written χ^r, will be given solely in terms of this element's node values.

Element with Linear Deflections

Consider first the simplest case in which the function w is given in the triangle r with nodes i, j, and k, as a linear function of the values $\langle w_i \quad w_j \quad w_k \rangle$. This is of course equivalent to the constant-strain triangle in the plane stress analysis.

The interpolation for $\tilde{w}(\zeta)$ in r is given by

$$\tilde{w}(\zeta) = \langle w_i \quad w_j \quad w_k \rangle \begin{bmatrix} \zeta_1 \\ \zeta_2 \\ \zeta_3 \end{bmatrix}$$

or by using the definition for ϕ_1 from Eq. (8.65) and

$$\begin{aligned} w^T &= \langle w_i \quad w_j \quad w_k \rangle \\ \tilde{w} &= w^T \phi_1 = {\phi_1}^T w \end{aligned} \tag{8.182b}$$

The contributions to χ from triangle r, χ^r, must be given in terms of w. The derivatives of $\partial \chi^r / \partial w_n$ must also be given in terms of w for $n = i, j, k$ only. Let

$$\left[\frac{\partial \chi^r}{\partial w} \right]^T = \left\langle \frac{\partial \chi^r}{\partial w_i} \quad \frac{\partial \chi^r}{\partial w_j} \quad \frac{\partial \chi^r}{\partial w_k} \right\rangle \tag{8.182c}$$

Each of the derivatives in Eq. (8.182c) may be calculated by differentiation of Eq. (8.179) under the integral sign, the extent of the integration being over the triangle under consideration. That is,

$$\frac{\partial \chi^r}{\partial w_n} = \iint S \left[\frac{\partial \tilde{w}}{\partial x} \frac{\partial (\partial \tilde{w}/\partial x)}{\partial w_n} + \frac{\partial \tilde{w}}{\partial y} \frac{\partial (\partial \tilde{w}/\partial y)}{\partial w_n} \right] dx\, dy - \iint p \frac{\partial \tilde{w}}{\partial w_n} dx\, dy \tag{8.183}$$

The function $\tilde{w}$ is known within the triangle and

$$\begin{aligned} \frac{\partial \tilde{w}}{\partial x} &= \frac{\partial {\phi^T}_1}{\partial x} w = \frac{1}{2A} [b_1 \quad b_2 \quad b_3] w \\ \frac{\partial \tilde{w}}{\partial y} &= \frac{\partial {\phi_1}^T}{\partial y} w = \frac{1}{2A} [a_1 \quad a_2 \quad a_3] w \end{aligned} \tag{8.184}$$

Then,

$$\begin{aligned} \frac{\partial (\partial \tilde{w}/\partial x)}{\partial w_n} &= \frac{1}{2A} b_n \\ \frac{\partial (\partial \tilde{w}/\partial y)}{\partial w_n} &= \frac{1}{2A} a_n \end{aligned} \qquad n = 1, 2, 3, \quad \text{or} \quad i, j, k \tag{8.185}$$

From Eqs. (8.184) and (8.185), it is seen that

$$\frac{\partial \tilde{w}}{\partial x} \frac{\partial(\partial \tilde{w}/\partial x)}{\partial w_n} = \frac{1}{2A}[b_1 \quad b_2 \quad b_3]\begin{bmatrix} w_i \\ w_j \\ w_k \end{bmatrix} \frac{1}{2A} b_n$$

$$= \frac{1}{4A^2}[b_1 b_n \quad b_2 b_n \quad b_3 b_n]w$$

Similarly,

$$\frac{\partial \tilde{w}}{\partial y} \frac{\partial(\partial \tilde{w}/\partial y)}{\partial w_n} = \frac{1}{4A^2}[a_1 a_n \quad a_2 a_n \quad a_3 a_n]w$$

Then, since integration of the constant terms simply multiplies each by the area A,

$$\left[\frac{\partial \chi^r}{\partial w}\right] = \begin{bmatrix} \dfrac{\partial \chi^r}{\partial w_i} \\ \dfrac{\partial \chi^r}{\partial w_j} \\ \dfrac{\partial \chi^r}{\partial w_k} \end{bmatrix} = \frac{S}{4A}\left\{\begin{bmatrix} b_1^2 & b_1 b_2 & b_1 b_3 \\ b_2 b_1 & b_2^2 & b_2 b_3 \\ b_3 b_1 & b_3 b_2 & b_3^2 \end{bmatrix} + \begin{bmatrix} a_1^2 & a_1 a_2 & a_1 a_3 \\ a_2 a_1 & a_2^2 & a_2 a_3 \\ a_3 a_1 & a_3 a_2 & a_3^2 \end{bmatrix}\right\}\begin{bmatrix} w_i \\ w_j \\ w_k \end{bmatrix}$$

That is, by using the definitions for α and γ from Eq. (8.120),

$$\left[\frac{\partial \chi^r}{\partial w}\right] = \frac{S}{4A}[\alpha^T \alpha + \gamma^T \gamma]w \tag{8.186}$$

It remains now in Eq. (8.185) to calculate the load potential term

$$\iint p \frac{\partial \tilde{w}}{\partial w_n}\, dx\, dy$$

From Eq. (8.182*b*),

$$p\begin{bmatrix} \dfrac{\partial \tilde{w}}{\partial w_i} \\ \dfrac{\partial \tilde{w}}{\partial w_j} \\ \dfrac{\partial \tilde{w}}{\partial w_k} \end{bmatrix} = p\phi_1{}^T w$$

Hence,

$$\iint p\left[\frac{\partial \tilde{w}}{\partial \tilde{w}_n}\right] dx\, dy = \frac{pA}{3}\begin{bmatrix} 1 \\ 1 \\ 1 \end{bmatrix} \tag{8.187}$$

Thus in Eqs. (8.186) and (8.187) the contribution to nodes i, j, and k of element r for $\partial\chi^r/\partial w_n$ have been obtained.

When all contributions from all elements have been obtained to the derivatives $\partial\chi/\partial w_n$, where $n = 1, N$, the resulting N equations may be solved for the nodal displacements w. It will be necessary to apply boundary conditions $w = 0$ on s to the equations.

Application to the Torsion of Shafts

If the shaft is solid and of uniform material, the analogy with the membrane poses no problems. Then the equations solved can be written

$$\nabla^2\phi = -2 \tag{8.188}$$

The constant $G\theta$ is obtained from the relationship that the torque M_T is equal to twice the volume under the ϕ surface. That is, by using ϕ from Eq. (8.188),

$$M_T = 2\iint \phi\, dA\, G\theta$$

Hence

$$\theta = \frac{M_T}{2G\iint \phi\, dA} \tag{8.189}$$

Notice that in this case $\iint \phi\, dA = \Sigma$ (area of triangle $\times$ average height).

The shear stresses obtained from the ϕ derivatives calculated in Eq. (8.188) must be multiplied by $G\theta$. That is,

$$\begin{bmatrix} \tau_{zx} \\ \tau_{zy} \end{bmatrix} = G\theta \begin{bmatrix} \dfrac{\partial\phi}{\partial y} \\ -\dfrac{\partial\phi}{\partial x} \end{bmatrix} = \frac{M_T}{2A\iint \phi\, dA} \begin{bmatrix} a_1 & a_2 & a_3 \\ -b_1 & -b_2 & -b_3 \end{bmatrix} \phi \tag{8.190}$$

These shear stresses are constant over a given triangle. Node values may be obtained by averaging the values from triangles intersecting at each node. It will be evident that this constant-stress triangle will suffer a similar defect to the constant-strain triangle for plane stress analysis in that boundary values will be in error since they represent more nearly the stress conditions at the centroids of adjacent triangles.

Application to Hollow Shafts

The membrane analogous to the hollow shaft has a plate covering the opening, with the membrane covering the solid portion of the shaft attached to the plate and the external boundary. Thus the internal boundary has $\phi =$ constant.

Such a shaft is shown in Fig. 8.23. Notice elements must be used to cover the opening. For these elements the plate condition is realized by making their membrane tension S very large, that is $1/S = G$ very small. Then the shaft may be considered composed of two materials with the one filling the hole of very low modulus. It is seen in Eq. (8.183) that for S very large in a triangle, the contribution of the first integral $\int\int \partial\chi^2/\partial w_n$ should be multiplied by a factor such as $10^6 X$ (that used for the membrane triangles).

Element with Quadratic Displacements

In this case the interpolation function used is ϕ_2, and the six nodes i, j, k, l, m, n, that is, apices and midpoints of sides, are used as the defining points. Then

$$\tilde{w}(\zeta) = w^T\phi_2 = \phi_2{}^T w \tag{8.191}$$

where

$$\phi_2{}^T = \langle \zeta_1(2\zeta_1 - 1) \quad \zeta_2(2\zeta_2 - 1) \quad \zeta_3(2\zeta_3 - 1) \quad 4\zeta_1\zeta_2 \quad 4\zeta_2\zeta_3 \quad 4\zeta_3\zeta_1\rangle$$
$$w^T = \langle w_i \quad w_j \quad w_k \quad w_l \quad w_m \quad w_n\rangle$$

The method for generating the derivatives of χ with respect to the node parameters w follows the same scheme as for the linear displacement triangle. Now,

$$\frac{\partial\tilde{w}}{\partial x} = \frac{1}{2A}[b_1(4\zeta_1 - 1) \quad b_2(4\zeta_2 - 1) \quad b_3(4\zeta_3 - 1) \quad 4(\zeta_2 b_1 + \zeta_1 b_2) \quad 4(\zeta_3 b_2 + \zeta_2 b_3) \quad 4(\zeta_3 b_1 + \zeta_1 b_3)]w$$

$$\frac{\partial\tilde{w}}{\partial y} = \frac{1}{2A}[a_1(4\zeta_1 - 1) \quad a_2(4\zeta_2 - 1) \quad a_3(4\zeta_3 - 1) \quad 4(\zeta_2 a_1 + \zeta_1 a_2) \quad 4(\zeta_3 a_2 + \zeta_2 a_3) \quad 4(\zeta_3 a_1 + \zeta_1 a_3)]w$$

Then,

$$\frac{\partial(\partial\tilde{w}/\partial x)}{\partial w_n} = \frac{1}{2A}[b_n(4\zeta_n - 1)] \qquad \text{for} \qquad n = 1, 2, 3$$

$$= \frac{1}{2A}[4(\zeta_{m+1}b_m + \zeta_m b_{m+1})] \qquad n = 4, 5, 6 \quad m = n - 3$$

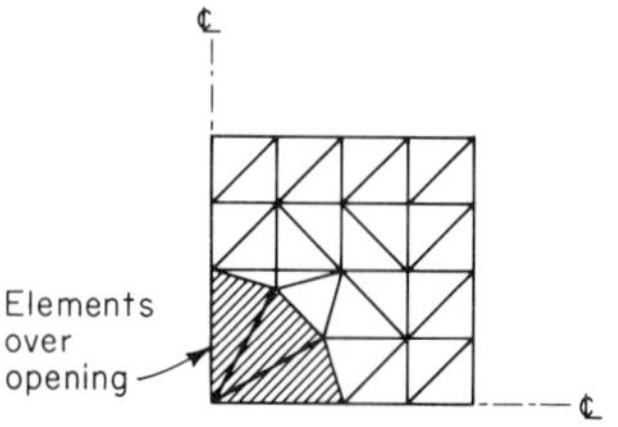

Figure 8.23 Torsion of hollow shaft.

In this expression $\zeta_4 = \zeta_1$, $b_4 = b_1$. Similar expressions are obtained for derivatives of $\partial\tilde{w}/\partial y$, with b replaced by a.

For the triangle, six derivatives with respect to the node displacements $w_i, w_j, \ldots, w_n$ are generated. Integration over the area of the triangle gives the contribution to the first term in Eq. (8.183). This may be written

$$\iint \begin{bmatrix} \dfrac{\partial \chi^r}{\partial w_i} \\ \dfrac{\partial \chi^r}{\partial w_j} \\ \dfrac{\partial \chi^r}{\partial w_k} \\ \dfrac{\partial \chi^r}{\partial w_l} \\ \dfrac{\partial \chi^r}{\partial w_m} \\ \dfrac{\partial \chi^r}{\partial w_n} \end{bmatrix} dA = \frac{1}{12A}[T_b + T_a]w \tag{8.192}$$

In Eq. (8.192), the 6×6 matrices T_b and T_a are generated by replacing $t = b$ and $t = a$ respectively in Eq. (8.193).

$$T = \begin{bmatrix} 3t_1^2 & -t_1t_2 & -t_1t_3 & 4t_1t_2 & 0 & 4t_1t_3 \\ & 3t_2^2 & -t_2t_3 & 4t_1t_2 & 4t_2t_3 & 0 \\ & & 3t_3^2 & 0 & 4t_2t_3 & 4t_1t_3 \\ & & & 8(t_1^2 + t_2^2 + t_1t_2) & 8t_1t_3 & 8t_2t_3 \\ & \text{symmetrical} & & & 8(t_2^2 + t_3^2 + t_2t_3) & 8t_1t_2 \\ & & & & & 8(t_3^2 + t_1^2 + t_1t_3) \end{bmatrix} \tag{8.193}$$

The load potential term is given by

$$\iint p\begin{bmatrix}\dfrac{\partial \tilde{w}}{\partial w_i}\\[2mm] \dfrac{\partial \tilde{w}}{\partial w_j}\\[2mm] \dfrac{\partial \tilde{w}}{\partial w_k}\\[2mm] \dfrac{\partial \tilde{w}}{\partial w_l}\\[2mm] \dfrac{\partial \tilde{w}}{\partial w_m}\\[2mm] \dfrac{\partial \tilde{w}}{\partial w_n}\end{bmatrix} dA = \iint p\begin{bmatrix}\zeta_1(2\zeta_1-1)\\ \zeta_2(2\zeta_2-1)\\ \zeta_3(2\zeta_3-1)\\ 4\zeta_1\zeta_2\\ 4\zeta_2\zeta_3\\ 4\zeta_3\zeta_1\end{bmatrix} dA = \frac{pA}{3}\begin{bmatrix}0\\0\\0\\1\\1\\1\end{bmatrix} \tag{8.194}$$

Calculation of the Torsion Constant and Shear Stresses

The procedure followed is identical with that given in Eqs. (8.188)ff. The integral $\iint \phi\, dA$ for the section is calculated

$$\iint \phi\, dA = \sum \text{(volume under one triangle)}$$

Since $\tilde{\phi}$ varies quadratically in this case, the volume under a triangle may be calculated from the interpolation function ϕ_2. That is,

$$\tilde{\phi} = \phi_2{}^T\phi$$

where

$$\phi^T = \langle \phi_i \quad \phi_j \quad \phi_k \quad \phi_l \quad \phi_m \quad \phi_n \rangle$$

Then,

$$\iint \tilde{\phi}\, dA = \frac{A}{3}(0 \quad 0 \quad 0 \quad 1 \quad 1 \quad 1)\phi = \frac{A}{3}(\phi_l + \phi_m + \phi_n)$$

Stresses vary linearly within each triangle, and are given by

$$\begin{bmatrix}\tau_{zx}\\ \tau_{zy}\end{bmatrix} = G\theta\begin{bmatrix}\dfrac{\partial\phi}{\partial y}\\[2mm] -\dfrac{\partial\phi}{\partial x}\end{bmatrix}$$

$$= \frac{M_T}{2A\iint\phi\, dA}\begin{bmatrix} a_1(4\zeta_1-1) & a_2(4\zeta_2-1) & a_3(4\zeta_3-1) & 4(\zeta_2a_1+\zeta_1a_2) & 4(\zeta_3a_2+\zeta_2a_3) & 4(\zeta_3a_1+\zeta_1a_3)\\ -b_1(4\zeta_1-1) & -b_2(4\zeta_2-1) & -b_3(4\zeta_3-1) & -4(\zeta_2b_1+\zeta_1b_2) & -4(\zeta_3b_2+\zeta_2b_3) & -4(\zeta_3b_1+\zeta_1b_3)\end{bmatrix}\phi \tag{8.195}$$

Node values are obtained by averaging values from contributing triangles.

9 NONLINEAR ASPECTS OF STRUCTURE BEHAVIOR

9.1 CAUSES OF NONLINEARITY IN THE LOAD–DEFLECTION RELATIONSHIP

It has been mentioned in Chap. 2, Fig. 2.1, that the linear elastic theory developed so far assumes that material properties are ideally elastic and that geometry changes do not alter member flexibility or stiffness or overall structure behavior. In Chaps. 4 and 6 expressions are developed for the nonlinear effects of axial force in a frame member on its bending flexibility and stiffness [see Eqs. (4.190) and (6.100)].

The analytic expressions were obtained for both axial tension and compression. Alternatively, an approximate expression is given [see Chap. 6, Eq. (6.106)] for the modification to member bending stiffness, assuming that the deflected shape is a cubic curve. Since the axial forces to be determined enter into the deflection calculations of the analysis, the procedure followed must be either a step by step or an iterative one.

When a structure, such as an arch or suspension cable, with a large dead load is given a load increment, it will be generally incorrect to base deflections solely on the elastic properties of the structure, even though deflections are small and the behavior over the load range is essentially linear. For example, the cable shown in Fig. 9.1a has an axial tension T of such a magnitude that for practical purposes the cable may be considered straight.

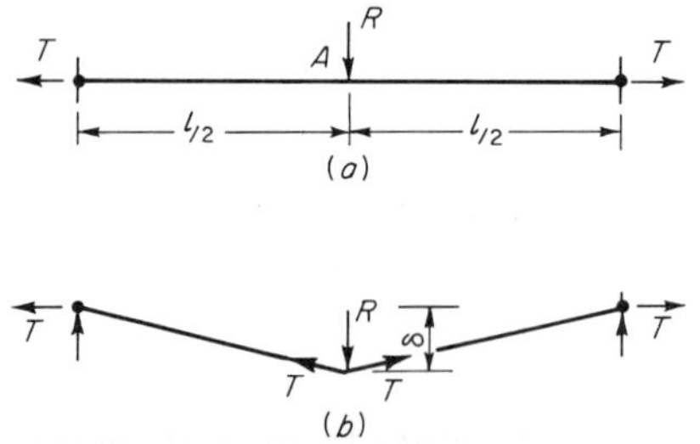

Figure 9.1 Deflected cable.

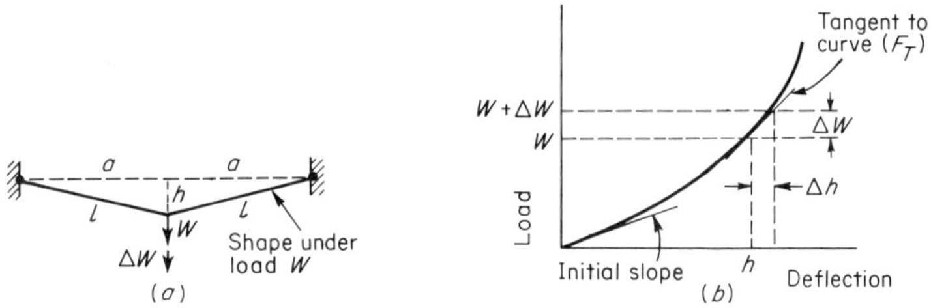

Figure 9.2 Tangent flexibility.

In Fig. 9.1*b* the deflected shape required to carry the transverse load R is shown, and if deflections are small,

$$R = \frac{4T}{l}\delta$$

Thus a linear relationship exists between deflection and transverse force even though the geometric configuration gives the elastic stiffness at A to be zero. When the cable is as shown in Fig. 9.2, then both elastic and geometric stiffness are available at A, *geometric stiffness* being the term used to describe the stiffness produced by the axial forces. For load increments at A both stiffnesses should be considered in order to obtain the increment of deflection. From Fig. 9.2*b*,

$$\Delta v = F_T \, \Delta W$$

where F_T is the tangent flexibility at load level W.

Changes in geometry of the structure are manifested in ways other than modification of member stiffness due to axial force. For example, suppose that the slender cantilever beam in Fig. 9.3*a* is composed of a material with a high-yield point so that when the load R is applied its deflected shape is that shown in Fig. 9.3*b*. Evidently in this case it would be a gross error to assume that from Fig. 9.3*a*,

$$M_x = R(l - x)$$

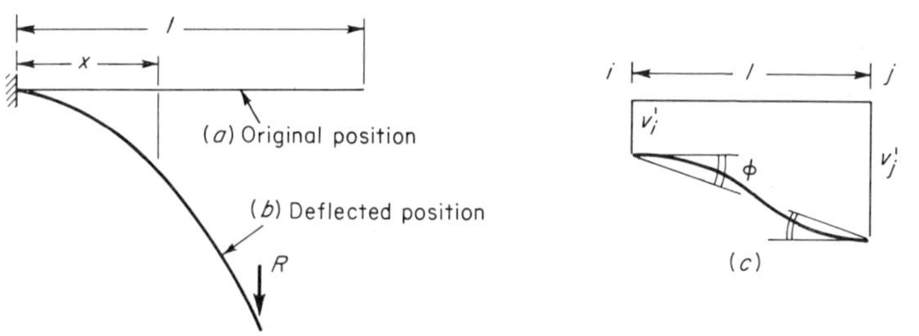

Figure 9.3 Beam deflection.

Also from Fig. 9.3*c* it should be evident that to apply in the case of large deflections v_i', v_j' the Eq. (3.68) for calculating the chord rotation ϕ will cause considerable error. That is, for large displacements,

$$\phi \neq \frac{1}{l}(v_j' - v_i')$$

The analysis is further complicated when the material yield point is exceeded at points of the structure. Some types of material behavior are shown in Fig. 9.4.

In Fig. 9.4*a* the curve given is that generally assumed for mild steel, with a perfectly elastic portion followed by a perfectly plastic plateau which is terminated when strain hardening commences. The curve in Fig. 9.4*b* has a perfectly elastic zone followed by a plastic zone in which linear strain hardening takes place. The curves *A* and *B* in Fig. 9.4*c* are for a rigid plastic material. The material *A* exhibits nonlinear strain hardening, whereas *B* is perfectly plastic. It should be noted that the rigid-plastic curve is that assumed in the limit theorems and their use for calculation of plastic-collapse loads of structures. The material of the type in Fig. 9.4*d* is nonlinear in its behavior, both in the elastic and plastic states. Some aspects of elastoplastic analysis will be considered in Sec. 9.4. In Secs. 9.2 and 9.3 only the effects of geometry changes on elastic performance are discussed.

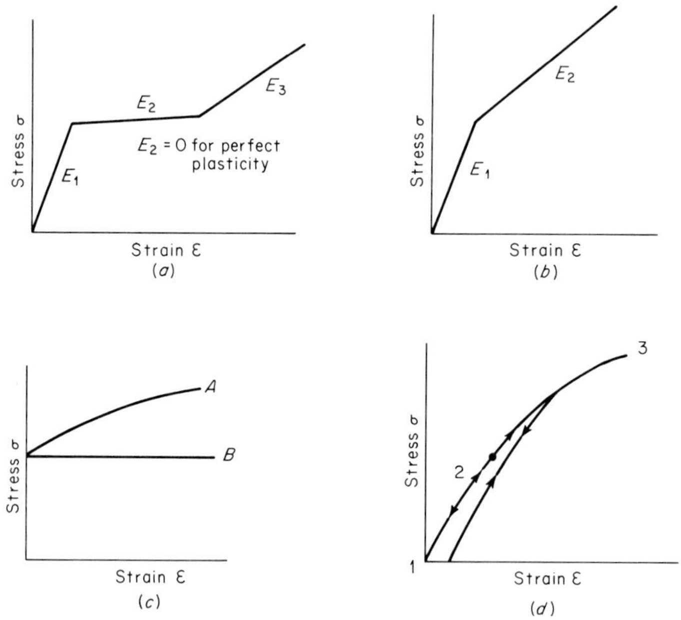

Figure 9.4 Material stress-strain curves.

9.2 NONLINEAR ANALYSIS BY THE FORCE METHOD; LARGE DISPLACEMENTS; AXIAL FORCES

The method used here (see Ref. 35) adopts the general linear equations by introducing initial member strains and fictitious node forces to allow for nonlinear effects. The general linear equations may be expressed for:

1 Member forces

$$S = bR + \beta H \tag{9.1}$$

In Eq. (9.1), R and H are the applied forces, and the initial member strains respectively [see Eq. (5.50) in which $\bar{v} \equiv H$]. The matrix β has been derived in Eq. (5.53) and is given by

$$\beta = -b_1 D_{11}^{-1} b_1^T \tag{9.2}$$

2 Node deflections

$$r = FR + b^T H \tag{9.3}$$

The matrices β, F, and b are all obtained as for the small deflection theory. It should be noted, however, that they may not be based on the original structure geometry, but rather on the geometry existing prior to the current load increment.

To allow for nonlinear effects, the fictitious (or *ersatz,* see Ref. 35) node forces R_i and initial member strains H_i are introduced. That is, the structure is analyzed with the original geometry and loads together with R_i and H_i. However, R_i and H_i are functions of the member forces and the node displacements, so that the modified equations are nonlinear. Thus Eqs. (9.1) and (9.3) are written

$$S = b(R + R_i) + \beta(H + H_i) \tag{9.4}$$

$$r = F(R + R_i) + b^T(H + H_i) \tag{9.5}$$

The nature of R_i and H_i may be determined by considering the simple two-bar structure shown in Fig. 9.5. The first effect to be ascertained is that of the change in geometry on the equilibrium of the joint A.

Suppose that S_o is the member force in the original position (calculated from R on the assumption of no geometry change) and that it remains constant as the member rotates through the angle $\Delta\phi$ as shown in Fig. 9.5*b*. The vector change ΔS_o in S_o is shown in Fig. 9.5*c* and thence,

$$\Delta S_o = S_o \, \Delta\phi = \frac{S_o r \cos\phi}{l} \tag{9.6}$$

Let

$$\cos\phi = c \qquad \sin\phi = s \qquad \text{and} \qquad \frac{r}{y} = \rho$$

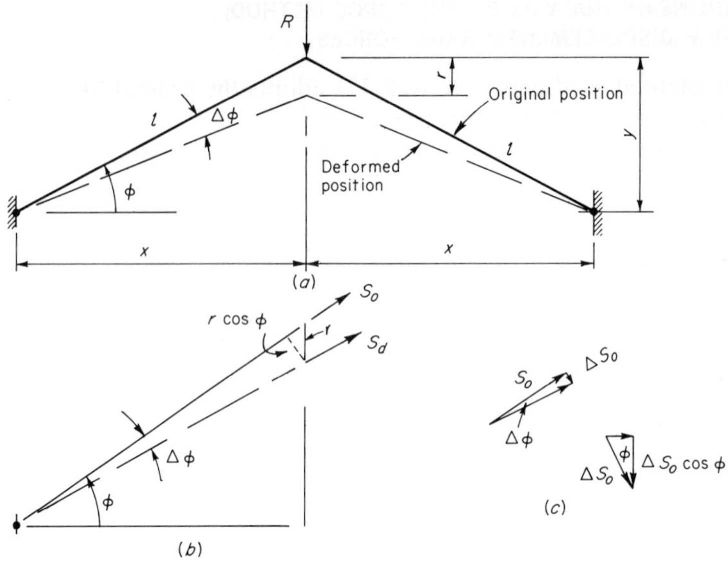

Figure 9.5 Large deflections. Forces.

Then

$$S_o = \frac{-1}{2s} R \qquad \text{hence} \qquad b = -\frac{1}{2s}$$

Substitution in Eq. (9.6) gives

$$\Delta S_o = -\frac{1}{2s}\frac{rc}{l} R = -\frac{1}{2}\rho c R$$

The vertical component of ΔS_o is written ΔR_i,

$$\Delta R_i = \Delta S_o \cos\phi = -\frac{1}{2}\rho c^2 R$$

For both members,

$$R_i = -\rho c^2 R \tag{9.7}$$

Alternatively,

$$R_i = 2\rho s c^2 S_o \tag{9.8}$$

In this derivation the same assumption has been made as in Fig. 9.1, that the deflection r is small, so that the above geometric approximations are valid. If large displacements occur for a given load, the structure must be analyzed for a number of load increments when this type of approximation is used.

Notice that in Eq. (9.7) or (9.8), R_i is a linear function of R or S_o and also of the geometry change r.

Consider now the shortening of the members in Fig. 9.5 and the effect on r when the small deflection transformation

$$r = b^T v$$

is no longer valid.

In Fig. 9.6 it is seen that when the member is shortened by the amount v, the vertical deflection r is produced by the rotation about end B. In order that the small-deflection equation may be used, the additional strain H_i must be added so that

$$r = b^T(v + H_i)$$

In Fig. 9.6, let the compressive strain $H_i = -X$, so that

$$(2l_d - X)X = (rc)^2$$

or

$$X \approx \frac{1}{2l_d} r^2c^2$$

Then

$$H_i \approx -\frac{r^2c^2}{2l_d} = \frac{y^2\rho^2c^2}{2l_d}$$

If the approximation $l_d \approx l$ is valid,

$$H_i \approx \frac{1}{2} ls^2\rho^2c^2 \tag{9.9}$$

It is seen now that for the displacement r relatively small, the initial strains H_i are quadratic functions of the geometry change. In matrix form, Eq. (9.8) will be written

$$R_i = a_i{}^T S \tag{9.10}$$

Then Eq. (9.4) becomes

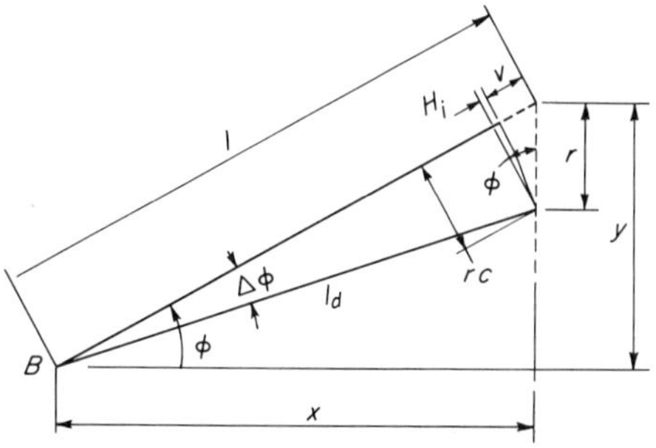

Figure 9.6 Large deflection. Distortions.

$$S = b(R + a_i^T S) + \beta(H + H_i) \tag{9.11}$$

Thence,

$$S = (I - ba_i^T)^{-1} bR + (I - ba_i^T)^{-1}\beta(H + H_i) \tag{9.12}$$

It is seen that

$$b_d = (I - ba_i^T)^{-1} b \tag{9.13}$$

is the force transformation matrix allowing for the R_i forces.

$$S = b_d R + (I - ba_i^T)^{-1}\beta(H + H_i) \tag{9.14}$$

Then from Eq. (9.10),

$$R_i = a_i^T(I - ba_i^T)^{-1}[bR + \beta(H + H_i)] \tag{9.15}$$

Introduction of Eq. (9.15) into Eq. (9.5) leads to a nonlinear matrix equation in the unknown displacement vector r,

$$r = FR + Fa_i^T(I - ba_i^T)^{-1}[bR + \beta(H + H_i)] + b^T(H + H_i)$$

or by using the expansion for the inverse of $(I - ba_i^T)^{-1}$,

$$r = FR + Fa_i^T[I + ba_i^T + (ba_i^T)^2 + \cdots] \times [bR + \beta(H + H_i)] + b^T(H + H_i) \tag{9.16}$$

In Eq. (9.16), let

$$a_i^T b = \phi_r \qquad \text{and} \qquad a_i^T \beta = \phi_H$$

Substitution for ϕ_r and ϕ_H in Eq. (9.16) gives

$$r = F(I + \phi_r + \phi_r + \cdots)R + [b^T + F(I + \phi_r + \phi_r^2 + \cdots)\phi_H](H + H_i) \tag{9.17}$$

Now $R_i = a_i^T S$ and $S = bR$, so that $a_i^T b$ is a matrix of fictitious loads resulting from the linear element forces produced by unit external loads. Furthermore, $a_i^T \beta$ is a matrix of fictitious loads resulting from the linear element forces produced by unit values of the element deformations. As an approximation, delete all powers of ϕ_r higher than the first and all products $\phi_r^n \phi_H$; then from Eq. (9.17)

$$r = F(I + \phi_r)R + (b^T + F\phi_H)(H + H_i) \tag{9.18}$$

The matrices $\phi_r R$ and $\phi_H H$ involve the deflections r; therefore, for the first approximation as in Eq. (9.7), assume that they are linear in r so that

$$\phi_r R = k_{ir} r \qquad \text{and} \qquad \phi_H H = k_{iH} r \tag{9.19}$$

Then by making these substitutions in Eq. (9.18),

$$r = FR + Fk_{ir} r + b^T(H + H_i) + Fk_{iH} r + F\phi_H H_i$$

Collecting terms involving r,

$$[I - F(k_{ir} + k_{iH})]r = (FR + b^T H) + (b^T + F\phi_H)H_i \tag{9.20}$$

For small deflections, $H_i = 0$, so that

$$r = [I - F(k_{ir} + k_{iH})]^{-1}(FR + b^T H) \tag{9.21}$$

When the matrix to be inverted tends to zero, the deflections tend to infinity. Hence the small-deflection buckling conditions and modes are defined by the equation

$$[I - F(k_{ir} + k_{iH})]r = 0 \tag{9.22}$$

The condition

$$(I - Fk_{ir})r = 0 \tag{9.23}$$

governs the load buckling, and

$$(I - Fk_{iH})r = 0 \tag{9.24}$$

the thermal buckling of the structure.

Example 9.1 Calculation of critical load for column

The column of length l with EI constant is shown in Fig. 9.7. The load P on the free end causes the member force $S = P$ in segments 1 and 2 of the column. The matrix F is the influence-coefficient matrix for transverse loads at nodes 1 and 2, and it is easily shown that

$$F = \frac{l^3}{24EI}\begin{bmatrix} 8 & 5/2 \\ 5/2 & 1 \end{bmatrix}$$

Now, k_{ir} is a matrix of fictitious loads resulting from unit deflections. From Fig. 9.7*b* it is seen that

$$k_{ir} = \frac{2P}{l}\begin{bmatrix} 1 & -1 \\ -1 & 2 \end{bmatrix}$$

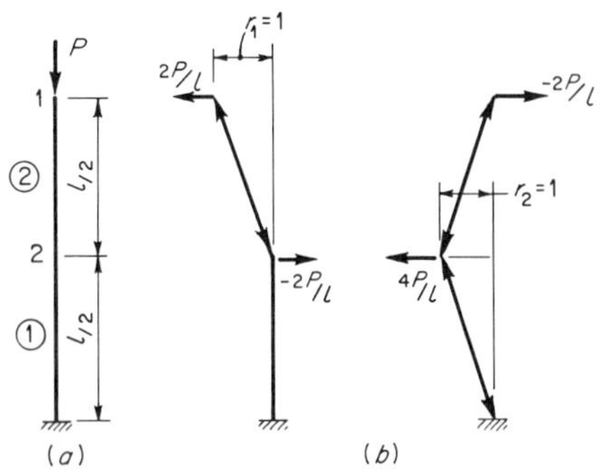

Figure 9.7 Column buckling.

This expression is derived as follows:

$$\begin{bmatrix} R_{1i} \\ R_{2i} \end{bmatrix} = \frac{2}{l}\begin{bmatrix} r_1 & -r_2 \\ -r_1 & 2r_2 \end{bmatrix}\begin{bmatrix} S_1 \\ S_2 \end{bmatrix} = \frac{2}{l}\begin{bmatrix} r_1 & -r_2 \\ -r_2 & 2r_2 \end{bmatrix}\begin{bmatrix} 1 \\ 1 \end{bmatrix} P$$

Substitution for F and k_{ir} in Eq. (9.23) gives

$$\left\{ I - \frac{Pl^2}{24EI}\begin{bmatrix} 11 & -6 \\ 3 & -1 \end{bmatrix} \right\} r = 0 \tag{i}$$

For $r \neq 0$ the determinant of the coefficient matrix must be zero, so that

$$\begin{vmatrix} \lambda - 11 & 6 \\ -3 & \lambda + 1 \end{vmatrix} = 0$$

where $\lambda = (24EI/Pl^2)$ and the values of λ are $\lambda = \pm 5\sqrt{18}$. It is seen that the largest value of λ gives the smallest value of P, so that

$$P_{cr} = \frac{2.61EI}{l^2}$$

An interesting alternative approach is available to set up the eigenvalue equation (9.23). Suppose that the column in Fig. 9.8, initially straight, has deflected under the axial force P, so that the deflections (r_1, r_2) are produced from moments due to P. From Fig. 9.8, the bending moments in members (1) and (2) are given by

$$\begin{bmatrix} M_{1i} \\ M_{1j} \\ M_{2i} \\ M_{2j} \end{bmatrix} = \begin{bmatrix} r_1 \\ r_1 - r_2 \\ r_1 - r_2 \\ 0 \end{bmatrix} P = \begin{bmatrix} 1 & 0 \\ 1 & -1 \\ 1 & -1 \\ 0 & 0 \end{bmatrix} rP = BrP$$

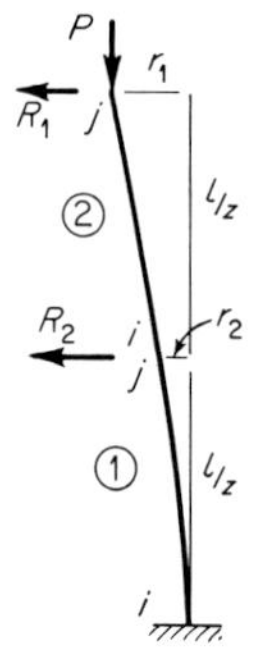

Figure 9.8 Column bending caused by axial load.

The expression for deflections r, calculated from the member bending moments is

$$r = b^T f S = (b^T f B r) P$$

Hence

$$[I - P(b^T f B)]r = 0 \tag{ii}$$

The b matrix is given by

$$\begin{bmatrix} M_{1i} \\ M_{1j} \\ M_{2i} \\ M_{2j} \end{bmatrix} = \begin{bmatrix} l & \frac{l}{2} \\ \frac{l}{2} & 0 \\ \frac{l}{2} & 0 \\ 0 & 0 \end{bmatrix} \begin{bmatrix} R_1 \\ R_2 \end{bmatrix}$$

$$f = \frac{l}{12EI} \begin{bmatrix} 2 & 1 & 0 & 0 \\ 1 & 2 & 0 & 0 \\ 0 & 0 & 2 & 1 \\ 0 & 0 & 1 & 2 \end{bmatrix}$$

Then,

$$b^T f = \frac{l^2}{24EI} \begin{bmatrix} 2 & 1 & 1 & 0 \\ 1 & 0 & 0 & 0 \end{bmatrix} \begin{bmatrix} 2 & 1 & 0 & 0 \\ 1 & 2 & 0 & 0 \\ 0 & 0 & 2 & 1 \\ 0 & 0 & 1 & 2 \end{bmatrix}$$

$$= \frac{l^2}{24EI} \begin{bmatrix} 5 & 4 & 2 & 1 \\ 2 & 1 & 0 & 0 \end{bmatrix}$$

Finally,

$$b^T f B = \frac{l^2}{24EI} \begin{bmatrix} 5 & 4 & 2 & 1 \\ 2 & 1 & 0 & 0 \end{bmatrix} \begin{bmatrix} 1 & 0 \\ 1 & -1 \\ 1 & -1 \\ 0 & 0 \end{bmatrix} = \frac{l^2}{24EI} \begin{bmatrix} 11 & -6 \\ 3 & -1 \end{bmatrix} \tag{iii}$$

When (*iii*) is substituted in Eq. (*ii*), it is seen that the same eigenvalue equation has been obtained. Notice also, that the bending moments may be obtained from the ersatz forces R_i as follows:

$$\begin{bmatrix} M_{1i} \\ M_{1j} \\ M_{2i} \\ M_{2j} \end{bmatrix} = \begin{bmatrix} l & \frac{l}{2} \\ \frac{l}{2} & 0 \\ \frac{l}{2} & 0 \\ 0 & 0 \end{bmatrix} \frac{2P}{l} \begin{bmatrix} r_1 - r_2 \\ -r_1 + r_2 \end{bmatrix} = P \begin{bmatrix} 2 & 1 \\ 1 & 0 \\ 1 & 0 \\ 0 & 0 \end{bmatrix} \begin{bmatrix} 1 & -1 \\ -1 & 2 \end{bmatrix} \begin{bmatrix} r_1 \\ r_2 \end{bmatrix}$$

On multiplication of the two matrices in this equation, the same result is obtained as previously.

Example 9.2 Calculation of fictitious node forces for a truss member with axial force

In the truss example in Fig. 9.5, the force R_i to be applied at joint A was found to be

$$R_i = -2\rho sc^2 S_o$$

When the member forms part of a truss system X and Y, force components will be generated at both ends of the member owing to the rotation of the member ($\delta\alpha$).

Consider the member ij shown in Fig. 9.9 whose end displacements are

$$r^T = \langle r_{ix} \quad r_{iy} \quad r_{jx} \quad r_{jy} \rangle \tag{9.25}$$

When rigid-body motions are subtracted from these displacements, the displacement of end j relative to end i will be given by the components $\langle \delta x \quad \delta y \rangle$ such that

$$\begin{bmatrix} \delta x \\ \delta y \end{bmatrix} = \begin{bmatrix} -1 & 0 & 1 & 0 \\ 0 & -1 & 0 & 1 \end{bmatrix} \begin{bmatrix} r_{ix} \\ r_{iy} \\ r_{jx} \\ r_{jy} \end{bmatrix} \tag{9.26}$$

From Fig. 9.9b, the rotation $\delta\alpha$ measured in a counterclockwise direction is found to be

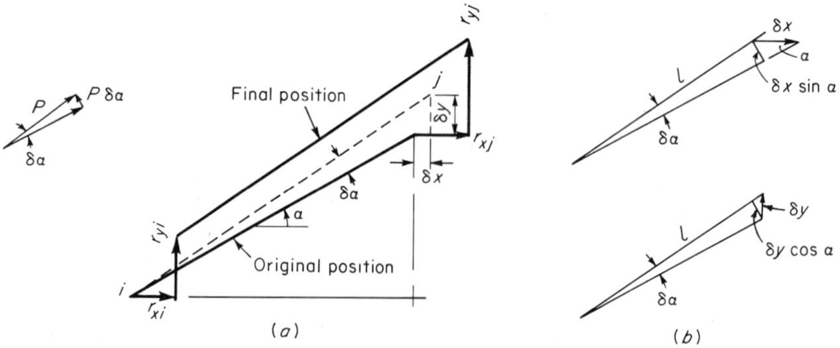

Figure 9.9 Member displacement and rotation. (a) End displacements of member; (b) rotation for displacements δx, δy.

$$\delta\alpha = \frac{1}{l}[-\sin\alpha \quad \cos\alpha]\begin{bmatrix}\delta x \\ \delta y\end{bmatrix} \tag{9.27}$$

Combining Eqs. (9.26) and (9.27),

$$\delta\alpha = \frac{1}{l}[\sin\alpha \quad -\cos\alpha \quad -\sin\alpha \quad \cos\alpha]r \tag{9.28}$$

Let P be the axial force in the member. The components of δP at end j for a rotation $\delta\alpha$ are

$$\begin{bmatrix}\Delta X \\ \Delta Y\end{bmatrix} = P\begin{bmatrix}-\sin\alpha \\ \cos\alpha\end{bmatrix}\delta\alpha \tag{9.29}$$

For ends i and j of the member,

$$\begin{bmatrix}\Delta X_i \\ \Delta Y_i \\ \Delta X_j \\ \Delta Y_j\end{bmatrix} = P\begin{bmatrix}\sin\alpha \\ -\cos\alpha \\ -\sin\alpha \\ \cos\alpha\end{bmatrix}\delta\alpha$$

When the value of $\delta\alpha$ is substituted from Eq. (9.28), it is found that the force increments are given by

$$\begin{bmatrix}\Delta X_i \\ \Delta Y_i \\ \Delta X_j \\ \Delta Y_j\end{bmatrix} = \frac{P}{l}\begin{bmatrix}s^2 & -cs & -s^2 & cs \\ -cs & c^2 & cs & -c^2 \\ -s^2 & cs & s^2 & -cs \\ cs & -c^2 & -cs & c^2\end{bmatrix}\begin{bmatrix}r_{ix} \\ r_{iy} \\ r_{jx} \\ r_{jy}\end{bmatrix} \tag{9.30}$$

In Eq. (9.30),

$$c = \cos\alpha \qquad s = \sin\alpha$$

The general application of Eqs. (9.30) in Eqs. (9.18), etc., is not pursued further here. It should be noted that Eq. (9.7) is obtained from Eq. (9.29) by the extraction of the appropriate row and column value of the coefficient matrix. The above equations are used in the analysis of the free cable given in Sec. 9.3.

9.3 ANALYSIS OF A PLANAR CABLE STRUCTURE

The cable, assumed planar, is shown in Fig. 9.10 under the influence of the loads R_1 and R_2. In the first instance the cable is assumed weightless. The self-weight may be accounted for by increasing the number of segments and applying segment dead loads as equivalent node forces.

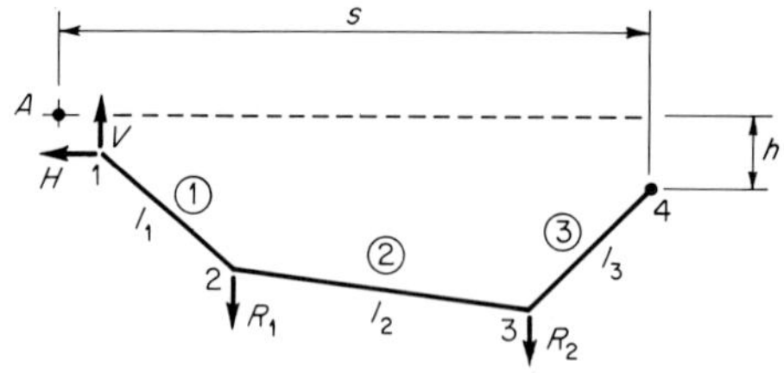

Figure 9.10 Cable analysis.

If the unstrained lengths l_{30}, l_{20}, etc., are known, and the end forces H and V are assumed, the position at end 1 may be calculated. If this differs from the required location of the support A, H and V must be modified in such a way as to bring 1 into coincidence with A. This is achieved in a manner that calculates flexibility coefficients at 1, including the effects of axial forces in the cable members.

The first problem then is to calculate the tangent flexibility of the cable member with axial force P and components $\langle X \quad Y\rangle$ for small variations $\langle \Delta X \quad \Delta Y\rangle$.

The member is shown in Fig. 9.11. That is, it is required to find F_T, the tangent-flexibility matrix, such that

$$\begin{bmatrix} \delta x \\ \delta y \end{bmatrix} = F_T \begin{bmatrix} \Delta X \\ \Delta Y \end{bmatrix}$$

Equation (9.27) may be written

$$[-\sin\alpha \quad \cos\alpha]\begin{bmatrix} -\sin\alpha \\ \cos\alpha \end{bmatrix}\delta\alpha = \frac{1}{l}[-\sin\alpha \quad \cos\alpha]\begin{bmatrix} \delta x \\ \delta y \end{bmatrix}$$

Hence

$$\begin{bmatrix} \delta x \\ \delta y \end{bmatrix}_\alpha = l\begin{bmatrix} -\sin\alpha \\ \cos\alpha \end{bmatrix}\delta\alpha \tag{9.31}$$

If both sides of Eq. (9.29) are multiplied by $[-\sin\alpha \quad \cos\alpha]$, the expression

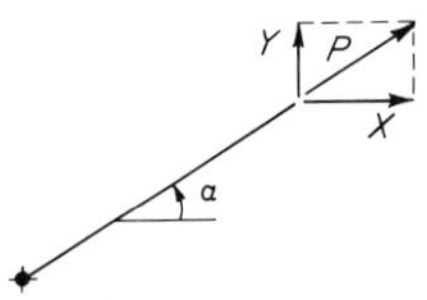

Figure 9.11 Cable element.

is obtained for $\delta\alpha$ in terms of the changes in the components of P. That is,

$$\delta\alpha = \frac{1}{P}[-\sin\alpha \quad \cos\alpha]\begin{bmatrix}\Delta X \\ \Delta Y\end{bmatrix} \tag{9.32}$$

Substitution for $\delta\alpha$ in Eq. (9.31) gives

$$\begin{bmatrix}\delta x \\ \delta y\end{bmatrix}_\alpha = \frac{l}{P}\begin{bmatrix}\sin^2\alpha & -\cos\alpha\sin\alpha \\ -\cos\alpha\sin\alpha & \cos^2\alpha\end{bmatrix}\begin{bmatrix}\Delta X \\ \Delta Y\end{bmatrix} \tag{9.33}$$

That is, if the cable rotates to cause $[\Delta X \quad \Delta Y]$ variation in the components of P, Eq. (9.33) gives the $[\delta x \quad \delta y]$ changes in the member projections.

The elastic flexibility may be calculated for the variation ΔP in P, caused by $[\Delta X \quad \Delta Y]$ based on original geometry.

The variation in member force P causes the change in length v,

$$v = \frac{l\,\Delta P}{EA}$$

The X and Y components are

$$\begin{bmatrix}\delta x \\ \delta y\end{bmatrix}_e = \begin{bmatrix}\cos\alpha \\ \sin\alpha\end{bmatrix} v = \frac{l}{EA}\begin{bmatrix}\cos\alpha \\ \sin\alpha\end{bmatrix}\Delta P \tag{9.34}$$

Now, the components of ΔP are

$$\Delta X = \Delta P\cos\alpha$$
$$\Delta Y = \Delta P\sin\alpha$$

Substitution in Eq. (9.34) gives

$$\begin{bmatrix}\delta x \\ \delta y\end{bmatrix}_e = \frac{l}{EA}\begin{bmatrix}\Delta X \\ \Delta Y\end{bmatrix} = \begin{bmatrix}\dfrac{l}{EA} & 0 \\ 0 & \dfrac{l}{EA}\end{bmatrix}\begin{bmatrix}\Delta X \\ \Delta Y\end{bmatrix} \tag{9.35}$$

The expression in Eq. (9.35) is the elastic flexibility relationship. The total effect on the changes in projections of l is obtained by adding the results of Eqs. (9.33) and (9.35). That is,

$$\begin{bmatrix}\delta x \\ \delta y\end{bmatrix} = \begin{bmatrix}\delta x \\ \delta y\end{bmatrix}_\alpha + \begin{bmatrix}\delta x \\ \delta y\end{bmatrix}_e = \begin{bmatrix}\dfrac{l}{P}\sin^2\alpha + \dfrac{l}{EA} & -\dfrac{l}{P}\cos\alpha\sin\alpha \\ -\dfrac{l}{P}\cos\alpha\sin\alpha & \dfrac{l}{P}\cos^2\alpha + \dfrac{l}{EA}\end{bmatrix}\begin{bmatrix}\Delta X \\ \Delta Y\end{bmatrix} \tag{9.36}$$

If, then, H and V are varied at node 1 and produce variations of components $[\Delta X \quad \Delta Y]$ in the cable members, the displacements at 1 will be, from Eq. (9.36),

$$\begin{bmatrix} \delta u \\ \delta v \end{bmatrix} = \sum_{1,\ \text{No. of members}} \begin{bmatrix} \delta x \\ \delta y \end{bmatrix}$$

Cable Forces P

If the values for H and V are assumed, a recurrence relationship can be established to determine the cable tensions P. Note that H is constant for every member of the cable. From Figs. 9.10 and 9.12 it is seen that the vertical components of force in cable members are given by

$$V_1 = V$$

and for $i \neq 1$,

$$V_i = V_{i-1} - R_i = V - \sum_{j=1,i-1} R_j \tag{9.37}$$

The cable tension for member i is

$$P = (H^2 + V_i^2)^{1/2} \tag{9.38}$$

The member position is determined from

$$\sin \alpha_i = \frac{V_i}{P_i} \qquad \cos \alpha_i = \frac{H}{P_i} \tag{9.39}$$

The strained length is

$$l'_i = l_i\left(1 + \frac{P_i}{EA}\right) \tag{9.40}$$

The projections on the coordinate axes are

$$\begin{aligned} x_i &= l'_i \cos \alpha_i \\ y_i &= l'_i \sin \alpha_i \end{aligned} \tag{9.41}$$

If H and V are correctly chosen,

$$s = \Sigma x_i$$
$$h = \Sigma y_i$$

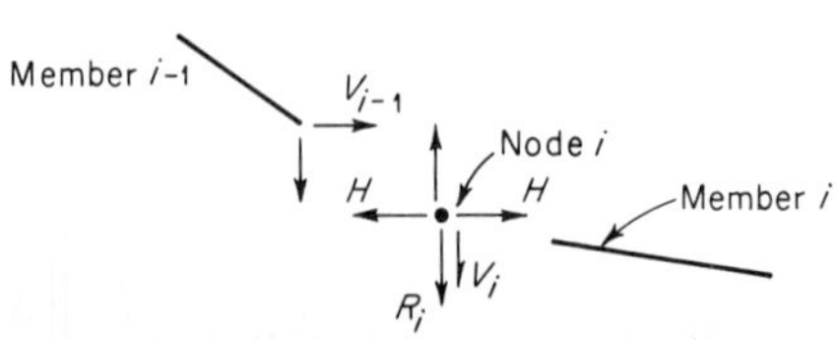

Figure 9.12 Forces at cable node.

If these equalities are not satisfied, an iteration procedure must be used to find the correct values of H and V. That is, let

$$x = s - \Sigma x_i$$

$$y = h - \Sigma y_i$$

Apply increments ΔH and ΔV to produce displacements of point 1, using the tangent flexibility at point 1.

$$F_T = \Sigma \begin{bmatrix} \frac{l}{P}s^2 + \frac{l}{EA} & -\frac{l}{P}cs \\ -\frac{l}{P}cs & \frac{l}{P}c^2 + \frac{l}{EA} \end{bmatrix} \tag{9.42}$$

Then for the first approximation,

$$\begin{bmatrix} x \\ y \end{bmatrix} = F_T \begin{bmatrix} \Delta H \\ \Delta V \end{bmatrix} \tag{9.43}$$

Hence

$$\begin{bmatrix} \Delta H \\ \Delta V \end{bmatrix} = F_T^{-1} \begin{bmatrix} x \\ y \end{bmatrix} \tag{9.44}$$

With the new H and V values, the process is repeated by starting with a re-calculation of member tensions.

Example 9.3 Cable analysis

A ⅝-in.-diameter cable of 1,000-ft span and an unstrained length of 1,100 ft has an area of 0.3 sq. in., and a modulus of elasticity of 30,000 in.-kip units. The cable is analyzed for:

1 A dead load of 20 lb/ft

2 A load of 2.0 kips applied in 20 equal increments (0.1 kip each) vertically upward at the center of the cable.

In the analysis, the cable has been divided into 20 equal lengths of 55 ft each, as shown in Fig. 9.13, so that the node loads for the first part of the analysis are each equal to 0.11 kip.

As an initial assumption, the trial values of H and V are

$$H = 1.67 \text{ kips}$$

$$V = 1.0 \text{ kip}$$

The horizontal tension may be obtained by assuming the initial curve to be circular or parabolic, hence fixing the center sag. When the load increment of $\Delta R_{11} = 0.10$ kip

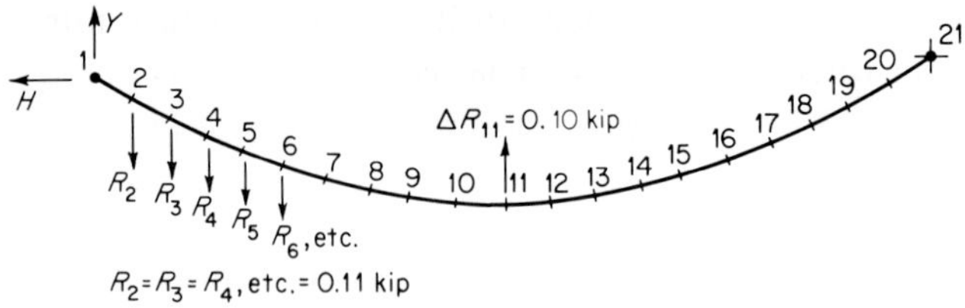

Figure 9.13 Cable structure.

is applied, the trial values of H and V that are used are taken as those at the end of the previous analysis.

The y coordinate of point 11 is tabulated in Table 9.1 for the successive load cases, together with the y cable coordinates for points 1 to 11 for the first and last load cases.

The values in Table 9.1 have been plotted in Figs. 9.14 and 9.15. In 9.14 the central deflection is shown, and it should be noted that the curve is highly nonlinear. That is,

TABLE 9.1 (*a*) **Cable deflections, ft**

Deflection at	*Dead load*	*Load case 20*
1	0.0	0.0
2	34.346	2.211
3	66.346	−0.980
4	95.702	−9.482
5	122.097	−23.062
6	145.199	−41.377
7	164.676	−64.010
8	180.215	−90.510
9	191.538	−120.427
10	198.426	−153.332
11	200.737	−188.830

(*b*) **Deflections at point 11, ft**

Load case increment No.	*y ordinate at 11*	*Deflection measured from dead-load position*
Dead load	200.738	0.0
2	188.830	11.908
4	169.424	31.314
6	138.036	62.702
8	91.427	109.311
10	32.070	168.668
12	−32.070	231.808
14	−91.427	292.165
16	−138.036	338.774
18	−169.424	370.162
20	−188.830	389.568

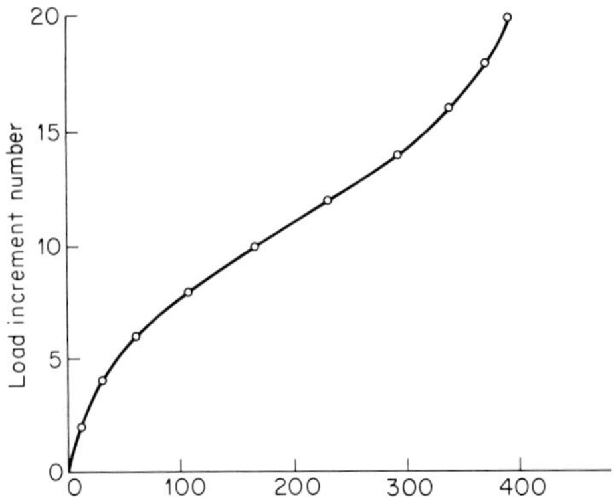

Figure 9.14 Deflection, in feet, of node 2 from dead-load position.

the tangent flexibility at node 11 changes significantly over the load increments. The cable profiles are plotted in Fig. 9.15.

9.4 NONLINEAR, LARGE–DISPLACEMENT ANALYSIS BY THE STIFFNESS METHOD

The approach followed is similar to that for the force method in Sec. 9.2. That is, the nonlinear effects are accounted for by the modification of the standard linear expressions. If the load is applied to the structure in small increments so that the displacements over any given load application are small, the only effect to be considered will be the change in stiffness due to axial forces in members. In Fig. 9.16, for example, the tangent-stiffness expression at the load level R may be written

$$\Delta R = K_T \, \Delta r \tag{9.45}$$

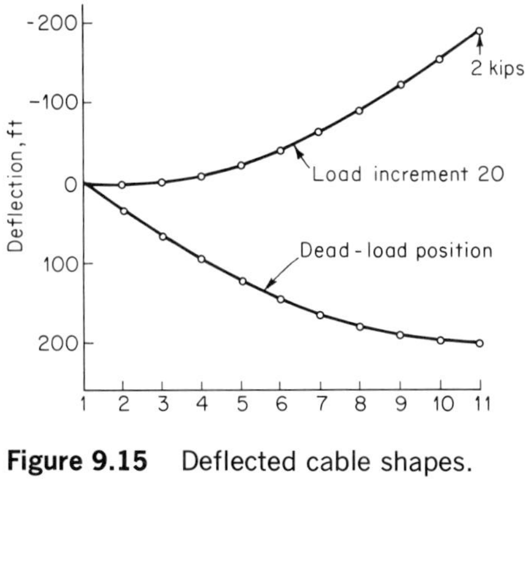

Figure 9.15 Deflected cable shapes.

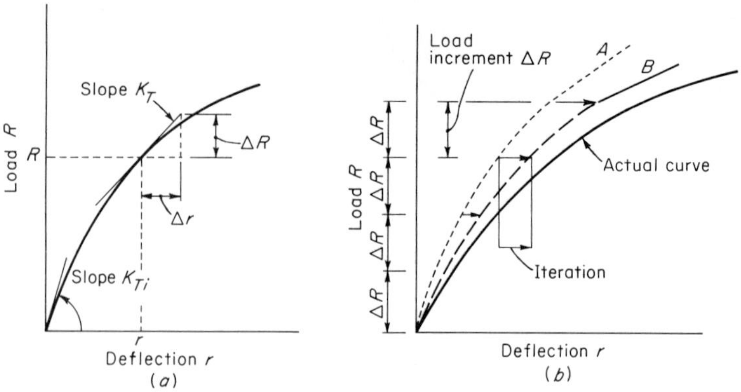

Figure 9.16 Nonlinear load-deflection curves.

In Eq. (9.45), K_T will be composed of two parts, an elastic stiffness K_E and a geometric stiffness K_G due to the presence of axial forces in the structure members at the load level and the effects of geometry changes on their components. Thus, the tangent stiffness K_T may be written

$$K_T = K_E + K_G \tag{9.46}$$

Equation (9.45) is then written

$$\Delta R = (K_E + K_G)\,\Delta r \tag{9.47}$$

For example, the cable shown in Fig. 9.1*a* has $K_E = 0$ and $K_G = 4T/l$ when in the undeflected position. When the situation shown in Fig. 9.2*a* is reached, K_E is no longer zero.

The process may be both iterative and incremental. Thus in Fig. 9.16*b* the load-deflection curve may be approximated by the broken line A for which K_G is calculated by using the forces in the members just prior to the load increment. A more accurate estimate of deflections for the increment ΔR will be obtained by using member forces that are intermediate between the newly found and old values. In this way iterations may be performed at each load step in order to make the curve A in Fig. 9.16*b* approach more nearly curve B, the actual behavior curve for the structure. It is important to realize that K_E must be based on the structure geometry just prior to the load increment under consideration. That is, although the deflections over the range of load ΔR are small, their aggregate effect is not. This will necessitate a carrying forward of structure node geometry to include node deflections and the subsequent recalculation of member stiffnesses. It may be necessary however to recalculate K_E, not after each load increment, but only after several cycles during each of which Δr remains relatively small. Alternatively expressions for K_E and K_G may be derived for the load increment Δr as nonlinear functions of the total deflection at load level R and the original geometry.

The nonlinear deflection effects will be discussed here in their relation to the direct stiffness method of analysis since this has proved such a useful tool for linear analysis. The treatment will be restricted to planar structures. It will be noticed that Eq. (9.47) leads to the calculation of buckling loads, since the criterion for buckling is that the structure will deflect without load increment. That is, for buckling,

$$(K_E + K_G)\,\Delta r = 0 \tag{9.48}$$

Now, K_G may be expressed as a linear function (for small deflections) of a single parameter λ such that

$$K_G = \lambda \bar{K}_G \tag{9.49}$$

For solutions to Eq. (9.48) to exist, the determinate of coefficients must vanish. That is,

$$|K_E + \lambda \bar{K}_G| = 0 \tag{9.50}$$

The transformations giving the stiffness matrix for the plane frame member in global coordinates when the basic elastic stiffness k' is known have been developed in Sec. 7.2 [see Eqs. (7.22)ff]. To recapitulate, two transformations are involved:

1 From the global coordinates to member coordinates

2 From the local (or member) coordinates to the basic member distortions

These transformations are shown in Fig. 9.17. The first, in Fig. 9.17*a*, is for the change in components due to rotation of coordinate axes and is given by

$$r' = L^T r \qquad R' = L^T R \tag{9.51}$$

or

$$r = Lr' \qquad R = LR'$$

where

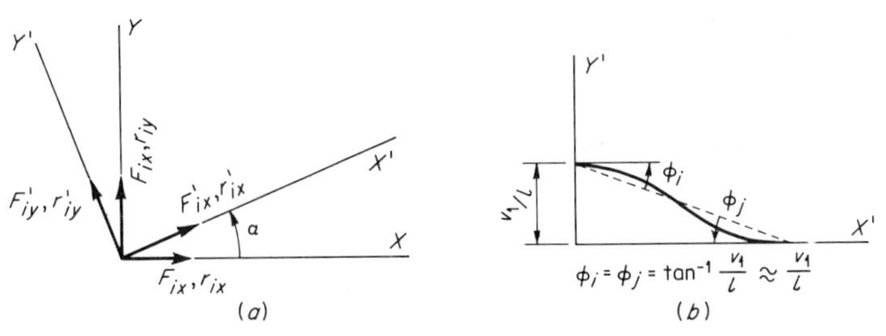

Figure 9.17 Transformation of forces and displacements.

$$L = \begin{bmatrix} \cos\alpha & -\sin\alpha & 0 \\ \sin\alpha & \cos\alpha & 0 \\ 0 & 0 & 1 \end{bmatrix}$$

The second transformation is constructed from unit-deflection diagrams such as that shown in Fig. 9.17*b*. For small deflections,

$$\begin{bmatrix} \phi_i \\ \phi_j \\ \delta \end{bmatrix} = \begin{bmatrix} 0 & \frac{1}{l} & 1 & 0 & -\frac{1}{l} & 0 \\ 0 & \frac{1}{l} & 0 & 0 & -\frac{1}{l} & 1 \\ -1 & 0 & 0 & 1 & 0 & 0 \end{bmatrix} \begin{bmatrix} r'_{ix} \\ r'_{iy} \\ \theta_i \\ r'_{jx} \\ r'_{jy} \\ \theta_j \end{bmatrix} \tag{9.52}$$

It is seen that Eq. (9.51) is correct regardless of the member displacements.[1] Equation (9.52), however, requires that displacements be small for the approximation that

$$\Delta\alpha = \frac{v_2 - v_1}{l}$$

When displacements may no longer be considered small, let

$$\begin{aligned} u &= r'_{jx} - r'_{ix} \\ w &= r'_{jy} - r'_{iy} \end{aligned} \tag{9.53}$$

Then

$$\delta = \sqrt{(l+u)^2 + w^2} - l \tag{9.54a}$$

$$\phi_i = \theta_i - \tan^{-1}\frac{w}{l+u} \tag{9.54b}$$

$$\phi_j = \theta_j - \tan^{-1}\frac{w}{l+u} \tag{9.54c}$$

These equations may be used to develop the required nonlinear transformations. In this first instance it will be assumed that for a given load increment ΔR, the deflections Δr are sufficiently small that Eq. (9.52) is still valid. In such a case it remains to calculate the effect of the axial forces on the stiffness matrix K_T. The similarity must be noticed between this approach and the calculation of the ersatz forces in the force method [see, for example, Eq. (9.30)].

[1] *Note:* If node geometry is corrected for deflections, the angle α will be measured for the next load increment in the deformed position.

In Sec. 9.5, the dependence of K_G on the quadratic strain tensor [see Sec. 2.8, Eqs. (2.177)ff] is shown and the geometric-stiffness matrix redeveloped for frame members. The theory is used then for the planar finite element.

The effects of large displacements on member performance have already been discussed in Sec. 6.5. It was found that axial force in a member causes a modification of bending stiffness that can be evaluated for the prismatic elastic member by using the analytic expression in Eq. (6.100) or alternatively approximated by the use of a cubic bowing function in Eq. (6.106). The simplicity of this approach over a wide range of problems makes its use very attractive.

It was found in Sec. 6.5, that the assumption of a cubic deflected shape leads to the modified bending stiffness given in Eq. (9.55).

$$\begin{bmatrix} M_i \\ M_j \end{bmatrix} = \left\{ \frac{2EI}{l} \begin{bmatrix} 2 & 1 \\ 1 & 2 \end{bmatrix} + \frac{P_o l}{30} \begin{bmatrix} 4 & -1 \\ -1 & 4 \end{bmatrix} \right\} \begin{bmatrix} \phi_i \\ \phi_j \end{bmatrix} \tag{9.55}$$

Since Eq. (9.52) is assumed to relate the member distortions to node displacements in the member coordinate frame, the modification to the stiffness matrix in these axes is given by

$$k_\phi = \begin{bmatrix} 0 & 0 \\ -\dfrac{1}{l} & -\dfrac{1}{l} \\ 1 & 0 \\ 0 & 0 \\ -\dfrac{1}{l} & -\dfrac{1}{l} \\ 0 & 1 \end{bmatrix} \frac{P_o l}{30} \begin{bmatrix} 4 & -1 \\ -1 & 4 \end{bmatrix} \begin{bmatrix} 0 & \dfrac{1}{l} & 1 & 0 & -\dfrac{1}{l} & 0 \\ 0 & \dfrac{1}{l} & 0 & 0 & -\dfrac{1}{l} & 1 \end{bmatrix}$$

That is, because of bowing,

$$k_\phi = \frac{Pl}{30} \begin{bmatrix} 0 & 0 & 0 & 0 & 0 & 0 \\ 0 & \dfrac{6}{l^2} & \dfrac{3}{l} & 0 & -\dfrac{6}{l^2} & \dfrac{3}{l} \\ 0 & \dfrac{3}{l} & 4 & 0 & -\dfrac{3}{l} & -1 \\ 0 & 0 & 0 & 0 & 0 & 0 \\ 0 & -\dfrac{6}{l^2} & -\dfrac{3}{l} & 0 & \dfrac{6}{l^2} & -\dfrac{3}{l} \\ 0 & \dfrac{3}{l} & -1 & 0 & -\dfrac{3}{l} & 4 \end{bmatrix} \tag{9.56}$$

This matrix must be added to that calculated below for the rotation of the member $\Delta\alpha$ and change in length δ.

Calculation of Geometric-stiffness Matrix for Frame Member

The straight prismatic member is shown in Fig. 9.18 in its local $X'Y'$ coordinate system. The direct force components in the plane on the ends i and j are $\langle F'_{ix} \quad F'_{iy} \quad F'_{jx} \quad F'_{jy} \rangle$.

For the rotation of axes from the $X'Y'$ axes to $\overline{X}\overline{Y}$ through the small angle $\Delta\alpha$, $\cos \Delta\alpha \simeq 1$, $\sin \Delta\alpha \simeq \Delta\alpha$, the transformation matrix for vector components is

$$L = \begin{bmatrix} \cos \Delta\alpha & -\sin \Delta\alpha \\ \sin \Delta\alpha & \cos \Delta\alpha \end{bmatrix} \simeq \begin{bmatrix} 1 & -\Delta\alpha \\ \Delta\alpha & 1 \end{bmatrix} \tag{9.57}$$

It is seen that in local displacement components [Eq. (9.28)],

$$\Delta\alpha = \frac{1}{l}[0 \quad -1 \quad 0 \quad 1] \begin{bmatrix} r'_{ix} \\ r'_{iy} \\ r'_{jx} \\ r'_{jy} \end{bmatrix} = Br' \tag{9.58}$$

In Eq. (9.58),

$$B = \frac{1}{l}[0 \quad -1 \quad 0 \quad 1] \tag{9.59}$$

Assume now that under the transformation, the axial force P_o and transverse shear S_o remain constant. Write force components in $\overline{X}\overline{Y}$ system by using the bar notation so that

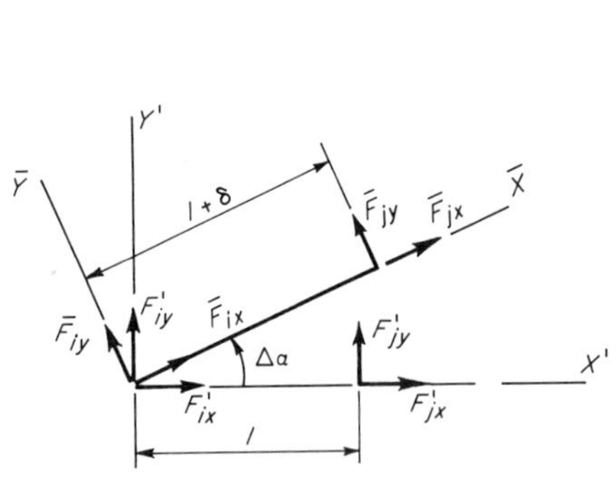

Figure 9.18 Member force. Local and displaced coordinates.

$$\begin{bmatrix} F'_{ix} \\ F'_{iy} \\ F'_{jx} \\ F'_{jy} \end{bmatrix} = \begin{bmatrix} \overline{F}_{ix} \\ \overline{F}_{iy} \\ \overline{F}_{jx} \\ \overline{F}_{jy} \end{bmatrix} = \begin{bmatrix} -1 & 0 \\ 0 & 1 \\ 1 & 0 \\ 0 & -1 \end{bmatrix} \begin{bmatrix} P_o \\ S_o \end{bmatrix} \tag{9.60}$$

Note that numerically $\overline{F} = F$ in the first instance. The components of $\overline{F}$ in the $X'Y'$ axes are given by using Eq. (9.51), and the variation in components ΔF is written

$$\Delta F' = \overline{F}' - F' = \begin{bmatrix} 1 & -Br' & 0 & 0 \\ Br' & 1 & 0 & 0 \\ 0 & 0 & 1 & -Br' \\ 0 & 0 & Br' & 1 \end{bmatrix} \begin{bmatrix} F'_{ix} \\ F'_{iy} \\ F'_{jx} \\ F'_{jy} \end{bmatrix} - \begin{bmatrix} F'_{ix} \\ F'_{iy} \\ F'_{jx} \\ F'_{jy} \end{bmatrix}$$

That is,

$$\Delta F' = \begin{bmatrix} 0 & -1 & 0 & 0 \\ 1 & 0 & 0 & 0 \\ 0 & 0 & 0 & -1 \\ 0 & 0 & 1 & 0 \end{bmatrix} F'Br'$$

Substitution for F' from Eq. (9.60) and for B from Eq. (9.59) gives

$$\Delta F' = \frac{1}{l} \begin{bmatrix} 0 & S_o & 0 & -S_o \\ 0 & P_o & 0 & -P_o \\ 0 & -S_o & 0 & S_o \\ 0 & -P_o & 0 & P_o \end{bmatrix} r'$$

Separating into P_o and S_o terms,

$$\Delta F' = \frac{P_o}{l} \begin{bmatrix} 0 & 0 & 0 & 0 \\ 0 & 1 & 0 & -1 \\ 0 & 0 & 0 & 0 \\ 0 & -1 & 0 & 1 \end{bmatrix} r' + \frac{S_o}{l} \begin{bmatrix} 0 & 1 & 0 & -1 \\ 0 & 0 & 0 & 0 \\ 0 & -1 & 0 & 1 \\ 0 & 0 & 0 & 0 \end{bmatrix} r' \tag{9.61}$$

It is seen in Eq. (9.61) that the first term gives the usual expression for the member with axial force alone. The second expression is unsymmetrical.

Symmetry is restored if account is taken of the change in length of the member and that the moments M_i and M_j producing the shear are constant rather than S_o. That is, owing to change in length δ, the product of shear times length is constant. If S_o is the initial shear force and ΔS_o its variation, then,

$$(S + \Delta S_o)(l + \delta) = S_o l$$

Neglecting the second-order term $\Delta S_o\,\delta$,

$$\Delta S_o = -\frac{S_o\,\delta}{l} \tag{9.62}$$

Again, from Eq. (9.26),

$$\delta = [-1 \quad 0 \quad 1 \quad 0]r' \tag{9.63}$$

The variation in F' force components due to ΔS_o will be designated $\Delta \overline{F}'$, and thence,

$$\Delta \overline{F}' = \begin{bmatrix} 1 & -Br' & 0 & 0 \\ Br' & 1 & 0 & 0 \\ 0 & 0 & 1 & -Br' \\ 0 & 0 & Br' & 1 \end{bmatrix} \begin{bmatrix} 0 \\ \Delta S_o \\ 0 \\ -\Delta S_o \end{bmatrix}$$

Neglecting the product $Br'\,\Delta S_o$ as a second-order term,

$$\Delta \overline{F}' = -\frac{S_o}{l} \begin{bmatrix} 0 \\ 1 \\ 0 \\ -1 \end{bmatrix} [-1 \quad 0 \quad 1 \quad 0]r'$$

That is,

$$\Delta \overline{F}' = \frac{S_o}{l} \begin{bmatrix} 0 & 0 & 0 & 0 \\ 1 & 0 & -1 & 0 \\ 0 & 0 & 0 & 0 \\ -1 & 0 & 1 & 0 \end{bmatrix} r' \tag{9.64}$$

Combining Eqs. (9.61) and (9.64) gives an expression for k_G that is symmetrical in the S_o term.

$$k'_G = \frac{P_o}{l}\begin{bmatrix} 0 & 0 & 0 & 0 \\ 0 & 1 & 0 & -1 \\ 0 & 0 & 0 & 0 \\ 0 & -1 & 0 & 1 \end{bmatrix} + \frac{S_o}{l}\begin{bmatrix} 0 & 1 & 0 & -1 \\ 1 & 0 & -1 & 0 \\ 0 & -1 & 0 & 1 \\ -1 & 0 & 1 & 0 \end{bmatrix} \tag{9.65}$$

When Eqs. (9.56) and (9.65) are combined so that both bowing and joint displacement effects are included, the geometric stiffness due to P_o is

$$k'_p = \frac{P_o l}{30}\begin{bmatrix} 0 & 0 & 0 & 0 & 0 & 0 \\ 0 & \frac{36}{l^2} & \frac{3}{l} & 0 & -\frac{36}{l^2} & \frac{3}{l} \\ 0 & \frac{3}{l} & 4 & 0 & -\frac{3}{l} & -1 \\ 0 & 0 & 0 & 0 & 0 & 0 \\ 0 & -\frac{36}{l^2} & -\frac{3}{l} & 0 & \frac{36}{l^2} & -\frac{3}{l} \\ 0 & \frac{3}{l} & -1 & 0 & -\frac{3}{l} & 4 \end{bmatrix} \tag{9.66}$$

This is a well-known expression (see, for example, Ref. 22). It is used in Eq. (9.50) for estimation of critical loads and is found to give very accurate results.

The second term in Eq. (9.65) involving S_o/l may be written in terms of the end moments $\langle M_i \quad M_j \rangle$, since

$$S_o = \frac{M_i + M_j}{l}$$

Then

$$k'_M = \frac{M_i}{l^2}\begin{bmatrix} 0 & 1 & 0 & -1 \\ 1 & 0 & -1 & 0 \\ 0 & -1 & 0 & 1 \\ -1 & 0 & 1 & 0 \end{bmatrix} + \frac{M_j}{l^2}\begin{bmatrix} 0 & 1 & 0 & -1 \\ 1 & 0 & -1 & 0 \\ 0 & -1 & 0 & 1 \\ -1 & 0 & 1 & 0 \end{bmatrix} \tag{9.67}$$

The geometric-stiffness matrix in Eq. (9.66), gives the relationship

$$\Delta R' = k_p \, \Delta r'$$

in member coordinates. Since $\Delta R' = L^T \Delta R$ and $\Delta r' = L^T \Delta r$, it follows that in global coordinates,

$$k_p = Lk'_pL^T \tag{9.68}$$

This is the expression for k_p that has been given in several references, for example, Refs. 10 and 22.

Calculation of Buckling Loads

It has been shown that the criterion for elastic instability is that the deflection Δr may occur without load increment. That is,

$$(K_E + K_G)\,\Delta r = 0$$

Solutions exist when the determinant of coefficients vanishes.

$$|K_E + \lambda \overline{K}_G| = 0$$

The parameter λ is a measure of the magnitude of the axial forces in the members.

For example, consider the cantilever column shown in Fig. 9.19, subject to the axial force P. The end of the column has the displacement freedoms $\langle v_2 \quad \theta_2 \rangle$, and thus the elastic-stiffness matrix is

$$K_E = \frac{2EI}{l^3}\begin{bmatrix} 6 & -3l \\ -3l & 2l^2 \end{bmatrix}$$

The geometric-stiffness matrix for end 2 is obtained from Eq. (9.66) by extracting the coefficients from the last two rows and columns. That is,

$$K_G = \frac{-P}{30l}\begin{bmatrix} 36 & -3l \\ -3l & 4l^2 \end{bmatrix}$$

The negative sign applies since P is compression. Hence, for buckling,

$$\left\{\frac{2EI}{l^3}\begin{bmatrix} 6 & -3l \\ -3l & 2l^2 \end{bmatrix} - \frac{P}{30l}\begin{bmatrix} 36 & -3l \\ -3l & 4l^2 \end{bmatrix}\right\}\Delta r = 0$$

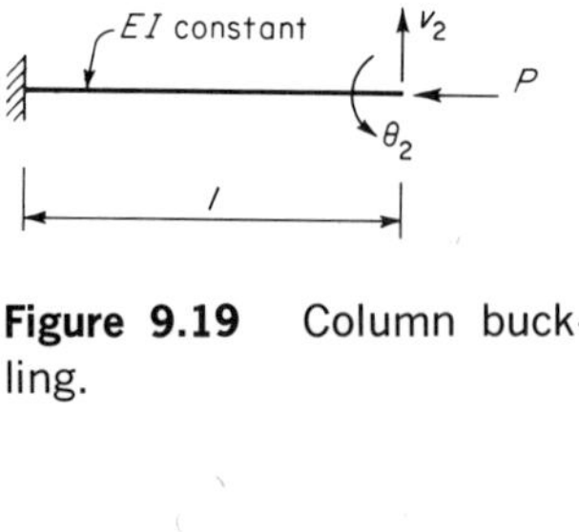

Figure 9.19 Column buckling.

Let $Pl^2/60EI = \lambda$; then

$$\begin{vmatrix} 6 - 36\lambda & -3l + 3l\lambda \\ -3l + 3l\lambda & 2l^2 - 4l^2\lambda \end{vmatrix} = 0$$

Expanding the determinant gives

$$4(1 - 8\lambda + 12\lambda^2) - 3(\lambda^2 - 2\lambda + 1) = 0$$

The smallest value of λ gives the P_{critical}. That is,

$$\lambda = 0.04143$$

and,

$$P_{\text{critical}} = \frac{2.486EI}{l^2} \tag{i}$$

The analytic solution

$$P_{\text{critical}} = \frac{\pi^2 EI}{4l^2} = \frac{2.467EI}{l^2}$$

The error in (i) is only 0.8 percent.

Development of Nonlinear Transformation Equations

Thus far, the theory developed considers the effect of displacements on structure stiffness in terms of the geometric-stiffness matrix determined as a linear function of the member loads. For example, Eq. (9.55) is really an incremental equation that should be written

$$\Delta\begin{bmatrix} M_i \\ M_j \end{bmatrix} = \left\{\frac{2EI}{l}\begin{bmatrix} 2 & 1 \\ 1 & 2 \end{bmatrix} + \frac{P_o l}{30}\begin{bmatrix} 4 & -1 \\ -1 & 4 \end{bmatrix}\right\}\Delta\begin{bmatrix} \phi_i \\ \phi_j \end{bmatrix}$$

In this expression the change in bending stiffness is given as due to the presence of the axial force P_o. No account is taken of the effect of moments on the axial stiffness.

Equations (9.54) are now examined to give the exact nonlinear displacement transformation to replace Eq. (9.52). From Eqs. (9.54),

$$\delta = \sqrt{(l + u)^2 + w^2} - l$$

$$\phi_i = \theta_i - \tan^{-1}\frac{w}{l + u}$$

$$\phi_j = \theta_j - \tan^{-1}\frac{w}{l + u}$$

Incremental values of member distortions $\Delta\langle\delta \quad \phi_i \quad \phi_j\rangle$ due to increments $\Delta\langle u \quad w \quad \theta_i \quad \theta_j\rangle$ are obtained by differentiating the above equations (9.54) partially with respect to the variables. Thus,

$$\Delta\delta = \frac{\partial \delta}{\partial u} du + \frac{\partial \delta}{\partial w} dw$$

That is,

$$\Delta\delta = \frac{l+u}{l+\delta} du + \frac{w}{l+\delta} dw \qquad (i)$$

For increments in end rotations,

$$\Delta\phi_i = \frac{\partial \phi_i}{\partial \theta_i} d\theta_i + \frac{\partial \phi_i}{\partial u} du + \frac{\partial \phi_j}{\partial w} dw$$

Hence,

$$\Delta\phi_i = d\theta_i + \frac{w}{(l+\delta)^2} du - \frac{l+u}{(l+\delta)^2} dw \qquad (ii)$$

$$\Delta\phi_j = d\theta_j + \frac{w}{(l+\delta)^2} du - \frac{l+u}{(l+\delta)^2} dw \qquad (iii)$$

In matrix form, the combined equations (i), (ii), and (iii) are written

$$\begin{bmatrix} \Delta\delta \\ \Delta\phi_i \\ \Delta\phi_j \end{bmatrix} = \begin{bmatrix} \dfrac{l+u}{l+\delta} & \dfrac{w}{l+\delta} & 0 & 0 \\ \dfrac{w}{(l+\delta)^2} & \dfrac{-(l+u)}{(l+\delta)^2} & 1 & 0 \\ \dfrac{w}{(l+\delta)^2} & \dfrac{-(l+u)}{(l+\delta)^2} & 0 & 1 \end{bmatrix} \begin{bmatrix} du \\ dw \\ d\theta_i \\ d\theta_j \end{bmatrix} \qquad (9.69)$$

This equation will be written

$$\Delta v = A\, \Delta \bar{r} \qquad (9.70)$$

Consider a set of forces,

$$\bar{R}^T = \langle R \quad S \quad M_i \quad M_j \rangle$$

that correspond to the displacements $\Delta\bar{r}$. It is seen that

$$\langle R \quad S \rangle = \langle F'_{jx} \quad F'_{jy} \rangle$$

For a set of virtual displacements defined by Eq. (9.70), the work done must be the same for both sets of forces (contragradient law), so that

$$\bar{R}^T \Delta\bar{r} = s^T \Delta v = s^T A\, \Delta\bar{r}$$

Hence

$$\bar{R} = A^T s \qquad (9.71)$$

The member forces,

$$s^T = \langle P \quad M_i \quad M_j \rangle$$

Let $\Delta\bar{R}$ and Δs be the variation in these force components. The transformation from member increment Δs to the forces $\Delta\bar{R}$ is obtained (remember that terms in A are functions of displacements) by differentiating Eq. (9.71) partially with respect to both force and displacement components. The result, after rearrangement of terms, is written

$$\Delta\bar{R} = A^T\,\Delta s + D\,\Delta\bar{r} \tag{9.72}$$

The matrix D contains terms involving R and S.

The matrix D is obtained by differentiating A with respect to u and w and rearranging terms. That is,

$$dA^T = \frac{\partial A^T}{\partial u}\,du + \frac{\partial A^T}{\partial w}\,dw$$

Firstly, write Eq. (9.71) in full by using Eq. (9.69).

$$\begin{bmatrix} R \\ S \\ M_i \\ M_j \end{bmatrix} = \begin{bmatrix} \dfrac{l+u}{l+\delta} & \dfrac{w}{(l+\delta)^2} & \dfrac{w}{(l+\delta)^2} \\ \dfrac{w}{l+\delta} & \dfrac{-(l+u)}{(l+\delta)^2} & \dfrac{-(l+u)}{(l+\delta)^2} \\ 0 & 1 & 0 \\ 0 & 0 & 1 \end{bmatrix} \begin{bmatrix} P_o \\ M_i \\ M_j \end{bmatrix}$$

Write the first two of these equations as

$$(l+\delta)^2 R = (l+\delta)(l+u)P_o + w(M_i + M_j)$$
$$(l+\delta)^2 S = w(l+\delta)P_o - (l+u)(M_i + M_j)$$

That is,

$$\begin{bmatrix} (l+\delta)(l+u) & w \\ w(l+\delta) & -(l+u) \end{bmatrix} \begin{bmatrix} P_o \\ M_i + M_j \end{bmatrix} = (l+\delta)^2 \begin{bmatrix} R \\ S \end{bmatrix}$$

Solve for $\langle P_o \quad (M_i + M_j)\rangle$, so that

$$\begin{bmatrix} P_o \\ M_i + M_j \end{bmatrix} = \frac{1}{l+\delta} \begin{bmatrix} l+u & w \\ w(l+\delta) & -(l+u)(l+\delta) \end{bmatrix} \begin{bmatrix} R \\ S \end{bmatrix}$$

It can be shown after some algebra that

$$D = \begin{bmatrix} d_{11} & d_{12} & 0 & 0 \\ d_{21} & d_{22} & 0 & 0 \\ 0 & 0 & 0 & 0 \\ 0 & 0 & 0 & 0 \end{bmatrix} \tag{9.73}$$

In Eq. (9.73),

$$d_{11} = \frac{1}{(l+\delta)^4}\{-w^2(l+u)R + [2(l+u)^2 + w^2]wS\} \tag{9.74a}$$

$$d_{12} = d_{21} = \frac{1}{(l+\delta)^4}[-w^3R - (l+u)^3S] \tag{9.74b}$$

$$d_{44} = \frac{1}{(l+\delta)^4}\{(l+u)[(l+u)^2 + 2w^2]R - (l+u)^2wS\} \tag{9.74c}$$

The derivation of the term d_{11} is given below. It is seen that

$$\frac{d(l+\delta)}{du} = \frac{l+u}{l+\delta}$$

$$\frac{d(l+u/l+\delta)}{du} = \frac{w^2}{(l+\delta)^3}$$

$$\frac{d[w/(l+\delta)^2]}{du} = \frac{-2w(l+u)}{(l+\delta)^4}$$

By using these derivatives, the variation ΔR due to du will give d_{11}, so that

$$d_{11} = \left[\frac{w^2}{(l+\delta)^3}P_o - \frac{2w(l+u)}{(l+\delta)^4}(M_i + M_j)\right]$$

Substitution for P_o and $M_i + M_j$ gives

$$d_{11} = \left\{\frac{w^2}{(l+\delta)^2}\frac{1}{l+\delta}[(l+u)R + wS] - \frac{2w(l+u)}{(l+\delta)^4}\frac{1}{l+\delta}[w(l+\delta)R - (l+u)(l+\delta)S]\right\}$$

On simplification,

$$d_{11} = \frac{1}{(l+\delta)^4}\{-(l+u)w^2R + w[2(l+u)^2 + w^2]S\}$$

The other terms are calculated in a similar manner.

It remains now to calculate the tangent-stiffness matrix $\bar{k}$ relating the force increment Δs to the displacement increment Δv. It should be noted that the above equations use the original structure geometry, together with total deflections in the incremental equations. From Chap. 6, Eqs. (6.104) and (6.106),

$$P = EA\left[\frac{\delta}{l} + \frac{1}{30}(2\phi_i^2 - \phi_i\phi_j + 2\phi_j^2)\right] \tag{9.75a}$$

$$\begin{bmatrix} M_i \\ M_j \end{bmatrix} = \frac{2EI}{l}\begin{bmatrix} 2 & 1 \\ 1 & 2 \end{bmatrix} + \frac{Pl}{30}\begin{bmatrix} 4 & -1 \\ -1 & 4 \end{bmatrix}\begin{bmatrix} \phi_i \\ \phi_j \end{bmatrix} \tag{9.75b}$$

When each of these expressions is differentiated with respect to δ, ϕ_i, and ϕ_j [remember that P in Eq. (9.75*b*) is a function of these variables as given in (9.75*a*)], it is found that

$$\Delta s = \begin{bmatrix} \Delta P \\ \Delta M_i \\ \Delta M_j \end{bmatrix} = \begin{bmatrix} \frac{EA}{l} & \frac{EA}{30}(4\phi_i - \phi_j) & \frac{EA}{30}(-\phi_i + 4\phi_j) \\ & k_{\Delta ii} & k_{\Delta ij} \\ & \text{symmetrical} & k_{\Delta jj} \end{bmatrix} \begin{bmatrix} \Delta\delta \\ \Delta\phi_i \\ \Delta\phi_j \end{bmatrix} \tag{9.76}$$

This equation is written

$$\Delta s = k_\Delta \, \Delta v \tag{9.76a}$$

In Eq. (9.76),

$$k_{\Delta ii} = \frac{4EI}{l} + \frac{4EA\,\delta}{30} + \frac{EAl}{300}(8\phi_i^2 - 4\phi_i\phi_j + 3\phi_j^2) \tag{9.77a}$$

$$k_{\Delta ij} = \frac{2EI}{l} - \frac{EAe}{30} + \frac{EAl}{300}(-2\phi_i^2 + 6\phi_i\phi_j - 2\phi_j^2) \tag{9.77b}$$

$$k_{\Delta jj} = \frac{4EI}{l} + \frac{4EA\,\delta}{30} + \frac{EAl}{300}(3\phi_i^2 - 4\phi_i\phi_j + 8\phi_j^2) \tag{9.77c}$$

The difference should be noted between the use of Eq. (9.55) and Eq. (9.76) as the incremental (tangent) stiffness relationship. In the first instance the effect of bowing is calculated from the constant axial force P_o on assuming that the incremented deflected shape for the member is cubic in form. That is, Eq. (9.55) is an incremental relationship as is the expanded form of k_p in Eq. (9.66).

In Eq. (9.76) it has been assumed that the initial state is that given by Eq. (9.55) [or Eq. (9.75*a*) and (9.75*b*)] and the variation is calculated from this position. Hence the appearance of the quadratic terms in Eqs. (9.77).

The tangent-stiffness matrix for the whole structure is now developed in the following manner. From Eqs. (9.53) it is seen that

$$\begin{bmatrix} u \\ w \end{bmatrix} = \begin{bmatrix} -1 & 0 & 1 & 0 \\ 0 & -1 & 0 & 1 \end{bmatrix} \begin{bmatrix} r'_{ix} \\ r'_{iy} \\ r'_{jx} \\ r'_{jy} \end{bmatrix} \tag{i}$$

Furthermore, from Eq. (9.51), these local displacements are expressed in terms of the global components,

$$\begin{bmatrix} r'_{ix} \\ r'_{iy} \\ r'_{jx} \\ r'_{jy} \end{bmatrix} = \begin{bmatrix} \cos\alpha & \sin\alpha & 0 & 0 \\ -\sin\alpha & \cos\alpha & 0 & 0 \\ 0 & 0 & \cos\alpha & \sin\alpha \\ 0 & 0 & -\sin\alpha & \cos\alpha \end{bmatrix} \begin{bmatrix} r_{ix} \\ r_{iy} \\ r_{jx} \\ r_{jy} \end{bmatrix} \tag{ii}$$

Combining Eqs. (*i*) and (*ii*),

$$\begin{bmatrix} u \\ w \end{bmatrix} = \begin{bmatrix} -\cos\alpha & -\sin\alpha & \cos\alpha & \sin\alpha \\ \sin\alpha & -\cos\alpha & -\sin\alpha & \cos\alpha \end{bmatrix} \begin{bmatrix} r_{ix} \\ r_{iy} \\ r_{jx} \\ r_{jy} \end{bmatrix}$$

Thence,

$$\begin{bmatrix} u \\ w \\ \theta_i \\ \theta_j \end{bmatrix} = \begin{bmatrix} -\cos\alpha & -\sin\alpha & 0 & \cos\alpha & \sin\alpha & 0 \\ \sin\alpha & -\cos\alpha & 0 & -\sin\alpha & \cos\alpha & 0 \\ 0 & 0 & 1 & 0 & 0 & 0 \\ 0 & 0 & 0 & 0 & 0 & 1 \end{bmatrix} \begin{bmatrix} r_{ix} \\ r_{iy} \\ \theta_i \\ r_{jx} \\ r_{jy} \\ \theta_j \end{bmatrix} \tag{9.78}$$

This is the equation

$$\bar{r} = Tr \tag{9.79}$$

Thence

$$\Delta\bar{r} = T\,\Delta r \tag{9.80}$$

And from Eq. (9.70),

$$\Delta v = AT\,\Delta r$$

Also, from Eq. (9.76*a*),

$$\Delta s = k_\Delta\,\Delta v$$

Therefore

$$\Delta s = k_\Delta AT\,\Delta r \tag{9.81}$$

Furthermore, applying the contragradient law to forces and displacements in

Eq. (9.80),

$$\Delta R = T^T \, \Delta \bar{R} \tag{9.82}$$

That is, from Eq. (9.72),

$$\Delta R = T^T(A^T \, \Delta s + D \, \Delta \bar{r})$$

Substituting for Δs from Eq. (9.81) in this expression gives

$$\Delta R = T^T(A^T k_\Delta A T + DT) \, \Delta r$$

That is,

$$\Delta R = T^T(A^T k_\Delta A + D) T \, \Delta r \tag{9.83}$$

Equation (9.83) may be used to build up the direct stiffness equations in the usual manner, and for a given load increment ΔR, solution of the equations gives the corresponding deflection increment Δr, where the vectors ΔR and Δr now indicate all the node force and deflection components respectively.

Example 9.4 Use of large-displacement equations

The cantilever beam shown in Fig. 9.20 has been analyzed by using the simple joint-translation geometric-stiffness matrix given in Eq. (9.65) in a program that recalculates the elastic stiffness in the deformed position, and also by using the refined approach with the bowing function included [Eqs. (9.69)ff]. In Fig. 9.20, the latter is referred to as *Jennings' (cubic deflected member)*. The curve plotted and labeled *Jennings' (ordinary)* is similar to the geometric stiffness [Eq. (9.65)] except that it uses nonlinear expressions based on the initial geometry to calculate the tangent stiffness. The geometric stiffness [Eq. (9.65)] has been run for one and five subdivisions of the beam. It will be noticed that for five subdivisions of the beam, both methods give results that are very close to the analytic solution calculated by the large-deflection theory for beams as given by Frisch-Fay (Ref. 11). Note that whereas the vertical deflection calculated from Jennings' (cubic) is in best agreement with the analytic solution, for the horizontal deflection u, the simple geometric stiffness is closest.

In Fig. 9.21, bending moments are given at the one-fifth points along the cantilever. The higher values given by Jennings' (cubic) reflect the smaller deflection u given in this method and shown in the plot $(L - u)/L$ in Fig. 9.20.

9.5 GEOMETRIC–STIFFNESS MATRICES FOR THE LINE–ELEMENT AND FINITE–ELEMENT PLANE TRIANGLES BY USING THE QUADRATIC STRAIN TENSOR[1]

In Chap. 2, it has been shown that the strain tensor may be expressed as the sum of the first-order strain tensor ε, which is used in the theory embodying small strains, and the quadratic strain tensor η, which is neglected in the classical theory of elasticity. It will be shown that the geometric-stiffness

[1] The treatment herein is similar to that in Ref. 10.

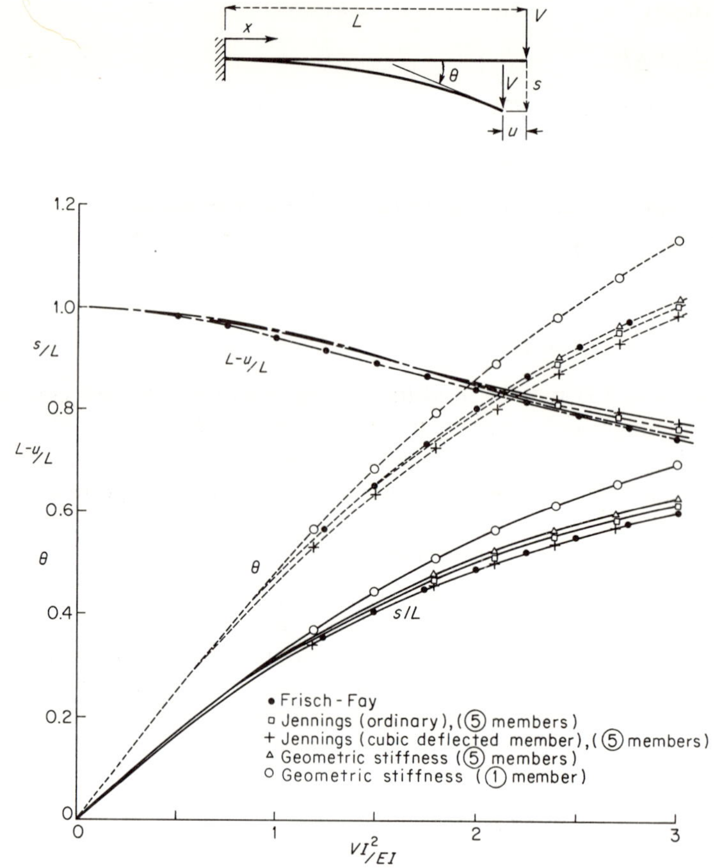

Figure 9.20 Cantilever deflections. Elastic nonlinear analyses (nondimensional plots).

matrix K_G may be related to the quadratic strain tensor η, and that η may be used in the generation of K_G.

The load-deflection relationship for a structure is shown in Fig. 9.22, in which the nonlinearity exhibited is due to geometric changes only. Consider the load-deflection state (R,r), and the increments ΔR and Δr. The tangent stiffness K_T will differ from the initial elastic stiffness K_E because of the presence of axial forces, so that

$$K_T = K_E + K_G \tag{9.84}$$

The slopes K_T and K_E are shown in Fig. 9.22.

For the increment in loads and deflections, the external work done is given by

$$\Delta W = R^T \Delta r + \tfrac{1}{2}\Delta r^T K_T \Delta r$$

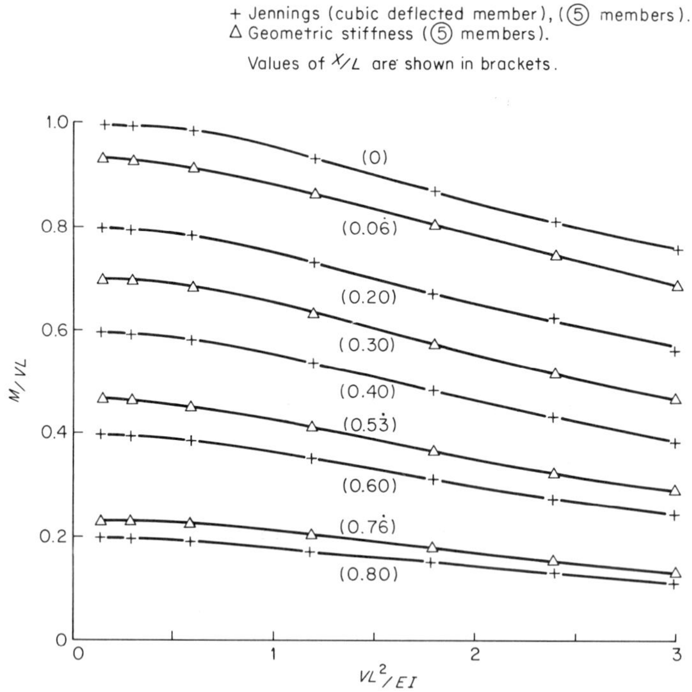

Figure 9.21 Cantilever-beam bending moments; elastic nonlinear analyses (nondimensional).

or, by using Eq. (9.84),

$$\Delta W = R^T \Delta r + \underbrace{\tfrac{1}{2}\Delta r^T K_E \Delta r}_{\text{elastic}} + \underbrace{\tfrac{1}{2}\Delta r^T K_G \Delta r}_{\text{geometric}} \tag{9.85}$$

In Eq. (9.85) the first term simply corresponds to the shaded rectangle in Fig. 9.22. The second term gives the linear elastic increment ΔU_E, and the third term gives the geometric increment ΔU_G. This external work done may be equated with the internal work produced by the total strains and the stress values. The strain tensor will be written

$$E_{ij} = \varepsilon_{ij} + \eta_{ij} \tag{9.86}$$

where, as shown in Chap. 2,

$$\varepsilon_{ij} = \tfrac{1}{2}(v_{i,j} + v_{j,i}) \tag{9.87}$$

$$\eta_{ij} = \tfrac{1}{2}(v_{k,i}v_{k,j}) \tag{9.88}$$

In (9.87) and (9.88), the comma in the subscripts indicates differentiation. That is,

$$v_{1,2} = \frac{\partial v_1}{\partial y} \qquad \text{etc.}$$

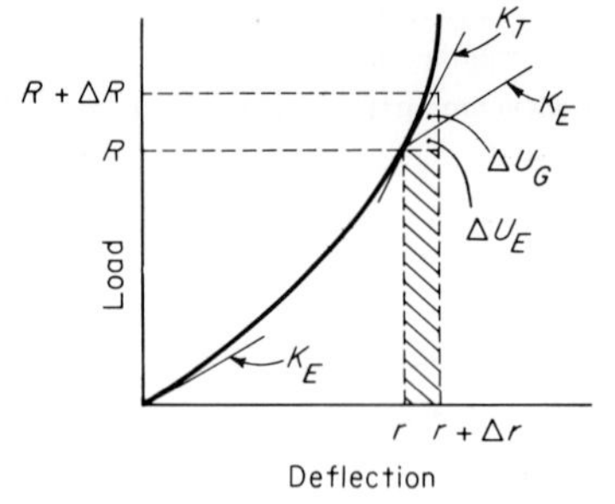

Figure 9.22 Strain energy.

The strain field corresponding to Δr will be

$$\Delta E_{ij} = \Delta \varepsilon_{ij} + \Delta \eta_{ij}$$

Let σ_{ij} be the stress field corresponding to state (R,r) and $\Delta\sigma_{ij}$ the increment due to $(\Delta R,\Delta r)$. The increment to the strain energy will be

$$\Delta U = \int_{\text{vol}} \sigma_{ij}\, \Delta E_{ij}\, dV + ½\int_{\text{vol}} \Delta\sigma_{ij}\, \Delta E_{ij}\, dV$$

Substitution for ΔE_{ij} gives

$$\Delta U = \int_{\text{vol}} \sigma_{ij}\, \Delta\varepsilon_{ij}\, dV + \int_{\text{vol}} \sigma_{ij}\, \Delta\eta_{ij}\, dV + ½\int_{\text{vol}} \Delta\sigma_{ij}\, \Delta\varepsilon_{ij}\, dV + ½\int_{\text{vol}} \Delta\sigma_{ij}\, \Delta\eta_{ij}\, dV$$

It is assumed that the last term is of second-order magnitude in comparison with the second term and hence will be neglected. Small deflections at the tangent are assumed, so that

$$R^T \Delta r = \int_{\text{vol}} \sigma_{ij}\, \Delta\varepsilon_{ij}\, dV$$

and hence,

$$\int_{\text{vol}} \sigma_{ij}\Delta\eta_{ij}\, dV = ½\Delta r^T K_G\, \Delta r \tag{9.89}$$

This expression may be used to calculate K_G. For convenience in the following theory, replace $\Delta\eta_{ij}$ with η_{ij} and Δr with r. That is, the incremental notation is dropped although it is still presumed that incremental displacements are being considered. In this new notation, Eq. (9.89) becomes

$$\int_{\text{vol}} \sigma_{ij}\eta_{ij}\, dV = ½ r^T K_G r \tag{9.90}$$

It is possible to use Eq. (9.90) to calculate K_G when the strain field E_{ij} is related to the displacement field, which is defined by the nodal values r, by the interpolation polynomials such as are developed in Chap. 8. For a given element in three dimensions, let the nodal displacement vector be

$$r^T = \langle U_1{}^T \quad U_2{}^T \quad U_3{}^T \rangle \tag{9.91}$$

The subvector U_1 refers to the node values in the direction of the cartesian coordinate 1, or x. The displacement function u_i within the element is assumed to be specified only by u_i, so that

$$\tilde{u}_i = \phi_i{}^T(\zeta_n) U_i$$

The quadratic strain tensor η_{ij} from Chap. 2 may be written in this notation as

$$2\eta_{ij} = M^T M = \begin{bmatrix} \tilde{u}_{1,1} & \tilde{u}_{2,1} & \tilde{u}_{3,1} \\ \tilde{u}_{1,2} & \tilde{u}_{2,2} & \tilde{u}_{3,2} \\ \tilde{u}_{1,3} & \tilde{u}_{2,3} & \tilde{u}_{3,3} \end{bmatrix} \begin{bmatrix} \tilde{u}_{1,1} & \tilde{u}_{1,2} & \tilde{u}_{1,3} \\ \tilde{u}_{2,1} & \tilde{u}_{2,2} & \tilde{u}_{2,3} \\ \tilde{u}_{3,1} & \tilde{u}_{3,2} & \tilde{u}_{3,3} \end{bmatrix} = \tilde{u}_{k,i}\tilde{u}_{k,j} \tag{9.92}$$

It is seen that

$$\tilde{u}_{i,k} = \phi_{i,k}^T U_i \tag{9.93}$$

Substitution in the matrix M for $\tilde{u}_{i,k}$ gives

$$M = \begin{bmatrix} \phi_{1,1}^T U_1 & \phi_{1,2}^T U_1 & \phi_{1,3}^T U_1 \\ \phi_{2,1}^T U_2 & \phi_{2,2}^T U_2 & \phi_{2,3}^T U_2 \\ \phi_{3,1}^T U_3 & \phi_{3,2}^T U_3 & \phi_{3,3}^T U_3 \end{bmatrix} \tag{9.94}$$

In the product $M^T M$, it is seen that η_{11}, for example, is given by

$$2\eta_{11} = U_1{}^T \phi_{1,1}\phi_{1,1}^T U_1 + U_2{}^T \phi_{2,1}\phi_{2,1}^T U_2 + U_3{}^T \phi_{3,1}\phi_{3,1}^T U_3$$

The general term is thus found to be

$$2\eta_{ij} = \sum_{k=1,3} U_k{}^T \phi_{k,i}\phi_{k,j}^T U_k \tag{9.95}$$

Instead of the summation given in Eq. (9.95), the expression for η_{ij} may be written in the matrix form of Eq. (9.96).

$$2\eta_{ij} = [U_1{}^T \quad U_2{}^T \quad U_3{}^T] \begin{bmatrix} \phi_{1,i}\phi_{1,j}^T & 0 & 0 \\ 0 & \phi_{2,i}\phi_{2,j}^T & 0 \\ 0 & 0 & \phi_{3,i}\phi_{3,j}^T \end{bmatrix} \begin{bmatrix} U_1 \\ U_2 \\ U_3 \end{bmatrix} \tag{9.96}$$

The stress field $\tilde{\sigma}_{ij}$ is also given by an interpolation formula expressed in terms of the nodal vector σ_{ij} by the interpolation function ϕ_σ, so that

$$\tilde{\sigma}_{ij} = \phi_\sigma{}^T \sigma_{ij} \tag{9.97}$$

The summation $2\eta_{ij}\tilde{\sigma}_{ij}$ is now given as

$$\underbrace{2\eta_{ij}\sigma_{ij}}_{\text{Sum on } i,j} = [U_1{}^T \quad U_2{}^T \quad U_3{}^T] \begin{bmatrix} \Sigma\Sigma\, \phi_{1,i}\phi_{1,j}^T\tilde{\sigma}_{ij} & 0 & 0 \\ 0 & \Sigma\Sigma\, \phi_{2,i}\phi_{2,j}^T\tilde{\sigma}_{ij} & 0 \\ 0 & 0 & \Sigma\Sigma\, \phi_{3,i}\phi_{3,j}\tilde{\sigma}_{ij} \end{bmatrix} \begin{bmatrix} U_1 \\ U_2 \\ U_3 \end{bmatrix} \tag{9.98}$$

A comparison of this expression with Eq. (9.90) shows that K_G is given by

$$K_G = \sum_{i=1,3} \sum_{j=1,3} \int_{\text{vol}} \begin{bmatrix} \phi_{1,i}\phi_{1,j}^T\tilde{\sigma}_{ji} & 0 & 0 \\ 0 & \phi_{2,i}\phi_{2,j}^T\tilde{\sigma}_{ij} & 0 \\ 0 & 0 & \phi_{3,i}\phi_{3,j}^T\tilde{\sigma}_{ij} \end{bmatrix} dV \tag{9.99}$$

This may be written as

$$K_G = \begin{bmatrix} K_{G11} & 0 & 0 \\ 0 & K_{G22} & 0 \\ 0 & 0 & K_{G33} \end{bmatrix} \tag{9.100}$$

It should be noted in this derivation of K_G that it has been assumed that $\tilde{u}_i$ is given entirely in terms of the nodal values U_i. This allows for the partitioning for K_G expressed in Eq. (9.100). The expression for K_G may be modified slightly by selecting a nodal system and interpolation functions for the displacement gradients $\tilde{u}_{i,k}$. Thus, let

$$\tilde{u}_{i,k} = \phi_{\varepsilon i}{}^T U_{i,k} \tag{9.101}$$

In Eq. (9.101), $U_{i,k}$ is a vector of the nodal values. Now,

$$U_{i,k} = \Phi_{ik} U_i \tag{9.102}$$

where Φ_{ik} is obtained from $\Phi_{i,k}$ evaluated at the nodal points. Thence

$$\tilde{u}_{i,k} = \phi_{\varepsilon i}{}^T \Phi_{ik} U_i \tag{9.103}$$

It is seen from a comparison of Eqs. (9.93) and (9.103) that

$$\phi_{i,k}^T = \phi_{\varepsilon i}{}^T \Phi_{ik} \tag{9.104}$$

Substitution in Eq. (9.99) gives

$$K_{Gnn} = \sum_{i=1,3} \sum_{j=1,3} \Phi_{ni}{}^T \int_{\text{vol}} \phi_{\varepsilon i}\phi_{\varepsilon j}{}^T \phi_\sigma{}^T \sigma_{ij}\, dV\, \Phi_{nj} \tag{9.105}$$

Either Eq. (9.99) or (9.105) may be used to calculate K_G. These will be used firstly for the spar, or line, elements treated in Sec. 9.4, and then finally for the plane-strain and linear-strain finite-element triangles.

Example 9.5 Calculation of geometric-stiffness matrix for a truss member

The geometric-stiffness matrix for the member with axial force P_o is given by the first term in Eq. (9.61). This expression is redeveloped in this example by using the quadratic strain expression. When the member is prismatic and subject to a constant axial force, the displacement field is necessarily a linear one that may be expressed in terms of the nodal displacements by the interpolation polynomial $\zeta^T = \langle \zeta_1 \quad \zeta_2 \rangle$ such that

$$\tilde{u} = \zeta^T U \tag{9.106a}$$

$$\tilde{v} = \zeta^T V \tag{9.106b}$$

where

$$U^T = \langle r'_{ix} \quad r'_{jx} \rangle \tag{9.107a}$$

$$V^T = \langle r'_{iy} \quad r'_{jy} \rangle \tag{9.107b}$$

From Eqs. (9.106a) and (9.106b), it is seen that the interpolation polynomials ϕ_1 and ϕ_2 are equal. That is,

$$\phi_1 = \phi_2 = \zeta \tag{9.108}$$

The stress tensor σ_{ij} contains only one nonzero value, the axial stress σ_1, and

$$\sigma_1 = \frac{P}{A}$$

Hence in Eqs. (9.98) and (9.99), only η_{11} need be considered in the quadratic strain tensor. Now,

$$2\eta_{11} = U_1{}^T \phi_{1,1} \phi_{1,1}^T U_1 + U_2{}^T \phi_{2,1} \phi_{2,1}^T U_2$$

and hence from Eq. (9.108), remembering

$$\phi_{1,1}^T = \frac{\partial \phi_1{}^T}{\partial x} = \frac{1}{l} \langle -1 \quad 1 \rangle$$

and similarly for $\phi_{2,1}$, it follows that

$$2\eta_{11} = U^T \frac{1}{l^2} \begin{bmatrix} -1 \\ 1 \end{bmatrix} [-1 \quad 1] U + V^T \frac{1}{l^2} \begin{bmatrix} -1 \\ 1 \end{bmatrix} [-1 \quad 1] V$$

That is,

$$2\eta_{11} = [U^T \quad V^T] \frac{1}{l^2} \begin{bmatrix} 1 & -1 & 0 & 0 \\ -1 & 1 & 0 & 0 \\ 0 & 0 & 1 & -1 \\ 0 & 0 & -1 & 1 \end{bmatrix} \begin{bmatrix} U \\ V \end{bmatrix}$$

From Eq. (9.99),

$$K_G = \int_{\text{vol}} \frac{P}{A l^2} \begin{bmatrix} 1 & -1 & 0 & 0 \\ -1 & 1 & 0 & 0 \\ 0 & 0 & 1 & -1 \\ 0 & 0 & -1 & 1 \end{bmatrix} dV = \frac{P}{l} \begin{bmatrix} 1 & -1 & 0 & 0 \\ -1 & 1 & 0 & 0 \\ 0 & 0 & 1 & -1 \\ 0 & 0 & -1 & 1 \end{bmatrix} \tag{9.109}$$

If shortening is neglected in Eq. (9.109),

$$K_G = \frac{P}{l} \begin{bmatrix} 0 & 0 & 0 & 0 \\ 0 & 0 & 0 & 0 \\ 0 & 0 & 1 & -1 \\ 0 & 0 & -1 & 1 \end{bmatrix}$$

This is the same result as obtained previously.

Example 9.6 Calculation of the geometric-stiffness matrix for a frame member

The frame member is shown in Fig. 9.23, and the same assumption is made here as in Eqs. (9.56)ff that the axial force and end moments together interact in a cubic distorted shape.

The axial force $P = F_{2X} = -F_{1X}$. The bending moment varies linearly along the column, so that

$$M = -M_1\zeta_1 + M_2\zeta_2 = \langle -\zeta_1 \quad \zeta_2 \rangle \begin{bmatrix} M_1 \\ M_2 \end{bmatrix} = \zeta^T M \tag{9.110}$$

The node displacement vector is

$$r^T = \langle u_1 \quad u_2 \quad v_1 \quad v_2 \quad \theta_1 \quad \theta_2 \rangle \tag{9.111}$$

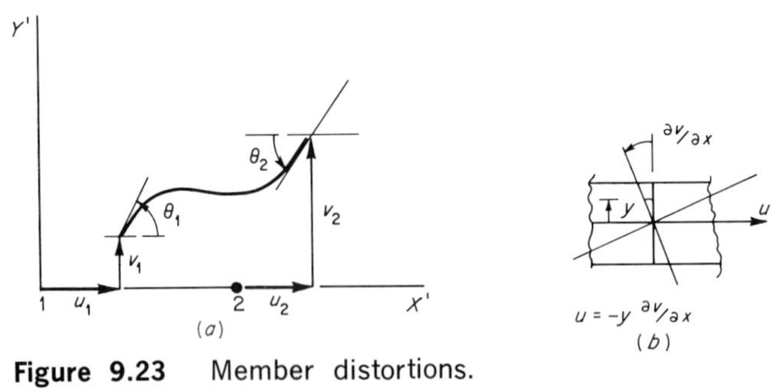

Figure 9.23 Member distortions.

The cubic curve for the deflection $\tilde{v}$ will be specified by the node values $v_1, v_2, \theta_1, \theta_2$. For this the orthogonal interpolating function has been found (see Chap. 6) to be

$$\tilde{v} = r^T \begin{bmatrix} 0 \\ 0 \\ \zeta_1^2(3 - 2\zeta_1) \\ \zeta_2^2(3 - 2\zeta_2) \\ l\zeta_1^2\zeta_2 \\ -l\zeta_1\zeta_2^2 \end{bmatrix} = r^T\phi_v \tag{9.112}$$

The displacement $\tilde{u}$ at a point in the cross section will be given not only as a linear function of $\langle u_1 \quad v_2 \rangle$ but also in terms of the tangent slope at the section (see Fig. 9.23*b*). That is,

$$\tilde{u} = u_1\zeta_1 + u_2\zeta_2 - y\frac{\partial \tilde{v}}{\partial x}$$

On differentiation of ϕ_v in Eq. (9.112) with respect to x, the expression for $\tilde{u}$ is given by

$$\tilde{u} = r^T \begin{bmatrix} \zeta_1 \\ \zeta_2 \\ \dfrac{6y\zeta_1(1 - \zeta_1)}{l} \\ \dfrac{-6y\zeta_2(1 - \zeta_2)}{l} \\ y\zeta_1(2 - 3\zeta_1) \\ y\zeta_2(2 - 3\zeta_2) \end{bmatrix} = r^T\phi_U \tag{9.113}$$

Notice that in this case $\tilde{u}$ depends on all the node values at i and j, and thence the form for η_{ij} will not be as in Eq. (9.96).

If shear distortions are neglected, the only contributing term to the quadratic strain tensor will be η_{11}, and

$$2\eta_{11} = \left(\frac{\partial \tilde{u}}{\partial x}\right)^2 + \left(\frac{\partial \tilde{v}}{\partial x}\right)^2 \tag{9.114}$$

The axial stress

$$\sigma_1 = \frac{P}{A} + (-M_1\zeta_1 + M_2\zeta_2)\frac{y}{I} \tag{9.115}$$

In this case K_G will be formed by integrating the various terms involved in the product $2\eta_{11}\sigma_1$.

Consider first the term $\partial \tilde{v}/\partial x$ in Eq. (9.114).

$$\frac{\partial \tilde{v}}{\partial x} = r^T \frac{\partial \phi_v}{\partial x} = r^T \begin{bmatrix} 0 \\ 0 \\ \dfrac{-6\zeta_1(1-\zeta_1)}{l} \\ \dfrac{6\zeta_2(1-\zeta_2)}{l} \\ \zeta_1(3\zeta_1 - 2) \\ \zeta_2(3\zeta_2 - 2) \end{bmatrix} = r^T \phi_{v,1} \tag{9.116}$$

Then

$$\left(\frac{\partial \tilde{v}}{\partial x}\right)^2 \sigma_1 = r^T \phi_{V,1} \phi_{V,1}^T \sigma_1 r \tag{9.117}$$

The contribution to K_G is given by

$$\int_{\text{vol}} \phi_{V,1}\phi_{V,1}^T \sigma_1 \, dV = \int_{\text{vol}} \begin{bmatrix} 0 \\ 0 \\ \dfrac{-6\zeta_1(1-\zeta_1)}{l} \\ \dfrac{6\zeta_2(1-\zeta_2)}{l} \\ \zeta_1(3\zeta_1 - 2) \\ \zeta_2(3\zeta_2 - 2) \end{bmatrix}$$
$$\times \left[0 \quad 0 \quad \frac{-6\zeta_1(1-\zeta_1)}{l} \quad \frac{6\zeta_2(1-\zeta_2)}{l} \quad \zeta_1(3\zeta_1 - 2) \quad \zeta_2(3\zeta_2 - 2) \right]$$
$$\times \left[\frac{P}{A} + \frac{y}{I}(-M_1\zeta_1 + M_2\zeta_2) \right] dV$$

On integration of the term involving P/A, it is found that

$$K_{Gv(P)} = \frac{P}{30} \begin{bmatrix} 0 & 0 & 0 & 0 & 0 & 0 \\ 0 & 0 & 0 & 0 & 0 & 0 \\ 0 & 0 & \dfrac{36}{l} & -\dfrac{36}{l} & 3 & 3 \\ 0 & 0 & -\dfrac{36}{l} & \dfrac{36}{l} & -3 & -3 \\ 0 & 0 & 3 & -3 & 4l & -l \\ 0 & 0 & 3 & -3 & -l & 3l \end{bmatrix} \tag{9.118}$$

A comparison of Eqs. (9.66) and (9.118) shows that the results obtained are identical.

The bending stresses do not contribute to the integral since $\int_{\text{area}} y\,dA = 0$. The remaining term contributing to η_{11} is $(\partial\tilde{u}/\partial x)^2$, and

$$\left(\frac{\partial \tilde{u}}{\partial x}\right)^2 \sigma_1 = r^T \phi_{u,1} \phi_{u,1}^T \sigma_1 r \tag{9.119}$$

In Eq. (9.119),

$$\phi_{u,1} = \frac{1}{l}\begin{bmatrix} -1 \\ 1 \\ \dfrac{6y(2\zeta_1 - 1)}{l} \\ \dfrac{6y(2\zeta_2 - 1)}{l} \\ 2y(3\zeta_1 - 1) \\ -2y(3\zeta_2 - 1) \end{bmatrix} \tag{9.120}$$

In the product $\phi_{u,1}\phi_{u,1}^T\sigma_1$, three terms appear corresponding to the stresses produced by P, M_1, and M_2. Write

$$K_{GU} = K_{GU}{}^P + K_{GU}{}^{M_1} + K_{GU}{}^{M_2} \tag{9.121}$$

On integration, the three matrices are found to be

$$K_{GU}{}^P = \frac{P}{A}\begin{bmatrix} \dfrac{A}{l} & -\dfrac{A}{l} & 0 & 0 & 0 & 0 \\ -\dfrac{A}{l} & \dfrac{A}{l} & 0 & 0 & 0 & 0 \\ 0 & 0 & \dfrac{12I}{l^3} & -\dfrac{12I}{l^3} & -\dfrac{6I}{l^2} & \dfrac{6I}{l^2} \\ 0 & 0 & -\dfrac{12I}{l^3} & \dfrac{12I}{l^3} & -\dfrac{6I}{l^2} & \dfrac{6I}{l^2} \\ 0 & 0 & -\dfrac{6I}{l^2} & -\dfrac{6I}{l^2} & \dfrac{4I}{l} & \dfrac{2I}{l} \\ 0 & 0 & \dfrac{6I}{l^2} & \dfrac{6I}{l^2} & \dfrac{2I}{l} & \dfrac{4I}{l} \end{bmatrix} \tag{9.122}$$

This term has usually been neglected in K_G.

$$K_{GU}{}^{M_1} = -\frac{M_1}{l^2}\begin{bmatrix} 0 & 0 & -1 & 1 & -l & 0 \\ 0 & 0 & 1 & -1 & 0 & l \\ -1 & 1 & & & & \\ 1 & -1 & & \text{zero if section is} & & \\ -l & 0 & & \text{doubly symmetrical} & & \\ 0 & l & & & & \end{bmatrix} \tag{9.123}$$

$$K_{GU}{}^{M_2} = \frac{M_2}{l^2}\begin{bmatrix} 0 & 0 & 1 & -1 & 0 & l \\ 0 & 0 & -1 & 1 & -l & 0 \\ 1 & -1 & & & & \\ -1 & 1 & & \text{zero if section is} & & \\ 0 & -l & & \text{doubly symmetrical} & & \\ l & 0 & & & & \end{bmatrix} \tag{9.124}$$

In Eqs. (9.123) and (9.124), the terms not shown involve $\int_{\text{area}} y^3\, dA$, which is zero only when the section is doubly symmetrical.

A comparison of Eqs. (9.67) with Eqs. (9.123) and (9.124) shows, on allowing for a rearrangement of terms, that these expressions for $K_{GU}{}^{M_1}$ and $K_{GU}{}^{M_2}$ agree with that found previously from the shear force S_o. It must be emphasized that the above derivations based on the quadratic strain tensor allow for the inclusion of all effects and are not based on an ad hoc selection of the member forces.

Example 9.7 Calculation of the geometric-stiffness matrix for plane stress triangular elements

It must be noted in Examples 9.5 and 9.6 that the geometric-stiffness matrix calculation will vary depending on the displacement functions chosen to define member distortions. Both Examples 9.5 and 9.6 give K_G stiffness that with K_E will approximate the tangent stiffness in Eq. (9.84). It will be seen that Example 9.6 leads to considerably more complex expressions than are in Example 9.5, and such will also be the case for the more elaborate planar triangular elements. Generally then, the type of K_G matrix developed will correspond to that for the line element of Example 9.5. If the plate element has displacements (u,v,w), where (u,v) are the in-plane and w the transverse displacements, K_G for w when coupled with the plate bending element may be used in Eq. (9.50) for calculation of buckling loads and modes.

In the following, the transverse deflection w is taken to be a function of x, y only. For the triangle, the nodal values will be

$$r^T = \langle U^T \quad V^T \quad W^T \rangle \tag{9.125}$$

The displacements at point $\langle \zeta_1 \quad \zeta_2 \quad \zeta_3 \rangle$ are given in Eq. (9.126). In this equation it is assumed that the same interpolation polynomial ϕ applies to $\tilde{u}$ and $\tilde{v}$, whereas the ϕ_w for w may be different.

$$\begin{bmatrix} \tilde{u} \\ \tilde{v} \\ \tilde{w} \end{bmatrix} = \begin{bmatrix} \phi^T & 0 & 0 \\ 0 & \phi^T & 0 \\ 0 & 0 & \phi_W{}^T \end{bmatrix} \begin{bmatrix} U \\ V \\ W \end{bmatrix} \tag{9.126}$$

The displacement gradients with respect to x and y are obtained from Eq. (9.126) by differentiation of ϕ^T and $\phi_W{}^T$. Thus, with respect to x,

$$\begin{bmatrix} \tilde{u}_X \\ \tilde{v}_X \\ \tilde{w}_X \end{bmatrix} = \begin{bmatrix} \phi_X{}^T & 0 & 0 \\ 0 & \phi_X{}^T & 0 \\ 0 & 0 & \phi_{WX}{}^T \end{bmatrix} \begin{bmatrix} U \\ V \\ W \end{bmatrix} \tag{9.127}$$

and with respect to y,

$$\begin{bmatrix} \tilde{u}_y \\ \tilde{v}_y \\ \tilde{w}_y \end{bmatrix} = \begin{bmatrix} \phi_y{}^T & 0 & 0 \\ 0 & \phi_y{}^T & 0 \\ 0 & 0 & \phi_{Wy}{}^T \end{bmatrix} \begin{bmatrix} U \\ V \\ W \end{bmatrix} \tag{9.128}$$

Furthermore,

$$u_z = v_z = w_z = 0$$

From Eqs. (9.127) and (9.128), nodal gradient vectors U_X, V_X, W_X, etc., are obtained, and from these the interpolation polynomials ϕ_ε and $\phi_{\varepsilon w}$ are calculated. That is,

$$\begin{bmatrix} \tilde{u}_X \\ \tilde{v}_X \\ \tilde{w}_X \end{bmatrix} = \begin{bmatrix} \phi_\varepsilon{}^T & 0 & 0 \\ 0 & \phi_\varepsilon{}^T & 0 \\ 0 & 0 & \phi_{W\varepsilon}{}^T \end{bmatrix} \begin{bmatrix} U_X \\ V_X \\ W_X \end{bmatrix} \tag{9.129}$$

$$\begin{bmatrix} \tilde{u}_y \\ \tilde{v}_y \\ \tilde{w}_y \end{bmatrix} = \begin{bmatrix} \phi_\varepsilon{}^T & 0 & 0 \\ 0 & \phi_\varepsilon{}^T & 0 \\ 0 & 0 & \phi_{W\varepsilon}{}^T \end{bmatrix} \begin{bmatrix} U_Y \\ V_Y \\ W_Y \end{bmatrix} \tag{9.130}$$

The expressions U_X, V_X, U_Y, etc., are obtained, as mentioned, in Eq. (9.127) and will be expressed in the form

$$\begin{bmatrix} U_X \\ V_X \\ W_X \end{bmatrix} = \begin{bmatrix} M & 0 & 0 \\ 0 & M & 0 \\ 0 & 0 & M_W \end{bmatrix} \begin{bmatrix} U \\ V \\ W \end{bmatrix} \tag{9.131}$$

$$\begin{bmatrix} U_y \\ V_y \\ W_y \end{bmatrix} = \begin{bmatrix} N & 0 & 0 \\ 0 & N & 0 \\ 0 & 0 & N_W \end{bmatrix} \begin{bmatrix} U \\ V \\ W \end{bmatrix} \tag{9.132}$$

The terms of the quadratic strain tensor are

$$2\eta_{XX} = U_X{}^T\phi_\varepsilon\phi_\varepsilon{}^T U_X + V_X{}^T\phi_\varepsilon\phi_\varepsilon{}^T V_X + W_X{}^T\phi_{\varepsilon W}\phi_{\varepsilon W}{}^T W_X$$
$$2\eta_{yy} = U_y{}^T\phi_\varepsilon\phi_\varepsilon{}^T U_y + V_y{}^T\phi_\varepsilon\phi_\varepsilon{}^T V_y + W_y{}^T\phi_{\varepsilon W}\phi_{\varepsilon W}{}^T W_y$$
$$\begin{aligned}2(\eta_{xy} + \eta_{yx}) = {} & U_X{}^T\phi_\varepsilon\phi_\varepsilon{}^T U_y + U_y{}^T\phi_\varepsilon\phi_\varepsilon{}^T U_X + V_X{}^T\phi_\varepsilon\phi_\varepsilon{}^T V_y \\ & + V_y{}^T\phi_\varepsilon\phi_\varepsilon{}^T V_X + W_X{}^T\phi_\varepsilon\phi_\varepsilon{}^T W_y + W_y{}^T\phi_\varepsilon\phi_\varepsilon{}^T W_X\end{aligned}$$

The in-plane stresses are given by

$$\begin{bmatrix}\tilde{\sigma}_x \\ \tilde{\sigma}_y \\ \tilde{\tau}_{xy}\end{bmatrix} = \begin{bmatrix}\phi_\sigma{}^T\sigma_x \\ \phi_\sigma{}^T\sigma_y \\ \phi_\sigma{}^T\sigma_{xy}\end{bmatrix}$$

The values $\langle\sigma_x \quad \sigma_y \quad \tau_{xy}\rangle$ are, as usual, the nodal stress vectors. The stresses $\tilde{\sigma}_x$, $\tilde{\sigma}_y$, $\tilde{\tau}_{xy}$ are of course components of the stress tensor, so that

$$\begin{bmatrix}\tilde{\sigma}_x & \tilde{\tau}_{xy} \\ \tilde{\tau}_{yx} & \tilde{\sigma}_y\end{bmatrix} = \begin{bmatrix}\sigma_{11} & \sigma_{12} \\ \sigma_{21} & \sigma_{22}\end{bmatrix}$$

The terms of K_G as given in Eq. (9.100) may now be generated from the substitution of the above expressions in Eq. (9.97). For the two-dimensional stress state, from Eq. (9.99) the value for K_{G11} is

$$\begin{aligned}K_{G11} = \int_{\text{vol}} [& \phi_{1,1}\phi_{1,1}^T(\phi_\sigma{}^T\sigma_{11}) + \phi_{1,1}\phi_{1,2}^T(\phi_\sigma{}^T\sigma_{12}) \\ & + \phi_{1,2}\phi_{1,1}^T(\phi_\sigma{}^T\sigma_{21}) + \phi_{1,2}\phi_{1,2}^T(\phi_\sigma{}^T\sigma_{22})]\end{aligned} \tag{9.133}$$

In the present rotation,

$$\phi_{1,1}^T = \phi_\varepsilon{}^T M$$
$$\phi_{2,1}^T = \phi_\varepsilon{}^T M$$
$$\phi_{1,2}^T = \phi_\varepsilon{}^T N$$
$$\phi_{2,2}^T = \phi_\varepsilon{}^T N$$

Then,

$$\begin{aligned}K_{G11} = \int_{\text{vol}} [& M^T\phi_\varepsilon\phi_\varepsilon{}^T M(\phi_\sigma{}^T\sigma_x) + M^T\phi_\varepsilon\phi_\varepsilon{}^T N(\phi_\sigma{}^T\tau_{xy}) \\ & + N^T\phi_\varepsilon\phi_\varepsilon{}^T M(\phi_\sigma{}^T\tau_{yx}) + N^T\phi_\varepsilon\phi_\varepsilon{}^T N(\phi_\sigma{}^T\sigma_y)]\end{aligned} \tag{9.134}$$

In matrix form,

$$K_{G11} = [M^T \quad N^T] \int_{\text{vol}} \begin{bmatrix}\phi_\varepsilon\phi_\varepsilon{}^T(\phi_\sigma{}^T\sigma_x) & \phi_\varepsilon\phi_\varepsilon{}^T(\phi_\sigma{}^T\tau_{xy}) \\ \phi_\varepsilon\phi_\varepsilon{}^T(\phi_\sigma{}^T\tau_{yx}) & \phi_\varepsilon\phi_\varepsilon{}^T(\phi_\sigma{}^T\sigma_y)\end{bmatrix}\begin{bmatrix}M \\ N\end{bmatrix} \tag{9.135}$$

All terms are now multiplied by the thickness $\tilde{h}$ given in terms of the nodal values

$$\tilde{h} = h_r\phi_\xi{}^T\xi \tag{9.136}$$

Write Eq. (9.135) in the form

$$K_{G11} = [M^T \quad N^T]\begin{bmatrix} J_{xx} & J_{xy} \\ J_{yx} & J_{yy} \end{bmatrix}\begin{bmatrix} M \\ N \end{bmatrix} \tag{9.137}$$

Then,

$$J_{xx} = h_r \int_{\text{area}} \phi_\varepsilon(\phi_\sigma{}^T\sigma_x)(\phi_\xi{}^T\xi)\phi_\varepsilon{}^T\, dA \tag{9.138}$$

Similar expressions are obtained for J_{xy} and J_{yy} by substitution for τ_{xy} and σ_y.

Consider now the constant stress triangle. The stress field σ_x, σ_y, τ_{xy} is constant, and so also is the strain field. Thence,

$$\phi_\varepsilon = \phi_\sigma = 1$$

If the thickness is constant,

$$J_{xx} = h_r \int_{\text{area}} \phi_\varepsilon(\phi_\sigma{}^T\sigma_x)(\phi_\xi{}^T\xi)\phi_\varepsilon{}^T\, dA = Ah\sigma_x$$

Similarly,

$$J_{xy} = Ah\tau_{xy}$$
$$J_{yy} = Ah\sigma_y$$

In this case, Eqs. (9.127) and (9.128) are given by

$$\begin{bmatrix} \tilde{u}_x \\ \tilde{v}_x \end{bmatrix} = \frac{1}{2A}\begin{bmatrix} b_1 & b_2 & b_3 & 0 & 0 & 0 \\ 0 & 0 & 0 & b_1 & b_2 & b_3 \end{bmatrix}\begin{bmatrix} u_i \\ u_j \\ u_k \\ v_i \\ v_j \\ v_k \end{bmatrix}$$

$$\begin{bmatrix} \tilde{u}_y \\ \tilde{v}_y \end{bmatrix} = \frac{1}{2A}\begin{bmatrix} a_1 & a_2 & a_3 & 0 & 0 & 0 \\ 0 & 0 & 0 & a_1 & a_2 & a_3 \end{bmatrix}\begin{bmatrix} u_i \\ u_j \\ u_k \\ v_i \\ v_j \\ v_k \end{bmatrix}$$

Since any point in the triangle may be used to describe the constant stress state, it is seen in Eqs. (9.131) that

$$M = \frac{1}{2A}[b_1 \quad b_2 \quad b_3]$$

$$N = \frac{1}{2A}[a_1 \quad a_2 \quad a_3]$$

Then,

$$K_{G11} = \frac{h}{4A}\begin{bmatrix} b_1 & a_1 \\ b_2 & a_2 \\ b_3 & a_3 \end{bmatrix}\begin{bmatrix} \sigma_x & \tau_{xy} \\ \tau_{yx} & \sigma_y \end{bmatrix}\begin{bmatrix} b_1 & b_2 & b_3 \\ a_1 & a_2 & a_3 \end{bmatrix} \tag{9.139}$$

It will be seen from Eqs. (9.131) and (9.132) that

$$K_{G12} = K_{G22}$$

For $K_{GW} = K_{G33}$ a similar expression is obtained with M and N replaced by M_W and N_W respectively. If $\phi_W = \phi$, that is, w is assumed to be given by the same interpolation formula as that for u and v,

$$K_{Gw} = K_{G11} = K_{G22}$$

For large-displacement theory of plate bending for which rotations from w are the contributing nonlinear effects,

$$K_G = \begin{bmatrix} 0 & 0 & 0 \\ 0 & 0 & 0 \\ 0 & 0 & K_{GW} \end{bmatrix} \tag{9.140}$$

Linear-strain Triangle with Constant Thickness From Eq. (9.137) for K_{G11},

$$J_{xx} = h\int_{\text{area}} \phi_\varepsilon(\phi_\sigma{}^T\sigma_x)\phi_\varepsilon{}^T\,dA$$

The displacements vary quadratically, so that displacement gradients (and strains) vary linearly. The displacement vectors will give values at the apices (i,j,k) and midpoints (l,m,n) of sides of the triangle. That is,

$$U^T = (u_i, v_j, u_k, u_l, u_m, u_n)$$
$$V^T = (v_i, v_j, v_k, v_l, v_m, v_n)$$

Then

$$\tilde{u} = \phi_2{}^T U$$
$$\tilde{v} = \phi_2{}^T V$$

in which

$$\phi_2{}^T = [\zeta_1(2\zeta_1 - 1) \quad \zeta_2(2\zeta_2 - 1) \quad \zeta_3(2\zeta_3 - 1) \quad 4\zeta_1\zeta_2 \quad 4\zeta_2\zeta_3 \quad 4\zeta_3\zeta_1]$$

The displacement gradients are given by

$$\tilde{u}_x = \phi_{2x}U \qquad \tilde{u}_y = \phi_{2y}U$$
$$\tilde{v}_x = \phi_{2x}V \qquad \tilde{v}_y = \phi_{2y}V$$

It is seen that

$$\phi_{2x}{}^T = \frac{1}{2A}[b_1(4\zeta_1 - 1) \quad b_2(4\zeta_2 - 1) \quad b_3(4\zeta_3 - 1) \quad 4(\zeta_2 b_1 + \zeta_1 b_2)$$
$$4(\zeta_3 b_2 + \zeta_2 b_3) \quad 4(\zeta_1 b_3 + \zeta_3 b_1)]$$

$$\phi_{2y}{}^T = \frac{1}{2A}[a_1(4\zeta_1 - 1) \quad a_2(4\zeta_2 - 1) \quad a_3(4\zeta_3 - 1) \quad 4(\zeta_2 a_1 + \zeta_1 a_2)$$
$$4(\zeta_3 a_2 + \zeta_2 a_3) \quad 4(\zeta_1 a_3 + \zeta_3 a_1)]$$

On substitution of the coordinates of i, j, and k in the expressions for $\tilde{u}_x$, $\tilde{u}_y$, etc., the matrices M and N are obtained, which relate nodal strains to nodal displacements. Thus

$$\begin{bmatrix} u_{xi} \\ u_{xj} \\ u_{xk} \end{bmatrix} = \frac{1}{2A}\begin{bmatrix} 3b_1 & -b_2 & -b_3 & 4b_2 & 0 & 4b_3 \\ -b_1 & 3b_2 & -b_3 & 4b_1 & 4b_3 & 0 \\ -b_1 & -b_2 & 3b_3 & 0 & 4b_2 & 4b_1 \end{bmatrix} U$$

Then,

$$M = \frac{1}{2A}\begin{bmatrix} 3b_1 & -b_2 & -b_3 & 4b_2 & 0 & 4b_3 \\ -b_1 & 3b_2 & -b_3 & 4b_1 & 4b_3 & 0 \\ -b_1 & -b_2 & 3b_3 & 0 & 4b_2 & 4b_1 \end{bmatrix} \tag{9.141}$$

Similarly,

$$\begin{bmatrix} u_{yi} \\ u_{yj} \\ u_{yk} \end{bmatrix} = \frac{1}{2A}\begin{bmatrix} 3a_1 & -a_2 & -a_3 & 4a_2 & 0 & 4a_3 \\ -a_1 & 3a_2 & -a_3 & 4a_1 & 4a_3 & 0 \\ -a_1 & -a_2 & 3a_3 & 0 & 4a_2 & 4a_1 \end{bmatrix} U$$

That is,

$$N = \frac{1}{2A}\begin{bmatrix} 3a_1 & -a_1 & -a_3 & 4a_2 & 0 & 4a_3 \\ -a_1 & 3a_2 & -a_3 & 4a_1 & 4a_3 & 0 \\ -a_1 & -a_2 & 3a_3 & 0 & 4a_2 & 4a_1 \end{bmatrix} \tag{9.142}$$

In terms of the nodal displacement gradients,

$$\tilde{u}_x = (\zeta_1, \zeta_2, \zeta_3)\begin{bmatrix} u_{xi} \\ u_{xj} \\ u_{xk} \end{bmatrix} \qquad \text{etc.}$$

That is,

$$\phi_\varepsilon{}^T = (\zeta_1, \zeta_2, \zeta_3) = \zeta^T$$

If the stresses vary linearly such that

$$\tilde{\sigma}_x = \phi_\sigma{}^T\sigma_x = \zeta^T\sigma_x$$

then

$$J_{xx} = h\int_{\text{area}} \zeta\zeta^T\sigma_x\zeta^T\,dA$$

On multiplication of the matrices under the integration sign and then integrating term by term, it is found that

$$J_{xx} = \frac{Ah}{60}\left\{\begin{bmatrix} 6 & 2 & 2 \\ 2 & 2 & 1 \\ 2 & 1 & 2 \end{bmatrix}\sigma_{xi} + \begin{bmatrix} 2 & 2 & 1 \\ 2 & 6 & 2 \\ 1 & 2 & 2 \end{bmatrix}\sigma_{xj} + \begin{bmatrix} 2 & 1 & 2 \\ 1 & 2 & 2 \\ 2 & 2 & 6 \end{bmatrix}\sigma_{xk}\right\} \tag{9.143}$$

The terms J_{xy} and J_{yy} are constructed in a similar manner. Then K_{G11} is constructed from Eq. (9.137) and is of course a 6×6 matrix. The block K_{G22} is the same as K_{G11}. The block K_{G33} for the transverse displacements will depend on the displacement function chosen for $\tilde{w}$. For example, suppose $\tilde{w}$ varies linearly so that $\phi_{\varepsilon W}$ is a constant. Then for linear stresses,

$$J_{xx} = h\int_{\text{area}} \zeta^T\sigma_x\,dA = \frac{Ah}{3}(\sigma_{xi} + \sigma_{xj} + \sigma_{xk}) \tag{9.144}$$

In this case,

$$M_W = \frac{1}{2A}[b_1 \quad b_2 \quad b_3]$$

$$N_W = \frac{1}{2A}[a_1 \quad a_2 \quad a_3]$$

When the variation in $\tilde{w}$ is assumed parabolic,

$$K_{G11} = K_{G22} = K_{G33}$$

9.6 ELASTOPLASTIC ANALYSIS OF STRUCTURES

The nonlinear structure behavior discussed so far has been associated with geometry changes that affect member stiffness as well as the joint equations of equilibrium. The analysis has been approached in an incremental form for which the tangent stiffness at the start of the load interval is calculated. This involves the concept of the geometric-stiffness matrix as well as the geometry change on the elastic stiffness. The structure analysis is further complicated if the material has a nonlinear stress-strain curve such as one of those shown in Fig. 9.4. If the material exhibits rigid plastic properties such as shown in Fig. 9.4*c*, curve *B*, the analysis proceeds by using the limit theorems, in which plastic mechanisms are investigated to obtain a lower-bound collapse load for the structure.

In elastoplastic analysis, the problem is further complicated by the dependence of plastic strain on the previous stress-strain history of the material. For example, if a mild-steel member in a frame is deformed into the plastic range and then straightened, its elastic moduli will be as for the original member, but the plastic behavior will be altered by the strain history. Again, a member may suffer stress reversal in a loading cycle so that in one stage it may be elastic and in a subsequent stage, plastic and then still later, elastic.

In the discussion that follows, it will be assumed that member stress-strain or load-deflection curves are known prior to the loading sequence. In such a case it should be evident that the incremental analysis may follow the same steps as for the large-displacement analysis. That is, the incremental load-deflection relationship is written

$$\Delta R = (K_{ET} + K_G)\,\Delta r \tag{9.145}$$

In Eq. (9.145), K_G is calculated as in the previous secs. 9.2 to 9.5, whereas K_{ET} is now the tangent stiffness based on the member properties at the current force level.

Alternatively, the stiffness method may be employed in an iterative procedure that calculates the increments of plasticity for load increment ΔR from the load state R_{n-1}. This method recognizes the fact that when a structure has been strained plastically from load level R_{n-1} to $R_n = R_{n-1} + \Delta R$, then unloaded to R_{n-1}, then reloaded, the presence of residual stresses produces elastic behavior until the stress levels of the first loading are exceeded (see Ref. 35). Thus, to the elastic analysis for ΔR, there must be added the forces from the initial plastic distortions of members. As these distortions are unknown, a process of iteration is used. That is, the elastic analysis is carried out and the plastic deformations for member forces calculated. A proportion of these strains is used as the initial lack of fit in the members for the next cycle of iteration. In this process, some care is required in selecting the proportion to be used so that the solution proceeds monotonically to the next equilibrium position rather than oscillate or diverge.

For the load increment ΔR, let v be the initial lack-of-fit vector for the member with stiffness k. The forces J required to completely constrain the member to the position of zero distortion (or that at load level R_{n-1}) will be

$$J = -kv \tag{9.146}$$

Member forces S_n are given by

$$S_n = S_{n-1} + ka\,\Delta r + J \tag{9.147}$$

The node forces to be used in the calculation of Δr must take account of the constraint forces J as well as the load increment ΔR. Then,

$$\Delta\bar{R} = \Delta R - a^T J \tag{9.148}$$

$$\Delta r = K^{-1}\,\Delta\bar{R} = K^{-1}(\Delta R - a^T J) \tag{9.149}$$

In the direct stiffness variation, J will be given in the global coordinates if k is the member stiffness in the global system, and then a is a unit matrix. This implies that in Eq. (9.147) the S_n forces are global components.

The two methods of elastoplastic analysis are shown diagrammatically in Fig. 9.24. That is, the analysis proceeds either:

1 By using the tangent stiffness $K_{ET} + K_G$

2 From the initial strain position A, by using the elastic stiffness $K_E + K_G$

The above procedures may be followed in the finite-element method of analysis of the continuum, where now, the material-property curves in Fig. 9.4 may be the equivalent stress-strain curves for a material with a yield criterion such as that given by the von Mises–Hencky theory.

In the analysis of framed structures for which the stress states can be considered generally rather simple, that is, direct tension or compression, these curves can be generalized to be load-deflection or moment-curvature diagrams. The application to the two- and three-dimensional stress states found in finite elements is not discussed herein.

The simplest material-property curve that may be considered is that given in Fig. 9.25, that is, an approximation based on two straight lines. It will be

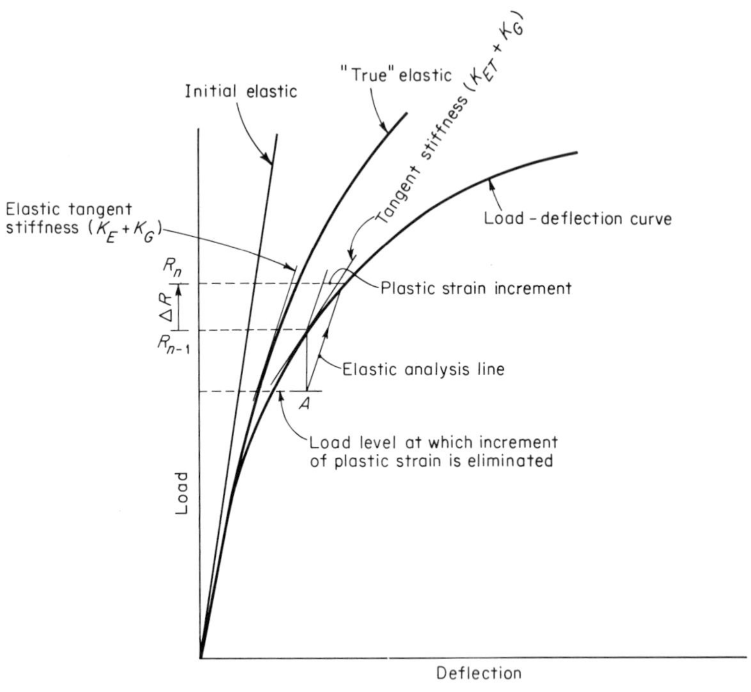

Figure 9.24 Load-deflection curve.

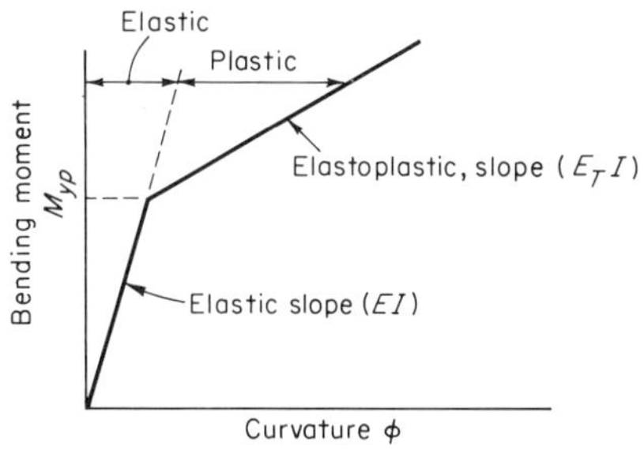

Figure 9.25 Moment-curvature diagram.

considered a moment-curvature diagram for a member in a plane frame structure. Axial forces are considered small, so that their influence on the moment curve is also small and may be neglected.

From Fig. 9.25, it is seen that the moment-curvature relationships are as follows:

1 When $M < M_{yp}$:

$$\phi = \frac{M}{EI} \tag{9.150a}$$

2 When $M > M_{yp}$:

$$\phi = \frac{M}{EI} + \left(\frac{1}{E_T I} - \frac{1}{EI}\right)(M - M_{yp}) \tag{9.150b}$$

Notice that even in the plastic range, elastic deformation occurs.

To consider the initial-strain method of analysis first, it will be necessary to calculate plastic strains along a member when the end moments $\langle M_i \quad M_j \rangle$ are given. Various situations arise depending on whether M_i and M_j exceed M_{yp} and on the signs of these moments. The possibilities are shown in Fig. 9.26.

From Fig. 9.26, the following plastic conditions may exist:

1 M_i, M_j of the same sign (Fig. 9.26*a*)

a If $|M_i| > M_{yp}$, $\quad p = \dfrac{(|M_i| - M_{yp})l}{|M_i + M_j|}$

b If $|M_i| \leqslant M_{yp}$, $\quad p = 0$

c If $|M_j| > M_{yp}$, $\quad q = \dfrac{(|M_i| - M_{yp})l}{|M_i + M_j|}$

d If $|M_j| \leqslant M_{yp}$, $\quad q = 0$

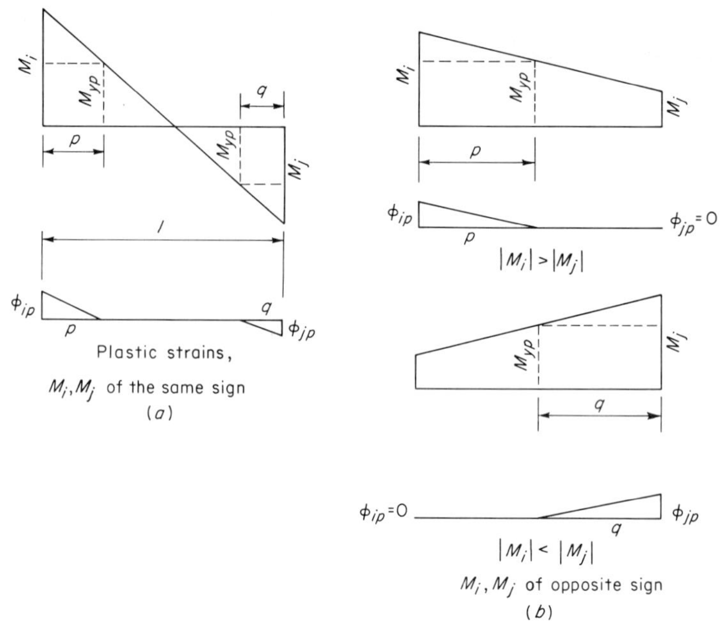

Figure 9.26 Plastic strains.

2 M_i, M_j of opposite sign (Fig. 9.26*b*)

e If $|M_i| > M_{yp}$ and $|M_j| > M_{yp}$,

$$p = q = l \qquad \text{fully plastic}$$

f If $|M_i| > M_{yp}$ and $|M_j| \leqslant M_{yp}$,

$$p = \frac{(|M_i| - M_{yp})l}{|M_i| - |M_j|} \qquad q = 0$$

g If $|M_i| \leqslant M_{yp}$ and $|M_j| > M_{yp}$,

$$p = 0 \qquad q = \frac{(|M_j| - M_{yp})l}{|M_j| - |M_i|}$$

From these zones of plasticity, defined by the distances p and q, and from the plastic curvatures ϕ_{ip} and ϕ_{jp} at the ends of the member, it is possible to calculate the end rotations υ_{ip} and υ_{jp} by the principle of virtual displacements, with the equilibrium force systems' being, in turn, $M_i = 1$ and $M_j = 1$. That is,

$$\begin{bmatrix} \upsilon_{ip} \\ \upsilon_{jp} \end{bmatrix} = \frac{1}{6l} \begin{bmatrix} p(3l - p) & -q^2 \\ -p^2 & q(3l - q) \end{bmatrix} \begin{bmatrix} \phi_{ip} \\ \phi_{jp} \end{bmatrix} \tag{9.151}$$

The plastic curvatures ϕ_{ip} and ϕ_{jp} at the member ends must be calculated with due regard to the signs of M_i and M_j. Let

$$\alpha_i = 1 - \frac{M_{yp}}{|M_i|} \qquad \alpha_j = 1 - \frac{M_{yp}}{|M_j|}$$

$$C = \frac{1}{E_T I} - \frac{1}{EI}$$

Then,

$$\begin{bmatrix} \phi_{ip} \\ \phi_{jp} \end{bmatrix} = C \begin{bmatrix} \alpha_i M_i \\ \alpha_j M_j \end{bmatrix}$$

Substitution in Eq. (9.151) gives the plastic rotations in terms of the end moments to be

$$\begin{bmatrix} v_{ip} \\ v_{jp} \end{bmatrix} = \frac{C}{6l} \begin{bmatrix} p(3l-p) & -q^2 \\ -p^2 & q(3l-q) \end{bmatrix} \begin{bmatrix} \alpha_i M_i \\ \alpha_j M_j \end{bmatrix} \tag{9.152}$$

The forces necessary to completely constrain the member ends elastically will be

$$\begin{bmatrix} M_{iF} \\ M_{jF} \end{bmatrix} = -\frac{2EI}{l} \begin{bmatrix} 2 & 1 \\ 1 & 2 \end{bmatrix} \begin{bmatrix} v_{ip} \\ v_{jp} \end{bmatrix} \tag{9.153}$$

The release forces to be applied to the structure at the nodes will be

$$R_F = -a^T J = a^T \bar{k} v = kv \tag{9.154}$$

In Eq. (9.154), $\bar{k}$ is the member-stiffness matrix given in Eq. (9.153), and a is the displacement transformation matrix relating member distortions to node displacements. The stiffness matrix k is the 6×6 matrix for member stiffness in terms of the global components at the ends of the member. Then,

$$v^T = \langle 0 \quad 0 \quad v_{ip} \quad 0 \quad 0 \quad v_{jp} \rangle \tag{9.155}$$

From Fig. 9.24 it is seen that the process of obtaining the load-deflection and load-member force relationships is both an incremental and iterative one. At the load level R_{n-1}, the plastic strains for the next load increment ΔR may in the first instance be taken as those for the previous increment. When the plastic strains are calculated from the current load increment, it will be usual to find that they are overestimated. To ensure convergence, only proportions of v_{ip} and v_{jp} are used. The proportion to be used may be found by experiment and may be as low as one-tenth or one-twentieth.

From Fig. 9.24 it is seen that in the elastic analysis from the initial strain position (point A in the figure), the tangent stiffness $K_G + K_E$ should be

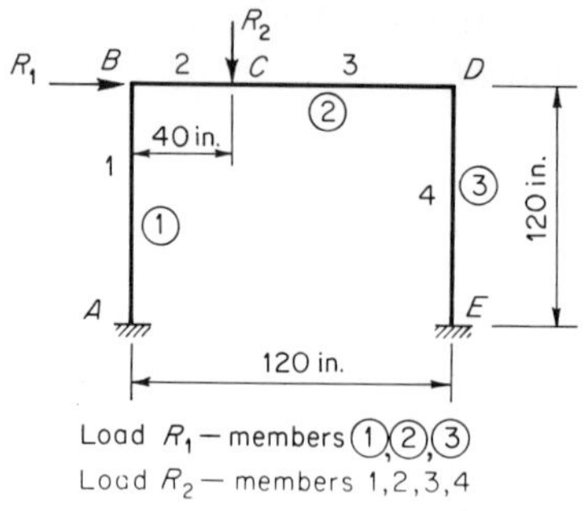

Figure 9.27 Portal-frame loads and dimensions.

used. The method of analysis [Eqs. (9.146) ff] may have difficulty in following the deflected shape of a structure that experiences snap through phenomena during which load decrements may occur.

Example 9.8 Elastoplastic analysis of a frame structure (small-displacement theory)

A single-bay portal frame with fixed bases is shown in Fig. 9.27. It is composed of a 10- by 4½-in. by 25-lb steel I beam whose yield moment is $M_{yp} = 687$ in.-kips and full plastic moment of resistance is $M_{\text{ult}} = 800$ in.-kips. The continuous-moment-curvature diagram is shown in Fig. 9.28, together with the two-line approximation for which $M_{yp} = 730$ in.-kips. The ratio of the slopes of the two lines is 100 to 1.

Two load cases have been analyzed to show different patterns of plastic deformation. The first consists of a horizontal load R_1 applied at the top of the columns, and the second of a vertical force R_2 downward on the beam at point C, a distance of 40 in. from the left-hand column. The resulting load-deflection curves are shown in Fig. 9.29*a* and *b*. The load–member-bending-moment curves are drawn for the critical sections in Fig. 9.30. Both the curves in Figs. 9.30 and 9.31 have been taken from values given in Table 9.2. Notice also, that even for a vertical load of 75 kips,

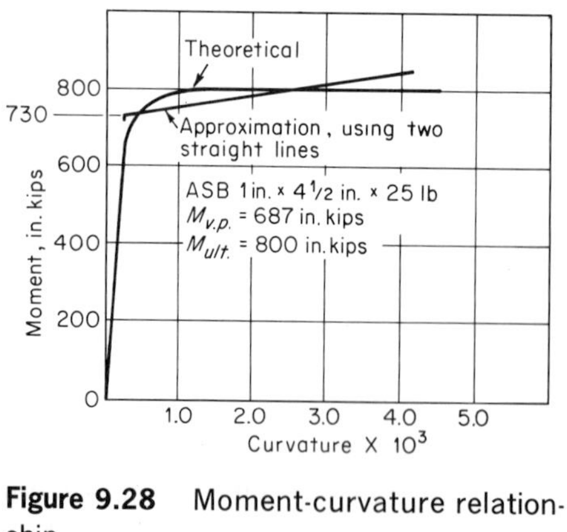

Figure 9.28 Moment-curvature relationship.

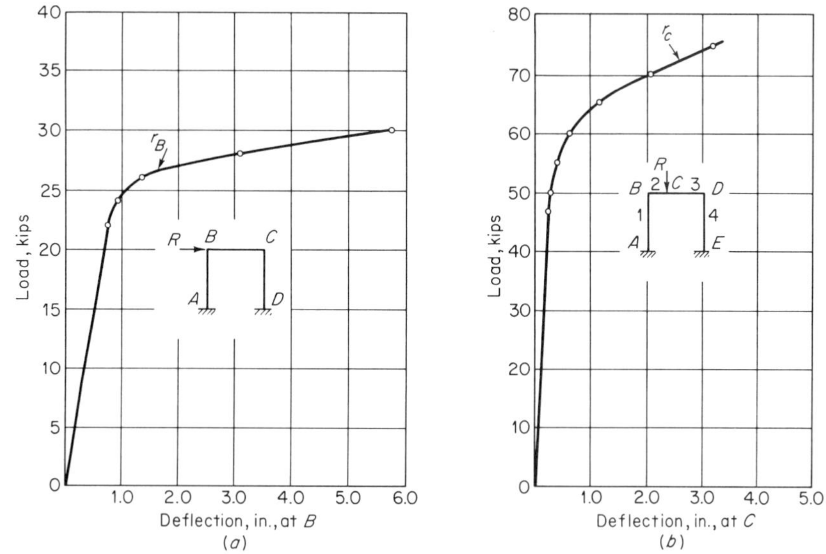

Figure 9.29 Load-deflection curves.

the maximum beam deflection is only 3.21 in. This would suggest that the geometry changes are not significant for this load pattern.

Notice in Fig. 9.30*b* that for the beam mechanism, considerably more plastic strain occurs before the plastic moments ($M_{\text{ult}} = 800$ in.-kips) are reached than for the sidesway type of loading in Fig. 9.30*a*. The presence of strain hardening does, of course, obscure the actual collapse mechanism.

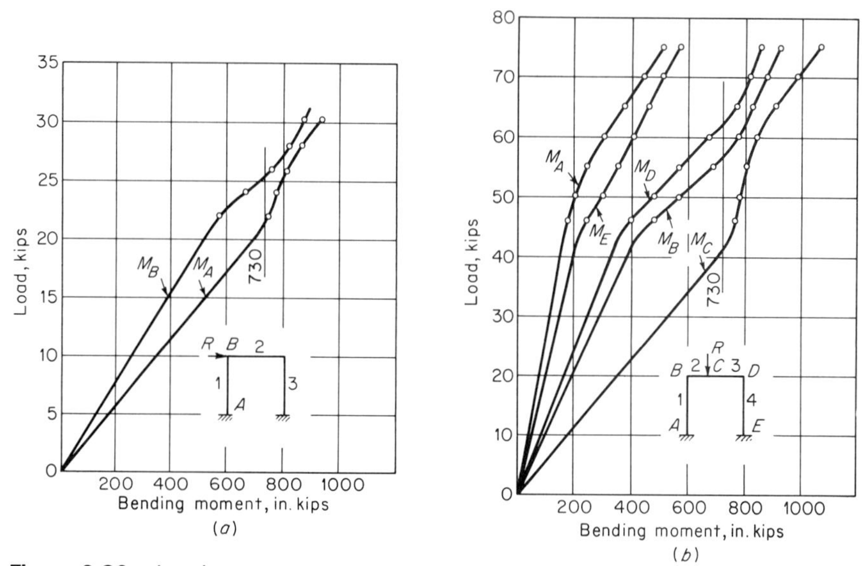

Figure 9.30 Load-moment curves.

TABLE 9.2 **(*a*) Horizontal load**

Elastoplastic analyses

Load, kips	Member 1 M_i	Member 1 M_j	Member 1 p	Member 1 q	Member 2 M_i	Member 2 M_j	Member 2 p	Member 2 q	Member 3 M_i	Member 3 M_j	Member 3 p	Member 3 q	δ_H, in.
22	747.9	573.7	1.60	...	−573.7	−571.7	...	...	571.7	746.8	...	1.50	0.762
24	772.9	668.4	3.55	...	−668.4	−666.3	...	...	666.3	772.3	...	3.50	0.960
26	801.5	758.9	5.48	2.20	−758.9	−758.5	2.26	2.23	758.5	801.1	2.17	5.45	1.379
28	865.8	815.2	9.67	6.06	−815.2	−814.6	6.25	6.21	814.6	864.4	6.03	9.58	3.095
30	933.3	866.9	13.53	9.11	−866.8	−866.7	9.45	9.45	866.7	933.1	9.10	13.52	5.739
Elastic analysis													
22	759.3	565.0	...	...	−565.0	−562.1	...	...	562.1	753.5	...	...	0.733

(*b*) Vertical load

Elastoplastic analyses

Load, kips	Member 1 M_i	Member 1 M_j	Member 1 p	Member 1 q	Member 2 M_i	Member 2 M_j	Member 2 p	Member 2 q	Member 3 M_i	Member 3 M_j	Member 3 p	Member 3 q	Member 4 M_i	Member 4 M_j	Member 4 p	Member 4 q	δ_v, in.
46	−186.9	−482.3	...	...	482.3	768.0	...	1.21	− 768.0	−411.5	2.56	...	411.5	257.7	...	...	0.263
50	−216.0	−576.5	...	...	576.5	788.4	...	1.70	− 788.4	−481.9	3.66	...	481.9	310.6	...	...	0.328
55	−253.9	−697.7	...	...	697.7	810.4	...	2.13	− 810.4	−573.2	4.63	...	573.2	378.4	...	...	0.415
60	−317.6	−781.8	...	5.62	781.8	850.7	1.26	2.95	− 850.7	−684.2	6.28	...	684.2	415.3	...	...	0.644
65	−390.2	−831.9	...	9.98	831.9	917.6	2.32	4.28	− 917.6	−783.4	8.81	2.50	783.4	438.7	5.21	...	1.196
70	−438.8	−887.0	...	14.19	887.0	999.5	3.32	5.71	− 999.5	−827.4	11.79	4.25	827.4	498.4	8.79	...	2.107
75	−507.6	−942.4	...	17.55	942.4	1082.1	4.19	6.95	−1082.1	−868.9	14.43	5.69	868.9	581.1	11.48	...	3.214
Elastic analysis																	
50	−191.0	−473.5	...	...	473.5	879.7	...	...	−879.7	−414.0	...	...	414.0	250.5	...	...	0.244

Frame or Beam Members with Strain Hardening Included

The same method of analysis as that given above may be used when the moment-curvature diagram is as given in Fig. 9.31. It must be realized that the two-line approximation given in Fig. 9.25 cannot be used for a material with no strain hardening in the plastic zone, because theoretically the plastic curvature strains would then be infinite. When strain hardening is included, as in Fig. 9.31, the intermediate slope E_1I can be made as flat as required.

For the member with end moments M_i and M_j the conditions arising when these values exceed the yield moment are shown in Fig. 9.32 and are very similar to those for the two-line case shown in Fig. 9.26.

The plastic end rotations are calculated in an identical manner with those for the two-line approximation, and hence,

$$\begin{bmatrix} v_{ip} \\ v_{jp} \end{bmatrix} = \frac{1}{6l}\begin{bmatrix} c(3l-c) & -d^2 \\ -c^2 & d(3l-d) \end{bmatrix}\begin{bmatrix} \phi_{s2i} \\ \phi_{s2j} \end{bmatrix} + \frac{1}{6l}\begin{bmatrix} (3l-c)(p+c)-p^2 & -q(q+d)-d^2 \\ -p(p+c)-c^2 & (3l-d)(q+d)-q^2 \end{bmatrix}\begin{bmatrix} \phi_{s1i} \\ \phi_{s1j} \end{bmatrix} \tag{9.156}$$

When $|M_i| < M_s$ and $|M_j| < M_s$, then $c = d = 0$, and then Eq. (9.156) reduces to the two-line approximation for which ϕ_{s1i} and ϕ_{s1j} are the intercepts of the M value on the line of slope E_1I.

The expressions used to calculate the distances p and q for the spread of plasticity along the member are given by the equations listed under a to g for the two-line case (see page 579). To calculate c and d, it is necessary only to replace M_{yp} and M_s in these same expressions. Having calculated v_{ip} and v_{jp}, the analysis proceeds in a fashion identical with that given previously [see Eqs. (9.153)ff].

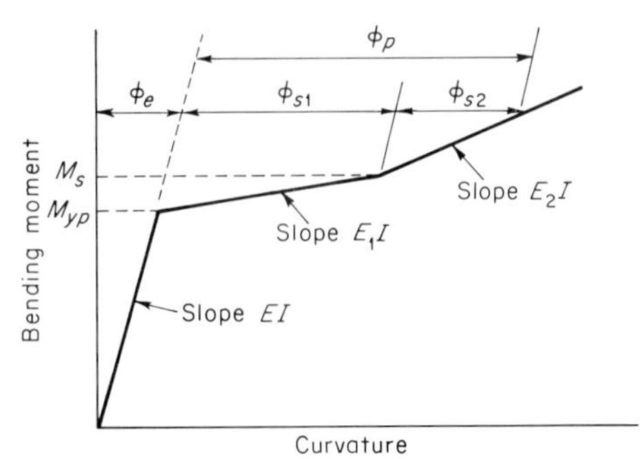

Figure 9.31 Three-line moment-curvature diagram.

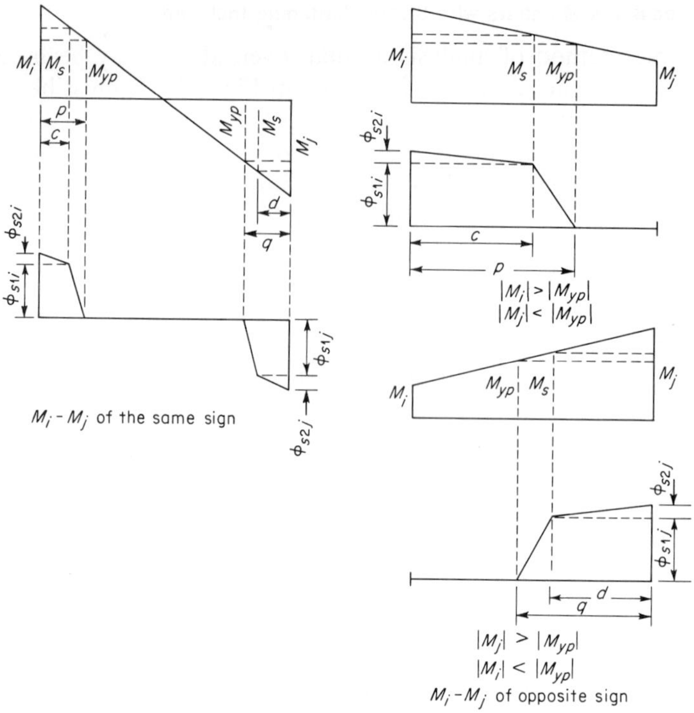

Figure 9.32 Plastic strains; material with three-line approximation to curvature diagram.

The General Moment-curvature Relationship

When the member under consideration has variable moment of inertia or the material-property curve is approximated by a general function instead of the series of straight lines, it will be found that the zones of plasticity along the member will not vary in the linear manner shown in Figs. 9.26 and 9.32. It may be inconvenient to develop explicit expressions for the end rotations v_{ip} and v_{jp}, and better, instead, to calculate these values by numerical quadrature. That is, the values of the plastic curvature strains are calculated at several defined points along the member and then Gauss' or Simpson's quadrature used to integrate these values with the unit-moment diagrams to obtain the end rotations.

Tangent Structure Stiffness

The incremental load-deflection relationship for a structure has been written in Eq. (9.145) as

$$\Delta R = (K_{ET} + K_G)\, \Delta r$$

When zones of plastic flow occur in which the member stiffness is reduced, the effect of these must be accounted for in K_{ET}. In the incremental approach, K_{ET} may be calculated by considering the conditions that exist at the beginning of the load increment ΔR. Alternatively, an iterative approach may be used in which the plastic zones over the load range ΔR are taken at some intermediate level between the previous values and the newly calculated ones. When the moment-curvature relationship of a beam element is approximated by two straight lines as in Fig. 9.25, the reduced stiffness for the member is readily calculated. If a more complicated analytic function is used to describe the M/ϕ curve, the stiffness $E_T I$ will vary over the zone of plasticity as a function of M, and numerical quadrature may be used to calculate the tangent stiffness in a manner similar to calculation of plastic end rotations for the same situation.

It is informative for the present discussion to extract the values of p and q (the extent of the zones of plasticity) from Table 9.2 for one of the load levels, for example, for the load $R_1 = 28$ kips for member 1, and from these to construct the stiffness matrix K_{ET}. In this way the deflection for the load increment of 2 kips can be calculated and compared with the value obtained in Table 9.2, δ_H, by using the initial-strain method.

Calculation of the Reduced Member Stiffness

The member whose M/ϕ curve is as in Fig. 9.25 is shown in Fig. 9.33 with the zones p and q of reduced stiffness $E_T I$.

For the increment of load, the lengths p and q are considered constant, so that the member-stiffness matrix can be calculated as the inverse of the flexibility matrix for the structure composed of the three members (1), (2), and (3). Now,

$$r = b^T f b R_1$$

Hence

$$K = F^{-1} = (b^T f b)^{-1}$$

see Eq. (6.2)

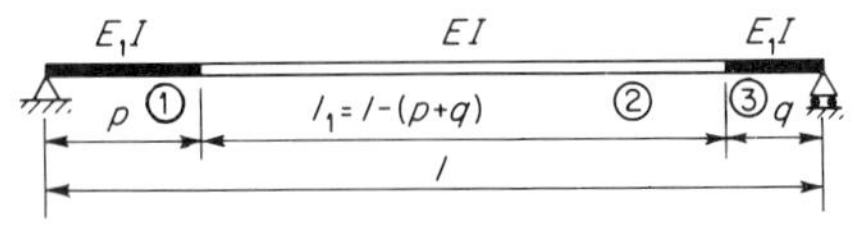

Figure 9.33 Change in member flexibility (stiffness).

In the present case,

$$b = \begin{bmatrix} 1 & 0 \\ 1 - \dfrac{p}{l} & \dfrac{p}{l} \\ 1 - \dfrac{p}{l} & \dfrac{p}{l} \\ \dfrac{q}{l} & 1 - \dfrac{q}{l} \\ \dfrac{q}{l} & 1 - \dfrac{q}{l} \\ 0 & 1 \end{bmatrix}$$

and

$$f = \begin{bmatrix} \dfrac{p}{6E_T I}\begin{bmatrix} 2 & 1 \\ 1 & 2 \end{bmatrix} & & \mathbf{0} \\ & \dfrac{l_1}{6EI}\begin{bmatrix} 2 & 1 \\ 1 & 2 \end{bmatrix} & \\ \mathbf{0} & & \dfrac{q}{6EI}\begin{bmatrix} 2 & 1 \\ 1 & 2 \end{bmatrix} \end{bmatrix}$$

From Table 9.2, the plastic zones for the columns AB and CD are taken at 28 kips to be

$p = 9.67$ in. and $q = 6.06$ in.

and those for the beam member BC,

$p = 6.25$ in. and $q = 6.21$ in.

By using the b and f matrices above, the reduced stiffnesses for the column and beam members are given as

$$k_{\text{column}} = \begin{bmatrix} 3413.7 & 396.7 \\ 396.7 & 5145.2 \end{bmatrix} \qquad k_{\text{beam}} = \begin{bmatrix} 521.0 & 408.5 \\ 408.5 & 5021.0 \end{bmatrix}$$

With these stiffness matrices, the frame is now analyzed by the standard stiffness method, in this case with axial distortions neglected. That is,

$$\Delta r = (a^T k a)^{-1}\, \Delta R$$

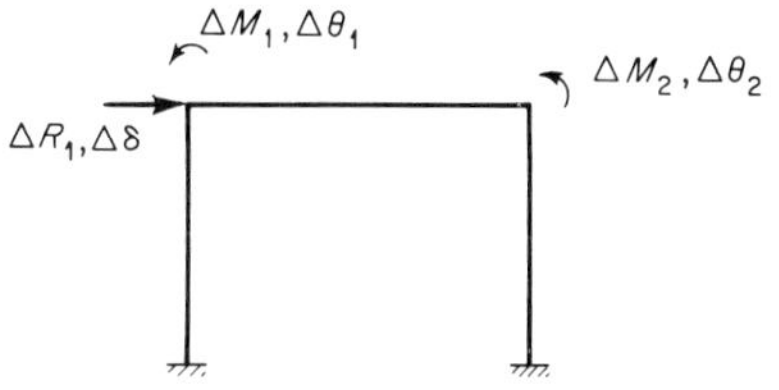

Figure 9.34 Node forces for tangent-stiffness analysis.

where k is composed of the matrices above as diagonal blocks, and a is given for $(\theta_1,\theta_2,\delta)$, shown in Fig. 9.34,

$$a = \begin{bmatrix} 0 & 0 & 0.00833 \\ 1 & 0 & 0.00833 \\ 1 & 0 & 0 \\ 0 & 1 & 0 \\ 0 & 1 & 0.00833 \\ 0 & 0 & 0.00833 \end{bmatrix}$$

It is found then that

$$\begin{bmatrix} \Delta M_1 \\ \Delta M_2 \\ \Delta R_1 \end{bmatrix} = \begin{bmatrix} 10166.2 & 408.5 & 46.164 \\ 408.5 & 10163.2 & 46.139 \\ 46.164 & 46.139 & 1.29768 \end{bmatrix} \begin{bmatrix} \Delta\theta_1 \\ \Delta\theta_2 \\ \Delta\delta \end{bmatrix}$$

When the above equations are solved for $\Delta\delta$, with $\Delta R_1 = 2$ kips and $\Delta M_1 = \Delta M_2 = 0$, it is found that the deflection is

$\Delta\delta = 2.237$ in.

From Table 9.2,

$\Delta\delta_H = 5.739 - 3.095 = 2.644$ in.

It is seen that the tangent stiffness has underestimated the deflection. However, from Table 9.2 it is seen that from 28 to 30 kips, there is a considerable spread of p and q along the members and this will account for the difference in the deflection values.

REFERENCES

1 Argyris, J. H.: Matrix Displacement Analysis of Plates and Shells, *Ing. Arch.*, vol. 25, 1966.

2 Argyris, J. H., and S. Kelsey: "Energy Theorems and Structural Analysis," Butterworth & Co. (Publishers), Ltd., London, 1960.

3 Bienzeno, C. B., and R. Grammel: "Engineering Dynamics," vol. 3, Blackie & Son, Ltd., Glasgow, 1955.

4 Busacher, R. G., and T. L. Saaty: "Finite Graphs and Networks," McGraw-Hill Book Company, New York, 1965.

5 Clough, R. W., E. L. Wilson, and I. P. King: Large Capacity Multi-story Frame Analysis Programs, *J. Struct. Div., ASCE,* vol. 89, August, 1963.

6 Denke, P. H.: A Computerized, Static and Dynamic Aircraft Structural Analysis System, *Soc. Automotive Engrs., S. P.* 264, December, 1964.

7 De Fries-Skene, A., and A. C. Scordelis: Direct Stiffness Solution for Folded Plates, *J. Struct. Div. Am. Soc. Civil Engrs.,* vol. 90, no. ST4, *Proc. Paper* 3994, August, 1964.

8 Ergatoudis, I., B. M. Irons, and O. C. Zienkiewicz: Curved, Isoparametric, "Quadrilateral" Elements for Finite Element Analysis, *Intern. J. Solids, Structures,* vol. 4., 1968.

9 Fenves, S. J., and F. H. Branin: Network Topological Formulation of Structural Analysis, *J. Struct. Div., ASCE,* vol. 89, no. ST4, August, 1963.

10 Felippa, C. A.: Refined Finite Element Analysis of Linear and Nonlinear Two Dimensional Structures, SESM 66-2A, Rept. to Natl. Sci. Found., NSF Grant GK7S, *Struct. Eng. Lab., Univ. Calif.,* Berkeley, October, 1966.

11 Frisch-Fay, R.: Flexible Bars, Butterworth & Co. (Publishers), Ltd., London, 1962.

12 Kani, Gaspar: "Analysis of Multi-story Frames," Frederick Ungar Publishing Co., New York, 1957.

13 Holand, Ivar, and Bell Kolbein: Finite Element Methods in Stress Analysis, Tapir, Tech. Univ. Norway, 1969.

14 Hrennikoff, A.: Solution of Problems by the Framework Method, *J. Appl. Mech., ASME,* December, 1941.

15 Irons, B. M.: Engineering Applications of Numerical Integrations in Stiffness Method, *AIAA J.*, vol. 4, no. 11, November, 1966.

16 Jennings, A.: The Free Cable, *The Engineer*, Dec. 28, 1962.

17 Jennings, A.: The Elastic Stability of Rigid Jointed Frames, *Intern. J. Mech. Sci.,* vol. 5, 1963.

18 Jennings, A.: Frame Analysis Including Change of Geometry, *J. Struct. Div., ASCE,* March, 1968.

19 Kantorovich and Krylov: "Approximate Methods of Higher Analysis," Erven P. Noordhoff, Ltd., Groningen, Netherlands, 1958.

20 Langhaar, H. L.: "Energy Methods in Applied Mechanics," John Wiley & Sons, Inc., New York, 1962.

21 Livesley, R. L.: "Matrix Methods of Structural Analysis," Pergamon Press, New York, 1964.

22 Martin, H. C.: On the Derivation of Stiffness Matrices for the Analysis of Large De-

flection and Stability Problems, *Proc. Conf. Matrix Methods in Struct. Mech.,* October, 1965. Air Force Flight Dynamics Lab., Air Force Inst. Tech.

23 Meessner, C. J., and R. S. Levy: Flexibility Method of Coupling Redundant Complex Structures, *J. Struct. Div., ASCE,* vol. 89, no. ST6, December, 1963.

24 Melosh, R. J., and R. M. Bainford: Efficient Solution of Load Deflection Equations, *J. Struct. Div., ASCE,* April, 1968.

25 Di Maggio, F. L.: Statical Indeterminacy and Stability of Structures, *J. Struct. Div., ASCE,* vol. 99, no. ST3, June, 1963.

26 Marcal, P. V., and I. King: Elastic-Plastic Analysis of Two Dimensional Stress Systems by the Finite Element Methods, *Intern. J. Mech. Sci.,* vol. 9, 1967.

27 Morley, L. S. D.: "Skew Plates and Structures," Pergamon Press, New York, 1963.

28 Norris, C. H., and J. B. Wilbur: "Elementary Structural Analysis," McGraw-Hill Book Company, New York, 1960.

29 Pipes, L. A.: "Matrix Methods for Engineering," Prentice-Hall, Inc., Englewood Cliffs, N.J., 1963.

30 Sokolnikoff, "Mathematical Theory of Elasticity," McGraw-Hill Book Company, New York, 1956.

31 Stricklin, J. A.: *AIAA J.,* vol. 6, no. 10, October, 1968.

32 Timoshenko, S. P., and S. Weinowsky-Krieger: "Theory of Plates and Shells," McGraw-Hill Book Company, New York, 1959.

33 Timoshenko, S. P., and J. N. Goodier: "Theory of Elasticity," McGraw-Hill Book Company, New York, 1951.

34 Tocher, J.: Selective Inversion of Stiffness Matrices, *J. Struct. Div., ASCE,* vol. 92, no. STI, February, 1966.

35 Veubeke, F. de: "Matrix Methods of Structural Analysis," The Macmillan Company, New York, 1964.

36 Wilson, E. L.: Matrix Analysis of Non-linear Structures, *Second Conf. Electronic Computation, ASCE.*

37 Yamada, Y., and N. Yoshimara: Plastic Stress-Strain Matrix and Its Application for the Solution of Elastic Plastic Problems by the Finite Element Method, *Intern. J. Mech. Sci.,* vol. 10, no. 5, 1968.

38 Zienkiewicz, O. C., and Y. K. Cheung: "The Finite Element Method in Structural and Continuum Mechanics," McGraw-Hill Book Company, New York, 1967.

39 Zienkiewicz, O. C., S. Valliappan, and I. P. King: Stress Analysis of Rock as a 'No Tension' Material, *Géotechnique,* vol. 18, no. 1, March, 1968.

APPENDIX A SMIS 3000

OPERATING INSTRUCTIONS FOR SYMBOLIC MATRIX INTERPRETATIVE PROGRAM

Purpose

The symbolic matrix interpretative system (SMIS) provides a means of performing matrix operations under the control of a sequence of punched cards. Matrices are designated by symbols that are selected by the user.

Restrictions

1 The size of the matrix will be limited by the availability of core storage. In the coding given, only 3,000 floating point numbers are available in the dynamic storage of the program. No more than 40 matrices may be retained in this store at one time. Each matrix is referred to by its name, one to six alphanumeric characters. If an attempt is made to load or generate a matrix with the same name as a previously defined matrix, the previous matrix is deleted before the operation is executed.

2 The program is written in Fortran IV for the General Electric Company 225 computer and is run under the normal Fortran monitor system. Overlays are used, so that a scratch tape is required on handler 3.

Usage

The input deck is composed of *control cards, data cards,* and *remark cards.* A control card designates the *operation* to be performed, along with its *calling sequence.* The calling sequence contains the names of matrices to be used in the operation, size of matrices, and other possible numbers. Data cards only follow the matrix LOAD operation. Information is punched on the control cards and data cards in free-field format, that is, each piece of information is separated from the next by one or more blank spaces. Remark cards are designated by placing an asterisk * in column 1. The remainder of a card beginning with column 2 is printed out and no further processing is carried out on the card. The character in column 2 can be used for carriage control purposes. If an asterisk is placed on a control card, information following the asterisk is printed only, and is not processed. Matrix names can contain any alphanumeric character, but must not begin with +, −, ., *, or the numbers 0 to 9. Only the first six characters in a matrix name are used to identify the matrix; any further characters are ignored. The standard end-of-file card is used to terminate processing.

Operations

The following is a list of possible operations and their calling sequences. The data-card layout is first shown, and a description of the operation follows. The letters N1, N2, N3, N4 refer to integer numbers, and these must *not* contain a decimal point. All other members may or may not contain a decimal point.

LOAD A N1 N2

The elements of matrix A of size N1 $\times$ N2 are loaded from the following data cards. The elements of the matrix must be punched by rows in free-field format.

PRINT A

Matrix A will be printed.

ADD A B

Elements of matrix A are added to corresponding elements of matrix B. Matrix A remains unaltered.

SUB A B

Elements of matrix A are subtracted from corresponding elements of matrix B. Matrix A remains unaltered.

SCALE A S1

Each element of matrix A is replaced by S1 times the element where S1 is the scale factor given as a floating point number.

MULT A B C

The product of matrix A times matrix B is generated and defined as matrix C.

INVERT A

Matrix A is replaced by the inverse of itself.

TRANS A B

The transpose of matrix A is generated and defined as matrix B.

START

This operation causes the elimination of all matrices in core storage. If an error is encountered in the input deck, the program automatically searches for the next START card, at which point the program is continued. The first card in any deck must be a START card.

DELETE A

Matrix A is eliminated from core storage.

ZERO A N1 N2

Matrix A of sizes N1 $\times$ N2 is generated as a null matrix.

STOSM A B N1 N2

Submatrix A is stored in matrix B with its first element on row N1 and column N2 of matrix B.

ADDSM A B N1 N2

Submatrix A is added to matrix B with its first element on row N1 and column N2 of matrix B.

RMVSM A B N1 N2 N3 N4

Submatrix A of size N1 $\times$ N2 is removed from matrix B such that the first element of submatrix A is in row N3 and column N4 of matrix B. Matrix B is unaltered.

STODG A B

Stores the matrix A on the diagonal of B. If A is of size M $\times$ N, then N must be an integer multiple of M and B must be of size N $\times$ N. A and B must both already exist and the elements of A replace the corresponding elements of B.

RMVDG A B N1

Matrix A consisting of N1 rows is removed from the diagonal of B. B is unaltered. The operation is similar to STODG except that A is created from B. B must be square and if it is of size N $\times$ N then N must be an integer multiple of N1. A is created with size N1 $\times$ N.

DUPL A B

Matrix B is created identical with matrix A.

SQREL A

Each element of matrix A is replaced by the square root of itself.

UNIT A N1

Matrix A is generated as a unit matrix of order N1.

```
$FORTRAN  MAIN PROGRAM SMIS 3000
C      PROGRAM-SMIS 3000
C      FORTRAN IV ---MAXIMUM OF 40 MATRICES AND/OR 3000 TOTAL ELEME
                                                        (CONT)NTS
C      MAIN PROGRAM
C      TO CARRY OUT MATRIX OPERATIONS USING A MATRIX INTERPRETIVE S
                                                        (CONT)CHEME
C
C      COMMON CARDS USED IN SUBROUTINES,
C   7  COMMON W1(50),W2(50)
C   6  COMMON A(3000)
C   5  COMMON XNAME(40),NROW(40),NCOL(40),NN(40)
C   4  COMMON OPLIST(30)
C   3  COMMON TABLE(40),IDENT(40)
C   2  COMMON NX,NR,NC,NNR,NNC,NEVEN
C   1  COMMON NSIZE,NUM,MAXSIZ,MAXNUM
C      NOTE  AN EVEN NUMBER OF INTEGER VARIABLES ON A COMMON CARD
C
       COMMON W1(50),W2(50)
       COMMON A(3000)
       COMMON XNAME(40),NROW(40),NCOL(40),NN(40)
       COMMON OPLIST(30)
       COMMON TABLE(40),IDENT(40)
       COMMON NX,NR,NC,NNR,NNC,NEVEN
       COMMON NSIZE,NUM,MAXSIZ,MAXNUM
C
C      LIST OF OPERATIONS
C      **** ** **********
C
C      1  LOAD
C      2  PRINT
C      3  ADD
C      4  SUB
C      5  SCALE
C      6  MULT
C      7  INVERT
C      8  TRANS
C      9  START
C      10 DELETE
C      11 ZERO
C      12 EIGEN
C      13 STOSM
C      14 ADDSM
C      15 RMVSM
C      16 STODG
C      17 RMVDG
C      18 DUPL
C      19 SQREL
C      20 UNIT
C      21 PRINTE
C
       DATA OPLIST/4HLOAD,5HPRINT,3HADD,3HSUB,5HSCALE, 4HMULT,6HINV
                                                        (CONT)ERT ,
     15HTRANS,5HSTART,6HDELETE,4HZERO,5HEIGEN,5HSTOSM,5HADDSM,5HRM
                                                        (CONT)VSM,
     25HSTODG,5HRMVDG,4HDUPL,5HSQREL,4HUNIT,
     36HPRINTE,9*1H /
       DATA MAXNUM,MAXSIZ/40,3000/
C
       CALL CARD(2)
       END

$FORTRAN SUBROUTINE SMIS FOR SMIS 3000

       SUBROUTINE SMIS
C
C      THIS SUBROUTINE REQUIRES COMMON CARDS--C1,C2,C3,C4,C5,C6,C7
C
       COMMON W1(50),W2(50)
       COMMON A(3000)
```

```
      COMMON XNAME(40),NROW(40),NCOL(40),NN(40)
      COMMON OPLIST(30)
      COMMON TABLE(40),IDENT(40)
      COMMON NX,NR,NC,NNR,NNC,NEVEN
      COMMON NSIZE,NUM,MAXSIZ,MAXNUM
C
C
 5    CALL CARD(0)
C
C     DETERMINATION OF OPERATION
C
      DO  40 I=1,30
      IF(OPLIST(I).EQ.TABLE(1))GO TO 20
 40   CONTINUE
       GO TO 800
 20   GO TO (110,120,160,170,180,190,200,210,260,370,380,280,220,3
                                                              (CONT)50,
     1230,400,410,150,310,450,120),I
C
C     MATRIX LOAD
C
C     CHECK  3 AND 4 FOR INTEGERS
 110  CALL CHECK(34)
      NR =TABLE(3)
      NC =TABLE(4)
      CALL MLIST(2)
      PRINT 102,NR,NC
 102  FORMAT(1I6,6H ROWS, 1I6,8H COLUMNS)
      CALL CARD(1)
      GO TO 5
C
C     MATRIX PRINT
C
 120  CALL MFIND(2)
      ITAG=I-2
      NA=NX
      NRA=NR
      NCA=NC
      CALL PRINT(A(NA),NRA,NCA,ITAG)
      GO TO 5
C
C     ADDITION OR SUBTRACTION OF MATRICES
C
 160  TAG=1.0
      GO TO 171
 170  TAG=-1.0
 171  CALL MFIND(2)
      NA=NX-1
      NNR=NR
      NNC=NC
      CALL MFIND(3)
      NB=NX-1
      IF(NR-NNR)173,172,173
 172  IF(NC-NNC)173,174,173
 173  CALL ERROR(1)
 174  NS=NR*NC
      NAB=NB
      IF(IDENT(4).EQ.0) GO TO 176
      IF(TABLE(4).EQ.TABLE(3))GO TO 176
      IF(TABLE(4).EQ.TABLE(2)) GO TO 175
      CALL MLIST(4)
      NAB=NX-1
      CALL MFIND(2)
      NA=NX-1
      CALL MFIND(3)
      NB=NX-1
      GO TO 176
 175  NAB=NA
 176  DO 177 I=1,NS
      J=I+NA
      K=I+NB
      KK=I+NAB
 177  A(KK)=A(K)+TAG*A(J)
      GO TO 5
```

```
C
C      SCALAR MULTIPLICATION
C
 180   IF(IDENT(3).LT.2) CALL ERROR(4)
       PRINT 103,TABLE(3)
 103   FORMAT(9H  SCALAR=,1E15.7)
       CALL MFIND(2)
       NA=NX
       NB=NX-1+NR*NC
       SCAL1=TABLE(3)
       DO 185 I=NA,NB
 185   A(I)=SCAL1*A(I)
       GO TO 5
C
C      MATRIX MULTIPLICATION
C
 190   CALL MFIND(2)
       NRA=NR
       NCA=NC
       CALL MFIND(3)
       NRB=NR
       NCB=NC
       IF(NCA-NRB)191,192,191
 191   CALL ERROR(1)
 192   NR=NRA
       CALL MLIST(4)
       ND=NX
       CALL MFIND(2)
       NA=NX
       CALL MFIND(3)
       IF(TABLE(4).EQ.TABLE(3).OR.TABLE(4).EQ.TABLE(2))CALL ERROR(4
                                                               (CONT))
       NB=NX
       CALL MULT(NRA,A(NA),NCA,A(NB),NCB,A(ND))
       GO TO 5
C
C      MATRIX INVERSION
C
 200   CALL MFIND(2)
       NRA=NR
       NA=NX
       IF(NR-NC)201,202,201
 201   CALL ERROR(10)
 202   IF(NR.GT.50) CALL ERROR(5)
       CALL INVERT(A(NA),NRA,W1,W2)
       GO TO 5
C
C      MATRIX TRANSPOSE
C
 210   CALL MFIND(2)
       NRA=NR
       NR=NC
       NC=NRA
       CALL MLIST(3)
       NB=NX
       CALL MFIND(2)
       NA=NX
       NCA=NC
       CALL TRANS(A(NA),A(NB),NRA,NCA)
       GO TO 5
C
C      START
C
C
 260   CALL START
       GO TO 5
C
C      EIGENVALUES AND EIGENVECTORS
C
C      EIGEN NOT INCLUDED IN THIS VERSION
 280   GO TO 800
C      CARDS  21370 TO 21530 NOT INCLUDED
C
C      DELETION OPERATION
C
```

```
 370   CALL DELETE(2)
       GO TO 5
C
C      ZERO OPERATION
C
C      CHECK 3 AND 4 FOR INTEGERS
 380   CALL CHECK(34)
       NR =TABLE(3)
       NC =TABLE(4)
       PRINT 102,NR,NC
       CALL MLIST(2)
       DO 381 I=NX,NSIZE
 381   A(I)=0.0
       GO TO 5
C
C      STORE SUBMATRIX
C
 220   NTAG=1
C
C      CHECK 4 AND 5 FOR INTEGER
 223   CALL CHECK(45)
       J=TABLE(4)
       K=TABLE(5)
       CALL MFIND(2)
       GO TO 235
C
C      REMOVE SUBMATRIX
C
C
C
C
C      CHECK 4 5 6 AND 7 FOR INTEGERS
 230   CALL CHECK(4567)
       J=TABLE(6)
       K=TABLE(7)
       NR=TABLE(4)
       NC=TABLE(5)
       NTAG=0
       CALL MLIST(2)
 235   NB=NX
       NRB=NR
       NCB=NC
       CALL MFIND(3)
       NA=NX
       NRA=NR
       NCA=NC
       CALL STOSM(A(NA),A(NB),NRA,NCA,NRB,NCB,J,K,NTAG)
       GO TO 5
C
C      ADD SUBMATRIX
C
 350   NTAG=-1
       GO TO 223
C
C      SQUARE ROOT OF EACH ELEMENT
C
 310   CALL MFIND(2)
       NA=NX
       NB=NA+NR*NC-1
       DO 315 K=NA,NB
 315   A(K)=SQRT(A(K))
       GO TO 5
C
C      STORES SUBMATRICES OF A ON DIAGONAL OF B
C
 400   CALL MFIND(2)
       NA=NX
       NRA=NR
       NCA=NC
       CALL MFIND(3)
       NB=NX
       NRB=NR
       NCB=NC
       IF((NCA/NRA*NRA).NE.NCA.OR.NCA.NE.NCB) CALL ERROR(1)
       IF(NCB.NE.NRB) CALL ERROR(10)
```

```
      IL=NCA/NRA
      NTAG=0
      DO 406 J=1,NRA
      K=1
      DO 405 I=1,IL
      CALL STOSM(A(NA),A(NB),NRA,NCA,1,NRA,J,K,NTAG)
      NB=NB+NRA*(NCB+1)
 405  K=K+NRA
 406  NB=NX+J*NCB
      GO TO 5
C
C     REMOVES SUBMATRICES FROM DIAGONAL OF B TO MATRIX A
C
 410  CALL CHECK(4)
      CALL MFIND(3)
      NRB=NR
      NR=TABLE(4)
      IF(NRB.NE.NC.OR.(NRB/NR*NR).NE.NRB) CALL ERROR(1)
      CALL MLIST(2)
      NA=NX
      NRA=NR
      CALL MFIND(3)
      NB=NX
      IL=NRB/NRA
      NTAG=0
      DO 416 J=1,NRA
      DO 415 I=1,IL
      M=NRA*(I-1)+1
      K=M+J-1
      CALL STOSM(A(NB),A(NA),NRB,NRB,1,NRA,K,M,NTAG)
 415  NA=NA+NRA
 416  CONTINUE
      GO TO 5
C
C     DUPLICATION OF A MATRIX
C
 150  CALL MFIND(2)
      CALL MLIST(3)
      NB=NX-1
      CALL MFIND(2)
      NA=NX-1
      NS=NR*NC
      DO 155 I=1,NS
      J=I+NB
      K=I+NA
 155  A(J)=A(K)
      GO TO 5
C
C     UNIT MATRIX
C
C     CHECK 3 FOR INTEGER
 450  CALL CHECK(3)
      NR=TABLE(3)
      NC=NR
      CALL MLIST(2)
      DO 451 I=NX,NSIZE
 451  A(I)=0.0
      K=NX
      IL=NX
      IH=NX+NR-1
      DO 452 I=IL,IH
      A(K)=1.0
 452  K=K+NC+1
      GO TO 5
 800  CALL ERROR(3)
       RETURN
      END

$FORTRAN          SUBROUTINE CHECK
      SUBROUTINE CHECK(J)
C     THIS PROGRAM REQUIRES COMMON CARDS --C1,C2,C3
      COMMON TABLE(40),IDENT(40)
      COMMON NX,NR,NC,NNR,NNC,NEVEN
      COMMON NSIZE,NUM,MAXSIZ,MAXNUM
      N=J
```

```
 10     I=N-(N/10)*10
        IF(I.LE.0)RETURN
        IF(IDENT(I).NE.2.OR.TABLE(I).GT.524287.)CALL ERROR(4)
        N=N/10
        GO TO 10
        END

$FORTRAN  SUBROUTINE CARD FOR SMIS 3000

        SUBROUTINE CARD(NM)
C
C       NORMAL USE IS WITH N=NM=0
C         READS CARD AND PRINTS IT. IF * IN COLUMN 1 IT READS AGAIN.
C         DECODES CARD PLACING EACH WORD IN THE ARRAY TABLE AND A CO
                                                            (CONT)RR-
C         ESPONDING IDENTIFIER IN THE ARRAY IDENT. THEN RETURNS.
C           IDENT=1  WORD CONTAINS FIRST 6 CHARACTERS OF AN ALPHANUM
                                                            (CONT)ERIC.
C                    ALPHANUMERIC STARTS WITH ANYTHING OTHER THAN +-
                                                            (CONT).0123
C                    456789
C           IDENT=2  WORD WAS AN INTEGER,STORED AS REAL.
C          IDENT=3   REAL NUMBER DETECTED.
C          IDENT=0   END OF TABLE
C         IF AN ERROR IS DETECTED A MESSAGE IS PRINTED AND START CARD
                                                            (CONT) IS
C         SEARCHED FOR.
C         DETECTION OF AN * TERMINATES DECODING OF THAT CARD.
C       N=NM=1  FILL AN ARRAY STARTING AT A(NA) OF SIZE NRA*NCA
C       N=NM=2  LOOK FOR START CARD AND CALL S/R START
C       END OF FILE CARD ENDS PROGRAM
C
C       THIS SUBROUTINE REQUIRES COMMON CARDS--C1,C2,C3,C4,C5,C6
C
        COMMON A(3000)
        COMMON XNAME(40),NROW(40),NCOL(40),NN(40)
        COMMON OPLIST(30)
        COMMON TABLE(40),IDENT(40)
        COMMON NX,NR,NC,NNR,NNC,NEVEN
        COMMON NSIZE,NUM,MAXSIZ,MAXNUM
        DIMENSION IA(6),IX(2),IN(80)
        EQUIVALENCE (X,IX)
        LOGICAL BEGUN
        N=NM
        NY=0
        NS=NR*NC
C
 10     IF (IEOFX(1))CALL ERROR(8)
        DO 5 I=1,40
 5      IDENT(I)=0
        READ 11,IN
 11     FORMAT (80A1)
        IF (N .EQ. 2)  GO TO 20
        IF (IN(1) .NE. 3H*  ) GO TO 20
        PRINT 11,(IN(I),I=2,80)
        GO TO 10
 20     BEGUN=.FALSE.
        J=1
        DO 900 I=1,80
        ICHAR=IN(I)/4096
        IF(ICHAR.EQ.3H00 ) GO TO 24
        IF(ICHAR.NE.3H00*) GO TO 25
        I=80
 24     IF (.NOT. BEGUN ) GO TO 900
        GO TO 300
 25     IF (BEGUN) GO TO 500
 100    BEGUN= .TRUE.
        X=0.
        F=10.
        G=1.
        IDENT(J)=2
        SIGN=1.
        IF (ICHAR .LE. 9) GO TO 120
        IF (ICHAR .EQ. 3H00+) GO TO 900
        IF (ICHAR .EQ. 3H00.) GO TO 140
```

```
      IF (ICHAR .NE. 3H00-) GO TO 160
      SIGN=-1.
      GO TO 900
 120  ADD=ICHAR
      X=X*F+ADD*G
      GO TO 900
 140  IDENT(J)=3
      F=1.
      GO TO 900
 160  IF (N .EQ. 1) GO TO 1000
      IDENT(J)=1
      DO 162 K=1,6
 162  IA(K)=3H00
      K=0
 170  K=K+1
      IF (K .GT. 6) GO TO 900
      IA(K)=ICHAR
      GO TO 900
C
C     END WORD SECTION
C
 300  IF (IDENT(J) .NE. 1) GO TO 350
      IX(1)=IA(3) + IA(2) * 64 + IA(1) * 4096
      IX(2)=IA(6) + IA(5) * 64 + IA(4) * 4096
      TABLE(J)=X
      GO TO 400
 350  X=X*SIGN
      IF (N .EQ. 1) GO TO 360
      TABLE(J)=X
      GO TO 400
 360  A(NX+NY)=X
      NY=NY+1
      IF (NY .GE. NS) RETURN
 400  J=J+1
      BEGUN=.FALSE.
      IF(J .NE. 2) GO TO 900
      IF (N .NE. 2) GO TO 900
      IF (IDENT(1) .NE. 1) GO TO 10
      IF (TABLE(1) .NE. 5HSTART) GO TO 10
      GO TO 900
C
C     CONTINUE WORD SECTION
C
 500  IF (IDENT(J) .EQ. 1) GO TO 170
      IF (IDENT(J) .EQ. 3) GO TO 550
      IF (ICHAR .LE. 9) GO TO 120
      IF (ICHAR .NE. 3H00.) GO TO 1010
      GO TO 140
 550  IF (ICHAR .GT. 9) GO TO 1010
      G=G/10.
      GO TO 120
 900  CONTINUE
      IF (BEGUN) GO TO 300
C     IDENT(J)=0  NOT NEEDED BECAUSE IDENT IS ZEROED AT 5
      IF (J .EQ. 1) GO TO 10
      IF (N .EQ. 1) GO TO 950
      IF (TABLE(1) .NE. 5HSTART) GO TO 920
      PRINT 919,IN
 919  FORMAT(1H1//X,80A1)
C
      CALL START
 920  PRINT 921,IN
 921  FORMAT(/X,80A1)
      RETURN
 950  IF (NY .LT. NS) GO TO 10
      RETURN
 1000 PRINT 1001,J
 1001 FORMAT(/15H****ERROR, WORD,I3,18H SHOULD BE NUMERIC)
      IF (J .NE. 1) GO TO 1020
      PRINT 1002,NY,NS
 1002 FORMAT (1H+,35X,5H,ONLY,I5,3H OF,I5,14H NUMBERS FOUND)
      GO TO 1020
 1010 PRINT 1011,J,IN(I),I
 1011 FORMAT(/15H****ERROR, WORD,I3,38H APPEARS TO BE A NUMERIC BU
                                                      (CONT)T INCL
```

```
     1UDES ,A1,10H IN COLUMN,I3)
1020 PRINT 921,IN
     N=2
     GO TO 10
     END
$FORTRAN SUBROUTINE MLIST
      SUBROUTINE MLIST(M)
C
C     SUBROUTINE TO LIST A MATRIX
C     THIS SUBROUTINE REQUIRES COMMON CARDS--C1,C2,C3,C4,C5
C
      COMMON XNAME(40),NROW(40),NCOL(40),NN(40)
      COMMON OPLIST(30)
      COMMON TABLE(40),IDENT(40)
      COMMON NX,NR,NC,NNR,NNC,NEVEN
      COMMON NSIZE,NUM,MAXSIZ,MAXNUM
      IF(IDENT(M).NE.1) CALL ERROR(4)
      XM=TABLE(M)
      I=1
 5    IF(NUM-I)30,10,10
 10   IF(XNAME(I)-XM)20,25,20
 20   I=I+1
      GO TO 5
 25   CALL DELETE(M)
 30   NUM=NUM+1
      IF(NUM.GT.MAXNUM) CALL ERROR(6)
      NROW(NUM)=NR
      NCOL(NUM)=NC
      XNAME(NUM)=XM
      NN(NUM)=NSIZE
      NX=NSIZE
      NSIZE=NSIZE+NR*NC
      IF((NSIZE-1).GT.MAXSIZ) CALL ERROR(5)
 60   RETURN
      END
$FORTRAN SUBROUTINE MFIND
      SUBROUTINE MFIND(M)
C
C     THIS SUBROUTINE REQUIRES COMMON CARDS--C1,C2,C3,C4,C5
C
      COMMON XNAME(40),NROW(40),NCOL(40),NN(40)
      COMMON OPLIST(30)
      COMMON TABLE(40),IDENT(40)
      COMMON NX,NR,NC,NNR,NNC,NEVEN
      COMMON NSIZE,NUM,MAXSIZ,MAXNUM
C
C     TO FIND LOCATION AND SIZE OF MATRIX XX
C
      IF(IDENT(M).NE.1) CALL ERROR(4)
      XM=TABLE(M)
      I=1
 5    IF(NUM-I)40,10,10
 10   IF (XNAME(I)-XM)20,30,20
 20   I=I+1
      GO TO 5
 30   NR=NROW(I)
      NC=NCOL(I)
      NX=NN(I)
      RETURN
 40   PRINT 41,XM
 41   FORMAT(1H+,9X,A6)
      CALL ERROR(2)
      END
$FORTRAN SUBROUTINE DELETE
      SUBROUTINE DELETE(M)
C
C     SUBROUTINE TO DELETE A MATRIX XM
C
C     THIS SUBROUTINE REQUIRES COMMON CARDS--C1,C2,C3,C4,C5,C6
C
      COMMON A(3000)
      COMMON XNAME(40),NROW(40),NCOL(40),NN(40)
```

```
      COMMON OPLIST(30)
      COMMON TABLE(40),IDENT(40)
      COMMON NX,NR,NC,NNR,NNC,NEVEN
      COMMON NSIZE,NUM,MAXSIZ,MAXNUM
      IF(IDENT(M).NE.1) CALL ERROR(4)
      XM=TABLE(M)
      I=1
 5    IF (NUM-I) 40,10,10
40    CALL ERROR(2)
10    IF (XNAME(I)-XM)20,30,20
20    I=I+1
      GO TO 5
30    NX=NN(I)
      NUM=NUM-1
      NS=NROW(I)*NCOL(I)
      NSIZE=NSIZE-NS
      DO 50 J=I,NUM
      XNAME(J)=XNAME(J+1)
      NROW(J)=NROW(J+1)
      NCOL(J)=NCOL(J+1)
50    NN(J)=NN(J+1)-NS
      DO 60 J=NX,NSIZE
      I=J+NS
60    A(J)=A(I)
      RETURN
      END

$FORTRAN SUBROUTINE PRINT
      SUBROUTINE PRINT (A,NR,NC,ITAG)
C
C     TO PRINT A MATRIX  ITAG NEQ 0 USE E FORMAT
C
      DIMENSION A(1),NCOL(8),IFMT(9)
      DATA IFMT/3H(1I,3H6,2,3HX, ,3HSCF,3HP 8,3HE14,3H,  ,3HDEC,3H
                                                       (CONT))  /
      DATA  MIN/3H-00/
      IF (ITAG.EQ.0) GO TO 10
      IFMT(4)=1
      IFMT(6)=3HE14
      IFMT(8)=7
      GO TO 34
10    IFMT(6)=3HF14
      NEL=NR*NC
      AMAX=0.
      DO 11 M=1,NEL
      AM=ABS(A(M))
      IF(AM.GT.AMAX)AMAX=AM
11    CONTINUE
      IF(AMAX.EQ.0.)AMAX=1.
      ILOG=ALOG10(AMAX)
      IF(AMAX.LT.1.)ILOG=ILOG-1
      IFMT(4)=0
      IFMT(8)=6
      IF(ILOG.GT.6) GO TO 12
      IF(ILOG.LT.(-3)) GO TO 12
      IFMT(8)=6-ILOG
      GO TO 34
12    M=IABS(ILOG)
      M=(M/10)*64+M-(M/10)*10
      IF(ILOG.GT.0)M=M+MIN
      IFMT(4)=M
      PRINT 13,ILOG
13    FORMAT(20X,I3,/21HMULTIPLY VALUES BY 10)
34    DO 50 M=1,NC,8
      NN=NC-M
      IF(NN-7)40,40,35
35    NN=7
40    MM=NN+1
      DO 45 N=1,MM
45    NCOL(N)=M-1+N
      PRINT 101,(NCOL(N),N=1,MM)
      DO 50 N=1,NR
      NL=M+(N-1)*NC
      NH=NL+NN
50    PRINT IFMT,(N,(A(I),I=NL,NH))
```

```
      RETURN
 101  FORMAT(3H0  ,8I14)
      END

$FORTRAN SUBROUTINE TRANS
      SUBROUTINE TRANS(A,B,NR,NC)
      DIMENSION A(1),B(1)
      K=1
      DO 100 I=1,NC
      L=I
      DO 100 J=1,NR
      B(K)=A(L)
      L=L+NC
 100  K=K+1
      RETURN
      END

$FORTRAN SUBROUTINE MULT
      SUBROUTINE MULT(NRA,A,NCA,B,NCB,C)
      DIMENSION A(1),B(1),C(1)
C
C     TO MULTIPLY TWO MATRICES
C
      K=1
      DO 200 I=1,NRA
      NN=(I-1)*NCA+1
      NH=NN-1+NCA
      DO 200 J=1,NCB
      MM=J
      C(K)=0.0
      DO 100 L=NN,NH
      C(K)=C(K)+A(L)*B(MM)
 100  MM=MM+NCB
 200  K=K+1
      RETURN
      END

$FORTRAN     SUBROUTINE STOSM
      SUBROUTINE STOSM(A,B,NRA,NCA,NRB,NCB,NR,NC,NTAG)
C
C     SUBROUTINE TO STORE OR REMOVE A SUBMATRIX
C
      DIMENSION A(1),B(1)
      IF(NRA+1-NR-NRB) 60,50,50
 50   IF(NCA+1-NC-NCB) 60,70,70
 60   CALL ERROR(1)
 70   DO 100 I=1,NRB
      K=(NC-1)+(NR+I-2)*NCA
      L=(I-1)*NCB
      DO 100 J=1,NCB
      N=K+J
      M=L+J
      IF(NTAG) 75,90,80
 75   A(N)=A(N)+B(M)
      GO TO 100
 80   A(N)=B(M)
      GO TO 100
 90   B(M)=A(N)
 100  CONTINUE
      RETURN
      END

$FORTRAN SUBROUTINE INVERT
      SUBROUTINE INVERT(A,N,M,C)
      DIMENSION A(1),M(1),C(1)
C
C     TO INVERT A MATRIX
C
      DO 90 I=1,N
 90   M(I)= -I
      DO 140 I=1,N
C
C     LOCATE LARGEST ELEMENT
C
      D=0.0
```

```
      DO 112 L=1,N
      IF(M(L))100,100,112
 100  J=L
      DO 110 K=1,N
      IF(M(K))103,103,108
 103  IF(ABS (D)-ABS (A(J)))105,105,108
 105  LD=L
      KD=K
      D=A(J)
 108  J=J+N
 110  CONTINUE
 112  CONTINUE
C
C     INTERCHANGE COLUMNS
C
      TEMP=-M(LD)
      M(LD)= M(KD)
      M(KD)= TEMP
      L=LD
      K=KD
      DO 114 J=1,N
      C(J)=A(L)
      A(L)=A(K)
      A(K)=C(J)
      L=L+N
 114  K=K+N
C
C     DIVIDE ROW BY LARGEST ELEMENT
C
      NR=(KD-1)*N+1
      NH=NR+N-1
      IF(ABS(D).LE.1.0E-30) CALL ERROR(9)
      DO 115 K=NR,NH
 115  A(K)=-A(K)/D
C
C     REDUCE REMAINING ROWS AND COLUMNS
C
      L=1
      DO 135 J=1,N
      IF(J-KD)130,125,130
 125  L=L+N
      GO TO 135
 130  DO 134 K=NR,NH
      A(L)=A(L)+C(J)*A(K)
 134  L=L+1
 135  CONTINUE
C
C     REDUCE COLUMN
C
      C(KD)=1.0
      J=KD
      DO 140 K=1,N
      A(J)=C(K)/D
 140  J=J+N
C
C     INTERCHANGE ROWS
C
      DO 200 I=1,N
      L=0
 150  L=L+1
      IF(M(L)-I)150,160,150
 160  K=(L-1)*N+1
      J=(I-1)*N+1
      M(L)=M(I)
      M(I)=I
      DO 200 LL=1,N
      TEMP=A(K)
      A(K)=A(J)
      A(J)=TEMP
      J=J+1
 200  K=K+1
      RETURN
      END
```

```
$FORTRAN SUBROUTINE START
      SUBROUTINE START
C
C     THIS SUBROUTINE REQUIRES COMMON CARDS--C1
C
      COMMON NSIZE,NUM,MAXSIZ,MAXNUM
      NSIZE=1
      NUM=0
      CALL SMIS
      END

$FORTRAN   SUBROUTINE ERROR
      SUBROUTINE ERROR(N)
C
C3    ERROR CALLS
C
      M=N
      CALL OVRLAY(1)
      GO TO (1,2,3,4,5,6,7,8,9,10),M
 1    PRINT 101
      GO TO 100
 2    PRINT 102
      GO TO 100
 3    PRINT 103
      GO TO 100
 4    PRINT 104
      GO TO  100
 5    PRINT 105
      GO TO 100
 6    PRINT 106
      GO TO 100
 7    PRINT 107
      GO TO 100
 8    PRINT 108
      CALL EXIT
 9    PRINT 109
      GO TO 100
 10   PRINT 110
      GO TO 100
C     LOOK FOR START CARD
 100  CALL CARD(2)
 101  FORMAT(27HMATRICES INCOMPATIBLE          )
 102  FORMAT(6HMATRIX,12X,9HUNDEFINED)
 103  FORMAT(27HOPERATION UNDEFINED          )
 104  FORMAT(27H*** ERROR IN COMMAND FORMAT                    )
 105  FORMAT(27HSTORAGE EXCEEDED        )
 106  FORMAT(27HOVER 40 MATRICES IN STORE        )
 107  FORMAT(28HEND OF FILE DETECTED ON TAPE    )
 108  FORMAT(/35HALL CARDS PROCESSED  END OF JOB
                                                  (CONT)   )
 109  FORMAT(15HSINGULAR MATRIX)
 110  FORMAT(17HNON-SQUARE MATRIX)
      END
```

APPENDIX B

PROGRAM FOR CALCULATION OF FIXED-END MOMENTS; STANDARD CASES

Description of Problems

This program calculates fixed-end moments for various load distributions on a uniform beam. Function subprograms and subroutines are used to minimize coding.

Method of Solution

Tables for the calculation of fixed-end moments for various load distributions are given in many texts on the theory of structures. Those used here are listed below, 13 common cases being considered.

Computer Program

The program is listed below. Figure A.1 defines the cases considered.

```
C          PROGRAM TO CALCULATE FIXED END MOMENTS
C          FOR UNIFORM BEAM FOR 13 COMMON CASES
    1002   READ 1,N,AL,W,A,B
           PRINT 3
    3      FORMAT(/44HTYPE   SPAN(L)   LOAD(W)   A   B  )
           PRINT 4,N,AL,W,A,B
    1      FORMAT(I2,4F10.2)
    4      FORMAT(I4,4F10.2)
           DO 5 I=1,13
           IF (I-N) 5,6,5
    5      CONTINUE
           PRINT 2
    2      FORMAT(20HOERROR IN TYPE LABEL)
           GO TO 1002
    6      GO TO (101,102,103,104,105,106,107,108,109,110,111,112,113),N
    101    FEMI=AMA(W,AL)
           FEMJ=-FEMI
           GO TO 1000
    102    FEMI=AMB(W,A,AL)
           FEMJ=-AMC(W,A,AL)
           GO TO 1000
```

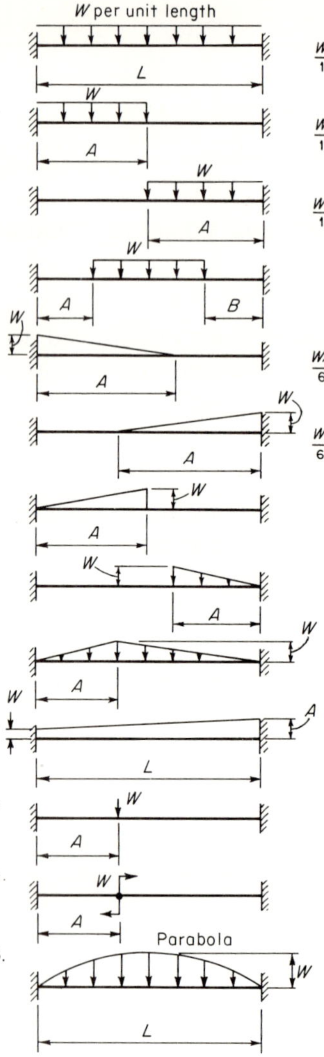

Case	Fixed-end moments, Left	Fixed-end moments, Right
1.	$\frac{WL^2}{12}$	$-\frac{WL^2}{12}$
2.	$\frac{WA^2}{12}\left[6-8\left(\frac{A}{L}\right)+3\left(\frac{A}{L}\right)^2\right]$	$-\frac{WA^2}{12}\left[4\left(\frac{A}{L}\right)-3\left(\frac{A}{L}\right)^2\right]$
3.	$\frac{WA^2}{12}\left[4\left(\frac{A}{L}\right)-3\left(\frac{A}{L}\right)^2\right]$	$-\frac{WA^2}{12}\left[6-8\left(\frac{A}{L}\right)+3\left(\frac{A}{L}\right)^2\right]$
4.	(1 - 2 - 3)	
5.	$\frac{WA^2}{60}\left[10-10\left(\frac{A}{L}\right)+3\left(\frac{A}{L}\right)^2\right]$	$-\frac{WA^2}{60}\left[5\left(\frac{A}{L}\right)-3\left(\frac{A}{L}\right)^2\right]$
6.	$\frac{WA^2}{60}\left[5\left(\frac{A}{L}\right)-3\left(\frac{A}{L}\right)^2\right]$	$-\frac{WA^2}{60}\left[10-10\left(\frac{A}{L}\right)+3\left(\frac{A}{L}\right)^2\right]$
7.	(2 - 5)	
8.	(3 - 6)	
9.	(1 - 5 - 6)	
10.	(5 + 6)	
11.	$\frac{WA(L-A)^2}{L^2}$	$-\frac{WA^2(L-A)}{L^2}$
12.	$\frac{W(L-A)(3A-L)}{L^2}$	$-\frac{WA(2L-3A)}{L^2}$
13.	$\frac{WL^2}{15}$	$-\frac{WL^2}{15}$

Figure A.1

```
103     FEMI=AMC(W,A,AL)
        FEMJ=-AMB(W,A,AL)
        GO TO 1000
104     FEMI=AMA(W,AL)-AMB(W,A,AL)-AMC(W,B,AL)
        FEMJ=-AMA(W,AL)+AMC(W,A,AL)+AMB(W,B,AL)
        GO TO 1000
105     FEMI=AMD(W,A,AL)
        FEMJ=-AME(W,A,AL)
        GO TO 1000
106     FEMI=AME(W,A,AL)
        FEMJ=-AMD(W,A,AL)
        GO TO 1000
107     FEMI=AMB(W,A,AL)-AMD(W,A,AL)
        FEMJ=-AMC(W,A,AL)+AME(W,A,AL)
        GO TO 1000
108     FEMI=AMC(W,A,AL)-AME(W,A,AL)
        FEMJ=-AMB(W,A,AL)+AMD(W,A,AL)
        GO TO 1000
109     B=AL-A
        FEMI=AMA(W,AL)-AMD(W,A,AL)-AME(W,B,AL)
        FEMJ=AMA(W,AL)-AME(W,A,AL)-AMD(W,B,AL)
        FEMJ=-FEMJ
        GO TO 1000
100     FEMI=AMD(W,AL,AL)+AME(A,AL,AL)
        FEMJ=AME(W,AL,AL)+AMD(A,AL,AL)
        FEMJ=-FEMJ
        GO TO 1000
11      CALL CONP(FEMI,FEMJ,W,A,AL)
        GO TO 1000
112     CALL CONM(FEMI,FEMJ,W,A,AL)
        GO TO 1000
113     FEMI=AMG(W,AL)
        FEMJ=-FEMI
1000    PRINT 10,FEMI,FEMJ
10      FORMAT(39H0FIXED END MOM LEFT FIXED END MOM RIGHT/
       1(6X,F10.2,14X,F10.2//))
        GO TO 1002
        END

        FUNCTION AMA(W,AL)
C       UNIFORMLY DISTRIBUTED LOAD WHOLE SPAN
        AMA=W*AL*AL/12.0
        RETURN
        END
```

```
      FUNCTION AMB(W,A,AL)
C     UNIFORMLY DISTRIBUTED LOAD OVER PORTION A FROM L.H.END
C     LEFT HAND F.E.M OR VICE-VERSA
      AMB=W*A*A*(6.0-8.0*(A/AL)+3.0*(A/AL)**2)/12.0
      RETURN
      END

      FUNCTION AMC(W,A,AL)
C     UNIFORM LOAD OVER PORTION A FROM L.H.END
C     RIGHT HAND F.E.M OR VICE-VERSA
      AMC=W*A*A*(4.0*(A/AL)-3.0*(A/AL)**2)/12.0
      RETURN
      END

      FUNCTION AMD(W,A,AL)
C     TRIANGULAR LOAD DISTRIBUTION OVER PORTION A FROM L.H.END
C     LEFT HAND F.E.M OR VICE-VERSA
C     MAXIMUM ORDINATE OF TRIANGLE AT LEFT SIDE
      AMD=W*A*A*(10.0-10.0*(A/AL)+3.0*(A/AL)**2)/60.0
      RETURN
      END

      FUNCTION AME(W,A,AL)
C     TRIANGULAR LOAD OVER PORTION A FROM L.H.END
C     RIGHT HAND F.E.M OR VICE-VERSA
C     MAXIMUM ORDINATE OF TRIANGLE AT LEFT
      AME=W*A*A*(5.0*(A/AL)-3.0*(A/AL)**2)/60.0
      RETURN
      END

      SUBROUTINE CONP(AMF,AMG,W,A,AL)
C     CONCENTRATED LOAD IN SPAN
      AMF=W*A*(AL-A)**2/(AL**2)
      AMG=-W*(AL-A)*A**2/(AL**2)
      RETURN
      END

      SUBROUTINE CONM(AMF,AMG,W,A,AL)
C     CONCENTRATED MOMENT IN SPAN
      AMF=(AL-A)*(3.0*A-AL)*W/(AL**2)
      AMG=A*(2.0*AL-3.0*A)*W/(AL**2)
      RETURN
      END
```

```
      FUNCTION AMG(W,AL)
C     PARABOLIC LOAD OVER WHOLE SPAN
      AMG=W*AL**2/15.0
      RETURN
      END
```

Data Input

For each case a card is required in (I2, 4F10.2) format containing in order TYPE, L, W, A, and B. A field may be left blank if it does not apply to the case being considered. W, A, and B are defined for each case in Fig. A.1.

APPENDIX C

PROGRAM FOR PLANE FRAME TRUSS ANALYSIS

The program listed below is based on the theory given in Chap. 2 (see Sec. 3.7A). In the compilation given, the number of joints allowed is 33, and the number of reactions 10. The program has been arranged to handle any number of load cases. All reaction components must be included as input. Thus, for example, a pin joint provides two components whose direction cosines in the global coordinates may be taken

	Cos α	*Sin α*
1	1.0	0.0
2	0.0	1.0

A roller support provides one reaction component whose direction cosines are those of the normal to the roller plane.

Input Details

Card (1) Format (3I5)	Number of joints, number of reactions, number of load cases.
Cards (Group 2) Format (2F10.3)	Joint coordinates arranged in ascending order, one pair of coordinates X, Y per card.
Cards (Group 3) Format (I5, 2F10.6)	Reaction, direction cosines. The first field on these cards gives the node number at which the reaction occurs. The remaining fields give the direction cosines of the reaction.
Cards (Group 4) Format (2I5)	These cards give the member-node numbers arranged in sequential order, one member per card.
Cards (Group 5) Format (2F10.2)	Joint loads. The joint loads R_X and R_Y are input for all joints, including those where reactions occur. Loads cases are arranged in groups.

Output Details

1 All data is ouput as echo check.

2 Member forces and reactions are listed for each load case under the appropriate headings.

PROGRAM FOR TRUSS ANALYSIS.

```
      DIMENSION X(40),Y(40),F(66),R(66),IP(66),
     1          IQ(66),NB(10),DCOS(10,2),M1(66),T2(66)
      COMMON B(66,66)
  999 READ 10,NJNTS,NRCT,NCASES
   10 FORMAT (3I5)
      PRINT 11,NJNTS,NRCT,NCASES
   11 FORMAT (1H1///
     1       18H NUMBER OF JOINTS=,I5/
     3       21H NUMBER OF REACTIONS=,I5/
     2       22H NUMBER OF LOAD CASES=,I5)
      NEQTNS=NJNTS*2
      NMBS=NEQTNS-NRCT
      READ 12,(X(N),Y(N),N=1,NJNTS)
   12 FORMAT(2F10.3)
C
      PRINT 13,(I,X(I),Y(I),I=1,NJNTS)
   13 FORMAT (/38H NODE NO   X COORDINATE     Y COORDINATE/
     1       (3X,I3,6X, F10.3,6X, F10.3))
      READ 15,(NB(I),DCOS(I,1),DCOS(I,2),I=1,NRCT)
   15 FORMAT (I5,2F10.6)
      PRINT16,(NB(I),DCOS(I,1),DCOS(I,2),I=1,NRCT)
   16 FORMAT (/8H NODE NO,5X,6HDCOS X,5X,6HDCOS Y/(I5,3X,2F11.6))
C
      READ 17,(IP(N),IQ(N),N=1,NMBS)
   17 FORMAT (2I5)
      PRINT 18,(I,IP(I),IQ(I),I=1,NMBS)
   18 FORMAT (23H0  NODE   END I   END J/(I7,2I8))
C
      DO 30 I=1,NEQTNS
      DO 30 J=1,NEQTNS
   30 B(I,J)=0.0
      DO 100 I=1,NMBS
      L=IP(I)
      K=IQ(I)
      D1=X(K)-X(L)
      D2=Y(K)-Y(L)
      AL=SQRT(D1**2+D2**2)
      C=D1/AL
      S=D2/AL
      M=2*L
      B(M-1,I)=-C
      B(M,I)=-S
      M=2*K
      B(M-1,I)= C
  100 B(M,I)= S
      DO 109 I=1,NRCT
      N=NMBS+I
      K=NB(I)
      M=2*K-2
      DO 109 L=1,2
      J=M+L
  109 B(J,N)=-DCOS(I,L)
C
      CALL INVERT(B,NEQTNS,66,M1,T2)
C
      DO 250 I=1,NCASES
      PRINT 203,I
  203 FORMAT (1H1///12H  LOAD CASE=,I5)
      READ 204,(R(J),J=1,NEQTNS)
  204 FORMAT (2F10.2)
      PRINT 202
  202 FORMAT (/28H  NODE     X  LOAD     Y  LOAD/)
      DO 300 J=1,NJNTS
      JX=2*J-1
      JY=2*J
  300 PRINT 205,J,R(JX),R(JY)
  205 FORMAT (I6,2F11.3)
C
      DO 210 M=1,NEQTNS
      F(M)=0.0
```

```
      DO 210 N=1,NEQTNS
  210 F(M)=F(M)+B(M,N)*R(N)
C
      PRINT 220,(N,F(N),N=1,NMBS)
  220 FORMAT(/28H    MEMBER NO    AXIAL   FORCE/(4X,I5,6X,F9.3))
      PRINT 230
  230 FORMAT(/21H   NODE NO   REACTION)
C
      DO 235 J=1,NRCT
      K=NMBS+J
      M=NB(J)
      PRINT 226,M,F(K)
  226 FORMAT (5X,I5,3X,F8.3)
  235 CONTINUE
  250 CONTINUE
      END
```

INVERSION SUB ROUTINE

```
      SUBROUTINE INVERT(A,NN,N,M,C)
      DIMENSION A(1),M(1),C(1)
      IF (NN-1) 80,70,80
   70 A(1)=1./A(1)
      GO TO 300
   80 DO 90 I=1,NN
   90 M(I)=-I
      DO 140 I=1,NN
C     LOCATE LARGEST ELEMENT
      D=0.0
      DO 112 L=1,NN
      IF (M(L)) 100,100,112
  100 J=L
      DO 110 K=1, NN
      IF (M(K)) 103,103,108
  103 IF (ABS(D)-ABS(A(J))) 105,105,108
  105 LD=L
      KD=K
      D=A(J)
  108 J=J+N
  110 CONTINUE
  112 CONTINUE
C     INTERCHANGE ROWS
      TEMP=-M(LD)
      M(LD)=M(KD)
      M(KD)=TEMP
      L=LD
      K=KD
      DO 114 J=1,NN
      C(J)=A(L)
      A(L)=A(K)
      A(K)=C(J)
      L=L+N
  114 K=K+N
C     DIVIDE COLUMN BY LARGEST ELEMENT
      NR=(KD-1)*N+1
      NH=NR+N-1
      DO 115 K=NR,NH
  115 A(K)=A(K)/D
C     REDUCE REMAINING ROWS AND COLUMNS
      L=1
      DO 135 J=1,NN
      IF (J-KD) 130,125,130
  125 L=L+N
      GO TO 135
  130 DO 134 K=NR,NH
      A(L)=A(L)-C(J)*A(K)
  134 L=L+1
  135 CONTINUE
C     REDUCE ROW
      C(KD)=-1.0
      J=KD
      DO 140 K=1,NN
      A(J)=-C(K)/D
  140 J=J+N
C     INTERCHANGE COLUMNS
```

```
      DO 200 I=1,NN
      L=0
150   L=L+1
      IF (M(L)-I) 150,160,150
160   K=(L-1)*N+1
      J=(I-1)*N+1
      M(L)=M(I)
      M(I)=I
      DO 200 L=1,NN
      TEMP=A(K)
      A(K)=A(J)
      A(J)=TEMP
      J=J+1
200   K=K+1
300   RETURN
      END
```

APPENDIX D

PROGRAM FOR CALCULATION OF MEMBER DEFLECTIONS

Description of Problem

Given the deflection and slope at the left-hand end of a straight nonprismatic member and the deflection at the right-hand end and that there may be a hinge at an arbitrary intermediate point, calculate deflections at the one-tenth points, using inertias and bending moments given also at the one-tenth points.

Method of Solution

See Sec. 4.11.

Computer Program

The following listing gives a suitable program of the theory in Sec. 4.11, in Fortran IV coding.

```
C         PROGRAM TO CALCULATE BEAM DEFLECTIONS GIVEN END
C         DEFLECTIONS
C         AND SLOPE AT LEFT HAND END
C         HINGE AT DISTANCE A FROM L.H.END
          DIMENSION AINERT(21),BMOM(21),DEFL(11),SLOPE(11),BINERT(11),
         1AMOM(11)
          KK=0
    108   PRINT 1
          KK=KK+1
    1     FORMAT(1H1//)
          PRINT 2,KK
    2     FORMAT(20H BEAM DEFLECTIONS NO,I10/)
          READ 3,ALTH,A.E
          PRINT 4
    4     FORMAT(33H   LENGTH   HINGE   YNGS MODULUS)
          PRINT 3,ALTH,A,E
    3     FORMAT(2F10.2,E12.2)
          PRINT 5
    5     FORMAT(19H0MOMENTS OF INERTIA)
          READ 6,BINERT
          PRINT 6, BINERT
```

```
6       FORMAT(6F10.1)
        CALL INTERP(AINERT,BINERT)
        PRINT 7
7       FORMAT(21H0BEAM BENDING MOMENTS)
        READ 6,AMOM
        PRINT 6,AMOM
        CALL INTERP(BMOM,AMOM)
        H=ALTH/20.0
        HA=H/3.0
        DO 100 I=1,21
100     BMOM(I)=BMOM(I)/(AINERT(I)*E)
        READ 8,U1,THETA1,U2
        PRINT 9
9       FORMAT(27H0END SLOPES AND DEFLECTIONS)
        PRINT 8,U1,THETA1,U2
8       FORMAT(3E12.3)
        SLOPE(1)=THETA1
        DEFL(1)=U1
        DO 101 I=3,21,2
        J=(I*1)/2
        SLOPE(J)=SLOPE(J-1)+HA*(BMOM(I-2)+4.0*
       1BMOM(I-1)+BMOM(I))
101     DEFL(J)=DEFL(J-1)+SLOPE(J-1)*2.0*H+HA*(BMOM(I-2)*
       12.0*H+4.0*BMOM(I-1)*H)
        IF(A) 102,102,103
103     Q=U2-DEFL(11)
        X=0.0
        XB=ALTH/10.0
        DO 104 I=1,11
        IF(A-X) 105,105,106
105     DEFL(I)=DEFL(I)+Q*(X-A)/(ALTH-A)
106     CONTINUE
104     X=X+XB
        GO TO 107
102     PRINT 10
10      FORMAT(21H0NO HINGE THIS MEMBER)
107     PRINT 11
11      FORMAT(17H0BEAM DEFLECTIONS)
        PRINT 12,(DEFL(I),I=1,11)
12      FORMAT(8F12.5)
        GO TO 108
        END
```

```
      SUBROUTINE INTERP(A,B)
      DIMENSION A(21),B(11)
      K=1
      DO 100 I=2,10.2
      A(K)=B(I-1)
      A(K+2)=B(I)
      A(K+4)=B(I+1)
      X=(B(I-1)-2.0*B(I)+B(I+1))/2.0
      Y=(B(I+1)-B(I-1))/2.0
      A(K+1)=((X/2.0-Y)/2.0)+B(I)
      A(K+3)=((X/2.0+Y)/2.0)+B(I)
100   K=K+4
      RETURN
      END
```

Input Details

Card 1 Format (2F10.2, E12.2)	Length, hinge distance, Young's modulus. If the field for hinge distance is zero, no hinge is present.
Cards 2 and 3 Format (6F10.1)	Moment of inertia (11 values) at one-tenth points.
Cards 5 and 6 Format (6F10.1)	Bending moments (11 values) at one-tenth points.
Card 7 Format (3E12.3)	Deflection end i, rotation end i, deflection end j.

APPENDIX E

STRUCTURE–CUTTER PROGRAM TO DEVELOP b_o AND b_1 MATRICES

This program accepts as input the equation of equilibrium a^T written to include both member forces and reactions, and it generates the b_o and b_1 matrices according to the theory given in Sec. 5.6 (see Example 5.15).

Input Data

Card 1 Format (2I5)	Number of equations, number of variables
Card 2 (and subsequent cards) Format (8F10.4)	The statics matrix of the joint equilibrium equations, read in row by row

Output

1 Input is printed as an echo check.

2 If the program fails to find a nonzero pivot value, it will output an error message giving the position reached when this has occurred.

3 If the reduction of all rows has been undertaken satisfactorily, the complete b_o and b_1 matrices will be output. The rearranged order that has taken place in the a^T matrix to achieve the reduction is also given.

PROGRAM– b_0–b_1 GENERATOR:

```
      COMMON    A(50,70),R(50),IROW(50),ICOL(70) ,X(70),STCOL(50)
C
C      NR= NO OF ROWS
C      NC=NO OF COLUMNS
      READ1,NR,NC
 1    FORMAT(2I5)
      PRINT 2
 2    FORMAT(20H1 ECHO CHECK ON DATA/20H NO ROWS  NO COLUMNS)
      PRINT 4,NR,NC
 4    FORMAT(I6,5X,I6)
      READ 3,((A(I,J),J=1,NC),I=1,NR)
      PRINT 5
 5    FORMAT(46H  STATICS  MATRIX  JOINT EQUILIBRIUM EQUATIONS)
      PRINT3,((A(I,J),J=1,NC),I=1,NR)
 3    FORMAT(8F10.4)
      DO 45 I=1,NR
 45   R(I)=0.0
      REWIND 7
      DO 50 I=1,NR
      IF(I.EQ.1) 49,48
```

```
 48    R(I-1)=0.0
 49    R(I)=1.0
       WRITE(7) (R(J),J=1,NR)
 50    CONTINUE
       N1=5
       N2=7
       DO 100 I=1,50
 100   IROW(I)=I
       DO 102 I=1,70
 102   ICOL(I)=I
       MP=NC-1
C      FIND LARGEST TERM IN MATRIX LEFT TO BE REDUCED
       DO 300 I=1,NR
       AC=ABS(A(I,I))
       IR =I
       IC=I
       DO 210 J=I,NR
       DO 210 L=I,NC
       IF(ABS(A(J,L)).LE.AC) GO TO 210
       IR=J
       IC=L
       AC=ABS(A(J,L))
 210   CONTINUE
       IF(IR.EQ.I) GO TO 240
C      INTERCHANGE ROWS OF A AND OF R
       DO 230 J=1,NC
       TEMP=A(I,J)
       A(I,J)=A(IR,J)
 230   A(IR,J)=TEMP
       ITEMP=IROW(I)
       IROW(I)=IROW(IR)
       IROW(IR)=ITEMP
C      TEST FOR COLUMN INTERCHANGE AND OPERATE IF NECESSARY
 240   IF(IC.EQ.I) GO TO 258
       DO 250 J=1,NR
       TEMP=A(J,I)
       A(J,I)=A(J,IC)
 250   A(J,IC)=TEMP
       ITEMP=ICOL(I)
       ICOL(I)=ICOL(IC)
       ICOL(IC)=ITEMP
C      NOW REDUCE ONE MORE ROW
 258   DIAG=A(I,I)
       IF(ABS(DIAG).LE.1.0E-30)    GO TO 305
       DIAG=1.0/DIAG
       DO 260 K=I,NC
 260   A(I,K)=A(I,K)*DIAG
       J=I+1
       IF(J.GT.NR) GO TO 279
       DO270 L=J,NR
       TEMP=A(L,I)
       STCOL(L)=TEMP
       DO 265 K=I,NC
 265   A(L,K)=A(L,K)-TEMP*A(I,K)
 270   CONTINUE
C      REDUCTION PROCESS ON LOADS
 279   REWIND 7
       REWIND 5
       IF(N1.EQ.5) 281,280
 281   N1=7
       N2=5
       GO TO 284
 280   N1=5
       N2=7
 284   CONTINUE
       DO 400 M=1,NR
 10    FORMAT(I20)
       READ(N1) (R(L),L=1,NR)
       IF(IR.EQ.I) GO TO 390
       TEMPA=R(I)
       R(I)=R(IR)
       R(IR)=TEMPA
 390   R(I)=R(I)*DIAG
       IF(J.GT.NR) GO TO 409
       DO 410 L=J,NR
```

```
  410   R(L)=R(L)-STCOL(L)*R(I)
  409   WRITE(N2) (R(L),L=1,NR)
  400   CONTINUE
  300   CONTINUE
C       BACK SUBSTITUTION
        REWIND 6
        REWIND 7
        REWIND 5
        NRA=NR-1
        DO 600 IM=1,NR
        READ(N2) (R(J),J=1,NR)
        X(NR)=R(NR)
        DO 610 I=1,NRA
        K=NR-I
        M=K+1
        X(K)=R(K)
        DO 612 L=M,NR
  612   X(K)=X(K)-A(K,L)*X(L)
  610   CONTINUE
        WRITE(6) (X(J),J=1,NR)
  600   CONTINUE
        REWIND 7
C       CALCULATE B1
        NRED=NR+1
        DO 700 M=NRED,NC
        X(NR)=A(NR,M)
        DO 710 I=1,NRA
        K=NR-I
        J=K+1
        X(K)=A(K,M)
        DO 712 L=J,NR
  712   X(K)=X(K)-A(K,L)*X(L)
  710   CONTINUE
        DO 713 L=1,NR
  713   X(L)=-X(L)
        WRITE(7) (X(J),J=1,NR)
  700   CONTINUE
        GO TO 310
  305   PRINT 20, IROW(I),ICOL(I)
  20    FORMAT(16H SINGULAR MATRIX,2I10)
        CALL EXIT
  310    PRINT 805
  805    FORMAT(25H0REARRANGED ROW NUMBERING)
        PRINT 11,(IROW(N),N=1,NR)
         PRINT 806
  806    FORMAT(29H0REARRANGED  COLUMN NUMBERING)
        PRINT 11,(ICOL(N),N=1,NC)
  11    FORMAT(30I4)
        REWIND 6
        PRINT 810
  810   FORMAT(1H1,20X,38H   B ZERO MATRIX  PRINTED  BY  COLUMNS)
        DO 860 IR=1,NR
        READ(6) (R(I),I=1,NR)
        DO 800 I=1,NC
  800   X(I)=0.0
        DO 801 J=1,NR
        K=ICOL(J)
  801   X(K)=R(J)
        PRINT 731, (X(I),I=1,NC)
  860   CONTINUE
        REWIND 7
        NRA=NC-NR
        PRINT 865
  865   FORMAT(1H1,20X,35H  B ONE MATRIX  PRINTED BY COLUMNS )
        DO 870 IR=1,NRA
        READ(7) (R(I),I=1,NR)
        DO 871 I=1,NC
  871   X(I)=0.
        DO 872 J=1,NR
        K=ICOL(J)
  872   X(K)=R(J)
        K=NR+IR
        J=ICOL(K)
        X(J)=1.0
```

```
      PRINT 731, (X(I),I=1,NC)
870   CONTINUE
731   FORMAT(15F8.3/)
      CALL EXIT
      END
```

APPENDIX F

SUBROUTINE TO SOLVE BANDED SYMMETRIC MATRIX BY CHOLESKY DECOMPOSITION

The subroutine is actually composed of two parts. The first reduces the coefficient matrix A to LL^T, and the second undertakes the back-substitution (see Sec. 7.5 for theory). The equations are given by

$$AX = B$$

The number of equations is given by NEQT and the band width by NDIF + 1. The diagonal terms are to be positive and greater than 10^{-30}.

1 Reduction Subroutine

```
      SUBROUTINE DECOM
      COMMON A(200,60), B(200), X(200),NEQT, NDIF
      DATA (XDG = 1.0E-30)
      NN = 0
      LIM = 0
      LIM = NDIF+1
 4    IF(NN+1.GE.NEQT) RETURN
      NN = NN+1
      DIAG = A(NN,1)
      IF(DIAG.LT.XDG) GO TO 20
      DIAG = SQRT(DIAG)
      DO 2 J = 1,LIM
 2    A(NN,J) = A(NN,J)/DIAG
      DO 6 J = 1,NDIF
      L = NN+J
      IF(L.GT.NEQT) GO TO 4
      AA = A(NN,J+1)
      IF(AA.EQ.0.0) GO TO 6
      DO 3 K = J,NDIF
      M=1+K-J
      A(L,M)=A(L,M)-AA*A(NN,K+1)
 3    CONTINUE
 6    CONTINUE
      GO TO 4
 20   PRINT 5, NN
```

```
5     FORMAT(10H SINGULAR,I5)
      CALL EXIT
      RETURN
      END
```

2 Back-substitution Subroutine

```
      SUBROUTINE RESOLV
      COMMON A(200,60), B(200), X(200), NEQT, NDIF
      NN = 0
1     IF(NN+1.GE.NEQT) GO TO 4
      NN = NN+1
      B(NN) = B(NN)/A(NN,1)
      DO 2 J = 1,NDIF
      L = NN+J
      IF(L.GT.NEQT) GO TO 1
      AA = A(NN,J+1)
      IF(AA.EQ.0.0) GO TO 2
      B(L) = B(L) - AA*B(NN)
2     CONTINUE
      GO TO 1
4     X(NEQT) = B(NEQT)/A(NEQT,1)
      NP=NEQT-1
      DO 5 L = 1,NP
      K = NEQT-L
      SUM = 0.0
      DO 6 J = 1,NDIF
      M = J+K
      IF(NEQT.LT.M) GO TO 5
6     SUM=SUM+A(K,J+1)*X(M)
5     X(K) = (B(K)-SUM)/A(K,1)
      RETURN
      END.
```

INDEX